THE METALS BLACK BOOK™

Ferrous Metals
3rd Edition

Volume 1
The Metals Data Book Series™

Published By:

CASTI **Publishing Inc.**
10566 - 114 Street
Edmonton, Alberta, T5H 3J7, Canada
Tel: (403) 424-2552 Fax: (403) 421-1308
E-Mail: castiadm@casti-publishing.com
Internet Web Site: http://www.casti-publishing.com

ISBN 1-894038-08-8
Printed in Canada

Canadian Cataloguing in Publication Data

Main entry under title:

The metals black book, ferrous metals

Includes bibliographical references and index.
ISBN 1-894038-08-8 (pbk.) -- ISBN 1-894038-07-X (CD-ROM)

1. Iron alloys--Metallurgy--Handbooks, manuals, etc.
2. Steel alloys--Metallury--Handbooks, manuals, etc. I. Bringas, John E.,
1953-
II. Wayman, Michael L. (Michael Lash), 1943-
TA464.M47 1998   669'.1   C98-910456-7

Important Notice

The material presented herein has been prepared for the general information of the reader and should not be used or relied upon for specific applications without first securing competent technical advice. Nor should it be used as a replacement for current complete engineering standards. In fact, it is highly recommended that current engineering standards be reviewed in detail prior to any decision-making. See the list of standards organizations, technical societies and associations in Appendix 7, many of which prepare engineering standards, to acquire the appropriate metal standards or specifications.

While the material in this book was compiled with great effort and is believed to be technically correct, *CASTI* Publishing Inc. and its staff do not represent or warrant its suitability for any general or specific use and assume no liability or responsibility of any kind in connection with the information herein.

Nothing in this book shall be construed as a defense against any alleged infringement of letters of patents, copyright, or trademark, or as defense against liability for such infringement.

First printing of 3rd Edition, July 1998
ISBN 1-894038-08-8   Copyright © 1992, 1995, 1998

CASTI PUBLICATIONS

THE METALS DATA BOOK SERIES™

Volume 1 - The Metals Black Book™ - Ferrous Metals
Volume 2 - The Metals Red Book™ - Nonferrous Metals
Volume 3 - The Metals Blue Book™ - Welding Filler Metals

The Practical Guidebook Series™

Volume 1 - The Practical Guide to ASME Section II - Materials Index
Volume 2 - The Practical Guide to ASME Section IX - Welding
      Qualifications
Volume 3 - The Practical Guide to ASME B31.3 - Process Piping
Volume 4 - The Practical Guide to ASME Section VIII - Pressure Vessels
Volume 5 - The Practical Guide to ASME B16 - Flanges, Fittings & Valves

The Practical Handbook Series™

Volume 1 - Practical Handbook of Cladding Technology
Volume 2 - Practical Handbook of Iron and Nickel Corrosion Resistant
      Alloys

The Practical Self-Study Series™

Volume 1 - Practical Self-Study Guide to Corrosion Control

Practical Engineering CD-ROM Series™

100 Best Engineering Shareware CD-ROM

Acknowledgments

CASTI Publishing Inc. gratefully acknowledges the assistance of John F. Grubb and Joseph Chivinsky, Allegheny Ludlum Steel, for their technical review of the metallurgy section, and W. Stasko, Crucible Materials Corp. for his review of the tool steel metallurgy chapter. These acknowledgments cannot, however, adequately express the publisher's appreciation and gratitude for their valued assistance.

A special thank you is extended to Christine Doyle, who entered all the data in the book with care and diligence.

Authors

Metallurgy chapters 1 and 8 were co-authored by Michael L. Wayman, Ph.D., P.Eng., Professor of Metallurgy, University of Alberta, Edmonton, Alberta, Canada, and John E. Bringas, P.Eng., Publisher and Metallurgical Engineer, *CASTI* Publishing Inc. Metallurgy chapters 2 to 7 were written solely by Michael L. Wayman, Ph.D., P.Eng.

The French definitions and the glossary of French, Spanish and German metallurgical terms, chapter 9, were translated by Denise Lamy, M.Sc., P.Eng. The Spanish terms in the glossary were reviewed and edited by Ana Benz, M.Sc., P.Eng. and the German terms were reviewed and edited by Richard Witzke, M.Sc., P.Eng.

The metals data section was researched, compiled and edited by John E. Bringas, P.Eng.

Dedication

The Metals Black Book™ - Ferrous Metals is dedicated to my mother, Mary Bringas, and her parents, my grandparents, Luigi and Regina Zorzit; whose dream they shared of having their son and grandson become an engineer, and my honour to have accomplished their dream.

Additionally this book is dedicated to the memory of my first metallurgy teacher, Mr. George Chirgwin, W.D. Lowe Technical School, Windsor, Ontario. Not only did Mr. Chirgwin encourage me to study metallurgy, but his unique way of teaching influenced me to also teach the wondrous science of metals.

John E. Bringas, P.Eng.
Edmonton, Alberta

We Would Like To Hear From You

Our mission at *CASTI* Publishing Inc. is to provide industry and educational institutions with practical technical books at low cost. To do so, the book must have a valuable topic and be current with today's technology. *The Metals Black Book*™ - Ferrous Metals, 3rd Edition, is the first volume in *The Metals Data Book Series*™, containing over 700 pages with more than 600,000 pieces of practical metals data. Since accurate data entry of more than 600,000 numbers is contingent on normal human error, we extend our apologies for any errors that may have occurred. However, should you find errors, we encourage you to inform us so that we may keep our commitment to the continuing quality of *The Metals Data Book Series*™.

If you have any comments or suggestions we would like to hear from you:

CASTI Publishing Inc., 15066 - 114 Street
Edmonton, Alberta, T5H 3J7 Canada
tel: (403) 424-2552, fax: (403) 421-1308
e-mail: castiadm@casti-publishing.com

Browse Through Our Books On The Internet

You can browse through our electronic bookstore and read the table of contents and selected pages from each of our books. You can find our home page at http://www.casti-publishing.com.

Contents

SECTION II METALS DATA (Continued)

SECTION III APPENDICES & INDEX

Chapter

1

INTRODUCTION TO THE METALLURGY OF FERROUS MATERIALS

Introduction

The expression *ferrous materials* is used to mean the metallic element iron and the entire range of iron-based metallic alloys. There are a great many different ferrous materials, but they can be divided into three basic categories, namely wrought iron, steel and cast iron.

Wrought iron, which is no longer commercially produced, is a relatively pure iron containing non-metallic slag inclusions. Modern wrought iron products are actually made of low carbon steel.

Steels are iron-based alloys whose most important component element next to iron itself is carbon. The carbon contents of steels are low, usually below 1%, but the presence and amount of carbon in the steel have a major effect on its behavior in service. By far the most common type of steel is plain carbon steel, i.e. steel containing only iron and carbon plus small amounts of manganese and, usually, silicon or aluminum. The manganese, silicon and aluminum are added to compensate for the presence of the impurities sulfur, oxygen and nitrogen. Another important type of steel, the alloy steels, contain in addition to the above-mentioned elements, significant quantities of such elements as chromium, nickel and molybdenum, which distinguishes them from plain carbon steels. A specialized range of alloy steels, known as stainless steels, contain a minimum of 11.5% chromium. Tool steels, the final type to be considered here, are specialized carbon or alloy steels which are capable of functioning under the demanding service conditions associated with the working and shaping of metallic and non-metallic materials into desired forms. Some steel is used in the form of steel castings, but most steel objects are mechanically worked into their final forms and are thus categorized as wrought products.

Cast irons contain much higher carbon and silicon levels than steels, typically 3-5% carbon and 1-3% silicon. These comprise another category of ferrous materials, which are intended to be cast from the liquid state to the final desired shape.

Ferrous alloys dominate the world of construction materials. Their widespread applications are the result of a broad range of desirable material properties combined with favorable economics. Iron is the least expensive of all the metals and the second most abundant in nature.

This chapter supplies an introduction to the metallurgical aspects of ferrous materials, especially steels. Subsequent chapters provide data on many aspects of various ferrous materials. More details relating to the metallurgy of particular products are discussed in introductions to the sections on Carbon and Alloy Steels, Cast Steels, Cast Irons, Tool Steels and Stainless Steels.

Historical Aspects

Iron is one of the seven metals of antiquity, and is associated with the Roman god Mars and the planet of the same name. The first iron to have been used by humanity was probably meteoritic iron; this is readily identifiable because the so-called 'iron' meteorites are in fact iron-nickel alloys containing an average of about 8% nickel. Objects made from meteoritic iron are found among the archaeological artifacts left by many ancient cultures worldwide. Meteoritic iron, the metal from the sky, was used for utilitarian, decorative and ornamental purposes, and in some cases for objects with ceremonial functions.

However, most of the iron found in the archaeological record has been smelted from ores of iron, and the existence of this early material has led to the designation "Iron Age" for a particular stage of the evolution of societies, which began during the second millennium B.C. The first instances of iron smelting are not known, but it is possible that the earliest smelted iron was an inadvertent by-product of copper smelting operations. Here it was sometimes necessary to add iron oxide to the smelting furnace charge as a flux in order to lower the melting temperature of the silicate slag. Overly reducing conditions in the copper smelting furnace could have led to the subsequent reduction of metallic iron from the slag. Certainly smelted iron was in use by about 2000 B.C. and was relatively widespread by 1000 B.C. The original form in which smelted iron was used was wrought iron, a heterogeneous mixture of iron with silicate slag. Wrought iron was produced in bloomery furnaces by the solid state reduction of iron ore to metal, well below the melting temperature of iron. The product of this smelting operation was a bloom, a mixture of slag and metallic iron which was hot hammered to remove as much slag as possible.   With improvements in bellows and furnace technology, smelting temperatures increased until they were adequate to permit the production of liquid iron, in blast furnaces.   This iron, which contained a high carbon content, could then be cast directly to useful shapes, as cast iron.   Alternatively it could be converted to wrought iron by subjecting it to decarburization, initially by treating it in fining furnaces and later by the puddling process.   Cast iron came into use in the western world some time in the 11th-13th centuries A.D., although it had been used since about 500 B.C. in China, where higher temperature furnaces were available much earlier than in the west.

The intentional addition of carbon to wrought iron to make steel was being carried out during the first millennium B.C. but once again the precise chronological and geographical origins are not yet known.   A high production industrial version of the solid state carburization process, known as cementation, was widely used in the western world beginning early in the 17th century.   For severe service applications this was supplemented in the mid-18th century by crucible steel, which was made by remelting cementation steel to produce a higher quality material.   However, steel remained a relatively low volume, high cost product until the development of the mass production processes (the Bessemer process and the Siemens-Martin open hearth process) for producing steel from blast furnace iron in the mid-19th century.   Prior to these developments wrought iron and cast iron, rather than steel, were the predominant ferrous materials in use for structural applications.

Iron and Steel Production

Iron is one of the most abundant elements in the earth's crust, where it is a major constituent of many minerals including oxides, sulfides, silicates and carbonates.   Commercially viable ores are predominantly of the oxide or carbonate type, and metallic iron is reduced from such ores with relative ease.   Most of the iron produced goes directly for conversion into steel, with minor amounts being modified for use as cast iron.   Typical primary production operations involve the blast furnace reduction of iron ore to produce liquid pig iron, a metallic iron containing some 4% carbon.   Liquid pig iron is subsequently treated in steelmaking furnaces where carbon and impurities are removed by preferential oxidation.   The basic oxygen furnace is the technology most widely used for this purpose at present.   Some steel is also produced by remelting scrap or combinations of scrap and ore (e.g. pre-treated pellets) in electric arc furnaces.

All steel is produced in the liquid state, so that before further processing it must be allowed to solidify.   In commercial operations this solidification is carried out using two different practices. The molten steel may be teemed into tall rectangular molds to solidify as ingots, which are subsequently reheated and worked, usually by hot rolling, into semifinished products known as blooms, billets or slabs depending on their dimensions. More commonly, blooms, billets or slabs may be produced directly from liquid iron by continuous casting. In this process the liquid steel is poured into the top of an open-bottomed water-cooled mold while a strand of solid steel is withdrawn continuously from the bottom of the mold. Semifinished steel is converted into finished wrought products such as bars, sheets, strips, plates, structural shapes, wire, rails and tubular products using such operations as hot and cold rolling, drawing, forging and extruding. Many finished products require specific forms of heat treatment which, in combination with the forming operations, produce the specific combinations of properties desired for particular engineering applications.

Deoxidation and Desulfurization

Liquid steel coming from the steelmaking furnace contains high levels of dissolved oxygen which must be removed before the steel is cast, either in a continuous caster or as ingots. Typical deoxidants employed for this purpose are aluminum and silicon (in the form of ferrosilicon). The removal of oxygen from the steel is referred to as killing, thus the expressions "silicon-killed" and "aluminum-killed" steel. Fully killed steel is relatively homogeneous in its chemical composition and properties, and this practice is common in alloy steels and steels which are intended to be forged or carburized. Semi-killed steel has less deoxidant added than is the case for killed steel, and is typically used for low to medium carbon steels for structural applications. However some steels are not killed, and the oxygen remains in the steel where it reacts with carbon forming porosity (blowholes) of carbon oxide gases. These rimmed steels solidify with marked variations in chemical composition within the ingot. They have an outer rim of relatively pure iron, low in carbon, phosphorus and sulfur. These elements occur at higher than average levels in the center of the ingot, especially near the top. The higher purity outer rim makes these steels more suitable for the production of low carbon steel sheet with good surface quality. Capped steels are intermediate between rimmed and semi-killed steels, and are suitable for sheet, strip, wire and bars with carbon levels above 0.15%.

The sulfur in steel originates as impurities in coal, the material which, after it is converted to coke, is used as the fuel and reductant in the iron blast furnace. Conventionally, the addition of manganese compensates for the sulfur in steel, the sulfur being tied up as manganese sulfides, rather than iron sulfides which would be molten at hot rolling temperatures, causing the steel to be brittle or hot short.   However manganese sulfide inclusions tend to degrade the fracture toughness, so modern blast furnace iron and steels are subjected to various types of desulfurization involving, for example, the injection of agents such as calcium carbide or calcium silicide into the molten iron or steel to remove the sulfur.   Thus many modern steels have much lower sulfur contents than the levels which were acceptable several decades ago, and which are still reflected in the specifications for standard grades of steel.

Pure Iron and its Allotropy

Pure metallic iron is of very limited usefulness for engineering applications as it has a very low strength and poor resistance to corrosion. Its density (specific gravity 7.87) is slightly lower than that of copper, and its melting temperature is 1540°C (2804°F), slightly above that of nickel and well above copper and aluminum.   However the most significant characteristic of iron is its allotropy.   Between room temperature and its melting temperature pure solid iron undergoes two changes of crystal structure, known as allotropic phase transformations, so that it has one type of crystal structure at high and low temperatures and another at intermediate temperatures.   Below 912°C (1674°F) iron exists with a body-centered cubic (bcc) crystal structure, with its atoms packed as shown in Fig. 1.1. This material is known as alpha (α) iron.

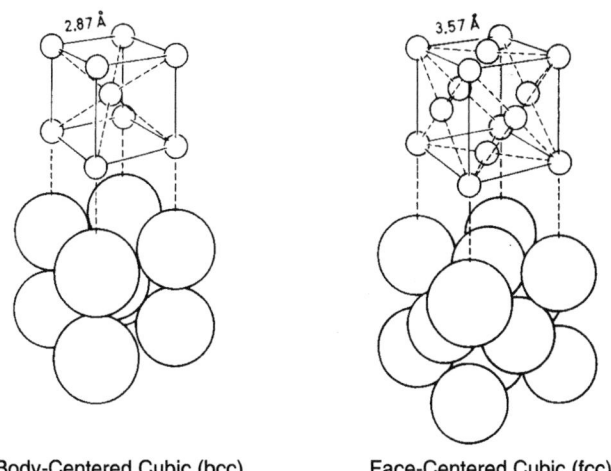

Body-Centered Cubic (bcc)   Face-Centered Cubic (fcc)

Figure 1.1 Crystal structures of iron.

At much higher temperatures, from 1395°C (2545°F) up to the melting temperature, iron also has the bcc crystal structure and in this temperature range it is referred to as delta (δ) iron.   However, at intermediate temperatures, between 912°C and 1395°C (1674 and 2545°F), the crystal structure of iron is face-centered cubic (fcc), with atom packing shown at the right side of Fig. 1.1; this is known as gamma (γ) iron. Thus if iron is slowly heated from room temperature to above its melting point, several phase changes (phase transformations) occur, and on subsequent slow cooling back to room temperature the reverse changes occur. Note that this applies only to slow cooling; if cooling rates are too high the behavior can be different as discussed below.

Iron-Carbon Alloys

The presence of carbon and alloying elements in iron make the allotropic behavior more complex, but in so doing they create the opportunity for an even wider range of microstructures and properties. The effects of carbon are the most significant in this respect. Carbon dissolves in the bcc iron forming a solid solution known as *ferrite* but the solubility of carbon is very low, with a maximum of only 0.025%C at 725°C (1337°F) in α-ferrite, and only 0.09%C at 1495°C (2725°F) in δ-ferrite. (Common practice is to use the expression ferrite synonymously with alpha (α) ferrite, and to use the full expression delta (δ) ferrite to refer to the high temperature phase.) On the other hand, in the fcc γ-iron the solubility of carbon is much greater, reaching a maximum of 2.1%C at a temperature of 1150°C (2100°F); the fcc solid solution of carbon in fcc iron is known as *austenite*. Both ferrite and austenite are interstitial solid solutions, that is, the carbon atoms dissolve in the iron by locating themselves in interstitial sites between the iron atoms which are arranged on the bcc or fcc crystal lattice. Fig. 1.2 shows the fcc structure of austenite with some of the lattice interstitial sites occupied by carbon atoms. Ferrite and austenite are referred to as phases since they are physically homogeneous and structurally distinct constituents of the microstructure of the alloy. Other phases which can occur in ferrous alloys will be discussed below.

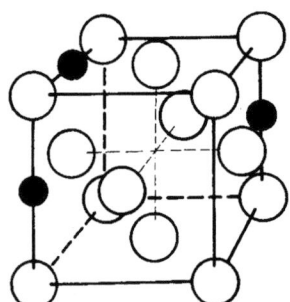

Figure 1.2 Carbon dissolved interstitially in fcc austenite. Carbon atoms are represented by black circles and iron atoms by white circles.

The temperature ranges over which the bcc and fcc crystal structures are stable are dramatically affected by the presence of carbon. Carbon lowers the lower limit of stability of the fcc phase from 912°C (1675°F) to as low as 725°C (1337°F) and raises the upper limit from 1395°C (2545°F) to as high as 1495°C (2725°F), as shown in Fig. 1.3, with the exact limits depending on the amount of carbon present.

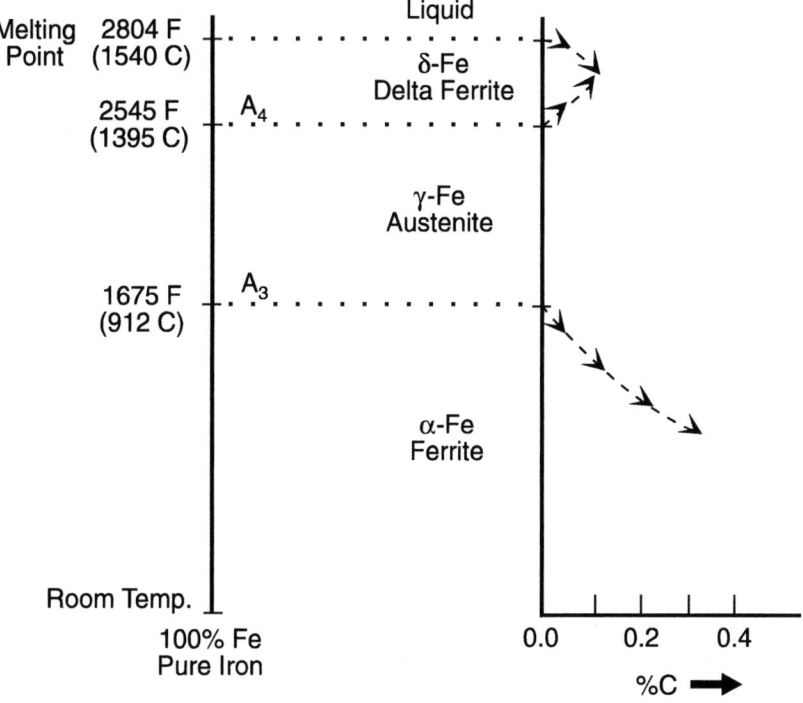

Figure 1.3 Effects of the addition of carbon to pure iron.

This complexity of phase behavior and phase stability can be dealt with most easily by making use of a phase diagram; Fig. 1.4 shows the phase diagram for the effect of carbon on iron. The left end side of this diagram represents pure iron, with increasing carbon content toward the right. Temperature increases toward the top of the diagram. Before considering this diagram in detail, it is important to be aware that phase diagrams apply to conditions of equilibrium, that is the thermodynamically stable state which is achieved only after there has been sufficient time for the atoms to move around and re-organize themselves into the stable phase or phases which are predicted by the diagram. Equilibrium can require long times to achieve, especially when temperatures are low since the diffusion process by which atoms move is slow at low temperatures where the thermal energy in the system is low.

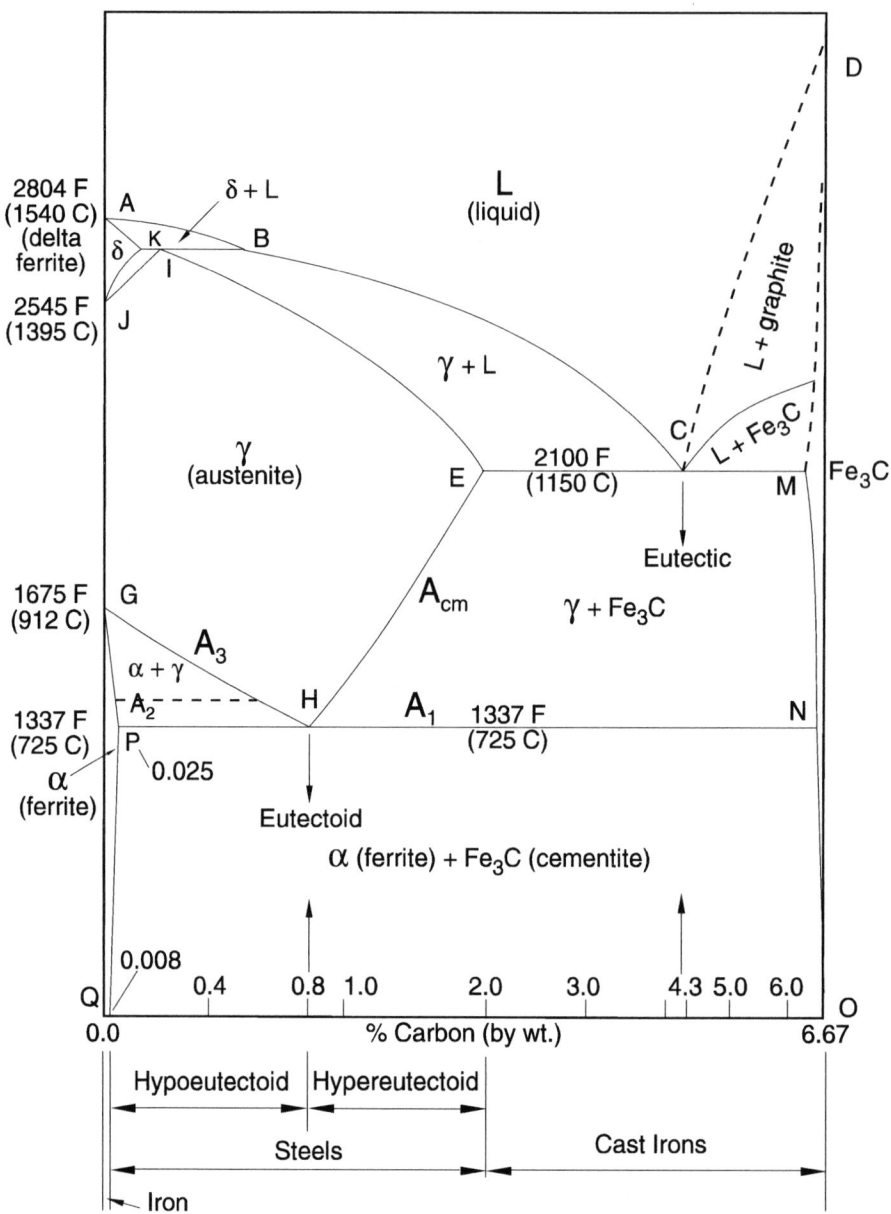

Figure 1.4 The iron-iron carbide phase diagram.

If the system is subjected to a rapid change of temperature the atoms may be unable to diffuse fast enough to keep up with any phase changes which are demanded by the phase diagram. As a result, during rapid temperature changes the phase diagram does not accurately predict the phase behavior; a different type of diagram is used for rapid changes of temperature, as discussed below. Understanding of all these diagrams is of great importance since steels are virtually always heat treated in some manner to develop their properties, and the diagrams allow the consequences of heat treatment to be predicted and understood.

The phase diagram is basically a map which predicts which phases are stable for any alloy with a given carbon content at a given temperature, i.e. as represented by a point on the phase diagram. Each such point lies either in a single-phase region, e.g. the austenite region, or in one of the two-phase regions which exist between the single-phase regions. The single-phase solid solutions are readily apparent on Fig. 1.4, with α-ferrite in the region GQPG and δ-ferrite in region AJKA. The borders of these regions (e.g. the lines QPG and JKA) represent the limits of solid solubility of carbon. This diagram also illustrates the fact that the carbon solubility is much greater in austenite than in ferrite, the limits of the austenite phase being JIEHGJ with a maximum of 2.1%C at point E.

There are other two phases shown on this phase diagram. One of these is the liquid solution of carbon in iron, which occurs at high temperatures, across the top of the diagram. The lower boundary of this region shows how the freezing temperature (or more accurately the liquidus temperature, the lowest temperature at which the entire material is liquid) of iron-carbon alloys changes with carbon content. For example it shows how at carbon contents in the cast iron range, between 3%C and 5%C, this temperature is low, reaching 1150°C (2100°F) at about 4.3%C. The low freezing temperature is a major reason why cast iron products can be produced with relative ease and low cost. The other phase shown on the phase diagram, at the right hand end within the boundary ONMDO, is *iron carbide* or *cementite*, a compound with the formula Fe_3C, corresponding to 6.67%C (note that percentages on phase diagrams are conventionally given in weight percent; Fe_3C contains 25 atomic percent carbon, which corresponds to 6.67 weight percent carbon). The crystal structure of cementite is orthorhombic, giving it a very high hardness, strength and brittleness in contrast to the soft ferrite and austenite phases.

Between any two single-phase regions, in a horizontal sense on the phase diagram, are two-phase regions, for example a steel with a carbon content of 0.4% at a temperature of 760°C (1400°F) lies in the two-phase region with ferrite on the left and austenite on the right. Thus the diagram predicts that this steel at this temperature would consist of a mixture of ferrite and austenite. By constructing an artificial horizontal line across the two-phase region at the temperature of interest, the relative amounts of ferrite and austenite, as well as their carbon contents can be predicted. Thus for the 0.4%C steel at 760°C, the left (ferrite) end of the horizontal line intersects the ferrite solubility limit line PG at about 0.01%C, read from the composition scale across the bottom of the diagram, so this is the predicted carbon content of the ferrite. Correspondingly, the right (austenite) end of the artificial horizontal line intersects the austenite solubility limit line HE at about 0.7%C, which is the predicted carbon content of the austenite. The horizontal position of the alloy composition, 0.4%C, relative to the two ends of the line, 0.01%C and 0.7%C, permits the relative amounts of ferrite and austenite to be determined; the closer the overall alloy composition is to the austenite end of the line, the more austenite is present in the steel's microstructure. In this case, there is somewhat more austenite than ferrite present because the alloy composition, 0.4%C, is somewhat closer to 0.7% than to 0.01%. In this manner, the phase diagram is used to predict the equilibrium phase structure of any alloy at any temperature, including which phases are present, their individual phase compositions (i.e. their carbon contents) and their relative amounts.

Phase diagrams can also be used to predict the changes (transformations) which occur during heating and cooling, as long as the temperature changes are slow as explained above. For example, one typical heat treatment given to a 0.2%C steel consists of slowly cooling from a temperature in the austenite region of the phase diagram, say 900°C (1650°F). In this case, the phase diagram predicts that when the austenite temperature falls below the line GH, about 865°C (1590°F), ferrite begins to form in the austenite. As the temperature continues to decrease, more and more ferrite forms so that by the time the steel reaches a temperature just above the horizontal 725°C (1337°F) boundary, line HP, about three-quarters of it has transformed to ferrite, while the rest remains austenite. On cooling through the 725°C (1337°F) temperature line, the ferrite remains unaffected, while all of the remaining austenite transforms to a mixture of ferrite and cementite. There is little change during further slow cooling to room temperature so that the final microstructure of the steel consists mainly of ferrite, with a small amount of cementite. The morphology of such a microstructure will be discussed below, along with further examples of the use of phase diagrams.

It is important to remember that all of these heat treatments which involve the cooling of austenite occur completely in the solid state. Austenite is a solid, as are its transformation products when it is cooled. This type of heat treatment is typically carried out after the material has been formed into its final or near-final shape.

This phase diagram in Fig. 1.4 shows the phase behavior of iron-carbon alloys with compositions up to 6.67%C; it is therefore called the iron-iron carbide (Fe-Fe$_3$C) phase diagram. It is equally possible to show a phase diagram which covers the full range of carbon contents from pure iron to pure carbon, however carbon contents of greater than 6.67% are not relevant to useful ferrous alloys. The ranges of carbon content applicable to plain carbon steels and to cast irons are shown in Fig. 1.4 below the bottom of the diagram.

In reality, the Fe-Fe$_3$C phase diagram shown in Fig. 1.4 is a metastable phase diagram rather than a true equilibrium phase diagram. Since graphite is thermodynamically more stable than iron carbide, an equilibrium phase diagram would have iron at one end and graphite at the other and iron carbide would not be shown. However iron carbide formation is kinetically favored over graphite formation, and the conversion of iron carbide to graphite is normally extremely slow. For example, iron carbide is present in steels which are more than 3000 years old. For this reason the Fe-Fe$_3$C system is applied to most ferrous alloys as if it were an equilibrium diagram. The boundaries of the iron-rich phases (α, γ, δ) are not appreciably different in the Fe-Fe$_3$C phase diagram and the Fe-C (graphite) phase diagram. However alloying elements can have a significant influence on whether Fe$_3$C or graphite forms; this is particularly important in cast irons as discussed in Chapter 4.

There are a number of other important features on the phase diagram in Fig. 1.4. The point at 0.8%C, 725°C (1337°F) is called the *eutectoid point*, while the eutectic point is at 4.3%C, 1150°C (2100°F). The latter point is important for cast irons, while the former applies to steels. Steels with the eutectoid composition, 0.8%C, are referred to as *eutectoid steels*, steels with less carbon are called *hypoeutectoid steels* while those with more carbon are called *hypereutectoid steels*. Most steels are hypoeutectoid; in particular, those with carbon contents below about 0.15%C are referred to as *mild steels*. Very important for steel heat treating is the line PN, called the *eutectoid temperature*, also referred to as the *lower critical temperature* or the *A$_1$ temperature*. This is the first transformation line reached on slowly heating steel from room temperature, i.e. the temperature at which austenite first begins to form during slow heating. The line GH represents the temperature at which the last ferrite disappears from hypoeutectoid steels on heating so that the entire microstructure then consists of austenite; this is referred to as the *A$_3$ temperature* or *upper critical temperature*. Note that unlike the A$_1$ temperature, the A$_3$ temperature depends on the carbon content of the steel. For hypereutectoid steels the corresponding temperature line is the line HE (known as the *A$_{cm}$ temperature*), the temperature at which the last cementite disappears on heating. The A$_2$ temperature is the *Curie* *temperature* of the ferrite, while the temperature given by line JI , the lowest temperature at which δ-ferrite is stable, is called the A_4 *temperature*.

In fact, for real heating and cooling processes, which occur at finite rates, and bearing in mind the need for sufficient time for atoms to diffuse as discussed above, these phase transformations do not occur at exactly the same temperatures on heating as they do on cooling. Specifically the transformation temperatures on heating are higher than on cooling, the extent of the discrepancy (or hysteresis) increasing at higher cooling or heating rates. For this reason they are distinguished in the terminology as *Ar* and *Ac temperatures*, the r and c referring to the French words *refroidissement* and *chauffage* for cooling and heating respectively. Thus for example the Ac_1 temperature is higher than the Ar_1 temperature. In this way slight departures from equilibrium are represented.

If steel having the eutectoid composition, 0.8%C, is heated within the austenite temperature range for a sufficient amount of time to form 100% austenite, a heat treatment called *austenitizing*, and then allowed to cool slowly, it remains completely austenitic until the temperature reaches the eutectoid temperature, 725°C (1337°F), at which time it transforms to a two-phase mixture of ferrite and cementite. If the carbon contents of the phases are considered, it can be seen on the phase diagram that this eutectoid transformation consists of austenite which contains a uniform carbon level of 0.8% transforming to ferrite which contains a very low (0.025%) carbon content plus cementite with a high (6.67%) carbon content. The actual morphology of the ferrite and cementite phases in this slowly cooled eutectoid steel are characteristically lamellar, that is they occur as alternating plates of ferrite and cementite; this lamellar microstructural constituent, shown in Fig. 1.5, is called *pearlite*. Note that pearlite is not a phase, rather it is made up of two phases, ferrite and cementite.

Fig. 1.6 shows how as the ferrite and cementite grow out into the austenite, carbon is rejected by the growing ferrite and accumulates as adjacent cementite plates. It is apparent that this transformation must require a great deal of carbon atom diffusion, hence pearlite formation can only occur if the cooling rate through the temperature range around 725°C (1337°F) is relatively slow. If the cooling rate is too rapid for sufficient carbon diffusion to occur, other microstructural constituents and phases which are not shown on the phase diagram can form, as discussed below.

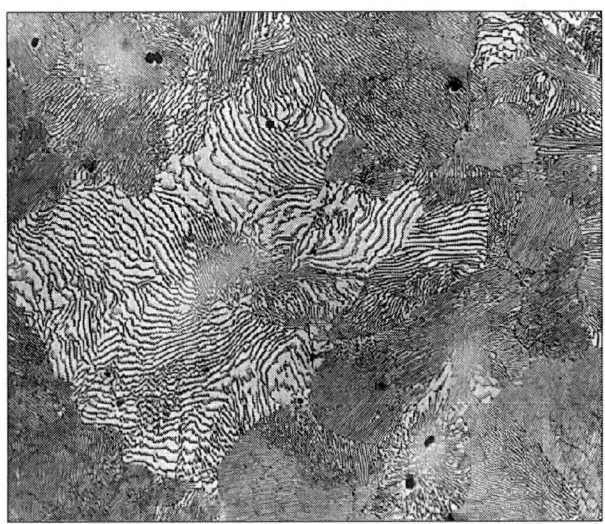

Figure 1.5 Showing the pearlite microstructure of
high carbon (1.07%) steel. Mag. 550X

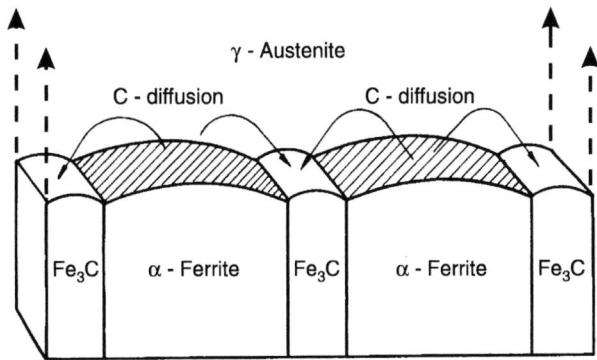

Figure 1.6 Schematic illustration of pearlite formation.

The thicknesses of the ferrite and cementite plates in the pearlite
strongly affect the mechanical properties of the steel. When pearlite
forms under conditions of very slow cooling, the individual ferrite and
cementite lamellae are thicker than if the steel is cooled more rapidly.
Pearlite can thus be coarse or fine, with *fine pearlite* having a higher
strength and lower ductility. This is only one of the many ways in which
cooling rate affects microstructure; further aspects of the heat treating of
steel, including the effects of slow cooling hypoeutectoid and
hypereutectoid steels, and of more rapid cooling, are now considered in
more depth.

Heat Treating of Steel -
The Effects of Carbon Content and Cooling Rate

The heat treating of steel normally begins with heating into the austentite temperature range and allowing the pre-existing microstructure to transform fully to austenite as required by the phase diagram. This austenitizing process may be carried out in any one of a number of atmospheres including air, inert gas, vacuum or molten salt. The hot austenitic steel is then cooled at some rate ranging from rapid (e.g. thousands of degrees per second by quenching in chilled brine) to slow (e.g. as little as a few degrees per hour by furnace cooling in a hot furnace which is allowed to cool with the steel inside). It is important to remember that the cooling rate is normally not uniform throughout the cross-section of the steel object, particularly at rapid cooling rates. The inside of a thick section can only cool by conducting its heat to the surface, where it is removed into the cooling medium; this is always a relatively slow process. The consequence is that if a thick section of steel is quenched, its surface undergoes a much higher cooling rate than its center, and therefore the surface and the center can have different microstructures and properties. Furthermore there will be residual stresses in the material associated with this situation. These effects can be beneficial or detrimental to the application of the material. However in the following discussion such complications will be avoided by considering only the cooling of a thin section, where it can be assumed that the cooling rate is constant throughout the cross-section.

During cooling the austenite becomes unstable, as predicted by the phase diagram, and decomposes or transforms to form a different microstructure, the characteristics of which depend on the austenitization conditions, the carbon content and the cooling rate. There are also effects due to the presence of other alloying elements as discussed below.

A description will first be given of the effects of slow cooling of hypoeutectoid (e.g. 0.4%C) steel from the temperature range where austenite is stable. Note that a finite austenitization time is required to dissolve all pre-existing carbides and to take all carbon into solid solution; this time depends on the thickness of the steel part and is frequently specified as one hour per inch of thickness. The higher the austenitizing temperature above the A_3 temperature (which for this 0.4%C steel is about 820°C or 1504°F), and the longer the time at the austenitizing temperature, the larger the austenite grain size will become. This austenite grain growth has detrimental effects on mechanical properties, so austenitization is generally carried out no more than 60°C (110°F) above the A_3 temperature.

Upon slow cooling from the austenitizing temperature, there is no change until the A_3 temperature is reached, at which time the fcc austenite begins to transform to the bcc ferrite. This occurs by the nucleation and growth of ferrite grains in the austenite, beginning at the austenite grain boundaries.   As cooling continues below the A_3 toward the A_1 temperature, the transformation of austenite to ferrite proceeds, by the continuing nucleation of new ferrite grains, and their continuing growth into the austenite.   This ferrite, which forms above the eutectoid temperature, is called proeutectoid ferrite.   Since the ferrite has a very low carbon content (maximum of about 0.025%C), the carbon content of the remaining (untransformed) austenite increases continuously as more and more of the austenite is replaced by ferrite.   By the time the temperature is just above the A_1 temperature, enough ferrite has formed that the carbon content of the remaining austenite has reached 0.8%C, the eutectoid composition.

As the slow cooling proceeds and the temperature drops below A_1, the untransformed austenite becomes unstable and transforms to pearlite. The microstructure is then a mixture of proeutectoid ferrite and pearlite, the pearlite being itself a mixture of ferrite and cementite. This is shown in Fig. 1.7, where the proeutectoid ferrite grains are the white areas and the pearlite is darker.

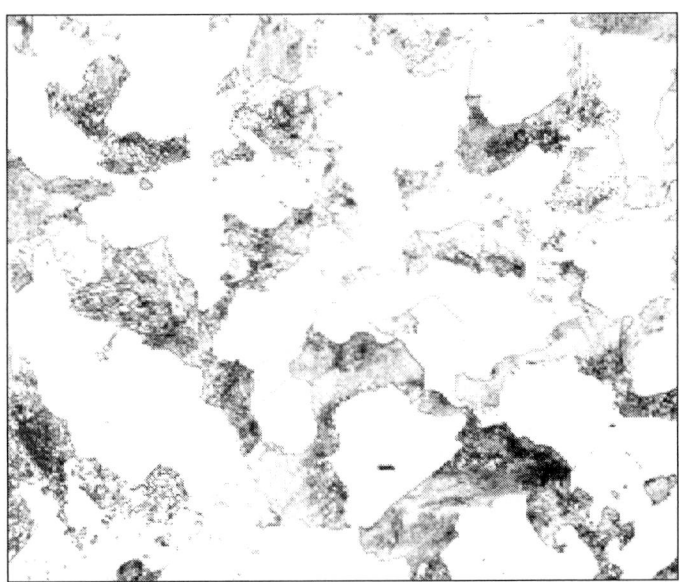

Figure 1.7 Showing a hypoeutectoid steel where the proeutectoid ferrite grains are the white areas and the pearlite is the darker areas.
Mag. 750X

Cooling from the A_1 temperature to room temperature produces virtually no further phase changes. The relative amounts, by weight, of ferrite and cementite in the final microstructure of this 0.4%C steel are predicted from the phase diagram to be approximately 94 parts ferrite to 6 parts cementite. The relative amounts of pearlite and proeutectoid ferrite can similarly be estimated by noting that they are the same as the relative amounts of austenite and ferrite which exist just above the A_1 temperature, i.e. about equal amounts of each. Slowly cooled hypoeutectoid steels with carbon contents less than 0.4%C will thus have microstructures containing more proeutectoid ferrite and less pearlite than the 0.4%C steel considered in this example. On the other hand the microstructures of slowly cooled steels containing between 0.4%C and the eutectoid composition, 0.8%C, will contain more pearlite than proeutectoid ferrite.

Cooling rates somewhat faster than those used for the above example can result in several differences in the behavior, the result of insufficient time being available for the diffusion necessary for complete equilibrium to be attained. One effect is that the proeutectoid ferrite which forms above the A_1 temperature does not grow as massive grains on the austenite grain boundaries. Instead it grows into the austenite in various crystallographic fashions after its nucleation on the austenite grain boundaries. One common proeutectoid ferrite morphology consists of crystallographic ferrite plates growing out into the austenite grains in several different orientations. This type of morphology, first observed in iron meteorites, is known as Widmanstätten ferrite. Another effect of more rapid cooling is that the pearlite forms at a temperature farther below the A_1. As a result it forms with thinner lamellae of ferrite and cementite, i.e. as finer as opposed to coarser pearlite.

Hypereutectoid steels behave in much the same way as the hypoeutectoid steels described above during slow cooling after austenitization, with the following differences. For full austenitization, it is necessary to austenitize above the A_{cm} temperature. The proeutectoid phase which forms during cooling between A_{cm} and A_1 is cementite which precipitates on the austenite grain boundaries. By the time the temperature has reached A_1, sufficient proeutectoid cementite has formed to lower the carbon content of the austenite to 0.8%C, the eutectoid composition. Hence the remaining austenite transforms to pearlite as the temperature falls below A_1. Since again there are no phase transformations below the A_1 temperature, the final room temperature microstructure consists of pearlite with a network of (proeutectoid) cementite along the prior austenite grain boundaries, as shown in Fig. 1.8. The phase diagram shows that for typical high carbon steels, which contain approximately

1%C, the total amount of proeutectoid cementite is small (i.e. very much smaller than the typical amount of proeutectoid ferrite in a hypoeutectoid steel).

Figure 1.8 A hypereutectoid steel (1.3% C) illustrating
pearlite with a network of (proeutectoid) cementite
along the prior austenite grain boundaries.

The above examples show the great usefulness of the iron-iron carbide phase diagram in predicting the changes which occur in the microstructure, and hence properties, of steel during certain heat treatments. However, it must be remembered that there are definite restrictions on the use of this diagram. For example, the presence of alloying elements distorts the diagram by shifting the positions of the various lines according to the particular alloying element and the amount added. Furthermore, as mentioned previously, faster rates of heating or cooling, such as those occurring during quenching or welding, greatly exceed the equilibrium rates so that the transformation reactions are shifted, delayed or simply do not have sufficient time to occur. The effects of more rapid cooling are now discussed.

The slowest rates of cooling correspond to furnace cooling, whereby a furnace containing the steel, at temperature, is switched off, or its heat source gradually reduced, so that the entire system, furnace and contents, cools to ambient temperature. This will always be a slow process, even if the heat source is suddenly removed, because of the large thermal inertia of the system, which consists typically of a mass of hot refractory brick. Faster cooling at a lower cost, can be obtained by air cooling, i.e. simply removing the steel from the furnace and allowing it to cool in the ambient atmosphere. Still higher cooling rates are obtained by quenching the steel, i.e. by removing it from the furnace and immediately immersing it in a cold medium with particular heat transfer characteristics. Common examples, listed in order of increasing cooling rate, include oil quenching, water quenching and brine (salt water) quenching. Alternatively a wide range of chemicals (typically polymers) is available which when added to water permit cooling rates to be controlled over a range from slower than water to faster. Quench media may be subjected to various degrees of agitation, which increases the cooling rate and permits more uniform cooling. A standard scale used to quantify cooling rate is the index of quench severity (called the H value), which has the value 1.0 for still water, as compared to as little as 0.25 for still oil and as much as 5.0 for agitated brine. In general it is desirable to utilize the minimum cooling rate necessary to achieve the desired microstructure, as more rapid cooling increases the magnitude of the residual stresses left in the quenched component, with consequent increased probability of distortion or cracking (quench cracking).

Several effects of increased cooling rate on the formation of ferrite-pearlite microstructures have already been alluded to, namely the different morphologies of proeutectoid ferrite, and the increasing fineness of the pearlite. However if cooling rates are increased still further, the limited time available during cooling is insufficient to permit the atom diffusion which is necessary for pearlite to form. As a result, microstructural constituents other than pearlite form when the austenite, which has become unstable below the A_1 temperature, transforms. These transformation products, including bainite and martensite, are non-equilibrium constituents which are therefore not present on the (equilibrium) phase diagram. Their formation occurs by processes which rely only partially (bainite), or alternatively not at all (martensite), on the diffusion of atoms. Thus martensite and bainite are able to form even at rapid cooling rates.

Bainite Formation

Bainite is a constituent which forms from austenite in a temperature range below about 535°C (1000°F) and above a critical temperature (the M_s temperature, discussed below) which depends on carbon content and is about 275°C (525°F) for eutectoid steel. Bainite forms together with pearlite in steels which are cooled somewhat too rapidly to permit full transformation to pearlite. Bainite is, like pearlite, a mixture of ferrite and iron carbide, but its morphology is different from that of pearlite, as its formation involves both atomic diffusion and a diffusionless shuffle of atoms referred to as shear. This latter characteristic enables bainite to form at cooling rates faster than the maximum at which pearlite can form. Furthermore the details of bainite formation depend strongly on the temperature at which the austenite transforms. At transformation temperatures in the upper part of the bainite formation range, upper bainite is formed. This is a rather feathery-appearing microstructural constituent, in contrast to lower bainite which forms at lower temperatures and is finer and more lenticular (lens-shaped). The distinction between upper and lower bainite is significant, and they can differ appreciably in mechanical properties. For the most part, a steel with a bainitic microstructure is harder, stronger and tougher at low temperatures than steels with ferrite-pearlite or fully pearlitic microstructures and equivalent carbon content. Unfortunately it can be extremely difficult to distinguish a steel microstructure as upper or lower bainite using the optical microscope, or to distinguish upper bainite from fine pearlite or lower bainite from martensite, a phase which will be discussed at length below. Examination using the electron microscope is needed to fully characterize bainites.

Martensite Formation

If austenite can be cooled to a sufficiently low temperature, for example by cooling very rapidly, its diffusion-controlled transformation to ferrite, pearlite or even bainite will not be possible. Instead, the austenite becomes so unstable that it is able to change its crystal structure by a diffusionless shearing transformation which moves blocks of atoms by small distances simultaneously. The transformation product is then *martensite*, a metastable phase which, like bainite, does not appear on the phase diagram since it does not exist under equilibrium conditions. The martensite structure is basically the result of the steel's attempt to transform from austenite (fcc) to ferrite (bcc), a process which is prevented by the presence in the austenite of a large amount of carbon, an amount far above the very low solubility limit of carbon in ferrite. This large supersaturation of carbon prevents a true bcc structure from forming so that the martensite is therefore a compromise structure. It can be thought of as a bcc structure which is highly distorted to accommodate the presence of the excessive amounts of carbon which are trapped at interstitial sites within the martensite structure. As a result, martensite possesses a crystal structure which is body-centered but not cubic; it is rather a body-centered tetragonal (bct) structure as shown in Fig. 1.9.

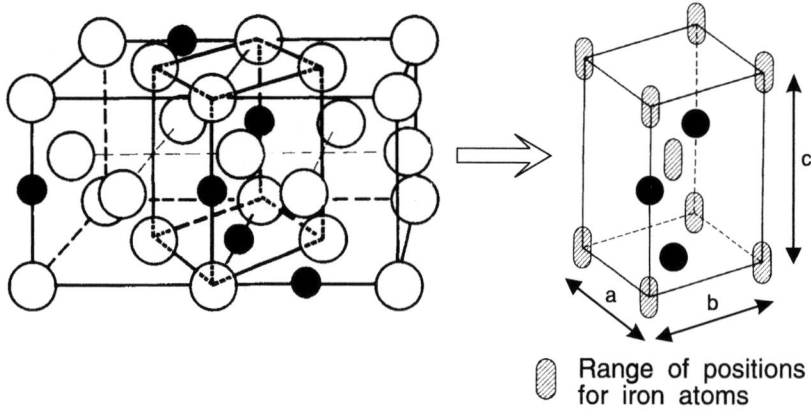

Range of positions
for iron atoms

Figure 1.9 Austenite (fcc) to martensite (bct) transformation.

The distortion of the crystal structure, and the associated residual stresses, are responsible for the very high hardness and strength and the extremely poor ductility and toughness of martensite. The amount of distortion depends on the carbon content, as shown in Fig. 1.10, since there is more carbon trapped in the martensites of higher carbon steels.

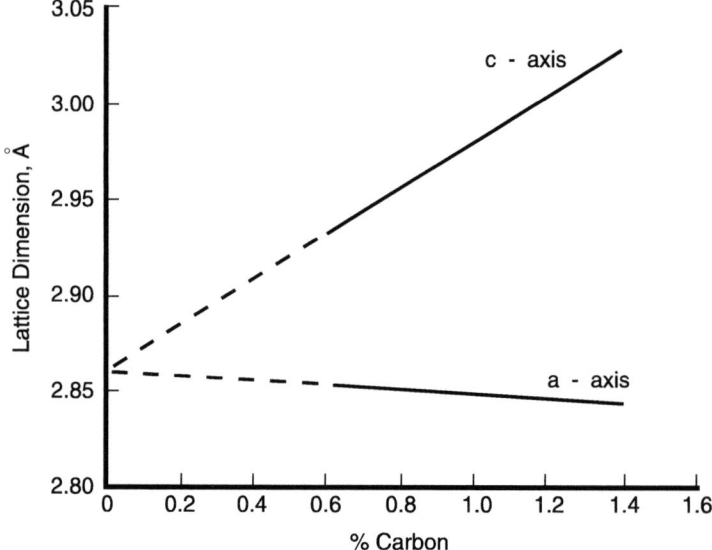

Figure 1.10 Distortion effect on bct structure due to carbon content. The
a–axis and c–axis are shown in Fig. 1.9.

This in turn causes the hardness of martensite to be directly related to carbon content (up to a limiting hardness at about 0.8%C), as shown in Fig. 1.11.

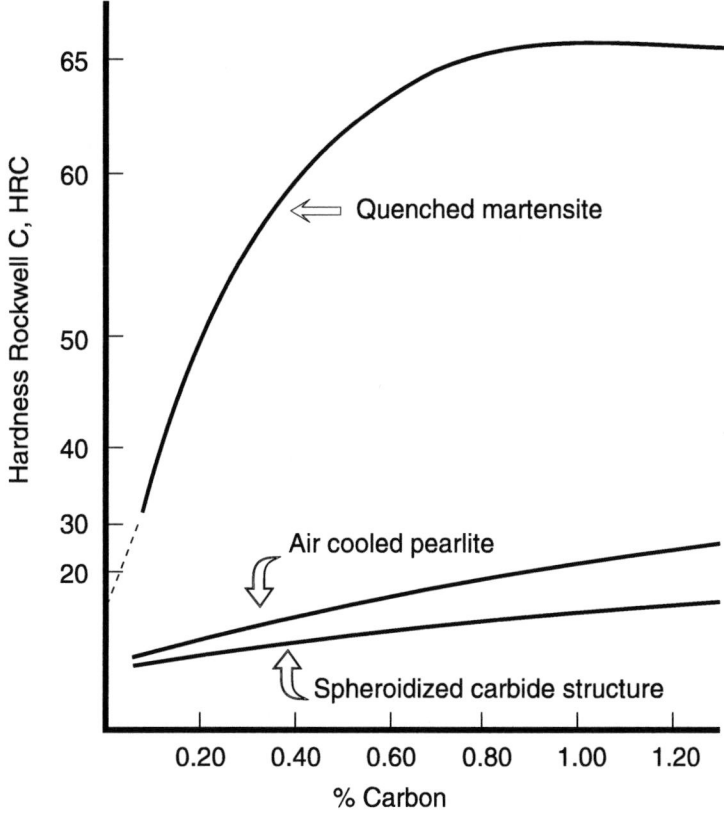

Figure 1.11 Effect of carbon on martensite hardness.

Because of the high strains associated with the martensite transformation, martensite forms in small plates or laths within the austenite grains.   The appearance of this microstructure is often described as *acicular*, meaning *needle-like*, as shown in Fig. 1.12. However, close examination shows that the appearance of martensite depends on its carbon content, with high carbon martensites exhibiting lenticular or plate-like morphology while low-carbon martensites are lath-like.   In the intermediate range of carbon content both lath and plate types are present. Any untransformed austenite, referred to as *retained austenite*, which remains in the microstructure appears between the martensite plates or between packets of martensite laths.

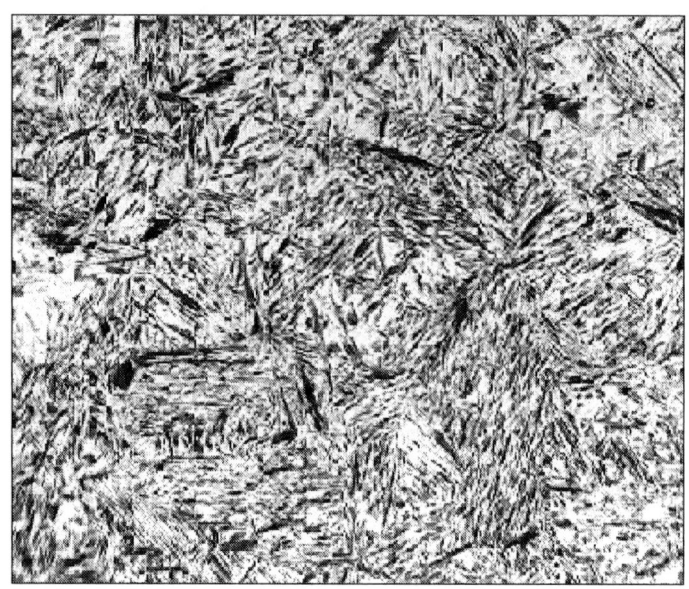

Figure 1.12 Acicular martensite showing needle-like appearance.
Mag. 475X

As mentioned above, in order for martensite to form, austenite must be brought to a sufficiently low temperature at a cooling rate which is rapid enough that the formation of ferrite, pearlite and bainite are avoided. Thus there is a critical cooling rate which must be exceeded in order to form martensite. At cooling rates slightly less than this critical rate, mixtures of martensite with the other transformation products, usually pearlite, are formed. This is described as a slack-quenched microstructure. The austenite in steels of higher carbon content requires more carbon diffusion in order to form the ferrite, pearlite or bainite constituents, hence martensite can be formed more easily, i.e. the critical cooling rate is lower for higher carbon steels. In low carbon steels, it is very difficult to form martensite as prohibitively high cooling rates are required. This is especially true in thick sections since, as mentioned above, cooling is never rapid in the center of a thick section.

During rapid cooling of austenite, martensite begins to form only when the austenite has reached a sufficiently low temperature, called the martensite start or M_s temperature, and the austenite-to-martensite transformation is not complete until the temperature has fallen below an even lower temperature, the martensite finish or M_f temperature. Since higher carbon martensites are more highly distorted, higher carbon austenite must be cooled further below the eutectoid temperature, i.e. to a lower temperature, in order that it becomes sufficiently unstable to form martensite. Thus the M_s and M_f temperatures decrease with increasing carbon content of the steels. Alloying elements (other than Co) also lower the M_s and M_f temperatures. The M_f temperatures of medium and high carbon steels can be below room temperature, with the consequence that if such a steel is austenitized and quenched to room temperature, its microstructure can consist of a mixture of martensite along with some retained austenite. This retained austenite can affect mechanical properties by lowering the strength and increasing the toughness, but is generally considered to be undesirable since it can cause problems by transforming to (brittle) martensite during subsequent stages of heat treatment or service.

Tempered Martensite

Although martensite is a very hard, strong, wear resistant material it lacks ductility and toughness, so much so that in all but low carbon steels brittle failure of martensite is so easily initiated that its strength cannot normally even be measured. Thus a steel through-hardened (transformed to martensite throughout its thickness) is not a satisfactory engineering material for most applications. However, a surface layer of martensite on a tougher ferrite-pearlite base can provide useful properties as discussed below. Furthermore, and even more usefully, martensite can be heat treated by tempering to obtain a tempered martensite microstructure with properties which are appropriate for industrial purposes. This tempering heat treatment allows a limited amount of carbon diffusion in the distorted bct martensite structure, so as to allow some degree of structural change (e.g. limited carbide formation). This reduces the distortion of the martensite and its internal stresses, with a consequent increase in ductility and toughness at some expense to hardness, strength and wear resistance. The extent of tempering and hence the mechanical properties can be controlled by varying the tempering time and temperature.

As predicted by the iron-iron carbide phase diagram, the stable structure for a carbon steel at ambient temperature is a mixture of ferrite and cementite. Thus during tempering martensite, a non-equilibrium (metastable) phase, tends to decompose in the direction of ferrite and cementite. At normal ambient temperatures, the mobility of carbon in martensite is too low for such changes to occur, and examples of martensite are known which date from more than 3000 years ago. Thus for engineering purposes, martensite can be considered stable. However if carbon is given the chance to diffuse, by tempering even at relatively low temperatures (e.g. hours at as little as 150°C or 300°F) then the amount of carbon diffusion can be sufficient to permit important changes in the martensite. At low tempering temperatures and/or short tempering times, decomposition of the martensite is minimal and the martensite remains hard and strong with slight increases in ductility and toughness. In the extreme, after tempering at high temperatures (still below the eutectoid temperature of 725°C or 1337°F) for long times, decomposition of the martensite can be so complete that its mechanical properties approach those of ferrite (i.e. soft, ductile, low strength). Thus the tempering temperature and time at temperature can be selected so as to achieve the specified mechanical properties, i.e. the balance between strength and toughness necessary for the intended application. This range of obtainable properties make quenched-and-tempered (Q & T) steels among the most versatile and useful engineering materials.

The actual changes that occur in the martensite during tempering are complex and depend on the tempering time and temperature. Among the effects are the precipitation of several types of iron carbides, including cementite, and the decomposition of retained austenite. During very heavy tempering at high temperatures the martensite decomposes to cementite and ferrite, as predicted by the phase diagram, however the cementite here is in the form of coarse spheroidal particles (Fig. 1.13), quite unlike the normal lamellar structure of pearlite.

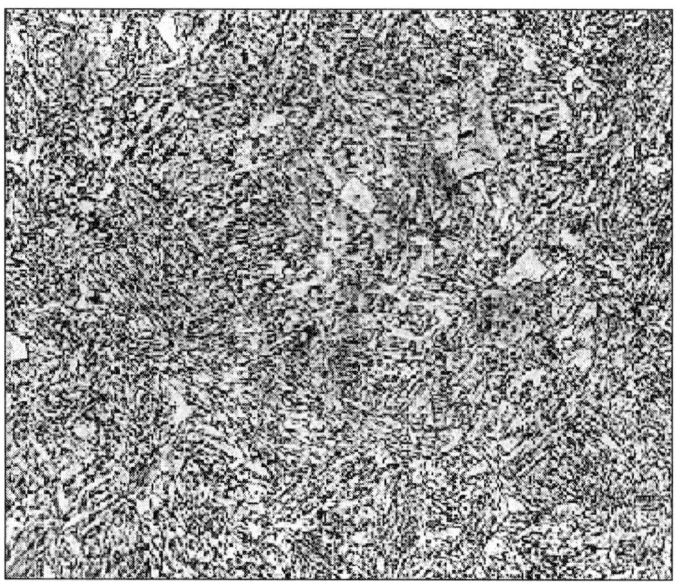

Figure 1.13 Tempered martensite exhibiting
spheroidized cementite in a matrix of ferrite.

In fact this spheroidized morphology of the ferrite-cementite microstructure is more energetically stable than lamellar pearlite, so if pearlite is held at high temperatures (below but close to 725°C [1337°F]) for long times either during heat treatment or in service, its cementite lamellae will gradually spheroidize also, giving a microstructure similar to that of very heavily tempered martensite.

Alloying elements, such as chromium and molybdenum, affect the tempering of martensite in several respects.   For example, alloying elements generally reduce the rate at which martensite tempers, as well as changing the type of carbide which precipitates during tempering from an iron carbide to an alloy carbide. Furthermore the combined presence of alloying elements and impurities can be responsible for the occurrence of temper embrittlement, a loss of room temperature ductility when alloy steels are tempered in the range 350-525°C (660-980°F). However the most important effect of alloying elements is on the hardenability, and this is discussed in the following section.

Hardenability

It is important to distinguish clearly between the terms "hardness" and "hardenability". Hardness is the resistance of a surface to being indented by an indenter under standard conditions, such as in the Rockwell or Brinell hardness tests. The hardness of a steel is determined by its composition and its microstructure (i.e. its thermo-mechanical processing). Hardenability, on the other hand, refers to the ability of a steel to harden, i.e. to form martensite to depth. This corresponds to the steel having a low critical cooling rate, i.e. having the ability to form martensite at low cooling rates. Steels with low hardenability are those which form only a thin surface layer of martensite when quenched from the austenite. As mentioned above, this is related to the inherent slowness of the removal of heat from the center of a thick section, so that the center of a thick section always cools relatively slowly. Thus typically below some depth beneath the surface of a quenched steel the critical cooling rate will no longer be exceeded, and some non-martensitic product such as pearlite will be present in the microstructure, lowering the hardness as shown schematically in Fig. 1.14. It must be stressed again that in most cases the reason for forming martensite is so that it can be tempered to form tempered martensite, with its highly controllable and desirable combination of strength and toughness.

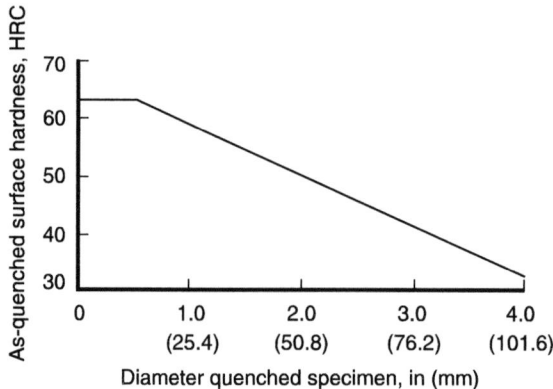

Figure 1.14 Section size effect on surface hardness of 0.54%C steel, water
quenched from 830°C (1525°F).

Low carbon steels are characterized by low hardenability, with critical
cooling rates only exceeded in thin sections. In higher carbon steels, the
hardenability is greater, but the very high hardness and brittleness of
high carbon martensite is normally undesirable. In alloy steels, however,
the situation is such that a high hardenability can be combined with a
good combination of strength and toughness. This is because in alloy
steels the interactions between the atoms of alloying element (e.g. Cr, Mo)
and carbon atoms slow the rate of carbon diffusion and thus increase the
time required for the formation of diffusion-controlled austenite
transformation products such as ferrite and pearlite. In other words,
alloying elements increase the hardenability, permitting the attainment
of a martensite microstructure, which is suitable for tempering, at lower
cooling rates. This high hardenability is one of the major reasons for the
use of alloy steels. In fact highly alloyed steels can form martensite at
rates slow enough to be equivalent to cooling in oil or even in air, hence
the designations oil-hardening and air-hardening steels. The use of
slower cooling rates to produce martensite is beneficial since it permits
martensite formation at a greater depth beneath the surface.
Furthermore, slower cooling reduces the magnitude of the residual
stresses which are present in the steel after quenching. These residual
stresses are caused by a combination of thermal contraction and the
volume expansion (2-4%) which accompanies the transformation of
austenite to martensite. When these volume changes occur at different
times at the surface and at the center of a piece of steel, high levels of
residual stresses can result, with consequent distortion and the potential
for quench cracking of the steel.

Because the cooling rate will vary with depth, there will also be a
hardness gradient from the surface to the center of a quenched bar, the

details of which depend on the bar diameter, cooling rate and hardenability. A standard test, the Jominy end-quench test (ASTM A 255, SAE J406), based on this effect is used to quantitatively evaluate hardenability. Furthermore, the effect can be used in a deliberate way by quenching a steel of limited hardenability to obtain a surface layer (case) of hard brittle martensite on a tough ferrite-pearlite bulk (core) microstructure, as discussed in more detail below.

IT, TTT and CCT Diagrams

Two types of diagram are used to display the hardenability characteristics of steels graphically. These are the *isothermal transformation* (IT) or *time-temperature-transformation* (TTT) *diagram*, and the *continuous cooling transformation* (CCT) *diagram*. The detailed appearance of these diagrams depends on the steel's composition (carbon and alloy content) and its austenite grain size (i.e. austenitization conditions), and thus every steel will have its own diagram for a given set of austenitization conditions. Both types of diagram are designed to predict in detail the transformation characteristics of a particular steel after austenitization, by showing which transformation product microstructures (and in some cases the as-transformed hardnesses) are obtained by various cooling conditions for that steel. The diagrams can also be used in the reverse sense for predicting the cooling conditions necessary to obtain a given microstructure and hardness.

IT diagrams are plots of temperature versus (log) time. They are determined by evaluating the microstructures of austenite which has been quenched from the austenitizing temperature into baths which are at fixed temperatures (hence the name isothermal) below the eutectoid temperature, and held for various amounts of time to allow the austenite to transform. They therefore give a graphical illustration of the time necessary for austenite to transform to its various transformation products at different temperatures.

For example the times necessary for the beginning and the end of the formation of pearlite and bainite can thus be determined as can the M_s and M_f temperatures. These IT curves for plain carbon steels have a very characteristic shape, as shown in Fig. 1.15 which is an IT diagram for 0.8%C steel with an austenite grain size of ASTM No. 6 (the ASTM grain size measurement system is discussed in a section near the end of this chapter). Here the two heavy lines represent the beginning and completion of the transformation of austenite to pearlite and bainite, with an intermediate line showing an estimate of the time necessary for 50% of the austenite to have transformed. Above the nose of the curve the austenite transforms to pearlite while below the nose only bainite forms.

The M_s temperature is shown along with the temperature at which various percentages (50% and 90%) of the austenite have transformed to martensite. The fact that the diagram illustrates the effects of time means that it does not represent equilibrium conditions, but the situation at the far right of the diagram (very long time) must approach the equilibrium state represented by the phase diagram.

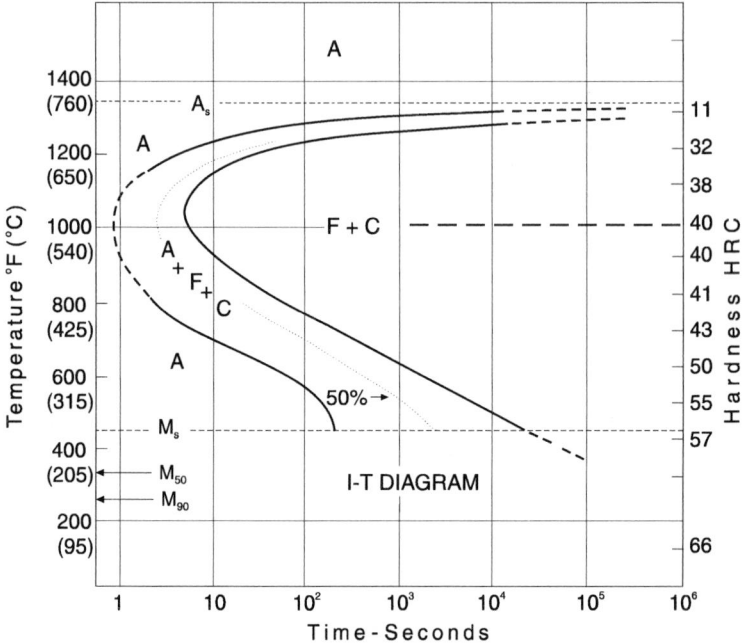

Figure 1.15 Showing an IT diagram for 0.8%C steel
with an austenite grain size of ASTM No. 6.

Although this type of diagram is determined by allowing austenite to transform at a fixed temperature (isothermally), it can give some indications of the effects of cooling rate on the austenite transformation. For example the closeness of the nose of the curve to the vertical axis (i.e. the "zero" time axis) gives an indication of the critical cooling rate necessary to transform austenite completely to martensite. However for the more usual industrial heat treatments which involve the continuous cooling of austenite, for example air cooling or water quenching, an alternative called the CCT diagram is more useful. This diagram also shows temperature plotted against time (on a log scale), but differs from the IT diagram in that it is determined by experimentally cooling steels at different rates, rather than allowing them to transform at fixed temperatures. The CCT diagrams give predictions which are roughly similar to those of IT diagrams but differ in some important respects such as a general shift of the beginning of the austenite transformation to lower temperatures and longer times. As well, the formation of bainite is more prevalent on CCT diagrams, which also give data on hardness after cooling to room temperature, as a function of cooling rates shown on the diagram, as in Fig. 1.16.

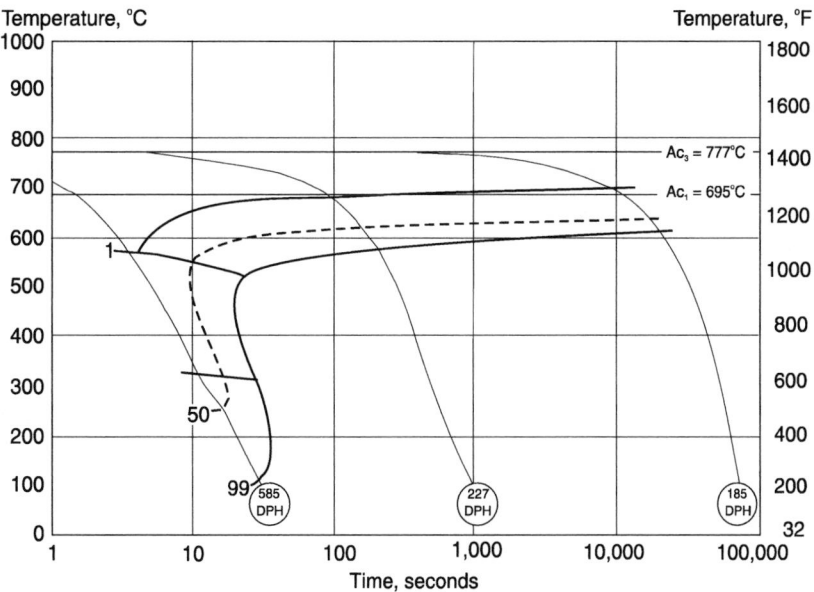

Figure 1.16 CCT diagram of an alloyed steel.

An alternative representation of CCT behavior is also available in the form of temperature plotted versus bar diameter, rather than time. Since cooling rate is a function of depth below the surface when a steel is cooled, it is possible to convert cooling rate to the diameter of a bar whose center undergoes a particular rate of cooling when the bar is quenched in a given cooling medium. Hence along the horizontal axis of this type of CCT curve is shown, typically, the bar diameter for air-cooled, oil-quenched and water-quenched steel of the specific composition for which the diagram is applicable (Fig. 1.17). A vertical line drawn on this diagram predicts the microstructure present at the center of a bar for a selected diameter and cooling medium. This gives useful representations of industrial heat treatment behavior.

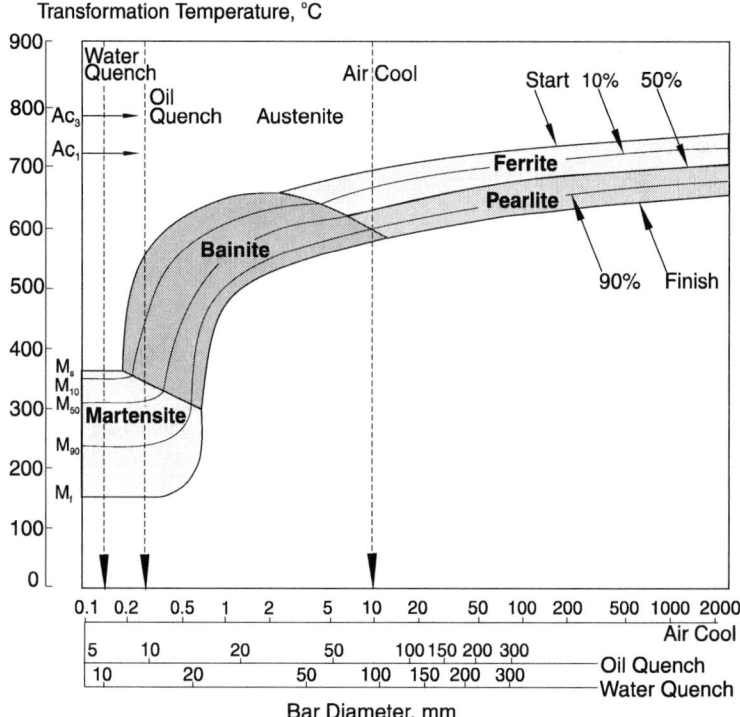

Figure 1.17 Temperature versus bar diameter CCT diagram
for a medium carbon steel (0.38%C).

The Roles of Alloying Elements

As mentioned in the preceding discussion, the alloying elements in steels play many different roles in determining the microstructure, properties and response to processing including heat treatment. Alloying elements generally lower the eutectoid carbon content and can raise or lower the eutectoid temperature. Some elements are austenite stabilizers, while others are ferrite stabilizers. Some are strong carbide formers while others prefer, to a greater or lesser extent, to dissolve in solution in ferrite. The presence of several elements together can give synergistic effects, or their individual effects can be fully or partially cancelled.

Carbon is the most important alloying element in steel. It is an austenite stabilizer, and the much increased solubility of carbon in austenite, as compared to ferrite, provides the basis for all the heat treating processes. In ferrite, carbon solubility is very low so that carbon exists in steels mainly as carbides (in plain carbon steels as the iron carbide, cementite). Increasing the carbon content of steels tends to increase their strengths (by increasing the amount of carbide present) and hardenabilities, while ductility, toughness, workability, weldability and machinability are reduced.

Manganese is normally present in all commercial steels. It combines with sulfur impurities to form particles of manganese sulfide (or iron-manganese sulfide), thereby avoiding the possibility of the formation of the detrimental iron sulfide phase (iron sulfides are brittle, and furthermore they melt at low temperatures causing the steel to be hot short). Manganese is an austenite stabilizer and it is not a strong carbide former. It contributes to strength by solution strengthening the ferrite and refining the pearlite. Another major role of manganese is in strongly increasing hardenability, especially when present in amounts greater than 0.8%. It also acts a deoxidizer, and oxygen is frequently present in its sulfides which are then referred to as oxysulfides.

Manganese contents much in excess of 2% tend to severely embrittle steel, however high carbon austenitic steels containing about 12%Mn display toughness combined with a high work-hardening rate, which make them useful industrial alloys for wear resistant applications, especially when the wear is accompanied by impact loading. These alloys, known as Hadfield's Manganese Steels, find applications in the mining industry (e.g. jaw crushers) and in severe service rail applications.

Silicon is present in all steels in minor amounts. It is a ferrite stabilizer, and is not a carbide former, but dissolves in the ferrite. Its major role is as a deoxidizer, since it combines readily with dissolved oxygen in molten steel to form silicates. This removes the possibility of porosity (blow holes) upon subsequent solidification, ensuring the production of dense sound steel. In cast irons the presence of silicon promotes graphite formation and provides resistance to attack by corrosive acids.

Silicon is added to steels designed for electrical and magnetic applications such as motors and transformers since it reduces eddy current losses in alternating current magnetization. It is also added to nickel-chromium-manganese steels where it increases strength and toughness. It increases hardenability especially in high carbon steels, and improves the castability of steels.

Sulfur is almost always a deleterious impurity in steels. It segregates strongly in steel castings and ingots, and degrades surface quality. It tends to combine with iron to form iron sulfides which are hard and have low melting points, hence they cause cracking during both hot and cold working. Thus the presence of sulfur in steel must be compensated for by additions of manganese or other strong sulfide-forming elements such as calcium and the rare earths. Intentional sulfur additions are made in the free-machining AISI/SAE 11xx and 12xx steels where iron sulfide chip breakers are desired for improved machinability.

Phosphorus has both detrimental and beneficial effects on the properties of steels. Like sulfur, it segregates strongly in ingots and castings, and it forms a brittle iron phosphide phase which reduces toughness. Phosphorus segregation during heat treatment or service of alloy steels is responsible for embrittlement effects known as temper embrittlement. On the other hand phosphorus is a potent hardener of iron. In recent years, re-phosphorized steels have been utilized as high strength sheet for applications such as automotive body panels. It is also intentionally added to the low-carbon free-machining AISI/SAE 12xx steels, and can have beneficial effects on sea water corrosion behavior of some steels.

Aluminum is widely used for deoxidation and for control of grain size. As a deoxidizer it readily combines with oxygen in liquid steel to form aluminum oxides, which can be removed into the slag. When added to steel in small concentrations it also combines with oxygen and nitrogen in the solid state to form aluminum oxides and nitrides which can have beneficial effects in pinning grain boundaries hence in controlling austenite grain growth in reheated steels. Furthermore in sheet steel the aluminum nitride precipitation can be controlled during annealing so as to give improved grain shape control with consequent improved sheet formability. Aluminum is not a carbide former.

Aluminum is used as an alloying addition in steels designed for nitriding, where it is present in amounts near 1%. The extremely high hardness of the nitrided case is due to the formation of a hard stable aluminum nitride surface layer. When present at high concentrations (above 1%), aluminum substantially increases the oxidation resistance of iron. This is applied in the iron aluminide alloys (Fe_3Al, FeAl) and in aluminum bearing stainless steels.

Chromium is a strong ferrite stabilizer and carbide former; as well it provides some solution strengthening of ferrite. Chromium also has a strong effect on increasing hardenability, especially in medium carbon steel, and it imparts both strength and oxidation resistance to steels at high temperatures. In high carbon steels it also provides resistance to abrasion and wear.

When steels with chromium contents above 11.5% are exposed to oxidizing environments, the chromium atoms in the steel react with oxygen to create a protective surface layer which gives corrosion and oxidation resistance; this is the basis for the stainless steels.

Nickel, like manganese, is an austenite stabilizer and does not form carbides, rather it acts as a ferrite strengthener and toughener in solution.   It also increases hardenability, especially in medium carbon steels.   Nickel produces a significant increase in notch toughness, even in the low temperature range, and is therefore added for increased toughness in case-hardening, heat-treatable and low temperature steels. Nickel is often used together with chromium in alloy steels to provide high hardenability, high impact strength and improved fatigue resistance.

The ability of nickel to stabilize austenite is utilized in stainless steels, where nickel contents of greater than 7% are able to impart austenitic structure even below room temperature.   In austenitic stainless steels nickel also provides improved resistance to environments which are not strongly oxidizing.

Molybdenum is a strong carbide former, like chromium, and it also contributes to solution strengthening of the ferrite.   Molybdenum is usually present in steels in combination with other alloying elements such as chromium and nickel.   It is a ferrite stabilizer, and is capable of acting as a grain refiner.   Molybdenum is most effective in improving hardenability, especially in high carbon steels.   It can induce secondary hardening during tempering of quenched steels, and it enhances the creep resistance of low alloy steels at elevated temperatures.

The presence of molybdenum in solution in steels gives improved resistance to pitting corrosion and therefore it is frequently added to high alloy chromium steels and to austenitic stainless steels.   It also acts to reduce susceptibility to temper embrittlement in alloy steels.

Vanadium is a stronger carbide former than chromium or molybdenum; it also promotes grain refinement and contributes to secondary hardening during tempering.   Its complex carbides and carbonitrides provide wear resistance, edge holding quality and high temperature strength. Vanadium also dissolves to some extent in ferrite, imparting strength and toughness.   It is used primarily in high speed, hot forming and creep resistant steels.

Titanium and **zirconium** are such strong deoxidizers that they can only be used in fully killed steels.   They are strong carbide formers and also strong nitride formers.   They show some solubility in ferrite and act to stabilize the ferrite, as well as serving as effective grain refining agents.

Titanium and zirconium, as well as calcium and the rare earths, notably cerium, are used to control the shape of manganese sulfide inclusions, causing them to maintain a shape which is spherical rather than elongated after the steel is worked; this gives improved fracture toughness, and more isotropic properties.

Titanium is used widely in stainless steels as a stabilizer against intergranular corrosion (sensitization), a role it is able to play because its carbides are more stable than those of chromium.

Niobium (columbium) and **tantalum** are often used together, both being strong carbide formers and stabilizers of ferrite.   Niobium strengthens ferrite; as well it is an effective grain refiner and as such acts to improve toughness. In addition, like titanium it is added to stainless steels to counteract sensitization (grain boundary corrosion due to intergranular chromium carbide precipitation).

Boron is added to steel only in very small quantities (between 0.0005 and 0.003%), specifically to improve hardenability, especially in fully deoxidized steels. Its presence degrades weldability.

Selenium is used in some free-machining steel to improve machinability. **Lead** additions in the range 0.015 to 0.035% are also made to improve machinability; in recent years **bismuth** has been used as a substitute for lead.

Industrial Heat Treatments

The above discussion shows clearly that the heat treatment which a steel receives is critically important in determining its microstructure and hence its properties.   Here the industrial heat treatments commonly given to steel will be summarized and examples given of industrial applications.

Normalizing consists of austenitizing above the A_3 temperature (typically 35 to 60°C or 65 to 110°F above) followed by air cooling. This is carried out to provide grain refinement, and to give a finer and more uniform carbide size than that which is present after hot working. Normalizing can be carried out as a batch process or in a continuous furnace.

Full annealing is carried out to develop a coarse pearlite microstructure, for example to provide optimum machining or forming properties. Typically for steels, the term *annealing* is used for the more formal description of full annealing. Full annealing involves austenitizing at a relatively high temperature to fully dissolve all carbides, then cooling slowly so as to allow transformation to pearlite at a relatively high temperature (not far below 725°C or 1337°F). Full annealing is common for sheet and strip, and can be carried out by a batch process (*box annealing*) or continuously (*continuous strip annealing*). Alternatively a coarse pearlite microstructure can be obtained by isothermal annealing whereby a sample is quenched from the austenite to below the eutectoid temperature, and held there at a constant (isothermal) temperature until it has transformed. This is carried out in a salt or lead bath or in a continuous furnace.

In medium to high carbon steels, coarse pearlite microstructures are too hard for optimum machinability or formability so in these steels a spheroidized pearlite is preferred. This can be developed by holding pearlitic (e.g. normalized) steel for long times just below the eutectoid temperature and is called *spheroidization annealing*. Alternatively, spheroidized pearlite can be obtained directly by transforming austenite in which not all the carbides have been taken into solution. One way of achieving this is by austenitization under restricted time-temperature conditions so that not all the pre-existing carbides dissolve, followed by slow cooling.

Process annealing or *sub-critical annealing* is carried out below the eutectoid temperature. This is done in order to soften steel which has been hardened by cold working for example in the processing of cold-rolled low carbon steel sheet or coiled strip. Here the annealing can be carried out either in batches (box annealing) or continuously. Sub-critical annealing is also performed in order to reduce residual stresses, in which case it is referred to as s*tress-relief annealing*.

In some low carbon steels austenitization is carried out between the eutectoid and the A_3 temperature, the process being referred to as *intercritical annealing*. The result is a proeutectoid ferrite matrix containing small uniform islands of austenite, which transform to martensite on subsequent cooling, giving a product known as dual-phase steel which has a good combination of strength and ductility.

Isothermal transformation of austenite to bainite is carried out by austenitizing, then quenching into a bath (usually molten salt) held at a temperature below the nose of the IT curve and above the M_s temperature. This process is similar to the isothermal annealing described above, but in the production of bainite it is referred to as *austempering*.

Quenching from the austenite range into a molten salt bath is also utilized in the *martempering process*. In this case the salt bath is held just above the M_s temperature, and the quenched steel is held in the bath only until the center of its cross-section has reached the bath temperature. Before the bainite transformation has had time to begin, the material is air cooled to room temperature so that the steel transforms to martensite. The advantage of martempering is that the martensite formation occurs during the relatively slow air cooling rather than during a rapid quench, thus the severe temperature gradients which are responsible for the high residual stresses in conventionally quenched martensite are eliminated.

Surface Hardening

For wear-resistance and other specialized applications it can be desirable to have a steel with a high hardness layer (case) on the surface of a high toughness base material (core). This can be accomplished with or without the case and core having different chemical compositions. If the chemical compositions of the case and core are the same, heat treatment is carried out so as to obtain different case and core microstructures. Typically, a high toughness core is first obtained by normalizing or quenching and tempering. This material is then subjected to localized intense surface heating so that only the surface layer becomes hot enough to form austenite, and the steel is then cooled rapidly so that the case transforms to martensite without the core having been markedly affected. The surface heating techniques include the direct impingement of a flame (*flame hardening*), surface heating by a high-frequency induction coil (*induction hardening*) or heating by high intensity light sources. Alternatively, laser beams can be used, especially for hardening of localized surface regions (*laser hardening*).

Processes in which the chemical composition of the surface is changed in order to permit the surface to be given a high hardness are known collectively as *case-hardening* processes. These involve the addition of either or both of the elements carbon and nitrogen, leading to the specific processes carburizing, nitriding and carbonitriding. The processes themselves involve the diffusion of carbon and/or nitrogen into the surface of the steel at high temperature.

In *carburizing* and *carbonitriding*, the material is subjected to a quench after the surface composition is changed so that a surface layer of martensite is formed in the high carbon, high hardenability surface while the low carbon, low hardenability case transforms to a tougher ferrite-pearlite microstructure. Carburizing has the advantage over the previously described surface hardening heat treatments in that the hardness of martensite is directly related to its carbon content, so carburizing provides a harder surface layer.

Several different industrial processes are utilized for these purposes. *Pack carburizing* consists of sealing the steel in a box along with carbonaceous solids and heating externally to permit carbon to diffuse into the surface of the steel. Temperatures in the austenite range are necessary and many hours of heating are generally required even for case depths as little as 1 mm (0.040 in). Alternatively the steel can be *gas carburized* by heating in an atmosphere of carburizing gases such as a mixture of methane with carbon monoxide. If dissociated ammonia is included in the gas mixture, higher contents of both nitrogen and carbon develop in the surface layer; this process is known as *carbonitriding*. For shallow cases, a liquid carburizing bath such as molten salt containing cyanide can be employed; the presence of nitrogen in the cyanide causes nitrogen to enter the steel surface along with carbon, so that this is also in fact a carbonitriding process. The presence of nitrogen increases the hardenability of the surface layer, allowing oil quenching rather than water quenching; it also increases resistance to softening during tempering.

With gas or liquid carburizing, the steel is directly quenched from the carburizing environment. Pack carburized steel is air cooled, extracted from the pack, then reheated to the austenite range and quenched.

Nitriding consists of holding the steel well below the eutectoid temperature, typically 510-538°C (950-1000°F) in a nitrogen-rich environment, typically ammonia gas. This results in a shallow surface layer rich in iron nitrides, with a very high hardness. The fact that the temperatures used are not high enough to austenitize the steel means that no martensite is formed, and also that residual stresses and consequent distortion are less than in carburized steel.

Microstructure

The term *microstructure* refers to the basic structural makeup of materials as viewed in the magnification range 10X to 10,000X using a light optical microscope or an electron microscope. In this magnification range individual atoms are not visible, nor are the geometrical atomic arrangements (unit cells) which characterize the different crystal structures or differentiate crystalline from non-crystalline (amorphous) materials. Rather what is seen is the material structure at the level of grains or particles of the different phases (matrix, second-phase particles, non-metallic inclusions, porosity, microcracks) which constitute the material. The microstructure of a specific piece of material is determined by both its chemical composition (the types and numbers of specific elements present in the material) and its entire thermo-mechanical history (the mechanical processing and heat treatments to which the material has been subjected). The microstructure is of great importance, along with chemical composition, in determining the mechanical properties (strength, ductility, toughness) of the material, so that microstructural characterization is absolutely essential in understanding material behavior and in the development of materials with improved properties.

Microstructure cannot normally be observed directly on an unprepared surface; rather, special specimen preparation techniques are employed prior to microscopic examination, processes which collectively are referred to as metallography. The specimen must be prepared so as to exhibit a smooth flat surface which represents a planar section through the microstructure and contains no distortions caused by the preparation techniques. A typical metallographic specimen preparation consists of first cutting an appropriate section through the material to obtain a specimen of size such that it can be readily handled (e.g. displaying a face of area between 0.1 and 1 cm^2 or ½ in.^{2}), then embedding the specimen in a polymer mount (e.g. a cylindrical mount 3 cm or 1.25 in. diameter and about 1 cm or ½ in. high) with the surface of interest revealed on one end face of the mount. This face is then subjected to a series of abrasive grinding operations, typically using successively finer abrasives in the range from 80 to 600 grit silicon carbide papers with water lubrication. This is followed by polishing with abrasives such as diamond paste or water-based slurries of aluminum or magnesium oxide on cloth polishing wheels to a final grade of 1 micron or less. Variations on this procedure include the use of diamond abrasives in the grinding stages, and the use of lapping machines where the sample is ground and polished on metal platens with abrasive slurries. Various degrees of mechanization are possible.

The final result of this specimen preparation is a smooth shiny planar surface through the material. Microscopic examination at this stage is useful for revealing such microstructural features as inclusions, some second phase particles, porosity and cracks. Chemical microanalysis, using electron microbeam techniques such as the electron probe microanalyzer or a scanning electron microscope equipped with a fluorescent x-ray analyzer are conveniently applied at this stage to determine, quantitatively or qualitatively, the nature of the different phases visible in the microstructure.

However a full understanding of the microstructure can only be obtained by examining the specimen after subjecting the polished surface to etching, using an acidic or basic solution. The etchant attacks some aspect of the microstructure in such a manner as to affect the surface topography or surface chemistry so that upon examination in a microscope some aspects of the microstructure become apparent. Typically this means that in the optical microscope the surface then reflects light in such a way as to provide contrast (gray level differences) among different microstructural features, while in the electron microscope the contrast is provided by changes in electron emission and/or the detection of emitted electrons. In many cases, etching with several different etchants may be necessary in order to obtain a full understanding of the material's microstructure. Etching typically reveals individual grains, either by providing contrast between adjacent grains based on their differing crystallographic orientations, or else by delineating the network of grain boundaries which mark the limits of the individual grains. In a similar manner, second phase particles and non-metallic inclusions are revealed by differences in gray level and/or color. Porosity in general appears dark. The contrast created in the optical microscope is best understood by thinking in terms of the scattering of incident light from the specimen surface. Regions of the surface which are smooth reflect light back into the objective lens of the microscope and such regions appear bright. On the other hand, regions which are roughened, grooved or pitted by the etchant scatter light in all directions, so that only some of the light is collected by the objective aperture of the microscope, hence these regions appear darker. This is illustrated in Figure 1.18.

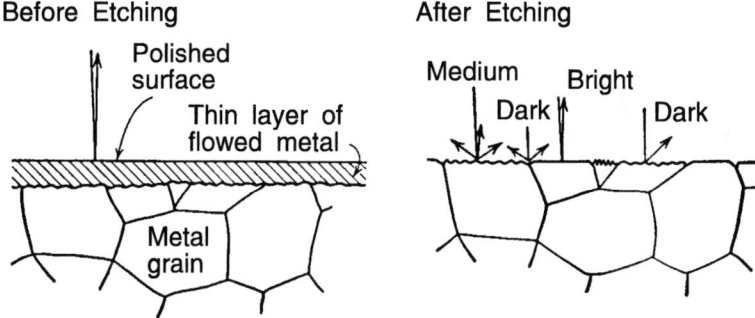

Figure 1.18 Etching effects on optical microscopy
of grains and grain boundaries.

The size, shape and distribution of porosity are useful microstructural parameters. In castings, the porosity often has a characteristic shape, reflecting the fact that it represents the space left as the result of solidification shrinkage as the last regions solidify. Since the shapes of solid grains as they solidify are normally dendritic (tree-like), the shrinkage porosity is generally located between the dendrite arms, and is hence termed *interdendritic porosity*. Porosity with this shape and distribution is thus characteristic of as-cast microstructures. If castings are mechanically worked the porosity will be completely or partially closed up, with its shape altered in ways geometrically compatible with the working geometry.

Metallographic examination of material containing cracks can provide information on the crack path (e.g. intergranular or transgranular), fracture mode (cleavage, microvoid coalescence, metal fatigue, creep, stress-corrosion cracking) and reveal the relationship between microstructure and cracking including the existence of unexpected or special microstructures in the vicinity of crack tips. The presence and characteristics of corrosion can also be studied and interpreted in this manner. Thus metallography is useful both for materials development and for failure analysis.

The great usefulness of metallographic examination cannot be overstated. The determination of the microstructure of the material can permit an understanding of the material properties and their relationships to manner in which the material was processed or used; alternatively metallographic examination can provide information on the processing of a material whose history is uncertain. It is widely used in materials research and development, in production (quality control) and in failure analysis.

Grain Size

Among the microstructural parameters which are most useful in characterizing materials is the *grain size*. This is most straightforward in simple single phase materials such as ferrite or austenite, but is still applicable to the matrix phase in multiphase materials. In materials where a second phase is of sufficient amount, size or continuity to be significant, its grain size may be reported separately. Grain size determination in single phase materials is carried out on polished and etched metallographic specimens using ASTM Standard E112, Test Method for Determining the Average Grain Size. This standard lists three methods for determining grain size, namely the Comparison Procedure, the Planimetric (Jeffries') Procedure, and the Intercept Procedure. Because of their purely geometric bases, these are quite independent of the metal concerned and may also be used for comparable measurements in non-metallic materials. It must be remembered that the grain size and shape as seen in a metallographic specimen represent the projection of the true microstructure on a planar section taken randomly through the specimen. A metal grain is a three dimensional shape, and within any microstructure is a range of true grain sizes and shapes. The grain cross-section as observed in a metallographic specimen is dependent on where the plane of observation cut through each individual grain. Thus no two fields of observation are identical.

The Comparison Method is popular since it is simple and straightforward. This method involves viewing grains in a microscope and comparing them at the same magnification (100X or 75X) with charts defined in ASTM E112; two examples are given in Fig. 1.19.

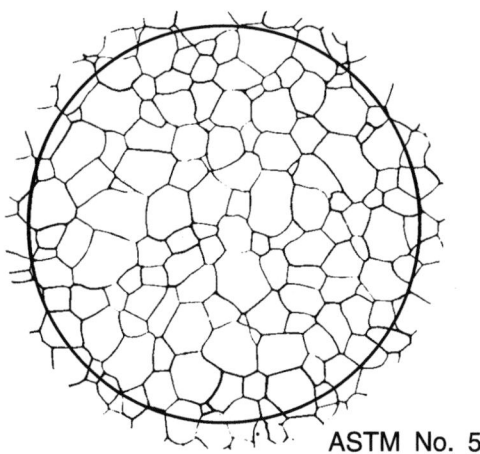

ASTM No. 5

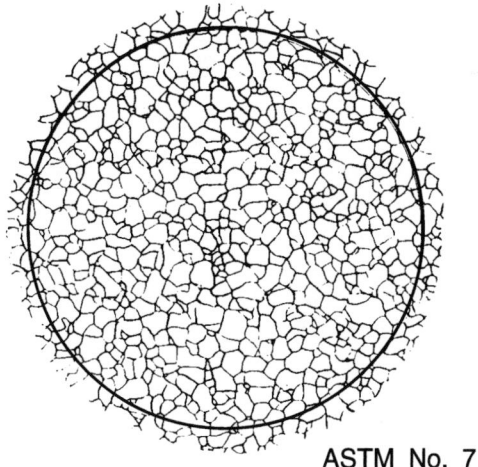

ASTM No. 7 1/2

Figure 1.19 Examples of ASTM E 112 comparison procedure grain size charts for No. 5 and No. 7½.

The ASTM grain size number corresponds to a certain number of grains per unit area of the image at the specified magnification, as shown in Table 1.1.

Table 1.1 ASTM No, Grains/in^2, grains/mm^2

ASTM No.	Grains/in^2 at 100X	grains/mm^2 at 1X
0	0.5	8
1	1	16
2	2	31
3	4	62
4	8	124
5	16	248
6	32	496
7	64	992
8	128	1980
9	256	3970
10	512	7940
11	1024	15,870
12	2048	31,700

The relationships between the grain size number and the number of grains per unit area are given by the expressions:

$$N = 2^{(n-1)}$$

where N= number of grains per square inch at 100X magnification, and n = ASTM grain size number, or, for SI units, and

$$N = 2^{(n+3)}$$

where N = number of grains per square millimeter (at 1X magnification), and n = ASTM grain size number (BS4490:1969).

In the Planimetric (Jeffries') Procedure a known area is inscribed in the observed field, and the grains within this area (minimum 50) are counted (including half the number of grains which intersect the perimeter of the field) and multiplied by Jeffries' multiplier. The product is the number of grains per square millimeter.

The Intercept Method has two procedures, the lineal (Heyn) procedure and the circular procedure. Both involve placing a grid pattern on the field of observation, and counting the number of grains at each grid intercept within a selected area.

In recent years, software and procedures have been developed to permit the automatic determination of grain size directly from images obtained in microscopes. A number of these computerized image analysis systems and procedures are available, generally from microscope manufacturers or suppliers.

The grain size is one of the most important of the microstructural parameters, as it has a strong effect on the strength and fracture resistance of materials. At high temperatures where creep is a problem, coarser grained materials have advantages. However at normal temperatures a finer-grained microstructure provides improved strength and fracture resistance (toughness) properties compared to a coarser-grained microstructure. For example the yield strength varies as the inverse square root of the average grain diameter. *Grain refinement* is the one process which improves both the yield strength and toughness simultaneously; other strengthening mechanisms, such as cold-working, tend to degrade toughness. Hence many proprietary line pipe steel specifications contain requirements on ferrite grain size to minimize the risk of brittle fracture.

The influence of grain size on fatigue resistance is profound, but not simple. Fine grained materials show greater resistance to fatigue crack initiation, but less resistance to fatigue crack propagation. Fine grained material often exhibits superior formability, particularly in cold bending. Even in high temperature service, fine grained materials often exhibit superior thermal fatigue resistance. Thus, for many elevated temperature service applications an intermediate grain size (e.g. ASTM grain size no. 3 to 8) will provide the best combination of formability, creep and thermal fatigue resistance.

Another good example of the importance of grain size concerns the heat treating of steel. During austenitization above the A_3 or A_{cm} temperatures, the previous microstructure of the steel disappears, pre-existing carbides gradually dissolve, and a new microstructure consisting of equiaxed austenite grains replaces the old one. At higher austenitizing temperatures and/or longer austenitizing times the austenite becomes more homogeneous and opportunity is provided in high carbon steels for more of the pre-existing carbides to dissolve. However, in addition to these effects, growth of the newly formed austenite grains also takes place. For hypoeutectoid steels which have been deoxidized with silicon the austenite grain size increases continuously and progressively as the austenitizing temperature is raised above the A_3 temperature, a behavior considered to be normal. *Coarser grained austenite* tends to transform to coarser grained products after cooling, with correspondingly poorer strength and toughness. For this reason austenitizing temperatures are limited to the region within 60°C (140°F) of the A_3 temperature, and austenitizing times are limited as well. Some steels are given small additions of such alloying elements as Al, Nb, V, Ti or Zr during steelmaking to inhibit *austenite grain growth*. This is accomplished by the formation of alloy carbides or carbonitrides which are stable at austenitizing temperatures and impede with the migration of austenite grain boundaries thus inhibiting grain coarsening. The result is *fine-grained austenite*, even after heating at normal austenitizing temperatures; the product is then called *fine-grained or grain refined steel*.

Grain coarsening of austenite is reversible in normal silicon-deoxidized steels by re-austenitization. During re-austenitization several new grains of austenite can be nucleated within the volume occupied by a single pre-existing austenite grain, and the size to which the new austenite grains grow is dependent on the re-austenitizing temperature and time. Thus the new austenite grain size will generally be smaller than the former grain size if the re-austenitizing temperature is lower then the previous one. However an acceptably small austenite grain size may not be recovered in a single re-austenitization heat treatment if the initial austenite grains are very coarse. In such a case several re-austenitizing heat treatments may be required to obtain a uniform and small final grain size.

The austenite grain size of normalized or annealed medium carbon steels can be readily observed because proeutectoid ferrite precipitates along the austenite grain boundaries during slow cooling. Thus bands of ferrite outline the prior austenite grain boundaries. However, it is not so easy to recognize the locations and paths of the prior austenite grain boundaries in low carbon steels when a large volume fraction of ferrite is present. Similarly, for quenched and tempered (martensitic) steels special etching techniques are required to reveal the prior austenite grain size. A picric acid etchant is suggested for revealing prior austenite grains in steels which have been fully hardened to martensite (see ASTM Standard E112, Appendix 3 for more details).

Grain size is not, however, the only microstructural parameter of importance. The shapes of grains are determined by thermo-mechanical history of the material just as grain sizes are. For example, the annealing of a cold-worked metal or alloy above about half of its melting temperature (in Kelvins) causes the material to undergo a series of processes (*recovery, recrystallization, grain growth*) which result in the creation of a new set of equiaxed grains, whose size depends on the previous cold-work and the annealing time and temperature. When annealed material is cold-worked (i.e. deformed plastically below the recrystallization temperature), its equiaxed grains are deformed in shape in a manner which follows the geometry of the bulk deformation. Thus in rolling, the equiaxed grains are deformed into pancakes which are flattened in the rolling plane and elongated in the rolling direction, while in wire drawing or rod rolling the grains are elongated in the rolling/drawing direction and proportionally compressed in the transverse directions. Furthermore, the non-metallic inclusions in the material may also be elongated by the working, provided that they are capable of plastic deformation at the working temperature. For example, manganese sulfide inclusions in steel are plastic at hot-rolling temperatures and hence are elongated by the deformation, whereas additions of rare earth elements such as cerium (mischmetal) render the inclusions non-deformable at these temperatures so that the sulfide inclusions remain spheroidal even during hot-rolling.

Chapter

2

WROUGHT CARBON & ALLOY STEEL METALLURGY

Carbon steels, the most important group of engineering materials in use today, are defined as steels for which there is no minimum specified content of the elements which are normally considered to be alloying elements, including Cr, Co, Mo, Nb (Cb), Ni, Ti, V, W and Zr. Carbon steels must also contain less than 1.65%Mn, less than 0.60%Si and less than 0.60%Cu. Alloy steels contain manganese, silicon and copper in excess of these limits and furthermore have specified ranges or minimums for one or more other alloying elements. *Low alloy steels* are those alloy steels with total alloying element contents, including carbon, of less than about 5%. Low alloy constructional steels are dealt with in this chapter, and the higher alloy stainless steels and tool steels in later chapters. Another type of steel, the *High Strength Low Alloy* (HSLA) steels, can be considered to be intermediate between carbon steels and alloy steels. The HSLA steels are designed to have higher mechanical properties than carbon steels but without the need for the quench and temper heat treatments which are associated with alloy steels.

Carbon and alloy steels are classified in a wide variety of ways. Most common is designation by chemical composition, as in the AISI/SAE/SAE system (specified by SAE J402), and the UNS system (specified by ASTM E 527 and SAE J1086). Alternatively, steels can be classified on the basis of mechanical properties such as strength, as in ASTM standards, or on the basis of quality descriptors such as forging quality or structural quality. Other bases for classification include the manufacturing method (e.g. basic oxygen furnace or electric furnace), the finishing method (hot rolling, cold rolling, etc.), the product form (bar, sheet, etc.), the deoxidation practice (rimmed, capped, semi-killed, killed), the microstructure (ferritic, pearlitic, martensitic), the heat treatment (annealed, normalized, quenched and tempered). More details describing the various North American metal designation systems are given in Chapter 7.

Carbon Steels

In this discussion, carbon steels will be considered in four categories, namely low carbon steels (sheet, and heat treatable steels of low carbon content 0.10 to 0.30%C), medium carbon steels (0.30 to 0.60%C) and high carbon steels (more than 0.60%C). These categories correspond approximately to divisions into non-heat treatable (the low carbon steels) and heat treatable (the higher carbon steels) categories, referring to their abilities to respond to quench and temper heat treatments. Carbon steels represent well over three quarters of steel production.

Carbon steels are further divided in groups by the amount of deoxidation they receive during the last stage of the steelmaking process, since liquid steel in the steel making furnace typically contains 400 to 800 ppm of oxygen. Deoxidation is performed by adding into the ladle measured amounts of ferromanganese, ferrosilicon and/or aluminum. The four deoxidation groups of carbon steel are: rimmed steel, capped steel, semi-killed steel, and fully-killed steel.

Rimmed steel usually is tapped from the furnace without adding deoxidizers to the steel in the furnace and only small amounts are added to the liquid steel in the ladle. This is done to have sufficient oxygen in the molten steel to give the desired gas evolution by reacting in the ingot with the carbon present in the steel. When the molten metal in the ingot begins to solidify, there is a lively effervescence of carbon monoxide, resulting in an outer ingot skin of relatively clean metal, i.e. low in carbon and other solutes. Rimmed steels are best suited for the manufacture of steel sheets, e.g. car body panels, sheet for home refrigerators, stoves and washing machines.

Capped steel is produced much like rimmed steel, except that the rimming action is stopped at the end of a minute or more by sealing the ingot mold with a cap. Capped steels made for sheet, strip, wire and bars generally have a carbon content greater than 0.15%.

Semi-killed steel is dexoidized less than fully-killed steel, with sufficient oxygen remaining free to react with carbon to form carbon monoxide to counterbalance solidification shrinkage. Semi-killed steel typically contains 0.15 to 0.30%C and is used widely in structural shapes.

Fully-killed steel, sometimes referred as killed steel, is 'fully' deoxidized to such an extent that there is no carbon monoxide or any other gas evolution during ingot solidification. This type of steel is generally used when a homogeneous microstruture is required in the final product, e.g. pressure vessel steels, pipeline steels, and forgings.

Low Carbon Steel

Non-heat treatable low carbon steels, sometimes called *mild steel*, are most strongly represented by the flat-rolled products, in particular the low carbon sheet steels which are by far the highest volume of all steel products. These steels are notable for their formability, and for this reason contain less than 0.10%C and less than 0.50%Mn. The sheet is produced in both hot-rolled and cold-rolled conditions and also in galvanized form, as tin plate and as base material for porcelain enamelling. The microstructures of these sheet steels consist predominantly of ferrite, with small amounts of spheroidized carbide. Both rimmed steel and aluminum-killed steel can be used for low carbon sheet, with the latter having become the norm. Low carbon sheet steel is also available as dual-phase steel which has a desirable combination of strength and formability due to processing which provides a microstructure consisting of islands of martensite in a ferrite matrix. Typical applications are in consumer products, notably automotive body panels and appliances.

Rolled structural steel plates and sections typically have higher carbon contents, up to 0.3% with manganese up to 1.5%. These are used for such applications as forgings, seamless tubing and boiler plate, usually in the as-rolled or normalized conditions.

Heat treatable (hardenable) low carbon steel products have carbon contents in the range 0.10-0.30%C. These have increased strength and reduced cold formability and although they can be directly quenched and tempered, they are more generally used in the carburized condition, with higher carbon contents used for thicker sections where increased hardenability is necessary.

Medium Carbon Steel

Medium carbon steels containing 0.30 to 0.60%C and 0.60 to 1.65%Mn are normally used in the quenched and tempered condition. Oil quenching can be used if section size is not too great. These steels, normally produced in the killed condition, are versatile since the balance between strength and ductility can be controlled by adjusting the tempering time and temperature. For example tempering a quenched AISI/SAE 1040 steel at 200°C (400°F) results in a tensile stress of about 830 MPa (120,00 psi) and a ductility of 18% elongation, whereas raising the tempering temperature to 650°C (1200°F) lowers the tensile strength to 650 MPa (94,000 psi) while raising the ductility to 28% elongation. Quenched and tempered medium carbon steels are used for such applications as automotive engine, transmission and suspension components, for example axles, gears and crankshafts. Railway applications include rails, railway wheels and axles.

High Carbon Steel

High carbon steel containing 0.6% to 1.0%C finds applications as springs and as high strength wires. These steels have lower ductility than the medium carbon steels as well as restricted formability and weldability. High carbon steels are normally processed by quenching and tempering, with oil quenching common except in heavy sections and for cutting edges.

High Strength Low Alloy (HSLA) Steels

HSLA steels offer improved mechanical properties and in some cases improved resistance to atmospheric corrosion in the as-rolled or normalized condition without the necessity of going to a quenched and tempered product. They are not characterized as alloy steels, rather they can be considered to be carbon steels with enhanced properties resulting from a combination of small alloying element additions and, usually, special processing methods. The HSLA steels are generally produced to mechanical property specifications (e.g. ASTM A 242), with quality descriptors (e.g. structural quality or pressure vessel quality) and less emphasis on chemical composition. They are capable of highly attractive combinations of strength and toughness at a reasonable cost. Many are described in SAE specification J410.

There are many categories of HSLA steels. One of the common ones is the microalloyed steels, so named because they contain alloying elements such as Nb (Cb), V, Ti or Mo in amounts rarely exceeding 0.1%. Manganese levels are generally high, in the vicinity of 1.5%. Another type is the acicular ferrite HSLA steels which contain less than 0.1% carbon, with additions of manganese, along with such elements as molybdenum and boron. This material finds wide application in linepipe for low temperature service.

These HSLA steels obtain their high strengths through a combination of mechanisms including extremely fine grain size, with precipitation hardening by carbide, nitride and carbonitride particles. To achieve the desired microstructure and properties it is necessary to carefully control the processing of these steels with emphasis on the temperature and deformation during the final stages of rolling, and the cooling conditions after finish rolling.

There are also dual-phase HSLA steels whose microstructures consist of small, uniformly distributed islands of high-carbon martensite in a ferrite matrix. Martensite typically accounts for about 20% of the volume. These materials have the excellent formability of low strength steel but can yield a high strength in the finished component, because they are greatly strengthened by the plastic deformation during forming.

Alloy Steels

Alloy steels are more expensive and can be more difficult to weld than carbon steels but have many attractive characteristics. Firstly, the alloying elements increase the hardenability so that thicker sections can be through hardened. Furthermore, the increased hardenability means that lower quench rates can be utilized, (e.g. oil quenching rather than water quenching) with consequent lower distortion and less susceptibility to quench cracking. The hardenability provided by the alloying elements also means that a lower carbon content can be used, giving a lower hardness martensite after quenching, further reducing the risk of quench cracking. In addition, the effectiveness of the alloying elements in retarding tempering permits the use of higher tempering temperatures with a consequent improvement in toughness. In some cases the alloying elements also provide resistance to environmental degradation under particular service conditions.

As mentioned in Chapter 1, it is important to distinguish the difference between hardness and hardenability. The principal purpose of adding alloying elements to make alloy steel is to increase hardenability. While carbon by itself controls the maximum attainable hardness for any standard steel (see discussion of martensite in Chapter 1), the carbon content has only a minor effect on hardenability.

Low carbon alloy steels (0.10 to 0.25%C) are used primarily as carburizing steels, the alloying elements giving improved mechanical properties in the core in comparison with corresponding carbon steels, as well as the other advantages discussed in the previous paragraph. Examples include alloy steels AISI/SAE 4023 and 5015 (UNS G40230 and G50150), which are used for such applications as shafts and other automotive components.

Low alloy higher carbon steels are used in (non-carburized) applications involving small sections and severe service conditions, such as high strength bolts and small machinery axles and shafts. Examples include manganese alloy steels such as AISI/SAE 1345 (UNS G13450), the molybdenum alloy steels AISI/SAE 4037 and 4047 (UNS G40370 and G40470) and the low nickel-chromium-molybdenum alloy steels AISI/SAE

8630-8650 (UNS G86300-G86500).   The higher strengths of these materials are also used to advantage in reducing the size and weight of components.

Higher levels of alloying element are used in carburizing steels where superior properties of case or core are necessary. The nickel-molybdenum alloy steel AISI/SAE 4620 (UNS G46200), the chromium steel AISI/SAE 5120 (UNS G51200) and the low nickel-chromium-molybdenum alloy steel AISI/SAE 8620 (UNS G86200) find applications as small hand tools, automotive gears and bearings for relatively severe service. The higher alloy nickel-molybdenum steel AISI/SAE 4815 (UNS G48150) and the nickel-chromium-molybdenum steels AISI/SAE 9310 and 94B17 (UNS G93100 and G94171) are used in severe service applications such as truck transmissions and differentials and in rock bit cutters. Non-carburizing high alloy constructional steels such as AISI/SAE 4340 and 86B45 (UNS G43400 and G86451) find applications in heavy aircraft and truck components where service conditions are severe and/or where quench distortion must be minimized.

Specialized alloy steels include the AISI/SAE 52100 (UNS G52986) ball bearing steel and the AISI/SAE 5155 and 5160 (UNS G51550 and 51600) spring steels.

There are also low carbon quenched and tempered constructional alloy steels which are not included in the AISI/SAE designation system. These have good toughness and weldability as a result of their low carbon levels. Included here are the T1 steels (UNS K11576, K11630, K11646) which find uses in pressure vessels, in mining equipment and as structural members in buildings.

CAST STEEL METALLURGY

Steel castings are produced by allowing molten steel to solidify in molds which are appropriately formed so that the solidified steel has a desired shape. Molds suitable for steel castings can be made from metal, ceramic, graphite, or any of a wide variety of types of sand, the choice of mold material being determined by the size, intricacy, surface finish and dimensional accuracy of the casting as well as cost. Castings are made in approximately the same steel compositions that are available as wrought products, and the properties can be expected to be similar, although the mechanical properties of castings are generally less directional. Castings are however susceptible to internal defects and surface imperfections, which can have a potentially serious effect on service performance.

In general, castings are made using fully-killed steel, most often aluminum-killed. Steel castings normally have sulfur and phosphorus limits of 0.06% and 0.05% respectively, slightly higher than those of wrought steels, and they also contain 0.30 to 0.65% silicon and 0.50 to 1.0% manganese. Steel castings are heat treated in much the same manner as are wrought products.

Castings are available in low, medium and high carbon steel as well as in low alloy steel. Low carbon cast steels typically contain between 0.16 and 0.19%C, and are typically either annealed or normalized after casting to refine the structure and relieve residual stresses. Some are quenched and tempered, and some are carburized for wear resistance. Free machining grades contain 0.08 to 0.30% sulfur. Applications include automotive and railway castings as well as furnace components and castings for electrical and magnetic equipment.

Medium carbon grades (0.20 to 0.50%C) are the most commonly produced cast steels. These are heat treated, typically by normalizing and tempering, to produce the desired mechanical properties, but this heat treatment also serves to relieve internal stresses and to refine the microstructure. Alternatively, quench and temper treatments can be used for maximum mechanical properties.

High carbon cast steels are most often heat treated by full annealing, but can be normalized and tempered or oil quenched and tempered.

Low alloy cast steels are utilized when higher strength requirements must be met, but there are other reasons why alloy steels are selected, including improved hardenability, wear resistance, impact resistance, machinability, strength at high or low temperatures, and resistance to oxidation or corrosion. Low alloy cast steels find applications in machine tools, steam turbines, valves and fittings, and in the transportation, excavating and chemical process industries.

Of the low alloy steels, manganese-molybdenum steels (AISI/SAE 80xx and 84xx), nickel steels (AISI/SAE 23xx) and manganese-nickel-chromium-molybdenum (AISI/SAE 95xx) steels are the most common. Less frequently specified are cast nickel-chromium (AISI/SAE 31xx and 33xx) steels, chromium (AISI/SAE 51xx), molybdenum (AISI/SAE 40xx), chromium-vanadium (AISI/SAE 61xx), nickel-molybdenum (AISI/SAE 46xx), and silicon (AISI/SAE 92xx) steels.

Despite the ability to classify cast steels on the basis of AISI/SAE chemical composition designations, most cast steels are produced to ASTM or SAE specifications which are based on mechanical properties although some have restrictions on composition to ensure mechanical properties and weldability. The SAE J435c specification includes three grades with specified hardenability requirements. For some industries, including the aerospace industry (SAE/AMS) and the railroad industry, manufacturers use their own, their industry association or military (MIL) specifications for cast steels.

Commonly selected grades include a medium carbon steel following ASTM A 27 grade 65-35 (or SAE J435 grade 0030) and a higher strength steel, which may be alloyed and heat treated, to the ASTM specification A 148 grade 105-85 (or SAE J435 grade 0105). Here the grade designations refer to the minimum tensile and yield strengths in ksi, thus grade 105-85 has a minimum tensile strength of 105 ksi and a minimum yield strength of 85 ksi. The ASTM A 27 steels have a chemical composition specification, and corresponding UNS numbers in the J series (grade 65-35 is UNS J03001). On the other hand there are no compositions in the specification for ASTM A 148 steels, other than maximums for sulfur and phosphorus, and so either the manufacturer or the purchaser can select the composition (in the latter case subject to agreement by the manufacturer).

Steel castings are produced in sizes varying from a few ounces to hundreds of tons. While only about two percent of total steel production is in the form of castings, many objects can be made much more readily by casting than by other processes such as mechanical working. This is true, for example, of turbine shells and diaphragms, valve bodies, exhaust manifolds and pump casings.

Chapter

4

CAST IRON METALLURGY

Cast irons are ferrous alloys which contain carbon contents in the 2-5% range, well above the normal carbon contents of steels. The other critical alloying element in cast irons is silicon, which is present at concentrations between 1 and 3%. Further alloying elements can be added as required to control specific properties such as resistance to abrasion, wear and corrosion. Like steels, perhaps to an even greater extent, the microstructures and properties of cast irons are determined not only by composition but also by the specific processing conditions which include the solidification process, the solidification rate, the cooling rate in the solid state, and the subsequent heat treating schedules.

The main applications of cast irons arise from a combination of their relatively low cost and wide ranges of properties. Their relatively low melting temperatures compared to steel permit lower cost casting processes and their compositions are such that sound and intricate castings can be obtained. Six basic types of cast iron are produced, namely gray cast iron (gray iron), white cast iron (white iron), ductile (nodular) cast iron, malleable cast iron, compacted graphite cast iron and high alloy cast iron. These cannot be clearly distinguished solely on the basis of chemical composition, since the thermal processing is at least as important as the composition in determining the type of cast iron produced. This is illustrated in Fig. 4.1, which shows the ranges of silicon and iron content for the first four types.

One feature of cast irons which is not encountered to a significant extent in steels is the presence of graphite in the microstructure. In carbon steels the carbon is mainly in the form of cementite (iron carbide) with only a small amount in solution in the ferrite. However, in cast irons (other than white iron) some or all of the carbon is in elemental form as graphite, the balance being mainly as iron carbide. The presence of particular alloying elements in iron stabilizes graphite at the expense of iron carbide, the most important of these graphite stabilizers being silicon and carbon (other less common ones include nickel, aluminum, copper, titanium and zirconium). Thus the higher the silicon and carbon contents, the more likely it is that graphite will form during a given heat treatment. Long holding time at high temperature and slow cooling also favor graphite formation, whereas more rapid cooling and the presence of such elements as manganese, sulfur and chromium stabilize the cementite phase at the expense of graphite (as do molybdenum, tungsten and vanadium). It is clear that a combination of chemical composition and thermal treatment determines the balance between graphite and carbide in a given cast iron.

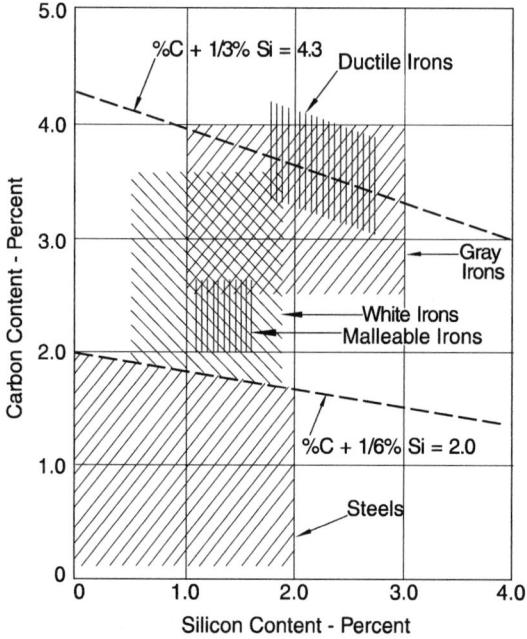

Figure 4.1 Showing the silicon and iron content of cast irons.

In Chapter 1 of this book attention was given to explaining equilibrium phase diagrams and their use in predicting the microstructures which would be present in an alloy of given composition which had been slowly cooled. Thus Fig. 1.4 might be expected to be useful for slowly cooled cast irons, however the presence of silicon modifies the situation in a number of important respects. One effect of silicon is to cause equilibrium eutectic solidification to occur over a narrow range of temperatures, unlike the case in the Fe-Fe$_3$C system where it occurs at one fixed temperature. In a similar manner it causes the eutectoid point of the Fe-Fe$_3$C system to become a narrow range of temperatures. This is shown in Fig. 4.2, which is an equilibrium phase diagram for the Fe-C system at a constant silicon content of 2%. The important region of this diagram for cast irons is the 2-5% carbon region, and here the most striking feature is the presence of the eutectic reaction (about 1160°C [2120°F], at a carbon content of 3.6%) where liquid solidifies over a narrow temperature range to a two-phase mixture of austenite (γ) and cementite. This eutectic constituent has a lamellar morphology and is known as ledeburite.

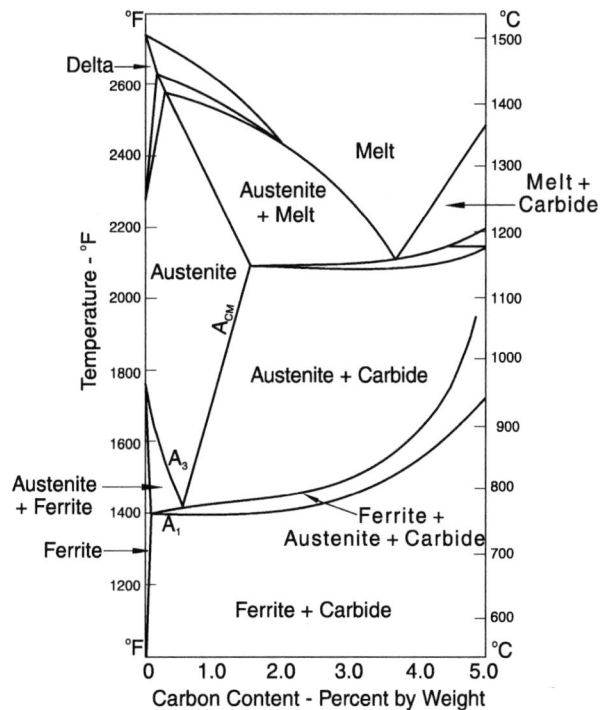

Figure 4.2 An equilibrium phase diagram for the Fe-C system
at a constant silicon content of 2%.

What is perhaps the most important effect of silicon on the iron-carbon system is that it stabilizes carbon in the form of graphite, rather than cementite, Fe_3C. Thus it is necessary to use a phase diagram which has carbon in the form of graphite rather than cementite, Fe_3C. This phase diagram looks much like Fig. 4.2 but with the cementite replaced by graphite. It is important to note that in different cast irons either or both of graphite and cementite can be present, the cementite often being in the form of pearlite, as in carbon steels. In other words, cast irons can solidify in the $Fe-Fe_3C$ system or the Fe-C system or a combination of both.

In addition to the effects discussed above, the presence of silicon progressively lowers the eutectic and eutectoid compositions by about one-third of a percent carbon for every percent of silicon in the iron so that, as shown in Fig. 4.2, a 2%Si cast iron has a eutectic composition of about 3.6%C in contrast to silicon-free iron carbon alloys which have a eutectic composition of 4.3%C (Fig. 1.4). In fact, instead of discussing the properties of cast irons in relation to carbon content, it is often more useful to relate them to a term which includes the effects of both carbon and silicon. This term, the *carbon equivalent* (CE) is given by the equation:

$$CE = \%C + 1/3\%Si$$

Other alloying elements in the cast iron also affect the eutectic composition, and with these taken into consideration, the equation becomes:

$$CE = \%C + 0.3\%Si + 0.33\%P + 0.4\%S - 0.027\%Mn$$

Since the eutectic composition always occurs at a carbon equivalent of 4.3, cast irons can be classified as hypoeutectic or hypereutectic depending on whether their carbon equivalents are respectively lower or higher than this value.

Other elements important to consider in discussing cast irons are sulfur, manganese and phosphorus. In cast irons as in steels, manganese and sulfur react together (and with oxygen) to form manganese sulfides (or oxysulfides), sometimes containing iron as well. These remain in the microstructure as non-metallic inclusions. Manganese is added intentionally, the optimum level being 1.7%S + 0.15%, since without manganese the sulfur (and oxygen) would react to form iron sulfides (or oxysulfides) which have low melting temperatures, thus they would cause hot shortness and hot cracking. Excess manganese which is not tied up in compounds acts to restrict graphitization and promote the formation of cementite (i.e. pearlite).

Phosphorus combines with iron and carbon to form a complex low-melting eutectic constituent called steadite, a mixture of ferrite, cementite and iron phosphide. Being hard and brittle, the presence of steadite (i.e. the presence of phosphorus in amounts above about 0.3%) increases the hardness and decreases the toughness of the cast iron, however it does increase the wear resistance, so it can be desirable in some applications.

With this general background in mind, the various types of cast iron will now be discussed.

Gray Iron

Gray iron is the most common of the cast irons as a result of its low cost and attractive properties. *Gray iron* obtains its name from the appearance of its fracture surface. The microstructure of gray iron consists of flakes of graphite in a metallic matrix. Since the fracture path lies to a great extent along the graphite flakes, a high fraction of the fracture surface consists of graphite, causing it to exhibit a grayish color.

The morphology of the graphite flakes is determined by composition and solidification conditions. Five types of graphite morphology are given by ASTM Specification A 247, as shown in Fig. 4.3.

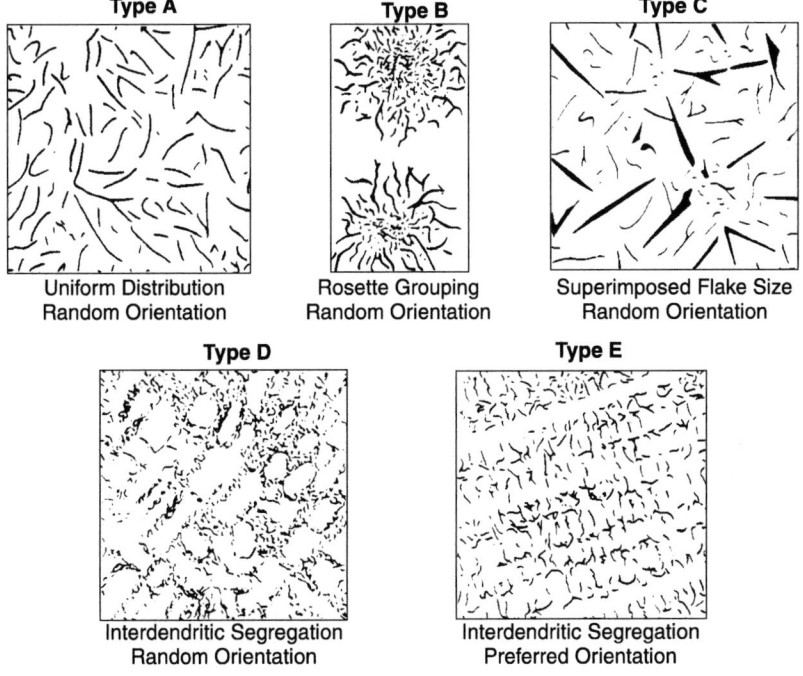

Figure 4.3 Illustrating the five types of gray iron graphite morphology.

Type A flakes with a random distribution and small size are normally preferred with regard to mechanical properties for most applications; they are found in inoculated (see below) irons cooled at moderate rates. Type B graphite is found in near-eutectic irons which have cooled more rapidly, for example in thin sections and on the surface of thicker castings. Type C graphite is found in hypereutectic irons which have been slowly cooled. They provide enhanced resistance to thermal shock, but degrade the strength and toughness. Types D and E are found in rapidly cooled irons, type D in eutectic or hypoeutectic compositions and type E in strongly hypoeutectic alloys. Type D flakes are normally associated with ferrite, which can cause soft spots in the casting, whereas type E flakes are more often associated with pearlite. The sizes of the graphite flakes are also important; Fig. 4.4 illustrates the standards.

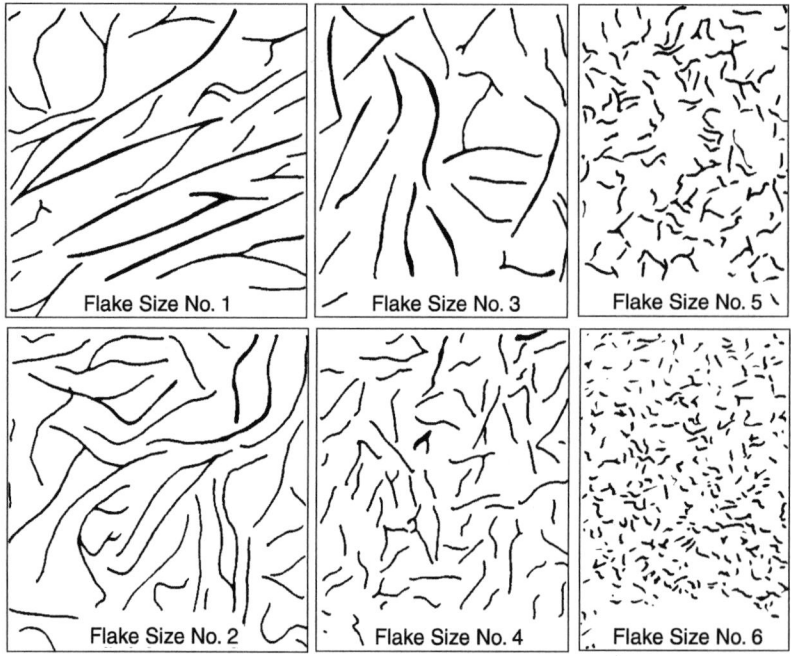

Figure 4.4 Exhibiting the standards used for describing the graphite flake size in gray cast irons. Mag. 100X

Solidification of hypereutectic gray cast iron begins with the formation of flakes of graphite (kish graphite) in the liquid iron. This is followed by solidification of the surrounding matrix to a eutectic mixture of graphite and austenite as cooling proceeds through the eutectic temperature range. The eutectic graphite is finer than the kish graphite, the two being shown in the type C illustration in Fig. 4.3. Care must be taken to avoid excessive amounts of graphite floating to the top of the casting, especially at high carbon contents and slow cooling rates.

If the composition is hypoeutectic, the first solid to form is austenite in dendritic form. This is followed by solidification of the austenite-graphite eutectic. This eutectic forms in growing spherical cells which ultimately meet and trap the last remaining liquid, which contains a high solute (e.g. phosphorus) content, between them. During growth the eutectic cells incorporate the proeutectic austenite dendrites, so that the microstructure upon completion of solidification consists of a dispersion of (eutectic) graphite flakes in an austenite matrix. During these processes the cooling rate, hence the solidification rate, is of utmost importance. Very slow cooling favors coarse randomly oriented graphite flakes (type A). More rapid solidification favors finer graphite, with types D and E morphology, or at very rapid cooling, possibly even the formation of iron carbide to form white cast iron. This latter case is referred to as chilling and is normally undesirable. However in some circumstances the chilling effect can be utilized in a casting by incorporating metal chills into the mold walls to achieve locally enhanced cooling rates; the result is localized regions of white iron on the surface of a gray iron casting at locations where, for example, extra wear resistance is desired. There are also situations where the cooling rate is such that a mixture of white and gray iron occurs within a single component. This mixture is referred to as mottled iron.

Inoculation of the liquid iron just before solidification is carried out using a ferrosilicon inoculant; this affects the microstructure by favoring type A graphite flakes and fine eutectic cells, especially in hypoeutectic gray irons.

The microstructure of the matrix can be either ferritic, a ferrite-pearlite mixture, or completely pearlitic (as shown in Fig. 4.5) depending on the cooling rate after solidification. During this cooling the austenite decomposes, as shown in the phase diagram, Fig. 4.2, first by precipitating some of its carbon as the solubility of carbon in austenite decreases along the A_{cm} line and then, at the eutectoid temperature, by undergoing complete transformation. At very slow cooling rates, there is sufficient time for carbon to diffuse from the decomposing austenite to deposit on the pre-existing eutectic and/or proeutectic graphite flakes so that the final microstructure consists only of graphite and ferrite. Higher silicon contents and higher carbon equivalents also favor this graphite formation. At faster cooling rates and lower carbon equivalents the austenite transforms too rapidly for this slow process to occur with the result that some of the carbon from the austenite ends up in the form of cementite through the formation of pearlite. Furthermore, the matrix microstructure, like the graphite flakes, is finer in irons cooled at higher cooling rates. Thus, for any given gray iron composition, the cooling rate from the eutectic temperature down to about 650°C (1200°F) controls the ratio of combined carbon (cementite) to graphitic carbon and thereby affects the strength of the iron. For a given cooling rate, lower carbon content leads to more cementite, less graphite and thus higher strength. At the same time, it must be remembered that thicker cross-sections cool more slowly than thin sections, with correspondingly coarser graphite, coarser pearlite, and a lower pearlite/ferrite ratio, thus lower strength.

The effects of section thickness on mechanical properties is much greater in cast irons than in steels, and the compositions necessary to obtain a given strength level are adjusted according to section thickness.

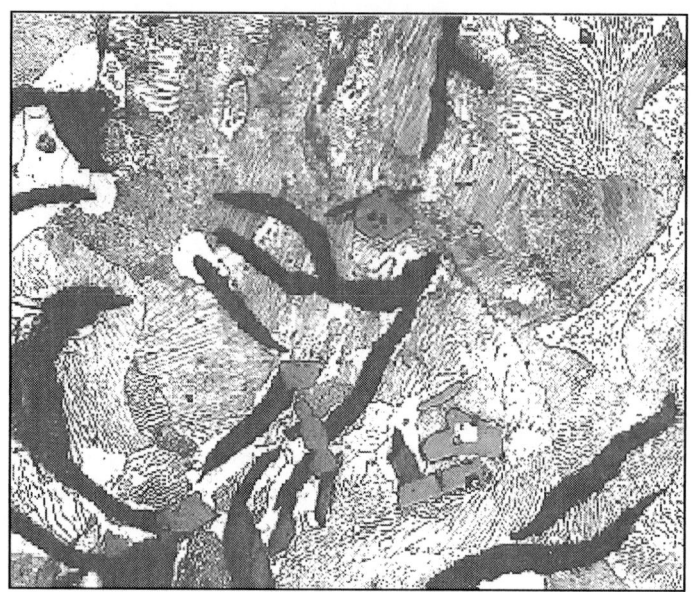

Figure 4.5 Gray iron with pearlitic matrix. Mag. 300X

Gray iron is normally used in the as-cast condition, but occasionally is stress relieved or annealed.   Gray irons are heat treatable by oil quenching and tempering although this is not normal practice.

Gray irons are classified according to ASTM Specifications A48 (gray iron castings) and A 278 (pressure-containing parts for temperatures up to 340°C or 650°F) into five classes based on minimum tensile strength. Class 20, the lowest class, has a minimum tensile strength of 20,000 psi (138 MPa), while the strength increases through classes 25, 30, and higher in increments of 5 with class 60, the highest class, having a strength of 60,000 psi (414 MPa). The corresponding UNS designations are of the form F1xxxx. Other ASTM cast iron specifications include A 159 (automotive), A 126 (valves), A 74 (soil pipe and fittings) and A 319 (non-pressure-containing parts for elevated temperature service). ASTM Specification A 436 covers austenitic gray irons and uses a different system of designations based on both tensile strength and hardness range; these have the UNS designations F4xxxx. SAE specification J431 covers grades of automotive gray cast iron.

Gray iron is a very versatile material as a result of not only its favorable economics but also its range of properties. It has good strength in compression, where the presence of the graphite flakes does not cause premature failure as is the case in tension. On average, a gray iron with a tensile strength of 20,000 psi (138 MPa) will have a compressive strength of 80,000 psi (552 MPa). Gray iron has good capability to damp vibrations and sound as a result of the presence of the graphite flakes which attenuate elastic waves in the iron structure. The graphite flakes also act as chip breakers, lubricants and oil reservoirs giving gray cast iron excellent machinability as well as the ability to resist sliding friction and galling. Its fluidity in the liquid state permits it to be cast to intricate shapes including thin sections.

Gray iron finds countless applications in many industries, including the automotive and machine tool industries and in general engineering use. The most commonly used gray irons for general use are classes 25, 30 and 35. Low strength grades, such as class 25 are used as brake drums and clutch plates as well as ingot molds, taking advantage of their superior resistance to heat checking. Low strength gray iron is also used in machine tools and other components subject to vibrations, as the lower strength material is more effective in damping vibrations. Furthermore both the machinability and the ability to be cast in thin sections are better in the lower strength classes. Properties which increase with increasing strength class include not only strength but also stiffness, wear resistance and the ability to be machined to a fine finish. Higher strength gray iron castings are used for heavy duty diesel engine blocks, heads and cylinder liners, gearboxes, pistons and flywheels.

White Iron

White irons contain no graphite; here virtually all the carbon is in the form of the hard brittle iron carbide, Fe_3C. The solidification process in these (hypoeutectic) irons begins with the formation of primary austenite, followed by the eutectic solidification reaction to form ledeburite, the lamellar austenite-cementite eutectic constituent. This process can be assisted by the use of inoculants, the carbide stabilizing elements tellurium, bismuth and sometimes vanadium.

After solidification the cooling austenite rejects carbon as the solubility of carbon falls along the A_{cm} line; this carbon is taken up by growth of the eutectic cementite. Then, in the eutectoid temperature range the remaining austenite transforms to ferrite and cementite, normally by forming a fine pearlite. A less common alternative is that the austenite can decompose forming ferrite accompanied by further growth of the existing eutectic cementite. The final microstructure is thus cementite and pearlite, as shown in Fig. 4.6, or less commonly, cementite and ferrite. The factors which cause this type of behavior as opposed to the graphite formation of gray cast iron are lower carbon and silicon contents and faster solidification rate.

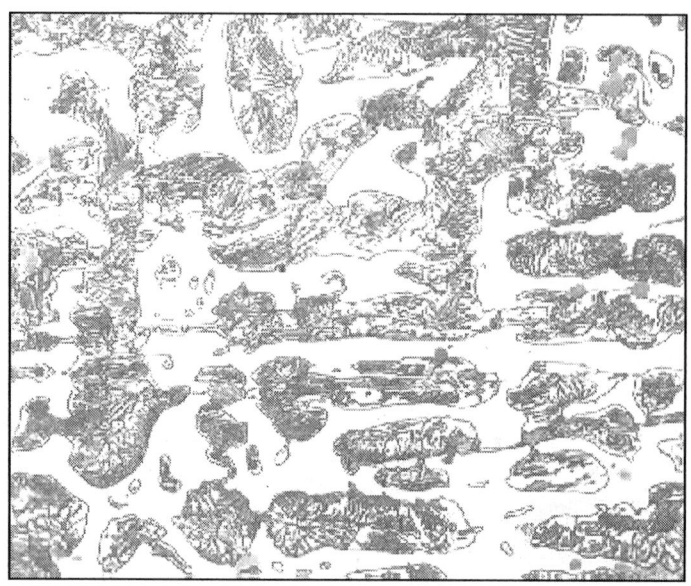

Figure 4.6 White cast iron showing pearlite (dark grains)
in a cementite (white background) matrix. Mag. 300X

Since the cementite is a hard brittle phase, white cast iron is also hard and brittle, and it fractures in a brittle manner creating a highly reflective fracture surface hence the name *white cast iron*. White iron is highly resistant to abrasive wear, however its brittleness limits its usefulness in bulk sections. It can be formed locally on the surface of gray iron castings if metal or graphite chills are incorporated into the walls of the mold. These act as heat sinks causing rapid cooling locally of the adjacent liquid iron, the result being that the casting surface adjacent to the chill will be white cast iron (sometimes referred to as chilled iron). This is done in order to control local surface properties such as wear resistance. White iron castings are also produced for conversion to malleable cast iron by subsequent heat treatment, as discussed below. Alloyed white cast irons are produced for use as abrasion- and wear-resistant materials, as discussed below in the section on alloy cast iron.

Ductile Iron

Ductile Iron is also commonly known as *nodular cast iron* and in Britain as *spherulitic graphite (SG) iron* although the accepted international term is ductile iron. This material contains graphite in the form of spheroids, rather than in the flake form which is present in gray irons. It is not the presence of graphite but its flake morphology which is responsible for the brittleness and low tensile strength of gray iron, whereas many of the beneficial effects that graphite imparts are not strongly dependent on its morphology, and are maintained in ductile iron. Thus ductile iron has a good range of yield strength, ductility, toughness and hot workability as well as excellent fluidity, castability, machinability and wear resistance. It combines the processing advantages of gray cast iron with the engineering advantages of steel. As a result ductile iron has found a steadily increasing range of applications, notably in the automotive, railroad and agricultural industries, since it was introduced in the late 1940s.

Ductile iron has composition similar to that of the higher carbon gray irons but with low levels of sulfur and phosphorus. The reason why graphite forms as flakes in gray iron is that impurities, notably sulfur and oxygen, poison the growth of graphite in the liquid iron. The graphite would normally grow as spheroids but this poisoning inhibits the growth on particular planes in the graphite crystals so that they are forced to grow as flakes. In order to produce ductile iron, not only is the impurity content kept low but also the liquid iron is treated with small amounts of magnesium or cerium just prior to casting. These elements tie up the sulfur and oxygen in the liquid iron and prevent them from poisoning the graphite growth. The result is that graphite forms with a spheroidal (nodular) morphology. The amount of nodularizing inoculant, typically a nickel-magnesium alloy, which is added to the liquid iron is carefully controlled. It is desirable that the liquid iron take up a small amount of residual magnesium in excess of that required to tie up all the sulfur. The amount of magnesium inoculant required is thus:

$$Mg_{added} = \frac{0.75S_{in} + Mg_{residual}}{\eta}$$

where S_{in} is the initial sulfur level and η is the fractional recovery of magnesium in the particular inoculation process used. The optimum amount of residual magnesium required for the formation of spheroidal graphite is 0.03-0.05%, the precise value depending on cooling rate (less being needed at faster cooling rates). If the residual magnesium content is too low, the graphite will have insufficient nodularity and a degradation of mechanical properties will result. On the other hand too much residual magnesium will promote carbide formation which is also deleterious.

In general ductile irons have finer microstructures and, because of the higher carbon contents, a higher density of graphite than gray irons. The compositions of ductile irons do not vary widely, with carbon contents between 3.6% and 3.9% and silicon contents between 2% and 2.8%. The mechanical properties are determined primarily by the matrix microstructures which in turn are determined by thermal processing treatments. Many ductile iron castings are given some form of heat treatment, either a stress relief treatment, an annealing treatment to produce a ferrite matrix, a normalize and temper treatment to produce a pearlitic or ferrite-pearlite matrix, or an oil quench and temper treatment. Martempering and austempering are also possible, as are surface hardening treatments including induction-, flame- and laser-hardening. The most common matrix microstructure, as shown in Fig. 4.7, consists of a rim of ferrite surrounding each graphite nodule (bulls-eye ferrite), with the balance of the matrix being pearlite.

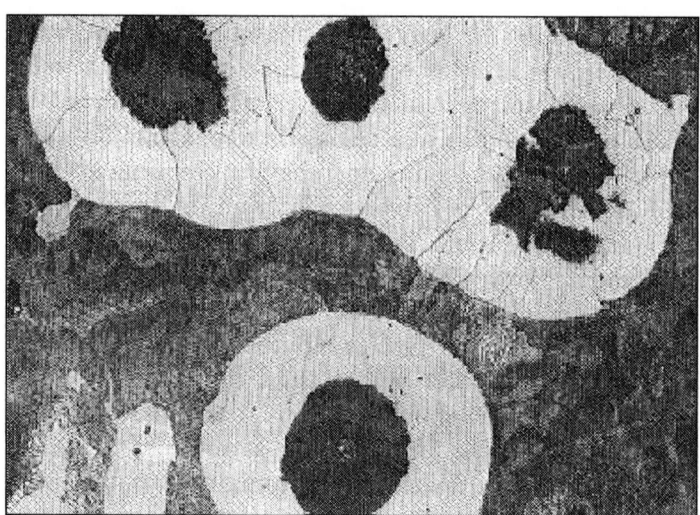

Figure 4.7 Microstructure of ductile iron showing a rim of ferrite (lighter area) surrounding each graphite (black) nodule in a pearlite matrix.
Mag. 450X

Most specifications for standard grades of ductile iron are based on properties. The specifications ASTM A 395, A 476 and A 536 for standard grades of ductile iron, and their ASME equivalents, provide grade designations which are made up of the minimum tensile strength in ksi, the minimum yield strength in ksi and the ductility in percent elongation, as for example grade 60-40-18. The SAE specifications J434 and AMS 5315C also cover these materials using a system based on microstructure and hardness. The ISO system, ISO 1083 uses a designation based on tensile strength (in MPa) and elongation percentage, for example grade 600-3. UNS designations follow the form F3xxxx.

Tensile strengths between 60 and 120 ksi are possible from different heat treatments for the standard grades of ductile iron. Low strength ferritic matrix grades such as 60-40-18 and 65-45-12 (note that these are much stronger than low strength gray irons) are used for applications such as pressure resistant valve and pump bodies, and machinery castings which must resist fatigue and shock loading. The medium strength ferrite-pearlite gray irons, either annealed 80-55-6 or normalized and tempered 100-70-03, are used for gears and other automotive and machine components, as well as general service applications; for more severe service the quenched and tempered martensitic grade 120-90-02 can be used. Austempered ductile irons, as covered by ASTM A 897 and its Metric version A 897M, have matrix microstructures consisting of mixtures of acicular (bainitic) ferrite and stabilized austenite. These provide higher strength grades ranging from 125-80-10 to 200-155-1 and even 230-185, which are used primarily for gear and wear resistance applications.

Malleable Iron

The many desirable properties possible in graphitic cast iron when the graphite exists as clusters, rather than flakes, were recognized long before the development of ductile iron in the 1940s. In fact there is evidence for its existence in ancient China more than 2000 years ago. This material is produced by forming the graphite clusters through the decomposition of iron carbide in the solid state; cast iron formed in this manner is called malleable iron. Here the initial castings are made in white iron with microstructures consisting of cementite in a pearlite matrix. These castings are then given a lengthy annealing (malleablizing) heat treatment to promote the decomposition of the cementite and the formation of graphite (temper graphite or temper carbon). The final microstructure consists of clusters of graphite in a matrix which can be ferritic, ferrite-pearlite or pearlitic as determined by the remaining dissolved carbon and the cooling rate after annealing.

The compositions of malleable irons must be low in silicon and carbon, both of which are graphite stabilizers and would inhibit the formation of a white iron casting. On the other hand if the contents of these elements are too low extra time would be required in the malleablizing anneal, thus the compositions are carefully controlled. Typical malleablizing anneals lie in the 40-90 hour range at temperatures between 870 and 950°C (1600 and 1750°F), after slow controlled heating to the malleablizing temperature. During the anneal the transformation of iron carbide to graphite occurs on nuclei formed during the heating period. Carbide-forming elements such as chromium delay the process which is completed when all the carbide has disappeared, leaving a microstructure of austenite and graphite. The iron is then cooled at a controlled rate in order to control the decomposition of the austenite, hence the matrix microstructure and the properties of the casting. Slow cooling through the austenite transformation temperature range permits all the carbon from the austenite to diffuse to the existing graphite clusters so that the final matrix microstructure is ferritic. Faster cooling rates, e.g. slow cooling from the malleablizing temperature to about 870°C (1600°F), then air cooling, give ferrite-pearlite mixtures or fully pearlitic matrix microstructures. Alternatively the malleablizing can be followed by oil quenching and tempering to give a tempered martensite matrix. The terminology used for these different microstructures is only partially accurate: ferritic malleable iron has a ferritic matrix, as shown in Fig. 4.8, but pearlitic malleable iron can have a ferrite-pearlite, fully pearlitic or martensitic matrix.

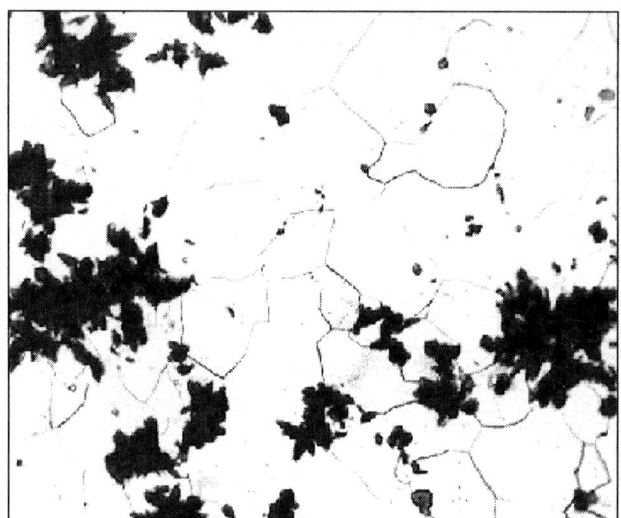

Figure 4.8 Ferritic malleable cast iron microstructure showing graphite flakes (black areas) in a ferrite (white areas) matrix. Mag. 300X

As in the case of other cast irons, malleable irons are normally specified by tensile properties. The appropriate ASTM Specifications are A 47, A 197, A 220, A 338 and A 602; and the SAE Standard is J158. ASTM specifications provide a five digit system for designations based on tensile strength and ductility with the first three digits giving the minimum tensile strength in psi divided by 100 and the final two digits giving the elongation percentage. Thus alloy 32510 has a tensile strength of 32,500 psi (225 MPa) and an elongation of 10%. This is a ferritic malleable iron, such as would be used for decorative architectural applications, hardware and automotive components. Pearlitic malleable irons range from 40010 to 90001, the highest strength irons being used for severe service applications such as diesel engine components and truck parts.

The applications of malleable iron are similar to those for ductile iron, lying mainly in the agricultural, railroad and automotive industries, including components of power trains, suspensions, frames and wheels. The choice between ductile iron and malleable iron for a given application can be difficult, as their properties are similar. Ductile iron tends to cost less, as there is no need for the lengthy heat treatments. There is a maximum thickness limitation on malleable iron castings since initially a white iron casting must be produced. However because all malleable iron castings are heat treated their homogeneity of properties, especially machinability, is excellent. The other properties which are responsible for their applications include shock resistance, ductility, toughness and castability.

Compacted Graphite Iron

Another type of cast iron, *compacted graphite iron* (CG iron) is in many respects intermediate between gray iron and ductile iron. In practice, CG iron has been inadvertently produced on occasion by undertreating ductile iron with the nodularizing inoculants, but it has now become an accepted form of cast iron. As in ductile iron, the nodularizing inoculant is normally magnesium or cerium although calcium, titanium and aluminum serve the same purpose. The inoculating treatment is carried out so as to leave a residual sulfur content in the range 0.01-0.02%. This is in contrast to ductile iron, in which the sulfur is more than compensated for and there is a residual magnesium content. The graphite morphology in CG iron consists of interconnected clusters which appear thicker and shorter than flakes. In the ASTM graphite shape classification system (Fig. 4.9) these are classified as type IV graphite. In an acceptable CG iron at least 80% of the graphite is of this form with no more than 20% spheroidal and no flake graphite present.

The morphology of the graphite is such that it does not dominate the strength properties as is the case in gray iron, thus the matrix microstructure can be controlled to control mechanical properties, as in ductile iron. There is a strong tendency in these alloys for the matrix to be ferritic, and additions of copper, tin, molybdenum or aluminum can be used to stimulate a higher pearlite/ferrite ratio. As in all cast irons the section thickness, through its affect on the cooling rate, has important effects on the microstructure. The pearlite/ferrite ratio increases with decreasing section thickness while at the same time the graphite becomes more nodular and the possibility of some chilling (carbide formation) increases.

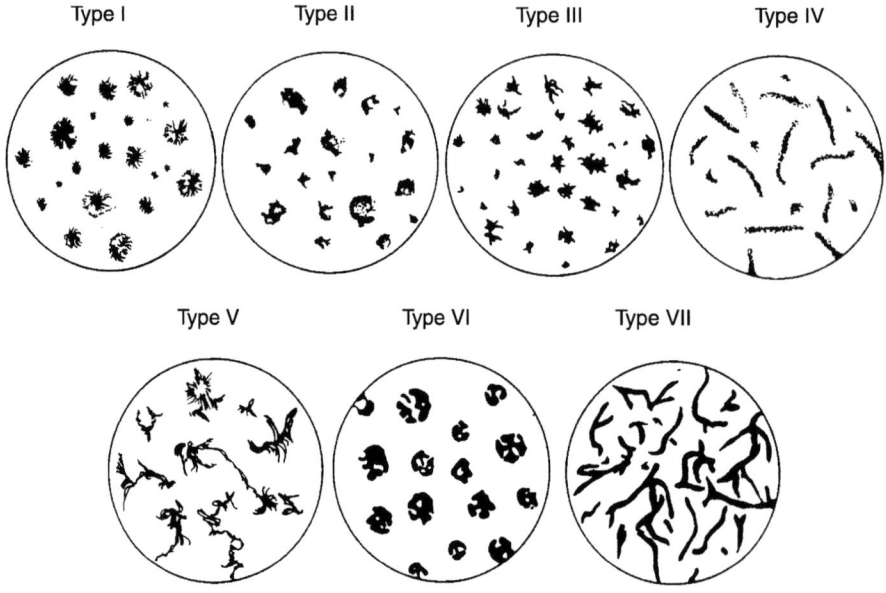

Figure 4.9 ASTM graphite shape classification system.

CG irons have somewhat poorer strength properties than ductile iron, but better castability, thermal conductivity, resistance to thermal shock, damping capability, and machinability. Compared to gray iron, CG irons have higher tensile strength at the same carbon equivalent, much higher ductility and toughness, and better high temperature stability. They can be substituted for gray iron when higher strength is desired in a complex casting, as in many automotive applications such as cylinder heads and especially manifolds which operate at elevated temperatures where the high thermal conductivity and resistance to warping and cracking give CG irons an advantage. The same properties are responsible for CG iron's extensive use in large ingot molds.

Alloy Cast Iron

Alloy cast irons have alloying elements added in amounts greater than those used for inoculation and these are intended to affect the occurrence, properties, or distribution of constituents in the microstructure. These alloying elements are used specifically to enhance resistance to abrasion, corrosion or high temperatures, and are discussed here in these three categories. High alloy contents are required for the most significant improvements in these properties, but lower levels of alloying elements can be effective.

Abrasion-resistant alloy cast iron obtains its abrasion resistance from its microstructure, which is often in the form of a hard abrasive-resistant surface layer of white cast iron above a softer core of gray cast iron. This can be achieved with cast iron of a fixed composition by casting against a chill made of metal or graphite on the inner wall of the mold, so that the solidifying cast iron cools rapidly. Castings produced in this manner are sometimes called *chilled iron castings*, as mentioned above. The microstructure of the chilled surface is then a fine mixture of dendritic pearlite with interdendritic cementite, whereas in the core the microstructure consists of graphite flakes in a ferrite-pearlite or fully pearlitic matrix. The depth of the chilled portion depends on the composition of the iron, the extent of inoculation and the pouring temperature.

Alternatively the entire cross-section can have a white iron microstructure. This requires that the carbon equivalent is low or else that the alloy content is high so that even in the center of the section, where cooling rates are low, no graphite forms. In chilled iron, the surface grains are fine and have directionality perpendicular to the surface, while bulk white iron has coarse, randomly oriented grains throughout. Both can be effective abrasion resistant materials.

Chilled iron and unalloyed white iron can be cheap and effective abrasion resistant materials but alloy cast irons give improved properties, with chromium being the most common alloying element. Chromium in the 1-4% range is used to increase hardness and abrasion resistance, while at the 12-35% level it increases corrosion and oxidation resistance as well. Nickel-chromium alloy cast irons are also of use in this regard. Nickel is a graphite stabilizer while chromium stabilizes cementite, so their net effects on phase stability can be minimized while they provide improvements in strength, and in oxidation and corrosion resistance. Up to about 2%Ni and 1%Cr a white cast iron with a refined pearlitic matrix is produced, while slightly above these levels the matrix is martensitic, giving higher strength. These are known commercially as the *Ni-Hard cast irons*. In general, martensitic white irons have considerably better abrasion resistance than pearlitic or austenitic white irons so most commercial abrasion-resistant alloy irons are of this type. Typical applications include components of grinding mills and materials handling equipment in the cement industry and the mineral extraction and processing industries.

Corrosion-Resistant Alloy Cast Irons

Corrosion-resistant alloy cast irons obtain their resistance to attack from their high alloy contents but their microstructures are also important. Depending on which of three alloying elements - silicon, chromium or nickel - dominates the composition, the microstructure of the material can be ferritic, pearlitic, martensitic or austenitic. Furthermore, the composition, inoculation and cooling rate determine whether the iron is white, gray or nodular in form, and the morphology and distribution of the carbon. From these possibilities come the four basic types of corrosion-resistant alloy cast irons: high-silicon iron, high-chromium iron, high-nickel gray iron and high-nickel ductile iron.

The high-silicon irons contain 12-18%Si, which gives them excellent resistance to organic acids and other corrosive acids. Above 14.5%Si they resist boiling sulfuric acid. These irons have poor mechanical properties including low thermal and mechanical shock resistance. Furthermore they are hard and brittle, difficult to cast and virtually unmachinable. However their corrosion resistance is such that they find applications such as drain pipe in chemical plants and laboratories, and as equipment for use in the explosives and fertilizer industries. They are also used in the paper, electroplating and dyestuffs industries for pumps, valves, nozzles and tank outlets, as well as in aggressive environments as anodes in impressed current cathodic protection systems.

The high-chromium irons are white irons containing chromium in the 15-35%Cr range. These high chromium contents allow some of the chromium to be tied up as carbides while leaving sufficient chromium remaining in solution in the ferritic matrix to give a protective surface layer and hence corrosion resistance. High-chromium irons find applications because of their resistance to oxidation in oxidizing acids (particularly nitric acid), in weak acids under oxidizing conditions, and in many organic and salt solutions and marine and industrial atmospheres. They are abrasion resistant, but have poor machinability especially at high carbon contents. Their mechanical properties are better than those of the high-silicon irons, and they can be heat treated if the chromium and carbon contents are properly balanced. They are used in the chemical and pulp and paper industries.

The high-nickel irons, containing 14-30%Ni, generally referred to as Ni-resist cast irons, are widely used. These alloys are particularly useful in alkaline environments at high temperatures, but also have resistance to weakly oxidizing acid conditions. Some of these materials, the high-nickel gray irons, are austenitic with flake graphite. These are castable, machinable and tough, although their strengths are not high. Other high-nickel corrosion resistant irons have graphite in nodular form and are produced using ductile iron practice. These high-nickel ductile irons have much improved mechanical properties compared with the high-nickel gray irons, with no loss of corrosion resistance.

Heat-Resistant Alloy Cast Irons

Heat-resistant alloy cast irons are produced as both gray and ductile irons with additions of silicon, chromium, nickel, molybdenum or aluminum, to improve their high-temperature properties. A special problem with cast irons in service above 425°C (800°F), caused either by prolonged heating or by cyclic heating and cooling, is a permanent change in volume called growth. This results from the expansion caused by graphitization of cementite and the internal oxidation of the iron. In addition, above this temperature range the mechanical properties deteriorate and surface oxidation becomes important, particularly if the oxide scale which forms is porous or non-adherent. Again, as for the corrosion-resistant irons, these heat-resistant alloys can be classified according to whether their principal alloying element is chromium, silicon or nickel.

Chromium provides a strong protective surface oxide and stable carbides in the microstructure. Low-chromium irons (up to 2%Cr) have good growth resistance, but in long service at high temperatures the pearlite can decompose to ferrite and its carbides can spheroidize. The high-chromium alloys have excellent resistance to both growth and oxidation up to nearly 1000°C (1830°F), however they have white iron microstructures and thus good strength but limited toughness and only fair machinability.

Silicon contents below 3.5% increase growth by promoting graphitization. However in the range 4.5-8%Si both oxidation and growth are reduced, the former because of the formation of a protective surface oxide. Ferritic high-silicon iron has poor toughness and thermal shock resistance at room temperature but is superior to ordinary gray iron above 260°C (500°F). A high-silicon high-nickel austenitic gray iron known as nicrosilal (5%Si-18%Ni-2-5%Cr) exhibits much higher toughness and thermal shock resistance and good scaling resistance. High-silicon heat-resistant ferritic ductile alloy cast irons are also available.

High-nickel irons containing 18-36%Ni with austenitic microstructures and up to 7%Cu and 2-4%C find applications where both heat and corrosion resistance are required. These alloys, called the Ni-resist alloys as are the corresponding corrosion-resisting alloy cast irons, resist both high temperature scaling and growth up to 815°C (1500°F) although they cannot be used above 540°C (1000°F) in sulfur-bearing atmospheres. Despite their limited strength, these alloys have superior shock resistance and toughness in comparison with the other heat-resisting irons. Ductile high-nickel heat-resistant irons with nodular graphite have improved strength and toughness.

Low-silicon chromium-molybdenum gray cast iron is used for automotive disk and drum brakes. Here the required properties are high heat capacity, good thermal conductivity, high thermal emissivity, high temperature strength, and resistance to thermal shock and to growth.

Chapter

5

TOOL STEEL METALLURGY

Tool steels are ferrous alloys which are intended for use in the working and shaping of metallic and non-metallic materials. They are heat-treatable alloys, ranging from simple carbon steels of high carbon content to complex alloy steels. Applications include drills, dies for extrusion, casting, forming, thread rolling and deep drawing, shear blades, punches, and cutting tools such as chisels, lathe tools and milling cutters. These are demanding applications, and consequently tool steels are high quality materials produced in relatively small quantities with good quality control in compliance with strict specifications. They are supplied in the annealed condition, the purchaser being responsible for forming the steel to the desired shape, then heat treating it by austenitizing, quenching and tempering to obtain the desired properties. Depending on the application, the required properties after heat treatment include strength (hardness), resistance to deformation, to wear and to high temperatures, dimensional stability and edge retention.

Although tool steels were traditionally bar materials ("long products"), much flat bar is now produced as plate ("flat product") and is sawed to size by the distributor.

Tool steels are also employed for structural applications in severe service conditions and in machinery components including high temperature springs, fasteners, valves and bearings.

Austenitization is carried out at a temperature between 770 and 1300°C (1420-2370°F) depending on the alloy. In some cases this involves complete dissolution of pre-existing carbides so that during the subsequent tempering an entirely new dispersion of carbide particles can form. Other cases require only partial carbide dissolution so that the ultimate heat treated microstructure includes both a coarse dispersion of primary carbides and a fine dispersion of secondary carbides, precipitated during tempering, to provide wear resistance and strength respectively. In some cases, such as the high speed steels, carbides can account for up to 40% of the microstructure in the annealed condition and undissolved carbides can make up 10% or more of the microstructure after austenitization.

It is important that surface carburization and decarburization be avoided during heat treatment as the surface properties are critically important to tool performance in service. Slow or multi-step heating (preheating) is often required to avoid distortion during heat treatment. With the exception of the type W tool steels, the hardenability is sufficiently high that drastic water quenching can be avoided, and step cooling from austenitization, for example into salt or oil and then in air, is frequently employed.

The high contents of carbon and alloying elements cause significant amounts of retained austenite to be present in the microstructure after quenching. Tempering is carried out immediately after quenching, in order to avoid delayed quench cracking.

During tempering, the slow formation of alloy carbides, in particular Mo_2C, W_2C and VC, is responsible for secondary hardening, which in some cases can bring the tempered hardness to a higher value than that of the as-quenched condition, along with better toughness and resistance to wear. Higher austenitizing temperatures can give higher hardness values because of higher amounts of dissolved primary carbides leading to higher amounts of carbide formed in tempering.

The particular carbides which form during solidification, annealing and tempering of the alloy tool steels depend on the alloy and thermal treatment, but are in general alloy carbides including:

- MC, where M is vanadium with tungsten and molybdenum,
- M_2C where M is a tungsten-molybdenum carbide containing some vanadium
- M_3C where M is mainly iron with some chromium
- $M_{23}C_6$ and M_6C in both of which M includes iron, chromium, molybdenum and tungsten.

At peak hardness during tempering the principal carbide is MC. Problems can arise from non-uniform distribution of carbides which results from carbon segregation in the original cast tool steel ingot, but these can be minimized by using powder metal manufacturing techniques as discussed below.

Multiple tempering treatments are necessary in many tool steels, as retained austenite which is present after quenching and remains during tempering can transform to untempered martensite during cooling from the tempering treatment. This brittle material must be tempered by a second tempering treatment, but any remaining retained austenite can transform to martensite during cooling from this second tempering, thus a third or even a fourth tempering may be required to completely eliminate untempered martensite.

Most tool steels are wrought products but cast and powder metallurgy products are also available. The powder products have several advantages and hence this has become a major process for the production of tool steels, in particular advanced high-speed tool steels but also improved cold-work and hot-work tool steels. Rapid powder solidification permits the production of alloys which cannot be produced by conventional ingot processing, thus for example alloys with 9-10%V and up to 15 vol% carbide become possible. Subsequent powder consolidation produces a product with little or no macrosegregation as well as a high volume fraction of carbide particles with a more uniform size distribution. The result is superior properties, including improved machinability and edge toughness in high-speed steels, and better toughness, ductility and wear resistance in hot-work and cold-work tool steels.

Tool Steel Classifications

The most common system for designating tool steels in North America is the AISI system (Table 5.1). This system classifies tool steels based on a combination of quenching method, composition and application, e.g. AISI type A2. A second system is the UNS designation system, which uses the prefix letter T followed by five digits to identify tool steels, e.g. UNS T30102, which corresponds to AISI type A2. There are also ASTM specifications for tool steels, including ASTM A 600 for tungsten and molybdenum high speed steels, A 681 for hot-work, cold-work, shock-resisting, special purpose and mold steels, and A 686 for water hardening tool steel. Many tool steels are identified by trade name, since products with the same AISI designation can vary in composition and performance from one supplier to another. More details regarding tool steel designation systems can be found in Chapter 7.

Table 5.1. AISI Classification of Tool Steels With UNS Number

AISI Type	Letter Symbol	UNS No.
Water-hardening tool steels	W	T723xx
Shock-resistant tool steels	S	T419xx
Cold-work tool steels		
Oil-hardening	O	T315xx
Medium alloy, air-hardening	A	T301xx
High carbon, high chromium	D	T304xx
Hot-work tool steels	H	T208xx
Chromium type	H1-H19	
Tungsten type	H20-H39	
Molybdenum type	H40-H59	
High speed tool steels		
Tungsten type	T	T120xx
Molybdenum type	M	T113xx
Special purpose low alloy tool steels	L	T612xx
Mold tool steels	P	T516xx

Water Hardening Tool Steels

Water hardening (AISI type W) tool steels, which include AISI types W1, W2 and W5 are the least expensive of the tool steels. AISI type W1 (UNS T72301) is a simple high carbon steel containing 0.6-1.4%C, giving only enough hardenability for a surface layer of martensite, even with a drastic quench, unless very thin material is employed. Small additions of vanadium are used in AISI type W2 (UNS T72302) to inhibit austenite grain growth giving increased toughness. These materials are used for different applications depending on carbon content; for example the low carbon material is more suitable for chisels and shear blades, with the higher carbon material being more useful for glass cutters, drills and dies. AISI type W5 (UNS T72305) has 1.1%C and 0.5%Cr to give increased hardenability, so that it can be used for large punches and heavier dies and rolls.

Shock-Resistant Tool Steels

Shock-resistant (AISI type S) tool steels are designed for resistance to repeated impact loading, as in rivet sets and shear blades, so that toughness becomes more important than hardness. High carbon is incompatible with high toughness, so a relatively low carbon content of the order of 0.5% is common in these AISI type S materials. Alloying elements include tungsten, molybdenum, chromium, manganese and silicon. Common examples include AISI type S5 (UNS T41905) which contains 0.4%Mo-0.8%Mn-2.0%Si, and type S7 (UNS T41907, 1.4%Mo-3.25%Cr) which has higher hardenability.

Cold-Work - Oil Hardening Tool Steels

Cold-work oil hardening (AISI type O) tool steels are used in applications where toughness and wear resistance are of primary importance. Also important are high hardenability, resistance to quench cracking in intricate sections, and ability to maintain a sharp (cutting) edge. The applications include dies for such purposes as blanking, stamping, trimming, bending and forming as well as reamers, broaches and knives. The AISI type O tool steels are not suitable for use at elevated temperatures, e.g. for hot working or for cutting at high speed. AISI type O1 (UNS T31501, 0.9%C-1.0%Mn-0.5%W-0.5%Cr) is a widely used example, where the manganese, tungsten and chromium provide the hardenability required for slow oil quenching so that distortion and cracking can be minimized.

Cold-Work - Air-Hardening Tool Steels

Cold-work medium alloy air-hardening (AISI type A) tool steels have applications similar to the O types, but the increased hardenability which permits martensite formation at even slower (air) quench rates gives the possibility of even less dimensional change, distortion and quench cracking than occurs in type O. Thus it is possible for example to produce intricate dies for such purposes as blanking, forming and drawing. Exceptional toughness and good abrasion resistance are important for these steels, a common example of which is AISI type A2 (UNS T30102, 1.0%C, 5.0%Cr and 1.0%Mo).

Cold-Work - High-Carbon High-Chromium Tool Steels

Cold-work high-carbon high-chromium (AISI type D) tool steels are utilized primarily as die steels where advantage is taken of their high strength and excellent wear resistance. The high chromium and carbon levels yield a greater volume of carbides, and a consequently higher wear resistance. The chromium also provides resistance to staining, and to high temperature oxidation. Typical AISI type D tool steels include D2 (UNS T30402, 1.5%C-12.0%Cr-1.0%Mo-1.0%V) and D3 (UNS T30403, 2.25%C-12.0%Cr). Additions of molybdenum increase hardenability and toughness. Vanadium is added for grain refinement, but it also decreases the amount of retained austenite and increases toughness at levels below 1 percent.

Hot Work Tool Steels

The hot-work (AISI type H) tool steels are used for forming processes which are carried out at high temperatures. Specific applications include dies and die inserts for extrusion, forging and die casting, as well as mandrels, punches and piercing tools. Tools steels for these applications must have much greater high temperature deformation resistance than is possible with carbon steels. They must also resist high temperature erosion and wear, thermal and mechanical shock, heat checking and warpage.

The hot-work tool steels are subdivided into three types, namely chromium-base, tungsten-base and molybdenum-base alloys. The chromium-base steels AISI H11, H12 and H13 all have 5%Cr-1.5%Mo-0.35%C, along with varying amounts of vanadium, silicon and tungsten. These steels have high hardenability which permits thick sections to be heat treated by air cooling, keeping distortion to a minimum. The tungsten-base H type tool steels H21, H22, H24, H25 and H26 (UNS T20821, T20822, T20824, T20825, T20826) contain 9-19%W, 2-5%Cr and up to 1%V. These are more resistant to high-temperature softening than the chromium-based H type steels, but are also more brittle. The molybdenum-based H type tool steels are similar to the tungsten-based versions but are lower cost and must be heat treated more carefully. AISI type H42 (UNS T20842) contains 6%W-5%Mo-4%Cr-2%V. All of these steels have carbon contents in the range 0.25 to 0.6% for enhanced shock resistance.

High Speed Tool Steels

The high-speed tool steels are based on substantial additions of the alloying elements tungsten (AISI type T) or molybdenum (AISI type M). Applications of these materials are in cutting or machining tools which are used on hard materials cut at fast cutting rates. Service of this type involves high temperatures, into the red-heat range at the tip of the cutting tool, and the tool material must be resistant to softening under these conditions. A material such as this is said to have red hardness. Wear resistance and the ability to maintain a sharp edge are also requirements for high-speed working. Alloying elements in high speed steels include tungsten and/or molybdenum which contribute to both red hardness and the formation of carbides which provide wear resistance. Vanadium contributes to abrasion resistance and cobalt to red hardness. Chromium gives high temperature oxidation resistance.

The AISI T type steels contain 12-20%W, 4%Cr and 1-5%V with some also containing cobalt in the 5-12% range. They are characterized by high red hardness and wear resistance and are deep hardening (up to about 75mm or 3 in. section thickness for quenching in molten salt). The general purpose tool steel in this class is AISI type T1 (UNS T12001, 18%W-4%Cr-1%V-0.75%C). AISI type T15 (12%W-4%Cr-5%V-5%Co-1.5%C) has a higher carbide content as a result of its high vanadium and carbon levels, and hence is the most wear resistant of this group. It can be hardened to about 67 HRC as compared to 64 HRC for most other high-speed steels.

The performance of the AISI type M tool steels is generally similar to that of the T type steels, except that they have slightly greater toughness at the same hardness level. However they dominate high-speed tool use in North America because of their lower cost compared to the T type steels. Typically the AISI type M tool steels contain 5-10%Mo, 1-7%W as well as the 4%Cr, 1-4%V and 5-12%Co which are also present in the T types. As with the T types, increased vanadium and carbon contents provide more carbides and hence better wear resistance, while cobalt provides improved red hardness. The general purpose tool steel in this class is AISI type M2 (UNS T11302, 5%Mo-6%W-4%Cr-2%V-1%C) whose applications include lathe and planer tools, milling cutters, drills, taps, and shaper tools. Tool steels in the M40 series (UNS T1134x, 4-10%Mo, 1-10%W, 4%Cr and 1-3%V with 1-1.5%C) are used for cutting tools for machining high-toughness high-strength steels. Molybdenum-based tool steels are more susceptible to decarburization during heat treatment than the tungsten-based types, so better control over heat treatment conditions is necessary.

Many high-speed tool steels are coated in order to increase service length. Surface treatments include oxide coating, chromium plating, carburizing (for dies but not cutting tools) and liquid nitriding. Physical vapor deposition of titanium nitride has become a common treatment which markedly reduces friction between tool and workpiece, thereby extending tool life by as much as 400%.

Low-Alloy Special Purpose Tool Steels

The low-alloy special purpose (L type) tool steels are used for machine parts which require good strength and toughness, such as arbors, cams, chucks and collets. They are generally oil-hardening steels except in thick sections where water quenching may be necessary. AISI type L2 (UNS T61202) contains 1%Cr and 0.2%V with several carbon contents available between 0.5 and 1.1%. The vanadium acts here as a grain refiner. AISI type L6 (UNS T61206) contains 1%Cr and 0.5%Mo as well as 1.5%Ni for toughness.

Mold Tool Steels

The low carbon mold (P type) tool steels find applications in die casting dies and molds, particularly for the forming of plastics.   They are produced to high quality soundness, cleanliness and hardenability standards, especially important for the large die blocks used in some plastic forming applications, but their resistance to softening at elevated temperatures is low-to-medium. Some, such as AISI types P2, P4 and P6 (UNS T51602, T51604, T51606), with up to 5%Cr-4%Ni-1%Mo-0.1%C, are high quality carburizing steels, with low hardness, so that the mold impression can be made by cold hubbing, followed by the carburizing, quenching and tempering treatment. AISI types P20 (UNS T51620, 1.5%Cr-0.5%Mo-0.35%C) and P21 (T51621, 0.5%Cr-4%Ni-0.2%V-1%Al-0.20%C) are normally supplied in the heat treated condition and are then machined to shape with no further heat treatment required for many applications.   AISI type P21, which is pre-hardened by precipitation-hardening by nickel-aluminum intermetallic compounds, is especially useful for molds where surface finish is critical.

Chapter

6

STAINLESS STEEL METALLURGY

Development of stainless steels began about 1910 in England and Germany. Since that time these alloys have become an integral part of the world materials scene, playing a vital role in many industries while accounting for only a few percent of total steel production. The early stainless steels were simple alloys of iron and chromium, with chromium contributing the corrosion resistance, oxidation resistance and pleasing appearance which are responsible for the applications of these alloys. When exposed to oxidizing environments, chromium in amounts greater than about 11.5% is responsible for the formation of an invisible adherent passive film on the surface of iron which protects it from further reaction with the environment. This layer, which can be considered to be a chromium-rich oxide, is self-healing if ruptured in oxidizing environments and as long as it is present the alloy can be considered "stainless". The protective layer allows many of these alloys to be useful in high temperature gaseous oxidizing environments as well as in lower temperature and ambient liquid media. In environments which destroy this passive layer, globally or locally, the stainless steel corrodes as if it were carbon steel, and the extra cost of the material has been wasted.

Stainless steels are available in the complete range of product forms. Wrought alloys are produced as plate, sheet, strip, foil, bar, wire, expanded metal mesh, piping and tubing. Cast products are also available, with applicable foundry practices similar to those for carbon and alloy steels. It is important to note the split between long product (bar and billet) producers and flat product (plate, sheet and strip) producers. There are no full-line (long plus flat product) stainless steel producers in the U.S. and there is relatively little co-ordination between the two product lines. Those alloys which are produced across the board often show significant compositional differences. However, many alloys, especially the ferritic stainless steels, are produced only in one of these forms. For example, free-machining alloys are produced as bar and occasionally as plate product but never as sheet or strip. High chromium stabilized ferritic alloys are widely produced as sheet and strip but rarely (type 446 being the significant exception) as bar or plate due to the influence of section size on the ductile-brittle transition.

Five classes of stainless steel are generally recognized. Ferritic stainless steels contain between 11 and 30% chromium and have a low carbon content. Martensitic steels contain 11-17% chromium and carbon contents up to the 1% range. Austenitic stainless steels contain 17-25% chromium and 7-35% nickel. Duplex stainless steels contain 23-30% chromium, 2-7% nickel and minor amounts of titanium or molybdenum. In addition, there are some stainless steels which are alloyed with small quantities of other elements to permit precipitation-hardening. Detailed discussions of these five classes of stainless steel, as well as cast stainless steels, are presented below.

The most important alloying elements in stainless steel, other than chromium, are nickel, carbon, molybdenum and nitrogen. Nickel stabilizes the austenite phase down to or below room temperature, giving to the alloys the high formability and ductility which is inherent in austenitic (face-centered cubic) crystal structures. Furthermore, the presence of nickel provides increased protection from corrosion in reducing, neutral or weakly oxidizing media. Manganese can be used as a partial replacement for nickel in austenitic stainless steels, although its effects on austenite are complex as discussed below.

The carbon contents of stainless steels lie in the range from less than 0.03% to about 1%. Increased carbon levels facilitate the production of martensitic stainless steels which can be strengthened by quench and temper treatments. The carbon also provides strength at high temperatures, however higher carbon content tends to degrade the toughness, as well as having a consistently deleterious effect on corrosion resistance.

Molybdenum combines with chromium to stabilize the passive surface film in the presence of chlorides which can otherwise destroy the chromium-rich film, rendering the alloy susceptible to localized corrosion in the forms of pitting and crevice corrosion.

Nitrogen is a powerful austenite stabilizer. Furthermore it enhances the resistance of austenitic stainless steels to pitting corrosion as well as increasing the strength, but, like carbon, it seriously degrades the toughness of ferritic stainless.

The effects of manganese are complex. At high temperatures its presence increases the nitrogen solubility and hence it acts as an indirect austenite stabilizer. At intermediate to ambient temperatures, it has about half the austenite stabilizing ability of nickel, while at cryogenic temperatures its austenite stabilizing power is very low.

Rare earth metals are sometimes added to stainless steels to improve oxidation resistance. Several stainless steels contain tungsten, either for enhanced corrosion resistance or for enhanced creep and tempering resistance.

Other alloying elements added to stainless steels for specific purposes (improvements to mechanical properties or to corrosion and oxidation resistance) include titanium, niobium (columbium), copper, silicon, aluminum, sulfur and selenium.

The Iron-Chromium Phase Diagram

Since all stainless steels, by definition, are iron alloys containing chromium, it is of importance to understand the equilibrium phase diagram for the iron-chromium system, as shown in Figure 6.1. This phase diagram demonstrates that the effect of chromium is to reduce the stability of the face-centered cubic austenite (γ or gamma) phase, so that it forms a gamma loop on the phase diagram. The gamma loop is surrounded by a region within which there are two stable phases, the austenite phase and the body-centered cubic ferrite (α or alpha) phase.

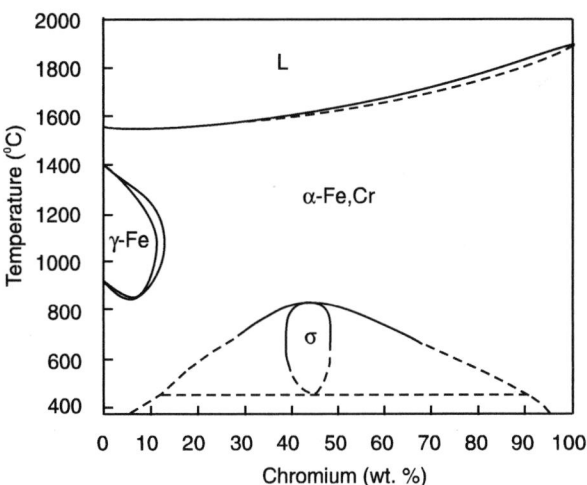

Figure 6.1 Fe-Cr phase diagram showing gamma (γ) loop.

Austenite is not stable outside these regions, i.e. above 17%Cr, and at any chromium level below 800°C (1470°F) or above 1400°C (2550°F). The existence of ferritic and martensitic stainless steels can thus be explained; ferritic stainless steels have compositions outside the gamma loop while martensitic stainless steels are developed by quenching alloys from inside the gamma loop to lower temperatures where the gamma phase becomes unstable and transforms to martensite. As in carbon and low-alloy steels, this quenching is customarily followed by tempering treatments. The addition of carbon to the simple iron-chromium alloys expands the gamma loop to higher chromium contents, and this is used to obtain martensitic structures at higher chromium levels in the higher carbon martensitic stainless steels.

Another noteworthy feature of the iron-chromium phase diagram is the existence of the intermetallic compound FeCr, called the σ (sigma) phase, which forms at temperatures below about 820°C (1508°F). This is a hard, brittle phase and its formation is responsible for embrittlement of alloys containing 15-70%Cr which are heated in the 500-800°C (930-1470°F) range for prolonged periods.

It is apparent from the shape of the phase diagram in Fig. 6.1 that the austenitic structure is not stable at room temperature in binary iron-chromium alloys. In order for stainless steels to have austenitic microstructures at ambient temperatures, it is necessary therefore for them to contain alloying elements which stabilize the austenite phase; by far the most important of these is nickel. The stability of austenite in austenitic stainless steels depends strongly on their nickel contents, with higher-nickel austenites being more stable. At room temperature, the common austenitic type 304 stainless, with 8% nickel, is stable whereas the lower nickel type 301 is not completely stable, as discussed below. The relative stabilities of the austenite, ferrite and martensite phases can be predicted for any stainless steel by considering the balance between the amounts of alloying elements which are ferrite stabilizers, i.e. chromium, molybdenum, silicon and niobium, and those which are austenite stabilizers, i.e. nickel, carbon, nitrogen, manganese and copper. This leads to the concept of the chromium equivalent and the nickel equivalent for any stainless steel, and several different equations have been developed to account for these "equivalent" values in terms of the relative stabilizing powers of the various alloying elements. For example:

$$Ni_{eq} = \%Ni + 30\%C + 30\%N + 0.5\%Mn$$

$$Cr_{eq} = \%Cr + \%Mo + 1.5\%Si + 0.5\%Nb$$

The chromium and nickel equivalents can then be related graphically to predict which phases are stable in an alloy of given composition, using a diagram such as the Schaeffler diagram, Fig. 6.2. This type of diagram was originally developed for use in predicting effects in air-cooled arc weld deposits, but it is qualitatively applicable to bulk microstructures of stainless steels.   Quantitative predictions are not possible because the phase stability depends on cooling rate.   Solidification under typical casting conditions, i.e. slower cooling than arc welds, produces a quite different diagram, as does more rapid solidification (e.g. laser or electron-beam welding).   Thus, for example, types 304 and 316 austenitic stainless steels, although usually balanced to provide about 5% ferrite in the as-welded condition, typically contain little or no (less than 0.5%) ferrite in the wrought plus annealed state.

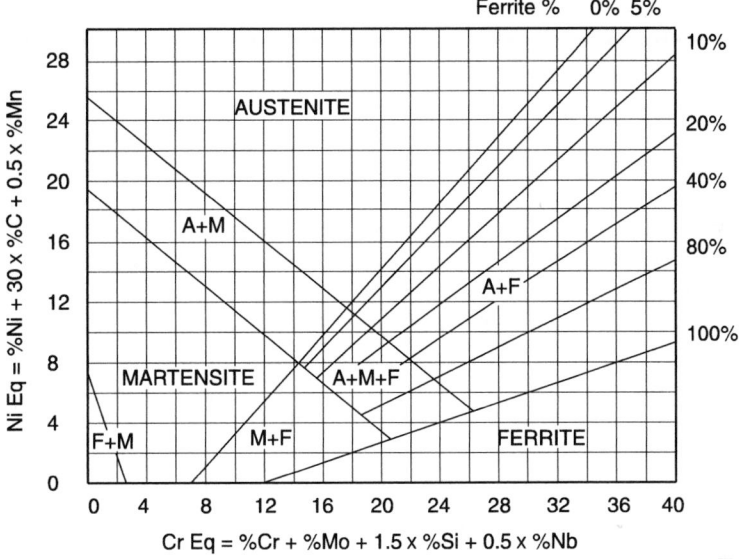

Figure 6.2 Schaeffler diagram.

Carbides are present within the microstructures of most stainless steels. In general there is a tendency for alloy carbides, notably chromium carbides, to form in preference to iron carbides, but mixed chromium-iron carbides are common.   At low chromium contents the cementite-type carbide $(Fe,Cr)_3C$ is prevalent, while at higher chromium contents, the carbides $(Fe,Cr)_7C_3$ and $(Fe,Cr)_{23}C_6$ predominate.   In the presence of other alloying elements, such as molybdenum, niobium (columbium) and vanadium, alloy carbides form which are referred to as M_7C_3 and $M_{23}C_6$ with M representing a mixture of alloying elements.   Titanium always produces MC carbide, which is usually also seen in niobium (columbium)-bearing alloys.   These MC carbides are often carbon deficient, and their compositions may approach M_2C.   Vanadium, if present in sufficient concentration, usually forms V_4C_3 carbide, but few if any stainless steels contain enough vanadium for this to occur.

Ferritic Stainless Steels

Ferritic stainless steels are basically iron-chromium alloys with compositions to the right of the gamma loop, i.e. greater than about 17% chromium. Commercial ferritic stainlesses exist with chromium contents as high as 30% although a few have chromium contents as low as 11%. The carbon content is kept low, typically less than 0.2% to optimize toughness, ductility and formability and to minimize susceptibility to intergranular sensitization (as discussed below), but the low carbon content also prevents these alloys from having high strengths. Most ferritic stainlesses are not heat-treatable by quenching and tempering (the exceptions being grades with around 17%Cr which can be heat treated by water quenching from a temperature in the 900-1200°C [(1650-2190°F)] range, giving islands of martensite in a ferrite matrix). The microstructures of most ferritic stainless steels consist of grains of the ferrite phase (body-centered cubic iron containing chromium in solid solution) with carbide precipitates. These alloys are ferromagnetic, and can have good ductility and formability, but relatively poor strength at high temperatures. Low temperature toughness can be poor. Higher chromium levels often indirectly degrade toughness and ductility as a result of the interaction of chromium with carbon. This is not always the case however, as demonstrated by the high toughness and ductility of properly processed 26%Cr-1%Mo alloys (e.g. UNS S44627). Some grades of ferritic stainless have extra low levels of the interstitial elements, carbon and nitrogen, for improved weldability. Machinability can be excellent in the free machining grades such as 430F, but not as good in the high chromium alloys such as type 446.

Ferritic stainless steels have good resistance to general liquid corrosion and high temperature oxidation, and to pitting and stress-corrosion cracking. Corrosion resistance of ferritic stainless increases with increasing chromium content, and is best for alloys in the annealed condition. Ferritic stainless steels are embrittled and rendered increasingly susceptible to corrosion if they are subjected to elevated temperatures. This problem, which can seriously affect alloys which have been welded, heat treated or cast, is due to chromium carbide and nitride precipitation in the ferrite grain boundaries. The degradation of corrosion resistance called *sensitization*, is due to the creation of a chromium-depleted zone adjacent to grain boundaries as chromium is removed into the growing intergranular chromium-rich carbides. When the chromium content of these depleted regions falls below a critical value, the passive protective surface film cannot form, and the alloy becomes susceptible to intergranular corrosion. This critical value is approximately 11%Cr but varies with the corrosive medium. For example the two standard test media for sensitization according to ASTM A 262 give different critical chromium levels.   Other alloying elements, including molybdenum, may also change the critical chromium level at which sensitization is detected.

Sensitization is circumvented in two ways: by the use of stainless steels with extra low levels of carbon and nitrogen, or by the use of grades which contain additions of titanium or niobium (columbium) at approximately 5-6 times the interstitial content.   These titanium or niobium additions stabilize the stainless steels against sensitization by promoting the formation of titanium or niobium carbides or nitrides, so that chromium is not depleted near the grain boundaries. Sensitization is a problem with most stainless steels, but occurs more rapidly in ferritic than in austenitic stainlesses because of the higher diffusion rate of carbon in ferrite.

Ferritic stainless steels are also susceptible to both sigma phase embrittlement and to 475°C embrittlement (caused by the precipitation of the alpha prime [α'] phase) if the steels spend prolonged periods at temperatures in the range 500-1000°C (930-1830°F) and 400-540°C (750-1000°F) respectively.

Ferritic stainless steels are used in general construction applications where special resistance to heat or corrosion is required.   They are cheaper than the nickel-bearing austenitic stainlesses, and have comparable corrosion resistance, but in general have somewhat less ductility and toughness. The most common ferritic stainless steels are type 430 (UNS S43000), (with its relatives, 430F and 430F-Se [UNS S43020 and UNS S43023]) which have 17%Cr, type 434 (UNS S43400, 17%Cr-1%Mo), type 439 (UNS S43035, 18%Cr + Ti), type 444 (UNS S44400, 18%Cr-2%Mo stabilized with Nb +Ti) and the lower chromium types 405 (UNS S40500) and 409 (UNS S40900). Type 409, a titanium-stabilized 11%Cr material, has become the second highest volume stainless steel produced (after type 304) as a result of its wide use in automotive exhaust systems. As performance requirements and service conditions become more severe, type 434 stainless has largely replaced type 430 for automotive trim applications, although type 430 remains more widely used in general service, for example as furnace and heater components. Type 446 (UNS S44600, 25%Cr) is declining in usage and is now commercially unobtainable as flat-rolled product, although it is still produced as bar. Type 446 is being replaced by alloys such as UNS S44627 (26%Cr-1%Mo). Type 444, after two decades of development and use, is now experiencing a significant increase in usage and production. Type 439 also deserves mention as it is finding use for both automotive and chemical process applications.

The higher chromium ferritic stainlesses are primarily used in high temperature service; notable among these are the aluminum-bearing oxidation resistant alloys. Alloys commonly used for electrical resistance heating element applications are covered by ASTM B 603 Specification for Drawn or Rolled Iron-Chromium-Aluminum Alloys for Electrical Heating Elements. Two common grades within this standard are UNS K91470 (13%Cr-3%Al) and UNS K92500 (22%Cr-5.5%Al).

Martensitic Stainless Steels

Martensitic stainless steels have chromium contents in the 11-17% range, along with 0.1-1.1% carbon. The presence of these carbon levels enlarges the gamma loop so that the alloy can be more readily austenitized within the gamma loop followed by quenching to a martensitic microstructure and tempering to obtain the desired properties. This type of stainless steel, which is ferromagnetic, is produced primarily for its high strength, and its corrosion resistance is relatively poor. Other alloying elements must be restricted to no more than a few percent as their presence lowers the M_s temperature, thus leading to retained austenite in the microstructure at room temperature. At a carbon content of 0.1% only those steels having less than about 13%Cr can be fully hardened (quenched to a fully martensitic microstructure); increasing the carbon to the 1% range permits a fully martensitic microstructure up to about 18%Cr.

For optimum hardness, austenitizing of these steels is carried out in the temperature range 980-1090°C (1800-2000°F). The high chromium content provides sufficient hardenability so that water quenching is not necessary; air quenching is sufficient to develop the martensitic microstructure. However, despite this, accelerated cooling, e.g. oil quenching, can produce a harder as-quenched material by preventing carbide precipitation. The martensite can then be tempered to increase ductility and toughness. The balance between strength and toughness is determined by the selection of tempering time and temperature. When tempered to equivalent strength, the quenched materials typically exhibit greater toughness than material which forms martensite by air cooling. The final microstructure consists of a martensitic matrix with a distribution of carbides - some primary carbides which are present after the quench, and a dispersion of finer carbides which precipitate during the tempering. The carbides provide wear resistance and allow the production and retention of good cutting edges. Tempering in the range between about 400 and 600°C (750 and 1100°F) causes some reduction in both toughness and corrosion resistance, particularly for the lower carbon alloys.

The most widely used of the martensitic stainless steels is type 410 (UNS S41000, 12.5%Cr, 0.15%maxC), which is used for machine parts, cutlery, hardware and engine components. Type 416 (UNS S41000), the free-machining sulfur-bearing modification of type 410, is one of the most widely used bar products. It is commonly substituted for type 410 because it has similar properties, however it is important to note that it is inferior to type 410 in its resistance to stress corrosion cracking and its low temperature notch toughness. Type 414 (UNS S41400) is a higher strength version of type 410, also containing 12.5%Cr, along with 1.8%Ni. Higher strength martensitic stainless steels include type 431 (UNS S43100, 16%Cr, 0.2%maxC+ 1.8%Ni), type 440A (UNS S44002, 17%Cr, 0.72%C), type 440B (UNS S44003, 17%Cr, 0.85%C) and type 440C (UNS S44004, 17%Cr, 1.07%C). Here the higher chromium is necessary to provide for the chromium content of the carbides while maintaining the minimum 11% necessary in the matrix for the passive protective surface film. The strength increases through this series with increasing carbon content, and toughness correspondingly decreases. Type 440C is used for the highest quality cutting implements as well as for ball bearings and races. However the 440C alloy has about 8%Cr tied up in the $M_{23}C_6$ carbides and thus only about 9%Cr remaining in the matrix. As a consequence, this alloy is not really stainless and must be treated carefully if rusting is to be avoided. The high carbon martensitic stainlesses are not recommended for cold forming, and they are more difficult to machine than lower carbon types. Martensitic stainless steels are the most difficult of all the stainless steels to weld, especially the three type 440s.

Austenitic Stainless Steels

Austenitic stainless steels make up about two-thirds of all stainless production because of their combination of high formability and good corrosion resistance. These are basically iron-chromium-nickel alloys containing 16-25%Cr and 7-35%Ni. In some austenitic alloys some of the nickel is replaced by manganese, which along with carbon and nitrogen contributes to the stability of the austenite phase. Some alloys, such as type 301, contain insufficient nickel for the austenite to be fully stable, and in these alloys plastic deformation at room temperature can cause a partial transformation of the austenite to deformation-induced martensite. These alloys are sometimes referred to as metastable austenitic stainless steels.

The microstructures of these materials consist mainly of single phase austenite grains but carbides can be present under some circumstances, with generally deleterious effects on corrosion properties.

Types 302 (UNS S30200) and 304 (UNS S30400) are the most widely used of all the stainless steels, with applications at both ambient and elevated temperatures. These grades contain approximately 18%Cr and 8%Ni and are used for such purposes as food-handling and chemical process equipment, tanks, appliances, trim, cryogenic vessels, railway cars and jewelry. Type 316 (UNS S31600), with 16%Cr and 12%Ni has a 2.5% molybdenum addition for added resistance to corrosive environments and high temperatures. It is used in more severe service conditions in the chemical, pulp and paper, food handling industries and as surgical implants. Type 309 (UNS S30900) has added chromium (23%Cr with 13%Ni) for increased resistance to high temperatures and is used in heat-treating equipment, heat exchangers and aircraft heaters. For more severe high temperature applications, such as combustion chambers and gas turbine components, higher alloy austenitics such as type 310 (UNS S31000, 25%Cr, 20%Ni) are preferred.

Type 304, although more stable than type 301, is also metastable. With sufficient deformation (e.g. cold working such as bending) type 304 will transform, at least partially, to martensite and become ferromagnetic. Increased alloy content, such as is found in type 305 stainless steel (UNS S30500, 18%Cr and 11%Ni) and also in types 309, 310 and 316 is needed to prevent this. However, this transformation can be useful in producing high strength products, such as the temper rolled products described in ASTM A 666, or in enhancing ductility, allowing more severe cold stretching. This occurs because the formation of deformation induced martensite results in a higher rate of strain hardening and retards the onset of necking instability.

The slight magnetism produced by the formation of deformation induced martensite explains why wrought 304 stainless steel can at times mislead a person using a magnet to distinguish between carbon steel and austenitic stainless.

Some austenitic stainless steels (the 2xx series) contain nitrogen as a solution strengthener, and have manganese replacing some of the nickel; the manganese also increases the solubility of nitrogen to give increased solution strengthening. Nitrogen-strengthened austenitic stainless steels have high strengths, excellent low temperature properties and corrosion resistance, and higher elevated temperature strength in comparison with 3xx steels. AISI types 201 and 202 (UNS S20100 and S20200), containing about 18% chromium, 5% nickel and 6-9% manganese, are general purpose austenitic stainless steels which are used for such applications as flatware, trim, kitchen equipment, automotive hub caps and historic railway cars (Fig. 6.3). Non-standard nitrogen-strengthened austenitics include the family of alloys, UNS S24100, S24000, S21904, S20910 and S21800 which contain 17-22%Cr, 5-13%Mn, 3-13%Ni, 0.1-0.4%N.

Figure 6.3 Canadian railway cars made of UNS S20100 stainless steel that have been in operation for over 40 years. Recent repairs were made with UNS S30100 and UNS S30400 stainless steels.

The austenitic stainless steels in general have the best overall corrosion resistance of all the stainless steels, and the most resistance to industrial atmospheres. For more severe corrosion conditions, alloy contents higher than that of the basic type 304 stainless become necessary. The addition of molybdenum (e.g. the 2.5%Mo in type 316) increases resistance to pitting corrosion, with higher nickel and molybdenum levels necessary in more aggressive pitting media such as those of high chloride content. Stress corrosion cracking in chloride-containing environments is a persistent problem with austenitic stainless steels. This can occur in the absence of general corrosion, as long as stress, applied or residual, and chloride ions (even in trace amounts) are present.   The result is transgranular fracture with little associated plastic deformation, and the consequences can be catastrophic. Alloys containing around 8% nickel, the most common concentration in austenitic stainless steels, are most susceptible to this phenomenon.   Types 304 and 316 are the most sensitive to chloride stress-corrosion cracking; higher alloyed austenitic stainless steels can be resistant to stress-corrosion cracking in sodium chloride but remain susceptible in magnesium chlorides.

Sensitization, the intergranular corrosion phenomenon described above in the section on ferritic stainless steels, is also a problem when austenitic stainless is slowly cooled through the carbide precipitation range between 850 and 400°C (1560 and 750°F). Stainless steels are therefore annealed above this temperature range, between 1050 and 1120°C (1920 and 2050°F), and then cooled rapidly to prevent chromium carbide precipitation. In welding, however, it is not always possible to control the cooling rate and sensitization can then occur. For these situations stabilized austenitic stainless steels have been developed which are comparable to type 304 but with additions of titanium (type 321, UNS S32100) or niobium (columbium) (types 347 and 348, UNS S34700 and S34800) as the stabilizing carbide-forming element. Alternatively, extra-low carbon stainless steels can be used to minimize sensitization, for example in the types 304L and 316L (UNS S30403 and 31603) which have a maximum carbon content of 0.03% compared to a maximum of 0.08% in normal types 304 and 316. These alloys are employed in welded components subject to severe corrosive environments, such as aircraft exhaust manifolds and stacks, tank cars, and boiler shells.

Historically the trend in austenitic stainless steels has been toward lower carbon contents, with the original type 302 having a specified maximum of 0.12%C, the improved alloy type 304 having 0.08%C maximum and the sensitization-resistant alloy type 304L having 0.03%C maximum. Type 304L still commands a price premium over the standard grade, type 304, because in the absence of the extra carbon in solution, increased nickel is required to produce a fully austenitic alloy. For type 316, where increased nickel is not required, the L-grade does not command a premium price since the argon-oxygen decarburization (AOD) refining technology has become widespread. The lower carbon content may facilitate mill production and much material sold as type 316 is actually 316L. This situation becomes even more involved when considering the high temperature alloys. The original type 310 alloy has a maximum of 0.25%C, the later type 310S has 0.08%C maximum and the corrosion-resistant type 310L contains 0.03%C maximum. The older type 310 alloy would have always contained carbide precipitates and was almost always heavily sensitized. This may not have been a problem for high temperature service, but did make mill processing and fabrication more difficult. The 310S grade substantially eliminated these problems. Later after the 310L grade became common, it was learned that the absence of carbide precipitates compromised creep resistance. The ASME design codes recognized a distinction between design creep properties for the low carbon L-grades and higher carbon standard grades. To provide end users a way to ensure that alloys ordered for high temperature service had this high carbon content, the H grades were introduced. These typically have between 0.04 and 0.11%C. The case of type 310 materials is particularly interesting in this regard, as the high carbon H-grade has a lower maximum permissible carbon content than the standard type 310 stainless steel.

The austenitic stainlesses are not hardenable by heat treatment, but can be cold worked for added strength. The tendency of some of these austenitic stainlesses to partially transform to martensite during cold work, as described above, is responsible for considerable increases in strength. Their austenitic structure gives them excellent toughness at low temperature, and they have found service as tanks for liquid oxygen, liquid nitrogen and liquid hydrogen, and in many other cryogenic applications. Wrought austenitic stainlesses are non-magnetic (unless cold worked which can impart some ferromagnetic attraction) and are the most weldable of all the stainless steels. Types 304 and 305 (UNS S30500) have the best formability, for example for deep drawing applications, with lower formability characteristics present in the lower nickel alloys such as type 301 (UNS S30100) and in the stabilized types 321 and 347/348. Sulfur or selenium can be added to austenitic stainless for improved machinability.

Superaustenitic Stainless Steels

Superaustenitic stainless steels are a relatively new group of stainless alloys. These have compositions which lie between the older austenitic stainless steels and traditional nickel base alloys. This family includes alloy 20 (e.g. 20Cb-3, UNS N08020, 20%Cr-35%Ni-2.5%Mo-3.5%Cu) developed for sulfuric acid service, alloy 904L (UNS N08904, 20%Cr-25%Ni-4.5%Mo-1.5%Cu), developed for resistance to a variety of chemical process environments, and the so-called "6-Mo" alloys. The latter (UNS S31254, N08367, N08925) contain approximately 20%Cr, 6%Mo, 0.2%N and 18-25%Ni. These alloys exhibit almost total immunity to pitting corrosion in ambient temperature (0-30°C or 32-86°F) sea water and similar high-chloride environments.   They also show greater stress-corrosion cracking resistance than the traditional austenitic stainless steels. These alloys have found extensive utilization in marine-related applications including sea water handling equipment (condensers, heat exchangers, offshore oil production equipment, etc.), pulp and paper production, chemical processing (typically involving chloride-bearing fluids) and pollution control equipment.

Precipitation-Hardening Stainless Steels

Precipitation-hardening stainless steels, with attractive combinations of strength, formability and corrosion resistance, can be classified as austenitic, semiaustenitic and martensitic types. The austenitic types (e.g. UNS S66286) are solution treated around 1200°C (2200°F), quenched, then aged at 700-800°C (1300-1470°F). These alloys contain sufficient nickel to remain austenitic even after the precipitation of nickel-bearing intermetallic compounds during aging. They are weldable, but are normally welded prior to heat treatment.

The more widely used semiaustenitic type involves chemical compositions which provide a careful balance between austenite-stabilizing elements and those which stabilize ferrite. The result is an alloy which is austenitic, or an austenite-ferrite mixture, in the annealed (solution treated) condition, but can transform to martensite as a result of thermal or thermo-mechanical treatment. Typically the alloy is fabricated in the solution treated condition, then given a "conditioning" treatment at high temperature; during this treatment chromium carbides precipitate from the austenite, lowering its chromium and carbon contents. This destabilizes the austenite so that on cooling from the conditioning treatment, martensite forms. The conditioning is followed by a precipitation-hardening heat treatment. For example a 17-7PH stainless steel (UNS S17700, 17%Cr-7%Ni-1.2%Al) can be solution treated at 1060°C (1940°F) ("Condition A"), fabricated, conditioned for 90 minutes at 760°C (1400°F), air cooled to room temperature and held for 30 minutes ("Condition T"), and finally aged at 565°C (1050°F) for 90 minutes. During aging the precipitation of NiAl and Ni_3Al occurs, strengthening the material. Other common semiaustenitics are PH15-7Mo (UNS S15700, 15%Cr-7%Ni-2.2%Mo-1.2%Al) and AM-350 (UNS S35000, 17%Cr-4%Ni-3%Mo). These alloys have carbon and nitrogen contents in the order of 0.1% and are readily weldable in the annealed condition. The semiaustenitic alloy AM 355 (UNS35500, 16%Cr-4%Ni-3%Mo) is used as end fittings on welded AM-350 bellows. These two alloys display better weldability than is exhibited by the other precipitation-hardening alloys.

Among the semiaustenitic precipitation-hardening alloys, the highest strengths are present in alloys with the "CH" temper. Here the solution annealed material, typically in sheet or wire form, is heavily cold worked. This cold working creates deformation induced martensite, which can be aged at a relatively low temperature to an extremely high strength.

Due to the relatively high contents of delta ferrite in the semiaustenitic precipitation-hardenable stainless steels, they are most suitable for use as sheet, strip or wire which experience extensive hot and/or cold working during processing. Plate, bar, billet and especially cast forms of the semiaustenitic alloys are usually less satisfactory, and here the martensitic precipitation-hardenable materials are usually preferable.

Martensitic type precipitation-hardening stainless steels are more widely used than the semiaustenitics or the austenitics. Here the balance between ferrite stabilizers and austenite stabilizers is such that after solution treatment and cooling to room temperature the alloys are in the martensitic condition. Solution treatment is followed by an aging treatment for precipitation-hardening. For example the alloy 17-4PH (UNS S17400, 16%Cr-4%Ni-3.4%Cu-0.25%Nb[Cb]) is solution annealed at

1040°C (1900°F) and aged in the range 450-510°C (850-950°F) to permit precipitation of a highly dispersed copper phase. This alloy is available and used in cast form (UNS J92150 and J92200) and as a powder metallurgy product, the latter typically for aerospace-related applications. Other alloys of this type include Custom 450 (UNS S45000, 15%Cr-6%Ni-0.8%Mo-1.5%Cu-0.3%Nb[Cb]) and higher strength variants such as PH13-8Mo (UNS S13800, 13%Cr-8%Ni-2.2%Mo-1.1%Al). The carbon contents are kept low (<0.1%) to maintain toughness.

The martensitic alloys are more widely used as bar, rod, plate and forgings than as sheet, because their relatively high strength after solution treatment and the distortion which accompanies the martensitic transformation make for difficulties in obtaining flat sheet. For applications in which thick (bar, plate, etc.) and thin (sheet, strip, wire) sections of precipitation-hardenable stainless steels must be joined, it is often possible to join semiaustenitic and martensitic alloys into a composite structure. The proper heat treatment can age both components of the part to high strengths.

Duplex Stainless Steels

The final category of stainless steels comprises the duplex ferrite-austenite stainless steels. Here the microstructures consists either of ferrite grains in an austenite matrix or vice-versa. In comparison with austenitic stainlesses, these alloys have much higher strength, comparable corrosion resistance, improved resistance to transgranular stress-corrosion cracking, good weldability, reasonable ductility and formability and relative freedom from susceptibility to intergranular sensitization. In light gages some duplex stainless steels can be made to exhibit superplastic behavior at elevated temperature. They are, however, not as readily hot-worked and, like the ferritic stainless steels, are susceptible to several types of embrittlement, the alpha prime (α') (i.e. 475°C or 885°F) embrittlement in the 300-550°C (575-1025°F) range and sigma (σ) phase embrittlement in the 500-800°C (930-1470°F) range.

The yield strengths of these alloys increase with increasing ferrite contents up to about 80% ferrite, but steels which have ferrite matrices undergo the low temperature ductile-to-brittle transition typical of ferritic steels, especially above 60% ferrite. Welding characteristics are similar to those of austenitic stainless, although care must be taken to avoid the embrittlement phenomena. Unlike the austenitics, these alloys are ferromagnetic.

Newer duplex stainless steels have roughly equal amounts of austenite and ferrite, and a low carbon content (0.03%max). These give improved chloride stress-corrosion cracking resistance combined with high strength and toughness and an ease of fabrication comparable to the austenitics. Among the important alloys in this category is type 2304 (UNS S32304, 23%Cr-4.5%Ni), a low alloy duplex stainless steel with higher strength than conventional austenitic stainless steels; this material is used for example for heat exchangers in the process industries and tensioning systems in offshore rigs. The highest volume duplex stainless steel is type 2205 (UNS S31803, 22%Cr-5%Ni-3%Mo), which is widely used in environments containing chlorides or hydrogen sulfide, such as in refineries and petroleum production, and in operations where brackish or chloride-bearing water is used as a cooling medium. There is also a group of higher chromium 'superduplex' stainless steels which contain 25%Cr-7%Ni-4%Mo; this group includes UNS S32750, S32760 and S32550. These high alloy duplex stainless steels are especially suitable for service in aggressive chloride-bearing environments, for example in the oil and gas and pulp and paper industries, in seawater heat exchangers, desalination plants and in applications involving high stresses in seawater and other chloride-containing environments.

Another advantage of the duplex stainless steels results from their thermal expansion properties. The austenitic stainless steels have relatively high coefficients of thermal expansion and low thermal conductivity, as compared to plain carbon steels and nickel-based alloys. This may lead to distortion and/or thermal fatigue problems. The ferritic and martensitic stainlesses have lower coefficients of thermal expansion and higher thermal conductivity, usually slightly changed than that of plain carbon steel, as shown in Table 6.1. In this regard duplex stainless steels are intermediate between ferritic and austenitic stainlesses, which may provide them with applications in situations where austenitic stainless is unsuitable.

Table 6.1 Typical physical properties of wrought stainless steels in the annealed condition.

Property	Units	Carbon Steel	Austenitic (Cr-Ni)	Ferritic (Cr)	Martensitic (Cr)	Precipitation -Hardening
Density	Mg/m^3	7.8	7.8-8.0	7.8	7.8	7.8
	$lb/in.^3$	0.28	0.28-0.29	0.28	0.28	0.28
Elastic	GPa	200	193-200	200	200	200
Modulus	10^6 psi	29	28-29	29	29	29
Coeff. of	10^{-6} m/m/°C	11.7	17.0-19.2	11.2-12.1	11.6-12.1	11.9
Thermal Exp.[a]	10^{-6} in./in./°F	6.5	9.4-10.7	6.2-6.7	6.4-6.7	6.6
Thermal	W/(m•K)	60	18.7-22.8	24.4-26.3	28.7	21.8-23
Conductivity[b]	Btu/hr•ft•°F	34.7	10.8-12.8	15.0-15.8	16.6	12.6-13.1
Melting	°C	1538	1400-1450	1480-1530	1480-1530	1400-1440
Range	°F	2800	2550-2650	2700-2790	2700-2790	2560-2625

a. Mean value from 0°C to 538°C (32°F to 1000°F)

b. Thermal conductivity at 100°C (212°F)

Cast Stainless Steels

The above discussion has focused primarily on wrought products, but stainless steel castings are also important industrial materials. Corrosion resistance of cast and wrought stainlesses are comparable, however there are some differences in mechanical properties. Cast austenitic stainless has increased resistance to stress-corrosion cracking, due partly to the presence of a small volume fraction ferrite in the austenite and partly to the higher silicon contents which are made primarily to improve the fluidity and castability of the liquid metal. However the heterogeneity of composition (alloy segregation) which is inherent in castings renders their mechanical properties generally inferior.

Cast stainless steels are especially important for service at elevated temperatures where carbon and alloy steels do not provide sufficient corrosion resistance and/or strength. Stainless steel castings are classified according to whether they are designed to resist corrosion (in aqueous environments below 650°C [1200°F]) or to resist heat (above 650°C [1200°F]). This distinction is based primarily on carbon content.

The designation system for cast stainless is that of the High Alloy Product Group of the Steel Founders Society of America. This group has replaced the Alloy Casting Institute, ACI, which was formerly responsible for the system which is still referred to as the ACI system. Details of this designation system can be found in Chapter 7.

C-type (corrosion-resistant) stainless steel castings include several alloy groups, namely the chromium steels, the chromium-nickel steels and the nickel-chromium steels. They are generally low in carbon, which is detrimental to corrosion resistance, and have more than 11%Cr and up to 30%Ni, although the majority of products fall in the range 18-22%Cr and 8-12%Ni. As for wrought stainless, the nickel improves resistance to corrosion in reducing, neutral or weakly oxidizing environments, and also increases the toughness. Molybdenum is added for improved resistance to crevice corrosion and pitting in seawater and other chloride-bearing environments. The most common alloys are the chromium-nickel alloys CF-8 and CF-8M (UNS J92600 and J92900), which are similar but not identical to wrought types 304 and 316. Among the higher nickel alloys, the most important is CN-7M (28%Ni-20%Cr-3.5%Cu-2.5%Mo), which has improved corrosion resistance in highly oxidizing environments and to hot, concentrated weakly oxidizing media. These alloys find wide applications in the chemical processing and power generating industries.

The microstructures of the cast C-type stainless steels may be ferritic, martensitic, austenitic or duplex ferritic-austenitic, depending on composition. Martensitic grades include CA-15 (UNS J91540), and its higher carbon heat-treatable modification CA-40 (UNS J91153); these are utilized where strength is more important than corrosion resistance. Among the austenitic grades are CH-20 (UNS J94302), CK-20 (UNS J94202) and CN-7M (UNS J95150). Ferritic grades include CB-30 (UNS J91803) which is resistant to many acids and is used in chemical processing and oil refinery applications. The duplex grades contain between 5% and 40% ferrite in austenite and, like the wrought duplex alloys, offer superior strength, corrosion resistance and weldability. Most important among the duplex alloys, particularly in the power generating industry, are the CF grades. Cast austenitic alloys cannot be strengthened by cold work as the wrought austenitic alloys can, or by heat treatment as the martensitic alloys can, or by carbide precipitation because of its deleterious effects on corrosion resistance. However strengthening by incorporating ferrite is highly effective, thus cast austenite-matrix alloys such as CF-8 are designed to contain a small volume fraction of ferrite. This has beneficial effects, including better strength, weldability and corrosion resistance, but it also increases the susceptibility to embrittlement phenomena at elevated temperatures. A common cast duplex alloy is CD-4MCu (UNS J93370) which, with its high amounts of ferrite and low nickel contents, exhibits improved resistance to stress-corrosion cracking.

The duplex cast alloys are easier to cast than the fully austenitic alloys because the presence of delta ferrite greatly reduces the tendency to undergo hot tearing to which austenitic stainless steels are highly susceptible. This also eases the task of weld repairing defective castings. Since castings are not usually subjected to significant hot deformation, the hot workability-related limitations on delta ferrite content, which restrict ferrite levels in nominally austenitic wrought products to less than about 10%, do not apply and highly duplex cast alloys are normal. Once again, as mentioned above, it is necessary to be aware that the presence of delta ferrite, which is ferromagnetic, can at times mislead a person who is using a magnet to distinguish between carbon steel and as-cast austenitic stainless steel.

There are also several precipitation-hardening C-type stainless steel casting alloys including ASTM A 747 types CB-7Cu-1 (UNS J92180) and CB-7Cu-2 (UNS J92110). These are considered where resistance to erosion and wear is required in addition to some degree of corrosion resistance.

H-type (heat-resistant) stainless steel castings are those which are capable of functioning in service at elevated temperatures. For better high temperature strength these alloys contain more carbon than the C-type stainless steel castings, and thus significant amounts of carbides are precipitated in the microstructure. The straight iron-chromium ferritic alloys contain 10 to 30%Cr giving good resistance to oxidation but low strength at elevated temperatures. Among this group are the ferritic alloys HA, HC and HD. Alloy HA (9%Cr, 1%Mo) has good resistance to oxidation up to 650°C (1200°F) and is widely used in refinery applications. Alloy HC (UNS J92605, 28%Cr) is useful to higher temperatures but its high temperature strength is poor; alloy HD (UNS J93005, 28%Cr, 5%Ni) is better in this regard.

Iron-chromium-nickel H-type alloys containing more than 13%Cr and more than 7%Ni are austenitic and have greater strength and ductility than the iron-chromium alloys. They find application in oxidizing and reducing sulfur-bearing gases. Included in this category are the widely used HF (UNS J92603, 20%Cr, 10%Ni) and HH (UNS J93503, 26%Cr, 12%Ni) alloys. Two grades of HH can be obtained, the type I (partially ferritic) and type II (fully austenitic). The latter has good strength at elevated temperatures and reasonable oxidation resistance. Alloy HI (UNS J94003, 28%Cr, 15%Ni) has higher levels of nickel and chromium and better oxidation resistance. This material has found application in vacuum retorts operating at 1175°C (2150°F). Alloy HK (UNS J94224, 26%Cr, 20%Ni) has good resistance to hot gases, oxidizing, reducing and sulfur-bearing, as well as being one of the strongest of the H-type alloys above 1040°C (1900°F). It is widely used in structural applications up to 1150°C (2100°F).

Iron-nickel-chromium cast H-type alloys containing more than 25%Ni and more than 10%Cr are also austenitic. These find applications in oxidizing and reducing atmospheres not containing sulfur, particularly in conditions of fluctuating temperature. The most widely used of these alloys are HP (35%Ni, 20%Cr), HT (UNS J94605, 35%Ni, 17%Cr), HU (39%Ni, 18%Cr), HW (60%Ni, 12%Cr) and HX (66%Ni, 17%Cr). Applications include trays for heat treating furnaces. The latter four of these alloys exhibit good corrosion resistance in molten salts and metals. The microstructures of all five consist of carbides in an austenite matrix.

Chapter

7

NORTH AMERICAN METAL STANDARD DESIGNATION SYSTEMS

Introduction

In the world of standardization, metals pioneered the way at the turn of this century. In 1895, the French government assigned a commission to formulate standard methods of testing materials of construction. Later that year, the European member countries of the International Association for Testing Materials (IATM) held their first conference in Zurich and standardization of metals began.

Today, there are numerous national, continental, and international standards each with its own cryptic designation system to identify metals and their alloys. The evolution of the metals industry has left us with numerous designation systems, even within an individual standards organization, and these have become blurred and less meaningful as new generations of technical personnel are passed the torch to carry on the task of standardization.

By reviewing some examples of the more prominent metals designation systems, a direction is offered to assist those who use metal standards as a part of their work or study. This chapter is not all inclusive. The amount of information on this topic could easily make up a complete book.

American Metal Standard Organizations

There are many metals standards organizations in the United States, a few of the more prominent ones are listed as follows:

AA	The Aluminum Association
AISI	American Iron and Steel Institute
ANSI	American National Standards Institute
AMS	Aerospace Material Specifications (SAE)
ASME	American Society of Mechanical Engineers
ASTM	American Society for Testing and Materials
AWS	American Welding Society
CSA	Canadian Standards Association
SAE	Society of Automotive Engineers

For each North American organization issuing metal specifications and standards, there is a designation system used to identify various metal and alloys. These designation systems grew according to the history of each group, and generally identify a metal by use of a coded number or alphanumeric designator. In some cases, numbers and letters were assigned in a sequential order by the respective listing organization, while in other cases they were given in a manner which directly identified chemical composition or mechanical properties. Some of the more popular North American designation systems for metals are presented below, with descriptive examples given.

American Society for Testing and Materials (ASTM)

The first complete book of ASTM Standards was published in 1915. Today there are 69 ASTM books of standards contained in 15 sections on various subjects. For the most part, the metals related standards are found in Section 1 - Iron and Steel Products (7 volumes), Section 2 - Nonferrous Metal Products (5 volumes), and Section 3 - Metals Test Methods and Analytical Procedures (6 volumes). These standards are revised yearly, as an example, from 1992 to 1993, 256 of the 631 standards were revised in Section 1 - Iron and Steel Products. Some standards (e.g. ASTM A 240) change several times a year and letter suffixes (a, b, c, etc.) are used to track mid-year revisions. This represents changes in 40% of these standards, not including the new standards that were issued that year. Consequently, it is an understatement to say that metal standards are very dynamic documents.

ASTM Specification System

Steel products are catagorized according to designation systems such as the AISI/SAE system or the UNS system described below, and also according to specification systems. These are statements of requirements, technical and commercial, that a product must meet, and therefore they can be used for purposes of procurement. One widely used system of specifications has been developed by the ASTM. The designation consists of a letter (A for ferrous materials) followed by an arbitrary serially-assigned number. These specifications often apply to specific products, for example A 548 is applicable to cold-heading quality carbon steel wire for tapping or sheet metal screws. Metric ASTM specifications have a suffix letter M. Some ASTM specifications (e.g. bars, wires and billets for forging) incorporate AISI/SAE designations for composition while others (e.g. plates and structural shapes) specify composition limits and ranges directly. Such requirements as strength levels, manufacturing and finishing methods and heat treatments are frequently incorporated into the ASTM product specifications.

Ferrous Metal Definition

Prior to 1993 the ASTM definition for ferrous metals was based on nominal chemical composition, where an iron content of 50% or greater determined the alloy to be ferrous. Consequently, these standards begin with the letter "A". If the iron content was less than 50%, then the next abundant element would determine the type of nonferrous alloy. Generally these standards begin with the letter "B". For example, should nickel be the next predominant element then the metal would be a nickel alloy.

Currently, ASTM has adopted the European definition of steel described in the EuroNorm Standard CEN EN10020 - Definition and Classification of Steel, which defines steel as:

> "A material which contains by weight more iron than any single element, having a carbon content generally less than 2% and containing other elements. A limited number of chromium steels may contain more than 2% of carbon, but 2% is the usual dividing line between steel and cast iron."

The CEN committee responsible for this standard has suggested changing the term "by weight" to "by mass" in order to stay consistent with the International System of Units.

ASTM Steels

Examples of the ASTM ferrous metal designation system, describing its use of specification numbers and letters, are as follows.

ASTM A 516/A 516M - 90 Grade 70 - Pressure Vessel Plates, Carbon Steel, for Moderate- and Lower-Temperature Service:

- the "A" describes a ferrous metal, but does not sub-classify it as cast iron, carbon steel, alloy steel or stainless steel.
- 516 is simply a sequential number without any direct relationship to the metal's properties.
- the "M" indicates that the standard A 516M is written in SI units (as a soft conversion) (the "M" comes from the word "Metric"), hence together A 516/A 516M utilizes both inch-pound and SI units.
- 90 indicates the year of adoption or revision.
- Grade 70 indicates the minimum tensile strength in ksi, i.e. 70 ksi (70,000 psi) minimum.

In the steel industry, the terms Grade, Type and Class have specific meaning. "Grade" is used to describe chemical composition, "Type" is used to define deoxidation practice, and "Class" is used to indicate other characteristics such as strength level or surface finish. However, within ASTM standards these terms were adapted for use to identify a particular metal within a metal standard and are used without any "strict" definition, but essentially mean the same thing. Some rules-of-thumb do exist, with a few examples as follows.

ASTM A 106 - 91 Grade A, Grade B, Grade C - Seamless Carbon Steel Pipe for High-Temperature Service:
- typically an increase in alphabet (such as the letters A, B, C) results in higher strength (tensile or yield) steels, and if it is an unalloyed carbon steel, an increase in carbon content.
- in this case: Grade A - 0.25%C (max.), 48 ksi tensile strength (min.); Grade B - 0.30%C (min.), 60 ksi tensile strength (min.); Grade C - 0.35%C 70 ksi tensile strength (min.).

ASTM A 48 - Class No. 20A, 25A, 30A - Gray Iron Castings:
- Class No. 20A describes this cast iron material as having a minimum tensile strength of 20 ksi (20,000 psi).
- similarly Class No. 25A has a minimum tensile strength of 25 ksi and Class No. 30A has a minimum tensile strength of 30 ksi.

ASTM A 276 Type 304, 316, 410 - Stainless and Heat-Resisting Steel Bars and Shapes:
- types 304, 316, 410 and others are based on the AISI designation system for stainless steels (see AISI description that follows).

Some ASTM standards will use more than one term to describe an individual metal within a group of metals from one standard, as shown in the following example.

ASTM A 193/193M-94 - Alloy Steel and Stainless Steel Bolting Materials for High Temperature Service:
- uses the terms "Type", "Identification Symbol", "Grade" and "Class" to describe bolting materials.
- example, Type: Austenitic steel, Identification Symbol: B8, Grade: Unstabilized 18 Chromium - 8 Nickel (AISI Type 304), is available in four different Classes: 1, 1A, 1D, and 2.

The ASTM designation system for cast stainless steels was adopted from the Alloy Casting Institute (ACI) system. According to this system, the designation consists of two letters followed by two digits and then optional suffix letters. The first letter of the designation is "C", if the alloy is intended for liquid corrosion service, or "H", for high temperature service. A second letter refers to the chromium and nickel contents of the alloy, increasing with increasing nickel content. The two letters are then followed by a number which gives the carbon content in hundredths of a percent, and in some cases a suffix letter or letters to indicate the presence of other alloying elements. It is important to note that the various casting grades of these stainless steels have a unique designation system different from that of their wrought counterparts.

For example, the designation "cast 304" stainless steel does not exist within the ASTM (ACI) system and is appropriately called grade CF8. Other examples are as follows.

ASTM A 351 Grade CF8M, Grade HK40 - Castings, Austenitic, Austenitic-Ferritic (Duplex), for Pressure Containing Parts:
- the "C" in CF8M indicates a Corrosion resistant metal and the "H" in HK40 indicates a Heat resistant metal.
- the numeric portion of the corrosion resistant designations represent the *maximum* carbon content multipled by 100, and those of the heat resistant designations represent its *nominal* carbon content multiplied by 100. For example: the maximum carbon content of grade CF8M is 0.08% C and the nominal carbon content of grade HK40 is 0.40%C (its actual carbon content range is 0.35-0.45%C).
- the "M" after the number represents an intentional addition of Molybdenum.

An interesting use of ASTM grade designators is found in pipe, tube and forging products, where the first letter "P" refers to pipe, "T" refers to tube, "TP" may refer to tube or pipe, and "F" refers to forging. Examples are found in the following ASTM specifications:

- ASTM A 335/A 335M - 91 grade P22 - Seamless Ferritic Alloy-Steel Pipe for High-Temperature Service.

- ASTM A 213/A 213M - 91 grade T22 - Seamless Ferritic and Austenitic Alloy-Steel Boiler, Superheater, and Heat-Exchanger Tubes.
- ASTM A 269 - 90 grade TP304 - Seamless and Welded Austenitic Stainless Steel Tubing for General Service.
- ASTM A 312/A 312M - 91 grade TP304 - Seamless and Welded Austenitic Stainless Steel Pipes.
- ASTM A 336/A 336M - 89 class F22 - Steel Forgings, Alloy, for Pressure and High-Temperature Parts.

ASTM Reference Standards and Supplementary Requirements

ASTM Standards contain a section known as "Reference Documents" that lists other ASTM Standards, that either become a part of the original standard or its supplementary requirements.   Supplementary requirements are listed at the end of the ASTM Standards and do not apply unless specified in the order, i.e. they are optional.

American Society of Mechanical Engineers (ASME) Designation System

The ASME Boiler and Pressure Vessel Code, Section II Parts A and B contain the material standards adopted for its use. These standards are based on the ASTM Standards and in most cases are identical to ASTM Standards, however some of the ASTM Standards have been edited to accommodate the ASME Boiler and Pressure Vessel Code.   When an ASTM standard has been adopted by the ASME Code, an "S" is placed in front of the ASTM standard identification number to indicate an ASME Code material.

ASME Code Section II Part A contains the ferrous metals and Part B contains the nonferrous metals. The following example illustrates the relationship between ASME and ASTM designation systems.

ASME SA-516/SA-516M-90 Grade 70 - Pressure Vessel Plates, Carbon Steel, for Moderate- and Lower-Temperature Service:
- identical to ASTM A 516/516M-90 Grade 70, except it has the ASME "SA" designator, indicating it has been adopted by the ASME Code.

In most American states and Canadian provinces, use of the ASME Boiler and Pressure Vessel Codes is mandated by law. ASME Code Cases allow the use of materials which have not been accepted by ASTM and do not have SA or SB designations. For more details regarding ASME Code Case Materials refer to *The Practical Guide Book to ASME Section II -* Materials Index, published by *CASTI* Publishing Inc.

AISI/SAE Designation Systems

Carbon and Alloy Steels

The AISI/SAE system (administered by the SAE) uses a basic four-digit system to designate the chemical composition of carbon and alloy steels. Throughout the system, the last two digits give the carbon content in hundredths of a percent. For carbon steels the first digit is 1. Plain carbon steels are designated 10xx therefore, for example, plain carbon steel containing 0.45% carbon is designated 1045 in this system. Resulfurized carbon steels are designated within the series 11xx, resulphurized and rephosphorized carbon steels 12xx and steels having manganese contents between 0.9 and 1.5% but no other alloying elements are designated 15xx. Composition ranges for manganese and silicon, and maxima for sulfur and phosphorus are also specified. For constructional and automotive alloy steels the first two digits of the AISI/SAE system describe the major alloying elements present in the material, the first digit giving the alloy group. For example the 43xx series steels contain 1.65-2.00%Ni, 0.50-0.80%Cr and 0.20-0.30%Mo, along with composition ranges for manganese and silicon and maximums for sulfur and phosphorus. Additional letters added between the first and last pairs of digits include "B" when boron is added (between 0.0005 and 0.003%) for enhanced hardenability, and "L" when lead is added (between 0.15 and 0.35%) for enhanced machinability. The prefix "M" is used to designate merchant quality steel (the least restrictive quality descriptor for hot-rolled steel bars used in noncritical parts of structures and machinery). The prefix "E" (electric furnace steel) and the suffix "H" (hardenability requirements) are mainly applicable to alloy steels. The full series of classification groups is shown in Table 7.1.

Table 7.1　Types and Identifying Elements in Standard Carbon and Alloy Steels

	AISI/SAE Classification
Carbon Steels	**Description**
10XX	Nonresulfurized, 1.00 manganese maximum
11XX	Resulfurized
12XX	Rephosphorized and resulfurized
15XX	Nonresulfurized, over 1.00 manganese maximum
Alloy Steels	
13XX	1.75 manganese
40XX	0.20 or 0.25 molybdenum or 0.25 molybdenum and 0.042 sulfur
41XX	0.50, 0.80, or 0.95 chromium and 0.12, 0.20, or 0.30 molybdenum

AISI/SAE Classification (Continued)

Alloy Steels	Description
43XX	1.83 nickel, 0.50 to 0.80 chromium, and 0.25 molybdenum
46XX	0.85 or 1.83 nickel and 0.20 or 0.25 molybdenum
47XX	1.05 nickel, 0.45 chromium, 0.20 or 0.35 molybdenum
48XX	3.50 nickel and 0.25 molybdenum
51XX	0.80, 0.88, 0.93, 0.95, or 1.00 chromium
51XXX	1.03 chromium
52XXX	1.45 chromium
61XX	0.60 or 0.95 chromium and 0.13 or 0.15 vanadium minimum
86XX	0.55 nickel, 0.50 chromium, and 0.20 molybdenum
87XX	0.55 nickel, 0.50 chromium, and 0.25 molybdenum
88XX	0.55 nickel, 0.50 chromium, and 0.35 molybdenum
92XX	2.00 silicon or 1.40 silicon and 0.70 chromium
50BXX	0.28 or 0.50 chromium
51BXX	0.80 chromium
81BXX	0.30 nickel, 0.45 chromium, and 0.12 molybdenum
94BXX	0.45 nickel, 0.40 chromium, and 0.12 molybdenum

Another common specification system used for steels is the Aerospace Materials Specifications (AMS), published by SAE. These specifications include chemical composition as well as mechanical property requirements and processing requirements are also common. In general these specifications are more restrictive than those for steels intended for less critical service applications.

Stainless Steels

The designation system for standard wrought stainless steels, originally developed by the AISI, consists of three digits followed by, in some cases, suffix letters. The first digit gives the alloy class, with the 2xx and 3xx series being austenitic stainless steels, while ferritic and martensitic stainless are designated in the 4xx series. However here the logic ends. The second and third digits are not related to chemical composition in a regular sequential manner, and furthermore within the 4xx series there is no regularity to the classification of martensitic versus ferritic stainless steels. For example type 430 and 446 are ferritic, while types 431 and 440 are martensitic. The suffix letters and the corresponding final two digits of the UNS system are described in Table 7.2:

Table 7.2 AISI and UNS Designators for Stainless Steels

AISI Suffix Designator	UNS No.	Description
xxxL	xxx01	low carbon (<0.03% as compared to the normal <0.08%) for improved resistance to intergranular corrosion as discussed below
xxxS	xxx08	low carbon (<0.08% as compared to standard <0.2% or higher)
xxxN	xxx51	added nitrogen for increased strength
xxxLN	xxx53	low carbon (<0.03%) plus added nitrogen
xxxF	xxx20	higher sulfur and phosphorus for improved machinability
xxxSe	xxx23	added selenium for better machined surfaces
xxxB	xxx15	added silicon to increase scaling resistance
xxxH	xxx09	wider allowable range of carbon content
xxxCu	xxx30	added copper

There are many stainless steels which are not covered by the AISI designation system. These non-standard stainless steels include the precipitation hardening stainless steels, and most of the duplex and alloy stainless steels. Here proprietary names are in widespread use. Some of the high alloy stainless steels, notably those with iron contents less than 50%, can be considered to be nickel-based alloys, and these are discussed in the chapter on nickel alloys in *The Metals Red Book* - Nonferrous Metals.

Numbering System (UNS) For Metals And Alloys

SAE and ASTM began discussion of a Unified Numbering System (UNS) for metals and alloys in 1967. The intent of the program was three-fold:

1. Simplify the many differing designation systems, independently established over the previous 60 years.
2. Give trade names a generic number, especially helpful when several companies produced the same alloy under different trade names.
3. Make a new system computer friendly by developing a consistent designation.

In 1969 the U.S. Army sponsored a study by SAE and ASTM into the development of a Unified Numbering System. The first published edition of the UNS for Metals and Alloys was in 1975, co-published by SAE and ASTM, and has since been updated about every 3 years. The UNS is available as a hard copy book and in a software version. The computer software program is known as UNSearch. Details of the Unified

Numbering System are specified in ASTM E 527: Practice for Numbering Metals and Alloys (UNS).

UNS Designation System

The UNS is an alphanumeric designation system consisting of a letter followed by five numbers. This system only represents chemical composition for an individual metal or alloy and is not a metal standard or specification. For the most part, existing systems, such the AISI/SAE designations, were incorporated into the UNS so that some familiarity was given to the system where possible. For example, the UNS prefix letter for carbon and alloy steels is "G", and the first four digits are the AISI/SAE designation, e.g. G10400. The intermediate letters "B" and "L" of the AISI/SAE system are replaced by making the fifth digit of the UNS designation 1 and 4 respectively, while the prefix letter "E" for electric furnace steels is designated in UNS system by making the fifth digit "6". The AISI/SAE steels which have a hardenability requirement indicated by the suffix letter "H" are designated by the Hxxxxx series in the UNS system. Carbon and alloy steels not referred to in the AISI/SAE system are categorized under the prefix letter "K".

Where possible, the first letter in the system denotes the metal group, for instance "S" designates stainless steels. Of the five digits of the UNS designation for stainless steels, the first three are the AISI alloy classification, e.g. S304xx. The final two digits are equivalent to the various modifications represented by suffix letters in the AISI system as given in the list of suffixes in Table 7.2. The UNS designations for ferrous metals and alloys are described in Table 7.3.

Table 7.3. UNS Designations for Ferrous Metals and Alloys

UNS Descriptor	Ferrous Metals
Dxxxxx	Specified mechanical properties steels
Fxxxxx	Cast irons
Gxxxxx	AISI and SAE carbon and alloy steels (except tool steels)
Hxxxxx	AISI H-steels
Jxxxxx	Cast steels
Kxxxxx	Miscellaneous steels and ferrous alloys
Sxxxxx	Heat and corrosion resistant (stainless) steels
Txxxxx	Tool steels

UNS Descriptor	Welding Filler Metals
Wxxxxx	Welding filler metals, covered and tubular electrodes classified by weld deposit composition.

Canadian Standards Association (CSA)

The Canadian Standards Association (CSA) has established metal standards for structural steels (CSA G40.20/40.21), pipeline steels (CSA Z245.1), corrugated steel pipe (G401), wire products (CSA G4, G12, G30.x, G279.2, G387), sprayed metal coatings (G189), and welding consumables (CSA W48.x).

Most CSA material standards use SI units, although some are available in both SI and Imperial units (e.g. CSA G4), while others are available in both units but published separately (e.g. CSA G40.20/G40.21-M92 (SI) and G40.20/G40.21-92 (Imperial)). When a CSA standard designation is followed by the letter "M", it uses SI units, and if the letter "M" is not present, it may use both units or use only Imperial units. The type of measurement units adopted in CSA standards are specific industry driven, with some industries moving faster towards the exclusive use of SI units than others, and thus the reason for these differences.

Welding filler metal specification under the CSA W48.x-M group of standards is a good example of a CSA standard using both units of measurement. These standards are based on the American Welding Society (AWS) system, but includes SI units of measurement followed by Imperial units in brackets. Examples are as follows.

- based on the AWS A5.1 Classification E7018, except it has the CSA W48.1-M designation and some minor specification differences.
- the first three digits in the classification designation, 480, represent minimum tensile strength in MPa, rather than ksi as in the AWS system.
- the fourth digit, and the fourth and fifth digits together, are used in the same manner as in the AWS standards, namely, position and type of covering, respectively.

CSA W48.2-M Classification E316L - Chromium and Chromium-Nickel Steel Covered Electrodes for Shielded Metal Arc Welding:
- based on the AWS A5.4 Classification E316L, except it has the CSA W48.2-M designation and some minor specification differences.
- the classification designation, 316L, is identical to AWS A5.4.

During the 1992 revision of CSA G4, Steel Wire Rope for General Purpose and for Mine Hoisting and for Mine Haulage, reference was made to standards published by the International Organization for Standardization (ISO) under the auspices of the Canadian Advisory Committee for ISO/TC105, Steel Wire Ropes. As far as practicable, rationalization with relevant ISO standards has been achieved in CSA

G4. In a similar light, the 1993 edition of CSA Z245.1-93, Steel Line Pipe, differs from its predecessor issued in 1990, by including requirements for ISO wire penetrameters.

American National Standards Institute (ANSI)

ANSI is an impartial organization that does not develop standards, but rather validates the general acceptability of standards developed by organizations such as ASTM, ASME, AWS, etc. It guarantees that the standards writing group used democratic procedures that gave everyone who will be "directly and materially" affected by the use of the standard an opportunity to participate in the development work or to comment on the document's provisions. It assures users that consensus was achieved on the standard's provisions and does not conflict with or unnecessarily duplicate other national standards.

The designations of American National Standards reflect ANSI's role as national coordinator of voluntary standards activities and as approval organization and clearinghouse for consensus standards. The standards are identified by the publication number or alphanumeric designator (where one exits) of the organization that administered the development work, preceded by ANSI's acronym and the organization's. If the organization does not have a designation system, ANSI assigns an identifier. Examples are as follows.

- ANSI/ASME B31.3 -1993 - Chemical Plant and Petroleum Refinery Piping.
- ANSI/AWS A5.1 -91 - Specification for Carbon Steel Arc Welding Electrodes.

EUROPEAN STANDARD (CEN)
STEEL DESIGNATION SYSTEM

Introduction

The European Committee for Standardization (CEN) is an association of the national standards organizations of 18 countries of the European Union and of the European Free Trade Association. The principal task of CEN is to prepare and issue *European Standards (EN),* defined as a set of technical specifications established and approved in collaboration with the parties concerned in the various member countries of CEN. They are established on the principle of consensus and adopted by the votes of weighted majority. Adopted standards must be implemented in their entirety as national standards by each member country, regardless of the way in which the national member voted, and any conflicting national standards must be withdrawn.

The identification of European Standards in each member country begin with the reference letters of the country's national standards body, e.g. BS for BSI in the United Kingdom, DIN for DIN in Germany, NF for AFNOR in France, etc., followed by the initials EN and a sequential number of up to five digits, e.g. BS EN 10025, DIN EN 10025, or NF EN 10025. An EN Standard may contain one document or may be made up of several parts, e.g. EN 10028 Parts 1 through 8, where each Part specifies a particular characteristic of the steel product, and may not include the word Part in the designation, but rather replace it with a hyphen, e.g. EN 10028-1, meaning Part 1. The prefix "pr" attached to the EN designation identifies the document as a draft standard that has not yet been approved, e.g. prEN 10088-1.

Other important documents issued by CEN are Harmonization Documents (HD) and European Prestandards (ENV). *Harmonization Documents* are prepared and adopted in the same manner as EN Standards but their application is more flexible so that particular national technical conditions pertaining to some countries can be accommodated.   As a minimum, National members are required to publicly announce approved Harmonization Document numbers and titles and to withdraw any conflicting national standards, in the same manner as EN Standards. Identification of Harmonization Documents is similar to that of EN Standards, with the exception that the letters EN are replaced with HD.

European Prestandards can be prepared as prospective standards for provisional application in the areas of technology where there is a high level of innovation or where an urgent need for guidance is felt, and where the safety of persons and goods is not involved. The time required for its preparation is therefore reduced.   Once adopted, a European Prestandard is subject to a trial period of up to three years, with the intent to transform it into a European Standard or Harmonization Document. The availability of the ENV document at the national level is at the discretion of the member countries. However, it should be made available in an appropriate form and its existence announced in the same manner as European Standards and Harmonization Documents. Identification of European Prestandards is similar to EN Standards, with the exception that the letters EN are replaced with ENV.

European Standards (EN) must not be confused with the older Euronorms, which were issued by the European Coal and Steel Community and are now in the process of being revised and transformed into European Standards. A number of Euronorms, e.g. EU 25, EU 28, EU 27, have already gone through this process.

As of 1995, approximately 40% of the European Standards issued each year are directly using International Organization for Standardization (ISO) Standards. The member countries of CEN are also members of ISO and the cooperation between them contributes significantly to their common work. The Vienna Agreement between CEN and ISO is a formal agreement of cooperation between the two parties, that presently covers about 800 projects of standardization.

European Standard Designation System for Steels - EN 10027

The European Standard designation system for steels is standardized in EN 10027, which is published in two parts:

- Part 1 - *Steel names*
- Part 2 - *Steel numbers*

A typical reference to a steel includes both name and number. This is done to minimize the danger of confusion between steels as a result of errors in the designation, an also to overcome the difficulties likely to occur when, in the future, it becomes necessary to modify a steel number or name. However, steel names and steel numbers are not necessarily interchangeable. A steel name can be a complete abbreviated identification of a steel product, whereas a steel number only identifies the steel (see discussion below). For example, the steel number a coated steel strip refers only to the substrate steel and not to the coated steel product as a whole.

Steel Names

The *steel name* is a combination of letters and numbers as described by EN 10027 Part 1 and supplemented by the CEN Report CR 10260 (formerly ECISS IC 10). Within this system, steel names are classified in two groups as described below. The system is similar in some respects to, but not identical with, that outlined in an ISO technical report (ISO TR 4949:1989 (E) "Steel names based on letter symbols".

Steel Names - Group 1

Group 1 within EN 10027 Part 1 refers to steels which are designated according to their application and mechanical or physical properties. These have names which consist of one or more letters, related to the application, followed by a number related to properties. For example, the name for structural steels begins with the letter S, for linepipe steels with the letter L, steels for pressure purposes with the letter P and rail steels with the letter R (see Table 8.1).

Table 8.1 Steel Names - Group 1 List	
S	Structural steels
P	Pressure purpose steels
L	Linepipe steels
E	Engineering steels
B	Steels for reinforcing concrete
Y	Steels for prestressing concrete
R	Rail steels or steels in the form of rails
H	Cold rolled flat products of high strength steels for cold forming
D	Flat products for cold forming
T	Tinmill products (steel products for packaging)
M	Electric Steels

This letter is followed by a value of some property, the specific property depending on the application. Thus in structural steels, the S is followed by a number which is the specified minimum yield strength in N/mm^2 for the smallest thickness range, while in rail steels the R is followed by the specified minimum tensile strength in N/mm^2. An example is given below.

- EN 10025 S185, where the S identifies this material to be a structural steel and 185 specifies the minimum yield strength in N/mm^2.

Steel Names - Group 2

Group 2 steel names are used for steels which are designated according to their chemical composition, and are further divided into four sub-groups depending on alloy content. The first sub-group consists of non-alloy steels (except high speed steels) with an average manganese content less than 1 percent. These have names consisting of the letter C followed by a number which is 100 times the specified average percentage carbon content.

The second sub-group includes non-alloy steels with an average manganese content equal to or greater than 1 percent, non-alloy free-cutting steels, and alloy steels (except high speed steels) where the content by weight of every alloy element is less than 5 percent. For this sub-group, the name consists of a number which is 100 times the specified average percentage carbon content, followed by chemical symbols representing the alloy elements that characterize the steel (in decreasing order of the values of their contents), followed by numbers indicating the values of contents of alloy elements. These latter numbers represent, respectively, the average percentage content of the element indicated, multiplied by a factor which depends on the element, as shown in Table 8.2.

Table 8.2 Alloying Element Factors for Steels	
Element	**Factor**
Cr, Co, Mn, Ni, Si, W	4
Al, Be, Cu, Mo, Nb, Pb, Ta, Ti, V, Zr	10
Ce, N, P, S	100
B	1000

The numbers referring to the different elements are rounded to the nearest integer and separated by hyphens. An example is shown below.

- EN 10028 Part 2, 13CrMo4-5
  nominally contains 0.13% C, 1% Cr, and 0.5% Mo.

The third sub-group of steel names based on chemical composition applies to alloy steels (except high speed steels) where the content by weight of at least one alloy element is greater than 5 percent. For this category, the name begins with the letter X, followed by a number which is 100 times the specified average percentage carbon content, followed by chemical symbols representing the alloying elements that characterize the steel (in decreasing order of the values of their contents), followed by numbers indicating the values of contents of these alloy elements. Here the number represent, respectively, the average percentage content of the element indicated, rounded to the nearest integer and separated by hyphens. An example is given as follows.

- EN 10088 Part 1, X2CrNi18-9
  nominally contains 0.02% C, 18% Cr, and 9% Ni.

The fourth and final sub-group of steel names based on chemical composition refers to high speed steels. Here the name consists of the letters HS followed by numbers indicating the values of percentage contents of alloy elements indicated in the order tungsten, molybdenum, vanadium, cobalt. Each number represents the average percentage content of the respective element, rounded to the nearest integer, with the numbers referring to the different elements separated by hyphens.

For both Group 1 (specified by application and properties) and Group 2 (specified by composition) steels, if the name as described above is preceded by the letter G then the steel is specified to be a casting.

Steel Numbers

EN 10027 Part 2 describes the system used for assigning steel numbers, which are complementary to the steel names described above. The number consists of a fixed number of digits and is hence more suitable than the name for data processing purposes. The number is of the form 1.XXXX, where the 1. refers to steel; other numbers may be allocated in the future to other industrial materials if this numbering system is expanded. The first two digits following the "1." give the steel group number. Examples for assigning these two digits are shown in Table 8.3. It can be seen that for this purpose, steels are divided into non-alloy and alloy steels, with subdivisions into base steels, quality steels and special steels, as defined below.

The final digits in the steel number are assigned sequentially. At present only two digits are used, but provision is made for expansion to a system using four digits in the future if required by an increase in the number of steel grades.

Definition and Classification of Steels - EN 10020

Non-Alloy Steels

The terms *base steels*, *quality steels* and *special steels* as used in the
second and third digits of the steel number in EN 10027-2 are quality
classes defined by main property or application characteristics. These are
explained in European Standard EN 10020, Definition and Classification
of Steel, which also defines steel and classifies steel grades into non-alloy
and alloy steels on the basis of chemical composition.

Table 8.3 Examples of Steel Numbers		
Non-alloy steel		
Base steel	1.00XX - base steels	
Quality steels	1.01XX - general structural steels with R_m < 500 N/mm^2	
Special steels	1.11XX - structural, pressure vessel and engineering steels with C < 0.50%	
Alloy steels		
Quality steels	1.08XX - steels with special physical properties	
Special steels		
	Tool steels	1.23XX - Cr-Mo, Cr-Mo-V or Mo-V steels
	Miscelaneous steels	1.35XX - bearing steels
	Stainless and heat resisting steels	1.46XX - chemical resistant and high temperature Ni alloys
	Structural, pressure vessel and engineering steels	1.51XX - Mn-Si or Mn-Cr steels

A boundary content of each alloy element is specified, so that a steel
containing any element in excess of this boundary value is designated an
alloy steel (with provision for the effects of particular combinations of
certain alloy elements). According to this standard, *non-alloy base steels*
are manufactured by normal steelmaking operations, do not require heat
treatment, meet certain property specifications but no other quality
requirements (such as suitability for cold drawing, etc.) and have no alloy
element requirements other than manganese and silicon.

Non-alloy quality steels are steels with more stringent requirements than
base steels for such parameters as fracture toughness, grain size or
formability. However quality steels do not have specified requirements

for response to heat treatment or for cleanliness in terms of non-metallic inclusions.

Non-alloy special steels are cleaner than base steels with regard to non-metallic inclusions. They are normally intended for quenching and tempering or surface hardening. Precise control of chemical composition and care in manufacture and process control are used to obtain improved properties (e.g. yield strength, hardenability) so that they meet one or more of a number of exacting requirements such as specified impact strength, hardness penetration depth, or surface hardness in particular heat-treated conditions, or particularly low contents of non-metallic inclusions, or phosphorus or sulfur.

Alloy Steels

European Standard EN 10020 also distinguishes between the two classes of alloy steels: alloy quality steels and alloy specialty steels. *Alloy quality steels* are used in applications similar to those of non-alloy quality steels but require additions of alloy elements to meet the specified property requirements. They are not intended for quenching and tempering or for surface hardening. This class includes some weldable fine-grained structural steels, some electrical steels, and alloy steels for severe cold forming applications.

The *alloy special steels* class includes stainless steels, heat and creep resisting steels, steels for bearings or tools, engineering steels, special structural steels and steels with special physical properties. These are characterized by precise control over chemical composition and manufacturing and process variables. Properties are frequently specified in combination and within closely controlled limits.

Other European Steel Standards

Steel *product forms* are defined according to EN 10079, Definition of Steel Products, which provides formal definitions for all products including flat products, coated and uncoated (sheet, strip, electrical steels, etc.), long products (rod, wire, bar, etc.), and other products such as forgings and powder metallurgy products.

Heat treatment terms can be found in EN 10052, Vocabulary of Heat Treatment Terms for Ferrous Products. These terms are defined in English only, but are listed in a glossary form in English, French, and German.

Chapter

9

METALLURICAL TERMS DEFINITIONS & GLOSSARY

English/French Definitions

A

A₁ temperature. The eutectoid temperature of a steel.
température A₁. La température eutectoïde d'un acier.

A₂ temperature. Curie temperature, where bcc iron upon reaching this temperature, 1420°F (770°C), becomes nonmagnetic.
température A₂. La température de Curie, température à laquelle le fer bcc, lorsqu'atteignant cette température, 1420°F (770°C), devient non magnétique.

A₃ temperature. The temperature at which proeutectoid ferrite begins to separate from austenite under conditions of slow cooling.
température A₃. La température à laquelle la ferrite proeutectoïde commence à se séparer de l'austénite sous des conditions de refroidissement lent.

A$_{cm}$ temperature. The temperature at which proeutectoid cementite begins to separate from austenite under conditions of slow cooling.
température A$_{cm}$. La température à laquelle la cémentite proeutectoïde commence à se séparer de l'austénite sous des conditions de refroidissement lent.

abrasion. The process of grinding or wearing away using abrasives.
abrasion. Le procédé consistant à moudre ou à user en utilisant des abrasifs.

abrasive. A substance capable of grinding away another material.

abrasif. Une substance capable de moudre un autre matériau.

age hardening. Hardening by aging, usually after rapid cooling or cold working.
durcissement par vieillissement. Durcissement par vieillissement, habituellement après refroidissement rapide ou après écrouissage.

aging. A change in properties that occurs at ambient or moderately elevated temperatures after hot working or a heat treating operation (quench aging in ferrous alloys), or after a cold working operation (strain aging). The change in properties is often, but not always, due to a phase change (precipitation), but does not involve a change in chemical composition.
vieillissement. Un changement de propriétés qui se produit à la température ambiante ou à des températures modérément élevées après formage à chaud ou après une opération de traitement thermique (vieillissement après trempe pour les alliages ferreux), ou après une opération d'écrouissage (vieillissement sous tension). Le changement de propriétés est souvent, mais pas toujours, dû à un changement de phase (précipitation), mais n'implique pas de changement de la composition chimique.

allotriomorph. A particle of a phase that has no regular external shape.
allotriomorphe. Une particule d'une phase qui n'a pas de forme externe régulière.

allotropy. The property whereby certain elements may exist in more than one crystal structure. See *polymorphism*.
allotropie. La propriété par laquelle certains éléments peuvent exister sous plus qu'une structure de cristal. Voir *polymorphisme*.

alloy. A substance having metallic properties and composed of two or more chemical elements of which at least one is a metal.
alliage. Une substance ayant des propriétés métalliques et composée de deux éléments chimiques ou plus dont au moins un est un métal.

alloy steel. Steel containing significant quantities of alloying elements (other than carbon and the commonly accepted amounts of manganese, silicon, sulfur and phosphorus) added to effect changes in mechanical or physical properties. Those containing less than 5% total metallic alloying elements tend to be termed low-alloy steels, and those containing more than 5% tend to be termed high-alloy steels.
acier allié. Acier contenant des quantités significatives d'éléments d'alliage (autre que le carbone et les quantités communément acceptées de manganèse, de silicium, de soufre, et de phosphore) ajoutés pour

effectuer des changements sur les propriétés mécaniques ou physiques. Ceux contenant moins que 5% au total en éléments d'alliage métalliques sont nommés acier faiblement alliés, et ceux contenant plus que 5% sont nommés aciers fortement alliés.

alloying element. An element added to a metal, and remaining in the metal, that effects changes in structure and properties.
élément d'alliage. Un élément ajouté à un métal, et demeurant dans le métal, qui produit des changements de structure et de propriétés.

angstrom unit (Å). A unit of linear measure equal to 10^{-10} m, or 0.1 nm; not an accepted Si unit, but still sometimes used for small distances such as interatomic distances and some wavelengths.
unité angström (Å). Une unité de mesure linéaire égale à 10^{-10} m, ou 0.1 nm; une unité qui n'est pas acceptée dans le SI, mais qui est encore quelquefois utilisée pour de petites distances telles que les distances interatomiques et certaines longueurs d'onde.

annealing. A general term denoting treatment of metal that involves heating to a suitable temperature followed by cooling at a suitable rate to produce discrete changes in microstructure and properties. See full *annealing, homogenizing annealing, spheroidizing annealing and subcritical annealing.* Full annealing is implied when the term "annealing" is used without qualification.
recuit. Un terme général indiquant le traitement d'un métal qui implique le chauffage jusqu'à une température appropriée suivi par un refroidissement à une vitesse appropriée afin de produire des changements distincts de microstructure et de propriétés. Voir *recuit complet, recuit d'homogénéisation, recuit de globulisation et recuit subcritique.* Le recuit complet est impliqué quand le terme "recuit" est utilisé sans restriction.

annealing twin. A twin formed in a metal during an annealing heat treatment.
macle de recuit. Une macle formée dans un métal lors d'un traitement thermique de recuit.

arc welding. A group of welding processes wherein the metal or metals being joined are coalesced by heating with an arc, with or without the application of pressure and with or without the use of filler metal.
soudage à l'arc. Un groupe de procédés de soudage par lesquels le métal ou les métaux à assembler sont fusionnés par chauffage avec un arc, avec ou sans application de pression et avec ou sans l'utilisation d'un métal d'apport.

artifact. In microscopy, a false structure introduced during preparation of a specimen.

artefact. En microscopie, une fausse structure introduite durant la préparation d'un spécimen.

austempering. Cooling an austenitized steel at a rate high enough to suppress formation of high-temperature transformation products, then holding the steel at a temperature below that for pearlite formation and above that for martensite formation until transformation to an essentially bainitic structure is complete.

trempe étagée bainitique. Refroidir un acier austénitisé à une vitesse suffisamment élevée afin de supprimer la formation de produits de transformation de haute température, ensuite maintenir l'acier à une température au-dessous de la température de formation de la perlite et au-dessus de celle de la formation de la martensite jusqu'à ce que la transformation en une structure essentiellement bainitique soit complète.

austenite. Generally, a solid solution of one or more alloying elements in the fcc polymorph of iron (γ-Fe). Specifically, in carbon steels, the interstitial solid solution of carbon in γ-Fe.

austénite. Généralement, une solution solide d'un ou plusieurs éléments d'alliage dans le fer polymorphe cfc (Fe-γ). Spécifiquement, dans les aciers au carbone, la solution solide interstitielle du carbone dans le Fe-γ.

austenitic grain size. The size of the grains in steel heated into the austenitic region.

grosseur de grain austénitique. La grosseur des grains de l'acier chauffé dans la région austénitique.

austenitizing. Forming austenite by heating a steel to between A_1 and A_3 or between A_1 and A_{cm} (partial austenitizing), or above A_1 or A_{cm} (complete austenitizing). When used without qualification, the term implies complete austenitizing.

austénitisation. Formation d'austénite en chauffant un acier entre A_1 et A_3 ou entre A_1 et A_{cm} (austénitisation partielle), ou au-delà de A_1 ou A_{cm} (austénitisation complète). Lorsqu'utilisé sans restriction, le terme implique l'austénitisation complète.

autoradiograph. A radiograph recorded photographically by radiation spontaneously emitted by radioisotopes that are produced in, or added to, the material. This technique identifies the locations of the radioisotopes.

autoradiographie. Une radiographie enregistrée photographiquement par la radiation émise spontanément par des radio-isotopes qui sont produits (ou ajoutés) dans le matériau.   Cette technique identifie l'emplacement des radio-isotopes.

autotempering. Tempering occurring immediately after martensite has formed, either as the martensite cools from the M_s temperature to room temperature or at room temperature. Also known as self tempering.
auto-revenu. Revenu qui se produit immédiatement après que la martensite se soit formée, soit lorsque la martensite se refroidie de la température M_s à la température de la pièce ou soit à la température de la pièce.

B

bainite. A eutectoid transformation product of ferrite and a fine dispersion of carbide, generally formed at temperatures below 840 to 930°F (450 to 500°C): upper bainite is an aggregate containing parallel lath-shape units of ferrite, produces the so-called "feathery" appearance in optical microscopy, and is formed at temperatures above about 660°F (350°C); lower bainite consists of individual plate-shape units and is formed at temperatures below about 660°F (350°C).
bainite. Un produit de la transformation eutectoïde de la ferrite et d'une dispersion fine de carbure, généralement formée à des températures sous 840 à 930°F (450 à 500°C): la bainite supérieure est un agrégat contenant des unités parallèles de ferrite en forme de latte, produit une apparence "plumeuse" sous le microscope optique, et est formée à des températures au-delà d'environ 660°F (350°C); la bainite inférieure consiste en unités individuelles en forme de plaque et est formée à des températures au-dessous d'environ 660°F (350°C).

bamboo grain structure. A structure in wire or sheet in which the boundaries of the grains tend to be aligned normal to the long axis and to extend completely through the thickness.
structure de grain en bambou. Une structure dans un fil ou une tôle dans laquelle les joints de grain tendent à s'aligner perpendiculairement à l'axe long et à s'étendre complètement à travers l'épaisseur.

banding. Inhomogeneous distribution of alloying elements or phases aligned in filaments or plates parallel to the direction of working. See *ferrite-pearlite banding* and *segregation banding.*
bande. Distribution hétérogène des éléments d'alliage ou des phases alignés en filaments ou en plaques parallèles à la direction de formage. Voir *bande de ferrite-perlite* et *ségrégation en bande.*

billet. A solid, semifinished steel round or square product that has been hot worked by forging, rolling or extrusion; usually smaller than a bloom. *billette (ou larget).* Un produit d'acier solide semi-fini rond ou carré qui a été formé à chaud par forgeage, par laminage ou par extrusion; habituellement plus petit qu'un bloom (brame).

blister steel. High-carbon steel produced by carburizing wrought iron. The bar, originally smooth, is covered with small blisters when removed from the cementation (carburizing) furnace. *acier comportant des soufflures (ou "pailles").* Acier à haut carbone produit par cémentation du fer corroyé. La barre, originellement lisse, est couverte par de petites ampoules lorsqu'elle est retirée du four de cémentation.

bloom. (1) Ancient: iron produced in a solid condition directly by the reduction of ore in a primitive furnace. The carbon content is variable but usually low. Also known as bloomery iron. The earliest ironmaking process, but still used in underdeveloped countries. (2) Modern: a semifinished hot rolled steel product, rectangular in section, usually produced on a blooming mill but sometimes made by forging. *bloom (brame). (1)* Ancien: fer produit directement en une forme solide par la réduction du minerai dans un four primitif. Le contenu en carbone est variable mais habituellement bas. Aussi connu sous le nom de fer de bloomerie. Le premier procédé de fabrication du fer, mais encore utilisé dans les pays sous-développés. *(2)* Moderne: un produit semi-fini en acier laminé à chaud, de section rectangulaire, habituellement produit dans un laminoir dégrossisseur mais quelquefois fabriqué par forgeage.

bloomery. A primitive furnace used for direct reduction of ore to iron. *bloomerie.* Un four primitif utilisé pour la réduction directe du minerai de fer.

box annealing. Annealing a metal or alloy in a sealed container under conditions that minimize oxidation. *recuit en caisse.* Recuit d'un métal ou d'un alliage dans un récipient scellé sous des conditions qui minimisent l'oxydation.

braze welding. A group of welding processes in which metals are joined by a filler metal that has a liquidus temperature below the solidus of the parent metal, but above 840°F (450°C). The filler metal is not distributed by capillary action. Compare brazing. *soudo-brasage.* Un groupe de procédés de soudage par lesquels les métaux sont joints par un métal d'apport qui a une température de liquidus sous le solidus du métal parent, mais au-dessus de 840°F

(450°C). Le métal d'apport n'est pas distribué par l'action capillaire. Comparer avec *brasage*.

brazing.   A welding process in which a third metal or alloy of comparatively low melting point is melted and caused, sometimes with the assistance of a flux, to wet the surfaces to be joined.
brasage.   Un procédé de soudage par lequel un troisième métal ou alliage, ayant une température de fusion comparativement basse, est mis en fusion et forcé, quelquefois avec l'assistance d'un flux, à mouiller les surfaces à joindre.

brittle fracture.   Fracture preceded by little or no plastic deformation.
rupture fragile.   Rupture précédée par peu ou pas de déformation plastique.

brittleness.   The tendency of a material to fracture without first undergoing significant plastic deformation.
fragilité.   La tendance d'un matériau à se rupturer sans d'abord passer par une déformation plastique significative.

buffer.   A substance added to aqueous solutions to maintain a constant hydrogen-ion concentration, even in the presence of acids or alkalis.
tampon.   Une substance ajoutée aux solutions aqueuses afin de maintenir constante la concentration d'ion hydrogène, même en présence d'acides ou d'alcalis.

burning. **(1)** During austenitizing, permanent damage of a metal or alloy by heating to cause incipient melting; see *overheating*. **(2)** During subcritical annealing, particularly in continuous annealing, production of a severely decarburized and grain-coarsened surface layer that results from heating to an excessively high temperature for an excessively long time.
brûlure. **(1)** Durant l'austénitisation, dommage permanent d'un métal ou d'un alliage par chauffage causant un début de fusion; voir *surchauffe*. **(2)** Durant le recuit subcritique, particulièrement lors de recuit continu, production d'une couche de surface, sévèrement décarburée et à gros grains, qui résulte du chauffage à une température excessivement haute pour une durée excessivement longue.

C

capped steel.   Semikilled steel cast in a bottle-top mold and covered with a cap fitting into the top of the mold to cause the top metal to solidify. Pressure increases in the sealed-in molten metal, resulting in a surface condition in the ingot similar to that of rimmed steel.

acier coiffé. Acier semi-calmé, moulé dans un moule à tête en forme de bouteille et couvert avec un larget de fermeture s'ajustant à la tête du moule, forçant le métal à la tête à se solidifier. La pression augmente dans le métal fondu scellé, résultant en une condition de surface du lingot similaire à celle de l'acier effervescent.

carbide. A compound of carbon with one or more metallic elements.
carbure. Un composé du carbone avec un ou plusieurs éléments métalliques.

carbon equivalent (for rating of weldability). A value that takes into account the equivalent additive effects of carbon and other alloying elements on a particular characteristic of a steel. For rating of weldability, a formula commonly used is:
$$CE = C + (Mn/6) + [(Cr + Mo + V)/5] + [(Ni + Cu)/15].$$
équivalent carbone (pour estimer la soudabilité). Une valeur qui prend en compte les effets équivalents additifs du carbone et de d'autres éléments d'alliage sur une caractéristique particulière d'un acier. Pour estimer la soudabilité, une formule communément utilisée est:
$$EC = C + (Mn/6) + [(Cr + Mo + V)/5] + [(Ni + Cu)/15].$$

carbon potential. A measure of the capacity of an environment containing active carbon to alter or maintain, under prescribed conditions, the carbon concentration in a steel.
potentiel carbone. Une mesure de la capacité d'un environnement contenant du carbone actif à altérer ou maintenir, sous des conditions prescrites, la concentration en carbone d'un acier.

carbon restoration. Replacing the carbon lost in the surface layer during previous processing by carburizing this layer to the original carbon level.
traitement de recarburation. Remplacer le carbone perdu à la couche de surface durant les procédés antérieurs, par cémentation de cette couche, afin de restaurer le niveau original de carbone.

carbon steel. A steel containing only residual quantities of elements other than carbon, except those added for deoxidation or to counter the deleterious effects of residual sulfur. Silicon is usually limited to about 0.60%, and manganese to about 1.65%. Also termed plain carbon steel, ordinary steel, straight carbon steel.
acier au carbone. Un acier contenant seulement des quantités résiduelles d'éléments autres que le carbone, à l'exception de ceux ajoutés pour la désoxydation ou pour contrer les effets nuisibles du soufre résiduel. Le silicium est habituellement limité à environ 0.60% et le manganèse à environ 1.65%. Aussi appelé acier ordinaire.

carbonitriding. A case hardening process in which a suitable ferrous material is heated above the lower transformation temperature in a gaseous atmosphere having a composition that results in simultaneous absorption of carbon and nitrogen by the surface and, by diffusion, creates a concentration gradient. The process is completed by cooling at a rate that produces the desired properties in the workpiece.

carbonitruration. Un procédé de durcissement de surface par lequel un matériau ferreux approprié est chauffé au-dessus de la température inférieure de transformation dans une atmosphère gazeuse ayant une composition qui résulte en l'absorption simultanée de carbone et d'azote à la surface et, par diffusion, crée un gradient de concentration. Le procédé est achevé en refroidissant à une vitesse qui produit les propriétés désirées dans la pièce.

carburizing. A process in which an austenitized ferrous material is brought into contact with a carbonaceous atmosphere having sufficient carbon potential to cause absorption of carbon at the surface and, by diffusion, create a concentration gradient.

cémentation. Un procédé par lequel un matériau ferreux austénétisé est amené en contact avec une atmosphère carbonée ayant un potentiel carbone suffisant pour causer l'absorption de carbone à la surface et, par diffusion, créer un gradient de concentration.

case. In a ferrous alloy, the outer portion that has been made harder than the inner portion, or core (see *case hardening)*.

couche superficielle traitée. Dans un alliage ferreux, la partie extérieure qui a été rendue plus dure que la partie intérieure, ou noyau (voir *traitement de durcissement de "surface"*).

case hardening. A generic term covering several processes applicable to steel that change the chemical composition of the surface layer by absorption of carbon or nitrogen, or a mixture of the two, and, by diffusion, create a concentration gradient.

traitement de durcissement de surface (trempe superficielle par cémentation). Un terme générique couvrant plusieurs procédés applicables à l'acier qui changent la composition chimique de la couche de surface par absorption de carbone ou d'azote, ou un mélange des deux, et, par diffusion, créent un gradient de concentration.

cast iron. Iron containing more carbon than the solubility limit in austenite (about 2%).

fonte de moulage. Fer contenant plus de carbone que la limite de solubilité de l'austénite (environ 2%).

cast steel. Steel in the form of castings, usually containing less than 2% carbon.
acier moulé. Acier en forme de moulage, contenant habituellement moins que 2% de carbone.

casting. (1) An object at or near finished shape obtained by solidification of a substance in a mold. (2) Pouring molten metal into a mold to produce an object of desired shape.
moulage. (1) Un objet à forme finie ou presque finie obtenu par la solidification d'une substance dans un moule. *(2)* Action de verser le métal fondu dans un moule afin de produire un objet d'une forme désirée.

cementation. (1) Introduction of one or more elements into the outer layer of a metal object by means of diffusion at high temperature. (2) An obsolete process used to convert wrought iron to blister steel by carburizing. Wrought iron bars were packed in sealed chests with charcoal and heated at about 2000°F (1100°C) for 6 to 8 days. Cementation was the predominant method of manufacturing steels, particularly high-carbon tool steels, prior to the introduction of the bessemer and open-hearth methods.
cémentation. (1) Introduction d'un ou plusieurs éléments dans la couche extérieure d'un objet en métal au moyen de la diffusion à haute température. *(2)* Un procédé désuet utilisé pour convertir le fer corroyé en acier comportant des soufflures par cémentation. Les barres de fer corroyé sont emballées dans des caisses scellées avec du charbon de bois et chauffées à environ 2000°F (1100°C) pendant 6 à 8 jours. La cémentation était la méthode prédominante de fabrication de l'acier, particulièrement des aciers à outils à haut carbone, avant l'introduction des méthodes bessemer et à four Martin.

cementite. A metastable carbide, with composition Fe_3C and orthorhombic crystal structure, having limited substitutional solubility for the carbide-forming elements, notably manganese.
cémentite. Un carbure métastable, de composition Fe_3C et avec un cristal de structure orthorhombique, ayant une solubilité substitutionnelle limitée pour les éléments carburigènes, notablement le manganèse.

chafery. A charcoal-fired furnace used in early ironmaking processes to reheat a bloom of wrought iron for forging to consolidate the iron and expel entrapped slag.
chaferie. Un four chauffé au charbon de bois utilisé dans les anciens procédés de fabrication du fer pour réchauffer un bloom de fer corroyé pour forger afin de consolider le fer et expulser la scorie emprisonnée.

chemical polishing. Improving the specular reflectivity of a metal surface by chemical treatment.
polissage chimique. Amélioration de la réflexivité spéculaire de la surface d'un métal par traitement chimique.

cleavage. Fracture of a crystal by crack propagation across a crystallographic plane of low index.
clivage. Rupture d'un cristal par propagation d'une fissure à travers un plan cristallographique à bas index.

cleavage fracture. Fracture of a grain, or most of the grains, in a polycrystalline metal by cleavage, resulting in bright reflecting facets.
rupture par clivage. Rupture d'un grain, ou de la plupart des grains, dans un métal polycristallin par clivage, résultant en des facette réfléchissant avec éclat.

cleavage plane. A characteristic crystallographic plane or set of planes in a crystal on which cleavage fracture occurs easily.
plan de clivage. Un plan ou ensemble de plans cristallographiques caractéristiques d'un cristal où la rupture par clivage se produit facilement.

columnar structure. A structure consisting of elongated grains whose tong axes are parallel.
structure basaltique. Une structure consistant en grains allongés dont les axes en tenaille sont parallèles.

constituent. A phase, or combination of phases, that occurs in a characteristic configuration in a microstructure.
constituant. Une phase, ou combinaison de phases, qui se produit en une configuration caractéristique dans une microstructure.

constitutional diagram. A graphical representation of the temperature and composition limits of phase fields in an alloy system as they actually exist under specific conditions of heating and cooling (synonymous with phase diagram). A constitutional diagram may be, or may approximate, an equilibrium diagram, or may represent metastable conditions or phases. Compare equilibrium diagram.
diagramme de constitution. Une représentation graphique des limites de température et de composition des champs de phase d'un système d'alliage tels qu'ils existent actuellement sous des conditions spécifiques de chauffage et de refroidissement (synonyme de diagramme de phase). Un diagramme de constitution peut être ou peut approcher un diagramme d'équilibre ou peut représenter des conditions ou des phases métastables. Comparer avec *diagramme d'équilibre.*

continuous phase. In an alloy or portion of an alloy containing more than one phase, the phase that forms the background or matrix in which the other phase or phases are present as isolated volumes.
phase continue. Dans un alliage ou portion d'alliage contenant plus qu'une phase, la phase qui forme le fond ou matrice dans laquelle l'autre phase ou les autres phases sont présentes en tant que volumes isolés.

controlled rolling. A hot rolling process in which the temperature of the steel is closely controlled, particularly during the final rolling passes, to produce a fine-grain microstructure.
laminage contrôlé. Un procédé de laminage à chaud dans lequel la température de l'acier est contrôlée de près, particulièrement durant les passes finales de laminage, afin de produire une microstructure à grains fins.

coring. A variation of composition between the center and surface of a unit of structure (such as a dendrite, a grain or a carbide particle) resulting from nonequilibrium growth over a range of temperature.
noyautage; cernes. Une variation de la composition entre le centre et la surface d'une unité de structure (telle qu'un dendrite, un grain ou une particule de carbure) résultant de la croissance hors d'équilibre sur une certaine étendue de température.

corrosion.   Deterioration of a metal by chemical or electrochemical reaction with its environment.
corrosion.   Détérioration d'un métal par réaction chimique ou électrochimique avec son environnement.

creep. Time-dependent strain occurring under stress.
fluage. Déformation dépendant de la durée et se produisant sous contrainte.

critical cooling rate. The limiting rate at which austenite must be cooled to ensure that a particular type of transformation product is formed.
vitesse de refroidissement critique. La vitesse limite à laquelle l'austénite doit être refroidie pour assurer qu'un type particulier de produit de transformation soit formé.

critical point. (1) The temperature or pressure at which a change in crystal structure, phase or physical properties occurs; same as *transformation temperature.* (2) In an equilibrium diagram, that specific combination of composition, temperature and pressure at which the phases of an inhomogeneous system are in equilibrium.

point critique (ou température de transformation). *(1)* La température ou pression à laquelle un changement de structure du cristal, de la phase, ou des propriétés physiques, se produit; même chose que *température de transformation.* *(2)* Dans un diagramme d'équilibre, cette combinaison spécifique de composition, température et pression à laquelle les phases d'un système hétérogène sont en équilibre.

critical strain. That strain which results in the formation of very large grains during recrystallization.
déformation critique. Cette déformation qui résulte en la formation de très gros grains durant la recristallisation.

critical temperature. Synonymous with *critical point* if pressure is constant.
température de transformation. Synonyme de *point critique* si la pression est constante.

cross rolling. A (hot) rolling process in which rolling reduction is carried out in a direction perpendicular to, as well as a direction parallel to, the length of the original slab.
laminage avec cylindres obliques. Un procédé de laminage à chaud par lequel la réduction de laminage est effectuée dans une direction perpendiculaire et aussi dans une direction parallèle à la longueur de la dalle d'origine.

crucible steel. High-carbon steel produced by melting blister steel in a covered crucible. Crucible steel was developed by Benjamin Huntsman in about 1750 and remained in use until the late 1940's.
acier au creuset. Acier à haut carbone produit par la fonte de l'acier à soufflures dans un creuset couvert. L'acier au creuset a été développé par Benjamin Huntsman vers 1750 et est demeuré en usage jusque vers la fin des années 1940.

crystalline fracture. A fracture of a polycrystalline metal characterized by a grainy appearance. Compare fibrous fracture.
rupture cristalline. Une rupture d'un métal polycristallin caractérisée par une apparence granuleuse. Comparer avec *rupture fibreuse.*

D

decarburization. Loss of carbon from the surface of a ferrous alloy as a result of heating in a medium that reacts with carbon.
décarburation. Perte du carbone à la surface d'un alliage ferreux résultant du chauffage dans un médium qui réagit avec le carbone.

decoration (of dislocations). Segregation of solute atoms to the line of a dislocation in a crystal. In ferrite, the dislocations may be decorated with carbon or nitrogen atoms.
garniture (de dislocations). Ségrégation d'atomes de soluté le long d'une dislocation dans un cristal. Dans la ferrite, les dislocations peuvent être garnies avec des atomes de carbone ou d'azote.

deformation bands. Generally, bands in which deformation has been concentrated inhomogeneously. See *kink band, microbands* and *shear bands* for specific types.
bandes de déformation. Généralement, des bandes dans lesquelles la déformation a été concentrée hétérogènement. Voir *bande de contorsion, microbandes et bandes de cisaillement* pour des types spécifiques.

degenerate structure. Usually refers to pearlite that does not have an ideally lamellar structure. The degree of degeneracy may vary from slight perturbations in the lamellar arrangement to structures that are not recognizably lamellar.
structure dégénérée. Réfère habituellement à la perlite qui n'a pas une structure lamellaire idéale. Le degré de dégénération peut varier de perturbations légères dans l'arrangement lamellaire à des structures dont on ne reconnaît plus les lamelles.

dendrite. A crystal that has grown in treelike branching mode.
dendrite. Un cristal qui a crû en une forme à l'allure de branches d'arbre.

dendritic segregation. Inhomogeneous distribution of alloying elements through the arms of dendrites.
ségrégation dendritique. Distribution hétérogène d'éléments d'alliage dans les bras des dendrites.

deoxidation. (1) Removal of oxygen from molten metals by use of suitable chemical agents. (2) Sometimes refers to removal of undesirable elements other than oxygen by the introduction of elements or compounds that readily react with them.
désoxydation. *(1)* Suppression de l'oxygène des métaux fondus par l'utilisation d'agents chimiques appropriés. *(2)* Réfère quelquefois à la suppression d'éléments indésirables autre que l'oxygène par l'introduction d'éléments ou de composés qui réagissent promptement avec eux.

diffusion. (1) Spreading of a constituent in a gas, liquid or solid, tending to make the composition of all parts uniform. (2) The spontaneous movement of atoms or molecules to new sites within a material.

diffusion. *(1)* Propagation d'un constituant dans un gaz, un liquide ou un solide, ayant tendance à rendre la composition de toutes les parties uniforme. *(2)* Le mouvement spontané des atomes ou des molécules vers de nouveaux sites dans un matériau.

dilatometer. An instrument for measuring the expansion or contraction of a solid metal resulting from heating, cooling, polymorphic changes, etc.

dilatomètre. Un instrument mesurant la dilatation ou le retrait d'un métal solide résultant du chauffage, du refroidissement, de changements polymorphes, etc.

dislocation. A linear defect in the structure of a crystal.

dislocation. Un défaut linéaire dans la structure d'un cristal.

ductility. The capacity of a material to deform plastically without fracturing.

ductilité. La capacité d'un matériau à se déformer plastiquement sans se rupturer.

E

elastic limit. The maximum stress to which a material may be subjected without any permanent strain remaining upon complete release of the stress.

limite d'élasticité. La contrainte maximum à laquelle un matériau peut être soumis sans qu'il reste de déformation permanente lors du relâchement complet de la contrainte.

elastic strain. Dimensional changes accompanying stress where the original dimensions are restored upon release of the stress.

déformation élastique. Changements dimensionnels accompagnant la contrainte où les dimensions originales sont restaurées lors du relâchement de la contrainte.

electron beam microprobe analyzer. An instrument for selective chemical analysis of a small volume of material. An electron beam bombards the area of interest and x-radiation thereby emitted is analyzed in a spectrometer.

analyseur par microsonde à rayon d'électrons. Un instrument pour l'analyse chimique sélective d'un petit volume de matériau. Un rayon d'électrons bombarde la région d'intérêt et des rayons-x sont ainsi émis et analysés dans un spectromètre.

electropolishing. Improving the specular reflectivity of a metal surface by electrochemical dissolution.
polissage électrolytique. Amélioration de la réflexivité spéculaire de la surface d'un métal par dissolution électrochimique.

elongation after fracture. In tensile testing, the increase in the gauge length measured after fracture of the specimen within the gauge length and usually expressed as a percentage of the original gauge length.
allongement après rupture. En essai de traction, l'augmentation de la longueur de la jauge mesurée après la rupture du spécimen à l'intérieur de la longueur de la jauge et habituellement exprimée en pourcentage de la longueur originelle de la jauge.

epitaxy. Induced orientation of the lattice of a crystal of a surface deposit by the lattice of the substrate crystal.
épitaxie. Orientation induite du réseau de cristal d'un dépôt de surface par le réseau du cristal substrat.

equiaxed structure. A structure in which the grains have approximately the same dimensions in all directions.
structure équiaxe. Une structure dans laquelle les grains ont approximativement les mêmes dimensions dans toutes les directions.

equilibrium diagram. A graphical representation of the temperature, pressure and composition limits of phase fields in an alloy system as they exist under conditions of thermodynamical equilibrium. In condensed systems, pressure is usually considered constant.
diagramme d'équilibre. Une représentation graphique de la température, de la pression et des limites de composition des champs de phase dans un système d'alliage tels qu'ils existent sous des conditions d'équilibre thermodynamique. Dans les systèmes condensés, la pression est habituellement considérée constante.

etchant. A chemical solution used to etch a metal to reveal structural details.
réactif d'attaque. Une solution chimique utilisée pour attaquer un métal afin de révéler les détails structuraux.

etching. Subjecting the surface of a metal to preferential chemical or electrolytic attack to reveal structural details.
attaque. Soumettre la surface d'un métal à l'attaque chimique ou électrolytique préférentielle afin de révéler les détails structuraux.

eutectoid. (1) An isothermal reversible transformation in which a solid solution is converted into two or more intimately mixed solids, the number of solids formed being the same as the number of components in the system.   **(2)** An alloy having the composition indicated by the eutectoid point on an equilibrium diagram.   **(3)** An alloy structure of intermixed solid constituents formed by a eutectoid transformation.

eutectoïde.   *(1)* Une transformation isotherme réversible dans laquelle une solution solide est convertie en deux ou plusieurs solides intimement mélangés, le nombre de solides formé étant le même que le nombre de composants dans le système.   *(2)* Un alliage ayant la composition indiquée par le point eutectoïde dans un diagramme d'équilibre.   *(3)* La structure d'un alliage de constituants solides entremélangés formés par une transformation eutectoïde.

F

fatigue.   The phenomenon leading to fracture under repeated or fluctuating stresses (having a maximum value less than the tensile strength of the material).

fatigue.   Le phénomène conduisant à la rupture sous des contraintes répétées ou variables (ayant une valeur maximale moindre que la résistance à la tension du matériau).

ferrite.   Generally, a solid solution of one or more alloying elements in the bcc polymorph of iron (α-Fe).   Specifically, in carbon steels, the interstitial solid solution of carbon in α-Fe.

ferrite.   Généralement, une solution solide d'un ou plusieurs éléments d'alliage du polymorphe cubique centré du fer (fer-α).   Spécifiquement, dans les aciers au carbone, la solution solide interstitielle du carbone dans le fer-α.

ferrite-pearlite banding.   Inhomogeneous distribution of ferrite and pearlite aligned in filaments or plates parallel to the direction of working.

bande ferrite-perlite.   Distribution hétérogène de la ferrite et de la perlite alignée en filaments ou plaques parallèles à la direction de formage.

ferritic grain size.   The grain size of the ferritic matrix of a steel.

grosseur de grain ferritique.   La grosseur de grain de la matrice ferritique d'un acier.

fiber.   **(1)** The characteristic of wrought metal that indicates directional properties.   It is revealed by etching a longitudinal section or manifested by the fibrous appearance of a fracture.   It is caused chiefly by extension of the constituents of the metal, both metallic and nonmetallic, in the direction of working. (2) The pattern of preferred orientation of metal crystals after a given deformation process. See *preferred orientation*.

fibre. *(1)* La caractéristique du métal corroyé qui indique les propriétés directionnelles. Elle est révélée en attaquant une section longitudinale, ou se manifeste par l'apparence fibreuse d'une rupture. Elle est causée principalement par l'extension des constituants du métal, autant métalliques que non métalliques dans la direction du formage. *(2)* Le motif d'orientation préférentielle des cristaux de métal après un procédé donné de déformation. Voir *orientation préférentielle*.

fibrous fracture. A fracture whose surface is characterized by a dull gray or silky appearance. Compare *crystalline fracture*.

rupture fibreuse. Une rupture dont la surface est caractérisée par un gris mat ou une apparence soyeuse. Comparer avec *rupture cristalline*.

filler metal. A third material that is melted concurrently with the parent metals during fusion or braze welding. It is usually, but not necessarily, of different composition from the parent metals.

métal d'apport. Un troisième matériau qui est fondu concurremment avec les métaux parents durant le soudage par fusion ou le soudo-brasage.   Il est habituellement, mais pas nécessairement, d'une composition différente de celle des métaux parents.

finery. A charcoal-fueled hearth furnace used in early processes for converting cast iron to wrought iron by melting and oxidizing it in an air blast, then repeatedly oxidizing the product in the presence of a slag. The carbon oxidizes more rapidly than the iron so that a wrought iron of low carbon content is produced.

affinerie. Un four à sole alimenté au charbon de bois utilisé dans les anciens procédés pour convertir la fonte de moulage en fer corroyé en la fondant et en l'oxydant dans un jet d'air, ensuite en oxydant le produit répétitivement en présence d'une scorie.   Le carbone s'oxyde plus rapidement que le fer de telle sorte qu'un fer corroyé d'un contenu peu élevé en carbone est produit.

finishing temperature. The temperature at which hot working is completed.

température de finissage. La température à laquelle le formage à chaud est complété.

flash. (1) In forging, the excess metal forced between the upper and lower dies. (2) In resistance butt welding, a fin formed perpendicular to the direction of applied pressure.

bavure. *(1)* En travail de forge, l'excès de métal forcé entre la matrice supérieure et inférieure. *(2)* En soudage en bout par résistance, une ailette formée perpendiculairement à la direction de la pression appliquée.

flow lines. (1) Texture showing the direction of metal flow during hot or cold working. Flow lines often can be revealed by etching the surface or a section of a metal part. (2) In mechanical metallurgy, paths followed by volume elements of metal during deformation.
lignes d'écoulement (lignes de Piobert-Lüders). *(1)* Texture montrant la direction de l'écoulement du métal durant le formage à chaud ou à froid. Les lignes d'écoulement peuvent souvent être révélées en attaquant la surface ou une section d'une pièce métallique. *(2)* En métallurgie mécanique, chemins suivis par les éléments du volume de métal durant la déformation.

flow stress. The uniaxial true stress required to cause plastic deformation at a specified value of strain.
contrainte d'écoulement. La vraie contrainte uniaxiale requise pour causer une déformation plastique suivant une valeur spécifiée de déformation.

flux. (1) In refining, a material used to remove undesirable substances as a molten mixture. It may also be used as a protective covering for molten metal. (2) In welding, a material used to prevent the formation of, or to dissolve and facilitate the removal of, oxides and other undesirable substances.
flux (fondant). *(1)* En affinage, un matériau utilisé pour enlever les substances indésirables sous forme de mélange fondu. Il peut aussi être utilisé comme couverture protectrice pour le métal fondu. *(2)* En soudage, un matériau utilisé pour prévenir la formation des oxydes et autres substances indésirables ou pour les dissoudre et en faciliter l'enlèvement.

forge welding. Welding hot metal by applying pressure or blows.
soudage à la forge. Le soudage d'un métal chaud en appliquant de la pression ou des coups.

forging. Plastically deforming metal, usually hot, into desired shapes with compressive force, with or without dies.
forgeage. Déformer le métal plastiquement, habituellement chaud, en formes désirées, avec une force compressive, avec ou sans matrices.

fractography. Descriptive treatment of fracture, especially in metals, with specific reference to photography of the fracture surface.

fractographie. Traitement descriptif d'une fracture, spécialement des métaux, avec une référence spécifique à la photographie de la surface de la rupture.

fragmentation.   The subdivision of a grain into small discrete crystallites outlined by a heavily deformed network of intersecting slip bands as a result of cold working.   These small crystals or fragments differ from one another in orientation and tend to rotate to a stable orientation determined by the slip systems.

fragmentation. La subdivision d'un grain en petits cristallites distincts soulignés par un réseau fortement déformé de bandes de glissement s'entrecoupant, résultant du formage à froid.   Ces petits cristaux ou fragments diffèrent l'un de l'autre par leur orientation et tendent à pivoter en une orientation stable déterminée par les systèmes de glissement.

free machining. Pertains to the machining characteristics of an alloy to which one or more ingredients have been introduced to produce small broken chips, low power consumption, better surface finish or longer tool life.

facilement   usinable   (décolletage   rapide).   A   trait   aux caractéristiques d'usinage d'un alliage auquel un ou plusieurs ingrédients ont été introduits pour produire de petits copeaux brisés, une faible consommation d'électricité, une meilleure finition de la surface ou une plus longue durée de l'outil.

full annealing (ferrous materials). An annealing treatment in which a steel is austenitized by heating to a temperature above the upper critical temperature (A_3 or A_{cm}) and then cooled slowly to room temperature. A typical cooling rate would be 210°F/h 100°C/h. Compare normalizing. Use of the term "annealing" without qualification implies full annealing.

recuit   complet   (matériaux   ferreux).   Un traitement de recuit par lequel un acier est austénétisé en chauffant à une température au-delà de la température critique supérieure (A_3 ou A_{cm}) et ensuite refroidi lentement jusqu'à la température de la pièce.   Une vitesse de refroidissement typique serait 210°F/h (100°C/h).   Comparer avec normalisation. L'utilisation du terme "recuit" sans restriction implique le recuit complet.

fusion welding. Any welding process in which fusion is employed to complete the weld.

soudage par fusion. Tout procédé de soudage où la fusion est employée pour compléter la soudure.

G

gas welding. Welding with a gas flame.
soudage au gaz. Soudage avec une flamme au gaz.

grain. An individual crystal in a polycrystalline metal or alloy, including twinned regions or subgrains if present.
grain. Un cristal individuel dans un métal ou un alliage polycristallin, incluant les régions maclées ou les sous-grains si présents.

grain-boundary liquation. An advanced stage of overheating in which material in the region of austenitic grain boundaries melts. Also known as burning.
liquation de joint de grain. Une étape avancée de surchauffe où le matériau dans la région des joints de grains austénitiques fond. Aussi appelé "brûlure".

grain-boundary sulfide precipitation. An intermediate stage of overheating in which sulfide inclusions are redistributed to the austenitic grain boundaries by partial solution at the overheating temperature and reprecipitation during subsequent cooling.
précipitation de sulfure aux joints de grain. Une étape intermédiaire de la surchauffe durant laquelle les inclusions de sulfure sont redistribuées aux joints de grains austénitiques par solution partielle à la température de surchauffe et reprécipitation durant le refroidissement qui s'ensuit.

grain coarsening. A heat treatment that produces excessively large austenitic grains.
grossissement du grain. Un traitement thermique qui produit des grains austénitiques excessivement gros.

grain flow. Fiberlike lines appearing on polished and etched sections of forgings, caused by orientation of the constituents of the metal in the direction of working during forging.
écoulement de grain. Lignes en forme de fibres apparaissant sur les sections polies et attaquées des pièces forgées, causées par l'orientation des constituants du métal dans la direction du formage durant le forgeage.

grain growth. An increase in the average size of the grains in polycrystalline metal, usually a result of heating at elevated temperature.
croissance du grain. Une augmentation de la grosseur moyenne des grains dans un métal polycristallin, habituellement un résultat du chauffage à température élevée.

grain size.   A measure of the areas or volumes of grains in a polycrystalline metal or alloy, usually expressed as an average when the individual sizes are fairly uniform.   In metals containing two or more phases, the grain size refers to that of the matrix unless otherwise specified.   Grain size is reported in terms of number of grains per unit area or volume, average diameter, or as a number derived from area measurements.

grosseur de grain.   Une mesure de la surface ou du volume des grains dans un métal ou un alliage polycristallin, habituellement exprimé en une moyenne quand les grosseurs individuelles sont relativement uniformes.   Dans les métaux contenant deux ou plusieurs phases, la grosseur de grain réfère à celle de la matrice à moins que spécifié autrement.   La grosseur de grain est reportée en termes de nombre de grains par unité de surface ou de volume, de diamètre moyen, ou en un nombre dérivé de mesures de surface.

granular fracture.   A type of irregular surface produced when metal fractures, characterized by a rough, grainlike appearance as differentiated from a smooth silky, or fibrous, type.   It can be subclassified into transgranular and intergranular forms.   This type of fracture is frequently called crystalline fracture, but the implication that the metal has crystallized is completely misleading.

rupture granuleuse.   Un type irrégulier de surface produite quand le métal se rupture, caractérisée par une apparence rugueuse, à apparence de grain et différentiée d'un type lisse soyeux, ou fibreux.   Elle peut être sous-classée en formes intragranulaire et intergranulaire.   Ce type de rupture est fréquemment appelée rupture cristalline, mais l'implication que le métal s'est cristallisé est complètement erronée.

graphite.   The polymorph of carbon with a hexagonal crystal structure.

graphite.   Le polymorphe du carbone ayant une structure à cristal hexagonal.

graphitization.   Formation of graphite in iron or steel.   Primary graphitization refers to formation of graphite during solidification.   Secondary graphitization refers to later formation during heat treatment.

graphitisation.   Formation de graphite dans le fer ou l'acier.   La graphitisation primaire réfère à la formation de graphite pendant la solidification.   La graphitisation secondaire réfère à la formation plus tard durant le traitement thermique.

grinding.   Removing material from a workpiece with a grinding wheel or abrasive belt.

meulage (rectification).   Enlèvement de matériau d'une pièce avec une meule ou une courroie abrasive.

H

hardenability. In ferrous alloys, the property that determines the depth and distribution of hardness induced by quenching.
trempabilité. Chez les alliages ferreux, la propriété qui détermine la profondeur et la distribution de la dureté induite par trempage.

hardening. Increasing hardness by suitable treatment, usually involving heating and cooling. When applicable, the following more specific terms should be used: age hardening, case hardening, flame hardening, induction hardening, precipitation hardening, quench hardening.
traitement de durcissement. Augmenter la dureté par un traitement approprié, impliquant habituellement chauffage et refroidissement. Lorsqu'applicable, les termes plus spécifiques qui suivent devraient être utilisés: durcissement par vieillissement, traitement de durcissement de surface, traitement de durcissement par trempe avec chauffage à la flamme, traitement de durcissement par trempe avec chauffage par induction, durcissement par précipitation, traitement de durcissement par trempe.

hardness (indentation). Resistance of a metal to plastic deformation by indentation. Various hardness tests such as Brinell, Rockwell and Vickers may be used. In the Vickers test, a diamond pyramid with an included face angle of 136° is used as the indenter.
dureté à la bille (empreinte). Résistance d'un métal à la déformation plastique par empreinte. Des essais variés de dureté tels que Brinell, Rockwell et Vickers peuvent être utilisés. Dans l'essai Vickers, un diamant en pyramide avec une face incluse à angle de 136° est utilisée pour faire l'empreinte.

heat-affected zone. That portion of the base metal which was not melted during brazing, cutting or welding, but within which microstructure and physical properties were altered by the treatment.
zone d'effet thermique. Cette portion du métal de base qui n'a pas fondu durant le brasage, le coupage ou le soudage, mais à l'intérieur de laquelle la microstructure et les propriétés physiques ont été altérées par le traitement.

heat tinting. Colouration of a metal surface through oxidation by heating to reveal details of structure.
coloration par chaleur. Coloration de la surface d'un métal due à l'oxydation par chauffage afin de révéler les détails de structure.

heat treatment. Heating and cooling a solid metal or alloy in such a way that desired structures, conditions or properties are attained. Heating for the sole purpose of hot working is excluded from the meaning of this term.

traitement thermique. Chauffage et refroidissement d'un métal ou d'un alliage solide de telle sorte que des structures, des conditions ou des propriétés désirées sont atteintes. Chauffer avec le seul but de former à chaud est exclu de la signification de ce terme.

hematite. The oxide of iron of highest valency which has a composition close to the stoichiometric composition Fe_2O_3.

hématite. L'oxyde de fer de la plus haute valence qui a une composition près de la composition stoechiométrique Fe_2O_3.

homogenizing annealing. An annealing treatment carried out at a high temperature, approaching the solidus temperature, for a sufficiently long time that inhomogeneous distributions of alloying elements are reduced by diffusional processes.

recuit d'homogénéisation. Un traitement de recuit accompli à haute température, approchant la température du solidus, pour une durée suffisamment longue pour que la distribution hétérogène des éléments d'alliage soit réduite par des procédés de diffusion.

hot working.   Deformation under conditions that result in recrystallization.

formage à chaud. Déformation sous des conditions qui résultent en recristallisation.

hypereutectoid alloy. In a eutectoid system, any alloy containing more than the eutectoid concentration of solute.

alliage hypereutectoïde. Dans un système eutectoïde, tout alliage contenant plus que la concentration eutectoïde de soluté.

hypoeutectic alloy. In a eutectic system, any alloy containing less than the eutectic concentration of solute.

alliage hypoeutectique. Dans un système eutectique, tout alliage contenant moins que la concentration eutectique de soluté.

I

idiomorph. A particle of a phase that has a regular external shape.

idiomorphe. Une particule d'une phase qui a une forme extérieure régulière.

impact test. A test for determining the behaviour of materials when subjected to high rates of loading under conditions designed to promote fracture, usually in bending, tension or torsion. The quantity measured is the energy absorbed when the specimen is broken by a single blow.
essai de choc (de résilience). Un essai pour déterminer le comportement des matériaux quand ils sont soumis à de hauts taux de charge sous des conditions désignées pour provoquer la rupture, habituellement par pliage, traction ou torsion. La quantité mesurée est l'énergie absorbée quand le spécimen est cassé en un seul coup (essai).

impurities. Elements or compounds whose presence in a material is undesired.
impuretés. Éléments ou composés dont la présence dans un matériau est non désirée.

inclusion. A nonmetallic material in a solid metallic matrix.
inclusion. Un matériau non métallique dans une matrice métallique solide.

ingot. A casting suitable for hot working or remelting.
lingot. Un moulage approprié pour le formage à chaud ou pour la refusion.

ingot iron. Commercially pure iron.
fer lingot. Fer commercialement pur.

intercrystalline. Between crystals, or between grains. Same as intergranular.
intergranulaire. Entre les cristaux, ou entre les grains.

internal oxidation. Formation of oxides beneath the surface of a metal.
oxydation interne. Formation d'oxydes sous la surface d'un métal.

interstitial solid solution. A solid solution in which the solute atoms occupy (interstitial) positions between the atoms in the structure of the solvent. See also *substitutional solid solution.*
solution solide interstitielle. Une solution solide dans laquelle les atomes de soluté occupent des positions (interstitielles) entre les atomes dans la structure du solvant. Voir aussi *solution solide substitutionnelle.*

intracrystalline. Within or across crystals or grains. Same as *transcrystalline* and *transgranular.*
intragranulaire. A l'intérieur ou à travers les cristaux ou grains.

iron. An element that has an average atomic number of 55.85 and that always, in engineering practice, contains small but significant amounts of carbon. Thus iron-carbon alloys containing less than about 0.1% C may be referred to as irons. Alloys with higher carbon contents are always termed steels.

fer. Un élément qui a un nombre atomique moyen de 55.85 et qui contient toujours, en ingénierie, une quantité de carbone petite mais significative. Ainsi les alliages de fer-carbone contenant moins qu'environ 0.1% C peuvent être appelés fers. Les alliages avec un contenu en carbone plus élevé sont toujours appelés aciers.

isothermal transformation (IT) diagram. A diagram that shows the isothermal time required for transformation of austenite to commence and to finish as a function of temperature. Same as *time-temperature-transformation* (TTT) *diagram* or S-curve.

diagramme de transformation en conditions isothermes (diagramme TI). Un diagramme qui montre la durée isotherme requise pour que la transformation de l'austénite commence et finisse, en fonction de la température. Même chose que *diagramme temps-température-transformation (TTT)* ou courbe en S.

K

killed steel. Steel deoxidized with a strong deoxidizing agent, such as silicon or aluminum, to reduce the oxygen content to such a level that no reaction occurs between carbon and oxygen during solidification.

acier calmé. Acier désoxydé avec un agent désoxydant puissant, tel que le silicium ou l'aluminium, pour réduire le contenu en oxygène à un niveau tel qu'aucune réaction ne se produit entre le carbone et l'oxygène pendant la solidification.

kink band (deformation). In polycrystalline materials, a volume of crystal that has rotated physically to accommodate differential deformation between adjoining parts of a grain while the band itself has deformed homogeneously. This occurs by regular bending of the slip lamellae along the boundaries of the band.

bande de contorsion (déformation). Dans les matériaux polycristallins, un volume de cristal qui a pivoté physiquement pour accommoder la déformation différentielle entre les parties adjointes d'un grain alors que la bande elle-même s'est déformée homogénéiquement. Ceci se produit par pliage régulier des lamelles de glissement le long des joints de la bande.

L

lamellar tear. A system of cracks or discontinuities aligned generally parallel to the worked surface of a plate. Usually associated with a fusion weld in thick plate.
déchirure lamellaire. Un système de criques ou discontinuités généralement alignées parallèlement à la surface travaillée d'une tôle. Habituellement associée avec une soudure par fusion dans une tôle épaisse.

lamination. An abnormal structure resulting in a separation or weakness aligned generally parallel to the worked surface of the metal.
feuilletage. Une structure anormale résultant en une séparation ou faiblesse généralement alignée parallèlement à la surface travaillée du métal.

lath martensite. Martensite formed, partly in steels containing less than about 1.0% C and solely in steels containing less than about 0.5% C, as parallel arrays or packets of lath-shape units about 0.1 to 0.3 mm thick, and having a habit plane that is close to {111}.
martensite massive. Martensite formée, partiellement chez les aciers contenant moins qu'environ 1.0% C et exclusivement chez les aciers contenant moins qu'environ 0.5% C, en arrangements parallèles ou paquets d'unités en forme de latte d'environ 0.1 à 0.3 μm d'épaisseur, et ayant un plan matrice qui est près de {111}.

liquation. Partial melting of an alloy.
liquation. Fusion partielle d'un alliage.

liquidus. In a constitutional diagram, the locus of points representing the temperatures at which various components commence freezing on cooling or finish melting on heating.
liquidus. Dans un diagramme de constitution, le lieu des points représentant les températures auxquelles les différents composants commencent à se solidifier lors du refroidissement ou finissent de fondre lors du chauffage.

Luders lines or bands. Elongated surface markings or depressions caused by localized plastic deformation that results from discontinuous (inhomogeneous) yielding.
lignes ou bandes de Luders. Marques de surface allongées ou dépressions causées par la déformation plastique localisée qui résulte de l'écoulement discontinu (hétérogène).

M

M_f temperature. The temperature at which martensitic transformation is essentially complete during cooling after austenitization.
température M_f La température à laquelle la transformation martensitique est essentiellement complète durant le refroidissement après l'austénitisation.

M_s temperature. The temperature at which a martensitic transformation starts during cooling after austenitization.
température M_s. La température à laquelle la transformation martensitique débute durant le refroidissement après l'austénitisation.

machinability. The capacity of a material to be machined easily.
usinabilité. La capacité d'un matériau à être facilement usiné.

macroetching. Etching of a metal surface with the objective of accentuating gross structural details, for observation by the unaided eye or at magnifications not exceeding ten diameters.
macroattaque. Attaque de la surface d'un métal avec l'objectif d'accentuer les détails structuraux grossiers pour observation à l'oeil nu ou à des grossissements n'excédant pas dix diamètres.

macrograph. A graphic reproduction of a prepared surface of a specimen at a magnification not exceeding ten diameters. When photographed, the reproduction is known as a photomacrograph (not a macrophotograph).
macrographie. Une reproduction graphique d'une surface préparée d'un spécimen à un grossissement n'excédant pas dix diamètres. Lorsque photographiée, la reproduction est appelée une photomacrographie (et non pas macrophotographie).

macrostructure. The structure of a metal as revealed by examination of the etched surface at a magnification not exceeding ten diameters.
macrostructure. La structure d'un métal telle que révélée par examen de la surface attaquée à un grossissement n'excédant pas dix diamètres.

magnetite. The oxide of iron of intermediate valence which has a composition close to the stoichiometric composition Fe_3O_4.
magnétite. L'oxyde de fer de valence intermédiaire qui a une composition près de la composition stoechiométrique Fe_3O_4.

manual welding. Welding wherein the entire welding operation is performed and controlled by hand.

soudage manuel. Soudage où l'opération entière de soudage est accomplie et contrôlée à la main.

martempering. (1) A hardening procedure in which an austenitized ferrous material is quenched into an appropriate medium at a temperature just above the M_s temperature of the material, held in the medium until the temperature is uniform through-out - but not long enough for bainite to form - and then cooled in air. The treatment is frequently followed by tempering. (2) When the process is applied to carburized material, the controlling M_s temperature is that of the case. This variation of the process is frequently called marquenching.
trempe étagée martensitique. *(1)* Un procédé de durcissement par lequel un matériau ferreux austénitisé est trempé dans un milieu approprié à une température juste au-dessus de la température M_s du matériau, maintenu dans le milieu jusqu'à ce que la température soit uniforme partout - mais pas suffisamment longtemps pour que la bainite se forme - et ensuite refroidi à l'air. Le traitement est fréquemment suivi par un revenu. *(2)* Quand le procédé est appliqué à un matériau cémenté, la température de contrôle M_s est celle de la surface. Cette variation du procédé est fréquemment appelée "marquenching".

martensite. In steel, a metastable transition phase with a body-centred-tetragonal crystal structure formed by diffusionless transformation of austenite generally during cooling between the M_s and M_f, temperatures.
martensite. Dans l'acier, une phase de transition métastable avec une structure à cristal tétragonal centré formée par la transformation sans diffusion de l'austénite, généralement durant le refroidissement entre les températures M_s et M_f.

martensite range. The interval between the M_s, and M_f temperatures.
domaine martensitique. L'intervalle entre les températures M_s et M_f.

matrix. The principal phase or aggregate in which another constituent is embedded.
matrice. La phase principale, ou agrégat, dans laquelle un autre constituant est encastré.

mechanical twin. A twin formed in a metal during plastic deformation by simple shear of the structure.
macle mécanique. Une macle formée dans un métal durant la déformation plastique par simple cisaillement de la structure.

mechanical polishing. A method of producing a specularly reflecting surface by use of abrasives.

polissage mécanique. Une méthode produisant une surface spéculairement réfléchissante par l'utilisation d'abrasifs.

melting point. The temperature at which a pure metal, compound or eutectic changes from solid to liquid; the temperature at which the liquid and the solid are in equilibrium.
température de fusion. La température à laquelle un métal pur, un composé, ou une substance eutectique change de solide à liquide; la température à laquelle le liquide et le solide sont en équilibres.

metal. An opaque, lustrous, elemental substance that is a good conductor of heat and electricity and, when polished, a good reflector of light. Most metals are malleable and ductile and are, in general, denser than other substances.
métal. Une substance élémentaire opaque, lustrée, qui est bonne conductrice de la chaleur et de l'électricité et qui, lorsque polie, est un bon réflecteur de lumière. La plupart des métaux sont malléables et ductiles et sont, en général, plus denses que les autres substances.

metallograph. An optical instrument designed for both visual observation and photomicrography of prepared surfaces of opaque materials at magnifications ranging from about 25 to about 1500 diameters.
métallographe. Un instrument optique conçu tant pour l'observation visuelle que pour la photomicrographie des surfaces préparées de matériaux opaques à des grossissements allant d'environ 25 à environ 1500 diamètres.

metastable. Possessing a state of pseudo-equilibrium that has a free energy higher than that of the true equilibrium state but from which a system does not change spontaneously.
métastable. Posséder un état de faux équilibre qui a une énergie libre plus élevée que celle du vrai état d'équilibre mais duquel un système ne change pas spontanément.

microbands (deformation). Thin sheetlike volumes of constant thickness in which cooperative slip occurs on a fine scale. They are an instability which carry exclusively the deformation at medium strains when normal homogeneous slip is precluded. The sheets are aligned at ± 55° to the compression direction and are confined to individual grains, which usually contain two sets of bands. Compare *shear bands*.
microbandes (déformation). Volumes minces en forme de feuille d'épaisseur constante dans lesquels un glissement coopératif se produit à une échelle fine. Elles consistent en une instabilité qui supporte exclusivement la déformation sous des contraintes moyennes quand le

glissement normal homogène est prévenu. Les feuilles sont alignées à ± 55° par rapport à la direction de compression et sont confinées à des grains individuels, lesquels contiennent habituellement deux séries de bandes. Comparer avec *bandes de cisaillement*.

microcrack. A crack of microscopic size.
microfissure. Une fissure de taille microscopique.

micrograph. A graphic reproduction of the prepared surface of a specimen at a magnification greater than ten diameters. When photographed, the reproduction is known as a photomicrograph (not a microphotograph).
micrographie. Une reproduction graphique de la surface préparée d'une spécimen à un grossissement de plus de dix diamètres. Lorsque photographiée, la reproduction est appelée photomicrographie (et non pas microphotographie).

microsegregation. Segregation within a grain, crystal or small particle. See coring.
microségrégation. Ségrégation à l'intérieur d'un grain, d'un cristal ou d'une petite particule. Voir *"noyautage; cernes" (coring)*.

microstructure. The structure of a prepared surface of a metal as revealed by a microscope at a magnification greater than ten diameters.
microstructure. La structure d'une surface préparée d'un métal telle que révélée par un microscope à un grossissement de plus de dix diamètres.

mild steel. Carbon steel containing a maximum of about 0.25% C.
acier doux. Acier au carbone contenant un maximum d'environ 0.25% C.

N

natural aging. Spontaneous aging of a supersaturated solid solution at room temperature. See *aging*.
vieillissement naturel. Vieillissement spontané d'une solution solide sursaturée, à la température de la pièce. Voir *vieillissement*.

necking. Local reduction of the cross-sectional area of metal by stretching.
striction. Réduction locale de la surface de la section d'un métal par étirement.

Neumann band. A mechanical (deformation) twin in ferrite.

bande de Neumann. Une macle mécanique (déformation) dans la ferrite.

nodular pearlite. Pearlite that has grown as a colony with an approximately spherical morphology.
perlite nodulaire. Perlite qui a grandi en une colonie ayant une morphologie approximativement sphérique.

normalizing. Heating a ferrous alloy to a suitable temperature above A_3 or A_{cm}. and then cooling in still air to a temperature substantially below A_1. The cooling rate usually is in the range 900 to 1800°F/h (500 to 1000°C/h).
traitement de normalisation. Chauffer un alliage ferreux jusqu'à une température adéquate au-dessus de A_3 ou A_{cm} et ensuite refroidir à l'abri des courants d'air jusqu'à une température substantiellement au-dessous de A_1. La vitesse de refroidissement est habituellement comprise entre 900 et 1800°F/h (500 à 1000°C/h).

notch brittleness. A measure of the susceptibility of a material to brittle fracture at locations of stress concentration. For example, in a notch tensile test a material is said to be "notch brittle" if its notch strength is less than its tensile strength; otherwise, it is said to be "notch ductile".
fragilité à l'entaille. Une mesure de la susceptibilité d'un matériau à la rupture fragile aux endroits de concentration de la contrainte. Par exemple, dans un essai de traction avec entaille un matériau est dit être "fragile à l'entaille" si sa résistance à l'entaille est moindre que sa résistance à la traction; autrement, il est dit être "ductile à l'entaille".

notch ductility. See *notch brittleness*.
ductilité à l'entaille. Voir *fragilité à l'entaille*.

notch sensitivity. A measure of the reduction in strength of a metal caused by the presence of stress concentration. Values can be obtained from static, impact or fatigue tests.
sensibilité à l'entaille. Une mesure de la réduction de la résistance d'un métal causée par la présence de la concentration de contrainte. Les valeurs peuvent être obtenues par des essais statiques, des essais de choc ou des essais de fatigue.

nucleation. Initiation of a phase transformation at discrete sites, the new phase growing from nuclei. See *nucleus* (1).
germination. Commencement d'une transformation de phase à des sites distincts, la nouvelle phase grandissant à partir des germes. Voir *germe (1)*.

nucleus. (1) The first structurally stable particle capable of initiating recrystallization of a phase or the growth of a new phase, and separated from the matrix by an interface. (2) The heavy central core of an atom, in which most of the mass and the total positive electrical charge are concentrated.

germe. *(1)* La première particule stable structurellement et capable d'initier la recristallisation d'une phase ou la croissance d'une nouvelle phase, séparée de la matrice par une interface. *(2)* Le noyau central et lourd d'un atome, où la presque totalité de la masse et la charge électrique positive totale sont concentrées.

O

orientation (crystal). Directions in space of the axes of the lattice of a crystal with respect to a chosen reference or coordinate system. See also *preferred orientation.*

orientation (cristal). Directions dans l'espace des axes du réseau d'un cristal par rapport à une référence choisie ou à un système de coordonnées. Voir aussi *orientation préférée.*

overaging. Aging under conditions of time and temperature greater than those required to obtain maximum change in a certain property. See aging.

(survieillissement, "overaging"). Vieillissement sous des conditions de durée et de température plus grandes que celles requises pour obtenir le maximum de changement d'une certaine propriété. Voir *vieillissement.*

overheating. Heating a metal or alloy to such a high temperature that its properties are impaired. When the original properties cannot be restored by further heat treating, by mechanical working or by a combination of working and heat treating, the overheating is known as burning.

surchauffe. Chauffer un métal ou un alliage à une température tellement élevée que ses propriétés sont compromises. Quand les propriétés originales ne peuvent pas être rétablies par d'autres traitements thermiques, par formage mécanique ou par une combinaison de formage et de traitements thermiques, la surchauffe est appelée brûlure.

oxidation. (1) A reaction in which there is an increase in valence resulting from a loss of electrons. (2) Chemical combination with oxygen to form an oxide.

oxydation. *(1)* Une réaction dans laquelle il y a une augmentation de la valence résultant d'une perte d'électrons. *(2)* Combinaison chimique avec l'oxygène pour former un oxyde.

oxidized surface. A surface having a thin, tightly adhering oxidized skin.

surface oxydée. Une surface ayant une peau oxydée mince et étroitement adhérente.

P

pack rolling. Hot rolling a pack of two or more sheets of metal; scale prevents the sheets from being welded together.

laminage en paquet. Laminer à chaud un paquet de deux ou plusieurs tôles de métal; la calamine empêche les tôles de se souder ensemble.

pancake grain structure. A structure in which the lengths and widths of individual grains are large compared to their thicknesses.

structure de grain en crêpe. Une structure dans laquelle la longueur et la largeur des grains individuels est grande comparée à leur épaisseur.

pass. (1) A single transfer of metal through a stand of rolls. (2) The open space between two grooved rolls through which metal is processed. (3) The weld metal deposited in one run along the axis of a weld.

passe. *(1)* Un transport unique du métal à travers un train de cylindres. *(2)* L'espace ouvert entre deux cylindres à rainure à travers lequel le métal est traité. *(3)* La soudure métallique déposée en un parcours le long de l'axe d'une soudure.

patenting. A heat treatment applied to medium and high-carbon steel prior to cold drawing to wire. The treatment involves austenitization followed by isothermal transformation at a temperature that produces a microstructure of very fine pearlite.

patentage. Un traitement thermique appliqué à l'acier à carbone moyen et haut avant l'étirement à froid en fil. Le traitement implique l'austénitisation suivie par transformation isotherme à une température qui produit une microstructure de perlite très fine.

pattern welding. A process in which strips or other small sections of iron or steel are twisted together and then forge welded. Homogeneity and toughness are thereby improved. A regular decorative pattern can be developed in the final product. Commonly used for making swords as early as the 3rd century A.D.

soudage en motif. Un procédé par lequel des bandes ou autres petites sections de fer ou d'acier sont tordues ensemble et ensuite soudées à la forge. L'homogénéité et la ténacité sont de ce fait améliorées. Un motif décoratif régulier peut être développé dans le produit fini.

Communément utilisé pour fabriquer des épées aussi tôt que le 3e siècle A.C.

pearlite. A eutectoid transformation product of ferrite and cementite that ideally has a lamellar structure but that is always degenerate to some extent.
perlite. Un produit de la transformation eutectoïde de la ferrite et de la cémentite ayant idéalement une structure lamellaire mais qui est toujours dégénérée jusqu'à un certain point.

peritectic. An isothermal reversible reaction in which a liquid phase reacts with a solid phase to produce another solid phase.
péritectique. Une réaction isotherme réversible par laquelle une phase liquide réagit avec une phase solide pour produire une autre phase solide.

phase. A physically homogeneous and distinct portion of a material system.
phase. Une partie physiquement homogène et distincte du système d'un matériau.

phase diagram. Synonymous with *constitutional diagram*.
diagramme de phase. Synonyme de *diagramme constitutionnel*.

photomacrograph. See *macrograph*.
photomacrographie. Voir *macrographie*.

photomicrograph. See *micrograph*
photomicrographie. Voir *micrographie*.

physical properties. Properties, other than mechanical properties, that pertain to the physical nature of a material; e.g., density, electrical conductivity, thermal expansion, reflectivity, magnetic susceptibility, etc.
propriétés physiques. Propriétés, autres que les propriétés mécaniques, qui appartiennent à la nature physique d'un matériau; ex.: densité, conductivité électrique, expansion thermique, réflexivité, susceptibilité magnétique, etc.

pig iron. (1) High-carbon iron made by reduction of iron ore in the blast furnace. (2) Cast iron in the form of pigs.
fonte brute. *(1)* Fer à carbone élevé fabriqué par la réduction du minerai de fer dans le haut fourneau. *(2)* Fonte de moulage en forme de gueuses.

piling. A process in which several bars are stacked and hot rolled together, with the objective of improving the homogeneity of the final product. Used in primitive ironmaking.
empilage. Un procédé par lequel plusieurs barres sont empilées et laminées à chaud ensemble, avec pour objectif d'améliorer l'homogénéité du produit final. Utilisé dans la fabrication primitive du fer.

plastic deformation. Deformation that remains, or will remain, permanent after release of the stress that caused it.
déformation plastique. Déformation qui demeure, ou qui demeurera, permanente après le relâchement de la contrainte qui l'a causée.

plasticity. The capacity of a metal to deform nonelastically without rupturing.
plasticité. La capacité d'un métal à se déformer inélastiquement sans se rompre.

plate. A flat-rolled metal product of some minimum thickness and width arbitrarily dependent on the type of metal.
plaque. Un produit métallique laminé plat d'une certaine épaisseur minimum et d'une certaine largeur minimum dépendant arbitrairement du type de métal.

plate martensite. Martensite formed, partly in steels containing more than about 0.5% C and solely in steels containing more than about 1.0% C, as lenticular-shape plates on irrational habit planes that are near $\{225\}_A$, or $\{259\}_A$ in very-high-carbon steels.
martensite en plaque. Martensite formée, partiellement chez les aciers contenant plus qu'environ 0.5% C et exclusivement chez les aciers contenant plus qu'environ 1% C, en plaques à forme lenticulaire sur des plans matrices "irrationnels" qui sont près de $\{225\}_A$, ou $\{259\}_A$ chez les aciers à très haut carbone.

polishing. Producing a specularly reflecting surface.
polissage. Produire une surface spéculairement réfléchissante.

polycrystalline. Comprising an aggregate of more than one crystal, and usually a large number of crystals.
polycristallin. Comprend un agrégat de plus d'un cristal, et habituellement un grand nombre de cristaux.

polymorphism. The property whereby certain substances may exist in more than one crystalline form, the particular form depending on the conditions of crystallization - e.g., temperature and pressure. Among elements, this phenomenon is also called *allotropy*.

polymorphisme. La propriété par laquelle certaines substances peuvent exister sous plus qu'une forme cristalline, la forme particulière dépendant des conditions de cristallisation -ex.: température et pression. Chez les éléments, ce phénomène est aussi appelé allotropie.

preferred orientation. A condition of a polycrystalline aggregate in which the crystal orientations are not random.

orientation préférentielle. Un état d'un agrégat polycristallin dans lequel les orientations du cristal ne sont pas au hasard.

primary crystal. The first type of crystal that separates from a melt during solidification.

cristal primaire. Le premier type de cristal qui se sépare du liquide durant la solidification.

proeutectoid (phase). Particles of a phase that precipitate during cooling after austenitizing but before the eutectoid transformation takes place.

proeutectoïde (phase). Particules d'une phase qui précipitent durant le refroidissement après l'austénitisation mais avant que la transformation eutectoïde ne se produise.

proof stress. See *yield strength*.

limite conventionnelle d'élasticité. Voir *résistance à l'écoulement*.

puddling process. A process for making wrought iron in which cast iron is melted in a hearth furnace and rabbled with slag and oxide until a pasty mass is obtained. This process was developed by Henry Cort about 1784 and remained in use until 1957, although on a very small scale during the present century.

procédé de puddlage. Un procédé de fabrication du fer corroyé par lequel la fonte de moulage est fondue dans un four à sole et agitée avec de la scorie et de l'oxyde jusqu'à ce qu'une masse pâteuse soit obtenue. Ce procédé a été développé par Henry Cort aux environs de 1784 et est demeuré en utilisation jusqu'en 1957, quoiqu'à une très petite échelle au siècle courant.

Q

quench aging. Aging that occurs after quenching following solution heat treatment.

"trempe de vieillissement". Vieillissement qui se produit après la trempe qui suit le traitement de mise en solution.

quench hardening. Hardening by austenitizing and then cooling at a rate such that a substantial amount of austenite is transformed to martensite.
durcissement par trempe. Durcissement par austénitisation et ensuite refroidissement à une vitesse telle qu'une quantité substantielle d'austénite est transformée en martensite.

quenching. Rapid cooling.
trempe. Refroidissement rapide.

R

recarburizing. (1) Increasing the carbon content of molten cast iron or steel by adding carbonaceous material, high-carbon pig iron or a high-carbon alloy. (2) Carburizing a metal part to return surface carbon lost in processing.
re-cémentation (recarburation). *(1)* Augmenter le contenu en carbone de la fonte de moulage fondue ou de l'acier en ajoutant un matériau carboné, de la fonte brute à haut carbone ou un alliage à haut carbone. *(2)* Recarburation d'une pièce de métal afin de retourner à la surface le carbone perdu lors de la préparation.

recovery. Reduction or removal of work-hardening effects, without motion of large-angle grain boundaries.
restauration. Réduire ou enlever les effets du durcissement par écrouissage, sans déplacement des joints de grain à grand angle.

recrystallization. (1) A change from one crystal structure to another, such as that occurring on heating or cooling through a critical temperature. (2) Formation of a new, strain-free grain, structure from the structure existing in cold worked metal.
recristallisation. *(1)* Un changement d'une structure cristalline en une autre, comme celui qui se produit lors du chauffage ou du refroidissement au-delà d'une température de transformation. *(2)* Formation d'une nouvelle structure, à grains sans déformation, à partir de la structure existant chez le métal formé à froid.

recrystallization annealing. Annealing cold worked metal to produce a new grain structure without a phase change.
traitement de recristallisation. Recuire le métal formé à froid afin de produire une nouvelle structure de grain, sans changement de phase.

recrystallization temperature. The approximate minimum temperature at which complete recrystallization of a cold worked metal occurs within a specified time.

température de recristallisation.   La température minimale approximative à laquelle la recristallisation complète d'un métal formé à froid se produit dans une durée de temps spécifiée.

residual elements.   Small quantities of elements unintentionally present in an alloy.
éléments résiduels.   Petites quantités d'éléments, présentes inintentionnellement dans un alliage.

residual stress.   Stress present in a body that is free of external forces or thermal gradients.
contrainte résiduelle.   Contrainte présente dans un corps qui est libre de forces extérieures ou de gradients thermiques.

resistance welding.   Welding with electrical resistance heating and pressure, the work being part of an electrical circuit.
soudage (électrique) par résistance.   Soudage avec chauffage à résistance électrique et pression, la pièce consistant en une partie du circuit électrique.

resolution.   The capacity of an optical or radiation system to separate closely spaced forms or entities; also, the degree to which such forms or entities can be discriminated.
résolution.   La capacité d'un système optique, ou à radiation, à séparer des formes ou des entités étroitement espacées; également, le degré auquel de telles formes ou entités peuvent être différentiées.

resulfurized steel.   Steel to which sulfur has been added in controlled amounts after refining. The sulfur is added to improve machinability.
acier re-sulfuré.   Acier auquel du soufre a été ajouté en quantités contrôlées après le raffinage.   Le soufre est ajouté pour améliorer l'usinage.

rimmed steel.   Low-carbon steel containing sufficient iron oxide to produce continuous evolution of carbon monoxide during ingot solidification, resulting in a case or rim of metal virtually free of voids.
acier effervescent.   Acier à bas carbone contenant suffisamment d'oxyde de fer pour produire l'évolution continue de l'oxyde de carbone pendant la solidification du lingot, résultant en une couche superficielle ou bordure de métal virtuellement dépourvue de cavités.

rolling.   Reducing the cross-sectional area of metal stock, or otherwise shaping metal products, through the use of rotating rolls.
laminage.   Réduire la surface de la section d'un bloc de métal, ou autrement former les produits du métal, par l'usage de cylindres rotatifs.

S

scale. A layer of oxidation products formed on a metal at high temperature.
calamine. Une couche de produits d'oxydation formés sur un métal à haute température.

scarf joint. A butt joint in which the plane of the joint is inclined with respect to the main axes of the members.
joint en enture. Un joint bout à bout où le plan du joint est incliné par rapport aux axes principaux des éléments.

segregation. Nonuniform distribution of alloying elements, impurities or phases.
ségrégation. Distribution non uniforme des éléments d'alliage, des impuretés ou des phases.

segregation banding. Inhomogeneous distribution of alloying elements aligned in filaments or plates parallel to the direction of working.
ségrégation en bande. Distribution hétérogène des éléments d'alliage alignés en filaments ou plaques parallèles à la direction du formage.

self diffusion. The spontaneous movement of an atom to a new site in a crystal of its own species.
autodiffusion. Le mouvement spontané d'un atome vers un nouveau site dans un cristal de sa propre espèce.

semikilled steel. Steel that is incompletely deoxidized and contains sufficient dissolved oxygen to react with the carbon to form carbon monoxide and thus offset solidification shrinkage.
acier demi-calmé. Acier qui est incomplètement désoxydé et qui contient suffisamment d'oxygène dissout pour réagir avec le carbone et former de l'oxyde de carbone et ainsi neutraliser le rétrécissement dû à la solidification.

shear. That type of force that causes or tends to cause two contiguous parts of the same body to slide relative to each other in a direction parallel to their plane of contact.
cisaillement. Ce type de force qui cause ou tend à causer le glissement de deux composantes contiguës d'un même corps l'une par rapport à l'autre dans une direction parallèle à leur plan de contact.

shear bands (deformation). Bands in which deformation has been concentrated inhomogeneously in sheets that extend across regional groups of grains. Usually only one system is present in each regional group of grains, different systems being present in adjoining groups. The bands are noncrystallographic and form on planes of maximum shear stress (55° to the compression direction). They carry most of the deformation at large strains. Compare *microbands*.

bandes de cisaillement (déformation). Bandes dans lesquelles la déformation a été concentrée hétérogènement en feuilles qui s'étendent au-delà de groupes "régionaux" de grains. Habituellement un seul système est présent dans chaque groupe régional de grains, différents systèmes étant présents dans des groupes contigus. Les bandes sont non cristallines et se forment sur des plans de contrainte de cisaillement maximum (55° par rapport à la direction de compression). Elles supportent la plus grande partie de la déformation sous de grandes contraintes. Comparer avec *microbandes*.

shear steel. Steel produced by forge welding together several bars of blister steel, providing a more homogeneous product.

acier de cisaillement. Acier produit en soudant ensemble à la forge plusieurs barres d'acier à soufflures, fournissant un produit plus homogène.

sheet. A flat-rolled metal product of some maximum thickness and minimum width arbitrarily dependent on the type of metal. Sheet is thinner than plate.

tôle. Produit métallique laminé plat d'une certaine épaisseur maximum et largeur minimum dépendant arbitrairement de la sorte de métal. Une tôle est plus mince qu'une plaque.

shortness. A form of brittleness in metal. It is designated as "cold", "hot", and "red", to indicate the temperature range in which the brittleness occurs.

friabilité. Une forme de fragilité du métal. Elle est aussi appelée "froide", "chaude" et "rouge", pour indiquer le domaine de température à laquelle la friabilité se produit.

sintering. Bonding of adjacent surfaces of particles in a mass of metal powders, or in a compact, by heating.

frittage. Liaison des surfaces adjacentes des particules dans une masse de poudres de métal, ou dans une masse compactée, par chauffage.

slab. A piece of metal, intermediate between ingot and plate, at least twice as wide as it is thick.

dalle (brame). Une pièce de métal, intermédiaire entre lingot et plaque, au moins deux fois aussi large qu'épaisse.

slag. A nonmetallic product resulting from mutual dissolution of flux and nonmetallic impurities in smelting and refining operations.
scories (laitier). Un produit non métallique résultant de la dissolution mutuelle du flux et des impuretés non métalliques dans les opérations de fonte et d'affinage.

slip. Plastic deformation by irreversible shear displacement of one part of a crystal relative to another in a definite crystallographic direction and on a definite crystallographic plane.
glissement. Déformation plastique par déplacement irréversible en cisaillement d'une partie d'un cristal par rapport à une autre dans une direction cristallographique définie et sur un plan cristallographique défini.

slip direction. The crystallographic direction in which translation of slip takes place.
direction de glissement. La direction cristallographique dans laquelle la translation du glissement se produit.

slip line. Trace of a slip plane on a viewing surface.
ligne de glissement. Tracé d'un plan de glissement sur une surface d'observation.

slip plane. The crystallographic plane on which slip occurs in a crystal.
plan de glissement. Le plan cristallographique sur lequel le glissement se produit dans un cristal.

solid solution. A solid crystalline phase containing two or more chemical species in concentrations that may vary between limits imposed by phase equilibrium.
solution solide. Une phase cristalline solide contenant deux espèces chimiques ou plus à des concentrations qui peuvent varier entre les limites imposées par l'équilibre de phase.

solidus. In a constitutional diagram, the locus of points representing the temperatures at which various components finish freezing on cooling or begin to melt on heating.
solidus. Dans un diagramme de constitution, le lieu des points représentant les températures auxquelles les composants variés finissent de se solidifier lors du refroidissement ou commencent à fondre lors du chauffage.

solute. The component of either a liquid or solid solution that is present to the lesser or minor extent; the component that is dissolved in the solvent.

soluté. Le composant d'une solution liquide ou solide qui est présent en plus petite quantité ou en quantité mineure; le composant qui est dissout dans le solvant.

solution heat treatment. A heat treatment in which an alloy is heated to a suitable temperature, held at that temperature long enough to cause one or more constituents to enter into solid solution, and then cooled rapidly enough to hold these constituents in solution.
traitement de mise en solution et trempe. Un traitement thermique par lequel un alliage est chauffé à une température appropriée, est tenu à cette température suffisamment longtemps pour permettre la mise en solution solide d'un ou plusieurs constituants et est ensuite refroidi suffisamment rapidement pour garder ces constituants en solution.

solvent. The component of either a liquid or solid solution that is present to the greater or major extent; the component that dissolves the solute.
solvant. Le composant d'une solution liquide ou solide qui est présent en plus grande quantité ou en quantité majeure; le composant qui dissout le soluté.

solvus. In a phase or equilibrium diagram, the locus of points representing the temperature at which solid phases with various compositions coexist with other solid phases; that is, the limits of solid solubility.
solvus (courbe limite de solubilité). Dans un diagramme de phase ou d'équilibre, le lieu des points représentant la température à laquelle les phases solides de compositions variées coexistent avec d'autres phases solides; c'est-à-dire, les limites de la solubilité solide.

spheroidized structure. A microstructure consisting of a matrix containing spheroidal particles of another constituent.
structure globulisée. Une microstructure consistant en une matrice contenant des particules globulisées d'un autre constituant.

spheroidizing. Heating and cooling to produce a spheroidal or globular form of carbide in steel.
globulisation. Chauffage et refroidissement pour produire une forme sphérique ou globulisée du carbure dans l'acier.

spheroidizing annealing. A subcritical annealing treatment intended to produce spheroidization of cementite or other carbide phases.
recuit de globulisation. Un traitement de recuit subcritique ayant pour but de produire la globulisation de la cémentite ou autres phases de carbure.

steel. An iron-base alloy usually containing carbon and other alloying elements. In carbon steel and low-alloy steel, the maximum carbon content is about 2.0%; in high-alloy steel, about 2.5%. The dividing line between low-alloy and high-alloy steels is generally regarded as the 5% level of total metallic alloying elements. Steel is differentiated from two general classes of "iron" - namely, cast irons, which have high carbon concentrations, and relatively pure irons, which have low carbon concentrations.

acier. Un alliage à base de fer contenant habituellement du carbone et d'autres éléments d'alliage. Dans l'acier au carbone et l'acier faiblement allié, le contenu maximum en carbone est environ 2.0%; dans l'acier fortement allié, environ 2.5%. La ligne de division entre aciers faiblement alliés et aciers fortement alliés est généralement considérée au niveau d'un total de 5% en éléments d'alliage métalliques. L'acier se différencie de deux catégories générales de "fer" - soit les fontes de moulage, qui ont de hautes concentrations en carbone et les fers relativement purs, qui ont de faibles concentrations en carbone.

strain. A measure of the relative change in the size of a body. Linear strain is the change per unit length of a linear dimension. True (or natural) strain is the natural logarithm of the ratio of the length at the moment of observation to the original gauge length. Shearing strain is the change in angle (expressed in radians) between two reference lines originally at right angles. When the term is used alone, it usually refers to linear strain in the direction of the applied stress.

déformation. Une mesure du changement relatif de la taille d'un corps. La déformation linéaire est le changement par unité de longueur d'une dimension linéaire. La vraie déformation (ou naturelle) est le logarithme naturel du rapport de la longueur au moment de l'observation par rapport à la longueur originale de la jauge. La déformation de cisaillement est le changement de l'angle (exprimé en radians) entre deux lignes de référence originellement à angle droit. Quand le terme est utilisé seul, il réfère habituellement à la déformation linéaire dans la direction de la contrainte appliquée.

strain aging. Aging induced by cold work. See *aging.*

(vieillissement sous tension). Vieillissement induit par le formage à froid. Voir *vieillissement.*

strain hardening. An increase in hardness and strength caused by plastic deformation at temperatures below the recrystallization range.

durcissement par écrouissage. Une augmentation de la dureté et de la résistance causée par la déformation plastique à des températures au-dessous du domaine de recristallisation.

stress. Force per unit area. True stress denotes stress determined by measuring force and area at the same time. Conventional stress, as applied to tension and compression tests, is force divided by original area. Nominal stress is stress computed by simple elasticity formulae.

contrainte. Force par unité de surface. La vraie contrainte dénote une contrainte déterminée en mesurant la force et la surface au même moment. La contrainte conventionnelle, telle qu'appliquée aux épreuves de traction et de compression, est la force divisée par la surface originale. La contrainte nominale est la contrainte calculée avec les formules simples d'élasticité.

stress-corrosion cracking. Failure by cracking under the combined action of corrosion and stress, either external (applied) or internal (residual). Cracking may be either intergranular or transgranular, depending on the metal and the corrosive medium.

fissuration par corrosion sous tension. Défaillance par fissuration sous l'action combinée de la corrosion et de contrainte soit externe (appliquée) ou interne (résiduelle). La fissuration peut être soit intergranulaire ou intragranulaire, dépendant du métal et du milieu corrosif.

stress relieving. Heating to a suitable temperature, holding long enough to reduce residual stresses and then cooling slowly enough to minimize the development of new residual stresses.

traitement de relaxation. Chauffer à une température appropriée, et maintenir suffisamment longtemps pour réduire les contraintes résiduelles et ensuite refroidir suffisamment lentement pour minimiser le développement de nouvelles contraintes résiduelles.

stretcher strains. Elongated markings that appear on the surfaces of some materials when they are deformed just past the yield point. These markings lie approximately parallel to the direction of maximum shear stress and are the result of localized yielding. See also *Luders lines*

vermiculures. Marques allongées qui apparaissent à la surface de certains matériaux quand ils sont déformés juste au-delà de la limite d'écoulement. Ces marques reposent approximativement parallèles à la direction de la contrainte de cisaillement maximum et sont le résultat d'écoulement localisé. Voir aussi *lignes de Luders*.

strip. A sheet of metal whose length is many times its width.

feuillard; acier en ruban. Une tôle de métal dont la longueur est plusieurs fois sa largeur.

sub-boundary structure (subgrain structure). A network of low-angle boundaries (usually with misorientations of less than one degree) within the main grains of a microstructure.
structure de sous-joint (structure de sous-grain). Un réseau de joints à angle faible (habituellement avec des misorientations de moins qu'un degré) à l'intérieur des grains principaux d'une microstructure.

subcritical annealing. An annealing treatment in which a steel is heated to a temperature below the A_1 temperature and then cooled slowly to room temperature.
recuit subcritique. Un traitement de recuit par lequel un acier est chauffé jusqu'à une température sous la température A_1 et ensuite refroidi lentement jusqu'à la température de la pièce.

subgrain. A portion of a crystal or grain slightly different in orientation from neighbouring portions of the same crystal. Generally, neighbouring subgrains are separated by low-angle boundaries.
sous-grain. Une portion d'un cristal ou grain légèrement différente en orientation de celle des portions voisines du même cristal. Généralement, les sous-grains voisins sont séparés par des joints à angle faible.

substitutional solid solution. A solid solution in which the solvent and solute atoms are located randomly at the atom sites in the crystal structure of the solution.
solution solide de substitution. Une solution solide dans laquelle les atomes de solvant et de soluté sont localisés au hasard dans les sites atomiques dans la structure du cristal de la solution.

substrate. The layer of metal underlying a coating, regardless of whether the layer is base metal.
substrat. La couche de métal au-dessous d'un revêtement, que la couche soit ou non le métal de base.

sulfide spheroidization. A stage of overheating in which sulfide inclusions are partly or completely spheroidized.
globulisation du sulfure. Une étape de surchauffe par laquelle les inclusions de sulfure sont partiellement ou complètement globulisées.

sulfur print. A macrographic method of examining distribution of sulfide inclusions.
empreinte de soufre (empreinte Baumann). Une méthode macrographique d'examen de la distribution des inclusions de sulfure.

supercooling. Cooling to a temperature below that of an equilibrium phase transformation without the transformation taking place.

sous-refroidissement. Refroidissement jusqu'à une température au-dessous de celle d'une transformation d'équilibre de phase sans que la transformation ne se produise.

superheating. (1) Heating a phase to a temperature above that of a phase transformation without the transformation taking place. **(2)** Heating molten metal to a temperature above the normal casting temperature to obtain more complete refining or greater fluidity.

surchauffe. (1) Chauffer une phase jusqu'à une température au-dessus de celle d'une transformation de phase sans que la transformation ne se produise. *(2)* Chauffer le métal fondu jusqu'à une température au-dessus de la température normale de moulage pour obtenir un raffinage plus complet ou une plus grande fluidité.

surface hardening. A generic term covering several processes applicable to a suitable ferrous alloy that produce, by quench hardening only, a surface layer that is harder or more wear resistant than the core. There is no significant alteration of the chemical composition of the surface layer. The processes commonly used are induction hardening, flame hardening and shell hardening. Use of the applicable specific process name is preferred.

traitement de durcissement superficiel. Un terme générique couvrant plusieurs procédés applicables à un alliage ferreux approprié qui produisent, par durcissement par trempe seulement, une couche de surface qui est plus dure ou plus résistante à l'usure que le coeur. Il n'y a pas d'altération importante de la composition chimique de la couche de surface. Les procédés communément utilisés sont le durcissement par trempe avec chauffage par induction, le durcissement par trempe avec chauffage à la flamme et le durcissement par trempe après chauffage superficiel. Il est préférable d'utiliser le nom du procédé spécifique qui s'applique.

T

taper section. A section made at an acute angle to a surface of interest, thereby achieving a geometrical magnification of depth. A sectioning angle of 5°43' achieves a depth magnification of 10: 1.

section conique. Une section faite à angle aigu par rapport à la surface d'intérêt, atteignant ainsi un grossissement géométrique de la profondeur. Un angle de section de 5°43' atteint un grossissement de la profondeur de 10:1.

teeming. Pouring molten metal from a ladle into ingot molds. The term applies particularly to the specific operation of pouring either iron or steel into ingot molds.

coulage. Verser le métal fondu d'une poche dans des moules à lingot. Le terme s'applique particulièrement à l'opération spécifique du coulage soit du fer ou de l'acier dans des moules à lingot.

temper brittleness. A reversible increase in the ductile-brittle transition temperature in steels heated in, or slowly cooled through, the temperature range from about 700 to 1100°F (375 to 575°C).

fragilité de revenu. Une augmentation réversible de la température de transition ductile-fragile des aciers chauffés (ou refroidis lentement) dans le domaine de température s'étendant d'environ 700 à 1100°F (375 à 575°C).

temper rolling. Light cold rolling of sheet steel. The operation is performed to improve flatness, to minimize the formation of stretcher strains, and to obtain a specified hardness or temper.

laminage de revenu. Léger laminage à froid d'une tôle d'acier. L'opération est exécutée afin d'améliorer l'aplatissement, ou pour minimiser la formation de vermiculures et pour obtenir une dureté ou un revenu spécifiés.

tempering. In heat treatment, reheating hardened steel to some temperature below the A_1 temperature for the purpose of decreasing hardness and/or increasing toughness. The process also is sometimes applied to normalized steel.

revenu. En traitement thermique, re-chauffer l'acier durci jusqu'à une certaine température sous la température A_1 dans le but de diminuer la dureté et/ou d'augmenter la ténacité. Le procédé est aussi quelquefois appliqué à l'acier normalisé.

tensile strength. In tensile testing, the ratio of the maximum force sustained to the original cross-sectional area.

résistance à la traction. En épreuve de traction, le rapport de la force maximum supportée par rapport à la surface de la coupe originelle.

texture. In a polycrystalline aggregate, the state of distribution of crystal orientations. In the usual sense, it is synonymous with preferred orientation, in which the distribution is not random.

texture. Dans un agrégat polycristallin, l'état de distribution de l'orientation des cristaux. Dans le sens habituel, elle est synonyme d'orientation préférentielle, dans laquelle la distribution n'est pas au hasard.

thermal analysis. A method of studying transformations in metal by measuring the temperatures at which thermal arrests occur.
analyse thermique. Une méthode d'étude des transformations dans le métal en mesurant les températures auxquelles les arrêts thermiques se produisent.

time-temperature-transformation (TTT) diagram. See *isothermal transformation (IT) diagram.*
diagramme de transformation en conditions isothermes (TTT). Voir *diagramme de transformation isotherme (TI).*

toughness. Capacity of a metal to absorb energy and deform plastically before fracturing.
ténacité. Capacité d'un métal à absorber l'énergie et à se déformer plastiquement avant de se casser.

transformation ranges (transformation temperature ranges). Those ranges of temperature within which austenite forms during heating and transforms during cooling. The two ranges are distinct, sometimes overlapping but never coinciding. The limiting temperatures of these ranges depend on the composition of the alloy and on the rate of change of temperature, particularly during cooling. See *transformation temperature.*
domaines de transformation (domaines de température de transformation). Ces domaines de température à l'intérieur desquels l'austénite se forme durant le chauffage et se transforme durant le refroidissement. Les deux domaines sont distincts, quelquefois se chevauchant mais ne coïncidant jamais. Les températures limites de ces domaines dépendent de la composition de l'alliage et de la vitesse de changement de température, particulièrement durant le refroidissement. Voir *température de transformation.*

transformation temperature. The temperature at which a change in phase occurs. The term is sometimes used to denote the limiting temperature of a transformation range. The following symbols are used:
A_1 - The temperature of the eutectoid transformation.
A_3 - The temperature at which proeutectoid ferrite begins to separate from austenite under conditions of slow cooling.
A_{cm} - The temperature at which proeutectoid cementite begins to separate from austenite under conditions of slow cooling.
M_f - The temperature at which transformation of austenite to martensite finishes during cooling.
M_s - The temperature at which transformation of austenite to martensite starts during cooling.

température de transformation.   La température à laquelle un changement de phase se produit. Le terme est quelquefois utilisé pour dénoter la température limite du domaine de transformation.   Les symboles suivant sont utilisés:

A_1 - La température de la transformation eutectoïde.

A_3 - La température à laquelle la ferrite proeutectoïde commence à se séparer de l'austénite sous des conditions de refroidissement lent.

A_{cm} - La température à laquelle la cémentite proeutectoïde commence à se séparer de l'austénite sous des conditions de refroidissement lent.

M_f - La température à laquelle la transformation de l'austénite en martensite se termine durant le refroidissement.

M_s - La température à laquelle la transformation de l'austénite en martensite commence durant le refroidissement.

transition temperature (ductile-brittle transition temperature). An arbitrarily defined temperature that lies within the temperature range in which metal fracture characteristics (as usually determined by tests of notched specimens) change rapidly, such as from primarily fibrous (shear) to primarily cleavage.

température de transition (température de transition ductile-fragile).   Une température définie arbitrairement qui se trouve à l'intérieur du domaine de température dans lequel les caractéristiques de rupture du métal (telles qu'habituellement déterminées par les épreuves de spécimens entaillés) changent rapidement, telles que principalement fibreuse (cisaillement) à principalement en clivage.

triple point.   The intersection of the boundaries of three adjoining grains, as observed in a section.

point triple.   L'intersection des joints de trois grains adjacents, telle qu'observée dans une coupe.

twin.   Two portions of a crystal having a definite orientation relationship; one may be regarded as the parent, the other as the twin. The orientation of the twin is either a mirror image of the orientation of the parent across a "twinning plane" or an orientation that can be derived by rotating the twin portion about a "twinning axis".

macle.   Deux portions d'un cristal ayant une relation d'orientation définie; l'une peut être considérée comme le parent et l'autre comme la macle.   L'orientation de la macle est soit une image-mirroir de l'orientation du parent à travers un "plan de maclage" ou une orientation qui peut être obtenue en tournant la portion macle autour d'un "axe de maclage".

twin, annealing. A twin produced as the result of heat treatment.

macle, recuit (macle de recuit). Une macle résultant d'un traitement thermique.

twin, crystal. A portion of a crystal in which the lattice is a mirror image of the lattice of the remainder of the crystal.
macle, cristal (grain maclé). Une portion d'un cristal dans laquelle le réseau est l'image-mirroir du réseau du reste du cristal.

twin, deformation. A twinned region produced by a shearlike distortion of the parent crystal structure during deformation. In ferrite, deformation twins form on {211} planes.
macle, déformation (macle de déformation). Une région maclée produite par une déformation en forme de cisaillement de la structure du cristal parent durant la déformation. Dans la ferrite, les macles de déformation se forment sur les plans {211}.

U

upper yield stress. See *yield point.*
limite supérieure d'élasticité. Voir limite *supérieure d'écoulement..*

upset. (1) The localized increase in cross-sectional area resulting from the application of pressure during mechanical fabrication or welding. (2) That portion of welding cycle during which the cross-sectional area is increased by the application of pressure.
"upset" (dérangement). (1) L'augmentation locale de la surface de la section résultant de l'application de pression durant la fabrication mécanique ou le soudage. *(2)* Cette portion du cycle de soudage pendant lequel la surface de la section est augmentée par l'application de pression.

V

vacancy. A type of structural imperfection in which an individual atom site is temporarily unoccupied.
lacune. Un type d'imperfection structurale dans laquelle un site d'atome individuel est temporairement inoccupé.

veining. A type of sub-boundary structure that can be delineated because of the presence of a greater-than-average concentration of precipitate or solute atoms.
veinage (marbrure). Un type de structure de sous-joint qui peut être soulignée par la présence d'une concentration plus grande que normale de précipités ou d'atomes de soluté.

W

Waloon process. An early two-hearth process for making wrought iron by refining cast iron. The conversion proper was carried out in a hearth furnace known as a finery; reheating for forging was carried out in a second hearth furnace known as a chafery.
procédé Waloon. Un procédé ancien à deux soles de fabrication du fer corroyé par raffinage de la fonte de moulage. La conversion propre était effectuée dans un four à sole appelé affinerie; le re-chauffage pour le forgeage était effectué dans un second four à sole appelé chaferie.

weld. A union made by welding.
soudure. Un joint fait par soudage.

weld bead. A deposit of filler metal from a single welding pass.
cordon de soudure. Un dépôt de métal d'apport en une passe unique de soudage.

weldability. Suitability of a metal for welding under specific conditions.
soudabilité. Aptitude d'un métal à être soudé sous des conditions spécifiques.

welding. Joining two or more pieces of material by applying heat or pressure, or both, with or without filler metal, to produce a localized union through fusion or recrystallization across the interface.
soudage. Joindre deux ou plusieurs pièces de matériau en appliquant de la chaleur ou de la pression, ou les deux, avec ou sans métal d'apport, produisant un joint localisé par la fusion ou la recristallisation à travers l'interface.

wetting agent. A surface-active agent that produces wetting by decreasing the cohesion within the liquid.
agent mouillant. Un agent à action de surface qui produit le mouillage en diminuant la cohésion à l'intérieur du liquide.

Widmanstatten structure. A structure characterized by a geometric pattern resulting from the formation of a new phase on certain crystallographic planes in the parent phase. The orientation of the lattice in the new phase is related crystallographically to the orientation of the lattice in the parent phase.
structure de Widmannstätten. Une structure caractérisée par un motif géométrique résultant de la formation d'une nouvelle phase sur certains plans cristallographiques de la phase parente. L'orientation du réseau de la nouvelle phase est reliée cristallographiquement à l'orientation du réseau de la phase parente.

wootz. A carbon steel containing 1 to 1.6% C produced by melting a bloomery iron or an inhomogeneous steel with charcoal in a crucible. The process originated in India as early as the 3rd century A.D.
wootz. Un acier au carbone contenant de 1 à 1.6% C produit en fondant un fer de bloom ou un acier hétérogène avec du charbon de bois dans un creuset. Le procédé a pris naissance aux Indes dès le 3e siècle.

wrought iron. An iron produced by direct reduction of ore or by refining molten cast iron under conditions where a pasty mass of solid iron with included slag is produced. The iron has a low carbon content.
fer corroyé. Un fer produit par réduction directe du minerai ou en affinant la fonte de moulage fondue sous des conditions où une masse pâteuse de fer solide est produite avec de la scorie incluse. Le fer a un contenu en carbone bas.

wustite. The oxide of iron of lowest valence which exists over a wide range of compositions that do not quite include the stoichiometric composition FeO.
wustite. L'oxyde de fer de plus basse valence qui existe sur une grande étendue de compositions mais n'incluant pas tout-à-fait la composition stoechiométrique FeO.

Y

yield point. The first stress in a material less than the maximum obtainable stress at which an increase in strain occurs without an increase in stress. Also known as upper *yield stress.*
limite d'écoulement. Dans un matériau, la première contrainte moindre que la contrainte maximum atteignable à laquelle une augmentation de la déformation se produit sans augmentation de la contrainte. Aussi appelée limite supérieure d'écoulement.

yield strength. The stress at which a material exhibits a specified limiting deviation from the proportionality of stress to strain. The deviation is expressed in terms of strain. Also known as *proof stress.*
résistance à l'écoulement. La contrainte à laquelle un matériau démontre une déviation limite spécifiée de la proportionnalité de la contrainte à la déformation. La déviation est exprimée en termes de déformation. Aussi appelée "limite conventionnelle d'élasticité" (proof stress).

GLOSSARY / LEXIQUE / LÉXICO / WÖRTERVERZEICHNIS

ENGLISH	FRENCH	SPANISH	GERMAN
A$_1$ temperature	température A$_1$	temperatura A$_1$	A$_1$ Temperatur
A$_2$ temperature	température A$_2$	temperatura A$_2$	A$_2$ Temperatur
A$_3$ temperature	température A$_3$	temperatura A$_3$	A$_3$ Temperatur
A$_{cm}$ temperature	température A$_{cm}$	temperatura A$_{cm}$	A$_{cm}$ Temperatur
abrasion	abrasion	abrasión	Schleifwirkung; Abnutzung
abrasive	abrasif	abrasivo	Schleifmittel; Abriebeigenschaften
age hardening	durcissement par vieillissement	endurecimiento por envejecimiento	Aushärtung; Vergütung
aging	vieillissement	envejecimiento	Alterung
allotriomorph	allotriomorphe	allotriomorfico	Allotriomorph
allotropy	allotropie	allotropía	Allotropie
alloy	alliage	aleación	Legierung
alloy steel	acier allié	acero aleado	legierter Stahl
alloying element	élément d'alliage	elemento de aleación	Legierungselement
angstrom unit (A)	unité angström (Å)	unidad angstrom (Å)	Ångström (Å)
annealing	recuit	recocido	Glühen
annealing twin	macle de recuit	macla por recocido	Rekristallisationszwilling
arc welding	soudage à l'arc	soldadura por arco	Lichtbogenschweißen
artifact	artefact	artefacto	künstlich
austempering	trempe étagée bainitique	atemperación escalonada (tratamiento con transformación isotérmica de austenita a bainita)	Zwischenstufen-Vergütung
austenite	austénite	austenita	Austenit
austenitic grain size	grosseur de grain austénitique	dimensión de grano austenítico	Austenitkorngröße
austenitizing	austénitisation	austenización	Austenitisieren
autoradiograph	autoradiographie	autorradiografía	Autoradiographie

GLOSSARY / LEXIQUE / LÉXICO / WÖRTERVERZEICHNIS

ENGLISH	FRENCH	SPANISH	GERMAN
autotempering	auto-revenu	autorevenido	Autoanlaßen
bainite	bainite	bainita	Bainit
bamboo grain structure	structure de grain en bambou	estructura de grano de bambú	Bambus Kornstruktur
banding	bande	bandas	Zeilenbildung
billet	billette (ou larget)	palanquilla	Walzblock; Barren; Knüppel
blister steel	acier comportant des soufflures (ou "pailles")	acero ampollado	Zement[ations]stahl
bloom	bloom (brame)	bloom; desbaste	Vorblock; Schmiedeblock; Blume
bloomery	bloomerie	planta para fabricar desbastes	Rennherd
box annealing	recuit en caisse	recocido en caja	Kistenglühen
braze welding	soudo-brasage	soldadura con metal amarillo	Schweißlötung
brazing	brasage	soldadura amarilla	Löten; Hartlöten
brittle fracture	rupture fragile	fractura frágil	Sprödbruch
brittleness	fragilité	fragilidad	Sprödigkeit
buffer	tampon	regulador; amortiguador	Pufferschicht
burning	brûlure	quemado	Verbrennung
capped steel	acier coiffé	acero efervescente	gedeckelter Stahl
carbide	carbure	carburo	Carbid
carbon equivalent (for rating of weldability)	équivalent carbone (pour estimer la soudabilité)	carbono equivalente	Kohlenstoffäquivalent (für Schweißbarkeit Schätzung)
carbon potential	potentiel carbone	potencial de carbono	Kohlenstoffpegel, C-Pegel
carbon restoration	traitement de recarburation	tratamiento de recementación	Wiederaufkohlung
carbon steel	acier au carbone	acero al carbono	unlegierter Stahl
carbonitriding	carbonitruration	carbonitruración	Karbonitrieren
carburizing	cémentation	cementación; carburación	Aufkohlen
case	couche superficielle traitée	capa superficial	Einsatz[härte]schicht

GLOSSARY / LEXIQUE / LÉXICO / WÖRTERVERZEICHNIS

ENGLISH	FRENCH	SPANISH	GERMAN
case hardening	trempe superficielle; durcissement superficiel	cementación en caja	Einsatzhärtung; Oberflächenhärten
cast iron	fonte de moulage	arrabio; hierro colado	Gußeisen
cast steel	acier moulé	acero moldeado; vaciado	Stahlguß
casting	moulage	pieza fundida; pieza de fundición; colada del acero líquido; fundición	Gußerzeugnisse
cementation	cémentation	cementación	Zementation
cementite	cémentite	cementita	Zementit
chafery	chaferie	rozamiento	chaferie
chemical polishing	polissage chimique	pulido químico	chemisches Polieren
cleavage	clivage	exfoliación o despegue	Spaltbruch; Spaltung
cleavage fracture	rupture par clivage	fractura transcristalina a través de planos de deslizamiento	Trennbruch; Sprödbruch
cleavage plane	plan de clivage	plano de despegue	Spaltfläche
columnar structure	structure basaltique	estructura columnar	Stengelgefüge
constituent	constituant	constituyente	Konstituent; Bestandteil
constitutional diagram	diagramme de constitution	diagrama de fases	Zustandsschaubild
continuous phase	phase continue	fase continua	kontinuierliches Phase
controlled rolling	laminage contrôlé	laminado controlado	Regelrollen
coring	noyautage; cernes	microsegregación o segregación intragranular	Kristallseigerung; Mikroseigerung
corrosion	corrosion	corrosión	Korrosion
creep	fluage	fluencia	Kriechen
critical cooling rate	vitesse de refroidissement critique	velocidad crítica de enfriamiento	kritische Abkühlungsgeschwindigkeit
critical point	point critique (ou température de transformation)	punto crítico; temperatura de transformación	Umwandlungstemperatur; Umwandlungspunkt

GLOSSARY / LEXIQUE / LÉXICO / WÖRTERVERZEICHNIS

ENGLISH	FRENCH	SPANISH	GERMAN
critical strain	déformation critique	deformación crítica	kritische Verformung
critical temperature	température de transformation	temperatura crítica	Umwandlungstemperatur
cross rolling	laminage avec cylindres obliques	laminación transversal	schrägwalzen
crucible steel	acier au creuset	acero al crisol	Tiegelstahlblock
crystalline fracture	rupture cristalline	fractura cristalina	kristallin[ischer Bruch
decarburization	décarburation	decarburación	Entkohlung
decoration (of dislocations)	garniture (de dislocations)	(guarnición de dislocaciones)	Dekoration (an Verschiebung)
deformation bands	bandes de déformation	bandas de deformación	Verformungszeilen
degenerate structure	structure dégénérée	estructura degenerada	entarteter Struktur
dendrite	dendrite	dendrita	Dendrit
dendritic segregation	ségrégation dendritique	segregación dendrítica	dendritisches Seigerung
deoxidation	désoxydation	desoxidación	Desoxidieren; Reduktion
diffusion	diffusion	difusión	Diffusion
dilatometer	dilatomètre	dilatómetro	Dilatometer
dislocation	dislocation	dislocación	Verschiebung; Versetzung
ductility	ductilité	ductilidad	Duktilität; Dehnbarkeit
elastic limit	limite d'élasticité	límite de elasticidad	Elastizitätsgrenze
elastic strain	déformation élastique	deformación elástica;	elastische Verformung
electron beam microprobe analyzer	analyseur par microsonde électronique	bombardeo de electrones	Elektronenstrahlmikrosonde analysator
electropolishing	polissage electrolytique	pulido electrolítico	elektrolytisch polieren
elongation after fracture	allongement après rupture	elongación a consecuencia de fractura	Bruchdehnung
epitaxy	épitaxie	epitaxia	Epitaxie
equiaxed structure	structure équiaxe	estructura equiaxial	gleichachsig Struktur
equilibrium diagram	diagramme d'équilibre	diagrama de equilibrio	Zustandsschaubild

GLOSSARY / LÉXIQUE / LÉXICO / WÖRTERVERZEICHNIS

ENGLISH	FRENCH	SPANISH	GERMAN
etchant	réactif d'attaque	ácido para grabar; reactivo para ataque	Ätzflüssigkeit
etching	attaque	ataque químico	Ätzen; Anätzung
eutectoid	eutectoïde	eutectoide	Eutektoid
fatigue	fatigue	fatiga	Ermüdung
ferrite	ferrite	ferrita	Ferrit
ferrite-pearlite banding	bande ferrite-perlite	bandas de ferrita y perlita	Ferrit-Perlit zeilenbildung
ferritic grain size	grosseur de grain ferritique	tamaño del grano ferrítico	ferritischer Korngrösse
fiber	fibre	fibra	Faser
fibrous fracture	rupture fibreuse	fractura fibrosa	Holzfaserbruch
filler metal	métal d'apport	metal de aportación	Schweißzusatzwerstoff; Zusastzwerkstoff
finery	affinerie	horno de afino	Affinerie
finishing temperature	temperature de finissage	temperatura de acabado; temperatura final	Endtemperatur des Werkstückes
flash	bavure	rebarba	Grat; Gratbildung
flow lines	lignes d'écoulement (lignes de Piobert-Lüders)	líneas de fluencia; líneas de Piobert-Lüders; líneas de flujo	Fließlinien; Schlieren
flow stress	contrainte d'écoulement	esfuerzo de fluencia	Fließspannung
flux	flux (fondant)	fundente	Flußmittel
forge welding	soudage à la forge	soldeo por calda a la forja; soldadura de forja	Hammerschweißen
forging	forgeage	forja; forjadura	Schmiedestück; Schmieden
fractography	fractographie	estudio de fracturas	Fraktographie
fragmentation	fragmentation	fragmentación	Zertrümmerung
free machining	facilement usinable (décolletage rapide)	de fácil maquinado; torneado rápido	spanbar

GLOSSARY / LEXIQUE / LÉXICO / WÖRTERVERZEICHNIS

ENGLISH	FRENCH	SPANISH	GERMAN
full annealing (ferrous materials)	recuit complet (matériaux ferreux)	recocido completo; recocido a fondo (materiales ferrosos)	Hochglühen (über A₃)
fusion welding	soudage par fusion	soldadura a fusión	Schmelzschweißen
gas welding	soudage au gaz	soldadura oxiacetilénica	Gasschweißen
grain	grain	grano	Korn
grain-boundary liquation	liquation de joint de grain	licuefacción de los límites de los granos	Korngrenzenangriff; Korngrenzeseigerung
grain-boundary sulfide precipitation	précipitation de sulfure aux joints de grain	(precipitación de sulfuro en los límites de granos	KorngrenzenSulfidausscheidung
grain coarsening	grossissement du grain	crecimiento del grano	Kornvergröberung
grain flow	écoulement de grain	fluencia de granos	Kornfluß
grain growth	croissance du grain	crecimiento del grano	Kornwachstum
grain size	grosseur de grain	tamaño de grano	Korngrösse
granular fracture	rupture granuleuse	fractura granulosa	körniger Bruch; kristallin[ischler Bruch
graphite	graphite	grafito	Graphit
graphitization	graphitisation	grafitización	Graphitisation
grinding	meulage (rectification)	rectificado; amolado	Schleifen
hardenability	trempabilité	templablidad; facultad de endurecimiento	Härtbarkeit
hardening	traitement de durcissement	endurecimiento; temple; cementación	Härten
hardness (indentation)	dureté à la bille (empreinte)	dureza (indentación)	Härte (Eindruck)
heat-affected zone (HAZ)	zone affectée thermiquement (ZAT)	zona térmicamente afectada; ZTA	Wärmeeinflußzone (WEZ)
heat tinting	coloration par chaleur	termocoloración	Anlaufen
heat treatment	traitement thermique	tratamiento térmico; termotratamiento	Wärmebehandlung

GLOSSARY / LÉXIQUE / LÉXICO / WÖRTERVERZEICHNIS

ENGLISH	FRENCH	SPANISH	GERMAN
hematite	hématite	hematita	Roteisenstein; roter Hämatit
homogenizing annealing	recuit d'homogénéisation	recocido de homogeneización	Diffusionsglühen
hot working	formage à chaud	formación o trabajo en caliente	Warmumformen
hypereutectoid alloy	alliage hypereutectoïde	aleación hipereutectoide	übereutektoidisch Legierung
hypoeutectic alloy	alliage hypoeutectique	aleación hipoeutéctica	untereutektoidisch Legierung
idiomorph	idiomorphe	idiomorfo	idiomorph
impact test	essai de choc (de résilience)	prueba de impacto; prueba de choque; prueba de resiliencia	Kerbschlagbiegeversuch
impurities	impuretés	impurezas	Unreinheit; Verunreinigung
inclusion	inclusion	inclusión	Einschluß
ingot	lingot	lingote	Blockmetall; Massel
ingot iron	fer lingot	hierro ARMCO	entkohlter Stahl; Bessemerstahl
intercrystalline	intercristalline	intercristalina	interkristallin
internal oxidation	oxydation interne	oxidación interna	innere Oxidation
interstitial solid solution	solution solide interstitielle	solución sólida intersticial	Einlagerungsmischkristall
intracrystalline	intragranulaire	intracristalina	intrakristallin
iron	fer	hierro	Eisen
isothermal transformation (IT) diagram	diagramme de transformation en conditions isothermes (diagramme TI)	diagrama de transformación isotérmica	isothermisches Glühen Schaubild
killed steel	acier calmé	acero desoxidado	beruhigter Stahl
kink band (deformation)	bande de contorsion (déformation)	banda de ensortijamiento	Zeilenverschlingung
lamellar tear	déchirure lamellaire	desgarre laminar	Terrassenbruch
lamination	feuilletage	separación en láminas	Dopplung
lath martensite	martensite massive	martensita de placas finas	Lattenmartensit
liquation	liquation	licuación; segregación	Seigerung
liquidus	liquidus	línea de líquido	Liquidus

GLOSSARY / LEXIQUE / LÉXICO / WÖRTERVERZEICHNIS

ENGLISH	FRENCH	SPANISH	GERMAN
Luder's lines or bands	lignes ou bandes de Lüder	líneas o bandas de Lüder	Fließfiguren; Fließlinien; Lüderssche Linien
M_f temperature	température M_f	temperatura M_f	M_f Temperatur
M_s temperature	température M_s	temperatura M_s	M_s Temperatur
machinability	usinabilité	maquinabilidad	Zerspanbarkeit
macroetching	macroattaque	macroataque químico	Grobätzung
macrograph	macrographie	macrografía	Makrograph; Grobgefügebild
macrostructure	macrostructure	macroestructura	Makrogefüge; Grobgefüge; Makrostruktur
magnetite	magnétite	magnetita	Magnetit; Magneteisenstein
manual welding	soudage manuel	soldeo manual	Handschweißung
martempering	trempe étagée martensitique	temple escalonado martensítico	Warmbadhärten
martensite	martensite	martensita	Martensit
martensite range	domaine martensitique	dominio martensítico	Martensitbereich
matrix	matrice	matriz	Grundmasse; Matrix
mechanical twin	macle mécanique	macla mecánica	mechanische Zwilling
mechanical polishing	polissage mécanique	pulido mecánico	Maschinenpolieren
melting point	température de fusion	punto de fusión	Schmelzpunkt
metal	métal	metal	Metall
metallograph	métallographe	metalografo; microscopio metalúrgico	Metallograph
metastable	métastable	metaestable	metastabil
microbands (deformation)	microbandes (déformation)	microbandas (deformación)	Mikrozeilen (Verformung)
microcrack	microfissure	microgrieta; microfissura	Mikroriß; mikroskopischer Riß
micrograph	micrographie	micrografía	Mikrobild; Schliffbild
microsegregation	microségrégation	microsegregación	Mikroseigerung

GLOSSARY / LEXIQUE / LÉXICO / WÖRTERVERZEICHNIS

ENGLISH	FRENCH	SPANISH	GERMAN
microstructure	microstructure	microestructura	Mikrogefüge
mild steel	acier doux	acero dulce; acero dúctil; acero suave	weicher unlegierte Stahl
natural aging	vieillissement naturel	envejecimiento natural	Kaltaushärtung; Kaltauslagern
necking	striction	estricción	Querschnittsverminderung; Einschnürung
Neumann band	bande de Neumann	banda de Neumann	NeumannStreifung
nodular pearlite	perlite nodulaire	perlita nodular	kugeliger Perlit
normalizing	traitement de normalisation	normalizado	normalglühen; normalisieren
notch brittleness	fragilité à l'entaille	fragilidad de entalla	Kerbsprödigkeit
notch ductility	ductilité à l'entaille	ductilidad de entalla	Kerbzähigkeit
notch sensitivity	sensibilité à l'entaille	sensibilidad de entalla	Kerbempfindlichkeit
nucleation	germination	nucleación; germinación	Keimbildung
nucleus	germe	núcleo	Keim; Kern
orientation (crystal)	orientation (cristal)	orientación (cristal)	Lagerung; Orientierung
overaging	(survieillissement, "overaging")	sobreenvejecimiento	überalterung; übervergütung
overheating	surchauffe	recalentamiento; sobrecalentado	Überhitzung
oxidation	oxydation	oxidación	Oxidation
oxidized surface	surface oxydée	superficie oxidada	oxidiertes Oberfläche
pack rolling	laminage en paquet	laminado en paquete	Paketwalzen
pancake grain structure	structure de grain en crêpe	estructura de grano de panqueque	Flachkornstruktur
pass	passe	paso; pasada	Schweißraupe
patenting	patentage	patentado	patentieren
pattern welding	soudage en motif	soldeo de motivo; configuración; forma	Muster Schweißen
pearlite	perlite	perlita	Perlit
peritectic	péritectique	punto peritéctico	peritektisch

GLOSSARY / LEXIQUE / LÉXICO / WÖRTERVERZEICHNIS

ENGLISH	FRENCH	SPANISH	GERMAN
phase	phase	fase	Phase
phase diagram	diagramme de phase	diagrama de fases	Zustandsschaubild
photomacrograph	photomacrographie	fotomacrografía	Lupenphotographie
photomicrograph	photomicrographie	fotomicrografía	Mikrophotographie
physical properties	propriétés physiques	propiedades físicas	physikalische Eigenschaften
pig iron	fonte brute	arrabio	Roheisen
plastic deformation	déformation plastique	deformación plástica	plastische Verformung
plasticity	plasticité	plasticidad	Plastizität
plate	plaque	placa	Platte
plate martensite	martensite en plaque	martensita en placa	Plattenmartensit
polishing	polissage	pulido; pulimiento	Polieren
polycrystalline	polycristallin	policristalino	polykristallin
polymorphism	polymorphisme	polimorfismo	Polymorphie
preferred orientation	orientation préférentielle	orientación preferencial	Vorzugsorientierung
primary crystal	cristal primaire	cristal primario	Primärkristall
proeutectoid (phase)	proeutectoïde (phase)	proeutectoide (fase)	voreutektoidisch (Phase)
proof stress	limite conventionnelle d'élasticité	prueba de capacidad	Dehngrenze
puddling process	procédé de puddlage	procedimiento de pudelado	Puddelnverarbeitung
quench aging	"trempe de vieillissement"	envejecimiento por temple	Abschreckalterung
quench hardening	durcissement par trempe	endurecimiento por temple	Abschreckhärten
quenching	trempe	temple	Abschrecken
recarburizing	re-cémentation (recarburation)	re-carburación	Rückkohlung
recovery	restauration	recuperación	Erholung
recrystallization	recristallisation	recristalización	Rekristallisation
recrystallization annealing	traitement de recristallisation	recocido para recristalizar	Rekristallisationsglühen
recrystallization temperature	température de recristallisation	temperatura de recristalización	Rekristallisationstemperatur
residual elements	éléments résiduels	elementos residuales	verbleibend Element

GLOSSARY / LEXIQUE / LÉXICO / WÖRTERVERZEICHNIS

ENGLISH	FRENCH	SPANISH	GERMAN
residual stress	contrainte résiduelle	tensiones residuales	Eigenspannungen
resistance welding	soudage (électrique) par résistance	soldadura por resistencia	Widerstandsschweißung
resolution	résolution	resolubilizar	Auflösung
resulfurized steel	acier re-sulfuré	acero con azufre adicional	rückschwefeltes Stahl
rimmed steel	acier effervescent	acero efervescente	Randstahl; Massenstahl
rolling	laminage	laminación	Walzen; Rollen
scale	calamine	cascarilla; laminilla; escama	Zunder; Sinter
scarf joint	joint en enture (à onglet)	empalme en los costados biselados del fleje	Schrägstoß
segregation	ségrégation	segregación	Seigerung
segregation banding	ségrégation en bande	segregación en bandas	Seigerungszeilen
self diffusion	autodiffusion	autodifusión	Selbstdiffusion
semikilled steel	acier demi-calmé	acero semicalmado	Halbberuhigtstahl
shear	cisaillement	cizallamiento	Scherung
shear bands (deformation)	bandes de cisaillement (déformation)	bandas de cizallamiento (deformación)	Scherzeilen (Verformung)
shear steel	acier de cisaillement	acero de cizallamiento	Scher-Stahl
sheet	tôle	hoja; chapa fina; plancha; lámina	Blech
shortness	friabilité	fragilidad	Sprödigkeit
sintering	frittage	fritaje; sinterizar;aglomerar en caliente	Zusammenbacken; sintern
slab	dalle (brame)	llantón; plancha; desbaste plano	Bramme
slag	scories (laitier)	escoria	Schlacke
slip	glissement	deslizamiento	Gleitung
slip direction	direction de glissement	dirección de deslizamiento	Gleitrichtung
slip line	ligne de glissement	línea de deslizamiento	Gleitlinie
slip plane	plan de glissement	plano de deslizamiento	Gleitebene

GLOSSARY / LEXIQUE / LÉXICO / WÖRTERVERZEICHNIS

ENGLISH	FRENCH	SPANISH	GERMAN
solid solution	solution solide	solución sólida	feste Lösung
solidus	solidus	solidus	Solidus
solute	soluté	soluto	gelöster Komponente
solution heat treatment	traitement de mise en solution et trempe	tratamiento térmico de solución	Lösungsglühen
solvent	solvant	solvente	Lösungsmittel
solvus	solvus (courbe limite de solubilité)	solvus; línea de solubilidad sólida	Löslichkeitskurve
spheroidized structure	structure globulisée	estructura esferoidizada	Kugelgefüge
spheroidizing	globulisation	esferoidización	Kugeligglühen
spheroidizing annealing	recuit de globulisation	recocido para esferoidizar	Weichglühen
steel	acier	acero	Stahl
strain	déformation	deformación	Verformung
strain aging	(vieillissement sous tension)	envejecimiento por deformación	Reckaltern
strain hardening	durcissement par écrouissage	endurecimiento por deformación	Kalthärtung; Kaltverfestigung
stress	contrainte	esfuerzo	Spannung
stress-corrosion cracking	fissuration par corrosion sous tension	rotura por corrosión bajo tensiones	Spannungskorrosionsriß
stress relieving	relaxation des contraintes	eliminación de tensiones	Entspannung
stretcher strains	vermiculures	marcas de esfuerzo por tracción	Fließfiguren
strip	feuillard; acier en ruban	fleje; banda; cinta; tira	Blechstreifen; Bandblech
sub-boundary structure (subgrain structure)	structure de sous-joint (structure de sous-grain)	estructura dentro de los límites de grano (intragranular)	Untergrenze Struktur; (Unterkorn Struktur)
subcritical annealing	recuit subcritique	recocido subcrítico	Rekristallisationsglühen
subgrain	sous-grain	subgrano	Subkorn; Unterkorn
substitutional solid solution	solution solide de substitution	solución solida de substitución	Substitutionsmischkristall
substrate	substrat	substrato	Substrat
sulfide spheroidization	globulisation du sulfure	esferoidización del sulfuro	Einformung des Sulfid

GLOSSARY / LEXIQUE / LÉXICO / WÖRTERVERZEICHNIS

ENGLISH	FRENCH	SPANISH	GERMAN
sulfur print	empreinte de soufre (empreinte Baumann	impresión de azufre	Baumannabdruck
supercooling	sous-refroidissement	superenfriamiento	Unterkühlung
superheating	surchauffe	sobrecalentamiento	Überhitzung
surface hardening; case hardening	traitement de durcissement superficiel	endurecimiento superficial	Oberflächenhärtung
taper section	section conique	sección terminada en punta	Schrägkante
teeming	coulage	vaciado	Gießen; Abguß
temper brittleness	fragilité de revenu	fragilidad de revenido	Anlaßsprödigkeit
temper rolling	laminage de revenu	laminado de endurecimiento	Nachwalzen; Dressieren
tempering	revenu	revenido	Anlassen
tensile strength	résistance à la traction	resistencia a la tracción	Zugfestigkeit
texture	texture	textura	Textur
thermal analysis	analyse thermique	análisis térmico	thermische Analyse; Thermoanalyse
time-temperature-transformation (TTT) diagram	diagramme de transformation en conditions isothermes (TTT)	diagrama de transformación isotérmica; curvas S; curvas TTT	Zeit-Temperatur-Umwandlungs-Schaubild, ZTU-Kurve
toughness	ténacité	tenacidad	Zähigkeit
transcrystalline	transgranulaire	transcristalina	transkristallin
transformation ranges (transformation temperature ranges)	domaines de transformation (domaines de température de transformation	dominios de transformación; dominios de las temperatura de transformación	Umwandlungsbereich
transformation temperature	température de transformation	temperatura de transformación	Umwandlungstemperatur
transgranular	intragranulaire	intracristalina	Transkristallin
transition temperature (ductile-brittle transition temperature)	température de transition (température de transition ductile-fragile)	temperatura de transición temperatura de transición de dúctil a frágil	Übergangstemperatur
triple point	point triple	punto triple	Tripelpunkt

GLOSSARY / LEXIQUE / LÉXICO / WÖRTERVERZEICHNIS

ENGLISH	FRENCH	SPANISH	GERMAN
twin	macle	macla	Zwilling
twin, annealing	macle, recuit (macle de recuit)	macla, recocido; macla de recocido	Rekristallisationszwilling
twin, crystal	macle, cristal (grain maclé)	macla, cristal	Zwillingskristall
twin, deformation	macle, déformation (macle de déformation)	macla, deformación	Verformungszwilling
upper yield stress	limite supérieure d'élasticité	lǐmite de fluencia superior	obere Streckgrenze
upset	"upset" (dérangement)	recalcado	stauchen
vacancy	lacune	vacante	Leerstelle
veining	veinage (marbrure)	veteado; jaspeado	Formriß; Blattrippe
Waloon process	procédé Waloon	proceso Waloon	Waloon Prozess
weld	soudure	soldadura	Schweißnaht
weld bead	cordon de soudure	cordón de soldadura	Schweißraupe
weldability	soudabilité	soldabilidad	Schweißbarkeit
welding	soudage	soldadura	Schweißen
wetting agent	agent mouillant	agente humectante	Klarspüler
Widmannstätten structure	structure de Widmannstätten	estructura de Widmannstätten	Widmannstätten - Gefüge
wootz	wootz	wootz	Wootz
work hardening	durcissement par écrouissage	endurecimiento del acero por trabajo en frío	Kalthärtung; Kaltverfestigung
wrought iron	fer corroyé.	hierro forjado; hierro pudelado; hierro dulce; acero suave	Schmiedeeisern
wustite	wustite	wurtzita	Wüstit
yield point	limite d'écoulement	límite aparente de elasticidad	Fliesspunkt
yield strength	résistance à l'écoulement	esfuerzo de fluencia	konventionelle o. technische Streckgrenze; 0,2% Dehngrenze; (proof stress: Ersatzstreckgrenze)

LEXIQUE / GLOSSARY / LÉXICO / WÖRTERVERZEICHNIS

FRENCH	ENGLISH	SPANISH	GERMAN
abrasif	abrasive	abrasivo	Schleifmittel; Abriebeigenschaften
abrasion	abrasion	abrasión	Schleifwirkung; Abnutzung
acier	steel	acero	Stahl
acier allié	alloy steel	acero aleado	legierter Stahl
acier au carbone	carbon steel	acero al carbono	unlegierter Stahl
acier au creuset	crucible steel	acero al crisol	Tiegelstahlblock
acier calmé	killed steel	acero desoxidado	beruhigter Stahl
acier coiffé	capped steel	acero efervescente	gedeckelter Stahl
acier comportant des soufflures (ou "pailles")	blister steel	acero ampollado	Zement[ations]stahl
acier de cisaillement	shear steel	acero de cizallamiento	Scher-Stahl
acier demi-calmé	semikilled steel	acero semicalmado	Halbberuhigtstahl
acier doux	mild steel	acero dulce; acero dúctil; acero suave	weicher unlegierte Stahl
acier effervescent	rimmed steel	acero efervescente	Randstahl; Massenstahl
acier moulé	cast steel	acero moldeado; vaciado	Stahlguß
acier re-sulfuré	resulfurized steel	acero con azufre adicional	rückschwefeltes Stahl
affinerie	finery	horno de afino	Affinerie
agent mouillant	wetting agent	agente humectante	Klarspüler
alliage	alloy	aleación	Legierung
alliage hypereutectoïde	hypereutectoid alloy	aleación hipereutectoide	übereutektoidisch Legierung
alliage hypoeutectique	hypoeutectic alloy	aleación hipoeutéctica	untereutektoidisch Legierung
allongement après rupture	elongation after fracture	elongación a consecuencia de fractura	Bruchdehnung
allotriomorphe	allotriomorph	allotriomorfico	Allotriomorph
allotropie	allotropy	allotropía	Allotropie
analyse thermique	thermal analysis	análisis térmico	thermische Analyse; Thermoanalyse

LEXIQUE / GLOSSARY / LÉXICO / WÖRTERVERZEICHNIS

FRENCH	ENGLISH	SPANISH	GERMAN
analyseur par microsonde électronique	electron beam microprobe analyzer	bombardeo de electrones	Elektronenstrahlmikrosonde analysator
artefact	artifact	artefacto	künstlich
attaque	etching	ataque químico	Ätzen; Anätzung
austénite	austenite	austenita	Austenit
austénitisation	austenitizing	austenización	Austenitisieren
auto-revenu	autotempering	autorevenido	Autoanlaßen
autodiffusion	self diffusion	autodiffusión	Selbstdiffusion
autoradiographie	autoradiograph	autorradiografía	Autoradiographie
bainite	bainite	bainita	Bainit
bande	banding	bandas	Zeilenbildung
bande de contorsion (déformation)	kink band (deformation)	banda de ensortijamiento	Zeilenverschlingung
bande de Neumann	Neumann band	banda de Neumann	NeumannStreifung
bande ferrite-perlite	ferrite-pearlite banding	bandas de ferrita y perlita	Ferrit-Perlit zeilenbildung
bandes de cisaillement (déformation)	shear bands (deformation)	bandas de cizallamiento (deformación)	Scherzeilen (Verformung)
bandes de déformation	deformation bands	bandas de deformación	Verformungszeilen
bavure	flash	rebarba	Grat; Gratbildung
billette (ou larget)	billet	palanquilla	Walzblock; Barren; Knüppel
bloom (brame)	bloom	bloom; desbaste	Vorblock; Schmiedeblock; Blume
bloomerie	bloomery	planta para fabricar desbastes	Rennherd
brasage	brazing	soldadura amarilla	Löten; Hartlöten
brûlure	burning	quemado	Verbrennung
calamine	scale	cascarilla; laminilla; escama	Zunder; Sinter
carbonitruration	carbonitriding	carbonitruración	Karbonitrieren
carbure	carbide	carburo	Carbid

LEXIQUE / GLOSSARY / LÉXICO / WÖRTERVERZEICHNIS

FRENCH	ENGLISH	SPANISH	GERMAN
cémentation	carburizing	cementación; carburación	Aufkohlen
cémentation	cementation	cementación	Zementation
cémentite	cementite	cementita	Zementit
chaferie	chafery	rozamiento	chaferie
cisaillement	shear	cizallamiento	Scherung
clivage	cleavage	exfoliación o despegue	Spaltbruch; Spaltung
coloration par chaleur	heat tinting	termocoloración	Anlaufen
constituant	constituent	constituyente	Konstituent; Bestandteil
contrainte	stress	esfuerzo	Spannung
contrainte d'écoulement	flow stress	esfuerzo de fluencia	Fließspannung
contrainte résiduelle	residual stress	tensiones residuales	Eigenspannungen
cordon de soudure	weld bead	cordón de soldadura	Schweißraupe
corrosion	corrosion	corrosión	Korrosion
couche superficielle traitée	case	capa superficial	Einsatz[härte]schicht
coulage	teeming	vaciado	Gießen; Abguß
cristal primaire	primary crystal	cristal primario	Primärkristall
croissance du grain	grain growth	crecimiento del grano	Kornwachstum
dalle (brame)	slab	llantón; plancha; desbaste plano	Bramme
décarburation	decarburization	decarburación	Entkohlung
déchirure lamellaire	lamellar tear	desgarre laminar	Terrassenbruch
déformation	strain	deformación	Verformung
déformation critique	critical strain	deformación crítica	kritische Verformung
déformation élastique	elastic strain	deformación elástica;	elastische Verformung
déformation plastique	plastic deformation	deformación plástica	plastische Verformung
dendrite	dendrite	dendrita	Dendrit
dérangement "upset"	upset	recalcado	stauchen
désoxydation	deoxidation	desoxidación	Desoxidieren; Reduktion

The Metals Black Book – 3rd Edition

LEXIQUE / GLOSSARY / LÉXICO / WÖRTERVERZEICHNIS

FRENCH	ENGLISH	SPANISH	GERMAN
diagramme d'équilibre	equilibrium diagram	diagrama de equilibrio	Zustandsschaubild
diagramme de constitution	constitutional diagram	diagrama de fases	Zustandsschaubild
diagramme de phase	phase diagram	diagrama de fases	Zustandsschaubild
diagramme de transformation en conditions isothermes (diagramme TI)	isothermal transformation (IT) diagram	diagrama de transformación isotérmica	isothermisches Glühen Schaubild
diagramme de transformation en conditions isothermes (TTT)	time-temperature-transformation (TTT) diagram	diagrama de transformación isotérmica; curvas S; curvas TTT	Zeit-Temperatur-Umwandlungs-Schaubild, ZTU-Kurve
diffusion	diffusion	difusión	Diffusion
dilatomètre	dilatometer	dilatómetro	Dilatometer
direction de glissement	slip direction	dirección de deslizamiento	Gleitrichtung
dislocation	dislocation	dislocación	Verschiebung; Versetzung
domaine martensitique	martensite range	dominio martensítico	Martensitbereich
domaines de transformation (domaines de température de transformation	transformation ranges (transformation temperature ranges)	dominios de transformación; dominios de las temperatura de transformación	Umwandlungsbereich
ductilité	ductility	ductilidad	Duktilität; Dehnbarkeit
ductilité à l'entaille	notch ductility	ductilidad de entalla	Kerbzähigkeit
durcissement par écrouissage	strain hardening	endurecimiento por deformación	Kalthärtung; Kaltverfestigung
durcissement par écrouissage	work hardening	endurecimiento del acero por trabajo en frío	Kalthärtung; Kaltverfestigung
durcissement par trempe	quench hardening	endurecimiento por temple	Abschreckhärten
durcissement par vieillissement	age hardening	endurecimiento por envejecimiento	Aushärtung; Vergütung
dureté à la bille (empreinte)	hardness (indentation)	dureza (indentación)	Härte (Eindruck)
écoulement de grain	grain flow	fluencia de granos	Kornfluß
élément d'alliage	alloying element	elemento de aleación	Legierungselement

LEXIQUE / GLOSSARY / LÉXICO / WÖRTERVERZEICHNIS

FRENCH	ENGLISH	SPANISH	GERMAN
éléments résiduels	residual elements	elementos residuales	verbleibend Element
empreinte de soufre (empreinte Baumann	sulfur print	impresión de azufre	Baumannabdruck
épitaxie	epitaxy	epitaxia	Epitaxie
équivalent carbone (pour estimer la soudabilité)	carbon equivalent (for rating of weldability)	carbono equivalente	Kohlenstoffäquivalent (für Schweißbarkeit Schätzung)
essai de choc (de résilience)	impact test	prueba de impacto; prueba de choque; prueba de resiliencia	Kerbschlagbiegeversuch
eutectoïde	eutectoid	eutectoide	Eutektoid
facilement usinable (décolletage rapide)	free machining	de fácil maquinado; torneado rápido	spanbar
fatigue	fatigue	fatiga	Ermüdung
fer	iron	hierro	Eisen
fer corroyé.	wrought iron	hierro forjado; hierro pudelado; hierro dulce; acero suave	Schmiedeeisern
fer lingot	ingot iron	hierro ARMCO	entkohlter Stahl; Bessemerstahl
ferrite	ferrite	ferrita	Ferrit
feuillard; acier en ruban	strip	fleje; banda; cinta; tira	Blechstreifen; Bandblech
feuilletage	lamination	separación en láminas	Dopplung
fibre	fiber	fibra	Faser
fissuration par corrosion sous tension	stress-corrosion cracking	rotura por corrosión bajo tensiones	Spannungskorrosionsriß
fluage	creep	fluencia	Kriechen
flux (fondant)	flux	fundente	Flußmittel
fonte brute	pig iron	arrabio	Roheisen
fonte de moulage	cast iron	arrabio; hierro colado	Gußeisen
forgeage	forging	forja; forjadura	Schmiedestück; Schmieden

LEXIQUE / GLOSSARY / LÉXICO / WÖRTERVERZEICHNIS

FRENCH	ENGLISH	SPANISH	GERMAN
formage à chaud	hot working	formación o trabajo en caliente	Warmumformen
fractographie	fractography	estudio de fracturas	Fraktographie
fragilité	brittleness	fragilidad	Sprödigkeit
fragilité à l'entaille	notch brittleness	fragilidad de entalla	Kerbsprödigkeit
fragilité de revenu	temper brittleness	fragilidad de revenido	Anlaßsprödigkeit
fragmentation	fragmentation	fragmentación	Zertrümmerung
friabilité	shortness	fragilidad	Sprödigkeit
frittage	sintering	fritaje; sinterizar;aglomerar en caliente	Zusammenbacken; sintern
garniture (de dislocations)	decoration (of dislocations)	(guarnición de dislocaciones)	Dekoration (an Verschiebung)
germe	nucleus	núcleo	Keim; Kern
germination	nucleation	nucleación; germinación	Keimbildung
glissement	slip	deslizamiento	Gleitung
globulisation	spheroidizing	esferoidización	Kugeligglühen
globulisation du sulfure	sulfide spheroidization	esferoidización del sulfuro	Einformung des Sulfid
grain	grain	grano	Korn
graphite	graphite	grafito	Graphit
graphitisation	graphitization	grafitización	Graphitisation
grosseur de grain	grain size	tamaño de grano	Korngrösse
grosseur de grain austénitique	austenitic grain size	dimensión de grano austenítico	Austenitkorngröße
grosseur de grain ferritique	ferritic grain size	tamaño del grano ferrítico	ferritischer Korngrösse
grossissement du grain	grain coarsening	crecimiento del grano	Kornvergröberung
hématite	hematite	hematita	Roteisenstein; roter Hämatit
idiomorphe	idiomorph	idiomorfo	idiomorph
impuretés	impurities	impurezas	Unreinheit; Verunreinigung
inclusion	inclusion	inclusión	Einschluß
intercristalline	intercrystalline	intercristalina	interkristallin

LEXIQUE / GLOSSARY / LÉXICO / WÖRTERVERZEICHNIS

FRENCH	ENGLISH	SPANISH	GERMAN
intragranulaire	intracrystalline	intracristalina	intrakristallin
intragranulaire	transgranular	intracristalina	Transkristallin
joint en enture (à onglet)	scarf joint	empalme en los costados biselados del fleje	Schrägstoß
lacune	vacancy	vacante	Leerstelle
laminage	rolling	laminación	Walzen; Rollen
laminage avec cylindres obliques	cross rolling	laminación transversal	schrägwalzen
laminage contrôlé	controlled rolling	laminado controlado	Regelrollen
laminage de revenu	temper rolling	laminado de endurecimiento	Nachwalzen; Dressieren
laminage en paquet	pack rolling	laminado en paquete	Paketwalzen
ligne de glissement	slip line	línea de deslizamiento	Gleitlinie
lignes d'écoulement (lignes de Piobert-Lüders)	flow lines	líneas de fluencia; líneas de Piobert-Lüders; líneas de flujo	Fließlinien; Schlieren
lignes ou bandes de Lüder	Luder's lines or bands	líneas o bandas de Lüder	Fließfiguren; Fließlinien; Lüderssche Linien
limite conventionnelle d'élasticité	proof stress	prueba de capacidad	Dehngrenze
limite d'écoulement	yield point	límite aparente de elasticidad	Fliesspunkt
limite d'élasticité	elastic limit	límite de elasticidad	Elastizitätsgrenze
limite supérieure d'élasticité	upper yield stress	lcmite de fluencia superior	obere Streckgrenze
lingot	ingot	lingote	Blockmetall; Massel
liquation	liquation	licuación; segregación	Seigerung
liquation de joint de grain	grain-boundary liquation	licuefacción de los límites de los granos	Korngrenzenangriff; Korngrenzeseigerung
liquidus	liquidus	línea de líquido	Liquidus
macle	twin	macla	Zwilling
macle de recuit	annealing twin	macla por recocido	Rekristallisationszwilling

LEXIQUE / GLOSSARY / LÉXICO / WÖRTERVERZEICHNIS

FRENCH	ENGLISH	SPANISH	GERMAN
macle mécanique	mechanical twin	macla mecánica	mechanische Zwilling
macle, cristal (grain maclé)	twin, crystal	macla, cristal	Zwillingskristall
macle, déformation (macle de déformation)	twin, deformation	macla, deformación	Verformungszwilling
macle, recuit (macle de recuit)	twin, annealing	macla, recocido; macla de recocido	Rekristallisationszwilling
macroattaque	macroetching	macroataque químico	Grobätzung
macrographie	macrograph	macrografía	Makrograph; Grobgefügebild
macrostructure	macrostructure	macroestructura	Makrogefüge; Grobgefüge; Makrostruktur
magnétite	magnetite	magnetita	Magnetit; Magneteisenstein
martensite	martensite	martensita	Martensit
martensite en plaque	plate martensite	martensita en placa	Plattenmartensit
martensite massive	lath martensite	martensita de placas finas	Lattenmartensit
matrice	matrix	matriz	Grundmasse; Matrix
métal	metal	metal	Metall
métal d'apport	filler metal	metal de aportación	Schweißzusatzwerstoff; Zusastzwerkstoff
métallographe	metallograph	metalógrafo; microscopio metalúrgico	Metallograph
métastable	metastable	metaestable	metastabil
meulage (rectification)	grinding	rectificado; amolado	Schleifen
microbandes (déformation)	microbands (deformation)	microbandas (deformación)	Mikrozeilen (Verformung)
microfissure	microcrack	microgrieta; microfissura	Mikroriß; mikroskopischer Riß
micrographie	micrograph	micrografía	Mikrobild; Schliffbild
microségrégation	microsegregation	microsegregación	Mikroseigerung
microstructure	microstructure	microestructura	Mikrogefüge

LEXIQUE / GLOSSARY / LÉXICO / WÖRTERVERZEICHNIS

FRENCH	ENGLISH	SPANISH	GERMAN
moulage	casting	pieza fundida; pieza de fundición; colada del acero líquido; fundición	Gußerzeugnisse
noyautage; cernes	coring	microsegregación o segregación intragranular	Kristallseigerung; Mikroseigerung
orientation (cristal)	orientation (crystal)	orientación (cristal)	Lagerung; Orientierung
orientation préférentielle	preferred orientation	orientación preferencial	Vorzugsorientierung
oxydation	oxidation	oxidación	Oxidation
oxydation interne	internal oxidation	oxidación interna	innere Oxidation
passe	pass	paso; pasada	Schweißraupe
patentage	patenting	patentado	patentieren
péritectique	peritectic	punto peritéctico	peritektisch
perlite	pearlite	perlita	Perlit
perlite nodulaire	nodular pearlite	perlita nodular	kugeliger Perlit
phase	phase	fase	Phase
phase continue	continuous phase	fase continua	kontinuierliches Phase
photomacrographie	photomacrograph	fotomacrografía	Lupenphotographie
photomicrographie	photomicrograph	fotomicrógrafía	Mikrophotographie
plan de clivage	cleavage plane	plano de despegue	Spaltfläche
plan de glissement	slip plane	plano de deslizamiento	Gleitebene
plaque	plate	placa	Platte
plasticité	plasticity	plasticidad	Plastizität
point critique (ou température de transformation)	critical point	punto crítico; temperatura de transformación	Umwandlungstemperatur; Umwandlungspunkt
point triple	triple point	punto triple	Tripelpunkt
polissage	polishing	pulido; pulimiento	Polieren
polissage chimique	chemical polishing	pulido químico	chemisches Polieren
polissage électrolytique	electropolishing	pulido electrolítico	elektrolytisch polieren

LEXIQUE / GLOSSARY / LÉXICO / WÖRTERVERZEICHNIS

FRENCH	ENGLISH	SPANISH	GERMAN
polissage mécanique	mechanical polishing	pulido mecánico	Maschinenpolieren
polycristallin	polycrystalline	policristalino	polykristallin
polymorphisme	polymorphism	polimorfismo	Polymorphie
potentiel carbone	carbon potential	potencial de carbono	Kohlenstoffpegel, C-Pegel
précipitation de sulfure aux joints de grain	grain-boundary sulfide precipitation	(precipitación de sulfuro en los límites de granos	KorngrenzenSulfidausscheidung
procédé de puddlage	puddling process	procedimiento de pudelado	Puddelnverarbeitung
procédé Waloon	Waloon process	proceso Waloon	Waloon Prozess
proeutectoïde (phase)	proeutectoid (phase)	proeutectoide (fase)	voreutektoidisch (Phase)
propriétés physiques	physical properties	propiedades físicas	physikalische Eigenschaften
re-cémentation (recarburation)	recarburizing	re-carburación	Rückkohlung
réactif d'attaque	etchant	ácido para grabar; reactivo para ataque	Ätzflüssigkeit
recristallisation	recrystallization	recristalización	Rekristallisation
recuit	annealing	recocido	Glühen
recuit complet (matériaux ferreux)	full annealing (ferrous materials)	recocido completo; recocido a fondo (materiales ferrosos)	Hochglühen (über A₃)
recuit d'homogénéisation	homogenizing annealing	recocido de homogeneización	Diffusionsglühen
recuit de globulisation	spheroidizing annealing	recocido para esferoidizar	Weichglühen
recuit en caisse	box annealing	recocido en caja	Kistenglühen
recuit subcritique	subcritical annealing	recocido subcrítico	Rekristallisationsglühen
relaxation des contraintes	stress relieving	eliminación de tensiones	Entspannung
résistance à l'écoulement	yield strength	esfuerzo de fluencia	konventionelle o. technische Streckgrenze; 0,2% Dehngrenze; (proof stress: Ersatzstreckgrenze)
résistance à la traction	tensile strength	resistencia a la tracción	Zugfestigkeit
résolution	resolution	resolubilizar	Auflösung

LEXIQUE / GLOSSARY / LÉXICO / WÖRTERVERZEICHNIS

FRENCH	ENGLISH	SPANISH	GERMAN
restauration	recovery	recuperación	Erholung
revenu	tempering	revenido	Anlassen
rupture cristalline	crystalline fracture	fractura cristalina	kristallin[ischer Bruch
rupture fibreuse	fibrous fracture	fractura fibrosa	Holzfaserbruch
rupture fragile	brittle fracture	fractura frágil	Sprödbruch
rupture granuleuse	granular fracture	fractura granulosa	körniger Bruch; kristallin[ischler Bruch
rupture par clivage	cleavage fracture	fractura transcristalina a través de planos de deslizamiento	Trennbruch; Sprödbruch
scories (laitier)	slag	escoria	Schlacke
section conique	taper section	sección terminada en punta	Schrägkante
ségrégation	segregation	segregación	Seigerung
ségrégation dendritique	dendritic segregation	segregación dendrítica	dendritisches Seigerung
ségrégation en bande	segregation banding	segregación en bandas	Seigerungszeilen
sensibilité à l'entaille	notch sensitivity	sensibilidad de entalla	Kerbempfindlichkeit
solidus	solidus	solidus	Solidus
soluté	solute	soluto	gelöster Komponente
solution solide	solid solution	solución sólida	feste Lösung
solution solide de substitution	substitutional solid solution	solución sólida de substitución	Substitutionsmischkristall
solution solide interstitielle	interstitial solid solution	solución sólida intersticial	Einlagerungsmischkristall
solvant	solvent	solvente	Lösungsmittel
solvus (courbe limite de solubilité)	solvus	solvus; línea de solubilidad sólida	Löslichkeitskurve
soudabilité	weldability	soldabilidad	Schweißbarkeit
soudage	welding	soldadura	Schweißen
soudage (électrique) par résistance	resistance welding	soldadura por resistencia	Widerstandsschweißung

LEXIQUE / GLOSSARY / LÉXICO / WÖRTERVERZEICHNIS

FRENCH	ENGLISH	SPANISH	GERMAN
soudage à l'arc	arc welding	soldadura por arco	Lichtbogenschweißen
soudage à la forge	forge welding	soldeo por calda a la forja; soldadura de forja	Hammerschweißen
soudage au gaz	gas welding	soldadura oxiacetilénica	Gasschweißen
soudage en motif	pattern welding	soldeo de motivo; configuración; forma	Muster Schweißen
soudage manuel	manual welding	soldeo manual	Handscheißung
soudage par fusion	fusion welding	soldadura a fusión	Schmelzschweißen
soudo-brasage	braze welding	soldadura con metal amarillo	Schweißlötung
soudure	weld	soldadura	Schweißnaht
sous-grain	subgrain	subgrano	Subkorn; Unterkorn
sous-refroidissement	supercooling	superenfriamiento	Unterkühlung
striction	necking	estricción	Querschnittsverminderung; Einschnürung
structure basaltique	columnar structure	estructura columnar	Stengelgefüge
structure de grain en bambou	bamboo grain structure	estructura de grano de bambú	Bambus Kornstruktur
structure de grain en crêpe	pancake grain structure	estructura de grano de panqueque	Flachkornstruktur
structure de sous-joint (structure de sous-grain	sub-boundary structure (subgrain structure)	estructura dentro de los límites de grano (intragranular)	Untergrenze Struktur; (Unterkorn Struktur)
structure de Widmannstätten	Widmannstatten structure	estructura de Widmannstätten	Widmannstätten - Gefüge
structure dégénérée	degenerate structure	estructura degenerada	entarteter Struktur
structure équiaxe	equiaxed structure	estructura equiaxial	gleichachsig Struktur
structure globulisée	spheroidized structure	estructura esferoidizada	Kugelgefüge
substrat	substrate	substrato	Substrat
surchauffe	overheating	recalentamiento; sobrecalentado	Überhitzung
surchauffe	superheating	sobrecalentamiento	Überhitzung
surface oxydée	oxidized surface	superficie oxidada	oxidiertes Oberfläche

LEXIQUE / GLOSSARY / LÉXICO / WÖRTERVERZEICHNIS

FRENCH	ENGLISH	SPANISH	GERMAN
survieillissement, "overaging"	overaging	sobreenvejecimiento	überalterung; übervergütung
tampon	buffer	regulador; amortiguador	Pufferschicht
température A_1	A_1 temperature	temperatura A_1	A_1 Temperatur
température A_2	A_2 temperature	temperatura A_2	A_2 Temperatur
température A_3	A_3 temperature	temperatura A_3	A_3 Temperatur
température A_{cm}	A_{cm} temperature	temperatura A_{cm}	A_{cm} Temperatur
température de finissage	finishing temperature	temperatura de acabado; temperatura final	Endtemperatur des Werkstückes
température de fusion	melting point	punto de fusión	Schmelzpunkt
température de recristallisation	recrystallization temperature	temperatura de recristalización	Rekristallisationstemperatur
température de transformation	critical temperature	temperatura crítica	Umwandlungstemperatur
température de transformation	transformation temperature	temperatura de transformación	Umwandlungstemperatur
température de transition (température de transition ductile-fragile)	transition temperature (ductile-brittle transition temperature)	temperatura de transición temperatura de transición de dúctil a frágil	Übergangstemperatur
température M_f	M_f temperature	temperatura M_f	M_f Temperatur
température M_s	M_s temperature	temperatura M_s	M_s Temperatur
ténacité	toughness	tenacidad	Zähigkeit
texture	texture	textura	Textur
tôle	sheet	hoja; chapa fina; plancha; lámina	Blech
traitement de durcissement	hardening	endurecimiento; temple; cementación	Härten
traitement de durcissement superficiel	surface hardening; case hardening	endurecimiento superficial	Oberflächenhärtung
traitement de mise en solution et trempe	solution heat treatment	tratamiento térmico de solución	Lösungsglühen

LEXIQUE / GLOSSARY / LÉXICO / WÖRTERVERZEICHNIS

FRENCH	ENGLISH	SPANISH	GERMAN
traitement de normalisation	normalizing	normalizado	normalglühen; normalisieren
traitement de recarburation	carbon restoration	tratamiento de recementación	Wiederaufkohlung
traitement de recristallisation	recrystallization annealing	recocido para recristalizar	Rekristallisationsglühen
traitement thermique	heat treatment	tratamiento térmico; termotratamiento	Wärmebehandlung
transgranulaire	transcrystalline	transcristalina	transkristallin
trempabilité	hardenability	templabilidad; facultad de endurecimiento	Härtbarkeit
trempe	quenching	temple	Abschrecken
trempe de vieillissement	quench aging	envejecimiento por temple	Abschreckalterung
trempe étagée bainitique	austempering	atemperación escalonada (tratamiento con transformación isotérmica de austenita a bainita)	Zwischenstufen-Vergütung
trempe étagée martensitique	martempering	temple escalonado martensítico	Warmbadhärten
trempe superficielle; durcissement superficiel	case hardening	cementación en caja	Einsatzhärtung; Oberflächenhärten
unité angström (Å)	angstrom unit (Å)	unidad angstrom (Å)	Ångström (Å)
usinabilité	machinability	maquinabilidad	Zerspanbarkeit
veinage (marbrure)	veining	veteado; jaspeado	Formriß; Blattrippe
vermicullures	stretcher strains	marcas de esfuerzo por tracción	Fließfiguren
vieillissement	aging	envejecimiento	Alterung
vieillissement naturel	natural aging	envejecimiento natural	Kaltaushärtung; Kaltauslagern
vieillissement sous tension	strain aging	envejecimiento por deformación	Reckalterung
vitesse de refroidissement critique	critical cooling rate	velocidad crítica de enfriamiento	kritische Abkühlungsgeschwindigkeit
wootz	wootz	wootz	Wootz
wustite	wustite	wurtzita	Wüstit
zone affectée thermiquement (ZAT)	heat-affected zone (HAZ)	zona térmicamente afectada; ZTA	Wärmeeinflußzone (WEZ)

LÉXICO / GLOSSARY / LEXIQUE / WÖRTERVERZEICHNIS

SPANISH	ENGLISH	FRENCH	GERMAN
abrasión	abrasion	abrasion	Schleifwirkung; Abnutzung
abrasivo	abrasive	abrasif	Schleifmittel; Abriebeigenschaften
acero	steel	acier	Stahl
acero al carbono	carbon steel	acier au carbone	unlegierter Stahl
acero al crisol	crucible steel	acier au creuset	Tiegelstahlblock
acero aleado	alloy steel	acier allié	legierter Stahl
acero ampollado	blister steel	acier comportant des souffiures (ou "pailles")	Zement[ations]stahl
acero con azufre adicional	resulfurized steel	acier re-sulfuré	rückschwefeltes Stahl
acero de cizallamiento	shear steel	acier de cisaillement	Scher-Stahl
acero desoxidado	killed steel	acier calmé	beruhigter Stahl
acero dulce; acero dúctil; acero suave	mild steel	acier doux	weicher unlegierte Stahl
acero efervescente	capped steel	acier coiffé	gedeckelter Stahl
acero efervescente	rimmed steel	acier effervescent	Randstahl; Massenstahl
acero moldeado; vaciado	cast steel	acier moulé	Stahlguß
acero semicalmado	semikilled steel	acier demi-calmé	Halbberuhigtstahl
ácido para grabar; reactivo para ataque	etchant	réactif d'attaque	Ätzflüssigkeit
agente humectante	wetting agent	agent mouillant	Klarspüler
aleación	alloy	alliage	Legierung
aleación hipereutectoide	hypereutectoid alloy	alliage hypereutectoïde	übereutektoidisch Legierung
aleación hipoeutéctica	hypoeutectic alloy	alliage hypoeutectique	untereutektoidisch Legierung
allotriomorfico	allotriomorph	allotriomorphe	Allotriomorph
allotropia	allotropy	allotropie	Allotropie
análisis térmico	thermal analysis	analyse thermique	thermische Analyse; Thermoanalyse

LÉXICO / GLOSSARY / LEXIQUE / WÖRTERVERZEICHNIS

SPANISH	ENGLISH	FRENCH	GERMAN
arrabio	pig iron	fonte brute	Roheisen
arrabio; hierro colado	cast iron	fonte de moulage	Gußeisen
artefacto	artifact	artefact	künstlich
ataque químico	etching	attaque	Ätzen; Anätzung
atemperación escalonada (tratamiento con transformación isotérmica de austenita a bainita)	austempering	trempe étagée bainitique	Zwischenstufen-Vergütung
austenita	austenite	austénite	Austenit
austenización	austenitizing	austénitisation	Austenitisieren
autodifusión	self diffusion	autodiffusion	Selbstdiffusion
autorevenido	autotempering	auto-revenu	Autoanlaßen
autorradiografía	autoradiograph	autoradiographie	Autoradiographie
bainita	bainite	bainite	Bainit
banda de ensortijamiento	kink band (deformation)	bande de contorsion (déformation)	Zeilenverschlingung
banda de Neumann	Neumann band	bande de Neumann	NeumannStreifung
bandas	banding	bande	Zeilenbildung
bandas de cizallamiento (deformación)	shear bands (deformation)	bandes de cisaillement (déformation)	Scherzeilen (Verformung)
bandas de deformación	deformation bands	bandes de déformation	Verformungszeilen
bandas de ferrita y perlita	ferrite-pearlite banding	bande ferrite-perlite	Ferrit-Perlit zeilenbildung
bloom; desbaste	bloom	bloom (brame)	Vorblock; Schmiedeblock; Blume
bombardeo de electrones	electron beam microprobe analyzer	analyseur par microsonde électronique	Elektronenstrahlmikrosonde analysator
capa superficial	case	couche superficielle traitée	Einsatz[härte]schicht
carbonitruración	carbonitriding	carbonitruration	Karbonitrieren
carbono equivalente	carbon equivalent (for rating of weldability)	équivalent carbone (pour estimer la soudabilité)	Kohlenstoffäquivalent (für Schweißbarkeit Schätzung)

LÉXICO / GLOSSARY / LEXIQUE / WÖRTERVERZEICHNIS

SPANISH	ENGLISH	FRENCH	GERMAN
carburo	carbide	carbure	Carbid
cascarilla; laminilla; escama	scale	calamine	Zunder; Sinter
cementación	cementation	cémentation	Zementation
cementación en caja	case hardening	trempe superficielle; durcissement superficiel	Einsatzhärtung; Oberflächenhärten
cementación; carburación	carburizing	cémentation	Aufkohlen
cementita	cementite	cémentite	Zementit
cizallamiento	shear	cisaillement	Scherung
constituyente	constituent	constituant	Konstituent; Bestandteil
cordón de soldadura	weld bead	cordon de soudure	Schweißraupe
corrosión	corrosion	corrosion	Korrosion
crecimiento del grano	grain growth	croissance du grain	Kornwachstum
crecimiento del grano	grain coarsening	grossissement du grain	Kornvergröberung
cristal primario	primary crystal	cristal primaire	Primärkristall
de fácil maquinado; torneado rápido	free machining	facilement usinable (décolletage rapide)	spanbar
decarburación	decarburization	décarburation	Entkohlung
deformación	strain	déformation	Verformung
deformación crítica	critical strain	déformation critique	kritische Verformung
deformación elástica;	elastic strain	déformation élastique	elastische Verformung
deformación plástica	plastic deformation	déformation plastique	plastische Verformung
dendrita	dendrite	dendrite	Dendrit
desgarre laminar	lamellar tear	déchirure lamellaire	Terrassenbruch
deslizamiento	slip	glissement	Gleitung
desoxidación	deoxidation	désoxydation	Desoxidieren; Reduktion
diagrama de equilibrio	equilibrium diagram	diagramme d'équilibre	Zustandsschaubild
diagrama de fases	constitutional diagram	diagramme de constitution	Zustandsschaubild

LÉXICO / GLOSSARY / LEXIQUE / WÖRTERVERZEICHNIS

SPANISH	ENGLISH	FRENCH	GERMAN
diagrama de fases	phase diagram	diagramme de phase	Zustandsschaubild
diagrama de transformación isotérmica	isothermal transformation (IT) diagram	diagramme de transformation en conditions isothermes (diagramme TI)	isothermisches Glühen Schaubild
diagrama de transformación isotérmica; curvas S; curvas TTT	time-temperature-transformation (TTT) diagram	diagramme de transformation en conditions isothermes (TTT)	Zeit-Temperatur-Umwandlungs-Schaubild, ZTU-Kurve
difusión	diffusion	diffusion	Diffusion
dilatómetro	dilatometer	dilatomètre	Dilatometer
dimensión de grano austenítico	austenitic grain size	grosseur de grain austénitique	Austenitkorngröße
dirección de deslizamiento	slip direction	direction de glissement	Gleitrichtung
dislocación	dislocation	dislocation	Verschiebung; Versetzung
dominio martensítico	martensite range	domaine martensitique	Martensitbereich
dominios de transformación; dominios de las temperatura de transformación	transformation ranges (transformation temperature ranges)	domaines de transformation (domaines de température de transformation	Umwandlungsbereich
ductilidad	ductility	ductilité	Duktilität; Dehnbarkeit
ductilidad de entalla	notch ductility	ductilité à l'entaille	Kerbzähigkeit
dureza (indentación)	hardness (indentation)	dureté à la bille (empreinte)	Härte (Eindruck)
elemento de aleación	alloying element	élément d'alliage	Legierungselement
elementos residuales	residual elements	éléments résiduels	verbleibend Element
eliminación de tensiones	stress relieving	relaxation des contraintes	Entspannung
elongación a consecuencia de fractura	elongation after fracture	allongement après rupture	Bruchdehnung
empalme en los costados biselados del fleje	scarf joint	joint en enture (à onglet)	Schrägstoß
endurecimiento del acero por trabajo en frío	work hardening	durcissement par écrouissage	Kalthärtung; Kaltverfestigung

LÉXICO / GLOSSARY / LEXIQUE / WÖRTERVERZEICHNIS

SPANISH	ENGLISH	FRENCH	GERMAN
endurecimiento por deformación	strain hardening	durcissement par écrouissage	Kalthärtung; Kaltverfestigung
endurecimiento por envejecimiento	age hardening	durcissement par vieillissement	Aushärtung; Vergütung
endurecimiento por temple	quench hardening	durcissement par trempe	Abschreckhärten
endurecimiento superficial	surface hardening; case hardening	traitement de durcissement superficiel	Oberflächenhärtung
endurecimiento; temple; cementación	hardening	traitement de durcissement	Härten
envejecimiento	aging	vieillissement	Alterung
envejecimiento natural	natural aging	vieillissement naturel	Kaltaushärtung; Kaltauslagern
envejecimiento por deformación	strain aging	vieillissement sous tension	Reckalterung
envejecimiento por temple	quench aging	trempe de vieillissement	Abschreckalterung
epitaxia	epitaxy	épitaxie	Epitaxie
escoria	slag	scories (laitier)	Schlacke
esferoidización	spheroidizing	globulisation	Kugeligglühen
esferoidización del sulfuro	sulfide spheroidization	globulisation du sulfure	Einformung des Sulfid
esfuerzo	stress	contrainte	Spannung
esfuerzo de fluencia	flow stress	contrainte d'écoulement	Fließspannung
esfuerzo de fluencia	yield strength	résistance à l'écoulement	konventionelle o. technische Streckgrenze; 0,2% Dehngrenze; (proof stress: Ersatzstreckgrenze)
estricción	necking	striction	Querschnittsverminderung; Einschnürung
estructura columnar	columnar structure	structure basaltique	Stengelgefüge
estructura de grano de bambú	bamboo grain structure	structure de grain en bambou	Bambus Kornstruktur
estructura de grano de panqueque	pancake grain structure	structure de grain en crêpe	Flachkornstruktur
estructura de Widmannstätten	Widmannstatten structure	structure de Widmannstätten	Widmannstätten - Gefüge
estructura degenerada	degenerate structure	structure dégénérée	entarteter Struktur

LÉXICO / GLOSSARY / LEXIQUE / WÖRTERVERZEICHNIS

SPANISH	ENGLISH	FRENCH	GERMAN
estructura dentro de los límites de grano (intragranular)	sub-boundary structure (subgrain structure)	structure de sous-joint (structure de sous-grain)	Untergrenze Struktur; (Unterkorn Struktur)
estructura equiaxial	equiaxed structure	structure équiaxe	gleichachsig Struktur
estructura esferoidizada	spheroidized structure	structure globulisée	Kugelgefüge
estudio de fracturas	fractography	fractographie	Fraktographie
eutectoide	eutectoid	eutectoide	Eutektoid
exfoliación o despegue	cleavage	clivage	Spaltbruch; Spaltung
fase	phase	phase	Phase
fase continua	continuous phase	phase continue	kontinuierliches Phase
fatiga	fatigue	fatigue	Ermüdung
ferrita	ferrite	ferrite	Ferrit
fibra	fiber	fibre	Faser
fleje; banda; cinta; tira	strip	feuillard; acier en ruban	Blechstreifen; Bandblech
fluencia	creep	fluage	Kriechen
fluencia de granos	grain flow	écoulement de grain	Kornfluß
forja; forjadura	forging	forgeage	Schmiedestück; Schmieden
formación o trabajo en caliente	hot working	formage à chaud	Warmumformen
fotomacrografía	photomacrograph	photomacrographie	Lupenphotographie
fotomicrografía	photomicrograph	photomicrographie	Mikrophotographie
fractura cristalina	crystalline fracture	rupture cristalline	kristallin[ischler Bruch
fractura fibrosa	fibrous fracture	rupture fibreuse	Holzfaserbruch
fractura frágil	brittle fracture	rupture fragile	Sprödbruch
fractura granulosa	granular fracture	rupture granuleuse	körniger Bruch; kristallin[ischler Bruch
fractura transcristalina a través de planos de deslizamiento	cleavage fracture	rupture par clivage	Trennbruch; Sprödbruch
fragilidad	brittleness	fragilité	Sprödigkeit

LÉXICO / GLOSSARY / LEXIQUE / WÖRTERVERZEICHNIS

SPANISH	ENGLISH	FRENCH	GERMAN
fragilidad	shortness	friabilité	Sprödigkeit
fragilidad de entalla	notch brittleness	fragilité à l'entaille	Kerbsprödigkeit
fragilidad de revenido	temper brittleness	fragilité de revenu	Anlaßsprödigkeit
fragmentación	fragmentation	fragmentation	Zertrümmerung
fritaje; sinterizar;aglomerar en caliente	sintering	frittage	Zusammenbacken; sintern
fundente	flux	flux (fondant)	Flußmittel
grafitización	graphitization	graphitisation	Graphitisation
grafito	graphite	graphite	Graphit
grano	grain	grain	Korn
guarnición de dislocaciones	decoration (of dislocations)	garniture (de dislocations)	Dekoration (an Verschiebung)
hematita	hematite	hématite	Roteisenstein; roter Hämatit
hierro	iron	fer	Eisen
hierro ARMCO	ingot iron	fer lingot	entkohlter Stahl; Bessemerstahl
hierro forjado; hierro pudelado; hierro dulce; acero suave	wrought iron	fer corroyé.	Schmiedeeisern
hoja; chapa fina; plancha; lámina	sheet	tôle	Blech
horno de afino	finery	affinerie	Affinerie
idiomorfo	idiomorph	idiomorphe	idiomorph
impresión de azufre	sulfur print	empreinte de soufre (empreinte Baumann	Baumannabdruck
impurezas	impurities	impuretés	Unreinheit; Verunreinigung
inclusión	inclusion	inclusion	Einschluß
intercristalina	intercrystalline	intercristalline	interkristallin
intracristalina	intracrystalline	intragranulaire	intrakristallin
intracristalina	transgranular	intragranulaire	Transkristallin
límite de fluencia superior	upper yield stress	limite supérieure d'élasticité	obere Streckgrenze

LÉXICO / GLOSSARY / LEXIQUE / WÖRTERVERZEICHNIS

SPANISH	ENGLISH	FRENCH	GERMAN
laminación	rolling	laminage	Walzen; Rollen
laminación transversal	cross rolling	laminage avec cylindres obliques	schrägwalzen
laminado controlado	controlled rolling	laminage contrôlé	Regelrollen
laminado de endurecimiento	temper rolling	laminage de revenu	Nachwalzen; Dressieren
laminado en paquete	pack rolling	laminage en paquet	Paketwalzen
licuación; segregación	liquation	liquation	Seigerung
licuefacción de los límites de los granos	grain-boundary liquation	liquation de joint de grain	Korngrenzenangriff; Korngrenzeseigerung
límite aparente de elasticidad	yield point	limite d'écoulement	Fliesspunkt
límite de elasticidad	elastic limit	limite d'élasticité	Elastizitätsgrenze
línea de deslizamiento	slip line	ligne de glissement	Gleitlinie
línea de líquido	liquidus	liquidus	Liquidus
líneas de fluencia; líneas de Piobert-Lüders; líneas de flujo	flow lines	lignes d'écoulement (lignes de Piobert-Lüders)	Fließlinien; Schlieren
líneas o bandas de Lüder	Luder's lines or bands	lignes ou bandes de Lüder	Fließfiguren; Fließlinien; Lüderssche Linien
lingote	ingot	lingot	Blockmetall; Massel
llantón; plancha; desbaste plano	slab	dalle (brame)	Bramme
macla	twin	macle	Zwilling
macla mecánica	mechanical twin	macle mécanique	mechanische Zwilling
macla por recocido	annealing twin	macle de recuit	Rekristallisationszwilling
macla, cristal	twin, crystal	macle, cristal (grain maclé)	Zwillingskristall
macla, deformación	twin, deformation	macle, déformation (macle de déformation)	Verformungszwilling
macla, recocido; macla de recocido	twin, annealing	macle, recuit (macle de recuit)	Rekristallisationszwilling
macroataque químico	macroetching	macroattaque	Grobätzung

LÉXICO / GLOSSARY / LEXIQUE / WÖRTERVERZEICHNIS

SPANISH	ENGLISH	FRENCH	GERMAN
macroestructura	macrostructure	macrostructure	Makrogefüge; Grobgefüge; Makrostruktur
macrografía	macrograph	macrographie	Makrograph; Grobgefügebild
magnetita	magnetite	magnétite	Magnetit; Magneteisenstein
maquinabilidad	machinability	usinabilité	Zerspanbarkeit
marcas de esfuerzo por tracción	stretcher strains	vermiculures	Fließfiguren
martensita	martensite	martensite	Martensit
martensita de placas finas	lath martensite	martensite massive	Lattenmartensit
martensita en placa	plate martensite	martensite en plaque	Plattenmartensit
matriz	matrix	matrice	Grundmasse; Matrix
metaestable	metastable	métastable	metastabil
metal	metal	métal	Metall
metal de aportación	filler metal	métal d'apport	Schweißzusatzwerkstoff; Zusastzwerkstoff
metalógrafo; microscopio metalúrgico	metallograph	métallographe	Metallograph
microbandas (deformación)	microbands (deformation)	microbandes (déformation)	Mikrozeilen (Verformung)
microestructura	microstructure	microstructure	Mikrogefüge
micrografía	micrograph	micrographie	Mikrobild; Schliffbild
microgrieta; microfissura	microcrack	microfissure	Mikroriß; mikroskopischer Riß
microsegregación	microsegregation	microségrégation	Mikroseigerung
microsegregación o segregación intragranular	coring	noyautage; cernes	Kristallseigerung; Mikroseigerung
normalizado	normalizing	traitement de normalisation	normalglühen; normalisieren
nucleación; germinación	nucleation	germination	Keimbildung
núcleo	nucleus	germe	Keim; Kern
orientación (cristal)	orientation (crystal)	orientation (cristal)	Lagerung; Orientierung

LÉXICO / GLOSSARY / LEXIQUE / WÖRTERVERZEICHNIS

SPANISH	ENGLISH	FRENCH	GERMAN
orientación preferencial	preferred orientation	orientation préférentielle	Vorzugsorientierung
oxidación	oxidation	oxydation	Oxidation
oxidación interna	internal oxidation	oxydation interne	innere Oxidation
palanquilla	billet	billette (ou larget)	Walzblock; Barren; Knüppel
paso; pasada	pass	passe	Schweißraupe
patentado	patenting	patentage	patentieren
perlita	pearlite	perlite	Perlit
perlita nodular	nodular pearlite	perlite nodulaire	kugeliger Perlit
pieza fundida; pieza de fundición; colada del acero líquido; fundición	casting	moulage	Gußerzeugnisse
placa	plate	plaque	Platte
plano de deslizamiento	slip plane	plan de glissement	Gleitebene
plano de despegue	cleavage plane	plan de clivage	Spaltfläche
planta para fabricar desbastes	bloomery	bloomerie	Rennherd
plasticidad	plasticity	plasticité	Plastizität
policristalino	polycrystalline	polycristallin	polykristallin
polimorfismo	polymorphism	polymorphisme	Polymorphie
potencial de carbono	carbon potential	potentiel carbone	Kohlenstoffpegel, C-Pegel
precipitación de sulfuro en los límites de granos	grain-boundary sulfide precipitation	précipitation de sulfure aux joints de grain	KorngrenzenSulfidausscheidung
procedimiento de pudelado	puddling process	procédé de puddlage	Puddelnverarbeitung
proceso Waloon	Waloon process	procédé Waloon	Waloon Prozess
proeutectoide (fase)	proeutectoid (phase)	proeutectoïde (phase)	voreutektoidisch (Phase)
propiedades físicas	physical properties	propriétés physiques	physikalische Eigenschaften
prueba de capacidad	proof stress	limite conventionnelle d'élasticité	Dehngrenze
prueba de impacto; prueba de choque; prueba de resiliencia	impact test	essai de choc (de résilience)	Kerbschlagbiegeversuch

LÉXICO / GLOSSARY / LEXIQUE / WÖRTERVERZEICHNIS

SPANISH	ENGLISH	FRENCH	GERMAN
pulido electrolítico	electropolishing	polissage électrolytique	elektrolytisch polieren
pulido mecánico	mechanical polishing	polissage mécanique	Maschinenpolieren
pulido químico	chemical polishing	polissage chimique	chemisches Polieren
pulido; pulimiento	polishing	polissage	Polieren
punto crítico; temperatura de transformación	critical point	point critique (ou température de transformation)	Umwandlungstemperatur; Umwandlungspunkt
punto de fusión	melting point	température de fusion	Schmelzpunkt
punto peritéctico	peritectic	péritectique	peritektisch
punto triple	triple point	point triple	Tripelpunkt
quemado	burning	brûlure	Verbrennung
re-carburación	recarburizing	re-cémentation (recarburation)	Rückkohlung
rebarba	flash	bavure	Grat; Gratbildung
recalcado	upset	dérangement "upset"	stauchen
recalentamiento; sobrecalentado	overheating	surchauffe	Überhitzung
recocido	annealing	recuit	Glühen
recocido completo; recocido a fondo (materiales ferrosos)	full annealing (ferrous materials)	recuit complet (matériaux ferreux)	Hochglühen (über A₃)
recocido de homogeneización	homogenizing annealing	recuit d'homogénéisation	Diffusionsglühen
recocido en caja	box annealing	recuit en caisse	Kistenglühen
recocido para esferoidizar	spheroidizing annealing	recuit de globulisation	Weichglühen
recocido para recristalizar	recrystallization annealing	traitement de recristallisation	Rekristallisationsglühen
recocido subcrítico	subcritical annealing	recuit subcritique	Rekristallisationsglühen
recristalización	recrystallization	recristallisation	Rekristallisation
rectificado; amolado	grinding	meulage (rectification)	Schleifen
recuperación	recovery	restauration	Erholung
regulador; amortiguador	buffer	tampon	Pufferschicht
resistencia a la tracción	tensile strength	résistance à la traction	Zugfestigkeit

LÉXICO / GLOSSARY / LEXIQUE / WÖRTERVERZEICHNIS

SPANISH	ENGLISH	FRENCH	GERMAN
resolubilizar	resolution	résolution	Auflösung
revenido	tempering	revenu	Anlassen
rotura por corrosión bajo tensiones	stress-corrosion cracking	fissuration par corrosion sous tension	Spannungskorrosionsriß
rozamiento	chafery	chaferie	chaferie
sección terminada en punta	taper section	section conique	Schrägkante
segregación	segregation	ségrégation	Seigerung
segregación dendrítica	dendritic segregation	ségrégation dendritique	dendritisches Seigerung
segregación en bandas	segregation banding	ségrégation en bande	Seigerungszeilen
sensibilidad de entalla	notch sensitivity	sensibilité à l'entaille	Kerbempfindlichkeit
separación en láminas	lamination	feuilletage	Dopplung
sobrecalentamiento	superheating	surchauffe	Überhitzung
sobreenvejecimiento	overaging	survieillissement, "overaging"	überalterung; übervergütung
soldabilidad	weldability	soudabilité	Schweißbarkeit
soldadura	welding	soudage	Schweißen
soldadura	weld	soudure	Schweißnaht
soldadura a fusión	fusion welding	soudage par fusion	Schmelzschweißen
soldadura amarilla	brazing	brasage	Löten; Hartlöten
soldadura con metal amarillo	braze welding	soudo-brasage	Schweißlötung
soldadura oxiacetilénica	gas welding	soudage au gaz	Gasschweißen
soldadura por arco	arc welding	soudage à l'arc	Lichtbogenschweißen
soldadura por resistencia	resistance welding	soudage (électrique) par résistance	Widerstandsschweißung
soldeo de motivo; configuración; forma	pattern welding	soudage en motif	Muster Schweißen
soldeo manual	manual welding	soudage manuel	Handscheißung
soldeo por calda a la forja; soldadura de forja	forge welding	soudage à la forge	Hammerschweißen

LÉXICO / GLOSSARY / LEXIQUE / WÖRTERVERZEICHNIS

SPANISH	ENGLISH	FRENCH	GERMAN
solidus	solidus	solidus	Solidus
solución sólida	solid solution	solution solide	feste Lösung
solución sólida de substitución	substitutional solid solution	solution solide de substitution	Substitutionsmischkristall
solución sólida intersticial	interstitial solid solution	solution solide interstitielle	Einlagerungsmischkristall
soluto	solute	soluté	gelöster Komponente
solvente	solvent	solvant	Lösungsmittel
solvus; línea de solubilidad sólida	solvus	solvus (courbe limite de solubilité)	Löslichkeitskurve
subgrano	subgrain	sous-grain	Subkorn; Unterkorn
substrato	substrate	substrat	Substrat
superenfriamiento	supercooling	sous-refroidissement	Unterkühlung
superficie oxidada	oxidized surface	surface oxydée	oxidiertes Oberfläche
tamaño de grano	grain size	grosseur de grain	Korngrösse
tamaño del grano ferrítico	ferritic grain size	grosseur de grain ferritique	ferritischer Korngrösse
temperatura A_1	A_1 temperature	température A_1	A_1 Temperatur
temperatura A_2	A_2 temperature	température A_2	A_2 Temperatur
temperatura A_3	A_3 temperature	température A_3	A_3 Temperatur
temperatura A_{cm}	A_{cm} temperature	température A_{cm}	A_{cm} Temperatur
temperatura crítica	critical temperature	température de transformation	Umwandlungstemperatur
temperatura de acabado; temperatura final	finishing temperature	température de finissage	Endtemperatur des Werkstückes
temperatura de recristalización	recrystallization temperature	température de recristallisation	Rekristallisationstemperatur
temperatura de transformación	transformation temperature	température de transformation	Umwandlungstemperatur
temperatura de transición	transition temperature	température de transition	Übergangstemperatur
temperatura de transición de dúctil a frágil	transition temperature (ductile-brittle transition temperature)	température de transition ductile-fragile)	
temperatura M_f	M_f temperature	température M_f	M_f Temperatur

LÉXICO / GLOSSARY / LEXIQUE / WÖRTERVERZEICHNIS			
SPANISH	**ENGLISH**	**FRENCH**	**GERMAN**
temperatura M_S	M_S temperature	température M_S	M_S Temperatur
templabilidad; facultad de endurecimiento	hardenability	trempabilité	Härtbarkeit
temple	quenching	trempe	Abschrecken
temple escalonado martensítico	martempering	trempe étagée martensitique	Warmbadhärten
tenacidad	toughness	ténacité	Zähigkeit
tensiones residuales	residual stress	contrainte résiduelle	Eigenspannungen
termocoloración	heat tinting	coloration par chaleur	Anlaufen
textura	texture	texture	Textur
transcristalina	transcrystalline	transgranulaire	transkristallin
tratamiento de recementación	carbon restoration	traitement de recarburation	Wiederaufkohlung
tratamiento térmico de solución	solution heat treatment	traitement de mise en solution et trempe	Lösungsglühen
tratamiento térmico; termotratamiento	heat treatment	traitement thermique	Wärmebehandlung
unidad angstrom (Å)	angstrom unit (Å)	unité angström (Å)	Ångström (Å)
vacante	vacancy	lacune	Leerstelle
vaciado	teeming	coulage	Gießen; Abguß
velocidad crítica de enfriamiento	critical cooling rate	vitesse de refroidissement critique	kritische Abkühlungsgeschwindigkeit
veteado; jaspeado	veining	veinage (marbrure)	Formriß; Blattrippe
wootz	wootz	wootz	Wootz
wurtzita	wustite	wustite	Wüstit
zona térmicamente afectada; ZTA	heat-affected zone (HAZ)	zone affectée thermiquement (ZAT)	Wärmeeinflußzone (WEZ)

WÖRTERVERZEICHNIS / GLOSSARY / LEXIQUE / LÉXICO

GERMAN	ENGLISH	FRENCH	SPANISH
A_1 Temperatur	A_1 temperature	température A_1	temperatura A_1
A_2 Temperatur	A_2 temperature	température A_2	temperatura A_2
A_3 Temperatur	A_3 temperature	température A_3	temperatura A_3
Abschreckalterung	quench aging	trempe de vieillissement	envejecimiento por temple
Abschrecken	quenching	trempe	temple
Abschreckhärten	quench hardening	durcissement par trempe	endurecimiento por temple
A_{cm} Temperatur	A_{cm} temperature	température A_{cm}	temperatura A_{cm}
Affinerie	finery	affinerie	horno de afino
Allotriomorph	allotriomorph	allotriomorphe	allotriomorfico
Allotropie	allotropy	allotropie	allotropia
Alterung	aging	vieillissement	envejecimiento
Ångström (Å)	angstrom unit (A)	unité angström (Å)	unidad angstrom (Å)
Anlassen	tempering	revenu	revenido
Anlaßprödigkeit	temper brittleness	fragilité de revenu	fragilidad de revenido
Anlaufen	heat tinting	coloration par chaleur	termocoloración
Ätzen; Anätzung	etching	attaque	ataque quimico
Ätzflüssigkeit	etchant	réactif d'attaque	ácido para grabar; reactivo para ataque
Aufkohlen	carburizing	cémentation	cementación; carburación
Auflösung	resolution	résolution	resolubilizar
Aushärtung; Vergütung	age hardening	durcissement par vieillissement	endurecimiento por envejecimiento
Austenit	austenite	austénite	austenita
Austenitisieren	austenitizing	austénitisation	austenización
Austenitkorngröße	austenitic grain size	grosseur de grain austénitique	dimensión de grano austenítico
Autoanlaßen	autotempering	auto-revenu	autorevenido
Autoradiographie	autoradiograph	autoradiographie	autorradiografia

WÖRTERVERZEICHNIS / GLOSSARY / LEXIQUE / LÉXICO

GERMAN	ENGLISH	FRENCH	SPANISH
Bainit	bainite	bainite	bainita
Bambus Kornstruktur	bamboo grain structure	structure de grain en bambou	estructura de grano de bambú
Baumannabdruck	sulfur print	empreinte de soufre (empreinte Baumann	impresión de azufre
beruhigter Stahl	killed steel	acier calmé	acero desoxidado
Blech	sheet	tôle	hoja; chapa fina; plancha; lámina
Blechstreifen; Bandblech	strip	feuillard; acier en ruban	fleje; banda; cinta; tira
Blockmetall; Massel	ingot	lingot	lingote
Bramme	slab	dalle (brame)	llantón; plancha; desbaste plano
Bruchdehnung	elongation after fracture	allongement après rupture	elongación a consecuencia de fractura
Carbid	carbide	carbure	carburo
chaferie	chafery	chaferie	rozamiento
chemisches Polieren	chemical polishing	polissage chimique	pulido químico
Dehngrenze	proof stress	limite conventionnelle d'élasticité	prueba de capacidad
Dekoration (an Verschiebung)	decoration (of dislocations)	garniture (de dislocations)	(guarnición de dislocaciones)
Dendrit	dendrite	dendrite	dendrita
dendritisches Seigerung	dendritic segregation	ségrégation dendritique	segregación dendrítica
Desoxidieren; Reduktion	deoxidation	désoxydation	desoxidación
Diffusion	diffusion	diffusion	difusión
Diffusionsglühen	homogenizing annealing	recuit d'homogénéisation	recocido de homogeneización
Dilatometer	dilatometer	dilatomètre	dilatómetro
Dopplung	lamination	feuilletage	separación en láminas
Duktilität; Dehnbarkeit	ductility	ductilité	ductilidad
Eigenspannungen	residual stress	contrainte résiduelle	tensiones residuales
Einformung des Sulfid	sulfide spheroidization	globulisation du sulfure	esferoidización del sulfuro
Einlagerungsmischkristall	interstitial solid solution	solution solide interstitielle	solución sólida intersticial

WÖRTERVERZEICHNIS / GLOSSARY / LEXIQUE / LÉXICO

GERMAN	ENGLISH	FRENCH	SPANISH
Einsatz[härte]schicht	case	couche superficielle traitée	capa superficial
Einsatzhärtung; Oberflächenhärten	case hardening	trempe superficielle; durcissement superficiel	cementación en caja
Einschluß	inclusion	inclusion	inclusión
Eisen	iron	fer	hierro
elastische Verformung	elastic strain	déformation élastique	deformación elástica;
Elastizitätsgrenze	elastic limit	limite d'élasticité	límite de elasticidad
elektrolytisch polieren	electropolishing	polissage électrolytique	pulido electrolítico
Elektronenstrahlmikrosonde analysator	electron beam microprobe analyzer	analyseur par microsonde électronique	bombardeo de electrones
Endtemperatur des Werkstückes	finishing temperature	température de finissage	temperatura de acabado; temperatura final
entarteter Struktur	degenerate structure	structure dégénérée	estructura degenerada
entkohlter Stahl; Bessemerstahl	ingot iron	fer lingot	hierro ARMCO
Entkohlung	decarburization	décarburation	decarburación
Entspannung	stress relieving	relaxation des contraintes	eliminación de tensiones
Epitaxie	epitaxy	épitaxie	epitaxia
Erholung	recovery	restauration	recuperación
Ermüdung	fatigue	fatigue	fatiga
Eutektoid	eutectoid	eutectoïde	eutectoide
Faser	fiber	fibre	fibra
Ferrit	ferrite	ferrite	ferrita
Ferrit-Perlit zeilenbildung	ferrite-pearlite banding	bande ferrite-perlite	bandas de ferrita y perlita
ferritischer Korngrösse	ferritic grain size	grosseur de grain ferritique	tamaño del grano ferrítico
feste Lösung	solid solution	solution solide	solución sólida
Flachkornstruktur	pancake grain structure	structure de grain en crêpe	estructura de grano de panqueque
Fliesspunkt	yield point	limite d'écoulement	límite aparente de elasticidad

WÖRTERVERZEICHNIS / GLOSSARY / LEXIQUE / LÉXICO

GERMAN	ENGLISH	FRENCH	SPANISH
Fließfiguren	stretcher strains	vermiculures	marcas de esfuerzo por tracción
Fließfiguren; Fließlinien; Lüderssche Linien	Luder's lines or bands	lignes ou bandes de Lüder	líneas o bandas de Lüder
Fließlinien; Schlieren	flow lines	lignes d'écoulement (lignes de Piobert-Lüders)	líneas de fluencia; líneas de Piobert-Lüders; líneas de flujo
Fließspannung	flow stress	contrainte d'écoulement	esfuerzo de fluencia
Flußmittel	flux	flux (fondant)	fundente
Formriß; Blattrippe	veining	veinage (marbrure)	veteado; jaspeado
Fraktographie	fractography	fractographie	estudio de fracturas
Gasschweißen	gas welding	soudage au gaz	soldadura oxiacetilénica
gedeckelter Stahl	capped steel	acier coiffé	acero efervescente
gelöster Komponente	solute	soluté	soluto
Gießen; Abguß	teeming	coulage	vaciado
gleichachsig Struktur	equiaxed structure	structure équiaxe	estructura equiaxial
Gleitebene	slip plane	plan de glissement	plano de deslizamiento
Gleitlinie	slip line	ligne de glissement	línea de deslizamiento
Gleitrichtung	slip direction	direction de glissement	dirección de deslizamiento
Gleitung	slip	glissement	deslizamiento
Glühen	annealing	recuit	recocido
Graphit	graphite	graphite	grafito
Graphitisation	graphitization	graphitisation	grafitización
Grat; Gratbildung	flash	bavure	rebarba
Grobätzung	macroetching	macroattaque	macroataque químico
Grundmasse; Matrix	matrix	matrice	matriz
Gußeisen	cast iron	fonte de moulage	arrabio; hierro colado
Gußerzeugnisse	casting	moulage	pieza fundida; pieza de fundición; colada del acero líquido; fundición

WÖRTERVERZEICHNIS / GLOSSARY / LEXIQUE / LÉXICO

GERMAN	ENGLISH	FRENCH	SPANISH
Halbberuhigtstahl	semikilled steel	acier demi-calmé	acero semicalmado
Hammerschweißen	forge welding	soudage à la forge	soldeo por calda a la forja; soldadura de forja
Handschweißung	manual welding	soudage manuel	soldeo manual
Härtbarkeit	hardenability	trempabilité	templabilidad; facultad de endurecimiento
Härte (Eindruck)	hardness (indentation)	dureté à la bille (empreinte)	dureza (indentación)
Härten	hardening	traitement de durcissement	endurecimiento; temple; cementación
Hochglühen (über A₃)	full annealing (ferrous materials)	recuit complet (matériaux ferreux)	recocido completo; recocido a fondo (materiales ferrosos)
Holzfaserbruch	fibrous fracture	rupture fibreuse	fractura fibrosa
idiomorph	idiomorph	idiomorphe	idiomorfo
innere Oxidation	internal oxidation	oxydation interne	oxidación interna
interkristallin	intercrystalline	intercristalline	intercristalina
intrakristallin	intracrystalline	intragranulaire	intracristalina
isothermisches Glühen Schaubild	isothermal transformation (IT) diagram	diagramme de transformation en conditions isothermes (diagramme TI)	diagrama de transformación isotérmica
Kaltaushärtung; Kaltauslagern	natural aging	vieillissement naturel	envejecimiento natural
Kalthärtung; Kaltverfestigung	strain hardening	durcissement par écrouissage	endurecimiento por deformación
Kalthärtung; Kaltverfestigung	work hardening	durcissement par écrouissage	endurecimiento del acero por trabajo en frío
Karbonitrieren	carbonitriding	carbonitruration	carbonitruración
Keim; Kern	nucleus	germe	núcleo
Keimbildung	nucleation	germination	nucleación; germinación
Kerbempfindlichkeit	notch sensitivity	sensibilité à l'entaille	sensibilidad de entalla

WÖRTERVERZEICHNIS / GLOSSARY / LEXIQUE / LÉXICO

GERMAN	ENGLISH	FRENCH	SPANISH
Kerbschlagbiegeversuch	impact test	essai de choc (de résilience)	prueba de impacto; prueba de choque; prueba de resiliencia
Kerbsprödigkeit	notch brittleness	fragilité à l'entaille	fragilidad de entalla
Kerbzähigkeit	notch ductility	ductilité à l'entaille	ductilidad de entalla
Kistenglühen	box annealing	recuit en caisse	recocido en caja
Klarspüler	wetting agent	agent mouillant	agente humectante
Kohlenstoffäquivalent (für Schweißbarkeit Schätzung)	carbon equivalent (for rating of weldability)	équivalent carbone (pour estimer la soudabilité)	carbono equivalente
Kohlenstoffpegel, C-Pegel	carbon potential	potentiel carbone	potencial de carbono
Konstituent; Bestandteil	constituent	constituant	constituyente
kontinuierliches Phase	continuous phase	phase continue	fase continua
konventionelle o. technische Streckgrenze; 0,2% Dehngrenze; (proof stress: Ersatzstreckgrenze)	yield strength	résistance à l'écoulement	esfuerzo de fluencia
Korn	grain	grain	grano
Kornfluß	grain flow	écoulement de grain	fluencia de granos
Korngrenzenangriff; Korngrenzeseigerung	grain-boundary liquation	liquation de joint de grain	licuefacción de los límites de los granos
KorngrenzenSulfidausscheidung	grain-boundary sulfide precipitation	précipitation de sulfure aux joints de grain	(precipitación de sulfuro en los límites de granos
Korngrösse	grain size	grosseur de grain	tamaño de grano
körniger Bruch; kristallin[isch|er Bruch	granular fracture	rupture granuleuse	fractura granulosa
Kornvergröberung	grain coarsening	grossissement du grain	crecimiento del grano
Kornwachstum	grain growth	croissance du grain	crecimiento del grano
Korrosion	corrosion	corrosion	corrosión
Kriechen	creep	fluage	fluencia

WÖRTERVERZEICHNIS / GLOSSARY / LEXIQUE / LÉXICO

GERMAN	ENGLISH	FRENCH	SPANISH
kristallin[ischer Bruch	crystalline fracture	rupture cristalline	fractura cristalina
Kristallseigerung; Mikroseigerung	coring	noyautage; cernes	microsegregación o segregación intragranular
kritische Abkühlungsgeschwindigkeit	critical cooling rate	vitesse de refroidissement critique	velocidad crítica de enfriamiento
kritische Verformung	critical strain	déformation critique	deformación crítica
Kugelgefüge	spheroidized structure	structure globulisée	estructura esferoidizada
kugeliger Perlit	nodular pearlite	perlite nodulaire	perlita nodular
Kugeliggluhen	spheroidizing	globulisation	esferoidización
künstlich	artifact	artefact	artefacto
Lagerung; Orientierung	orientation (crystal)	orientation (cristal)	orientación (cristal)
Lattenmartensit	lath martensite	martensite massive	martensita de placas finas
Leerstelle	vacancy	lacune	vacante
legierter Stahl	alloy steel	acier allié	acero aleado
Legierung	alloy	alliage	aleación
Legierungselement	alloying element	élément d'alliage	elemento de aleación
Lichtbogenschweißen	arc welding	soudage à l'arc	soldadura por arco
Liquidus	liquidus	liquidus	línea de líquido
Löslichkeitskurve	solvus	solvus (courbe limite de solubilité)	solvus; línea de solubilidad sólida
Lösungsglühen	solution heat treatment	traitement de mise en solution et trempe	tratamiento térmico de solución
Lösungsmittel	solvent	solvant	solvente
Löten; Hartlöten	brazing	brasage	soldadura amarilla
Lupenphotographie	photomacrograph	photomacrographie	fotomacrografía
Magnetit; Magneteisenstein	magnetite	magnétite	magnetita
Makrogefüge; Grobgefüge; Makrostruktur	macrostructure	macrostructure	macroestructura

WÖRTERVERZEICHNIS / GLOSSARY / LEXIQUE / LÉXICO

GERMAN	ENGLISH	FRENCH	SPANISH
Makrograph; Grobgefügebild	macrograph	macrographie	macrografía
Martensit	martensite	martensite	martensita
Martensitbereich	martensite range	domaine martensitique	dominio martensítico
Maschinenpolieren	mechanical polishing	polissage mécanique	pulido mecánico
mechanische Zwilling	mechanical twin	macle mécanique	macla mecánica
Metall	metal	métal	metal
Metallograph	metallograph	métallographe	metalógrafo; microscopio metalúrgico
metastabil	metastable	métastable	metaestable
M_f Temperatur	M_f temperature	température M_f	temperatura M_f
Mikrobild; Schliffbild	micrograph	micrographie	micrografía
Mikrogefüge	microstructure	microstructure	microestructura
Mikrophotographie	photomicrograph	photomicrographie	fotomicrografía
Mikroriß; mikroskopischer Riß	microcrack	microfissure	microgrieta; microfissura
Mikroseigerung	microsegregation	microségrégation	microsegregación
Mikrozeilen (Verformung)	microbands (deformation)	microbandes (déformation)	microbandas (deformación)
M_s Temperatur	M_s temperature	température M_s	temperatura M_s
Muster Schweißen	pattern welding	soudage en motif	soldeo de motivo; configuración; forma
Nachwalzen; Dressieren	temper rolling	laminage de revenu	laminado de endurecimiento
NeumannStreifung	Neumann band	bande de Neumann	banda de Neumann
normalglühen; normalisieren	normalizing	traitement de normalisation	normalizado
obere Streckgrenze	upper yield stress	limite supérieure d'élasticité	lǫmite de fluencia superior
Oberflächenhärtung	surface hardening; case hardening	traitement de durcissement superficiel	endurecimiento superficial
Oxidation	oxidation	oxydation	oxidación

The Metals Black Book – 3rd Edition

WÖRTERVERZEICHNIS / GLOSSARY / LEXIQUE / LÉXICO

GERMAN	ENGLISH	FRENCH	SPANISH
oxidiertes Oberfläche	oxidized surface	surface oxydée	superficie oxidada
Paketwalzen	pack rolling	laminage en paquet	laminado en paquete
patentieren	patenting	patentage	patentado
peritektisch	peritectic	péritectique	punto peritéctico
Perlit	pearlite	perlite	perlita
Phase	phase	phase	fase
physikalische Eigenschaften	physical properties	propriétés physiques	propiedades físicas
plastische Verformung	plastic deformation	déformation plastique	deformación plástica
Plastizität	plasticity	plasticité	plasticidad
Platte	plate	plaque	placa
Plattenmartensit	plate martensite	martensite en plaque	martensita en placa
Polieren	polishing	polissage	pulido; pulimiento
polykristallin	polycrystalline	polycristallin	policristalino
Polymorphie	polymorphism	polymorphisme	polimorfismo
Primärkristall	primary crystal	cristal primaire	cristal primario
Puddelnverarbeitung	puddling process	procédé de puddlage	procedimiento de pudelado
Pufferschicht	buffer	tampon	regulador; amortiguador
Querschnittsverminderung; Einschnürung	necking	striction	estricción
Randstahl; Massenstahl	rimmed steel	acier effervescent	acero efervescente
Reckalterung	strain aging	vieillissement sous tension	envejecimiento por deformación
Regelrollen	controlled rolling	laminage contrôlé	laminado controlado
Rekristallisation	recrystallization	recristallisation	recristalización
Rekristallisationsglühen	subcritical annealing	recuit subcritique	recocido subcrítico
Rekristallisationsglühen	recrystallization annealing	traitement de recristallisation	recocido para recristalizar
Rekristallisationstemperatur	recrystallization temperature	température de recristallisation	temperatura de recristalización
Rekristallisationszwilling	annealing twin	macle de recuit	macla por recocido

WÖRTERVERZEICHNIS / GLOSSARY / LEXIQUE / LÉXICO

GERMAN	ENGLISH	FRENCH	SPANISH
Rekristallisationszwilling	twin, annealing	macle, recuit (macle de recuit)	macla, recocido; macla de recocido
Rennherd	bloomery	bloomerie	planta para fabricar desbastes
Roheisen	pig iron	fonte brute	arrabio
Roteisenstein; roter Hämatit	hematite	hématite	hematita
Rückkohlung	recarburizing	re-cémentation (recarburation)	re-carburación
rückschwefeltes Stahl	resulfurized steel	acier re-sulfuré	acero con azufre adicional
Scher-Stahl	shear steel	acier de cisaillement	acero de cizallamiento
Scherung	shear	cisaillement	cizallamiento
Scherzeilen (Verformung)	shear bands (deformation)	bandes de cisaillement (déformation)	bandas de cizallamiento (deformación)
Schlacke	slag	scories (laitier)	escoria
Schleifen	grinding	meulage (rectification)	rectificado; amolado
Schleifmittel; Abriebeigenschaften	abrasive	abrasif	abrasivo
Schleifwirkung; Abnutzung	abrasion	abrasion	abrasión
Schmelzpunkt	melting point	température de fusion	punto de fusión
Schmelzschweißen	fusion welding	soudage par fusion	soldadura a fusión
Schmiedeeisern	wrought iron	fer corroyé.	hierro forjado; hierro pudelado; hierro dulce; acero suave
Schmiedestück; Schmieden	forging	forgeage	forja; forjadura
Schrägkante	taper section	section conique	sección terminada en punta
Schrägstoß	scarf joint	joint en enture (à onglet)	empalme en los costados biselados del fleje
schrägwalzen	cross rolling	laminage avec cylindres obliques	laminación transversal
Schweißbarkeit	weldability	soudabilité	soldabilidad
Schweißen	welding	soudage	soldadura
Schweißlötung	braze welding	soudo-brasage	soldadura con metal amarillo
Schweißnaht	weld	soudure	soldadura

WÖRTERVERZEICHNIS / GLOSSARY / LEXIQUE / LÉXICO

GERMAN	ENGLISH	FRENCH	SPANISH
Schweißraupe	weld bead	cordon de soudure	cordón de soldadura
Schweißraupe	pass	passe	paso; pasada
Schweißzusatzwerstoff; Zusastzwerkstoff	filler metal	métal d'apport	metal de aportación
Seigerung	liquation	liquation	licuación; segregación
Seigerung	segregation	ségrégation	segregación
Seigerungszeilen	segregation banding	ségrégation en bande	segregación en bandas
Selbstdiffusion	self diffusion	autodiffusion	autodifusión
Solidus	solidus	solidus	solidus
Spaltbruch; Spaltung	cleavage	clivage	exfoliación o despegue
Spaltfläche	cleavage plane	plan de clivage	plano de despegue
spanbar	free machining	facilement usinable (décolletage rapide)	de fácil maquinado; torneado rápido
Spannung	stress	contrainte	esfuerzo
Spannungskorrosionsriß	stress-corrosion cracking	fissuration par corrosion sous tension	rotura por corrosión bajo tensiones
Sprödbruch	brittle fracture	rupture fragile	fractura frágil
Sprödigkeit	brittleness	fragilité	fragilidad
Sprödigkeit	shortness	friabilité	fragilidad
Stahl	steel	acier	acero
Stahlguß	cast steel	acier moulé	acero moldeado; vaciado
stauchen	upset	dérangement "upset"	recalcado
Stengelgefüge	columnar structure	structure basaltique	estructura columnar
Subkorn; Unterkorn	subgrain	sous-grain	subgrano
Substitutionsmischkristall	substitutional solid solution	solution solide de substitution	solución solida de substitución
Substrat	substrate	substrat	substrato
Terrassenbruch	lamellar tear	déchirure lamellaire	desgarre laminar

WÖRTERVERZEICHNIS / GLOSSARY / LEXIQUE / LÉXICO

GERMAN	ENGLISH	FRENCH	SPANISH
Textur	texture	texture	textura
thermische Analyse; Thermoanalyse	thermal analysis	analyse thermique	análisis térmico
Tiegelstahlblock	crucible steel	acier au creuset	acero al crisol
Transkristallin	transgranular	intragranulaire	intracristalina
transkristallin	transcrystalline	transgranulaire	transcristalina
Trennbruch; Sprödbruch	cleavage fracture	rupture par clivage	fractura transcristalina a través de planos de deslizamiento
Tripelpunkt	triple point	point triple	punto triple
überalterung; übervergütung	overaging	survieillissement, "overaging"	sobreenvejecimiento
übereutektoidisch Legierung	hypereutectoid alloy	alliage hypereutectoïde	aleación hipereutectoide
Übergangstemperatur	transition temperature (ductile-brittle transition temperature)	température de transition (température de transition ductile-fragile)	temperatura de transición temperatura de transición de dúctil a frágil
Überhitzung	overheating	surchauffe	recalentamiento; sobrecalentado
Überhitzung	superheating	surchauffe	sobrecalentamiento
Umwandlungsbereich	transformation ranges (transformation temperature ranges)	domaines de transformation (domaines de température de transformation	dominios de transformación; dominios de las temperatura de transformación
Umwandlungstemperatur	critical temperature	température de transformation	temperatura crítica
Umwandlungstemperatur	transformation temperature	température de transformation	temperatura de transformación
Umwandlungstemperatur; Umwandlungspunkt	critical point	point critique (ou température de transformation)	punto crítico; temperatura de transformación
unlegierter Stahl	carbon steel	acier au carbone	acero al carbono
Unreinheit; Verunreinigung	impurities	impuretés	impurezas
untereutektoidisch Legierung	hypoeutectic alloy	alliage hypoeutectique	aleación hipoeutéctica

WÖRTERVERZEICHNIS / GLOSSARY / LEXIQUE / LÉXICO

GERMAN	ENGLISH	FRENCH	SPANISH
Untergrenze Struktur; (Unterkorn Struktur)	sub-boundary structure (subgrain structure)	structure de sous-joint (structure de sous-grain	estructura dentro de los límites de grano (intragranular)
Unterkühlung	supercooling	sous-refroidissement	superenfriamiento
verbleibend Element	residual elements	éléments résiduels	elementos residuales
Verbrennung	burning	brûlure	quemado
Verformung	strain	déformation	deformación
Verformungszeilen	deformation bands	bandes de déformation	bandas de deformación
Verformungszwilling	twin, deformation	macle, déformation (macle de déformation)	macla, deformación
Verschiebung; Versetzung	dislocation	dislocation	dislocación
Vorblock; Schmiedeblock; Blume	bloom	bloom (brame)	bloom; desbaste
voreutektoidisch (Phase)	proeutectoid (phase)	proeutectoïde (phase)	proeutectoide (fase)
Vorzugsorientierung	preferred orientation	orientation préférentielle	orientación preferencial
Waloon Prozess	Waloon process	procédé Waloon	proceso Waloon
Walzblock; Barren; Knüppel	billet	billette (ou larget)	palanquilla
Walzen; Rollen	rolling	laminage	laminación
Warmbadhärten	martempering	trempe étagée martensitique	temple escalonado martensitico
Wärmebehandlung	heat treatment	traitement thermique	tratamiento térmico; termotratamiento
Wärmeeinflußzone (WEZ)	heat-affected zone (HAZ)	zone affectée thermiquement (ZAT)	zona térmicamente afectada; ZTA
Warmumformen	hot working	formage à chaud	formación o trabajo en caliente
weicher unlegierte Stahl	mild steel	acier doux	acero dulce; acero dúctil; acero suave
Weichglühen	spheroidizing annealing	recuit de globulisation	recocido para esferoidizar
Widerstandsschweißung	resistance welding	soudage (électrique) par résistance	soldadura por resistencia
Widmannstätten - Gefüge	Widmannstatten structure	structure de Widmannstätten	estructura de Widmannstätten
Wiederaufkohlung	carbon restoration	traitement de recarburation	tratamiento de recementación

WÖRTERVERZEICHNIS / GLOSSARY / LEXIQUE / LÉXICO

GERMAN	ENGLISH	FRENCH	SPANISH
Wootz	wootz	wootz	wootz
Wüstit	wustite	wustite	wurtzita
Zähigkeit	toughness	ténacité	tenacidad
Zeilenbildung	banding	bande	bandas
Zeilenverschlingung	kink band (deformation)	bande de contorsion (déformation)	banda de ensortijamiento
Zeit-Temperatur-Umwandlungs-Schaubild, ZTU-Kurve	time-temperature-transformation (TTT) diagram	diagramme de transformation en conditions isothermes (TTT)	diagrama de transformación isotérmica; curvas S; curvas TTT
Zement(ations)stahl	blister steel	acier comportant des soufflures (ou "pailles")	acero ampollado
Zementation	cementation	cémentation	cementación
Zementit	cementite	cémentite	cementita
Zerspanbarkeit	machinability	usinabilité	maquinabilidad
Zertrümmerung	fragmentation	fragmentation	fragmentación
Zugfestigkeit	tensile strength	résistance à la traction	resistencia a la tracción
Zunder; Sinter	scale	calamine	cascarilla; laminilla; escama
Zusammenbacken; sintern	sintering	frittage	fritaje; sinterizar;aglomerar en caliente
Zustandsschaubild	equilibrium diagram	diagramme d'équilibre	diagrama de equilibrio
Zustandsschaubild	constitutional diagram	diagramme de constitution	diagrama de fases
Zustandsschaubild	phase diagram	diagramme de phase	diagrama de fases
Zwilling	twin	macle	macla
Zwillingskristall	twin, crystal	macle, cristal (grain maclé)	macla, cristal
Zwischenstufen-Vergütung	austempering	trempe étagée bainitique	atemperación escalonada (tratamiento con transformación isotérmica de austenita a bainita)

Chapter

10

CAST IRONS:

AMERICAN SPECIFICATION TITLES & DESIGNATIONS, CHEMICAL COMPOSITIONS, & MECHANICAL PROPERTIES

CAST IRONS

ASTM Spec.	Title
Gray Iron Castings	
A 48	Gray Iron Castings
A 48M	Gray Iron Castings [Metric]
A 159	Automotive Gray Iron Castings
A 278	Gray Iron Castings for Pressure-Containing Parts for Temperatures up to 650°F
A 278M	Gray Iron Castings for Pressure-Containing Parts for Temperatures Up to 350°C, [Metric]
A 319	Gray Iron Castings for Elevated Temperatures for Non-Pressure Containing Parts
A 436	Austenitic Gray Iron Castings
A 823	Statically Cast Permanent Mold Gray Iron Castings
Ductile Iron Castings	
A 395	Ferritic Ductile Iron Pressure-Retaining Castings for Use at Elevated Temperatures
A 395M	Ferritic Ductile Iron Pressure-Retaining Castings for Use at Elevated Temperatures [Metric]
A 439	Austenitic Ductile Iron Castings
A 476	Ductile Iron Castings for Paper Mill Dryer Rolls
A 476M	Ductile Iron Castings for Paper Mill Dryer Rolls [Metric]
A 536	Ductile Iron Castings
A 571	Austenitic Ductile Iron Castings for Pressure-Containing Parts Suitable for Low-Temperature Service
A 571M	Austenitic Ductile Iron Castings for Pressure-Containing Parts Suitable for Low-Temperature Service [Metric]
A 874	Ferritic Ductile Iron Castings Suitable for Low-Temperature Service
A 874M	Ferritic Ductile Iron Castings Suitable for Low-Temperature Service [Metric]
A 897	Austempered Ductile Iron Castings
A 897M	Austempered Ductile Iron Castings [Metric]
Malleable Iron Castings	
A 47	Ferritic Malleable Iron Castings
A 47M	Ferritic Malleable Iron Castings [Metric]
A 197	Cupola Malleable Iron
A 197M	Cupola Malleable Iron [Metric]

CAST IRONS (Continued)

ASTM Spec.	Title
Malleable Iron Castings (Continued)	
A 220	Pearlitic Malleable Iron
A 220M	Pearlitic Malleable Iron [Metric]
A 338	Malleable Iron Flanges, Pipe Fittings, and Valve Parts for Railroad, Marine, and Other Heavy Duty Service at Temperatures up to 650°F (345°C)
A 602	Automotive Malleable Iron Castings
White Iron Castings	
A 518	Corrosion-Resistant High-Silicon Iron Castings
A 518M	Corrosion-Resistant High-Silicon Iron Castings [Metric]
A 532/A 532M	Abrasion-Resistant Cast Irons
A 667/A 667M	Centrifugally Cast Dual Metal (Gray and White Cast Iron) Cylinders
A 748/A 748M	Statically Cast Chilled White Iron-Gray Iron Dual Metal Rolls for Pressure Vessel Use
A 942	Centrifugally Cast White Iron/Gray Iron Dual Metal Abrasion Resistant Roll Shells
Iron Pipe and Fittings	
A 74	Cast Iron Soil Pipe and Fittings
A 126	Gray Iron Castings for Valves, Flanges, and Pipe Fittings
A 338	Malleable Iron Flanges, Pipe Fittings, and Valve Parts for Railroad, Marine, and Other Heavy Duty Service at Temperatures Up to 650°F (345°C)
A 377	Ductile-Iron Pressure Pipe
A 674	Polyethylene Encasement for Ductile Iron Pipe for Water or Other Liquids
A 716	Ductile Iron Culvert Pipe
A 746	Ductile Iron Gravity Sewer Pipe
A 861	High-Silicon Iron Pipe and Fittings
A 888	Hubless Cast Iron Soil Pipe and Fittings for Sanitary and Storm Drain, Waste, and Vent Piping Applications
General	
A 644	Iron Castings
A 834	Common Requirements for Iron Castings for General Industrial Use
A 842	Compacted Graphite Iron Castings

CAST IRONS (Continued)

ASTM Spec.	Title
Methods of Testing Cast Iron	
A 247	Evaluating the Microstructure of Graphite in Iron Castings
A 327	Impact Testing of Cast Irons
A 327M	Impact Testing of Cast Irons [Metric]
A 367	Chill Testing of Cast Iron
A 438	Transverse Testing of Gray Cast Iron

CHEMICAL COMPOSITION OF AUSTENITIC GRAY IRON CASTINGS

ASTM Spec.	UNS	C	Mn	Si	S	Ni	Cr	Mo	Cu
A 436 Type 1	F41000	3.00	0.5-1.5	1.00-2.80	0.12	13.50-17.50	1.5-2.5	---	5.50-7.50
A 436 Type 1b	F41001	3.00	0.5-1.5	1.00-2.80	0.12	13.50-17.50	2.50-3.50	---	5.50-7.50
A 436 Type 2	F41002	3.00	0.5-1.5	1.00-2.80	0.12	18.00-22.00	1.5-2.5	---	0.50
A 436 Type 2b	F41003	3.00	0.5-1.5	1.00-2.80	0.12	18.00-22.00	3.00-6.00	---	0.50
A 436 Type 3	F41004	2.60	0.5-1.5	1.00-2.00	0.12	28.00-32.00	2.50-3.50	---	0.50
A 436 Type 4	F41005	2.60	0.5-1.5	5.00-6.00	0.12	29.00-32.00	4.50-5.50	---	0.50
A 436 Type 5	F41006	2.40	0.5-1.5	1.00-2.00	0.12	34.00-36.00	0.10	---	0.50
A 436 Type 6	F41007	3.00	0.5-1.5	1.50-2.50	0.12	18.00-22.00	1.00-2.00	1.00	3.50-5.50

Single values are maximums.

CHEMICAL COMPOSITION OF DUCITLE IRON CASTINGS

ASTM Spec.	UNS	Total C	Si	Mn	P	S	Ni	Cr	Other
A 395, A 395 M	F32800	3.00 min	2.50		0.08[a]	---	---	---	---
A 439 Type D-2	F43000	3.00	1.50-3.00	0.70-1.25	0.08	---	18.00-22.00	1.75-2.75	---
A 439 Type D-2B	F43001	3.00	1.50-3.00	0.70-1.25	0.08	---	18.00-22.00	2.75-4.00	---
A 439 Type D-2C	F43002	2.90	1.00-3.00	1.80-2.40	0.08	---	21.00-24.00	0.50	---
A 439 Type D-3	F43003	2.60	1.00-2.80	1.00	0.08	---	28.00-32.00	2.50-3.50	---

CHEMICAL COMPOSITION OF DUCTILE IRON CASTINGS (Continued)

ASTM Spec.		Total C	Si	Mn	P	S	Ni	Cr	Other
A 439 Type D-3A	F43004	2.60	1.00-2.80	1.00	0.08	---	28.00-32.00	1.00-1.50	---
A 439 Type D-4	F43005	2.60	5.00-6.00	1.00	0.08	---	28.00-32.00	4.50-5.50	---
A 439 Type D-5	F43006	2.40	1.00-2.80	1.00	0.08	---	34.00-36.00	0.10	---
A 439 Type D-5B	F43007	2.40	1.00-2.80	1.00	0.08	---	34.00-36.00	2.00-3.00	---
A 439 Type D-5S	-	2.30	4.90-5.50	1.00	0.08	---	34.00-37.00	1.75-2.25	---
A 476	F34100	3.0 min	3.0	---	0.08	0.05	---	---	3.8-4.5 CE
A 571	F43010	2.2-2.7	1.5-2.50	3.75-4.5	0.08	---	21.0-24.0	0.20	---
A 874	-	3.0-3.7	1.2-2.3	0.25	0.03	---	1.0	0.07	0.07 Mg, 0.1 Cu, 4.5 CE

a. For each reduction of 0.01% phosphorus below its maximum, an increase of 0.08% silicon above its specified maximum will be permitted up to 2.75% maximum silicon. CE - Carbon equivalent, see related standard for equation. Single values are maximums, unless otherwise designated.

CHEMICAL COMPOSITION OF WHITE IRON CASTINGS

ASTM Spec.	UNS	C	Mn	Si	P	S	Ni	Cr	Mo	Cu
A 518 Grade 1	---	0.65-1.10	1.50	14.20-14.75	---	---	---	0.50	0.50	0.50
A 518 Grade 2	---	0.75-1.15	1.50	14.20-14.75	---	---	---	3.25-5.00	0.40-0.60	0.50
A 518 Grade 3	---	0.70-1.10	1.50	14.20-14.75	---	---	---	3.25-5.00	0.20	0.50
A 532 Class I Type A	F45000	2.8-3.6	2.0	0.8	0.3	0.15	3.3-5.0	1.4-4.0	1.0	---
A 532 Class I Type B	F45001	2.4-3.0	2.0	0.8	0.3	0.15	3.3-5.0	1.4-4.0	1.0	---
A 532 Class I Type C	F45002	2.5-3.7	2.0	0.8	0.3	0.15	4.0	1.0-2.5	1.0	---
A 532 Class I Type D	F45003	2.5-3.6	2.0	2.0	0.10	0.15	4.5-7.0	7.0-11.0	1.5	---
A 532 Class II Type A	F45004	2.0-3.3	2.0	1.5	0.10	0.06	2.5	11.0-14.0	3.0	1.2
A 532 Class II Type B	F45005	2.0-3.3	2.0	1.5	0.10	0.06	2.5	14.0-18.0	3.0	1.2
A 532 Class II Type D	F45007	2.0-3.3	2.0	1.0-2.2	0.10	0.06	2.5	18.0-23.0	3.0	1.2
A 532 Class III Type A	F45009	2.0-3.3	2.0	1.5	0.10	0.06	2.5	23.0-30.0	3.0	1.2

Single values are maximums.

MECHANICAL PROPERTIES OF GRAY IRON CASTINGS

ASTM Spec.	UNS	Tensile Strength	
		ksi	MPa
A 48 Class 20 A, B, C, S	F11401	20	138
A 48 Class 25 A, B, C, S	F11701	25	172
A 48 Class 30 A, B, C, S	F12101	30	207
A 48 Class 35 A, B, C, S	F12401	35	241
A 48 Class 40 A, B, C, S	F12801	40	276
A 48 Class 45 A, B, C, S	F13101	45	310
A 48 Class 50 A, B, C, S	F13501	50	345
A 48 Class 55 A, B, C, S	F13801	55	379
A 48 Class 60 A, B, C, S	F14101	60	414
A 278[a] Class No. 20	F11401	20	---
A 278[a] Class No. 25	F11701	25	---
A 278[a] Class No. 30	F12101	30	---
A 278[a] Class No. 35	F12401	35	---
A 278[a] Class No. 40	F12803	40	---
A 278[a] Class No. 45	F13102	45	---
A 278[a] Class No. 50	F13502	50	---
A 278[a] Class No. 55	F13802	55	---
A 278[a] Class No. 60	F14102	60	---
A 278M[b] Class No. 150	---	---	150
A 278M[b] Class No. 175	---	---	175
A 278M[b] Class No. 200	---	---	200
A 278M[b] Class No. 225	---	---	225
A 278M[b] Class No. 250	---	---	250
A 278M[b] Class No. 275	---	---	275

MECHANICAL PROPERTIES OF GRAY IRON CASTINGS (Continued)

ASTM Spec.	UNS	Tensile Strength	
		ksi	MPa
A 278M[b] Class No. 300	---	---	300
A 278M[b] Class No. 325	---	---	325
A 278M[b] Class No. 350	---	---	350
A 278M[b] Class No. 380	---	---	380
A 278M[b] Class No. 415	---	---	415

a. Castings of all classes are suitable for use up to 450°F. For temperatures above 450°F and up to 650°F, only Class No. 40, 45, 50, 55, and 60 castings are suitable.

b. Castings of all classes are suitable for use up to 230°C. For temperatures above 230°C and up to 350°C, only Class No. 275, 300, 325, 350, 380, and 415 castings are suitable. UNS numbers are not assigned to ASTM A 278M.

Single values are minimums, unless otherwise designated.

MECHANICAL PROPERTIES OF AUSTENITIC GRAY IRON CASTINGS

ASTM Spec.	UNS	Tensile Strength, ksi	Tensile Strength, MPa	Hardness, HB
A 436 Type 1	F41000	25	172	131-183
A 436 Type 1b	F41001	30	207	149-212
A 436 Type 2	F41002	25	172	118-174
A 436 Type 2b	F41003	30	207	171-248
A 436 Type 3	F41004	25	172	118-159
A 436 Type 4	F41005	25	172	149-212
A 436 Type 5	F41006	20	138	99-124
A 436 Type 6	F41007	25	172	124-174

Single values are minimums, unless otherwise designated.

MECHANICAL PROPERTIES OF STATICALLY CAST PERMANENT MOLD GRAY IRON CASTINGS

ASTM A 823[a]	Tensile Strength, ksi	Tensile Strength, MPa	Hardness, HB
Uncored - Annealed			
Grade A-SA	30	207	163-207
Grade A-SB	25	172	163-207
Grade A-SC	20	138	163-207
Grade A-SS	18	124	143-207
Cored Annealed			
Grade A-CA	30	207	143-207
Grade A-CB	25	172	143-207
Grade A-CC	20	138	143-207
Uncored Normalized			
Grade N-SA	30	207	170-229
Grade N-SB	25	172	170-229
Grade N-SC	20	138	170-229
Grade N-SS	18	124	149-229
Cored Normalized			
Grade N-CA	30	207	170-229
Grade N-CB	25	172	170-229
Grade N-CC	20	138	170-229

a. UNS Numbers not assigned to ASTM A 823.

MECHANICAL PROPERTIES OF DUCTILE IRON CASTINGS

ASTM Spec.	UNS	Tensile Strength		Yield Strength		% Elongation	Hardness, HB
		ksi	MPa	ksi	MPa		
A 395	F32800	60	---	40	---	18	143-187
A 395M	---	---	415	---	275	18	143-187
A 439 Type D-2	F43000	58	400	30	207	8.0	139-202
A 439 Type D-2B	F43001	58	400	30	207	7.0	148-211
A 439 Type D-2C	F43002	58	400	28	193	20.0	121-171
A 439 Type D-3	F43003	55	379	30	207	6.0	139-202
A 439 Type D-3A	F43004	55	379	30	207	10.0	131-193
A 439 Type D-4	F43005	60	414	---	---	---	202-273
A 439 Type D-5	F43006	55	379	30	207	20.0	131-185
A 439 Type D-5B	F43007	55	379	30	207	6.0	139-193
A 439 Type D-5S	---	65	449	30	207	10	131-193
A 476 (1 in)	F34100	80	---	60	---	3.0	201
A 476 (3 in)	F34100	80	---	60	---	1.0	201
A 476 M (25 mm)	---	---	555	---	415	3.0	201
A 476 M (75 mm)	---	---	555	---	415	1.0	201
A 536 Gr 60-40-18	F32800	60	414	40	276	18	---
A 536 Gr 65-45-12	F33100	65	448	45	310	12	---
A 536 Gr 80-55-06	F33800	80	552	55	379	6.0	---
A 536 Gr 100-70-03	F34800	100	689	70	483	3.0	---
A 536 Gr 120-90-02	F36200	120	827	90	621	2.0	---
A 874	---	45	---	30	---	12	---
A 874 M	---	---	300	---	200	12	---

Single values are minimums, unless otherwise designated.

MECHANICAL PROPERTIES OF DUCTILE IRON CASTINGS WITH NOTCH TOUGHNESS REQUIREMENTS

ASTM Spec.	Tensile Strength		Yield Strength		% Elongation	Impact[a]	Hardness, HB[c]
	ksi	MPa	ksi	MPa			
A 571 Class 1	65	---	30	---	30	15/12 ft•lb[a]	121-171
A 571 Class 2	60	---	25	---	25	20/15 ft•lb[a]	111-171
A 571M Class 1	---	450	---	205	30	20/16 J[a]	121-171
A 571M Class 2	---	415	---	170	25	27/20 J[a]	111-171
A 897 Gr 125/80/10	125	---	80	---	10	75[b]	269-321
A 897 Gr 150/100/7	150	---	100	---	7	60[b]	302-363
A 897 Gr 175/125/4	175	---	125	---	4	45[b]	341-444
A 897 Gr 200/155/1	200	---	155	---	1	25[b]	388-477
A 897 Gr 230/185/-	230	---	185	---	---	---	444-555
A 897M Gr 850/550/10	---	850	---	550	10	100	269-321
A 897M Gr 1050/700/7	---	1050	---	700	7	80	302-363
A 897M Gr 1200/850/4	---	1200	---	850	4	60	341-444
A 897M Gr 1400/1100/1	---	1400	---	1100	1	35	388-477
A 897M Gr 1600/1300/-	---	1600	---	1300	---	---	444-555

a. Minimum Charpy V-notch impact energy required for average of each set of three specimens, followed by, minimum impact energy permitted for one specimen only of a set, test temperature to be agreed upon between manufacturer and purchaser.

b. Unnotched Charpy bars tested at 72°F ±7°F and measure in ft•lb. Values are minimum for the average of the highest three test values of the four tested samples.

c. Hardness is not mandatory and shown only for information.

Single values are minimums, unless otherwise designated.

MECHANICAL PROPERTIES OF MALLEABLE IRON CASTINGS

ASTM Spec.	UNS	Tensile Strength	Yield Strength	% Elongation	Hardness, HB
A 47 Gr 32510	F22200	50 ksi	32.5 ksi	10	156[a]
A 47M Gr 22010	---	340 MPa	220 MPa	10	156
A 197	F22000	40 ksi	30 ksi	5	---
A 197M	---	275 MPa	200 MP ksi a	5	---
A 220 Gr 40010	F22830	60 ksi	40 ksi	10	149-197[a]
A 220 Gr 45008	F23130	65 ksi	45 ksi	8	156-197[a]
A 220 Gr 45006	F23131	65 ksi	45 ksi	6	156-207[a]
A 220 Gr 50005	F23530	70 ksi	50 ksi	5	179-229[a]
A 220 Gr 60004	F24130	80 ksi	60 ksi	4	197-241[a]
A 220 Gr 70003	F24830	85 ksi	70 ksi	3	217-269[a]
A 220 Gr 80002	F25530	95 ksi	80 ksi	2	241-285[a]
A 220 Gr 90001	F26230	105 ksi	90 ksi	1	269-231[a]
A 220M Gr 280M10		400 MPa	280 MPa	10	149-197[a]
A 220M Gr 310M8		450 MPa	310 MPa	8	156-197[a]
A 220M Gr 310M6		450 MPa	310 MPa	6	156-207[a]
A 220M Gr 340M5		480 MPa	340 MPa	5	179-229[a]
A 220M Gr 410M4		550 MPa	410 MPa	4	197-241[a]
A 220M Gr 480M3		590 MPa	480 MPa	3	217-269[a]
A 220M Gr 550M2		650 MPa	550 MPa	2	241-285[a]
A 220M Gr 650M1		720 MPa	620 MPa	1	269-231[a]
A 602 Gr M3210[c]	F20000	---	---	---	156 max[b] (4.8 min)
A 602 Gr M4504[c]	---	---	---	---	163-217[b] (4.7-4.1)
A 602 Gr M5003[c]	---	---	---	---	187-241[b] (4.4-3.9)
A 602 Gr M5503[c]	---	---	---	---	187-241[b] (4.4-3.9)

MECHANICAL PROPERTIES OF MALLEABLE IRON CASTINGS (Continued)

ASTM Spec.	UNS	Tensile Strength	Yield Strength	% Elongation	Hardness, HB
A 602 Gr M7002[c]	---	---	---	---	229-269[b] (4.0-3.7)
A 602 Gr M8501[c]	---	---	---	---	269-302[b] (3.7-3.5)

a. Typical hardness values; at purchaser's option to specify.
b. Brinell impression diameter (BID) in mm, using a 10 mm ball at 3000 kg load.
c. See ASTM A 602 for heat treatment condition.

ABRASION-RESISTANT (WHITE) CAST IRON HARDNESS REQUIREMENTS[a]

ASTM A 532	UNS	Designation	As Cast &/or Stress Relieved, HB min	Hardened &/or Stress Relieved, min		Chill Cast[b]	
				Level 1, HB min	Level 2, HB min	HB min	Softened HB, max
Class I Type A	F45000	Ni-Cr-HiC	550	600	650	600	---
Class I Type B	F45001	Ni-Cr-LoC	550	600	650	600	---
Class I Type C	F45002	Ni-Cr-GB	550	600	650	600	400
Class I Type D	F45003	Ni-HiCr	500	600	650	550	---
Class II Type A	F45004	12% Cr	550	600	650	550	400
Class II Type B	F45005	15% Cr-Mo	450	600	650	---	400
Class III Type D	F45007	20% Cr-Mo	450	600	650	---	400
Class III Type A	F45009	25% Cr	450	600	650	---	400

a. Rockwell and Vickers hardness testing is also permitted, see ASTM A 532/A 532M for more details.
b. Non-chilled areas of casting shall meet minimum hardness or sand cast requirements.

Chapter

11

CASTINGS CARBON & ALLOY STEELS:

AMERICAN SPECIFICATION TITLES & DESIGNATIONS, CHEMICAL COMPOSITIONS & MECHANICAL PROPERTIES

CARBON AND ALLOY STEEL CASTINGS

ASTM Spec.	Title
General Applications, Structural Purposes	
A 27/A 27M	Steel Castings, Carbon, for General Application
A 148/A 148M	Steel Castings, High Strength, for Structural Purposes
A 732/A 732M	Castings, Investment, Carbon and Low Alloy Steel for General Application, and Cobalt Alloy for High Strength at Elevated Temperatures
A 781/A 781M	Castings, Steel and Alloy, Common Requirements, for General Industrial Use
High-Temperature/Pressure Service	
A 216/A 216M	Steel Castings, Carbon, Suitable for Fusion Welding, for High-Temperature Service
A 217/A 217M	Steel Castings, Martensitic Stainless and Alloy, for Pressure-Containing Parts, Suitable for High-Temperature Service
A 356/A 356M	Steel Castings, Carbon, Low Alloy, and Stainless Steel, Heavy-Walled for Steam Turbines
A 389/A 389M	Steel Castings, Alloy, Specially Heat-Treated, for Pressure Containing Parts, Suitable for High Temperature Service
A 426	Centrifugally Cast Ferritic Alloy Steel Pipe for High-Temperature Service
A 487/A 487M	Steel Castings, Suitable for Pressure Service
A 660	Centrifugally Cast Carbon Steel Pipe for High-Temperature Service
A 703/A 703M	Steel Castings, General Requirements, for Pressure-Containing Parts
Low-Temperature Service	
A 352/A 352M	Steel Castings, Ferritic and Martensitic, for Pressure-Containing Parts, Suitable for Low-Temperature Service
A 757/A 757M	Steel Castings, Ferritic and Martensitic, for Pressure-Containing and Other Applications, for Low-Temperature Service
Alloy Steel Castings	
A 128/A 128M	Steel Castings, Austenitic Manganese
A 915/A 915M	Steel Castings, Carbon and Alloy, Chemical Requirements Similar to Standard Wrought Grades

CHEMICAL COMPOSITION OF CARBON STEEL CASTINGS FOR GENERAL APPLICATIONS

ASTM Spec.[a]	UNS	C	Mn	Si	S	P	Condition
A 27 Gr N-1	J02500	0.25[b]	0.75[b]	0.80	0.06	0.05	Chemical analysis only
A 27 Gr N-2	J03500	0.35[b]	0.60[b]	0.80	0.06	0.05	Heat treated but not mechanically tested
A 27 Gr U60-30 [415-205]	J02500	0.25[b]	0.75[b]	0.80	0.06	0.05	Mechanically tested but not heat treated
A 27 Gr 60-30 [415-205]	J03000	0.30[b]	0.60[b]	0.80	0.06	0.05	Heat treated and mechanically tested
A 27 Gr 65-35 [450-240]	J03001	0.30[b]	0.70[b]	0.80	0.06	0.05	Heat treated and mechanically tested
A 27 Gr 70-36 [485-250]	J03501	0.35[b]	0.70[b]	0.80	0.06	0.05	Heat treated and mechanically tested
A 27 Gr 70-40 [485-275]	J02501	0.25[b]	1.20[b]	0.80	0.06	0.05	Heat treated and mechanically tested

a. Class 1 requires postweld heat treatment, Class 2 does not require postweld heat treatment. In either case, Class 1 or 2 must be specified in addition to the grade designation (see ASTM A 27 paragraph 9.2 for more details).
b. For each reduction of 0.01% C below the maximum specified, an increase of 0.04% Mn above the maximum specified is permitted to a maximum of 1.4% Mn for Grade 70-40 and 1.00% Mn for the other grades.00000
Single values are maximums, unless otherwise specified.

CHEMICAL COMPOSITION OF HIGH STRENGTH STEEL CASTINGS FOR STRUCTURAL PURPOSES

ASTM Spec.	C	Mn	Si	S	P	Condition
A 148 Gr 80-40 [550-275]	a	a	a	0.06	0.05	Composition and heat treatment necessary to achieve specified mechanical properties
A 148 Gr 80-50 [550-345]	a	a	a	0.06	0.05	Composition and heat treatment necessary to achieve specified mechanical properties
A 148 Gr 90-60 [620-415]	a	a	a	0.06	0.05	Composition and heat treatment necessary to achieve specified mechanical properties
A 148 Gr 105-85 [725-585]	a	a	a	0.06	0.05	Composition and heat treatment necessary to achieve specified mechanical properties
A 148 Gr 115-95 [795-655]	a	a	a	0.06	0.05	Composition and heat treatment necessary to achieve specified mechanical properties
A 148 Gr 130-115 [895-795]	a	a	a	0.06	0.05	Composition and heat treatment necessary to achieve specified mechanical properties
A 148 Gr 135-125 [930-860]	a	a	a	0.06	0.05	Composition and heat treatment necessary to achieve specified mechanical properties
A 148 Gr 150-135 [1035-930]	a	a	a	0.06	0.05	Composition and heat treatment necessary to achieve specified mechanical properties
A 148 Gr 160-145 [1105-1000]	a	a	a	0.06	0.05	Composition and heat treatment necessary to achieve specified mechanical properties
A 148 Gr 165-150 [1140-1035]	a	a	a	0.02	0.02	Composition and heat treatment necessary to achieve specified mechanical properties

CHEMICAL COMPOSITION OF HIGH STRENGTH STEEL CASTINGS FOR STRUCTURAL PURPOSES (Continued)

ASTM Spec.	C	Mn	Si	S	P	Condition
A 148 Gr 165-150L [1140-1035L]	a	a	a	0.02	0.02	Composition and heat treatment necessary to achieve specified mechanical properties
A 148 Gr 210-180 [1450-1240]	a	a	a	0.02	0.02	Composition and heat treatment necessary to achieve specified mechanical properties
A 148 Gr 210-180L [1450-1240L]	a	a	a	0.02	0.02	Composition and heat treatment necessary to achieve specified mechanical properties
A 148 Gr 260-210 [1795-1450]	a	a	a	0.02	0.02	Composition and heat treatment necessary to achieve specified mechanical properties
A 148 Gr 260-210L [1795-1450L]	a	a	a	0.02	0.02	Composition and heat treatment necessary to achieve specified mechanical properties

a. If not specified by the purchaser and agreed to by the manufacturer, the manufacturer may select the C, Mn, Si and alloying elements contents to obtain the required mechanical properties.
Single values are maximums, unless otherwise specified.

CHEMICAL COMPOSITION OF CARBON STEEL CASTINGS FOR HIGH TEMPERATURE SERVICE

ASTM Spec.	UNS	C	Mn	Si	S	P	Application
A 216 Gr WCA	J02502	0.25	0.70a	0.60	0.045	0.04	Weldable steel casting for high temperature service
A 216 Gr WCB	J03002	0.30	1.00a	0.60	0.045	0.04	Weldable steel casting for high temperature service
A 216 Gr WCC	J02503	0.25	1.20a	0.60	0.045	0.04	Weldable steel casting for high temperature service

a. For each reduction of 0.01% C below the maximum specified, an increase of 0.04% Mn above the maximum specified is permitted, up to a maximum of 1.10% Mn for Grade WCA, 1.28% Mn for Grade WCB, and 1.40% Mn for Grade WCC.
b. Specified maximum residual elements are 0.30% Cu, 0.50% Ni, 0.50% Cr, 0.20% Mo and 0.03% V, with total residual elements not exceeding 1.00%.
Single values are maximums, unless otherwise specified.

CHEMICAL COMPOSITION OF CARBON STEEL CASTINGS FOR PRESSURE CONTAINING PARTS

ASTM Spec.	UNS	C	Mn	Si	S	P	Application
A 352 Gr LCA[a,b]	J02504	0.25	0.70[a]	0.60	0.045	0.04	Low-temperature applications
A 352 Gr LCB[a,b]	J03003	0.30	1.00[a]	0.60	0.045	0.04	Low-temperature applications
A 352 Gr LCC[a,c]	J02505	0.25	1.20[a]	0.60	0.045	0.04	Low-temperature applications
A 356 Gr 1[a]	J03502	0.35	0.70[a]	0.60	0.030	0.035	Castings for cylinders (shells), valve chests, throttle valves, and other heavy-walled components for steam turbines
A 757 Gr A1Q[d]	J03002	0.30	1.00	0.60	0.025	0.025	Castings for pressure-containing applications at low temperatures
A 757 Gr A2Q[a,d]	J02503	0.25	1.20	0.60	0.025	0.025	Castings for pressure-containing applications at low temperatures

a. For each reduction of 0.01% C below the maximum specified, an increase of 0.04% Mn above the maximum specified is permitted, up to a maximum of 1.10% Mn for A 352 Gr LCA, 1.28% Mn for A 352 Gr LCB, 1.40% Mn for A 352 Gr LCC, 1.00% Mn for A 356 Gr 1, and 1.40% Mn for A 757 Gr A2Q. Specified maximum residual elements are 0.30% Cu, 0.50% Ni, 0.50% Cr, 0.20% Mo and 0.03% V, with total residual elements not exceeding 1.00%. b. Specified maximim residual elements are 0.50% Ni, 0.50% Cr, 0.20% Mo and 0.03% V, with total residual elements not exceeding 1.00%. d. Specified maximim residual elements are 0.50% Cu, 0.50% Ni, 0.40% Cr, 0.25% Mo and 0.03% V, with total residual elements not exceeding 1.00%. including phosphorus and sulfur. Single values are maximums, unless otherwise specified.

CHEMICAL COMPOSITION OF ALLOY STEEL CASTINGS FOR PRESSURE CONTAINING PARTS

ASTM Spec.	UNS	C	Mn	Si	S	P	Cr	Ni	Mo	Other
A 217 Gr WC1	J12524	0.25	0.50-0.80	0.60	0.045	0.04	---	---	0.45-0.65	a
A 217 Gr WC4	J12082	0.05-0.20	0.50-0.80	0.60	0.045	0.04	0.50-0.80	0.70-1.10	0.45-0.65	b
A 217 Gr WC5	J22000	0.05-0.20	0.40-0.70	0.60	0.045	0.04	0.50-0.90	0.60-1.00	0.90-1.20	b
A 217 Gr WC6	J12072	0.05-0.20	0.50-0.80	0.60	0.045	0.04	1.00-1.50	---	0.45-0.65	a
A 217 Gr WC9	J21890	0.05-0.18	0.40-0.70	0.60	0.045	0.04	2.00-2.75	---	0.90-1.20	a
A 217 Gr WC11	J11872	0.15-0.21	0.50-0.80	0.30-0.60	0.015	0.020	1.00-1.50	---	0.45-0.65	a
A 217 Gr C5	J42045	0.20	0.40-0.70	0.75	0.045	0.04	4.00-6.50	---	0.45-0.65	a
A 217 Gr C12	J82090	0.20	0.35-0.65	1.00	0.045	0.04	8.00-10.00	---	0.90-1.20	a

a. Total residual elements not to exceed 1.00%; see ASTM A 217 for more details. b. Total residual elements not to exceed 0.60%; see ASTM A 217 for more details. Single values are maximums, unless otherwise specified.

CHEMICAL COMPOSITION OF ALLOY STEEL CASTINGS FOR PRESSURE SERVICE

ASTM Spec.	UNS	C	Mn	Si	Cr	Ni	Mo	Other
A 487 Gr 1 Cl A, B, C	J13002	0.30	1.00	0.80	---	---	---	0.04-0.12 V[a, b]
A 487 Gr 2 Cl A, B, C	J13005	0.30	1.00-1.40	0.80	---	---	0.10-0.30	a, b
A 487 Gr 4 Cl A, B, C, D, E	J13047	0.30	1.00	0.80	0.40-0.80	0.40-0.80	0.15-0.30	b, c
A 487 Gr 6 Cl A, B	J13855	0.05-0.38	1.30-1.70	0.80	0.40-0.80	0.40-0.80	0.30-0.40	b, c
A 487 Gr 7 Cl A	J12084	0.05-0.20	0.60-1.00	0.80	0.40-0.80	0.70-1.00	0.40-0.60	0.03-0.10 V, 0.002-0.006 B, 0.15-0.50 Cu[b, c]
A 487 Gr 8 Cl A, B, C	J22091	0.05-0.20	0.50-0.90	0.80	2.00-2.75	---	0.90-1.10	b, c
A 487 Gr 9 Cl A, B, C, D, E	J13345	0.05-0.33	0.60-1.00	0.80	0.75-1.10	---	0.15-0.30	a, b
A 487 Gr 10 Cl A, B	J23015	0.30	0.60-1.00	0.80	0.55-0.90	1.40-2.00	0.20-0.40	b, c
A 487 Gr 11 Cl A, B	J12082	0.05-0.20	0.50-0.80	0.60	0.50-0.80	0.70-1.10	0.45-0.65	b, d
A 487 Gr 12 Cl A, B	J22000	0.05-0.20	0.40-0.70	0.60	0.50-0.90	0.60-1.00	0.90-1.20	b, d
A 487 Gr 13 Cl A, B	J13080	0.30	0.80-1.10	0.60	---	1.40-1.75	0.20-0.30	b, e
A 487 Gr 14 Cl A	J15580	0.55	0.80-1.10	0.60	---	1.40-1.75	0.20-0.30	b, e
A 487 Gr 16 Cl A	---	0.12[g]	2.10[g]	0.50	---	1.00-1.40	---	d, f

a. Total residual elements not to exceed 1.00%; see ASTM Spec. 487 for details.
b. Contains 0.040 P maximumand 0.045 S maximum.
c. Total residual elements not to exceed 0.60%; see ASTM A 487 for details.
d. Total residual elements not to exceed 0.50%; see ASTM A 487 for details.
e. Total residual elements not to exceed 0.75%; see ASTM A 487 for details.
f. Contains 0.02 P maximum and 0.02 S maximum.
g. For each reduction of 0.01% C below the maximum carbon content, an increase of 0.04% Mn above the maximum allowable is permitted up to a maximum of 2.30% Mn.
Single values are maximums, unless otherwise specified.

CHEMICAL COMPOSITION OF CARBON AND ALLOY STEEL CASTINGS FOR PRESSURE CONTAINING PARTS FOR LOW TEMPERATURE SERVICE

ASTM Spec.	UNS	C	Mn	Si	S	P	Cr	Ni	Mo	Other
A 352 Gr LCA	J02504	0.25[a]	0.70[a]	0.60	0.045	0.04	0.50	0.50	0.20	0.30 Cu, 0.03 V[b]
A 352 Gr LCB	J03003	0.30[a]	1.00[a]	0.60	0.045	0.04	0.50	0.50	0.20	0.30 Cu, 0.03 V[b]
A 352 Gr LCC	J02505	0.25[a]	1.20[a]	0.60	0.045	0.04	0.50	0.50	0.20	0.03 V[b]
A 352 Gr LC1	J12522	0.25	0.50-0.80	0.60	0.045	0.04	---	---	0.45-0.65	b
A 352 Gr LC2	J22500	0.25	0.50-0.80	0.60	0.045	0.04	---	2.00-3.00	---	b
A 352 Gr LC2-1	J42215	0.22	0.55-0.75	0.50	0.045	0.04	1.35-1.85	2.50-3.50	0.30-0.60	b
A 352 Gr LC3	J31550	0.15	0.50-0.80	0.60	0.045	0.04	---	3.00-4.00	---	b
A 352 Gr LC4	J41500	0.15	0.50-0.80	0.60	0.045	0.04	---	4.00-5.00	---	0.30 Cu[b]
A 352 Gr LC9	---	0.13	0.90	0.45	0.045	0.04	0.50	8.50-10.00	0.20	0.03 V[b]

a. For a reduction of 0.01% below the specified max. C content, an increase of 0.04% Mn above the specified max. is permitted up to a max. of 1.10% Mn for LCA, 1.28% Mn for LCB, and 1.40% Mn for LCC. b. Total residual elements not to exceed 1.00% max. Single values are max values, unless otherwise specified.

CHEMICAL COMPOSITION OF CARBON AND ALLOY STEEL CASTINGS SIMILAR TO STANDARD AISI WROUGHT GRADES

ASTM Spec.	Similar to AISI	C	Mn	Si	S	P	Cr	Ni	Mo
A 915 Gr SC 1020	1020	0.18-0.23	0.40-0.80	0.30-0.60	0.040	0.040	---	---	---
A 915 Gr SC 1025	1025	0.22-0.28	0.40-0.80	0.30-0.60	0.040	0.040	---	---	---
A 915 Gr SC 1030	1030	0.28-0.34	0.50-0.90	0.30-0.60	0.040	0.040	---	---	---
A 915 Gr SC 1040	1040	0.37-0.44	0.50-0.90	0.30-0.60	0.040	0.040	---	---	---
A 915 Gr SC 1045	1045	0.43-0.50	0.40-0.80	0.30-0.60	0.040	0.040	---	---	---
A 915 Gr SC 4130	4130	0.28-0.33	0.70-1.10	0.30-0.60	0.040	0.035	0.80-1.10	---	0.15-0.25
A 915 Gr SC 4140	4140	0.38-0.43	0.60-0.90	0.30-0.60	0.040	0.035	0.80-1.10	---	0.15-0.25
A 915 Gr SC 4330	4330	0.28-0.33	0.60-0.90	0.30-0.60	0.040	0.035	0.70-0.90	1.65-2.00	0.20-0.30
A 915 Gr SC 4340	4340	0.38-0.43	0.60-1.00	0.30-0.60	0.040	0.035	0.70-0.90	1.65-2.00	0.20-0.30
A 915 Gr SC 8620	8620	0.18-0.23	0.60-1.00	0.30-0.60	0.040	0.035	0.40-0.60	0.40-0.70	0.15-0.25
A 915 Gr SC 8625	8625	0.23-0.28	0.60-1.00	0.30-0.60	0.040	0.035	0.40-0.60	0.40-0.70	0.15-0.25
A 915 Gr SC 8630	8630	0.28-0.33	0.60-1.00	0.30-0.60	0.040	0.035	0.40-0.60	0.40-0.70	0.15-0.25

MECHANICAL PROPERTIES OF CARBON STEEL CASTINGS FOR GENERAL APPLICATIONS

ASTM Spec.	Tensile Strength		Yield Strength		% Elongation	% Reduction of Area
	ksi	MPa	ksi	MPa		
A 27 Gr N-1	a	a	a	a	a	a
A 27 Gr N-2	a	a	a	a	a	a
A 27 Gr U60-30 [415-205]	60	415	30	205	22	30
A 27 Gr 60-30 [415-205]	60	415	30	205	24	35
A 27 Gr 65-35 [450-240]	65	450	35	240	24	35
A 27 Gr 70-36 [485-250]	70	485	36	250	22	30
A 27 Gr 70-40 [485-275]	70	485	40	275	22	30

a. Not specified.
Single values are minimums, unless otherwise specified.

MECHANICAL PROPERTIES OF HIGH STRENGTH STEEL CASTINGS FOR STRUCTURAL PURPOSES

ASTM Spec.	Tensile Strength		Yield Strength		% Elongation	% Reduction of Area	Charpy ft-lb (J) @ Temp °F (°C)[a]
	ksi	MPa	ksi	MPa			
A 148 Gr 80-40 [550-275]	80	550	40	275	18	30	---
A 148 Gr 80-50 [550-345]	80	550	50	345	22	35	---
A 148 Gr 90-60 [620-415]	90	620	60	415	20	40	---
A 148 Gr 105-85 [725-585]	105	725	85	585	17	35	---
A 148 Gr 115-95 [795-655]	115	795	95	655	14	30	---
A 148 Gr 130-115 [895-795]	130	895	115	795	11	25	---
A 148 Gr 135-125 [930-860]	135	930	125	860	9	22	---
A 148 Gr 150-135 [1035-930]	150	1035	135	930	7	18	---
A 148 Gr 160-145 [1105-1000]	160	1105	145	1000	6	12	---
A 148 Gr 165-150 [1140-1035]	165	1140	150	1035	5	20	---
A 148 Gr 165-150L [1140-1035L]	165	1140	150	1035	5	20	20 (27) / 16 (22) @ -40 (-40)

Single values are minimums, unless otherwise specified.

MECHANICAL PROPERTIES OF HIGH STRENGTH STEEL CASTINGS FOR STRUCTURAL PURPOSES (Continued)

ASTM Spec.	Tensile Strength		Yield Strength		% Elongation	% Reduction of Area	Charpy ft-lb (J) @ Temp °F (°C)[a]
	ksi	MPa	ksi	MPa			
A 148 Gr 210-180 [1450-1240]	210	1450	180	1240	4	15	---
A 148 Gr 210-180L [1450-1240L]	210	1450	180	1240	4	15	15 (20) / 12 (16) @ -40 (-40)
A 148 Gr 260-210 [1795-1450]	260	1795	210	1450	3	6	---
A 148 Gr 260-210L [1795-1450L]	260	1795	210	1450	3	6	6 (8) / 4 (5) @ -40 (-40)

a. Minimum impact energy required for average of each set of three specimens in ft•lb (J), followed by, minimum impact energy permitted for one specimen only of a set in ft•lb (J), followed by the test temperature in °F (°C).
Single values are minimums, unless otherwise specified.

MECHANICAL PROPERTIES OF CARBON STEEL CASTINGS FOR PRESSURE CONTAINING PARTS

ASTM Spec.	Tensile Strength		Yield Strength		% Elongation	% Reduction of Area	Charpy ft-lb (J) @ Temp °F (°C)[a]
	ksi	MPa	ksi	MPa			
A 352 Gr LCA	60-85	415-585	30	205	24	35	13 (18) / 10 (14) @ - 25 (-32)
A 352 Gr LCB	65-90	450-620	35	240	24	35	13 (18) / 10 (14) @ - 50 (-46)
A 352 Gr LCC	70-95	485-655	40	275	22	35	15 (20) / 12 (16) @ -50 (-46)
A 356 Gr 1	70	485	36	250	20	35	---
A 757 Gr A1Q	65	450	35	240	24	35	13 (17) / 10 (14) @ -50 (-46)
A 757 Gr A2Q	70	485	40	275	22	35	15 (20) / 12 (16) @ -50 (-46)

a. Minimum impact energy required for average of each set of three specimens in ft•lb (J), followed by, minimum impact energy permitted for one specimen only of a set in ft•lb (J), followed by the test temperature in °F (°C).
Single values are minimums, unless otherwise specified.

MECHANICAL PROPERTIES OF CARBON STEEL CASTINGS FOR HIGH TEMPERATURE SERVICE

ASTM Spec.	Tensile Strength		Yield Strength		% Elongation	% Reduction of Area
	ksi	MPa	ksi	MPa		
A 216 Gr WCA	60-85	415-585	30	205	24	35
A 216 Gr WCB	70-95	485-655	36	250	22	35
A 216 Gr WCC	70-95	485-655	40	275	22	35

Single values are minimums, unless otherwise specified.

MECHANICAL PROPERTIES OF ALLOY STEEL CASTINGS FOR PRESSURE CONTAINING PARTS

ASTM Spec.	Tensile Strength		Yield Strength		% Elongation	% Reduction of Area
	ksi	MPa	ksi	MPa		
A 217 Gr WC1	65-90	450-620	35	240	24	35
A 217 Gr WC4	70-95	485-655	40	275	20	35
A 217 Gr WC5	70-95	485-655	40	275	20	35
A 217 Gr WC6	70-95	485-655	40	275	20	35
A 217 Gr WC9	70-95	485-655	40	275	20	35
A 217 Gr WC11	80-105	550-725	50	345	18	45
A 217 Gr C5	90-115	620-795	60	415	18	35
A 217 Gr C12	90-115	620-795	60	415	18	35

Single values are minimums.

MECHANICAL PROPERTIES OF ALLOY STEEL CASTINGS FOR PRESSURE CONTAINING PARTS FOR LOW TEMPERATURE SERVICE

ASTM Spec.	Tensile Strength ksi	Tensile Strength MPa	Yield Strength ksi	Yield Strength MPa	% Elongation	% Reduction of Area	Charpy ft-lb (J) @ Temp °F (°C)[a]
A 352 Gr LCA	60-85	415-585	30	205	24	35	13 (18) / 10 (14) @ - 25 (-32)
A 352 Gr LCB	65-90	450-620	35	240	24	35	13 (18) / 10 (14) @ - 50 (-46)
A 352 Gr LCC	70-95	485-655	40	275	22	35	15 (20) / 12 (16) @ - 50 (-46)
A 352 Gr LC1	65-90	450-620	35	240	24	35	13 (18) / 10 (14) @ - 75 (-59)
A 352 Gr LC2	70-95	485-655	40	275	24	35	15 (20) / 12 (16) @ - 100 (-73)
A 352 Gr LC2-1	105-130	725-895	80	550	18	30	30 (41) / 25 (34) @ -100 (-73)
A 352 Gr LC3	70-95	485-655	40	275	24	35	15 (20) / 12 (16) @ -150 (-101)
A 352 Gr LC4	70-95	485-655	40	275	24	35	15 (20) / 12 (16) @ -175 (-115)
A 352 Gr LC9	85	585	75	515	20	30	20 (27) / 15 (20) @ -320 (-196)

a. Minimum impact energy required for average of each set of three specimens in ft·lb (J), followed by, minimum impact energy permitted for one specimen only of a set in ft·lb (J), followed by the test temperature in °F (°C). Note: Single values are minimums, unless otherwise specified.

MECHANICAL PROPERTIES OF ALLOY STEEL CASTINGS FOR PRESSURE SERVICE

ASTM Spec.	Tensile Strength ksi	Tensile Strength MPa	Yield Strength ksi	Yield Strength MPa	% Elongation	% Reduction of Area	Hardness HRC[a], max
A 487 Gr 1 Cl A	85-110	585-760	55	380	22	40	---
A 487 Gr 1 Cl B	90-115	620-795	65	450	22	45	---
A 487 Gr 1 Cl C	90	620	65	450	22	45	22 (235)
A 487 Gr 2 Cl A	85-110	585-760	53	365	22	35	---
A 487 Gr 2 Cl B	90-115	620-795	65	450	22	40	---
A 487 Gr 2 Cl C	90	620	65	450	22	40	22 (235)
A 487 Gr 4 Cl A	90-115	620-795	60	415	18	40	---
A 487 Gr 4 Cl B	105-130	725-895	85	585	17	35	---
A 487 Gr 4 Cl C	90	620	60	415	18	35	22 (235)
A 487 Gr 4 Cl D	100	690	75	515	17	35	22 (235)

MECHANICAL PROPERTIES OF ALLOY STEEL CASTINGS FOR PRESSURE SERVICE (Continued)

ASTM Spec.	Tensile Strength		Yield Strength		% Elongation	% Reduction of Area	Hardness HRC[a], max
	ksi	MPa	ksi	MPa			
A 487 Gr 4 Cl E	115	795	95	655	15	35	---
A 487 Gr 6 Cl A	115	795	80	550	18	30	---
A 487 Gr 6 Cl B	120	825	95	655	12	25	---
A 487 Gr 7 Cl A	115	795	100	690	15	30	---
A 487 Gr 8 Cl A	85-110	585-760	55	380	20	35	---
A 487 Gr 8 Cl B	105	725	85	585	17	30	---
A 487 Gr 8 Cl C	100	690	75	515	17	35	22 (235)
A 487 Gr 9 Cl A	90	620	60	415	18	35	---
A 487 Gr 9 Cl B	105	725	85	585	16	35	---
A 487 Gr 9 Cl C	90	620	60	415	18	35	22 (235)
A 487 Gr 9 Cl D	100	690	75	515	17	35	22 (235)
A 487 Gr 9 Cl E	115	795	95	655	15	35	---
A 487 Gr 10 Cl A	100	690	70	485	18	35	---
A 487 Gr 10 Cl B	125	860	100	690	15	35	---
A 487 Gr 11 Cl A	70-95	485-655	40	275	20	35	---
A 487 Gr 11 Cl B	105-130	725-895	85	585	17	35	---
A 487 Gr 12 Cl A	70-95	485-655	40	275	20	35	---
A 487 Gr 12 Cl B	105-130	725-895	85	585	17	35	---
A 487 Gr 13 Cl A	90-115	620-795	60	415	18	35	---
A 487 Gr 13 Cl B	105-130	725-895	85	585	17	35	---
A 487 Gr 14 Cl A	120-145	825-1000	95	655	14	30	---
A 487 Gr 16 Cl A	70-95	485-655	40	275	22	35	---

a. Measured Brinell hardness (HB) in parentheses, with converted Rockwell hardness (HRC) number reported in accordance with ASTM E 140.
Single values are minimums, unless otherwise specified.

Chapter

12

WROUGHT CARBON & ALLOY STEELS:

AMERICAN SPECIFICATION TITLES & DESIGNATIONS, CHEMICAL COMPOSITIONS & MECHANICAL PROPERTIES

STRUCTURAL STEELS

ASTM Spec.	Title
A 6/A 6M	General Requirements for Rolled Structural Steel Bars, Plates, Shapes, and Sheet Piling
A 36/A 36M	Carbon Structural Steel
A 131/A 131M	Structural Steel for Ships
A 242/A 242M	High-Strength Low-Alloy Structural Steel
A 252	Welded and Seamless Steel Pipe Piles
A 283/A 283M	Low and Intermediate Tensile Strength Carbon Steel Plates, Shapes, and Bars
A 328/A 328M	Steel Sheet Piling
A 514/A 514M	High-Yield-Strength, Quenched and Tempered Alloy Steel Plate, Suitable for Welding
A 529/A 529M	High-Strength Carbon-Manganese Steel of Structural Quality
A 572/A 572M	High-Strength Low-Alloy Columbium-Vanadium Structural Steel
A 573/A 573M	Structural Carbon Steel Plates of Improved Toughness
A 588/A 588M	High-Strength Low-Alloy Structural Steel with 50 ksi [345 MPa] Minimum Yield Point to 4 in. [100 mm] Thick
A 633/A 633M	Normalized High-Strength Low-Alloy Structural Steel Plates
A 656/A 656M	Hot-Rolled, Structural Steel, High Strength Low-Alloy Plate with Improved Formability
A 673/A 673M	Sampling Procedure for Impact Testing of Structural Steel
A 678/A 678M	Quenched-and-Tempered Carbon and High-Strength Low-Alloy Structural Steel Plates
A 690/A 690M	High-Strength Low-Alloy Steel H-Piles and Sheet Piling for Use in Marine Environments
A 709/A 709M	Carbon and High-Strength Low-Alloy Structural Steel Shapes, Plates, and Bars and Quenched and Tempered-Alloy Structural Steel Plates for Bridges
A 710/A 710M	Age-Hardening Low-Carbon Nickel-Copper-Chromium-Molybdenum-Columbium Alloy Structural Steel Plates
A 769/A 769M	Carbon and High-Strength Electric Resistance Welded Steel Shapes
A 786/A 786M	Rolled Steel Floor Plates
A 808/A 808M	High-Strength Low-Alloy Carbon, Manganese, Columbium, Vanadium Steel of Structural Quality with Improved Notch Toughness
A 827/A 827M	Plates, Carbon Steel, for Forging and Similar Applications
A 829/A 829M	Alloy Structural Steel Plates
A 830/A 830M	Plates, Carbon Steel, Structural Quality, Furnished to Chemical Composition Requirements

STRUCTURAL STEELS (Continued)

ASTM Spec.	Title
A 852/A 852M	Quenched and Tempered Low-Alloy Structural Steel Plate with 70 ksi [485 MPa] Minimum Yield Strength to 4 in. [100 mm] Thick
A 857/A 857M	Steel Sheet Piling, Cold Formed, Light Gage
A 871/A 871M	Structural Steel Plate with Atmospheric Corrosion Resistance
A 898/A 898M	Straight Beam Ultrasonic Examination of Rolled Steel Structural Shapes
A 913/A 913M	High-Strength, Low-Alloy Steel Shapes of Structural Quality, Produced by Quenching and Self-Tempering Process (QST)
A 945/A 945M	High-Strength, Low-Alloy Structural Steel Plate with Low Carbon and Restricted Sulfur for Improved Weldability, Formability, and Toughness
A 950/A 950M	Fusion Bonded Epoxy-Coated Structural Steel H-Piles and Sheet Piling

CSA Std.	Title
G40.21	Structural Quality Steels

CARBON AND ALLOY STEEL PLATE, SHEET, STRIP

ASTM Spec.	Title
A 109	Steel, Strip, Carbon, Cold-Rolled
A 109M	Steel, Strip, Carbon, Cold-Rolled [Metric]
A 366/A 366M	Commercial Steel (CS), Sheet, Carbon (0.15 Maximum Percent), Cold-Rolled
A 414/A 414M	Steel, Sheet, Carbon, for Pressure Vessels
A 424	Steel, Sheet, for Porcelain Enameling
A 505	Steel, Sheet and Strip, Alloy, Hot-Rolled and Cold-Rolled, General Requirements for
A 506	Steel, Sheet and Strip, Alloy, Hot-Rolled and Cold-Rolled, Regular Quality and Structural Quality
A 507	Steel, Sheet and Strip, Alloy, Hot-Rolled and Cold-Rolled, Drawing Quality
A 568/A 568M	Steel, Sheet, Carbon, and High-Strength, Low-Alloy, Hot-Rolled and Cold-Rolled, General Requirements for
A 569/A 569M	Commercial Steel (CS), Sheet and Strip, Carbon (0.15 Maximum Percent), Hot-Rolled
A 570/A 570M	Steel, Sheet and Strip, Carbon, Hot-Rolled, Structural Quality
A 606	Steel, Sheet and Strip, High-Strength, Low-Alloy, Hot-Rolled and Cold-Rolled, with Improved Atmospheric Corrosion Resistance
A 607	Steel, Sheet and Strip, High-Strength, Low-Alloy, Columbium or Vanadium, or Both, Hot-Rolled and Cold-Rolled

CARBON AND ALLOY STEEL PLATE, SHEET, STRIP (Continued)	
ASTM Spec.	**Title**
A 611	Structural Steel (SS), Sheet, Carbon, Cold-Rolled
A 619/A 619M	Non-Killed Forming Steel (NKFS), Sheet, Carbon, Cold-Rolled
A 620/A 620M	Drawing Steel (DS), Sheet, Carbon, Cold-Rolled
A 621/A 621M	Forming Steel (FS), Sheet and Strip, Carbon, Hot-Rolled
A 622/A 622M	Drawing Steel (DS), Sheet and Strip, Carbon, Hot-Rolled
A 635/A 635M	Steel, Sheet and Strip, Heavy Thickness Coils, Carbon, Hot-Rolled
A 659/A 659M	Commercial Steel (CS), Sheet and Strip, Carbon (0.16 Maximum Percent), Hot-Rolled
A 682	Steel, Strip, High-Carbon, Cold-Rolled, Spring Quality, General Requirements for
A 682M	Steel, Strip, High-Carbon, Cold-Rolled, Spring Quality, General Requirements for [Metric]
A 684/A 684M	Steel, Strip, High-Carbon, Cold-Rolled
A 715	Steel, Sheet and Strip, High-Strength, Low-Alloy, Hot-Rolled, and Steel Sheet, Cold-Rolled, High Strength, Low-Alloy, with Improved Formability
A 749/A 749M	Steel, Strip, Carbon and High-Strength, Low-Alloy, Hot-Rolled, General Requirements for
A 794	Commercial Steel (CS), Sheet, Carbon (0.16 % Maximum to 0.25 % Maximum), Cold-Rolled
A 812/A 812M	Steel, Sheet, High-Strength, Low-Alloy, Hot-Rolled, for Welded Layered Pressured Vessels (Discontinued 1997)
A 873/A 873M	Steel, Sheet and Strip, Chromium-Molybdenum Alloy, for Pressure Vessels (Discontinued 1997)
A 907/A 907M	Steel, Sheet and Strip, Heavy Thickness Coils, Carbon, Hot-Rolled, Structural Quality

CARBON AND ALLOY STEEL BARS

ASTM Spec.	Title
A 29/A 29M	Steel Bars, Carbon and Alloy, Hot-Wrought and Cold-Finished, General Requirements for
A 31	Steel Rivets and Bars for Rivets, Pressure Vessels
A 108	Steel Bars, Carbon, Cold-Finished, Standard Quality
A 255	End-Quench Test for Hardenability of Steel
A 304	Carbon and Alloy Steel Bars, Subject to End-Quench Hardenability Requirements
A 311/A 311M	Cold-Drawn, Stress-Relieved Carbon Steel Bars, Subject to Mechanical Property Requirements
A 321	Steel Bars, Carbon, Quenched and Tempered
A 322	Steel Bars, Alloy, Standard Grades
A 331	Steel Bars, Alloy, Cold-Finished
A 355	Steel Bars, Alloy, for Nitriding
A 400	Steel Bars, Selection Guide, Composition, and Mechanical Properties
A 434	Steel Bars, Alloy, Hot-Wrought or Cold-Finished, Quenched and Tempered
A 499	Steel Bars and Shapes, Carbon, Rolled from "T" Rails
A 575	Steel Bars, Carbon, Merchant Quality, M-Grades
A 576	Steel Bars, Carbon, Hot-Wrought, Special Quality
A 663/A 663M	Steel Bars, Carbon, Merchant Quality, Mechanical Properties
A 675/A 675 M	Steel Bars, Alloy, Hot-Wrought, Special Quality, Mechanical Properties
A 695	Steel Bars, Carbon, Hot-Wrought, Special Quality, for Fluid Power Applications
A 696	Steel Bars, Carbon, Hot-Wrought or Cold-Finished, Special Quality, for Pressure Piping Components
A 702	Steel Fence Posts and Assemblies, Hot Wrought
A 739	Steel Bars, Alloy, Hot-Wrought, for Elevated Temperature or Pressure-Containing Parts, or Both
A 914/A 914M	Steel Bars, Subject to Restricted End-Quench Hardenability Requirements
A 920/A 920M	Steel Bars, Microalloy, Hot-Wrought, Special Quality, Mechanical Properties
A 921/A 921M	Steel Bars, Microalloy, Hot-Wrought, Special Quality, for Subsequent Hot Forging

CARBON AND ALLOY STEEL WIRES

ASTM Spec.	Title
A 227/A 227M	Steel Wire, Cold-Drawn for Mechanical Springs
A 228/A 228M	Steel Wire, Music Spring Quality
A 229/A 229M	Steel Wire, Oil-Tempered for Mechanical Springs
A 230/A 230M	Steel Wire, Oil-Tempered Carbon Valve Spring Quality
A 231/A 231M	Chromium-Vanadium Alloy Steel Spring Wire
A 232/A 232M	Chromium-Vanadium Alloy Steel Valve Spring Quality Wire
A 401/A 401M	Chromium-Silicon Alloy Steel Spring Wire
A 407	Steel Wire, Cold-Drawn, for Coiled-Type Springs
A 417	Steel Wire, Cold-Drawn, for Zig-Zag, Square-Formed, and Sinuous-Type Upholstery Spring Units
A 510	Wire Rods and Coarse Round Wire, Carbon Steel, General Requirements for
A 510M	Wire Rods and Coarse Round Wire, Carbon Steel, General Requirements for [Metric]
A 679/A 679M	Steel Wire, High Tensile Strength, Hard-Drawn
A 713	Steel Wire, High-Carbon Spring, for Heat-Treated Components
A 752	Wire Rods and Coarse Round Wire, Alloy Steel, General Requirements for
A 752M	Wire Rods and Coarse Round Wire, Alloy Steel, General Requirements for [Metric]
A 764	Steel Wire, Carbon, Drawn Galvanized and Galvanized at Size for Mechanical Springs
A 805	Steel, Flat Wire, Carbon, Cold-Rolled
A 853	Steel Wire, Carbon, for General Use
A 877/A 877M	Steel Wire, Chromium-Silicon Alloy Valve Spring Quality
A 878/A 878M	Steel Wire, Modified Chromium Vanadium Valve Spring Quality
A 899	Steel Wire, Epoxy-Coated
A 905	Steel Wire, Pressure Vessel Winding
B 415	Hard-Drawn Aluminum-Clad Steel Wire
B 501	Silver-Coated Copper-Clad Steel Wire for Electronic Application

CARBON AND ALLOY STEEL RAILS AND ACCESSORIES, WHEELS, AXLES, AND TIRES

ASTM Spec.	Title
A 1	Carbon Steel Tee Rails
A 2	Carbon Steel Girder Rails of Plain, Grooved, and Guard Types
A 3	Low, Medium, and High Carbon (Non-Heat-Treated) Steel Joint Bars
A 21	Carbon Steel Axles, Non-Heat-Treated and Heat-Treated, for Railway Use
A 49	Heat-Treated Carbon Steel Joint Bars, Microalloyed Joint Bars, and Forged Carbon Steel Compromise Joint Bars
A 65	Steel Track Spikes
A 66	Steel Screw Spikes
A 67	Low-Carbon and High-Carbon Hot-Worked Steel Tie Plates
A 183	Carbon Steel Track Bolts and Nuts
A 504	Wrought Carbon Steel Wheels
A 551	Steel Tires
A 583	Cast Steel Wheels for Railway Service
A 729	Alloy Steel Axles, Heat-Treated, for Mass Transit and Electric Railway Service
A 759	Carbon Steel Crane Rails

CARBON AND ALLOY STEELS FOR CONCRETE REINFORCEMENT AND PRESTRESSED CONCRETE

ASTM Spec.	Title
A 82	Steel Wire, Plain, for Concrete Reinforcement
A 184/A 184M	Fabricated Deformed Steel Bar Mats for Concrete Reinforcement
A 185	Steel Welded Wire Fabric, Plain, for Concrete Reinforcement
A 416/A 416M	Uncoated Seven-Wire Steel Strand for Prestressed Concrete
A 421	Uncoated Stress-Relieved Steel Wire for Prestressed Concrete
A 496	Steel Wire, Deformed, for Concrete Reinforcement
A 497	Welded Deformed Steel Wire Fabric for Concrete Reinforcement
A 615/A 615M	Deformed and Plain Billet-Steel Bars for Concrete Reinforcement

CARBON AND ALLOY STEELS FOR CONCRETE REINFORCEMENT AND PRESTRESSED CONCRETE (Continued)

ASTM Spec.	Title
A 616/A 616M	Rail-Steel Deformed and Plain Bars for Concrete Reinforcement
A 617/A 617M	Axle-Steel Deformed and Plain Bars for Concrete Reinforcement
A 648	Steel Wire, Hard Drawn for Prestressing Concrete Pipe
A 704/A 704M	Welded Steel Plain Bar or Rod Mats for Concrete Reinforcement
A 706/A 706M	Low-Alloy Steel Deformed and Plain Bars for Concrete Reinforcement
A 722A 722M	Uncoated High-Strength Steel Bar for Prestressing Concrete
A 767/A 767M	Zinc-Coated (Galvanized) Steel Bars for Concrete Reinforcement
A 775/A 775M	Epoxy-Coated Reinforcing Steel Bars
A 779	Steel Strand, Seven Wire, Uncoated, Compacted, Stress-Relieved for Prestressed Concrete
A 820	Steel Fibers for Fiber-Reinforced Concrete
A 821	Steel Wire, Hard-Drawn for Prestressing Concrete Tanks
A 864/A 864M	Steel Wire, Deformed, for Prestressed Concrete Railroad Ties
A 881/A 881M	Steel Wire, Deformed, Stress-Relieved or Low-Relaxation, for Prestressed Concrete Railroad Ties
A 882/A 882M	Epoxy-Coated Seven-Wire Prestressing Steel Strand
A 884/A 884M	Epoxy-Coated Steel Wire and Welded Wire Fabric for Reinforcement
A 886/A 886M	Steel Strand, Indented, Seven-Wire Stress-Relieved for Prestressed Concrete
A 910	Uncoated, Weldless, 2- and 3-Wire Steel Strand for Prestressed Concrete
A 911/A 911M	Uncoated, Stress-Relieved Steel Bars for Prestressed Concrete Ties
A 933/A 933M	Vinyl (PVC) Coated Steel Wire and Welded Wire Fabric for Reinforcement
A 934/A 934M	Epoxy-Coated Prefabricated Steel Reinforcing Bars
A 970	Welded Headed Bars for Concrete Reinforcement

CARBON AND ALLOY STEEL CHAINS

ASTM Spec.	Title
A 391/A 391M	Grade 80 Alloy Steel Chain
A 413/A 413M	Carbon Steel Chain
A 466/A 466M	Weldless Carbon Steel Chain
A 467/A 467M	Machine and Coil Chain
A 906/A 906M	Alloy Steel Chain Slings for Overhead Lifting
A 973/A 973M	Grade 100 Alloy Steel Chain

SPRING STEEL AND STEEL SPRINGS

ASTM Spec.	Title
A 125	Steel Springs, Helical, Heat-Treated
A 689	Carbon and Alloy Steel Bars for Springs

CARBON AND ALLOY STEELS FOR BEARINGS

ASTM Spec.	Title
A 295	High Carbon Anti-Friction Bearing Steel
A 485	High Hardenability Antifriction Bearing Steel
A 534	Carburizing Steels for Anti-Friction Bearings
A 535	Special-Quality Ball and Roller Bearing Steel
A 866	Medium Carbon Anti-Friction Bearing Steel

ASTM SPECIFICATIONS THAT USE AISI DESIGNATIONS

ASTM Spec.	Title
A 29	Steel bars, carbon and alloy, hot-wrought and cold-finished, general requirements for
A 108	Steel bars, Carbon, cold-finished, standard quality
A 295	High-carbon anti-friction bearing steel
A 304	Carbon and alloy steel bars subject to end-quench hardenability requirements
A 311	Cold-drawn, stress-relieved carbon steel bars subject to mechanical property requirements
A 322	Steel bars, alloy, standard grades
A 331	Steel bars, alloy, cold-finished
A 434	Steel bars, alloy, hot-wrought or cold-finished, quenched and tempered
A 505	Steel, sheet and strip, alloy, hot-rolled and cold-rolled, general requirements for
A 506	Steel, sheet and strip, alloy, hot-rolled and cold-rolled, regular quality and structural quality
A 507	Steel, sheet and strip, alloy, hot-rolled and cold-rolled, drawing quality
A 510	General requirements for wire rods and coarse round wire, carbon steel
A 534	Carburizing steels for anti-friction bearings
A 535	Special-quality ball and roller bearing steel
A 544	Scrapless nut quality carbon steel wire (Discontinued 1991)
A 545	Cold-heading quality carbon steel wire for machine screws (Discontinued 1991)
A 546	Cold-beading quality medium high carbon steel wire for hexagon-head bolts (Discontinued 1991)
A 547	Cold-heading quality alloy steel wire for hexagon-head bolts (Discontinued 1991)
A 548	Cold-heading quality carbon steel wire for tapping or sheet metal screws (Discontinued 1991)
A 549	Cold-heading quality carbon steel wire for wood screws (Discontinued 1991)
A 575	Steel bars, carbon, merchant quality, M-grades
A 576	Steel bars, carbon, hot-wrought, special quality
A 635	Steel, Sheet and Strip, Heavy Thickness Coils, Carbon, Hot-Rolled
A 659	Commercial steel (CS), sheet and strip, carbon (0.16 maximum to 0.25 maximum percent), hot-rolled
A 682	Steel, strip, high-carbon, cold-rolled, spring quality, general requirements for
A 684	Steel, strip, high-carbon, cold-rolled
A 689	Carbon and alloy steel bars for springs

ASTM SPECIFICATIONS THAT USE AISI DESIGNATIONS (Continued)

ASTM Spec.	Title
A 713	Steel wire, high-carbon spring, for heat-treated components
A 752	General requirements for wire rods and coarse round wire, alloy steel
A 794	Commercial steel (CS), sheet, carbon (0.16 % maximum to 0.25 % maximum), cold-rolled
A 827	Plates, carbon steel, for forging and similar applications
A 829	Alloy structural steel plates
A 830	Plates, carbon steel, structural quality, furnished to chemical composition requirements

AISI & UNS IDENTIFYING ELEMENTS

AISI	UNS	Carbon Steels
10xx	G10xx0	Nonresulfurized, Manganese 1.00% max
11xx	G11xx0	Resulfurized
12xx	G12xx0	Resulfurized & rephosphorized
13xx	G13xx0	Manganese Steels
23xx	G23xx0	Nickel Steels
25xx	G25xx0	Nickel Steels
31xx	G31xx0	Nickel-Chromium Steels
32xx	G32xx0	Nickel-Chromium Steels
33xx	G33xx0	Nickel-Chromium Steels
34xx	G34xx0	Nickel-Chromium Steels
40xx	G40xx0	Molybdenum Steels
41xx	G41xx0	Chromium-Molybdenum Steels
43xx	G43xx0	Nickel-Chromium-Molybdenum Steels
44xx	G44xx0	Molybdennum Steels
46xx	G46xx0	Nickel-Molybdenum Steels
47xx	G47xx0	Nickel-Chromium-Molybdemun Steels
48xx	G48xx0	Nickel-Molybdenum Steels

AISI & UNS IDENTIFYING ELEMENTS (Continued)

AISI	UNS	Carbon Steels
50xx	G50xx0	Chromium Steels
51xx	G51xx0	Chromium Steels
50xxx	G50xx6	Chromium Steels
51xxx	G51xx6	Chromium Steels
52xxx	G52xx6	Chromium Steels
61xx	G61xx0	Chromium-Vanadium Steels
71xxx	G71xx0	Tungsten-Chromium Steels
72xx	G72xx0	Tungsten-Chromium Steels
81xx	G81xx0	Nickel-Chromium-Molybdenum Steels
86xx	G86xx0	Nickel-Chromium-Molybdenum Steels
87xx	G87xx0	Nickel-Chromium-Molybdenum Steels
88xx	G88xx0	Nickel-Chromium-Motybdenum Steels
92xx	G92xx0	Silicon-Manganese Steels
93xx	G93xx0	Nickel-Chromium-Molybdenum Steels
94xx	G94xx0	Nickel-Chromium-Molybdenum Steels
97xx	G97xx0	Nickel-Chromium-Molybdenum Steels
98xx	G98xx	Nickel-Chromium-Molybdenum Steels
AISI	**UNS**	**Carbon & Alloy Steels**
xxBxx	Gxxxx1	B denotes Boron Steels[a]
xxLxx	Gxxxx4	L denotes Leaded Steels[b]
AISI	**UNS**	**Stainless Steels**
2xx	S2xxxx	Chromium-Nickel Steels
3xx	S3xxxx	Chromium-Nickel Steels
4xx	S4xxxx	Chromium Steels
5xx	S5xxxx	Chromium (Heat Resisting) Steels

a Boron added to a range of 0.0005-0.003% B. b Lead added to a range of 0.15-0.35 Pb.

CHEMICAL COMPOSITION OF CARBON STEELS[a, b]

AISI	UNS No.	C	Mn	P	S
1005	G10050	0.06	0.35	0.040	0.050
1006	G10060	0.08	0.25-0.40	0.040	0.050
1008	G10080	0.10	0.30-0.50	0.040	0.050
1010	G10100	0.08-0.13	0.30-0.60	0.040	0.050
1011	G10110	0.08-0.13	0.60-090	0.040	0.050
1012	G10120	0.10-0.15	0.30-0.60	0.040	0.050
1013	G10130	0.11-0.16	0.50-0.80	0.040	0.050
1015	G10150	0.13-0.18	0.30-0.60	0.040	0.050
1016	G10160	0.13-0.18	0.60-0.90	0.040	0.050
1017	G10170	0.15-0.20	0.30-0.60	0.040	0.050
1018	G10180	0.15-0.20	0.60-0.90	0.040	0.050
1019	G10190	0.15-0.20	0.70-1.00	0.040	0.050
1020	G10200	0.18-0.23	0.30-0.60	0.040	0.050
1021	G10210	0.18-0.23	0.60-0.90	0.040	0.050
1022	G10220	0.18-0.23	0.70-1.00	0.040	0.050
1023	G10230	0.20-.0.25	0.30-0.60	0.040	0.050
1025	G10250	0.22-0.28	0.30-0.60	0.040	0.050
1026	G10260	0.22-0.28	0.60-0.90	0.040	0.050
1029	G10290	0.25-0.31	0.60-0.90	0.040	0.050
1030	G10300	0.28-0.34	0.60-0.90	0.040	0.050
1033	G10330	0.29-0.36	0.70-1.00	0.040	0.050
1035	G10350	0.32-0.38	0.60-0.90	0.040	0.050
1037	G10370	0.32-0.38	0.70-1.00	0.040	0.050
1038	G10380	0.35-0.42	0.60-0.90	0.040	0.050
1039	G10390	0.37-0.44	0.70-1.00	0.040	0.050
1040	G10400	0.37-0.44	0.60-0.90	0.040	0.050

CHEMICAL COMPOSITION OF CARBON STEELS[a, b] (Continued)

AISI	UNS No.	C	Mn	P	S
1042	G10420	0.40-0.47	0.60-0.90	0.040	0.050
1043	G10430	0.40-0.47	0.70-1.00	0.040	0.050
1044	G10440	0.43-0.50	0.30-0.60	0.040	0.050
1045	G10450	0.43-0.50	0.60-0.90	0.040	0.050
1046	G10460	0.43-0.50	0.70-1.00	0.040	0.050
1049	G10490	0.46-0.53	0.60-0.90	0.040	0.050
1050	G10500	0.48-0.55	0.60-0.90	0.040	0.050
1053	G10530	0.48-0.55	0.70-1.00	0.040	0.050
1055	G10550	0.50-0.60	0.60-0.90	0.040	0.050
1059	G10590	0.55-0.65	0.50-0.80	0.040	0.050
1060	G10600	0.55-0.65	0.60-0.90	0.040	0.050
1064	G10640	0.60-0.70	0.50-0.80	0.040	0.050
1065	G10650	0.60-0.70	0.60-0.90	0.040	0.050
1069	G10690	0.65-0.75	0.40-0.70	0.040	0.050
1070	G10700	0.65-0.75	0.60-0.90	0.040	0.050
1074	G10740	0.70-0.80	0.50-0.80	0.040	0.050
1075	G10750	0.70-0.80	0.40-0.70	0.040	0.050
1078	G10780	0.72-0.85	0.30-0.60	0.040	0.050
1080	G10800	0.75-0.88	0.60-0.90	0.040	0.050
1084	G10840	0.80-0.93	0.60-0.90	0.040	0.050
1085	G10850	0.80-0.93	0.70-1.00	0.040	0.050
1086	G10860	0.80-0.93	0.30-0.50	0.040	0.050
1090	G10900	0.85-0.98	0.60-0.90	0.040	0.050
1095	G10950	0.90-1.03	0.30-0.50	0.040	0.050

CHEMICAL COMPOSITION OF CARBON STEELS[a, b] (Continued)

a. Standard carbon steels can be produced with a lead range of 0.15-0.35%. Such steels are identified by inserting the letter "L" between the second and third numerals of the grade designation, for example, 10L40. A cast or heat analysis is not determinable when lead is added to the ladle stream. b. When boron treatment for killed steels is specified, the steels can be expected to contain 0.0005-0.003% boron. If the usual titanium additive is not permitted, the steels can be expected to contain up to 0.005% boron. Single values are maximums, unless otherwise specified.

CHEMICAL COMPOSITION OF RESULFURIZED CARBON STEELS[a]

AISI	UNS	C	Mn	P	S
1108	G11080	0.08-0.13	0.50-0.80	0.040	0.08-0.13
1110	G11100	0.08-0.13	0.30-0.60	0.040	0.08-0.13
1116	G11160	0.14-0.20	1.10-1.40	0.040	0.16-0.23
1117	G11170	0.14-0.20	1.00-1.30	0.040	0.08-0.13
1118	G11180	0.14-0.20	1.30-1.60	0.040	0.08-0.13
1119	G11190	0.14-0.20	1.00-1.30	0.040	0.24-0.33
1123	G11230	0.20-0.27	1.20-1.50	0.040	0.06-0.09
1132	G11320	0.27-0.34	1.35-1.65	0.040	0.09-0.13
1137	G11370	0.32-0.39	1.35-1.65	0.040	0.08-0.13
1139	G11390	0.35-0.43	1.35-1.65	0.040	0.13-0.20
1140	G11400	0.37-0.44	0.70-1.00	0.040	0.08-0.13
1141	G11410	0.37-0.45	1.35-1.65	0.040	0.08-0.13
1144	G11440	0.40-0.48	1.35-1.65	0.040	0.24-0.33
1145	G11450	0.41-0.49	0.70-1.00	0.040	0.08-0.13
1146	G11460	0.42-0.49	0.70-1.00	0.040	0.08-0.13
1151	G11510	0.48-0.55	0.70-1.00	0.040	0.08-0.13

a. Standard carbon steels can be produced with a lead range of 0.15-0.35%. Such steels are identified by inserting the letter "L" between the second and third numerals of the grade designation, for example, 10L40. A cast or heat analysis is not determinable when lead is added to the ladle stream. Single values are maximums, unless otherwise specified.

CHEMICAL COMPOSITION OF REPHOSPHORIZED & RESULFURIZED CARBON STEELS[a]

AISI	UNS	C	Mn	P	S	Pb
1211	G12110	0.13	0.60-0.90	0.07-0.12	0.10-0.15	---
1212	G12120	0.13	0.70-1.00	0.07-0.12	0.16-0.23	---
1213	G12130	0.13	0.70-1.00	0.07-0.12	0.24-0.33	---
12L13	G12134	0.13	0.70-1.00	0.07-0.12	0.24-.33	0.15-0.35
12L14[a]	G12144	0.15	0.85-1.15	0.04-0.09	0.26-0.35	0.15-0.35
1215	G12150	0.09	0.75-1.05	0.04-0.09	0.26-0.35	---

a. Standard carbon steels can be produced with a lead range of 0.15-0.35%. Such steels are identified by inserting the letter "L" between the second and third numerals of the grade designation, for example, 10L40. A cast or heat analysis is not determinable when lead is added to the ladle stream. Single values are maximums, unless otherwise specified.

CHEMICAL COMPOSITION OF HIGH MANGANESE CARBON STEELS[a]

AISI	UNS	C	Mn	P max	S max
1513	G15130	0.10-0.16	1.10-1.40	0.040	0.050
1518	G15180	0.15-0.21	1.10-1.40	0.040	0.050
1522	G15220	0.18-0.24	1.10-1.40	0.040	0.050
1524	G15240	0.19-0.25	1.35-1.65	0.040	0.050
1525	G15250	0.23-0.29	0.80-1.10	0.040	0.050
1526	G15260	0.22-0.29	1.10-1.40	0.040	0.050
1527	G15270	0.22-0.29	1.20-1.50	0.040	0.050
1533	G15330	0.30-0.37	1.20-1.50	0.040	0.050
1534	G15340	0.30-0.37	1.20-1.50	0.040	0.050
1536	G15360	0.30-0.37	1.20-1.50	0.040	0.050
1541	G15410	0.36-0.44	1.35-1.65	0.040	0.050
1548	G15480	0.44-0.52	1.10-1.40	0.040	0.050
1551	G15510	0.45-0.56	0.85-1.15	0.040	0.050

CHEMICAL COMPOSITION OF HIGH MANGANESE CARBON STEELS[a] (Continued)

AISI	UNS	C	Mn	P max	S max
1552	G15520	0.47-0.55	1.20-1.50	0.040	0.050
1561	G15610	0.55-0.65	0.75-1.05	0.040	0.050
1566	G15660	0.60-0.71	0.85-1.15	0.040	0.050

a. Standard carbon steels can be produced with a lead range of 0.15-0.35%. Such steels are identified by inserting the letter "L" between the second and third numerals of the grade designation, for example, 10L40. A cast or heat analysis is not determinable when lead is added to the laddle stream.

CHEMICAL COMPOSITION OF CARBON H-STEELS[a]

AISI	UNS No.	C	Mn	P	S	Si
1038 H	H 10380	0.34-0.43	0.50-1.00	0.040	0.050	0.15-0.35
1045 H	H 10450	0.42-0.51	0.50-1.00	0.040	0.050	0.15-0.35
1522 H	H 15220	0.17-0.25	1.00-1.50	0.040	0.050	0.15-0.35
1524 H	H 15240	0.18-0.26	1.25-1.75	0.040	0.050	0.15-0.35
1526 H	H 15260	0.21-0.30	1.00-1.50	0.040	0.050	0.15-0.35
1541 H	H 15410	0.35-0.45	1.25-1.75	0.040	0.050	0.15-0.35
15B21 H[b]	H 15211[b]	0.17-0.24	0.70-1.20	0.040	0.050	0.15-0.35
15B35 H[b]	H 15351[b]	0.31-0.39	0.70-1.20	0.040	0.050	0.15-0.35
15B37 H[b]	H 15371[b]	0.30-0.39	1.00-1.50	0.040	0.050	0.15-0.35
15B41 H[b]	H 15411[b]	0.35-0.45	1.25-1.75	0.040	0.050	0.15-0.35
15B48 H[b]	H 15481[b]	0.43-0.53	1.00-1.50	0.040	0.050	0.15-0.35
15B62 H[b]	H 15621[b]	0.54-0.67	1.00-1.50	0.040	0.050	0.40-0.60

a. Standard H Steels can be produced with a lead range of 0.15-0.35%. Such steels are identified by inserting the letter "L" between the second and third numerals of the grade designation, for example, 15L22 H. Lead is generally reported as a range of 0.15-0.35%.
b. These steels can be expected to have 0.0005% minimum boron content.
Single values are maximums, unless otherwise specified.

CHEMICAL COMPOSITON OF ALLOY STEELS

AISI	UNS	C	Mn	Si	S	P	Ni	Cr	Mo
1330	G13300	0.28-0.33	1.60-1.90	0.15-0.35	0.040	0.035	---	---	---
1335	G13350	0.33-0.38	1.60-1.90	0.15-0.35	0.040	0.035	---	---	---
1340	G13400	0.38-0.43	1.60-1.90	0.15-0.35	0.040	0.035	---	---	---
1345	G13450	0.43-0.48	1.60-1.90	0.15-0.35	0.040	0.035	---	---	---
3140	G31400	0.38-0.43	0.70-0.90	0.15-0.35	0.040	0.040	1.10-1.40	0.55-0.75	---
4023	G40230	0.20-0.25	0.70-0.90	0.15-0.35	0.040	0.035	---	---	0.20-0.30
4024	G40240	0.20-0.25	0.70-0.90	0.15-0.35	0.035-0.050	0.035	---	---	0.20-0.30
4027	G40270	0.25-0.30	0.70-0.90	0.15-0.35	0.040	0.035	---	---	0.20-0.30
4028	G40280	0.25-0.30	0.70-0.90	0.15-0.35	0.035-0.050	0.035	---	---	0.20-0.30
4032	G40320	0.30-0.35	0.70-0.90	0.15-0.35	0.040	0.035	---	---	0.20-0.30
4037	G40370	0.35-0.40	0.70-0.90	0.15-0.35	0.040	0.035	---	---	0.20-0.30
4042	G40420	0.40-0.45	0.70-0.90	0.15-0.30	0.040	0.035	---	---	0.20-0.30
4047	G40470	0.45-0.50	0.70-0.90	0.15-0.35	0.040	0.035	---	---	0.20-0.30
4118	G41180	0.18-0.23	0.70-0.90	0.15-0.35	0.040	0.035	---	0.40-0.60	0.08-0.15
4130	G41300	0.28-0.33	0.40-0.60	0.15-0.35	0.040	0.035	---	0.80-1.10	0.15-0.25
4135	G41350	0.33-0.38	0.70-0.90	0.15-0.35	0.040	0.035	---	0.80-1.10	0.15-0.25
4137	G41370	0.35-0.40	0.70-0.90	0.15-0.35	0.040	0.035	---	0.80-1.10	0.15-0.25
4140	G41400	0.38-0.43	0.75-1.00	0.15-0.35	0.040	0.035	---	0.80-1.10	0.15-0.25
4142	G41420	0.40-0.45	0.75-1.00	0.15-0.35	0.040	0.035	---	0.80-1.10	0.15-0.25
4145	G41450	0.43-0.48	0.75-1.00	0.15-0.35	0.040	0.035	---	0.80-1.10	0.15-0.25
4147	G41470	0.45-0.50	0.75-1.00	0.15-0.35	0.040	0.035	---	0.80-1.10	0.15-0.25
4150	G41500	0.48-0.53	0.75-1.00	0.15-0.35	0.040	0.035	---	0.80-1.10	0.15-0.25
4161	G41610	0.56-0.64	0.75-1.00	0.15-0.35	0.040	0.035	---	0.70-0.90	0.25-0.35
4320	G43200	0.17-0.22	0.45-0.65	015-0.35	0.040	0.035	1.65-2.00	0.40-0.60	0.20-0.30
4340	G43400	0.38-0.43	0.60-0.80	0.15-0.30	0.040	0.035	1.65-2.00	0.70-0.90	0.20-0.30
E4340	G43406	0.38-0.43	0.65-0.85	0.15-0.35	0.025	0.025	1.65-2.00	0.70-0.90	0.20-0.30
4422	G44220	0.20-0.25	0.70-0.90	0.15-0.35	0.040	0.035	---	---	0.35-0.45

CHEMICAL COMPOSITON OF ALLOY STEELS (Continued)

AISI	UNS	C	Mn	Si	S	P	Ni	Cr	Mo
4427	G44270	0.24-0.29	0.70-0.90	0.15-0.35	0.040	0.035	---	---	0.35-0.45
4615	G46150	0.13-0.18	0.45-0.65	0.15-0.35	0.040	0.035	1.65-2.00	---	0.20-0.30
4617	G46170	0.15-0.20	0.45-0.65	0.15-0.35	0.040	0.035	1.65-2.00	---	0.20-0.30
4620	G46200	0.17-0.22	0.45-0.65	0.15-0.35	0.040	0.035	1.65-2.00	---	0.20-0.30
4626	G46260	0.24-0.29	0.45-0.65	0.15-0.35	0.040	0.035	0.70-1.00	---	0.15-0.25
4718	G47180	0.16-0.21	0.70-0.90	0.15-0.35	0.040	0.035	0.90-1.20	0.35-0.55	0.30-0.40
4720	G47200	0.17-0.22	0.50-0.70	0.15-0.35	0.040	0.035	0.90-1.20	0.35-0.55	0.15-0.25
4815	G48150	0.13-0.18	0.40-0.60	0.15-0.35	0.040	0.035	3.25-3.75	---	0.20-0.30
4817	G48170	0.15-0.20	0.40-0.60	0.15-0.35	0.040	0.035	3.25-3.75	---	0.20-0.30
4820	G48200	0.18-0.23	0.50-0.70	0.15-0.35	0.040	0.035	3.25-3.75	---	0.20-0.30
50B40[a]	G50401	0.38-0.42	0.75-1.00	0.15-0.35	0.040	0.035	---	0.40-0.60	---
50B44[a]	G50441	0.43-0.48	0.75-1.00	0.15-0.35	0.040	0.035	---	0.40-0.60	---
5046	G50460	0.43-0.50	0.75-1.00	0.15-0.35	0.040	0.035	---	0.20-0.35	---
50B46[a]	G50461	0.44-0.49	0.75-1.00	0.15-0.35	0.040	0.035	---	0.20-0.35	---
50B50[a]	G50501	0.48-0.53	0.75-1.00	0.15-0.35	0.040	0.035	---	0.40-0.60	---
5060	G50600	0.56-0.64	0.75-1.00	0.15-0.35	0.040	0.035	---	0.40-0.60	---
50B60[a]	G50601	0.56-0.64	0.75-1.00	0.15-0.35	0.040	0.035	---	0.40-0.60	---
5115	G51150	0.13-0.18	0.70-0.90	0.15-0.35	0.040	0.035	---	0.70-0.90	---
5117	G51170	0.15-0.20	0.70-0.90	0.15-0.35	0.040	0.035	---	0.70-0.90	---
---	G51190	0.16-0.21	1.00-1.30	0.15 max	0.040	0.035	---	0.70-0.90	---
5120	G51200	0.17-0.22	0.70-0.90	0.15-0.35	0.040	0.035	---	0.70-0.90	---
---	G51210	0.18-0.23	1.00-1.30	0.15-0.35	0.040	0.035	---	0.70-0.90	---
---	G51240	0.21-0.26	1.00-1.30	0.15-0.35	0.040	0.035	---	0.70-0.90	---
5130	G51300	0.28-0.33	0.70-0.90	0.15-0.35	0.040	0.035	---	0.80-1.10	---
5132	G51320	0.30-0.35	0.60-0.80	0.15-0.35	0.040	0.035	---	0.75-1.00	---
5135	G51350	0.33-0.38	0.60-0.80	0.15-0.35	0.040	0.035	---	0.80-1.05	---

CHEMICAL COMPOSITON OF ALLOY STEELS (Continued)

AISI	UNS	C	Mn	Si	S	P	Ni	Cr	Mo
5140	G51400	0.38-0.43	0.70-0.90	0.15-0.35	0.040	0.035	---	0.70-0.90	---
5147	G51470	0.46-0.51	0.70-0.95	0.15-0.35	0.040	0.035	---	0.85-1.15	---
5150	G51500	0.48-0.53	0.70-0.90	0.15-0.35	0.040	0.035	---	0.70-0.90	---
5155	G51550	0.51-0.59	0.70-0.90	0.15-0.35	0.040	0.035	---	0.70-0.90	---
5160	G51600	0.56-0.64	0.75-1.00	0.15-0.35	0.040	0.035	---	0.70-0.90	---
51B60[b]	G51601	0.56-0.64	0.75-1.00	0.15-0.35	0.040	0.035	---	0.70-0.90	---
E50100	G50986	0.95-1.10	0.25-0.45	0.15-0.35	0.025	0.025	---	0.40-0.60	---
E51100	G51986	0.98-1.10	0.25-0.45	0.15-0.35	0.025	0.025	---	0.90-1.15	---
E52100	G52986	0.98-1.10	0.25-0.45	0.15-0.35	0.025	0.025	---	1.30-1.60	---
6118[c]	G61180	0.16-0.21	0.50-0.70	0.15-0.35	0.040	0.035	---	0.50-0.70	---
6150[d]	G61500	0.48-0.53	0.70-0.90	0.15-0.35	0.040	0.035	---	0.80-1.10	---
8115	G81150	0.13-0.18	0.70-0.90	0.15-0.35	0.040	0.035	0.20-0.40	0.30-0.50	0.08-0.15
81B45[b]	G81451	0.43-0.48	0.75-1.00	0.15-0.35	0.040	0.035	0.20-0.40	0.35-0.55	0.08-0.15
8615	G86150	0.13-0.18	0.70-0.90	0.15-0.35	0.040	0.035	0.40-0.70	0.40-0.60	0.15-0.25
8617	G86170	0.15-0.20	0.70-0.90	0.15-0.35	0.040	0.035	0.40-0.70	0.40-0.60	0.15-0.25
8620	G86200	0.18-0.23	0.70-0.90	0.15-0.35	0.040	0.035	0.40-0.70	0.40-0.60	0.15-0.25
8622	G86220	0.20-0.25	0.70-0.90	0.15-0.35	0.040	0.035	0.40-0.70	0.40-0.60	0.15-0.25
8625	G86250	0.23-0.28	0.70-0.90	0.15-0.35	0.040	0.035	0.40-0.70	0.40-0.60	0.15-0.25
8627	G86270	0.25-0.30	0.70-0.90	0.15-0.35	0.040	0.035	0.40-0.70	0.40-0.60	0.15-0.25
8630	G86300	0.28-0.33	0.70-0.90	0.15-0.35	0.040	0.035	0.40-0.70	0.40-0.60	0.15-0.25
8637	G86370	0.35-0.40	0.75-1.00	0.15-0.35	0.040	0.035	0.40-0.70	0.40-0.60	0.15-0.25
8640	G86400	0.38-0.43	0.75-1.00	0.15-0.35	0.040	0.035	0.40-0.70	0.40-0.60	0.15-0.25
8642	G86420	0.40-0.45	0.75-1.00	0.15-0.35	0.040	0.035	0.40-0.70	0.40-0.60	0.15-0.25
8645	G86450	0.43-0.48	0.75-1.00	0.15-0.35	0.040	0.035	0.40-0.70	0.40-0.60	0.15-0.25
86B45[a]	G86451	0.43-0.48	0.75-1.00	0.15-0.35	0.040	0.035	0.40-0.70	0.40-0.60	0.15-0.25
8650	G86500	0.48-0.53	0.75-1.00	0.15-0.35	0.040	0.035	0.40-0.70	0.40-0.60	0.15-0.25

CHEMICAL COMPOSITON OF ALLOY STEELS (Continued)

AISI	UNS	C	Mn	Si	S	P	Ni	Cr	Mo
8655	G86550	0.51-0.59	0.75-1.00	0.15-0.35	0.040	0.035	0.40-0.70	0.40-0.60	0.15-0.25
8660	G86600	0.55-0.65	0.75-1.00	0.15-0.35	0.040	0.035	0.40-0.70	0.40-0.60	0.15-0.25
8720	G87200	0.18-0.23	0.70-0.90	0.15-0.35	0.040	0.035	0.40-0.70	0.40-0.60	0.20-0.30
8740	G87400	0.38-0.43	0.75-1.00	0.15-0.35	0.040	0.035	0.40-0.70	0.40-0.60	0.20-0.30
8822	G88220	0.20-0.25	0.75-1.00	0.15-0.35	0.040	0.035	0.40-0.70	0.40-0.60	0.30-0.40
9254	G92540	0.51-0.59	0.60-0.80	1.20-1.60	0.040	0.035	---	0.60-0.80	---
9260	G92600	0.56-0.64	0.75-1.00	1.80-2.20	0.040	0.035	---	---	---
E9310	G93106	0.08-0.13	0.45-0.65	0.15-0.35	0.025	0.025	3.00-3.50	1.00-1.40	0.08-0.15
94B15[b]	G94151	0.13-0.18	0.75-1.00	0.15-0.35	0.040	0.035	0.30-0.60	0.30-0.50	0.08-0.15
94B17[a]	G94171	0.15-0.20	0.75-1.00	0.15-0.35	0.040	0.035	0.30-0.60	0.30-0.50	0.08-0.15
94B30[a]	G94301	0.28-0.33	0.75-1.00	0.15-0.35	0.040	0.035	0.30-0.60	0.30-0.50	0.08-0.15

a. Contains boron 0.0005-0.003%.
b. Contains boron 0.0005 minimum.
c. Contains 0.10-0.15 vanadium.
d. Contains 0.15 minimum vanadium.
Single values are maximums, unless otherwise specified.

CHEMICAL COMPOSITION OF ALLOY H-STEELSᵇ

AISI	UNS No.ª	C	Mn	Si	Ni	Cr	Mo
1330 H	H 13300	0.27-0.33	1.45-2.05	0.15-0.35	---	---	---
1335 H	H 13350	0.32-0.38	1.45-2.05	0.15-0.35	---	---	---
1340 H	H 13400	0.37-0.44	1.45-2.05	0.15-0.35	---	---	---
1345 H	H 13450	0.42-0.49	1.45-2.05	0.15-0.35	---	---	---
4027 H	H 40270	0.24-0.30	0.60-1.00	0.15-0.35	---	---	0.20-0.30
4028 H	H 40280	0.24-0.30	0.60-1.00	0.15-0.35	---	---	0.20-0.30
4032 H	H 40320	0.29-0.35	0.60-1.00	0.15-0.35	---	---	0.20-0.30
4037 H	H 40370	0.34-0.41	0.60-1.00	0.15-0.35	---	---	0.20-0.30
4042 H	H 40420	0.39-0.46	0.60-1.00	0.15-0.35	---	---	0.20-0.30
4047 H	H 40470	0.44-0.51	0.60-1.00	0.15-0.35	---	---	0.20-0.30
4118 H	H 41180	0.17-0.23	0.60-1.00	0.15-0.35	---	0.30-0.70	0.08-0.15
4130 H	H 41300	0.27-0.33	0.30-0.70	0.15-0.35	---	0.75-1.20	0.15-0.25
4135 H	H 41350	0.32-0.38	0.60-1.00	0.15-0.35	---	0.75-1.20	0.15-0.25
4137 H	H 41370	0.34-0.41	0.60-1.00	0.15-0.35	---	0.75-1.20	0.15-0.25
4140 H	H 41400	0.37-0.44	0.65-1.10	0.15-0.35	---	0.75-1.20	0.15-0.25
4142 H	H 41420	0.39-0.46	0.65-1.10	0.15-0.35	---	0.75-1.20	0.15-0.25
4145 H	H 41450	0.42-0.49	0.65-1.10	0.15-0.35	---	0.75-1.20	0.15-0.25
4147 H	H 41470	0.44-0.51	0.65-1.10	0.15-0.35	---	0.75-1.20	0.15-0.25
4150 H	H 41500	0.47-0.54	0.65-1.10	0.15-0.35	---	0.75-1.20	0.15-0.25
4161 H	H 41600	0.55-0.65	0.65-1.10	0.15-0.35	---	0.65-0.95	0.25-0.35
4320 H	H 43200	0.17-0.23	0.40-0.70	0.15-0.35	1.55-2.00	0.35-0.65	0.20-0.30
4340 H	H 43400	0.37-0.44	0.55-0.90	0.15-0.35	1.55-2.00	0.65-0.95	0.20-0.30
E4340 Hᶜ	H 43406	0.37-0.44	0.60-0.95	0.15-0.35	1.55-2.00	0.65-0.95	0.20-0.30
4419 H	H 44190	0.17-0.23	0.35-0.75	0.15-0.35	---	---	0.45-0.60
4620 H	H 46290	0.17-0.23	0.35-0.75	0.15-0.35	1.55-2.00	---	0.20-0.30
4621 H	H 46210	0.17-0.23	0.60-1.00	0.15-0.35	1.55-2.00	---	0.20-0.30

| CHEMICAL COMPOSITION OF ALLOY H-STEELS[b](Continued) | | | | | | | |
AISI	UNS No.[a]	C	Mn	Si	Ni	Cr	Mo
4626	H 46260	0.23-0.29	0.40-0.70	0.15-0.35	0.65-1.05	---	0.15-0.25
4718 H	H 47180	0.15-0.21	0.60-0.95	0.15-0.35	0.85-1.25	0.30-0.60	0.30-0.40
4720 H	H 47200	0.17-0.23	0.45-0.75	0.15-0.35	0.85-1.25	0.30-0.60	0.15-0.25
4815 H	H 48150	0.12-0.18	0.30-0.70	0.15-0.35	3.20-3.80	---	0.20-0.30
4817 H	H 48170	0.14-0.20	0.30-0.70	0.15-0.35	3.20-3.80	---	0.20-0.30
4820 H	H 48200	0.17-0.23	0.40-0.80	0.15-0.35	3.20-3.80	---	0.20-0.30
50B40 H[d]	H 50401	0.37-0.44	0.65-1.10	0.15-0.35	---	0.30-0.70	---
50B44 H[d]	H 50441	0.42-0.49	0.65-1.10	0.15-0.35	---	0.30-0.70	---
5046 H	H 50460	0.43-0.50	0.65-1.10	0.15-0.35	---	0.13-0.43	---
50B46 H[d]	H 50461	0.43-0.50	0.65-1.10	0.15-0.35	---	0.13-0.43	---
50B50 H[d]	H 50501	0.47-0.54	0.65-1.10	0.15-0.35	---	0.30-0.70	---
50B60 H[d]	H 50601	0.55-0.65	0.65-1.10	0.15-0.35	---	0.30-0.70	---
5120 H	H 51200	0.17-0.23	0.60-1.00	0.15-0.35	---	0.60-1.00	---
5130 H	H 51300	0.27-0.33	0.60-1.10	0.15-0.35	---	0.75-1.20	---
5132 H	H 51320	0.29-0.35	0.50-0.90	0.15-0.35	---	0.65-1.10	---
5135 H	H 51350	0.32-0.38	0.50-0.90	0.15-0.35	---	0.70-1.15	---
5140 H	H 51400	0.37-0.44	0.60-1.00	0.15-0.35	---	0.60-1.00	---
5145 H	H 51450	0.42-0.49	0.60-1.00	0.15-0.35	---	0.60-1.00	---
5147 H	H 51470	0.45-0.52	0.60-1.05	0.15-0.35	---	0.80-1.25	---
5150 H	H 51500	0.47-0.54	0.60-1.00	0.15-0.35	---	0.60-1.00	---
5155 H	H 51550	0.50-0.60	0.60-1.00	0.15-0.35	---	0.60-1.00	---
5160 H	H 51600	0.55-0.65	0.65-1.10	0.15-0.35	---	0.60-1.00	---
51B60 H[d]	H 51601	0.55-0.65	0.65-1.10	0.15-0.35	---	0.60-1.00	---
6118 H[e]	H 61180	0.15-0.21	0.40-0.80	0.15-0.35	---	0.40-0.80	---
6150 H[f]	H 61500	0.47-0.54	0.60-1.00	0.15-0.35	---	0.75-1.20	---

CHEMICAL COMPOSITION OF ALLOY H-STEELS[b](Continued)

AISI	UNS No.[a]	C	Mn	Si	Ni	Cr	Mo
81B45 H[d]	H 81451	0.42-0.49	0.70-1.05	0.15-0.35	0.15-0.45	0.30-0.60	0.08-0.15
8617 H	H 86170	0.14-0.20	0.60-0.95	0.15-0.30	0.35-0.75	0.35-0.65	0.15-0.25
8620 H	H 86200	0.17-0.23	0.60-0.95	0.15-0.35	0.35-0.75	0.35-0.65	0.15-0.25
8622 H	H 86220	0.19-0.25	0.60-0.95	0.15-0.35	0.35-0.75	0.35-0.65	0.15-0.25
8625 H	H 86250	0.22-0.28	0.60-0.95	0.15-0.35	0.35-0.75	0.35-0.65	0.15-0.25
8627 H	H 86270	0.24-0.30	0.60-0.95	0.15-0.35	0.35-0.75	0.35-0.65	0.15-0.25
8630 H	H 86300	0.27-0.33	0.60-0.95	0.15-0.35	0.35-0.75	0.35-0.65	0.15-0.25
86B30 H[d]	H 86301	0.27-0.33	0.60-0.95	0.15-0.35	0.35-0.75	0.35-0.65	0.15-0.25
8637 H	H 86370	0.34-0.41	0.70-1.05	0.15-0.35	0.35-0.75	0.35-0.65	0.15-0.25
8640 H	H 86400	0.37-0.44	0.70-1.05	0.15-0.35	0.35-0.75	0.35-0.65	0.15-0.25
8642 H	H 86420	0.39-0.46	0.70-1.05	0.15-0.35	0.35-0.75	0.35-0.65	0.15-0.25
8645 H	H 86450	0.42-0.49	0.70-1.05	0.15-0.35	0.35-0.75	0.35-0.65	0.15-0.25
86B45 H[d]	H 86451	0.42-0.49	0.70-1.05	0.15-0.35	0.35-0.75	0.35-0.65	0.15-0.25
8650 H	H 86500	0.47-0.54	0.70-1.05	0.15-0.35	0.35-0.75	0.35-0.65	0.15-0.25
8655 H	H 86550	0.50-0.60	0.70-1.05	0.15-0.35	0.35-0.75	0.35-0.65	0.15-0.25
8660 H	H 86600	0.55-0.65	0.70-1.05	0.15-0.35	0.35-0.75	0.35-0.65	0.15-0.25
8720 H	H 87200	0.17-0.23	0.60-0.95	0.15-0.35	0.35-0.75	0.35-0.65	0.20-0.30
8740 H	H 87400	0.37-0.44	0.70-1.05	0.15-0.35	0.35-0.75	0.35-0.65	0.20-0.30
8822 H	H 88220	0.19-0.25	0.70-1.05	0.15-0.35	0.35-0.75	0.35-0.65	0.30-0.40
9260 H	H 92600	0.55-0.65	0.65-1.10	1.70-2.20	---	---	---
9310 H[c]	H 93100	0.07-0.13	0.40-0.70	0.15-0.35	2.95-3.55	1.00-1.45	0.08-0.15
94B15 H[d]	H 94151	0.12-0.18	0.70-1.05	0.15-0.35	0.25-0.65	0.25-0.55	0.08-0.15
94B17 H[d]	H 94171	0.14-0.20	0.70-1.05	0.15-0.35	0.25-0.65	0.25-0.55	0.08-0.15
94B30 H[d]	H 94301	0.27-0.33	0.70-1.05	0.15-0.35	0.25-0.65	0.25-0.55	0.08-0.15

CHEMICAL COMPOSITION OF ALLOY H-STEELS[b](Continued)

a. Designations established in accordance with Practice E 527 and SAE J1086. Recommended Practice for Numbering Metals and Alloys (UNS).

b. Sulfur content range is 0.035 to 0:050%.

c. Electric furnace steel.

d. These steels can be expected to have a boron content 0.0005-0.003%.

e. Vanadium content range is 0.10 to 0.15%.

f. Minimum vanadium content is 0.15%.

Notes:

Prosphorus and sulfur in open-hearth steel is 0.035%, and 0.040%, maximum respectively. Phosphorus and sulfur in electric-furnace steel (designated by the prefix letter "E") is 0.025%, maximum.

Small quantities of certain elements are present in alloy steels which are not specified or required. These elements are considered as incidental and may be present to the following maximum amounts: 0.35% copper, 0.25% nickel, 0.20% chromium; 0.06%molybdenum.

Chemical ranges and limits shown in this table are subject to the permissible variation for product analysis shown in ASTM A 29/A 29M.

Standard "H" Steels can be produced with a lead range of 0.15-0.35%. Such steels are identified by inserting the letter "L" between the second and third numerals of the grade designation, for example, 41L40H. Lead is generally reported as a range of 0.15-0.35%.

Single values are maximums, unless otherwise specified.

CHEMICAL COMPOSITON OF CSA G40.21 STRUCTURAL QUALITY STEELS - PLATES, FLOOR PLATES, BARS, SHAPES AND SHEET PILLING

Imperial	SI	C	Mn	P	S	Si	GRE[a]	Cr	Ni	Cu
33G	230G	0.26	1.20	0.05	0.05	0.40	0.10	---	---	---
50G	350G	0.28	1.65	0.04	0.05	0.40	0.10	---	---	---
60G	400G	0.28	1.65	0.04	0.05	0.40	0.10	---	---	---
38W	260W	0.20	0.50-1.50	0.04	0.05	0.40	0.10	---	---	---
44W	300W	0.22	0.50-1.50	0.04	0.05	0.40	0.10	---	---	---
50W	350W	0.23	0.50-1.50	0.04	0.05	0.40	0.10	---	---	---
55W	380W	0.23	0.50-1.50	0.04	0.05	0.40	0.10	---	---	---
60W	400W	0.23	0.50-1.50	0.40	0.05	0.40	0.10	---	---	---
70W	480W	0.26	0.50-1.50	0.04	0.05	0.40	0.10	---	---	---
38WT	260WT	0.20	0.80-1.50	0.03	0.04	0.15-0.40	0.10	---	---	---
44WT	300WT	0.22	0.80-1.50	0.03	0.04	0.15-0.40	0.10	---	---	---
50WT	350WT	0.22	0.80-1.50	0.03	0.04	0.15-0.40	0.10	---	---	---
60WT	400WT	0.22	0.80-1.50	0.03	0.04	0.15-0.40	0.10	---	---	---
70WT	480WT	0.26	0.80-1.50	0.03	0.04	0.15-0.40	0.10	---	---	---
50R	350R	0.16	0.75	0.05-0.15	0.04	0.75	0.10	0.30-1.25	0.90	0.20-0.60
50A	350A	0.20	0.75-1.35	0.03	0.04	0.15-0.50	0.10	0.70	0.90	0.20-0.60
60A	400A	0.20	0.75-1.35	0.03	0.04	0.15-0.50	0.10	0.70	0.90	0.20-0.60
70A	480A	0.20	1.00-1.60	0.025	0.035	0.15-0.50	0.12	0.70	0.25-0.50	0.20-0.60
50AT	350AT	0.20	0.75-1.35	0.03	0.04	0.15-0.50	0.10	0.70	0.90	0.20-0.60
60AT	400AT	0.20	0.75-1.35	0.03	0.04	0.15-0.50	0.10	0.70	0.90	0.20-0.60
70AT	480AT	0.20	1.00-1.60	0.025	0.035	0.15-0.50	0.12	0.70	0.25-0.50	0.20-0.60
100Q	700Q[b]	0.20	1.50	0.03	0.04	0.15-0.40	---	---	---	---
100QT	700QT[b]	0.20	1.50	0.03	0.04	0.15-0.40	---	---	---	---

a. GRE - grain refining elements. Aluminum, columbium (niobium), and vanadium may be used as grain refinement elements.   See CSA G40.21 for more details. b. Boron added 0.0005-0.005%.

Single values are maximums, unless otherwise speciifed.

TYPICAL MECHANICAL PROPERTIES OF CARBON STEELS

AISI	Heat Treatment	Austenitizing Temperature		Hardness HB	Tensile Strength		Yield Strength		% El	% RA	Izod Impact Strength	
		°C	°F		ksi	MPa	ksi	MPa			J	ft·lb
1015	As-rolled	---	---	126	61.0	420.6	45.5	313.7	39.0	61.0	110.5	81.5
	Normalized	925	1700	121	61.5	424.0	47.0	324.1	37.0	69.6	115.5	85.2
	Annealed	870	1600	111	56.0	386.1	41.3	284.4	37.0	69.7	115.0	84.8
1020	As-rolled			143	65.0	448.2	48.0	330.9	36.0	59.0	86.8	64.0
	Normalized	870	1600	131	64.0	441.3	50.3	346.5	35.8	67.9	117.7	86.8
	Annealed	870	1600	111	57.3	394.7	42.8	294.8	36.5	66.0	123.4	91.0
1022	As-rolled			149	73.0	503.3	52.0	358.5	35.0	67.0	81.3	60.0
	Normalized	925	1700	143	70.0	482.6	52.0	358.5	34.0	67.5	117.3	86.5
	Annealed	870	1600	137	62.3	429.2	46.0	317.2	35.0	63.6	120.7	89.0
1030	As-rolled			179	80.0	551.6	50.0	344.7	32.0	57.0	74.6	55.0
	Normalized	925	1700	149	75.5	520.6	50.0	344.7	32.0	60.8	93.6	69.0
	Annealed	845	1550	126	67.3	463.7	49.5	341.3	31.2	57.9	69.4	51.2
1040	As-rolled			201	90.0	620.5	60.0	413.7	25.0	50.0	48.8	36.0
	Normalized	900	1650	170	85.5	589.5	54.3	374.0	28.0	54.9	65.1	48.0
	Annealed	790	1450	149	75.3	518.8	51.3	353.4	30.2	57.2	44.3	32.7
1050	As-rolled			229	105.0	723.9	60.0	413.7	20.0	40.0	31.2	23.0
	Normalized	900	1650	217	108.5	748.1	62.0	427.5	20.0	39.4	27.1	20.0
	Annealed	790	1450	187	92.3	636.0	53.0	365.4	23.7	39.9	16.9	12.5
1060	As-rolled			241	118.0	813.6	70.0	482.6	17.0	34.0	17.6	13.0
	Normalized	900	1650	229	112.5	775.7	61.0	420.6	18.0	37.2	13.2	9.7
	Annealed	790	1450	179	90.8	625.7	54.0	372.3	22.5	38.2	11.3	8.3
1080	As-rolled			293	140.0	965.3	85.0	586.1	12.0	17.0	6.8	5.0
	Normalized	900	1650	293	146.5	1010.1	76.0	524.0	11.0	20.6	6.8	5.0
	Annealed	790	1450	174	89.3	615.4	54.5	375.8	24.7	45.0	6.1	4.5

TYPICAL MECHANICAL PROPERTIES OF CARBON STEELS (Continued)

AISI	Heat Treatment	Austenitizing Temperature °C	Austenitizing Temperature °F	Hardness HB	Tensile Strength ksi	Tensile Strength MPa	Yield Strength ksi	Yield Strength MPa	% EI	% RA	Izod Impact Strength J	Izod Impact Strength ft-lb
1095	As-rolled			293	140.0	965.3	83.0	572.3	9.0	18.0	4.1	3.0
	Normalized	900	1650	293	147.0	1013.5	72.5	499.9	9.5	13.5	5.4	4.0
	Annealed	790	1450	192	95.3	656.7	55.0	379.2	13.0	20.6	2.7	2.0
1117	As-rolled			143	70.6	486.8	44.3	305.4	33.0	63.0	81.3	60.0
	Normalized	900	1650	137	67.8	467.1	44.0	303.4	33.5	63.8	85.1	62.8
	Annealed	855	1575	121	62.3	429.5	40.5	279.2	32.8	58.0	93.6	69.0
1118	As-rolled			149	75.6	521.2	45.9	316.5	32.0	70.0	108.5	80.0
	Normalized	925	1700	143	69.3	477.8	46.3	319.2	33.5	65.9	103.4	76.3
	Annealed	790	1450	131	65.3	450.2	41.3	284.8	34.5	66.8	106.4	78.5
1137	As-rolled			192	91.0	627.4	55.0	379.2	28.0	61.0	82.7	61.0
	Normalized	900	1650	197	97.0	668.8	57.5	396.4	22.5	48.5	63.7	47.0
	Annealed	790	1450	174	84.8	584.7	50.0	344.7	26.8	53.9	49.9	36.8
1141	As-rolled			192	98.0	675.7	52.0	358.5	22.0	38.0	11.1	8.2
	Normalized	900	1650	201	102.5	706.7	58.8	405.4	22.7	55.5	52.6	38.8
	Annealed	815	1500	163	86.8	598.5	51.2	353.0	25.5	49.3	34.3	25.3
1144	As-rolled			212	102.0	703.3	61.0	420.6	21.0	41.0	52.9	39.0
	Normalized	900	1650	197	96.8	667.4	58.0	399.9	21.0	40.4	43.4	32.0
	Annealed	790	1450	167	84.8	584.7	50.3	346.8	24.8	41.3	65.1	48.0

TYPICAL MECHANICAL PROPERTIES OF ALLOY STEELS

AISI	Heat Treatment	Austenitizing Temperature °C	Austenitizing Temperature °F	Hardness HB	Tensile Strength ksi	Tensile Strength MPa	Yield Strength ksi	Yield Strength MPa	% El	% RA	Izod Impact Strength J	Izod Impact Strength ft-lb
1340	Normalized	870	1600	248	121.3	836.3	81.0	558.5	22.0	62.9	92.5	68.2
	Annealed	800	1475	207	102.0	703.3	63.3	436.4	25.5	57.3	70.5	52.0
3140	Normalized	870	1600	262	129.3	891.5	87.0	599.8	19.7	57.3	53.6	39.5
	Annealed	815	1500	197	100.0	689.5	61.3	422.6	24.5	50.8	46.4	34.2
4130	Normalized	870	1600	197	97.0	668.8	63.3	436.4	25.5	59.5	86.4	63.7
	Annealed	865	1585	156	81.3	560.5	52.3	360.6	28.2	55.6	61.7	45.5
4140	Normalized	870	1600	302	148.0	1020.4	95.0	655.0	17.7	46.8	22.6	16.7
	Annealed	815	1500	197	95.0	655.0	60.5	417.1	25.7	56.9	54.5	40.2
4150	Normalized	870	1600	321	167.5	1154.9	106.5	734.3	11.7	30.8	11.5	8.5
	Annealed	815	1500	197	105.8	729.5	55.0	379.2	20.2	40.2	24.7	18.2
4320	Normalized	895	1640	235	115.0	792.9	67.3	464.0	20.8	50.7	72.9	53.8
	Annealed	850	1560	163	84.0	579.2	61.6	424.7	29.0	58.4	109.8	81.0
4340	Normalized	970	1600	363	195.5	1279.0	125.0	861.8	12.2	36.3	15.9	11.7
	Annealed	810	1490	217	108.0	744.6	68.5	472.3	22.0	49.9	51.1	37.7
4620	Normalized	900	1650	174	83.3	574.3	53.1	366.1	29.0	66.7	132.9	98.0
	Annealed	855	1575	149	74.3	512.3	54.0	372.3	31.3	60.3	93.6	69.0
4820	Normalized	860	1580	229	109.5	750.0	70.3	484.7	24.0	59.2	109.8	81.0
	Annealed	815	1500	197	98.8	681.2	67.3	464.0	22.3	58.8	92.9	68.5
5140	Normalized	870	1600	229	115.0	792.9	68.5	472.3	22.7	59.2	38.0	28.0
	Annealed	830	1525	167	83.0	572.3	42.5	293.0	28.6	57.3	40.7	30.0
5150	Normalized	870	1600	255	126.3	870.8	76.8	529.5	20.7	58.7	31.5	23.2
	Annealed	825	1520	197	98.0	675.7	51.8	357.1	22.0	43.7	25.1	18.5
5160	Normalized	855	1575	269	138.8	957.0	77.0	530.9	17.5	44.8	10.8	8.0
	Annealed	815	1495	197	104.8	722.6	40.0	275.8	17.2	30.6	10.0	7.4

TYPICAL MECHANICAL PROPERTIES OF ALLOY STEELS (Continued)

AISI	Heat Treatment	Austenitizing Temperature		Hardness	Tensile Strength		Yield Strength		% El	% RA	Izod Impact Strength	
		°C	°F	HB	ksi	MPa	ksi	MPa			J	ft·lb
6150	Normalized	870	1600	269	136.3	939.8	89.3	615.7	21.8	61.0	35.5	26.2
	Annealed	815	1500	197	96.8	667.4	59.8	412.3	23.0	48.4	27.4	20.2
8620	Normalized	915	1675	183	91.8	632.9	51.8	357.1	26.3	59.7	99.7	73.5
	Annealed	870	1600	149	77.8	536.4	55.9	385.4	31.3	62.1	112.2	82.8
8630	Normalized	870	1600	187	94.3	650.2	62.3	429.5	23.5	53.5	94.6	69.8
	Annealed	845	1550	156	81.8	564.0	54.0	372.3	29.0	58.9	95.2	70.2
8650	Normalized	870	1600	302	148.5	1023.9	99.8	688.1	14.0	40.4	13.6	10.0
	Annealed	795	1465	212	103.8	715.7	56.0	386.1	22.5	46.4	29.4	21.7
8740	Normalized	870	1600	269	134.8	929.4	88.0	606.7	16.0	47.9	17.6	13.0
	Annealed	815	1500	201	100.8	695.0	60.3	415.8	22.2	46.4	40.0	29.5
9255	Normalized	900	1650	269	135.3	932.9	84.0	579.2	19.7	43.4	13.6	10.0
	Annealed	845	1550	229	112.3	774.3	70.5	486.1	21.7	41.1	8.8	6.5
9310	Normalized	890	1630	269	131.5	906.7	82.8	570.9	18.8	58.1	119.3	88.0
	Annealed	845	1550	241	119.0	820.5	63.8	439.9	17.3	42.1	78.6	58.0

TYPICAL MECHANICAL PROPERTIES OF QUENCHED & TEMPERED CARBON STEELS									
	Tempering Temp.		Hardness	Tensile Strength		Yield Strength			
AISI[a]	°C	°F	HB	ksi	MPa	ksi	MPa	% Elongation	% RA
1030[b]	205	400	495	123	848	94	648	17	47
	315	600	401	116	800	90	621	19	53
	425	800	302	106	731	84	579	23	60
	540	1000	255	97	669	75	517	28	65
	650	1200	207	85	586	64	441	32	70
1040[b]	205	400	514	130	896	96	662	16	45
	315	600	444	129	889	94	648	18	52
	425	800	352	122	841	92	634	21	57
	540	1000	269	113	779	86	593	23	61
	650	1200	201	97	669	72	496	28	68
1040	205	400	262	113	779	86	593	19	48
	315	600	255	113	779	86	593	20	53
	425	800	241	110	758	80	552	21	54
	540	1000	212	104	717	71	490	26	57
	650	1200	192	92	634	63	434	29	65
1050[b]	205	400	514	163	1124	117	807	9	27
	315	600	444	158	1089	115	793	13	36
	425	800	375	145	1000	110	758	19	48
	540	1000	293	125	862	95	655	23	58
	650	1200	235	104	717	78	538	28	65
1050	205	400	---	---	---	---	---	---	---
	315	600	321	142	979	105	724	14	47
	425	800	277	136	938	95	655	20	50
	540	1000	262	127	876	84	579	23	53
	650	1200	223	107	738	68	469	29	60

TYPICAL MECHANICAL PROPERTIES OF QUENCHED & TEMPERED CARBON STEELS (Continued)

AISI[a]	Tempering Temp.		Hardness	Tensile Strength		Yield Strength		% Elongation	% RA
	°C	°F	HB	ksi	MPa	ksi	MPa		
1060	205	400	321	160	1103	113	779	13	40
	315	600	321	160	1103	113	779	13	40
	425	800	311	156	1076	111	765	14	41
	540	1000	277	140	965	97	669	17	45
	650	1200	229	116	800	76	524	23	54
1080	205	400	388	190	1310	142	979	12	35
	315	600	388	189	1303	142	979	12	35
	425	800	375	187	1289	138	951	13	36
	540	1000	321	164	1131	117	807	16	40
	650	1200	255	129	889	87	600	21	50
1095[b]	205	400	601	216	1489	152	1048	10	31
	315	600	534	212	1462	150	1034	11	33
	425	800	388	199	1372	139	958	13	35
	540	1000	293	165	1138	110	758	15	40
	650	1200	235	122	841	85	586	20	47
1095	205	400	401	187	1289	120	827	10	30
	315	600	375	183	1262	118	813	10	30
	425	800	363	176	1213	112	772	12	32
	540	1000	321	158	1089	98	676	15	37
	650	1200	269	130	896	80	552	21	47
1137	205	400	352	157	1082	136	938	5	22
	315	600	285	143	986	122	841	10	33
	425	800	262	127	876	106	731	15	48
	540	1000	229	110	758	88	607	24	62
	650	1200	197	95	655	70	483	28	69

TYPICAL MECHANICAL PROPERTIES OF QUENCHED & TEMPERED CARBON STEELS (Continued)

AISI[a]	Tempering Temp.		Hardness	Tensile Strength		Yield Strength		% Elongation	% RA
	°C	°F	HB	ksi	MPa	ksi	MPa		
1137[b]	205	400	415	217	1496	169	1165	5	17
	315	600	375	199	1372	163	1124	9	25
	425	800	311	160	1103	143	986	14	40
	540	1000	262	120	827	105	724	19	60
	650	1200	187	94	648	77	531	25	69
1141	205	400	461	237	1634	176	1213	6	17
	315	600	415	212	1462	186	1282	9	32
	425	800	331	169	1165	150	1034	12	47
	540	1000	262	130	896	111	765	18	57
	650	1200	217	103	710	86	593	23	62
1144	205	400	277	127	876	91	627	17	36
	315	600	262	126	869	90	621	17	40
	425	800	248	123	848	88	607	18	42
	540	1000	235	117	807	83	572	20	46
	650	1200	217	105	724	73	503	23	55

a Heat treated specimens were oil quenched unless otherwise noted.
b Heat treated specimens were water quenched.

TYPICAL MECHANICAL PROPERTIES OF QUENCHED & TEMPERED ALLOY STEELS

AISI[a]	Tempering Temp.		Hardness	Tensile Strength		Yield Strength		% Elongation	% RA
	°C	°F	HB	ksi	MPa	ksi	MPa		
1330[b]	205	400	459	232	1600	211	1455	9	39
	315	600	402	207	1427	186	1282	9	44
	425	800	335	168	1158	150	1034	15	53
	540	1000	263	127	876	112	772	18	60
	650	1200	216	106	731	83	572	23	63
1340	205	400	505	262	1806	231	1593	11	35
	315	600	453	230	1586	206	1420	12	43
	425	800	375	183	1262	167	1151	14	51
	540	1000	295	140	965	120	827	17	58
	650	1200	252	116	800	90	621	22	66
4037	205	400	310	149	1027	110	758	6	38
	315	600	295	138	951	111	765	14	53
	425	800	270	127	876	106	731	20	60
	540	1000	247	115	793	95	655	23	63
	650	1200	220	101	696	61	421	29	60
4042	205	400	516	261	1800	241	1662	12	37
	315	600	455	234	1613	211	1455	13	42
	425	800	380	187	1289	170	1172	15	51
	540	1000	300	143	986	128	883	20	59
	650	1200	238	115	793	100	689	28	66
4130[b]	205	400	467	236	1627	212	1462	10	41
	315	600	435	217	1496	200	1379	11	43
	425	800	380	186	1282	173	1193	13	49
	540	1000	315	150	1034	132	910	17	57
	650	1200	245	118	814	102	703	22	64

AISI[a]	Tempering Temp.		Hardness	Tensile Strength		Yield Strength		% Elongation	% RA
	°C	°F	HB	ksi	MPa	ksi	MPa		
4140	205	400	510	257	1772	238	1641	8	38
	315	600	445	225	1551	208	1434	9	43
	425	800	370	181	1248	165	1138	13	49
	540	1000	285	138	951	121	834	18	58
	650	1200	230	110	758	95	655	22	63
4150	205	400	530	280	1931	250	1724	10	39
	315	600	495	256	1765	231	1593	10	40
	425	800	440	220	1517	200	1379	12	45
	540	1000	370	175	1207	160	1103	15	52
	650	1200	290	139	958	122	841	19	60
4340	205	400	520	272	1875	243	1675	10	38
	315	600	486	250	1724	230	1586	10	40
	425	800	430	213	1469	198	1365	10	44
	540	1000	360	170	1172	156	1076	13	51
	650	1200	280	140	965	124	855	19	60
5046	205	400	482	253	1744	204	1407	9	25
	315	600	401	205	1413	168	1158	10	37
	425	800	336	165	1138	135	931	13	50
	540	1000	282	136	938	111	765	18	61
	650	1200	235	114	786	95	655	24	66
50B46	205	400	560	---	---	---	---	---	---
	315	600	505	258	1779	235	1620	10	37
	425	800	405	202	1393	181	1248	13	47
	540	1000	322	157	1082	142	979	17	51
	650	1200	273	128	883	115	793	22	60

TYPICAL MECHANICAL PROPERTIES OF QUENCHED & TEMPERED ALLOY STEELS (Continued)

AISI[a]	Tempering Temp.		Hardness	Tensile Strength		Yield Strength		% Elongation	% RA
	°C	°F	HB	ksi	MPa	ksi	MPa		
50B60	205	400	600	---	---	---	---	---	---
	315	600	525	273	1882	257	1772	8	32
	425	800	435	219	1510	201	1386	11	34
	540	1000	350	163	1124	145	1000	15	38
	650	1200	290	130	896	113	779	19	50
5130	205	400	475	234	1613	220	1517	10	40
	315	600	440	217	1496	204	1407	10	46
	425	800	379	185	1275	175	1207	12	51
	540	1000	305	150	1034	136	938	15	56
	650	1200	245	115	793	100	689	20	63
5140	205	400	490	260	1793	238	1641	9	38
	315	600	450	229	1579	210	1448	10	43
	425	800	365	190	1310	170	1172	13	50
	540	1000	280	145	1000	125	862	17	58
	650	1200	235	110	758	96	662	25	66
5150	205	400	525	282	1944	251	1731	5	37
	315	600	475	252	1737	230	1586	6	40
	425	800	410	210	1448	190	1310	9	47
	540	1000	340	163	1124	150	1034	15	54
	650	1200	270	117	807	118	814	20	60
5160	205	400	627	322	2220	260	1793	4	10
	315	600	555	290	1999	257	1772	9	30
	425	800	461	233	1606	212	1462	10	37
	540	1000	341	169	1165	151	1041	12	47
	650	1200	269	130	896	116	800	20	56

TYPICAL MECHANICAL PROPERTIES OF QUENCHED & TEMPERED ALLOY STEELS (Continued)

TYPICAL MECHANICAL PROPERTIES OF QUENCHED & TEMPERED ALLOY STEELS (Continued)

AISI[a]	Tempering Temp.		Hardness HB	Tensile Strength		Yield Strength		% Elongation	% RA
	°C	°F		ksi	MPa	ksi	MPa		
51B60	205	400	600	---	---	---	---	---	---
	315	600	540	---	---	---	---	---	---
	425	800	460	237	1634	216	1489	11	36
	540	1000	355	175	1207	160	1103	15	44
	650	1200	290	140	965	126	869	20	47
6150	205	400	538	280	1931	245	1689	8	38
	315	600	483	250	1724	228	1572	8	39
	425	800	420	208	1434	193	1331	10	43
	540	1000	345	168	1158	155	1069	13	50
	650	1200	282	137	945	122	841	17	58
81B45	205	400	550	295	2034	250	1724	10	33
	315	600	475	256	1765	228	1572	8	42
	425	800	405	204	1407	190	1310	11	48
	540	1000	338	160	1103	149	1027	16	53
	650	1200	280	130	896	115	793	20	55
8630	205	400	465	238	1641	218	1503	9	38
	315	600	430	215	1482	202	1392	10	42
	425	800	375	185	1276	170	1172	13	47
	540	1000	310	150	1034	130	896	17	54
	650	1200	240	112	772	100	689	23	63
8640	205	400	505	270	1862	242	1669	10	40
	315	600	460	240	1655	220	1517	10	41
	425	800	400	200	1379	188	1296	12	45
	540	1000	340	160	1103	150	1034	16	54
	650	1200	280	130	896	116	800	20	62

TYPICAL MECHANICAL PROPERTIES OF QUENCHED & TEMPERED ALLOY STEELS (Continued)

AISI[a]	Tempering Temp. °C	Tempering Temp. °F	Hardness HB	Tensile Strength ksi	Tensile Strength MPa	Yield Strength ksi	Yield Strength MPa	% Elongation	% RA
86B45	205	400	525	287	1979	238	1641	9	31
	315	600	475	246	1696	225	1551	9	40
	425	800	395	200	1379	191	1317	11	41
	540	1000	335	160	1103	150	1034	15	49
	650	1200	280	131	903	127	876	19	58
8650	205	400	525	281	1937	243	1675	10	38
	315	600	490	250	1724	225	1551	10	40
	425	800	420	210	1448	192	1324	12	45
	540	1000	340	170	1172	153	1055	15	51
	650	1200	280	140	965	120	827	20	58
8660	205	400	580	---	---	---	---	---	---
	315	600	535	---	---	---	---	---	---
	425	800	460	237	1634	225	1551	13	37
	540	1000	370	190	1310	176	1213	17	46
	650	1200	315	155	1068	138	951	20	53
8740	205	400	578	290	1999	240	1655	10	41
	315	600	495	249	1717	225	1551	11	46
	425	800	415	208	1434	197	1358	13	50
	540	1000	363	175	1207	165	1138	15	55
	650	1200	302	143	986	131	903	20	60
9255	205	400	601	305	2103	297	2048	1	3
	315	600	578	281	1937	260	1793	4	10
	425	800	477	233	1606	216	1489	8	22
	540	1000	352	182	1255	160	1103	15	32
	650	1200	285	144	993	118	814	20	42

| TYPICAL MECHANICAL PROPERTIES OF QUENCHED & TEMPERED ALLOY STEELS (Continued) | | | | | | | | | | |
| AISI[a] | Tempering Temp. | | Hardness | Tensile Strength | | Yield Strength | | % Elongation | % RA |
	°C	°F	HB	ksi	MPa	ksi	MPa		
9260	205	400	600	---	---	---	---	---	---
	315	600	540	---	---	---	---	---	---
	425	800	470	255	1758	218	1503	8	24
	540	1000	390	192	1324	164	1131	12	30
	650	1200	295	142	979	118	814	20	43
94B30	205	400	475	250	1724	225	1551	12	46
	315	600	445	232	1600	206	1420	12	49
	425	800	382	195	1344	175	1207	13	57
	540	1000	307	145	1000	135	931	16	65
	650	1200	250	120	827	105	724	21	69

a Heat treated specimens were oil quenched unless otherwise noted.
b Heat treated specimens were water quenched.

MECHANICAL PROPERTIES OF CSA G40.21 STRUCTURAL QUALITY STEELS - PLATES, FLOOR PLATES, BARS, SHAPES AND SHEET PILLING

Grade		Tensile Strength		Yield Strength[a]		% Elongation[b]		Hardness, HB
Imperial	SI	ksi	MPa	ksi	MPa	Longitudinal	Transverse	
33G	230G	55-72	380-500	33	230	24	22	---
50G	350G	70-100	480-690	50	350	19	17	---
60G	400G	80-105	550-720	60	400	19	17	---
38W	260W	60-85	410-590	38	260	23	21	---
44W	300W	65-90	450-620	44	300	23	21	---
50W	350W	65-95	450-650	50	350	22	20	---
55W	380W	70-95	480-650	55	380	21	---	---
60W	400W	75-100	520-690	60	400	18	15	---
70W	480W	85-115	590-790	70	480	17	14	---
38WT	260WT	60-85	410-590	38	260	23	21	---
44WT	300WT	65-90	450-620	44	300	23	21	---
50WT	350WT	70-95	480-650	50	350	22	20	---
60WT	400WT	75-100	520-690	60	400	20	17	---
70WT	480WT	85-117	590-790	70	480	17	14	---
50R	350R	70-95	480-650	50	350	21	18	---
50A	350A	70-95	480-650	50	350	21	19	---
60A	400A	75-100	520-690	60	400	21	18	---
70A	480A	85-115	590-790	70	480	17	14	---
50AT	350AT	70-95	480-650	50	350	21	19	---
60AT	400AT	75-100	520-690	60	400	21	18	---
70AT	480AT	85-115	590-790	70	480	17	14	---
100Q	700Q	115-135	800-950	100	700	18	16	235-293
100QT	700QT	115-135	800-950	100	700	18	16	235-293

a. For thicknesses up to 2 ½ in. or 65 mm.
b. Elongation gage length is 2 in. or 50 mm.
Single values are minimum, unless otherwise specified.

Chapter

13

FORGINGS CARBON, ALLOY & STAINLESS STEELS:

AMERICAN SPECIFICATION TITLES & DESIGNATIONS, CHEMICAL COMPOSITIONS & MECHANICAL PROPERTIES

CARBON, ALLOY AND STAINLESS STEEL FORGINGS

Carbon and Alloy Steel Forgings

ASTM Spec.	Title
A 105/A 105M	Carbon Steel, Forgings, for Piping Applications
A 181/A 181M	Carbon Steel Forgings, for General-Purpose Piping
A 182/A 182M	Forged or Rolled Alloy-Steel Pipe Flanges, Forged Fittings, and Valves and Parts for High-Temperature Service
A 266/A 266M	Carbon Steel Forgings for Pressure Vessel Components
A 288	Carbon and Alloy Steel Forgings for Magnetic Retaining Rings for Turbine Generators
A 289	Alloy Steel Forgings for Nonmagnetic Retaining Rings for Generators
A 290	Carbon and Alloy Steel Forgings for Rings for Reduction Gears
A 291	Steel Forgings, Carbon and Alloy, for Pinions, Gears and Shafts for Reduction Gears
A 336/A 336M	Alloy Steel Forgings for Pressure and High-Temperature Parts
A 350/A 350M	Carbon and Low-Alloy Steel Forgings, Requiring Notch Toughness Testing for Piping Components
A 372/A 372M	Carbon and Alloy Steel Forgings for Thin-Walled Pressure Vessels
A 427	Wrought Alloy Steel Rolls for Cold and Hot Reduction
A 469	Vacuum-Treated Steel Forgings for Generator Rotors
A 470	Vacuum-Treated Carbon and Alloy Steel Forgings for Turbine Rotors and Shafts
A 471	Vacuum-Treated Alloy Steel Forgings for Turbine Rotor Disks and Wheels
A 508/A 508M	Quenched and Tempered Vacuum-Treated Carbon and Alloy Steel Forgings for Pressure Vessels
A 521	Steel, Closed-Impression Die Forgings for General Industrial Use
A 522/A 522M	Forged or Rolled 8 and 9 % Nickel Alloy Steel Flanges, Fittings, Valves, and Parts for Low-Temperature Service
A 541/A 541M	Quenched and Tempered Carbon and Alloy Steel Forgings for Pressure Vessel Components
A 579	Superstrength Alloy Steel Forgings
A 592/A 592M	High-Strength Quenched and Tempered Low-Alloy Steel Forged Fittings and Parts for Pressure Vessels
A 646	Premium Quality Alloy Steel Blooms and Billets for Aircraft and Aerospace Forgings
A 649/A 649M	Forged Steel Rolls Used for Corrugating Paper Machinery
A 668/A 668M	Steel Forgings, Carbon and Alloy, for General Industrial Use
A 694/A 694M	Carbon and Alloy Steel Forgings for Pipe Flanges, Fittings, Valves, and Parts for High-Pressure Transmission Service
A 707/A 707M	Forged, Carbon and Alloy Steel Flanges for Low-Temperature Service

CARBON, ALLOY AND STAINLESS STEEL FORGINGS (Continued)

ASTM Spec.	Title
Carbon and Alloy Steel Forgings (Continued)	
A 711	Steel Forging Stock
A 723/A 723M	Alloy Steel Forgings for High-Strength Pressure Component Application
A 727/A 727M	Carbon Steel Forgings for Piping Components with Inherent Notch Toughness
A 730	Forgings, Carbon and Alloy Steel, for Railway Use
A 765/A 765M	Carbon Steel and Low-Alloy Steel Pressure-Vessel-Component Forgings with Mandatory Toughness Requirements
A 788	Steel Forgings, General Requirements
A 836/A 836M	Titanium-Stabilized Carbon Steel Forgings for Glass-Lined Piping and Pressure Vessel Service
A 837	Steel Forgings, Alloy, for Carburizing Applications
A 859/A 859M	Age-Hardening Alloy Steel Forgings for Pressure Vessel Components
A 909	Steel Forgings, Microalloy, for General Industrial Use
A 940	Vacuum Treated Steel Forgings, Alloy, Differentially Heat Treated, for Turbine Rotors
A 952	Forged Grade 80 Alloy Steel Lifting Components and Welded Attachment Links
Stainless Steel Forgings	
A 314	Stainless Steel Billets and Bars for Forgings
A 473	Stainless Steel Forgings
A 565	Martensitic Stainless Steel Bars, Forgings, and Forging Stock for High-Temperature Service
A 638/A 638M	Precipitation Hardening Iron Base Superalloy Bars, Forgings, and Forging Stock for High-Temperature Service
A 705/A 705M	Age-Hardening Stainless Steel Forgings
A 768	Vacuum-Treated 12 % Chromium Alloy Steel Forgings for Turbine Rotors and Shafts
A 891	Precipitation Hardening Iron Base Superalloy Forgings for Turbine Rotor Disks and Wheels
A 965/A 965 M	Steel Forgings, Austenitic, for Pressure and High-Temperature Parts

CHEMICAL COMPOSITION OF CARBON STEEL FORGINGS

ASTM Spec.	UNS	C	Mn	Si	S	P	Other
A 105	K03504	0.35	0.60-1.05	0.35	0.050	0.040	0.40 Cu, 0.40 Ni, 0.30 Cr, 0.12 Mo, 0.05 V, 0.02 Cb
A 181 Cl 60,70	K03502	0.35	1.10	0.10-0.35	0.05	0.05	---
A 266 Gr 1, 2	K03506	0.35	0.40-1.05	0.15-0.35	0.025	0.025	---
A 266 Gr 3	K05001	0.45	0.50-0.90	0.35	0.025	0.025	---
A 266 Gr 4	K03017	0.30	0.80-1.35	0.15-0.35	0.025	0.025	---
A 288 Cl 1	K05002	0.50	0.60-1.00	0.15-0.30	0.025	0.025	---
A 290 Cl A, B	K04000	0.35-0.50	0.60-0.90	0.35	0.040	0.040	0.30 Ni, 0.25 Cr, 0.10 Mo, 0.06 V, 0.35 Cu
A 290 Cl C, D	K04500	0.40-0.50	0.60-0.90	0.35	0.040	0.040	0.30 Ni, 0.25 Cr, 0.10 Mo, 0.06 V, 0.35 Cu
A 291 Cl 1	K05500	0.55	0.60-0.90	0.35	0.040	0.040	0.30 Ni, 0.25 Cr, 0.10 Mo, 0.06 V, 0.35 Cu
A 291 Cl 2	K05000	0.50	0.40-0.90	0.35	0.040	0.040	0.10 V, 0.35 Cu
A 350 Gr LF1	K03009	0.35	0.60-1.35	0.15-0.30	0.040	0.035	0.40 Ni, 0.30 Cr, 0.12 Mo, 0.40 Cu, 0.02 Cb, 0.05 V
A 350 Gr LF2	K03011	0.35	0.60-1.35	0.15-0.30	0.040	0.035	0.40 Ni, 0.30 Cr, 0.12 Mo, 0.40 Cu, 0.02 Cb, 0.05 V
A 372 Gr A	K03002	0.30	1.00	0.15-0.35	0.025	0.025	---
A 372 Gr B	K04001	0.35	1.35	0.15-0.35	0.025	0.025	---
A 372 Gr C	K04801	0.48	1.65	0.15-0.35	0.025	0.025	---
A 469 Cl 1	K14501	0.45	0.90	0.15-0.35	0.015	0.015	0.03-0.12 V
A 470 Cl 1	K14501	0.45	0.90	0.15-0.35	0.025	0.025	0.03 min V
A 508 Gr 1	K13502	0.35	0.40-1.05	0.15-0.40	0.025	0.025	0.40 Ni, 0.25 Cr, 0.10 Mo, 0.05 V
A 508 Gr 1A	K13502	0.30	0.70-1.35	0.15-0.40	0.025	0.025	0.40 Ni, 0.25 Cr, 0.10 Mo, 0.05 V
A 541 Gr 1	K03506	0.35	0.40-0.90	0.15-0.35	0.025	0.025	0.40 Ni, 0.25 Cr, 0.10 Mo, 0.05 V
A 541 Gr 1A	---	0.30	0.70-1.35	0.15-0.40	0.025	0.025	0.40 Ni, 0.25 Cr, 0.10 Mo, 0.05 V
A 649 Cl 2	K05001	0.55	0.50-0.90	0.15-0.35	0.025	0.025	---
A 649 Cl 4	---	0.35	0.60-1.05	0.15-0.35	0.025	0.025	---
A 694 [a, c]	K03014	0.26	1.40	0.35	0.025	0.025	---
A 707 Gr L1[a]	K02302	0.20	0.60-1.50	0.35	0.030	0.030	0.30 Cr, 0.40 Ni, 0.12 Mo, 0.05 V, 0.40 Cu, 0.02 Cb
A 707 Gr L2[a]	K03301	0.30	0.60-1.35	0.35	0.030	0.030	0.30 Cr, 0.40 Ni, 0.12 Mo, 0.05 V, 0.40 Cu, 0.02 Cb

CHEMICAL COMPOSITION OF CARBON STEEL FORGINGS (Continued)

ASTM Spec.	UNS	C	Mn	Si	S	P	Application
A 727[a]	K02506	0.25	0.90-1.35	0.15-0.30	0.025	0.035	0.40 Ni, 0.30 Cr, 0.12 Mo, 0.40 Cu, 0.02 Cb, 0.05 V
A 765 Gr I [b]	K03046	0.30	0.60-1.05	0.15-0.35	0.020	0.020	0.50 Ni, 0.05 V, 0.05 Al, 0.40 Cr, 0.25 Mo, 0.35 Cu
A 765 Gr II [b]	K03047	0.30	0.60-1.35	0.15-0.35	0.020	0.020	0.50 Ni, 0.05 V, 0.05 Al, 0.40 Cr, 0.25 Mo, 0.35 Cu
A 765 Gr IV [b]	---	0.20	1.00-1.60	0.15-0.50	0.020	0.020	0.50 Ni, 0.06 V, 0.05 Al, 0.40 Cr, 0.25 Mo, 0.35 Cu

a. Heat analysis. b. Intentional additions of up to 0.40% Cr, 0.25% Mo and 0.50% Ni are permitted by the manufacturer. c. Chemistry listed is for all Grades of ASTM A 694. Single values are maximums, unless otherwise specified.

CHEMICAL COMPOSITION OF ALLOY STEEL FORGINGS

ASTM Spec.	UNS	C	Mn	Si	Cr	Ni	Mo	Other
A 182 Gr F1	K12822	0.28	0.60-0.90	0.15-0.35	---	---	0.44-0.65	---
A 182 Gr F2	K12122	0.05-0.21	0.30-0.80	0.10-0.60	0.50-0.81	---	0.44-0.65	---
A 182 Gr F5	K41545	0.15	0.30-0.60	0.50	4.0-6.0	0.50	0.44-0.65	---
A 182 Gr F5a	K42544	0.25	0.60	0.50	4.0-6.0	0.50	0.44-0.65	---
A 182 Gr F9	K90941	0.15	0.30-0.60	0.50-1.00	8.0-10.0	---	0.90-1.10	---
A 182 Gr F11 Cl 1	K11597	0.05-0.15	0.30-0.60	0.50-1.00	1.00-1.50	---	0.44-0.65	---
A 182 Gr F11 Cl 2, 3	K11572	0.10-0.20	0.30-0.80	0.50-1.00	1.00-1.50	---	0.44-0.65	---
A 182 Gr F12 Cl 1	K11562	0.05-0.15	0.30-0.60	0.50	0.80-1.25	---	0.44-0.65	---
A 182 Gr F12 Cl 2	K11564	0.10-0.20	0.30-0.80	0.10-0.60	0.80-1.25	---	0.44-0.65	---
A 182 Gr F21	K31545	0.05-0.15	0.30-0.60	0.50	2.7-3.3	---	0.80-1.06	---
A 182 Gr F3V	K31830	0.05-0.18	0.30-0.60	0.10	2.8-3.2	---	0.90-1.10	0.20-0.30 V, 0.015-0.035 Ti, 0.001-0.003 B
A 182 Gr F22 Cl 1, 3	K21590	0.05-0.15	0.30-0.60	0.50	2.00-2.50	---	0.87-1.13	---
A 182 Gr F22V	K31835	0.11-0.15	0.30-0.60	0.10	2.00-2.50	0.25	0.90-1.10	0.20 Cu, 0.25-0.35 V, 0.07 Cb, 0.002 B, 0.015 Ca, 0.030 Ti
A 182 Gr F91	K91560	0.08-0.12	0.30-0.60	0.20-0.50	8.0-9.5	0.40	0.85-1.05	0.06-0.10 Cb, 0.03-0.07 N, 0.04 Al, 0.18-0.25 V

CHEMICAL COMPOSITION OF ALLOY STEEL FORGINGS (Continued)

ASTM Spec.	UNS	C	Mn	Si	Cr	Ni	Mo	Other
A 182 Gr FR	K22035	0.20	0.40-1.06	---	---	1.60-2.24	---	0.75-1.25 Cu
A 290 Cl E, F	K14048	0.35-0.45	0.70-1.00	0.35	0.80-1.15	0.50	0.15-0.25	0.06 V, 0.35 Cu
A 290 Cl G,H,I,J,K,L	K24045	0.35-0.45	0.60-0.90	0.35	0.60-0.90	1.65-2.00	0.20-0.50	0.10 V, 0.35 Cu
A 290 Cl M, P	---	0.38-0.45	0.40-0.70	0.40	1.40-1.80	0.30	0.30-0.45	0.03 V, 0.35 Cu, 0.85-1.30 Al
A 291 Cl 3	K14507	0.45	0.40-0.90	0.35	1.25	0.50	0.15 min	0.50 V, 0.35 Cu
A 291 Cl 3A	K14557	0.45	0.40-0.90	0.35	1.50	1.00-3.00	0.15 min	0.10 V, 0.35 Cu
A 291 Cl 4 to 7	K24245	0.35-0.50	0.40-0.90	0.35	0.60 min	1.65 min	0.20-0.60	0.10 V, 0.35 Cu
A 291 Cl 8	---	0.38-0.45	0.40-0.70	0.40	1.40-1.80	0.30	0.30-0.45	0.03 V, 0.35 Cu, 0.85-1.30 Al
A 336 Gr F1	K12520	0.20-0.30	0.60-0.80	0.20-0.35	---	---	0.40-0.60	---
A 336 Gr F3V	---	0.10-0.15	0.30-0.60	0.10	2.7-3.3	---	0.90-1.10	0.20-0.30 V, 0.001-0.003 B, 0.015-0.035 Ti
A 336 Gr F5	K41545	0.15	0.30-0.60	0.50	4.0-6.0	0.50	0.45-0.65	---
A 336 Gr F5A	K42544	0.25	0.60	0.50	4.0-6.0	0.50	0.45-0.65	---
A 336 Gr F9	---	0.15	0.30-0.60	0.50-1.00	8.0-10.0	---	0.90-1.10	---
A 336 Gr F11 Cl 1	---	0.05-0.15	0.30-0.60	0.50-1.00	1.00-1.50	---	0.44-0.65	---
A 336 Gr F11 Cl 2, 3	K11572	0.10-0.20	0.30-0.80	0.50-1.00	1.00-1.50	---	0.45-0.65	---
A 336 Gr F12	K11564	0.10-0.20	0.30-0.80	0.10-0.60	0.80-1.10	---	0.45-0.65	---
A 336 Gr F21 Cl 1,3	K31545	0.05-0.15	0.30-0.60	0.50	2.7-3.3	---	0.80-1.06	---
A 336 Gr F22 Cl 1,3	K21590	0.05-0.15	0.30-0.60	0.50	2.00-2.50	---	0.90-1.10	---
A 336 Gr F22V	---	0.11-0.15	0.30-0.60	0.10	2.00-2.50	0.25	0.90-1.10	0.25-0.35 V, 0.07 Cb, 0.0020 B, 0.030 Ti, 0.20 Cu, 0.015 Ca
A 336 Gr F91	---	0.08-0.12	0.30-0.60	0.20-0.50	8.0-9.5	0.40	0.85-1.05	0.18-0.25 V, 0.06-0.10 Cb, 0.03-0.07 N, 0.04 Al
A 350 Gr LF3	K32025	0.20	0.90	0.20-0.35	0.30	3.3-3.7	0.12	0.40 Cu, 0.02 Cb, 0.03 V
A 350 Gr LF5	K13050	0.30	0.60-1.35	0.20-0.35	0.30	1.0-2.0	0.12	0.40 Cu, 0.02 Cb, 0.03 V

CHEMICAL COMPOSITION OF ALLOY STEEL FORGINGS (Continued)								
ASTM Spec.	UNS	C	Mn	Si	Cr	Ni	Mo	Other
A 350 Gr LF6	K12202	0.22	1.15-1.50	0.15-0.30	0.30	0.40	0.12	0.40 Cu,, 0.02 Cb, 0.04-0.11 V [a,b], 0.01-0.030 N
A 350 Gr LF9	K22036	0.20	0.40-1.06	---	0.30	1.60-2.24	0.12	0.75-1.25 Cu, 0.02 Cb, 0.03 V
A 350 Gr LF787	---	0.07	0.40-0.70	0.40	0.60-0.90	0.70-1.00	0.15-0.25	1.00-1.30 Cu, 0.02 Cb min, 0.03 V
A 372 Gr D	K14508	0.40-0.50	1.40-1.80	0.15-0.35	---	---	0.17-0.27	---
A 372 Gr E Cl 55, 65, 70	K13047	0.25-0.35	0.40-0.90	0.15-0.35	0.80-1.15	---	0.15-0.25	---
A 372 Gr F, Cl 55, 65, 70	G41350	0.30-0.40	0.70-1.00	0.15-0.35	0.80-1.15	---	0.15-0.25	---
A 372 Gr G, Cl 55, 65, 70	---	0.25-0.35	0.70-1.00	0.15-0.35	0.40-0.65	---	0.15-0.25	---
A 372 Gr H, Cl 55, 65, 70	K13547	0.30-0.40	0.75-1.05	0.15-0.35	0.40-0.65	---	0.15-0.25	---
A 372 Gr J Cl 55, 65, 70, 110[c]	K13548	0.35-0.50	0.75-1.05	0.15-0.35	0.80-1.15	---	0.15-0.25	---
A 372 Gr K	K31820	0.18	0.10-0.40	0.15-0.35	1.00-1.80	2.0-3.3	0.20-0.60	---
A 372 Gr L	K24055	0.38-0.43	0.60-0.80	0.15-0.35	0.70-0.90	1.65-2.00	0.20-0.30	---
A 372 Gr M Cl A, B	---	0.23	0.20-0.40	0.30	1.50-2.00	2.8-3.9	0.40-0.60	0.08 V
A 469 Cl 2	K22573	0.25	0.60	0.15-0.30	0.50	2.50 min	0.20-0.50	0.03 V min
A 469 Cl 3	K22773	0.27	0.60	0.15-0.30	0.50	2.50 min	0.20-0.50	0.03 V min
A 469 Cl 4	K32723	0.27	0.70	0.15-0.30	0.50	3.00 min	0.20-0.60	0.03 V min
A 469 Cl 5	K33125	0.31	0.70	0.15-0.30	0.50	3.00 min	0.20-0.70	0.05-0.15 V
A 469 Cl 6, 7, 8	K42885	0.28	0.60	0.15-0.30	1.25-2.00	3.25-4.00	0.30-0.60	0.05-0.15 V
A 470 Cl 2	K22578	0.25	0.20-0.60	0.15-0.30	0.75	2.50 min	0.25 min	0.03 V min
A 470 Cl 3, 4	K22878	0.28	0.20-0.60	0.15-0.30	0.75	2.50 min	0.25 min	0.03 V min
A 470 Cl 5, 6, 7	K42885	0.28	0.20-0.60	0.10	1.25-2.00	3.25-4.00	0.25-0.60	0.05-0.15 V
A 470 Cl 8	K23010	0.25-0.35	1.00	0.15-0.35	1.05-1.50	0.75	1.00-1.50	0.20-0.30 V
A 470 Cl 9	---	0.30	0.70	0.15-0.35	0.75	2.00 min	0.25 min	0.03-0.12 V
A 471 Cl 1, 2, 3	K32800	0.28	0.70	0.15-0.35	0.75-2.00	2.00-4.00	0.20-0.70	0.05 V min

CHEMICAL COMPOSITION OF ALLOY STEEL FORGINGS (Continued)

ASTM Spec.	UNS	C	Mn	Si	Cr	Ni	Mo	Other
A 471 Cl 4, 5	K32800	0.28	0.70	0.15-0.35	0.75-2.00	2.00-4.00	0.20-0.70	0.05 V min
A 471 Cl 6	K32800	0.28	0.70	0.15-0.35	0.75-2.00	2.00-4.00	0.20-0.70	0.05 V min
A 471 Cl 10	K23205	0.27-0.37	0.70-1.00	0.20 min	0.85-1.25	0.50	1.00-1.50	0.20-0.30 V
A 471 Cl 11, 12, 13	---	0.38-0.43	0.60-1.00	0.15-0.35	0.80-1.10	0.50	0.15 min	0.06 V
A 471 Cl 14	---	0.45	0.60-1.00	0.15-0.35	0.50-1.25	1.65-3.50	0.20 min	V optional
A 508 Gr 2	K12766	0.27	0.50-1.00	0.15-0.40	0.25-0.45	0.50-1.00	0.55-0.70	0.05 V
A 508 Gr 3	K12042	0.25	1.20-1.50	0.15-0.40	0.25	0.40-1.00	0.45-0.60	0.05 V
A 508 Gr 3V	K31830	0.10-0.15	0.30-0.60	0.10	2.8-3.3	---	0.90-1.10	0.20-0.30 V, 0.001-0.003 B, 0.015-0.035 Ti
A 508 Gr 4N	K22375	0.23	0.20-0.40	0.15-0.40	1.50-2.00	2.8-3.9	0.40-0.60	0.03 V
A 508 Gr 5	K42365	0.23	0.20-0.40	0.30	1.50-2.00	2.8-3.9	0.40-0.60	0.08 V
A 508 Gr 22	K21590	0.11-0.15	0.30-0.60	0.50	2.00-2.50	0.25	0.90-1.10	0.02 V
A 522 Type I[d]	K81340	0.13	0.90	0.15-0.30	---	8.5-9.5	---	---
A 522 Type II[d]	K71340	0.13	0.90	0.15-0.30	---	7.5-8.5	---	---
A 541 Gr 1C	K11800	0.18	1.30	0.15-0.35	0.15	0.25	0.05	0.02-0.12 V
A 541 Gr 2	K12765	0.27	0.50-0.90	0.15-0.35	0.25-0.45	0.50-1.00	0.55-0.70	0.05 V
A 541 Gr 3	K12045	0.25	1.20-1.50	0.15-0.35	0.25	0.40-1.00	0.45-0.60	0.05 V
A 541 Gr 3V	K31830	0.10-0.15	0.30-0.60	0.10	2.8-3.3	---	0.90-1.10	0.20-0.30 V, 0.015-0.035 Ti, 0.001-0.003 B
A 541 Gr 4N	K42343	0.23	0.20-0.40	0.30	1.25-2.00	2.8-3.9	0.40-0.60	0.03
A 541 Gr 5	K42348	0.23	0.20-0.40	0.30	1.25-2.00	2.8-3.9	0.40-0.60	0.08
A 541 Gr 11 Cl 4	K11572	0.10-0.20	0.30-0.80	0.50-1.00	1.00-1.50	0.50	0.45-0.65	0.05 V
A 541 Gr 22 Cl 3	K21390	0.11-0.15	0.30-0.60	0.50	2.00-2.50	0.25	0.90-1.10	0.02 V
A 541 Gr 22 Cl 4, 5	K21390	0.05-0.15	0.30-0.60	0.50	2.00-2.50	0.50	0.90-1.10	0.05 V
A 541 Gr 22V	---	0.11-0.15	0.30-0.60	0.10	2.00-2.50	0.25	0.90-1.10	0.25-0.35 V, 0.030 Ti, 0.0020 B, 0.20 Cu, 0.07 Cb, 0.015 Ca

CHEMICAL COMPOSITION OF ALLOY STEEL FORGINGS (Continued)								
ASTM Spec.	UNS	C	Mn	Si	Cr	Ni	Mo	Other
A 592 Gr A[d]	K11856	0.15-0.21	0.80-1.10	0.40-0.80	0.50-0.80	---	0.18-0.28	0.05-0.15 Zr, 0.0025 B
A 592 Gr E[d]	K11695	0.12-0.20	0.40-0.70	0.20-0.35	1.40-2.00	---	0.40-0.60	0.04-0.10 Ti, 0.20-0.40 Cu, 0.0015-0.005 B
A 592 Gr F[d]	K11576	0.10-0.22	0.60-1.00	0.15-0.35	0.40-0.65	0.70-1.00	0.40-0.60	0.03-0.08 V, 0.15-0.50 Cu, 0.002-0.006 B
A 646 Gr 4 (8620)	G86200	0.18-0.23	0.70-0.90	0.20-0.35	0.40-0.60	0.40-0.70	0.15-0.25	---
A 646 Gr 5 (4330 Modified)	K23080	0.28-0.33	0.75-1.00	0.20-0.35	0.70-0.95	1.65-2.00	0.35-0.50	0.05-0.10 V
A 646 Gr 7 (4340)	G43400	0.38-0.43	0.65-0.85	0.20-0.35	0.70-0.90	1.65-2.00	0.20-0.30	---
A 646 Gr 11 (4130)	G41300	0.28-0.33	0.40-0.60	0.20-0.35	0.80-1.10	---	0.15-0.25	---
A 646 Gr 12 (4140)	G41400	0.38-0.43	0.75-1.00	0.20-0.35	0.80-1.10	---	0.15-0.25	---
A 707 Gr L3	K12510	0.22	1.15-1.50	0.30	0.30	0.40	0.12	0.04-0.11 V, 0.010-0.030 N, 0.02 Cb, 0.20 Cu min.
A 707 Gr L4[d]	K12089	0.18	0.45-0.65	0.35	0.30	1.65-2.00	0.20-0.30	0.05 V, 0.40 Cu, 0.02 Cb
A 707 Gr L5[d]	K20934	0.07	0.40-0.70	0.35	0.60-0.90	0.70-1.00	0.15-0.25	0.05 V, 1.00-1.30 Cu, 0.03 Cb min
A 707 Gr L6[d]	K20902	0.07	1.85-2.20	0.15	0.30	0.40	0.25-0.35	0.05 V, 0.40 Cu, 0.06-0.10 Cb
A 707 Gr L7[d]	K32218	0.20	0.90	0.35	0.30	3.2-3.7	0.12	0.05 V, 0.40 Cu, 0.02 Cb
A 707 Gr L8[d]	K42247	0.20	0.20-0.40	0.35	1.50-2.00	2.8-3.9	0.40-0.60	0.05 V, 0.40 Cu, 0.02 Cb
A 723 Gr 1[d]	K23550	0.35	0.90	0.35	0.80-2.00	1.5-2.25	0.20-0.40	0.20 V
A 723 Gr 2[d]	K34035	0.40	0.90	0.35	0.80-2.00	2.3-3.3	0.30-0.50	0.20 V
A 723 Gr 3[d]	K44045	0.40	0.90	0.35	0.80-2.00	3.3-4.5	0.40-0.80	0.20 V
A 765 Gr III	K32026	0.20	0.90	0.15-0.35	0.20	3.3-3.8	0.06	0.05 V, 0.05 Al, 0.35 Cu
A 836	---	0.20	0.90	0.35	---	---	---	4 x %C min Ti, 1.00 Ti max
A 859	---	0.07	0.40-0.70	0.40	0.60-0.90	0.70-1.00	0.15-0.25	1.00-1.30 Cu, 0.02 Cb

CHEMICAL COMPOSITION OF ALLOY STEEL FORGINGS (Continued)

a. The sum of chromium and molybdenum shall not exceed 0.32% on heat analysis; by agreement, the heat analysis limit for V and Cb, or both, may be increased up to 0.10% and 0.05%, respectively; see ASTM A 350 for more details. b. The sum of copper, nickel, chromium and molybdenum shall not exceed 1.00% on heat analysis. c. ASTM A 372 Gr J Cl 110 - UNS No. G41370. d. Heat analysis. Single values are maximums, unless otherwise specified. Although sulfur and phosphorous contents are not listed in this table due to space limitation, they are specified, see the appropriate ASTM Specification for details.

MECHANICAL PROPERTIES OF CARBON STEEL FORGINGS

ASTM Spec.	Tensile Strength		Yield Strength		% El	% RA	Hardness HB	Impact Strength ft-lb (J) @ Temp °F (°C)[a]
	ksi	MPa	ksi	MPa				
A 105	70	485	36	250	30	30	187 max	---
A 181 Cl 60	60	415	30	205	22	35	---	---
A 181 Cl 70	70	485	36	250	18	24	---	---
A 266 Gr 1	60-85	415-585	30	205	23	38	121-170[b]	---
A 266 Gr 2, 4	70-95	485-655	36	250	20	33	137-197[b]	---
A 266 Gr 3	75-100	515-690	37.5	260	19	30	156-207[b]	---
A 288 Cl 1	70	485	45	310	18	40	---	15 (20)[c]
A 290 Cl A	80	550	45	310	22	45	163-202	10 (14) @ 70-80 (21-27)[d]
A 290 Cl B	---	---	---	---	---	---	163-202	---
A 290 Cl C	95	655	65	450	20	40	197-241	10 (14) @ 70-80 (21-27)[d]
A 290 Cl D	---	---	---	---	---	---	197-241	---
A 291 Cl 1[e]	85	585	50	345	22	45	170-223	Long. 10 (13)[b] Trans. 8 (11)[b]@70(21)
A 291 Cl 2[e]	95	655	70	485	20	45	201-241	Long. 10 (13)[b] Trans. 8 (11)[b]@70(21)
A 350 Gr LF1	60-85	415-585	30	205	25	38	197 max	13 (18) / 10 (14) @ -20 (-28.9)
A 350 Gr LF2	70-95	485-655	36	250	22	30	197 max	15 (20) / 12 (16) -50 (-45.6)
A 372 Gr A	60-85	415-585	35	240	20	---	121	---
A 372 Gr B	75-100	515-690	45	310	18	---	156	---
A 372 Gr C	90-115	620-795	55	380	15	---	187	---

MECHANICAL PROPERTIES OF CARBON STEEL FORGINGS (Continued)

ASTM Spec.	Tensile Strength		Yield Strength		% El	% RA	Hardness HB	Impact Strength ft-lb (J) @ Temp °F (°C)[a]
	ksi	MPa	ksi	MPa				
A 469 Cl 1	75	515	35	240	20	30	---	10 (13) room temperature
A 470 Cl 1	75	515	40	275	22	40	---	10 (13.6) room temperature
A 508 Gr 1, 1A	70-95	485-655	36	250	20	38	---	15 (20) / 10 (14) @ 40 (4.4)
A 541 Gr 1, 1A	70-95	480-660	36	250	20	38	---	15 (20) / 10 (14) @ 40 (4)
A 649 Cl 2	75	515	37.5	260	20	50	207-285[f]	---
A 649 Cl 4	60	415	30	205	22	55	---	---
A 694 Gr F42	60	415	42	290	20	---	---	---
A 694 Gr F46	60	415	46	315	20	---	---	---
A 694 Gr F48	62	425	48	330	20	---	---	---
A 694 Gr F50	64	440	50	345	20	---	---	---
A 694 Gr F52	66	455	52	360	20	---	---	---
A 694 Gr F56	68	470	56	385	20	---	---	---
A 694 Gr F60	75	515	60	415	20	---	---	---
A 694 Gr F65	77	530	65	450	20	---	---	---
A 694 Gr F70	82	565	70	485	18	---	---	---
A 707 L1 Cl 1	60	415	42	290	22	40	149-207	30 (41) / 24 (33) @ -20 (-29)
A 707 L2 Cl 1	60	415	42	290	22	40	149-207	30 (41) / 24 (33) @ -50 (-46)
A 707 L1 Cl 2	66	455	52	360	22	40	149-217	40 (54) / 32 (43) @ -20 (-29)
A 707 L2 Cl 2	66	455	52	360	22	40	149-217	40 (54) / 32 (43) @ -50 (-46)
A 707 L1 Cl 3	75	515	60	415	20	40	156-235	50 (68) / 40 (54) @ -20 (-29)
A 707 L2 Cl 3	75	515	60	415	20	40	156-235	50 (68) / 40 (54) @ -50 (-46)
A 707 L1 Cl 4	90	620	75	515	20	40	179-265	50 (68) / 40 (54) @ -20 (-29)
A 707 L2 Cl 4	90	620	75	515	20	40	179-265	50 (68) / 40 (54) @ -50 (-46)
A 727	60.0-85.0	415-585	36.0	250	22	30	187 max[g]	---
A 765 Gr I	60-85	415-585	30	205	25	38	---	13 (18) / 10 (14) @ -20 (-30)

MECHANICAL PROPERTIES OF CARBON STEEL FORGINGS (Continued)

ASTM Spec.	Tensile Strength		Yield Strength		% EI	% RA	Hardness HB	Impact Strength ft•lb (J) @ Temp °F (°C)[a]
	ksi	MPa	ksi	MPa				
A 765 Gr II	70-95	485-655	36	250	22	30	---	15 (20) / 12 (16) @ -50 (-45)
A 765 Gr IV	80-105	550-725	50	345	22	30	---	26 (35) / 20 (27) @ -20 (-30)[h]

a. Minimum impact energy required for average of each set of three specimens in ft•lb (J), followed by, minimum impact energy permitted for one specimen only of a set in ft•lb (J), followed by the test temperature in °F (°C); all values are for full size specimens 10 mm x 10 mm, for sub-size specimens see the related ASTM Specification. b. Supplementary requirement only. c. Number of test specimens shall be prescribed by the purchaser. Charpy V-notch testing performed at room temperature. d. Two set of impact tests shall be tested in accordance to ASTM A 370. e. Maximum 10 in. (250 mm) solid diameter or thickness. f. Surfaced hardened pressure roll body forgings shall have a hardness of 58-65 HRC. g. Hardness limit of 187 HB maximum only applies to liquid-quenched and tempered forgings. h. Optional upon agreement between the purchaser and the manufacturer. Single values are minimums, unless otherwise specified.

MECHANICAL PROPERTIES OF ALLOY STEEL FORGINGS

ASTM Spec.	Tensile Strength		Yield Strength		% EI	% RA	Hardness HB	Impact Strength ft•lb (J) @ Temp °F (°C)[a]
	ksi	MPa	ksi	MPa				
A 182 Gr F1	70	485	40	275	20	30	143-192	---
A 182 Gr F2	70	485	40	275	20	30	143-192	---
A 182 Gr F5	70	485	40	275	20	35	143-217	---
A 182 Gr F5a	90	620	65	450	22	50	187-248	---
A 182 Gr F9	85	585	55	380	20	40	179-217	---
A 182 Gr F91	85	585	60	415	20	40	248 max	---
A 182 Gr F11 Cl 1	60	415	30	205	20	45	121-174	---
A 182 Gr F11 Cl 2	70	485	40	275	20	30	143-207	---
A 182 Gr F11 Cl 3	75	515	45	310	20	30	156-207	---
A 182 Gr F12 Cl 1	60	415	32	220	20	45	121-174	---
A 182 Gr F12 Cl 2	70	485	40	275	20	30	143-207	---
A 182 Gr F21	75	515	45	310	20	30	156-207	---
A 182 Gr F3V	85-110	585-760	60	415	18	45	174-237	40 (54) / 35 (48) @ 0 (-18)

MECHANICAL PROPERTIES OF ALLOY STEEL FORGINGS (Continued)

ASTM Spec.	Tensile Strength		Yield Strength		% El	% RA	Hardness HB	Impact Strength ft·lb (J) @ Temp °F (°C)[a]
	ksi	MPa	ksi	MPa				
A 182 Gr F22 Cl 1	60	415	30	205	20	35	170 max	---
A 182 Gr F22 Cl 3	75	515	45	310	20	30	156-207	---
A 182 Gr F22V	85-110	585-780	60	415	18	45	174-237	40 (54) / 35 (48) @ 0 (-18)
A 182 Gr FR	63	435	46	315	25	38	197 max	---
A 291 Cl 3, 3A[b]	105	725	80	550	19	45	223-262	---
A 291 Cl 4[b]	120	825	95	655	16	40	248-293	---
A 291 Cl 5[b]	140	965	115	795	16	40	285-331	---
A 291 Cl 6[b]	145	1000	120	825	15	40	302-352	---
A 291 Cl 7[b]	170	1375	140	960	14	35	341-415	---
A 291 Cl 8[b]	120	825	85	585	15	40	255-302	---
A 336 Gr F1	70-95	485-660	40	275	20	40	---	---
A 336 Gr F3V	85-110	585-760	60	415	18	45	---	40 (54) / 35 (48) @ 0 (-18)
A 336 Gr F5	60-85	415-585	36	250	20	40	---	---
A 336 Gr F5A	80-105	550-725	50	345	19	35	---	---
A 336 Gr F6	85-110	585-760	55	380	18	35	---	---
A 336 Gr F9	85-110	585-760	55	380	20	40	---	---
A 336 Gr F11 Cl 1	60-85	415-585	30	205	20	45	---	---
A 336 Gr F11 Cl 2	70-95	485-660	40	275	20	40	---	---
A 336 Gr F11 Cl 3	75-100	515-690	45	310	18	40	---	---
A 336 Gr F12	70-95	485-660	40	275	20	40	---	---
A 336 Gr F21 Cl 1	60-85	415-585	30	205	20	45	---	---
A 336 Gr F21 Cl 3	75-100	515-690	45	310	19	40	---	---
A 336 Gr F22 Cl 1	60-85	415-585	30	205	20	45	---	---
A 336 Gr F22 Cl 3	75-100	515-690	45	310	19	40	---	---
A 336 Gr F22V	85-110	585-760	60	415	18	45	---	40 (54) / 35 (48) @ 0 (-18)

MECHANICAL PROPERTIES OF ALLOY STEEL FORGINGS (Continued)

ASTM Spec.	Tensile Strength		Yield Strength		% El	% RA	Hardness HB	Impact Strength ft·lb (J) @ Temp °F (°C)[a]
	ksi	MPa	ksi	MPa				
A 336 Gr F91	85-110	585-760	60	415	20	40	---	15 (20) / 12 (16) @ -150 (-101.1)
A 350 Gr LF3 Cl 2	70-95	485-655	37.5	260	22	35	197 max	15 (20) / 12 (16) @ -75 (-59)
A 350 Gr LF5 Cl 1	60-85	415-585	30	205	25	38	197 max	15 (20) / 12 (16) @ -75 (-59)
A 350 Gr LF5 Cl 2	70-95	485-655	37.5	260	22	35	197 max	15 (20) / 12 (16) @ -75 (-59.4)
A 350 Gr LF6 Cl 1	66-91	455-630	52	360	22	40	197 max	15 (20) / 12 (16) @ -60 (-50)
A 350 Gr LF6 Cl 2	75-100	515-690	60	415	20	40	197 max	20 (27) / 15 (20) @ -60 (-50)
A 350 Gr LF9	63-88	435-605	46	315	25	38	197 max	13 (18) / 10 (14) @ -100 (-73.3)
A 350 Gr LF787 Cl 2	65-85	450-585	55	380	20	45	197 max	15 (20) / 12 (16) @ -75 (-59)
A 350 Gr LF787 Cl 3	75-95	515-655	65	450	20	45	197 max	15 (20) / 12 (16) @ -100 (-73)
A 372 Gr D	105-130	725-895	65	450	15	---	217	---
A 372 Gr E, F, G, H, Cl 55	85-110	545-760	55	380	20	---	179	---
A 372 Gr E, F, G, H, Cl 65	105-130	725-895	65	450	19	---	217	---
A 372 Gr E, F, G, H, Cl 70	120-145	825-1000	70	485	18	---	248	---
A 372 Gr J Cl 110	135-160	930-1100	110	760	15	---	277	---
A 372 Gr K	100-125	690-860	80	550	20	---	207	---
A 372 Gr L	155-180	1070-1240	135	930	12	---	311	---
A 372 Gr M Cl A	105-130	725-895	85	585	18	---	217	---
A 372 Gr M Cl B	120-145	825-1000	100	690	16	---	248	---
A 469 Cl 2[c]	80	550	55	380	20	50	---	30 (41) @ room temperature
A 469 Cl 3[c]	90	620	70	485	20	50	---	30 (41) @ room temperature
A 469 Cl 4[c]	100	690	80	550	17	45	---	25 (34) @ room temperature
A 469 Cl 5[c]	110	760	90	620	15	40	---	15 (20) @ room temperature
A 469 Cl 6[c]	100	690	80	550	18	55	---	60 (81) @ room temperature
A 469 Cl 7[c]	110	760	90	620	17	50	---	50 (68) @ room temperature

MECHANICAL PROPERTIES OF ALLOY STEEL FORGINGS (Continued)								
	Tensile Strength		Yield Strength				Hardness	Impact Strength
ASTM Spec.	ksi	MPa	ksi	MPa	% El	% RA	HB	ft·lb (J) @ Temp °F (°C)[a]
A 469 Cl 8 [c]	120	825	100	690	16	45	---	40 (54) @ room temperature
A 470 Cl 2 [c]	80	550	55	380	22	50	---	28 (38.0) @ room temperature
A 470 Cl 3 [c]	90	620	70	483	20	48	---	25 (34.0) @ room temperature
A 470 Cl 4 [c]	105	725	85	585	17	45	---	20 (27.2) @ room temperature
A 470 Cl 5 [c]	90-110	620-760	70	483	20	52	---	50 (68.0) @ room temperature
A 470 Cl 6 [c]	105-125	725-860	85	585	18	52	---	45 (61.2) @ room temperature
A 470 Cl 7 [c]	120-135	825-930	95	655	18	52	---	40 (54.4) @ room temperature
A 470 Cl 8 [c]	105-125	725-860	85	585	17	43	---	6 (8.2) @ room temperature
A 470 Cl 9 [c]	95	655	70	485	20	40	---	12 (16) @ room temperature
A 471 Cl 1 [c]	100	690	80-100	550-690	20	50	---	50 @ room temperature
A 471 Cl 2 [c]	105	725	90-110	620-760	19	50	---	50 @ room temperature
A 471 Cl 3 [c]	110	760	100-120	690-825	18	47	---	45@ room temperature
A 471 Cl 4 [c]	120	830	110-130	760-895	17	45	---	45@ room temperature
A 471 Cl 5 [c]	130	900	120-140	825-965	16	43	---	40@ room temperature
A 471 Cl 6 [c]	140	965	130-150	895-1035	15	43	---	40@ room temperature
A 471 Cl 10 [c]	105	725	90-105	620-725	15	30	---	10@ room temperature
A 471 Cl 11	100	690	75-95	515-655	20.0	50.0	207-255	15 (20) @ room temperature
A 471 Cl 12	110	760	85-105	585-725	18.0	48.0	229-269	15 (20) @ room temperature
A 471 Cl 13	125	860	105-125	725-860	16.0	45.0	255-302	15 (20) @ room temperature
A 471 Cl 14	125	860	105-125	725-860	18.0	45.0	255-302	15 (20) @ room temperature
A 508 Gr 2 Cl 1, Gr 3 Cl 1	80-105	550-725	50	345	18	38	---	30 (41) / 25 (34) @ 40 (4.4)
A 508 Gr 2 Cl 2, Gr 3 Cl 2	90-115	620-795	65	450	16	35	---	35 (48) / 30 (41) @ 70 (21)
A 508 Gr 3V	85-110	585-760	60	415	18	45	---	40 (55) / 35 (50) @ 0 (-18)

MECHANICAL PROPERTIES OF ALLOY STEEL FORGINGS (Continued)

ASTM Spec.	Tensile Strength		Yield Strength		% El	% RA	Hardness HB	Impact Strength ft·lb (J) @ Temp °F (°C)[a]
	ksi	MPa	ksi	MPa				
A 508 Gr 4N Cl 1, Gr 5 Cl 1	105-130	725-895	85	585	18	45	---	35 (48) / 30 (41) @ -20 (-29)
A 508 Gr 4N Cl 2, Gr 5 Cl 2	115-140	795-965	100	690	16	45	---	35 (48) / 30 (41) @ -20 (-29)
A 508 Gr 4N Cl 3	90-115	620-795	70	485	20	48	---	35 (48) / 30 (41) @ -20 (-29)
A 508 Gr 22 Cl 3	85-110	585-760	55	380	18	45	---	40 (55) / 35 (50) @ 0 (-18)
A 541 Gr 2 Cl 1, Gr 3 Cl 1, 1C	80-105	550-720	50	340	18	38	---	30 (41) / 25 (34) @ 40 (4)
A 541 Gr 2 Cl 2, Gr 3 Cl 2	90-115	620-790	65	450	16	35	---	35 (47) / 30 (41) @ 70 (21)
A 541 Gr 3V, 22V	85-110	585-760	60	415	18	45	---	40 (55) / 35 (50) @ 0 (-18)
A 541 Gr 4N Cl 1, Gr 5 Cl 1	105-130	720-900	85	590	18	48	---	35 (47) / 30 (41) @ 40 (4)
A 541 Gr 4N Cl 2, Gr 5 Cl 2	115-140	790-1000	100	690	16	45	---	35 (47) / 30 (41) @ 40 (4)
A 541 Gr 4N Cl 3	90-115	620-790	70	480	20	48	---	35 (47) / 30 (41) @ 40 (4)
A 541 Gr 11 Cl 4	80-105	550-720	50	340	18	38	---	15 (20) / 10 (14) @ 40 (4)
A 541 Gr 22 Cl 3	85-110	585-760	55	380	18	45	---	40 (55) / 35 (50) @ 0 (-18)
A 541 Gr 22 Cl 4	105-130	720-900	85	590	16	45	---	35 (47) / 30 (41) @ 40 (4)
A 541 Gr 22 Cl 5	115-140	790-1000	100	690	15	40	---	25 (34) / 20 (27) @ 40 (4)
A 694 Gr F42[f]	60	415	42	290	20	---	---	---
A 694 Gr F46[f]	60	415	46	315	20	---	---	---
A 694 Gr F48[f]	62	425	48	330	20	---	---	---
A 694 Gr F50[f]	64	440	50	345	20	---	---	---
A 694 Gr F52[f]	66	455	52	360	20	---	---	---
A 694 Gr F56[f]	68	470	56	385	20	---	---	---
A 694 Gr F60[f]	75	515	60	415	20	---	---	---
A 694 Gr F65[f]	77	530	65	450	20	---	---	---
A 694 Gr F70[f]	82	565	70	485	18	---	---	---
A 707 Cl 1[d]	60	415	42	290	22	40	149-207	30 (41) / 24 (33)[e]

MECHANICAL PROPERTIES OF ALLOY STEEL FORGINGS (Continued)								
ASTM Spec.	**Tensile Strength**		**Yield Strength**		**% El**	**% RA**	**Hardness**	**Impact Strength**
	ksi	MPa	ksi	MPa			HB	ft·lb (J) @ Temp °F (°C)[a]
A 707 Cl 2[d]	66	455	52	360	22	40	149-217	40 (54) / 32 (43)[e]
A 707 Cl 3[d]	75	515	60	415	20	40	156-235	50 (68) / 40 (54)[e]
A 707 Cl 4[d]	90	620	75	515	20	40	179-265	50 (68) / 40 (54)[e]
A 723 Cl 1	115	795	100	690	16	50	---	35 (47) / 30 (41) @ 40 (4.5)
A 723 Cl 2	135	930	120	825	14	45	---	30 (41) / 25 (34) @ 40 (4.5)
A 723 Cl 2a	145	1000	130	895	12.5	43	---	28 (38) / 23 (31) @ 40 (4.5)
A 723 Cl 3	155	1070	140	965	13	40	---	25 (34) / 20 (27) @ 40 (4.5)
A 723 Cl 4	175	1205	160	1105	12	35	---	20 (27) / 15 (20) @ 40 (4.5)
A 723 Cl 5	190	1310	180	1240	10	30	---	12 (16) / 10 (14) @ 40 (4.5)
A 765 Gr III	70-95	485-655	37.5	260	22	35	---	15 (20) / 12 (16) @ -150 (-100)
A 909 Cl 60	75	515	60	415	18	---	167	---
A 909 Cl 80	95	655	80	550	15	---	201	---
A 909 Cl 100	125	860	100	690	10	---	269	---
A 909 Cl 120	150	1030	120	825	8	---	321	---

a. Min. impact energy required for average of each set of 3 specimens in ft·lb (J), followed by min. impact energy permitted for 1 specimen only of a set in ft·lb (J), followed by the test temperature in °F (°C); all values are for full size specimens 10 mm x 10 mm, for sub-size specimens see the related ASTM specification. b. Maximum 10 in. (250 mm) solid diameter or thickness. c. Transition temperature impact test required, see the related ASTM Specification for more details. d. The availability of a particular size of flange of a specific grade and class is limited only by the capability of the composition to meet the specific mechanical property requirements. However, current practice normally limits the following: Grades L1 and L7 to Classes 1 and 2, Grades L2, L3, and L4 to Classes 1, 2, and 3, and Grades L5, L6, and L8 are generally available in any class. e. Impact test temperature for Grade L1 is -20 °F (-29 °C); for Grades L2 and L3 is -50 °F (-46 °C); for Grades L4, L5, L6 is -80 °F (-62 °C) and for Grades L7 and L8 is -100 °F (-73 °C). f. Alloy steel shall conform to the requirements for Grade L5 of ASTM A 707/A 707M. Single values are minimums, unless otherwise specified.

MECHANICAL PROPERTIES OF MARTENSITIC STAINLESS STEEL FORGINGS

AISI Type (UNS No.) Product Form	ASTM Specification	Heat Treat Condition[a]	Tensile Strength		Yield Strength		% El	% RA	Hardness, max
			ksi	MPa	ksi	MPa			
Type 410 (UNS S41000)									
Flanges, Forged Fittings, Valves, Parts	A 182 Gr F 6a Cl 1	A, NT	70	485	40	275	18	35	143-187 HB
	A 182 Gr F 6a Cl 2	A, NT	85	585	55	380	18	35	167-229 HB
Type 410 (UNS S41000)									
Forging	A 182 Gr F 6a Cl 3	A, NT	110	760	85	585	15	35	235-302 HB
	A 182 Gr F 6a Cl 4	A, NT	130	895	110	760	12	35	263-321 HB
	A 473	A	70	485	40	275	20	45	223 HB
Type 416 (UNS S41600)									
Forging	A 473	A	70	485	40	275	20	45	223 HB
Type 440A (UNS S44002)									
Forging	A 473	A	---	---	---	---	---	---	269 HB
Type 440B (UNS S44003)									
Forging	A 473	A	---	---	---	---	---	---	269 HB
Type 440C (UNS S44004)									
Forging	A 473	A	---	---	---	---	---	---	269 HB

a. A - annealed, HF - hot finished, NT - normalized and tempered.
Single values are minimums, except for hardness values that are maximums, unless otherwise specified.

MECHANICAL PROPERTIES OF FERRITIC STAINLESS STEEL FORGINGS

Common Name (UNS No.) / Product Form	ASTM Specification	Heat Treat Condition[a]	Tensile Strength ksi	Tensile Strength MPa	Yield Strength ksi	Yield Strength MPa	% El	% RA	Hardness, max
AISI Type 410 (UNS S41000)									
Forging	A 336 Gr F6	A, NT	85-110	585-760	55	380	18	35	---
AISI Type 430 (UNS S43000)									
Flanges, Forged Fittings, Valves, Parts	A 182 Gr F 430	A	60	415	35	240	20.0	45.0	190 HB
Forging	A 473	A	70	485	35	240	20	45	217 HB
E-Brite, Type XM-27, 26-1 (UNS S44627)									
Flanges, Forged Fittings, Valves, Parts	A 182 Gr F XM-27Cb	A	60	415	35	240	20.0	45.0	190 HB

a. A - annealed. Note: Single values are minimums, except for hardness values that are maximums, unless otherwise specified.

MECHANICAL PROPERTIES OF AUSTENITIC STAINLESS STEEL FORGINGS[a]

Product Form	ASTM Specification	Grade	UNS No.	Heat Treat Condition[b]	Tensile Strength ksi	Tensile Strength MPa	Yield Strength ksi	Yield Strength MPa	% El	% RA
Flanges, Forged Fittings, Valves, Parts (≤ 5 in.)	A 182	F304	S30400	ST & Q	75	515	30	205	30	50
		F304L	S30403	ST & Q	70	485	25	170	30	50
		F310	S31000	ST & Q	75	515	30	205	30	50
		F316	S31600	ST & Q	75	515	30	205	30	50
		F316L	S31603	ST & Q	70	485	25	170	30	50
		F317	S31700	ST & Q	75	515	30	205	30	50
		F317L	S31703	ST & Q	70	485	25	170	30	50
		F321	S32100	ST & Q	75	515	30	205	30	50
		F347	S34700	ST & Q	75	515	30	205	30	50
		F348	S34800	---	75	515	30	205	30	50
Forging	A 336	F304	S30400	---	70	485	30	205	30	45

MECHANICAL PROPERTIES OF AUSTENITIC STAINLESS STEEL FORGINGS[a] (Continued)

Product Form	ASTM Specification	Grade	UNS No.	Heat Treat Condition[b]	Tensile Strength ksi	Tensile Strength MPa	Yield Strength ksi	Yield Strength MPa	% El	% RA
Forging (continued)	A 336	F304L	S30403	---	65	450	25	170	30	45
		F309H	S30909	---	70	---	30	---	30	45
		F310H	S31009	---	70	---	30	---	30	45
		F316	S31600	---	70	485	30	205	30	45
		F316L	S31603	---	65	450	25	170	30	45
		F321	S32100	---	70	485	30	205	30	45
		F347	S34700	---	70	485	30	205	30	45
		F348	S34800	---	70	485	30	205	30	45
Forging	A 473	304	S30400	A	75	515	30	205	40	50
		304L	S30403	A	65	450	25	170	40	50
		308	S30800	A	75	515	30	205	40	50
		310	S31000	A	75	515	30	205	40	50
		316	S31600	A	75	515	30	205	40	50
		316L	S31603	A	65	450	25	170	40	50
		317	S31700	A	75	515	30	205	40	50
		321	S32100	A	75	515	30	205	40	50
		347	S34700	A	75	515	30	205	40	50
		348	S34800	A	75	515	30	205	40	50
Pipe Flanges, Forged Fittings, Valves & Parts	B 462	20Cb-3	N08020	Stabilized A	80	551	35	241	30.0	50.0
		20Mo-4	N08024	A	80	551	35	241	30.0	50.0
		20Mo-6	N08026	Solution A	80	551	35	241	30.0	50.0
		AL-6XN	N08367	Solution A	95	655	45	310	30.0	50.0
Forging	B 564	AL-6XN	N08367	---	95	655	45	310	30	---

a. Fe-Ni-Cr-Mo alloys and their variations have been included in this table of "austenitic stainless steels". b. A - annealed, ST & Q - solution treated and quenched. Single values are minimums, unless otherwise specified.

MECHANICAL PROPERTIES OF DUPLEX STAINLESS STEEL FORGINGS

Common Name, UNS No.	ASTM Specification	Heat Treat Condition[a]	Tensile Strength ksi	Tensile Strength MPa	Yield Strength ksi	Yield Strength MPa	% El	% RA	Hardness, max
44LN, UNS S31200	A 182 Gr F50	ST & Q	100-130	690-900	65	450	25	50	---
2205, UNS S31803	A 182 Gr F51	ST & Q	90	620	65	450	25	45	---
UNS S39274	A 182 Gr F54	ST & Q	116	800	80	550	15	30	310 HB
2507, UNS S32750	A 182 Gr F53	ST & Q	116	800	80	550	15	---	310 HB
Zeron 100, UNS S32760	A 182 Gr F55	ST & Q	109-130	750-895	80	550	25.0	45	---
7-Mo PLUS, UNS S32950	A 182 Gr F52	ST & Q	100	690	70	485	15	---	---
UNS S39277	A 182 Gr F57	ST & Q	118	820	85	585	25	50	---
UNS S32520	A 182 Gr F59	ST & Q	112	770	80	550	25	40	---

a. A – annealed, ST & Q – solution treated and quenched.
Single values are minimums, except for hardness values that are maximums, unless otherwise specified.

MECHANICAL PROPERTIES OF PRECIPITATION-HARDENING STAINLESS STEEL FORGINGS

Common Names (UNS No.)	ASTM Spec.	Heat Treat Condition[a]	Tensile Strength ksi	Tensile Strength MPa	Yield Strength ksi	Yield Strength MPa	% El[b] L, T	% RA[b] L, T	Hardness, max	Charpy ft·lb/J
PH 13-8 Mo, XM-13 (UNS S13800)										
Forging	A 705	A	---	---	---	---	---	---	38 HRC/363 HB	---
		H950	220	1520	205	1420	10	45, 35	45 HRC/430 HB	---
		H1000	205	14200	190	1310	10	50, 40	43 HRC/400 HB	---
		H1025	185	1280	175	1210	11	50, 45	41 HRC/380 HB	---
		H1050	175	1210	165	1140	12	50, 45	40 HRC/372 HB	---
		H1100	150	1030	135	930	14	50	34 HRC/313 HB	---
		H1150	135	930	90	620	14	50	30 HRC/283 HB	---
		H1150M	125	860	85	585	16	55	26 HRC/259 HB	---

MECHANICAL PROPERTIES OF PRECIPITATION-HARDENING STAINLESS STEEL FORGINGS(Continued)

Common Names (UNS No.)	ASTM Spec.	Heat Treat Condition[a]	Tensile Strength ksi	Tensile Strength MPa	Yield Strength ksi	Yield Strength MPa	% El[b] L, T	% RA[b] L, T	Hardness, max	Charpy ft-lb/J
15-5 PH, XM-12 (UNS S15500)										
Forging	A 705	A	---	---	---	---	---	---	38 HRC/363 HB	---
		H900	190	1310	170	1170	10, 6	35, 15	40 HRC/388 HB	---
		H925	170	1170	155	1070	10, 7	38, 20	38 HRC/375 HB	5/6.8 L
		H1025	155	1070	145	1000	12, 8	45, 27	35 HRC/331 HB	15/20 L, 10/14 T
		H1075	145	1000	125	860	13, 9	45, 28	32 HRC/311 HB	20/27 L, 15/20 T
		H1100	140	965	115	795	14, 10	45, 29	31 HRC/302 HB	25/34 L, 15/20 T
		H1150	135	930	105	725	16, 11	50, 30	28 HRC/277 HB	30/41 L, 20/27 T
		H1150M	115	795	75	515	18, 14	55, 35	24 HRC255 HB	55/75 L, 35/47 T
PH 15-7 Mo, Type 632 (UNS S15700)										
Forging	A 705	A	---	---	---	---	---	---	100 HRB/269 HB	---
		RH950	200	1380	175	1210	7	25	415 HB	---
		TH1050	180	1240	160	1100	8	25	375 HB	---
17-4 PH, Type 630 (UNS S17400)										
Forging (≤ 3 in. - 75 mm)	A 564	A	---	---	---	---	---	---	38 HRC/363 HB	---
		H900	190	1310	170	1170	10	40	40 HRC/388 HB	---
		H925	170	1170	155	1070	10	44	38 HRC/375 HB	5/6.8
(≤ 8 in. - 200 mm)	A564	H1025	155	1070	145	1000	12	45	35 HRC/331 HB	15/20
		H1075	145	1000	125	860	13	45	32 HRC/311 HB	20/27
		H1100	140	965	115	795	14	45	31 HRC/302 HB	25/34
		H1150	135	930	105	725	16	50	28 HRC/277 HB	30/41
		H1150M	115	795	75	515	18	55	24 HRC/255 HB	55/75
Stainless W, Type 635 (UNS S17600)										
Forging	A 705	A	120	825	75	515	10	45	32 HRC or 302 HB	---
		H950	190	1310	170	1170	8	25	39 HRC/363 HB	---

MECHANICAL PROPERTIES OF PRECIPITATION-HARDENING STAINLESS STEEL FORGINGS (Continued)										
Common Names (UNS No.)	ASTM Spec.	Heat Treat Condition[a]	Tensile Strength ksi	Tensile Strength MPa	Yield Strength ksi	Yield Strength MPa	% El[b] L, T	% RA[b] L, T	Hardness, max	Charpy ft·lb/J
Forging (Continued)	A 705	H1000	180	1240	160	1100	8	30	37 HRC/352 HB	---
		H1050	170	1170	150	1035	10	40	35 HRC/331 HB	---
17-7 PH, Type 631 (UNS S17700)										
Forging	A 705	A	---	---	---	---	---	---	89 HRB/229 HB	---
		RH950	185	1280	150	1030	6	10	41 HRC/388 HB	---
		TH1050	170	1170	140	965	6	25	38 HRC/352 HB	---
AM-355, Type 634 (UNS S35500)										
Forging	A 705	A	---	---	---	---	---	---	363 HB	---
		H1000	170	1170	155	1070	12	25	37 HRC/341 HB	---
Custom 450, XM-25 (UNS S45000)										
Forging	A 705	A	125[c]	860[c]	95	655	10	40	33 HRC/311 HB	---
(≤ 8 in. - 200 mm)		H900	180	1240	170	1170	10	40	39 HRC/363 HB	---
		H950	170	1170	160	1100	10	40	37 HRC/341 HB	---
		H1000	160	1100	150	1030	12	45	36 HRC/331 HB	---
		H1025	150	1030	140	965	12	45	34 HRC/321 HB	---
		H1050	145	1000	135	930	12	45	34 HRC/321 HB	---
		H1100	130	895	105	725	16	50	30 HRC/285 HB	---
		H1150	125	860	75	515	15	50	26 HRC/262 HB	---
Custom 455, XM-16 (UNS S45500)										
Forging	A 705	A	---	---	---	---	---	---	36 HRC/331 HB	---
(≤ 6 in. - 150 mm)		H900	235	1620	220	1520	8	30	47 HRC/444 HB	---
		H950	220	1520	205	1410	10	40	44 HRC/415 HB	---
		H1000	205	1410	185	1280	10	40	40 HRC/363 HB	---

MECHANICAL PROPERTIES OF PRECIPITATION-HARDENING STAINLESS STEEL FORGINGS (Continued)

a Heat treatment condition A is solution treated. b. First value is longitudinal, second value is transverse, if single value is listed it represents both longitudinal and transverse tests. c. Tensile strengths of 130 to 165 ksi (895 to 1140 MPa) for sizes up to ½ in. (13 mm). Single values are minimums, unless otherwise specified.

Chapter

14

TUBULAR PRODUCTS CARBON, ALLOY & STAINLESS STEELS:

AMERICAN SPECIFICATION TITLES & DESIGNATIONS, CHEMICAL COMPOSITIONS & MECHANICAL PROPERTIES

CARBON, ALLOY AND STAINLESS STEEL PIPES

Carbon and Alloy Steel Pipes

ASTM Spec.	Title
A 53	Pipe, Steel, Black and Hot-Dipped, Zinc-Coated, Welded and Seamless
A 106	Seamless Carbon Steel Pipe for High-Temperature Service
A 134	Pipe, Steel, Electric-Fusion (Arc)-Welded (Sizes NPS 16 and Over)
A 135	Electric-Resistance-Welded Steel Pipe
A 139	Electric-Fusion (Arc)-Welded Steel Pipe (NPS 4 and Over)
A 252	Welded and Seamless Steel Pipe Piles
A 333/A 333M	Seamless and Welded Steel Pipe for Low-Temperature Service
A 335/A 335M	Seamless Ferritic Alloy-Steel Pipe for High-Temperature Service
A 369/A 369M	Carbon and Ferritic Alloy Steel Forged and Bored Pipe for High-Temperature Service
A 381	Metal-Arc-Welded Steel Pipe for Use with High-Pressure Transmission Systems
A 426	Centrifugally Cast Ferritic Alloy Steel Pipe for High-Temperature Service
A 523	Plain End Seamless and Electric-Resistance-Welded Steel Pipe for High-Pressure Pipe-Type Cable Circuits
A 524	Seamless Carbon Steel Pipe for Atmospheric and Lower Temperatures
A 530/A 530M	General Requirements for Specialized Carbon and Alloy Steel Pipe
A 587	Electric-Resistance-Welded Low-Carbon Steel Pipe for the Chemical Industry
A 589	Seamless and Welded Carbon Steel Water-Well Pipe
A 660	Centrifugally Cast Carbon Steel Pipe for High-Temperature Service
A 671	Electric-Fusion-Welded Steel Pipe for Atmospheric and Lower Temperatures
A 672	Electric-Fusion-Welded Steel Pipe for High-Pressure Service at Moderate Temperatures
A 691	Carbon and Alloy Steel Pipe, Electric-Fusion-Welded for High-Pressure Service at High Temperatures
A 714	High-Strength Low-Alloy Welded and Seamless Steel Pipe
A 795	Black and Hot-Dipped Zinc-Coated (Galvanized) Welded and Seamless Steel Pipe for Fire Protection Use

ANSI Std.	Title
B 36.10M	Welded and Seamless Wrought Steel Pipe

CARBON, ALLOY AND STAINLESS STEEL PIPES (Continued)

Carbon and Alloy Steel Pipes (Continued)

API Spec.	Title
2B	Specification for fabricated structural steel and pipe
5CT	Specification for casing and tubing
5D	Specification for drill pipe
5L	Specification for line pipe

CSA Std.	Title
CAN3-Z245.1	Steel line pipe

Stainless Steel Pipes

ASTM Spec.	Title
A 312/A 312M	Seamless and Welded Austenitic Stainless Steel Pipes
A 358/A 358M	Electric-Fusion-Welded Austenitic Chromium-Nickel Alloy Steel Pipe for High-Temperature Service
A 376/A 376M	Seamless Austenitic Steel Pipe for High-Temperature Central-Station Service
A 409/A 409M	Welded Large Diameter Austenitic Steel Pipe for Corrosive or High-Temperature Service
A 451	Centrifugally Cast Austenitic Steel Pipe for High-Temperature Service
A 452	Centrifugally Cast Austenitic Steel Cold-Wrought Pipe for High-Temperature Service
A 733	Welded and Seamless Carbon Steel and Austenitic Stainless Steel Pipe Nipples
A 778	Welded, Unannealed Austenitic Stainless Steel Tubular Products
A 790/A 790M	Seamless and Welded Ferritic/Austenitic Stainless Steel Pipe
A 813/A 813M	Single- or Double-Welded Austenitic Stainless Steel Pipe
A 814/A 814M	Cold-Worked Welded Austenitic Stainless Steel Pipe
A 872	Centrifugally Cast Ferritic/Austenitic Stainless Steel Pipe for Corrosive Environments
A 928/A 928M	Ferritic/Austenitic (Duplex) Stainless Steel Pipe Electric Fusion Welded with Addition of Filler Metal
A 943/A 943M	Spray-Formed Seamless Austenitic Stainless Steel Pipes
A 949/A 949M	Spray-Formed Seamless Ferritic/Austenitic Stainless Steel Pipe
A 954	Austenitic Chromium-Nickel-Silicon Alloy Steel Seamless and Welded Pipe

ANSI Std.	Title
B 36.19M	Stainless Steel Pipe

CARBON, ALLOY AND STAINLESS STEEL PIPES (Continued)

Stainless Steel Fittings

ASTM Spec.	Title
A 403/A 403M	Wrought Austenitic Stainless Steel Piping Fittings
A 774/A 774M	As-Welded Wrought Austenitic Stainless Steel Fittings for General Corrosion Service at Low and Moderate Temperatures
A 815/A 815M	Wrought Ferritic, Ferritic/Austenitic, and Martensitic Stainless Steel Piping Fittings

CARBON, ALLOY AND STAINLESS STEEL TUBES

Carbon and Alloy Steel Tubes

Boiler, Superheater, and Miscellaneous Tubes

ASTM Spec.	Title
A 178/A 178M	Electric-Resistance-Welded Carbon Steel and Carbon-Manganese Steel Boiler and Superheater Tubes
A 192/A 192M	Seamless Carbon Steel Boiler Tubes for High-Pressure Service
A 209/A 209M	Seamless Carbon-Molybdenum Alloy-Steel Boiler and Superheater Tubes
A 210/A 210M	Seamless Medium-Carbon Steel Boiler and Superheater Tubes
A 226/A 226M	Electric-Resistance-Welded Carbon Steel Boiler and Superheater Tubes for High-Pressure Service (Discontinued 1997)
A 250/A 250M	Electric-Resistance-Welded Ferritic Alloy-Steel Boiler and Superheater Tubes
A 254	Copper-Brazed Steel Tubing
A 334/A 334M	Seamless and Welded Carbon and Alloy-Steel Tubes for Low-Temperature Service
A 423/A 423M	Seamless and Electric-Welded Low-Alloy Steel Tubes
A 450/A 450M	General Requirements for Carbon, Ferritic Alloy, and Austenitic Alloy Steel Tubes
A 520	Supplementary Requirements for Seamless and Electric-Resistance-Welded Carbon Steel Tubular Products for High-Temperature Service Conforming to ISO Recommendations for Boiler Construction
A 556/A 556M	Seamless Cold-Drawn Carbon Steel Feedwater Heater Tubes
A 822	Seamless Cold-Drawn Carbon Steel Tubing for Hydraulic System Service
A 953	Austenitic Chromium-Nickel-Silicon Alloy Steel Seamless and Welded Tubing

CARBON, ALLOY AND STAINLESS STEEL TUBES (Continued)

Carbon and Alloy Steel Tubes (Continued)

Still Tubes

ASTM Spec.	Title
A 161	Seamless Low-Carbon and Carbon-Molybdenum Steel Still Tubes for Refinery Service
A 200	Seamless Intermediate Alloy-Steel Still Tubes for Refinery Service
A 271	Seamless Austenitic Chromium-Nickel Steel Still Tubes for Refinery Service

Fuel Line Tubing

A 539	Electric-Resistance-Welded Coiled Steel Tubing for Gas and Fuel Oil Lines

Heat-Exchanger and Condenser Tubes

A 179/A 179M	Seamless Cold-Drawn Low-Carbon Steel Heat-Exchanger and Condenser Tubes
A 214/A 214M	Electric-Resistance-Welded Carbon Steel Heat-Exchanger and Condenser Tubes

Mechanical Tubing

A 512	Cold-Drawn Buttweld Carbon Steel Mechanical Tubing
A 513	Electric-Resistance-Welded Carbon and Alloy Steel Mechanical Tubing
A 519	Seamless Carbon and Alloy Steel Mechanical Tubing
A 787	Electric-Resistance-Welded Metallic-Coated Carbon Steel Mechanical Tubing

Structural Tubing

A 500	Cold-Formed Welded and Seamless Carbon Steel Structural Tubing in Rounds and Shapes
A 501	Hot-Formed Welded and Seamless Carbon Steel Structural Tubing
A 595	Steel Tubes, Low-Carbon, Tapered for Structural Use
A 618	Hot-Formed Welded and Seamless High-Strength Low-Alloy Structural Tubing
A 847	Cold-Formed Welded and Seamless High-Strength, Low-Alloy Structural Tubing with Improved Atmospheric Corrosion Resistance

Stainless Steel Tubes

Stainless Steel Boiler, Superheater and Miscellaneous Tubes

A 213/A 213M	Seamless Ferritic and Austenitic Alloy-Steel Boiler, Superheater, and Heat-Exchanger Tubes
A 249/A 249M	Welded Austenitic Steel Boiler, Superheater, Heat-Exchanger, and Condenser Tubes
A 268/A 268M	Seamless and Welded Ferritic and Martensitic Stainless Steel Tubing for General Service
A 269	Seamless and Welded Austenitic Stainless Steel Tubing for General Service

CARBON, ALLOY AND STAINLESS STEEL TUBES (Continued)

Stainless Steel Tubes (Continued)

Stainless Steel Boiler, Superheater and Miscellaneous Tubes (Continued)

ASTM Spec.	Title
A 270	Seamless and Welded Austenitic Stainless Steel Sanitary Tubing
A 608	Centrifugally Cast Iron-Chromium-Nickel High-Alloy Tubing for Pressure Application at High Temperatures
A 632	Seamless and Welded Austenitic Stainless Steel Tubing (Small-Diameter) for General Service
A 688/A 688M	Welded Austenitic Stainless Steel Feedwater Heater Tubes
A 771/A 771M	Seamless Austenitic and Martensitic Stainless Steel Tubing for Liquid Metal-Cooled Reactor Core Components
A 789/A 789M	Seamless and Welded Ferritic/Austenitic Stainless Steel Tubing for General Service
A 803/A 803M	Welded Ferritic Stainless Steel Feedwater Heater Tubes
A 826/A 826M	Seamless Austenitic and Martensitic Stainless Steel Duct Tubes for Liquid Metal-Cooled Reactor Core Components
A 851	High-Frequency Induction Welded, Unannealed, Austenitic Stainless Steel Condenser Tubes
A 908	Stainless Steel Needle Tubing

Stainless Steel Still Tubes

A 271	Seamless Austenitic Chromium-Nickel Steel Still Tubes for Refinery Service

Stainless Steel Heat-Exchanger and Condenser Tubes

A 213/A 213M	Seamless Ferritic and Austenitic Alloy-Steel Boiler, Superheater, and Heat-Exchanger Tubes
A 249/A 249M	Welded Austenitic Steel Boiler, Superheater, Heat-Exchanger, and Condenser Tubes
A 498	Seamless and Welded Carbon, Ferritic, and Austenitic Alloy Steel Heat-Exchanger Tubes with Integral Fins
A 851	High-Frequency Induction Welded, Unannealed, Austenitic Stainless Steel Condenser Tubes

Stainless Steel Mechanical Tubing

A 511	Seamless Stainless Steel Mechanical Tubing
A 554	Welded Stainless Steel Mechanical Tubing

CHEMICAL COMPOSITION OF CARBON STEEL PIPES

ASTM Specification	Pipemaking Process	C	Mn	P	S	Si
A 53 (Type F)[e]	Furnace welded	0.30	1.20	0.05	0.045	---
A 53 (Type E) Gr A[e]	Electric-resistance-welded	0.25	0.95	0.05	0.045	---
A 53 (Type E) Gr B[e]	Electric-resistance-welded	0.30	1.20	0.05	0.045	---
A 53 (Type S) Gr A[e]	Seamless	0.25	0.95	0.05	0.045	---
A 53 (Type S) Gr B[e]	Seamless	0.30	1.20	0.05	0.045	---
A 106 Gr A	Seamless	0.25[a]	0.27-0.93	0.035	0.035	0.10
A 106 Gr B	Seamless	0.30[a]	0.29-1.06	0.035	0.035	0.10
A 106 Gr C	Seamless	0.35[a]	0.29-1.06	0.035	0.035	0.10
A 135 Gr A	Electric resistance welded	0.25	0.95	0.035	0.035	---
A 135 Gr B	Electric resistance welded	0.30	1.20	0.035	0.035	---
A 139 Gr A	Arc welded	---	1.00	0.035	0.035	---
A 139 Gr B	Arc welded	0.30	1.00	0.035	0.035	---
A 139 Gr C	Arc welded	0.30	1.20	0.035	0.035	---
A 139 Gr D	Arc welded	0.30	1.30	0.035	0.035	---
A 139 Gr E	Arc welded	0.30	1.40	0.035	0.035	---
A 252	Welded or seamless	---	---	0.050	---	---
A 333 Gr 1	Welded or seamless	0.30	0.40-1.06	0.025	0.025	---
A 333 Gr 6	Welded or seamless	0.30	0.29-1.06	0.025	0.025	0.10 min
A 381	Submerged arc welded	0.26	1.40	0.025	0.025	---
A 523 Gr A	Seamless	0.22	0.90	0.035	0.050	---
A 523 Gr A	Electric resistance welded	0.21	0.90	0.035	0.050	---
A 523 Gr B	Seamless	0.27	1.15	0.035	0.050	---
A 523 Gr B	Electric resistance welded	0.26	1.15	0.035	0.050	---
A 524 Gr I, II	Seamless	0.21	0.90-1.35	0.035	0.035	0.10-0.40
A 587 [b]	Electric resistance welded	0.15	0.27-0.63	0.035	0.035	---

CHEMICAL COMPOSITION OF CARBON STEEL PIPES (Continued)

ASTM Specification (Continued)	Pipemaking Process	C	Mn	P	S	Si
A 589	Welded or seamless	---	---	0.050	0.060	---
A 795 Gr A	Welded or seamless	0.25	0.95	0.035	0.035	---
A 795 Gr B	Welded or seamless	0.30	1.20	0.035	0.035	---
API Specification						
5CT Gr H40, J55, K55, N80	Welded or seamless	---	---	0.030	0.030	---
5CT Gr C95	Welded or seamless	0.45	1.90	0.030	0.030	0.45
5CT Gr P110	Welded or seamless	---	---	0.030	0.030	---
5L Gr A25, Class I	Seamless	0.21	0.30-0.60	0.030	0.030	---
5L Gr A25, Class I	Electric or continuous welded	0.21	0.30-0.60	0.030	0.030	---
5L Gr A25, Class II	Seamless	0.21	0.30-0.60	0.045-0-080	0.030	---
5L Gr A25, Class II	Electric or continuous welded	0.21	0.30-0.60	0.045-0.080	0.030	---
5L Gr A	Seamless	0.22	0.90	0.030	0.030	---
5L Gr A	Welded	0.21	0.90	0.030	0.030	---
5L Gr B	Seamless	0.27	1.15	0.030	0.030	---
5L Gr B	Welded	0.26	1.15	0.030	0.030	---
5L Gr X42	Seamless, nonexpanded	0.29	1.25	0.030	0.030	---
5L Gr X42	Welded	0.28	1.25	0.030	0.030	---
5L Gr X42, X46, X52	Seamless, cold expanded	0.29	1.25	0.030	0.030	---
5L Gr X46, X52	Seamless, nonexpanded	0.31	1.35	0.030	0.030	---
5L Gr X46, X52	Welded, nonexpanded	0.30	1.35	0.030	0.030	---
5L Gr X46, X52	Welded, cold expanded	0.28	1.25	0.030	0.030	---
5L Gr X56, X60	Seamless	0.26	1.35	0.030	0.030	---
5L Gr X56, X60	Welded	0.26	1.35	0.030	0.030	---
5L Gr X65	Welded	0.26	1.40	0.030	0.030	---
5L Gr X70	Welded	0.23	1.60	0.030	0.030	---
5L Gr X80[c]	Welded	0.18	1.80	0.030	0.030	---

CHEMICAL COMPOSITION OF CARBON STEEL PIPES (Continued)

CSA Specification

Z245.1 All Grades[d]	Seamless or welded	0.31	2.00	0.030	0.035	0.50

a. For each reduction of 0.01% C below the specified maximum, an increase of 0.06% Mn above the specified maximum will be permitted up to a maximum of 1.35% Mn.

b. Contains 0.02-0.100% Al.

c. For each reduction of 0.01% C below the specified maximum, an increase of 0.05% Mn above the specified maximum will be permitted up to a maximum of 2.0% Mn.

d. Other chemical composition limits (maximums): 0.11% Nb, 0.11% Ti, 0.11% V, 0.001% B. Carbon equivalent 0.40 maximum.

e. Combination of the following five elements shall not exceed 1.00%: Cu, Ni, Cr, Mo, V.

Single values are maximum unless otherwise specified.

CHEMICAL COMPOSITION OF CARBON STEEL TUBES					
ASTM Spec.[a]	C	Mn	P	S	Si
A 161 low carbon grade	0.10-0.20	0.30-0.80	0.035	0.035	0.25
A 178 Gr A	0.06-0.18	0.27-0.63	0.035	0.035	---
A 178 Gr C	0.35	0.80	0.035	0.035	---
A 178 Gr D	0.27	1.00-1.50	0.030	0.015	0.10 min
A 179	0.06-0.18	0.27-0.63	0.035	0.035	---
A 192	0.06-0.18	0.27-0.63	0.035	0.035	0.25
A 210 Gr A-1	0.27	0.93	0.035	0.035	0.10 min
A 210 Gr C	0.35	0.29-1.06	0.035	0.035	0.10 min
A 214	0.18	0.27-0.63	0.035	0.035	---
A 226	0.06-0.18	0.27-0.63	0.035	0.035	0.25
A 334 Gr 1[b]	0.30	0.40-1.06	0.025	0.025	---
A 334 Gr 6[b]	0.30	0.29-1.06	0.025	0.025	0.10 min
A 539	0.15	0.63	0.035	0.035	---
A 556 Gr A2	0.18	0.27-0.63	0.035	0.035	---
A 556 Gr B2	0.27	0.29-0.93	0.035	0.035	0.10 min
A 556 Gr C2	0.30	0.29-1.06	0.035	0.035	0.10 min

a. Also published as ASME specifications as SA XXX, except for A 161 and A 539.
b. For each reduction of 0.01% C below 0.30%, an increase of 0.05% Mn above 1.06% will be permitted to a maximum of 1.35% manganese.
Single values are maximums, unless otherwise specified.

CHEMICAL COMPOSITION OF ALLOY STEEL PIPE[f]								
ASTM Spec.	Form	C	Mn	Si	Cr	Ni	Mo	Others
A 333 Gr 3	W, S	0.19	0.31-0.64	0.18-0.37	---	3.18-3.82	---	---
A 333 Gr 4[a]	W, S	0.12	0.50-1.05	0.08-0.37	0.44-1.01	0.47-0.98	---	0.40-0.75 Cu
A 333 Gr 7	W, S	0.19	0.90	0.13-0.32	---	2.03-2.57	---	---
A 333 Gr 8	W, S	0.13	0.90	0.13-0.32	---	8.40-9.60	---	---
A 333 Gr 9	W, S	0.20	0.40-1.06	---	---	1.60-2.24	---	0.75-1.25 Cu
A 333 Gr 10[b]	W, S	0.20	1.15-1.50	0.10-0.35	0.15	0.25	0.05	0.15 Cu
A 335 Gr P1	S	0.10-0.20	0.30-0.80	0.10-0.50	---	---	0.44-0.65	---
A 335 Gr P2	S	0.10-0.20	0.30-0.61	0.10-0.30	0.50-0.81	---	0.44-0.65	---
A 335 Gr P5	S	0.15	0.30-0.60	0.50	4.00-6.00	---	0.45-0.65	---
A 335 Gr P5b	S	0.15	0.30-0.60	1.00-2.00	4.00-6.00	---	0.45-0.65	---
A 335 Gr P5c	S	0.12	0.30-0.60	0.50	4.00-6.00	---	0.45-0.65	0.70< Ti ≥ 4 x C or Cb 8-10 x C
A 335 Gr P9	S	0.15	0.30-0.60	0.25-1.00	8.00-10.00	---	0.90-1.10	---
A 335 Gr P11	S	0.05-0.15	0.30-0.60	0.50-1.00	1.00-1.50	---	0.44-0.65	---
A 335 Gr P12	S	0.05-0.15	0.30-0.61	0.50	0.80-1.25	---	0.44-0.65	---
A 335 Gr P15	S	0.05-0.15	0.30-0.60	1.15-1.65	---	---	0.44-0.65	---
A 335 Gr P21	S	0.05-0.15	0.30-0.60	0.50	2.65-3.35	---	0.80-1.06	---
A 335 Gr P22	S	0.05-0.15	0.30-0.60	0.50	1.90-2.60	---	0.87-1.13	---
A 335 Gr P91	S	0.08-0.12	0.30-0.60	0.20-0.50	8.00-9.50	0.40	0.85-1.05	0.18-0.25 V 0.030-0.070 N, 0.04 Al, 0.06-0.10 Cb
A 714 Gr I	W, S	0.22	1.25	---	---	---	---	0.20 Cu min
A 714 Gr II[c]	W, S	0.22	0.85-1.25	0.30	---	---	---	0.20 Cu min
A 714 Gr III[c]	W, S	0.23	1.35	0.30	---	---	---	0.20 Cu min

CHEMICAL COMPOSITION OF ALLOY STEEL PIPE[f] (Continued)

ASTM Spec. (Continued)	Form	C	Mn	Si	Cr	Ni	Mo	Others
A 714 Gr IV	W,S	0.10	0.60	---	0.80-1.20	0.20-0.50	---	0.25-0.45 Cu
A 714 Gr V	W,S	0.16	0.40-1.01	---	---	1.65 min	---	0.80 Cu min
A 714 Gr VI	W,S	0.15	0.50-1.00	---	0.30	0.40-1.10	0.10-0.20	0.30-1.00 Cu
A 714 Gr VII	W,S	0.12	0.20-0.50	0.25-0.75	0.30-1.25	0.65	---	0.25-0.55 Cu
A 714 Gr VIII[d]	W,S	0.19	0.80-1.25	0.30-0.65	0.40-0.65	0.40	---	0.25-0.40 Cu
API Specification								
5CT Gr L80 Type 1	W,S	0.43[e]	1.90	0.45	---	0.25	---	0.35 Cu
5CT Gr L80 Type 9 Cr	S	0.15	0.30-0.60	1.00	8.0-10.0	0.50	0.90-1.10	0.25 Cu
5CT Gr C90 Type 1	S	0.35	1.00	---	1.20	0.99	0.75	---
5CT Gr C90 Type 2	S	0.50	1.90	---	N.L.	0.99	N.L.	---
5CT Gr C95	W,S	0.45[e]	1.90	0.45	---	---	---	---
5CT Gr T95 Type 1	S	0.35	1.20	---	0.40-1.50	0.99	0.25-0.85	---
5CT Gr T95 Type 2	S	0.50	1.90	---	---	0.99	---	---
5CT Gr P-110	W,S	---	---	---	---	---	---	---
5CT Gr Q125 Type 1	W,S	0.35	1.00	---	1.20	0.99	0.75	---
5CT Gr Q125 Type 2	W,S	0.35	1.00	---	N.L.	0.99	N.L.	---
5CT Gr Q125 Type 3	W,S	0.50	1.90	---	N.L.	0.99	N.L.	---
5CT Gr Q125 Type 4	W,S	0.50	1.90	---	N.L.	0.99	N.L.	---

a. Contains 0.04-0.30% Al.
b. Contains 0.06% Al, 0.12% V, 0.05% Nb.
c. Contains 0.02% V minimum.
d. Contains 0.02-0.10% V.
e. Carbon content for Grade L80 Type 1 and Grade C95 may be increased to 0.50% C and 0.55% C respectively, if the product is oil quenched.
f. Although sulfur and phosphorous are not listed in this table due to limited space, they are specified; see appropriate material standard for more details.
S - Seamless; W - Welded; N.L. - No limit.
Single values are maximums, unless otherwise specified.

CHEMICAL COMPOSITION OF ALLOY STEEL TUBES

ASTM Spec.[a]	C	Mn	P	S	Si	Cr	Mo	Others
A 161 Gr T1, A 209 Gr T1, A 250 Gr T1	0.10-0.20	0.30-0.80	0.025	0.025	0.10-0.50	---	0.44-0.65	---
A 209 Gr T1a, A 250 Gr T1a	0.15-0.25	0.30-0.80	0.025	0.025	0.10-0.50	---	0.44-0.65	---
A 209 Gr T1b, A 250 Gr T1b	0.14	0.30-0.80	0.025	0.025	0.10-0.50	---	0.44-0.65	---
A 213 Gr T2, A 250 Gr T2	0.10-0.20	0.30-0.61	0.025	0.025	0.10-0.30	0.50-0.81	0.44-0.65	---
A 200 Gr T4	0.05-0.15	0.30-0.60	0.025	0.025	0.50-1.00	2.15-2.85	0.44-0.65	---
A 200 Gr T5, A 213 Gr T5	0.15	0.30-0.60	0.025	0.025	0.50	4.00-6.00	0.45-0.65	---
A 213 Gr T5b	0.15	0.30-0.60	0.025	0.025	1.00-2.00	4.00-6.00	0.45-0.65	---
A 213 T5c[b]	0.12	0.30-0.60	0.025	0.025	0.50	4.00-6.00	0.45-0.65	---
A 200 Gr T7	0.15	0.30-0.60	0.025	0.025	0.50-1.00	6.00-8.00	0.45-0.65	---
A 200 Gr T9, A 213 Gr T9	0.15	0.30-0.60	0.025	0.025	0.25-1.00	8.00-10.00	0.90-1.10	---
A 200 Gr T11, A 213 Gr T11, A 250 Gr T11	0.05-0.15	0.30-0.60	0.025	0.025	0.50-1.00	1.00-1.50	0.44-0.65	---
A 213 Gr T12, A 250 Gr T12	0.05-0.15	0.30-0.61	0.025	0.025	0.50	0.80-1.25	0.44-0.65	---
A 213 Gr T17	0.15-0.25	0.30-0.61	0.025	0.025	0.15-0.35	0.80-1.25	---	0.15 V min
A 200 Gr T21, A 213 Gr T21	0.05-0.15	0.30-0.60	0.025	0.025	0.50	2.65-3.35	0.80-1.06	---
A 200 Gr T22, A 213 Gr T22, A 250 Gr T22	0.05-0.15	0.30-0.60	0.025	0.025	0.50	1.90-2.60	0.87-1.13	---
A 213, Gr T91	0.08-0.12	0.30-0.60	0.020	0.010	0.20-0.50	8.00-9.50	0.85-1.05	0.18-0.25V, 0.06-0.1 Cb, 0.030-0.070 N, 0.40 Ni, 0.04 Al,
A 334 Gr 3	0.19	0.31-0.64	0.025	0.025	0.18-0.37	---	---	3.18-3.82 Ni
A 334 Gr 7	0.19	0.90	0.025	0.025	0.13-0.32	---	---	2.03-2.57 Ni
A 334 Gr 8	0.13	0.90	0.025	0.025	0.13-0.32	---	---	8.40-9.60 Ni
A 334 Gr 9	0.20	0.40-1.06	0.025	0.025	---	---	---	0.75-1.25 Cu, 1.60-2.24 Ni

CHEMICAL COMPOSITION OF ALLOY STEEL TUBES (Continued)

ASTM Spec.[a]	C	Mn	P	S	Si	Cr	Mo	Others
A 423 Gr 1	0.15	0.55	0.06-0.16	0.060	0.10 min	0.24-1.31	---	0.20-0.60 Cu, 0.20-0.70 Ni
A 423 Gr 2	0.15	0.50-1.00	0.04	0.05	---	---	0.10	0.30-1.00 Cu, 0.40-1.10 Ni

a. Also published as ASME specifications as SA XXX.
b. Grade 5Tc has a Ti content of not less than 4 times %C and not more than 0.70% C.
Single values are maximums, unless otherwise specified.

MECHANICAL PROPERTIES OF CARBON STEEL PIPES

ASTM Spec.	Form	Tensile Strength		Yield Strength		% Elongation	Other Tests
		ksi	MPa	ksi	MPa		
A 53 Type F	FBW, W	48	330	30	205	see A 53	---
A 53 Types E, S Gr A	ERW, S	48	330	30	205	see A 53	---
A 53 Types E, S Gr B	ERW, S	60	415	35	240	see A 53	---
A 106 Gr A	S	48	330	30	205	35 L, 25 T	---
A 106 Gr B	S	60	415	35	240	30 L, 16.5 T	---
A 106 Gr C	S	70	485	40	275	30 L, 16.5 T	---
A 135 Gr A	ERW	48	331	30	207	35	---
A 135 Gr B	ERW	60	414	35	241	30	---
A 139 Gr A	AW	48	330	30	205	35	---
A 139 Gr B	AW	60	415	35	240	30	---
A 139 Gr C	AW	60	415	42	290	25	---
A 139 Gr D	AW	60	415	46	315	23	---
A 139 Gr E	AW	66	455	52	360	22	---

MECHANICAL PROPERTIES OF CARBON STEEL PIPES (Continued)

ASTM Spec.	Form	Tensile Strength		Yield Strength		% Elongation	Other Tests
		ksi	MPa	ksi	MPa		
A 252 Gr 1	W, S	50	345	30	205	30	---
A 252 Gr 2	W, S	60	414	35	240	25	---
A 252 Gr 3	W, S	66	455	45	310	20	---
A 333 Gr 1	W, S	55	380	30	205	35 L, 25 T	13 (18) / 10 (14) @ -50 (-45)[a]
A 333 Gr 6	W, S	60	415	35	240	30 L, 16.5 T	13 (18) / 10 (14) @ -50 (-45)[a]
A 381 Cl Y35	SAW	60	415	35	240	26	---
A 381 Cl Y42	SAW	60	415	42	290	25	---
A 381 Cl Y46	SAW	63	435	46	316	23	---
A 381 Cl Y48	SAW	62	430	48	330	21	---
A 381 Cl Y50	SAW	64	440	50	345	21	---
A 381 Cl Y52	SAW	66	455	52	360	20	---
A 381 Cl Y56	SAW	71	490	56	385	20	---
A 381 Cl Y60	SAW	75	515	60	415	20	---
A 381 Cl Y65	SAW	77	535	65	450	20	---
A 523 Gr A	ERW, S	48	330	30	205	35	---
A 523 Gr B	ERW, S	60	415	35	240	30	---
A 524 Gr I	S	60-85	414-586	35	240	30 L, 16.5 T	---
A 524 Gr II	S	55-80	380-550	30	205	35 L, 25 T	---
A 587	ERW	48	331	30	207	40	---
A 589	FBW	48	330	30	205	see A 589	---
A 589 Gr A	ERW, S	48	331	30	207	see A 589	---
A 589 Gr B	ERW, S	60	413	35	241	see A 589	---
API Spec.							
5CT H-40	W, S	60	---	40-80	---	---	---
5CT J-55	W, S	75	---	55-80	---	---	---

MECHANICAL PROPERTIES OF CARBON STEEL PIPES (Continued)

API Spec.	Form	Tensile Strength ksi	Tensile Strength MPa	Yield Strength ksi	Yield Strength MPa	% Elongation	Other Tests
5CT K-55	W, S	95	---	55-80	---	---	---
5CT N-80	W, S	100	---	80-110	---	---	---
5CT L-80 Type 1	W, S	95	---	80-95	---	---	23 HRC, 241 HB max
5L A25	W, S	45	310	25	172	see 5L	---
5L A	W, S	48	331	30	207	see 5L	---
5L B	W, S	60	413	35	241	see 5L	---
5L X42	W, S	60	413	42	289	see 5L	---
5L X46	W, S	63	434	46	317	see 5L	---
5L X52	W, S	66	455	52	358	see 5L	---
5L X56	W, S	71	489	56	386	see 5L	---
5L X60	W, S	75	517	60	413	see 5L	---
5L X65	W, S	77	530	65	448	see 5L	---
5L X70	W, S	82	565	70	482	see 5L	---
5L X80	W, S	90-120	620-827	80	551	see 5L	---
CSA Spec.							
Z245.1 Gr 172	W	---	310	---	172	see Z245.1	27 J (< 457 mm OD), 40 J (≥ 457 mm OD)[b]
Z245.1 Gr 207	W	---	331	---	207	see Z245.1	27 J (< 457 mm OD), 40 J (≥ 457 mm OD)[b]
Z245.1 Gr 241	W	---	414	---	241	see Z245.1	27 J (< 457 mm OD), 40 J (≥ 457 mm OD)[b]
Z245.1 Gr 290	W	---	414	---	290	see Z245.1	27 J (< 457 mm OD), 40 J (≥ 457 mm OD)[b]
Z245.1 Gr 317	W	---	434	---	317	see Z245.1	27 J (< 457 mm OD), 40 J (≥ 457 mm OD)[b]
Z245.1 Gr 359	W	---	455	---	359	see Z245.1	27 J (< 457 mm OD), 40 J (≥ 457 mm OD)[b]
Z245.1 Gr 386	W	---	490	---	386	see Z245.1	27 J (< 457 mm OD), 40 J (≥ 457 mm OD)[b]
Z245.1 Gr 414	W	---	517	---	414	see Z245.1	27 J (< 457 mm OD), 40 J (≥ 457 mm OD)[b]
Z245.1 Gr 448	W	---	531	---	448	see Z245.1	27 J (< 457 mm OD), 40 J (≥ 457 mm OD)[b]

MECHANICAL PROPERTIES OF CARBON STEEL PIPES (Continued)

CSA Spec.	Form	Tensile Strength		Yield Strength		% Elongation	Other Tests
		ksi	MPa	ksi	MPa		
Z245.1 Gr 483	W	---	565	---	483	see Z245.1	27 J (< 457 mm OD), 40 J (≥ 457 mm OD)[b]
Z245.1 Gr 550	W	---	620	---	550	see Z245.1	27 J (< 457 mm OD), 40 J (≥ 457 mm OD)[b]

a. Minimum impact energy required for average of each set of three specimens in ft•lb (J), followed by, minimum impact energy permitted for one specimen only of a set in ft•lb (J), followed by the test temperature in °F (°C), full size specimen (10 mm x 10 mm).

b. Pipe notch toughness requirement for Category II and III pipe only (Category I pipe has no notch toughness requirement). Test temperature to be agreed upon between purchaser and supplier, and meet requirements listed in CSA Z245.1 paragraphs 8.4.2.2, 8.4.2.3, and 8.4.2.4.

FBW - Furnace butt welded;

ERW - Electric resistance welded;

W - Welded;

S - Seamless;

AW - Arc welded;

SAW - Submerged arc welded.

MECHANICAL PROPERTIES OF CARBON STEEL TUBES

ASTM Spec.	Form	Tensile Strength		Yield Strength		% Elongation	Other Tests
		ksi	MPa	ksi	MPa		
A 161 low carbon grade	S	47	324	26	179	35	125 HB[a], 137 HB[b] max
A 178 Gr C	ERW	60	415	37	255	30	---
A 178 Gr D	ERW	70	485	40	275	30	---
A 179	S	---	---	---	---	---	72 HRB max
A 192	S	---	---	---	---	---	137 HB, 77 HRB[c] max
A 210 Gr A-1	S	60	415	37	255	30	143 HB or 79 HRB max
A 210 Gr C	S	70	485	40	275	30	179 HB or 89 HRB max
A 214	ERW	---	---	---	---	---	72 HRB max
A 226	ERW	---	---	---	---	---	125 HB or 72 HRB max
A 334 Gr 1	S, W	55	380	30	205	35	163 HB or 85 HRB max 13 (18) / 10 (14) @ -50 (-45)[d]
A 334 Gr 6	S, W	60	415	35	240	30	190 HB or 90 HRB max 13 (18) / 10 (14) @ -50 (-45)[d]
A 539	ERW	45	310	35	241	21	---
A 556 Gr A2	S	47	320	26	180	35	72 HRB max
A 556 Gr B2	S	60	410	37	260	30	79 HRB max
A 556 Gr C2	S	70	480	40	280	30	89 HRB max

a. Hardness for cold finished tubes.
b. Hardness for hot rolled tubes.
c. Brinell hardness test performed on tubes 0.200 in. (5.1 mm) and over in wall thickness, Rockwell hardness test performed on tubes less than 0.200 (5.1 mm) in wall thickness.
d. Minimum impact energy required for average of each set of three specimens in ft•lb (J), followed by, minimum impact energy permitted for one specimen only of a set in ft•lb (J), followed by the test temperature in °F (°C).
S - Seamless; ERW - Electric resistance welded; W - welded.
All single values are minimum, unless otherwise specified.

MECHANICAL PROPERTIES OF ALLOY STEEL PIPES

ASTM Spec.	Form	Tensile Strength		Yield Strength		% Elongation	Other Tests
		ksi	MPa	ksi	MPa		
A 333 Gr 3	W, S	65	450	35	240	30 L, 20 T	13 (18) / 10 (14) @ -150 (-100)[a]
A 333 Gr 4	W, S	60	415	35	240	30 L, 16.5 T	13 (18) / 10 (14) @ -150 (-100)[a]
A 333 Gr 7	W, S	65	450	35	240	30 L, 22 T	13 (18) / 10 (14) @ -100 (-75)[a]
A 333 Gr 8	W, S	100	690	75	515	22 L	lateral expansion requirement
A 333 Gr 9	W, S	63	435	46	315	28 L	13 (18) / 10 (14) @ -100 (-75)[a]
A 333 Gr 10	W, S	80	550	65	450	22 L	13 (18) / 10 (14) @ -75 (-60)[a]
A 333 Gr 11	W, S	65	450	35	240	18 L	---
A 335 Gr P1, P2	S	55	380	30	205	30 L, 20 T	---
A 335 Gr P5, P5b, P5c, P9, P11, P15, P21, P22	S	60	415	30	205	30 L, 20 T	---
A 335 Gr P12	S	60	415	32	220	30 L, 20 T	---
A 335 Gr P91	S	85	585	60	415	20 L	---
A 714 Cl 2 Gr I	W, S	70	485	50	345	22	---
A 714 Cl 2Gr II	W, S	70	485	50	345	22	---
A 714 Cl 2 Gr III	W, S	65	450	50	345	20	---
A 714 Cl 4 Gr IV	W, S	58	400	36	250	see A 335	---
A 714 Cl 4 Gr V Type F	FBW	55	380	40	275	see A 335	---
A 714Cl 4 Gr V Type E, S	ERW, S	65	450	46	315	see A 335	---
A 714 Cl 4 Gr VI Type E, S	ERW, S	65	450	46	315	see A 335	---
A 714 Cl 4 Gr VII Type E, S	ERW, S	65	450	45	310	22	---
A 714 Cl 4 Gr VIII Type E, S	ERW, S	70	485	50	345	21	---
5CT Gr L80 Type 1	W, S	95	---	80-95	---	---	23 HRC, 241 HB[b] max
5CT Gr L80 Type 9 Cr	S	95	---	80-95	---	---	23 HRC, 241 HB[b] max
5CT Gr C90 Type 1, 2	S	100	---	90-105	---	---	25.4 HRC, 255 HB[b] max
5CT Gr C95	W, S	105	---	95-110	---	---	---

MECHANICAL PROPERTIES OF ALLOY STEEL PIPES (Continued)

API Spec.	Form	Tensile Strength		Yield Strength		% Elongation	Other Tests
		ksi	MPa	ksi	MPa		
5CT Gr T95 Type 1, 2	S	105	--	95-110	--	--	25.4 HRC, 255 HBᵇ max
5CT Gr P-110	W, S	125	--	110-140	--	--	--
5CT Gr Q125 Type 1,2,3,4	W, S	135	--	125-150	--	--	--

a. Minimum impact energy required for average of each set of three specimens in ft•lb (J), followed by, minimum impact energy permitted for one specimen only of a set in ft•lb (J), followed by the test temperature in °F (°C), full size specimen (10 mm x 10 mm).
b. In case of dispute, laboratory Rockwell C hardness tests shall be used as the referee method.
Single values are minimums, unless otherwise specified.

MECHANICAL PROPERTIES OF ALLOY STEEL TUBES

ASTM Spec.ᵃ	Tensile Strength		Yield Strength		% Elongation	Hardness, max
	ksi	MPa	ksi	MPa		
A 161 Gr T1	55	379	30	207	30	137 HBᵃ, 150 HBᵇ
A 200 Gr T4, T5, T11, T21, T22	60	414	25	172	30	163 HB
A 200 Gr T7, T9	60	414	25	172	30	179 HB
A 200 Gr T91	85	585	60	414	20	218 HB
A 209 Gr T1, A 250 Gr T1	55	380	30	205	30	146 HB, 80 HRBᶜ
A 209 Gr T1a, A 250 Gr T1a	60	415	32	220	30	153 HB, 81 HRBᶜ
A 209 Gr T1b, A 250 Gr T1b	53	365	28	195	30	137 HB, 77 HRBᶜ
A 213 Gr T5b, T7, T9	60	415	30	205	30	179HB/190HV (89 HRB)ᵈ
A 213 Gr T12	60	415	32	220	30	163 HB/170 HV (85 HRB)ᵈ
A 213 Gr T91	85	585	60	415	20	250 HB/265 HV (25 HRC)ᵈ
A 250 Gr T2, T11, T22	60	415	30	205	30	163 HB, 85 HRBᶜ
A 250 Gr T12	60	415	32	220	30	163 HB, 85 HRBᶜ

MECHANICAL PROPERTIES OF ALLOY STEEL TUBES (Continued)

ASTM Spec.[a]	Tensile Strength		Yield Strength		% Elongation	Hardness, max
	ksi	MPa	ksi	MPa		
A 334 Gr 3, 7	65	450	35	240	30	190 HB, 90 HRB[c]
A 334 Gr 8	100	690	75	520	22	---
A 334 Gr 9	63	435	46	315	28	---
A 423 Gr 1, 2	60	415	37	255	25	170 HB, 87 HRB[c]

a. Hardness for cold finished tubes.
b. Hardness for hot rolled tubes.
c. Brinell hardness test performed on tubes 0.200 in. (5.1 mm) and over in wall thickness, Rockwell hardness test performed on tubes less than 0.200 (5.1 mm) in wall thickness.
d. See ASTM E 140 for hardness scale conversion rules.
Single values are minimums unless other wise specified.

MECHANICAL PROPERTIES OF MARTENSITIC STAINLESS STEEL TUBES

Grade, UNS No.	ASTM Specification	Heat Treat Condition[a]	Tensile Strength		Yield Strength		% El	% RA	Hardness, max
			ksi	MPa	ksi	MPa			
TP405 UNS S40500	A 268	a	60	415	30	205	20	---	207 HB or 95 HRB
TP410 UNS S41000	A 268	a	60	415	30	205	20	---	207 HB or 95 HRB

a. Tubes shall be reheated to at least 1200°F (650°C) and cooled (as appropriate to the grade) to meet the requirements in the ASTM standard.
Single values are minimums, except that hardness values are maximums.

MECHANICAL PROPERTIES OF FERRITIC STAINLESS STEEL TUBES

Grade, UNS No.	ASTM Specification	Heat Treat Condition[a]	Tensile Strength		Yield Strength		% El	% RA	Hardness, max
TP409, S40900	A 268, A 803	A	55	380	30	205	20	---	207 HB or 95 HRB
TP430, S43000	A 268	A	60	415	35	240	20	---	190 HB or 90 HRB
TP439, S43035	A 268, A 803	A	60	415	30	205	20	---	207 HB or 95 HRB
TPXM-27, (E-Brite, 26-1), S44627	A 268, A 803	A	65	450	40	275	20	---	241 HB or 100 HRB
26-3-3, (SC-1), S44660	A 268	A	85	585	65	450	20	---	265 HB or 25 HRC
29-4C, S44735	A 268, A803	A	75	515	60	415	18	---	241 HB or 100 HRB
29-4-2, S44800	A 268, A 803	A	80	550	60	415	20	---	207 HB or 100 HRB

a. A - annealed.

Single values are minimums, except that hardness values are maximums.

MECHANICAL PROPERTIES OF AUSTENITIC STAINLESS STEEL TUBES AND PIPES

Product Form	ASTM Specification	Grade	UNS No.	Heat Treat Condition[a]	Tensile Strength		Yield Strength		% El
					ksi	MPa	ksi	MPa	
Tube	A 213	TP304	S30400	A	75	515	30	205	35
		TP304L	S30403	A	70	485	25	170	35
		TP304LN	S30453	A	75	515	30	205	35
		TP309H	S30909	A	75	515	30	205	35
		TP310H	S31009	A	75	515	30	205	35
		TP316	S31600	A	75	515	30	205	35
		TP316L	S31603	A	70	485	25	170	35
		TP316LN	S31653	A	75	515	30	205	35
		TP317	S31700	A	75	515	30	205	35
		TP317L	S31703	A	75	515	30	205	35
		TP321	S32100	A	75	515	30	205	35
		TP347	S34700	A	75	515	30	205	35
		TP348	S34800	A	75	515	30	205	35
Pipe	A 312	TP304	S30400	A	75	515	30	205	35
		TP304L	S30403	A	70	485	25	170	35
		TP304LN	S30453	A	75	515	30	205	35
		TP309H	S30909	A	75	515	30	205	35
		TP310H	S31009	A	75	515	30	205	35
		TP316	S31600	A	75	515	30	205	35
		TP316L	S31603	A	70	485	25	170	35
		TP316LN	S31653	A	75	515	30	205	35
		TP317	S31700	A	75	515	30	205	35
		TP317L	S31703	A	75	515	30	205	35
		TP321[b]	S32100	A	75	515	30	205	35
		TP347	S34700	A	75	515	30	205	35

| MECHANICAL PROPERTIES OF AUSTENITIC STAINLESS STEEL TUBES AND PIPES (Continued) | | | | | | | | |
Product Form	ASTM Specification	Grade	UNS No.	Heat Treat Condition[a]	Tensile Strength ksi	Tensile Strength MPa	Yield Strength ksi	Yield Strength MPa	% El
Pipe	A 312	TP348	S34800	A	75	515	30	205	35
	B 464, B 474	20Cb-3 20Mo-4 20Mo-6	N08020 N08024 N08026	A	80	551	35	241	30
	B 535, B 546, B 710	Type 330, RA-330	N08330	A	70	483	30	207	30
	B 675	AL-6XN	N08367	A	(> 3/16") 95	655	45	310	30
	B 673	Type 904L	N08904	SA	71	490	31	220	35
		1925 hMo, 25-6 Mo	N08926	SA	94	650	43	295	35
Tube	B 468	20Cb-3 20Mo-4 20Mo-6	N08020 N08024 N08026	A	80	550	35	240	30
	B 668	Sanicro 28	N08028	A	73	500	31	214	40
	B 739	Type 330, RA-330	N08330	A	70	483	30	207	30
	B 676 Cl 1, 2	AL-6XN	N08367	SA	(≤ 3/16") 100	690	45	310	30
	B 674	Type 904L	N08904	SA	71	490	31	220	35
		1925 hMo, 25-6 Mo	N08926	SA	94	650	43	295	35
Pipe and Tube	B 729	20Cb-3 20Mo-4 20Mo-6	N08020 N08024 N08026	A	80	550	35	240	30
	B 690	AL-6XN	N08367	A, HF or CF	(> 3/16") 95	655	45	340	30
	B 677	Type 904L	N08904	SA	71	490	31	220	35
		1925 hMo, 25-6 Mo	N08926	SA	94	650	43	295	35

a. Due to customer demand, Fe-Ni-Cr-Mo alloys and their variations have been included in this table of "austenitic stainless steels".
b. A - annealed, SA – solution annealed, HF - hot finished, CF - cold finished.
Single values are minimums.

MECHANICAL PROPERTIES OF DUPLEX STAINLESS STEEL TUBES AND PIPES									
Common Name (UNS No.)	ASTM	Heat Treat	Tensile Strength		Yield Strength				Hardness, max
Product Form	Specification	Condition[a]	ksi	MPa	ksi	MPa	% El	% RA	
44LN (UNS S31200)									
Tube	A 789	A	100	690	65	450	25	---	280 HB
Pipe	A 790	A	100	690	65	450	25	---	280 HB
DP-3 (UNS S31260)									
Tube	A 789	A	100	690	65	450	25	---	290 HB or 30.5 HRC
Pipe	A 790	A	100	690	65	450	25	---	---
3RE60 (UNS S31500)									
Tube	A 789	A	92	630	64	440	30	---	290 HB or 30.5 HRC
Pipe	A 790	A	92	630	64	440	30	---	290 HB or 30.5 HRC
2205 (UNS S31803)									
Tube	A 789	A	90	620	65	450	25	---	290 HB or 30.5 HRC
Pipe	A 790	A	90	620	65	450	25	---	290 HB or 30.5 HRC
Piping Fitting	A 815	A	90	620	65	450	20	---	290 HB
2304 (UNS S32304)									
Tube	A 789	A (≤ 1 in.)	100	690	65	450	25	---	---
		A (> 1 in.)	87	600	58	400	25	---	290 HB or 30.5 HRC
Pipe	A 790	A	87	600	58	400	25	---	290 HB or 30.5 HRC
Ferralium 255 (UNS S32550)									
Tube	A 789	A	110	760	80	550	15	---	297 HB or 31.5HRC
Pipe	A 790	A	110	760	80	550	15	---	297 HB or 31.5HRC
2507 (UNS S32750)									
Tube	A 789	A	116	800	80	550	15	---	310 HB or 32 HRC
Pipe	A 790	A	116	800	80	550	15	---	310 HB or 32 HRC
Type 329 (UNS S32900)									
Tube	A 789	A	90	620	70	485	20	---	271 HB or 28 HRC

MECHANICAL PROPERTIES OF DUPLEX STAINLESS STEEL TUBES AND PIPES (Continued)									
Common Name (UNS No.)	ASTM	Heat Treat	Tensile Strength		Yield Strength				
Product Form	Specification	Condition[a]	ksi	MPa	ksi	MPa	% El	% RA	Hardness, max
Type 329 (UNS S32900)									
Pipe	A 790	A	90	620	70	485	20	---	271 HB or 28 HRC
7-Mo PLUS (UNS S32950)									
Tube	A 789	A	100	690	70	480	20	---	290 HB or 30.5 HRC
Pipe	A 790	A	100	690	70	480	20	---	290 HB or 30.5 HRC
Piping Fitting	A 815	A	100	690	70	485	15.0	---	290 HB

a. A - annealed, HF - hot finished, CF - cold finished.
Single values are minimums, except for hardness values that are maximums, unless otherwise specified.

Chapter

15

BOILERS & PRESSURE VESSELS CARBON, ALLOY & STAINLESS STEELS:

AMERICAN SPECIFICATION TITLES & DESIGNATIONS, CHEMICAL COMPOSITIONS, MECHANICAL PROPERTIES, P NUMBERS & S NUMBERS

CARBON AND ALLOY STEEL PLATES FOR BOILERS AND PRESSURE VESSELS

ASTM Spec.	Title
A 20/A 20M	General Requirements for Steel Plates for Pressure Vessels
A 202/A 202M	Pressure Vessel Plates, Alloy Steel, Chromium-Manganese-Silicon
A 203/A 203M	Pressure Vessel Plates, Alloy Steel, Nickel
A 204/A 204M	Pressure Vessel Plates, Alloy Steel, Molybdenum
A 225/A 225M	Pressure Vessel Plates, Alloy Steel, Manganese-Vanadium-Nickel
A 285/A 285M	Pressure Vessel Plates, Carbon Steel, Low and Intermediate-Tensile Strength
A 299/A 299M	Pressure Vessel Plates, Carbon Steel, Manganese-Silicon
A 302/A 302M	Pressure Vessel Plates, Alloy Steel, Manganese-Molybdenum and Manganese-Molybdenum-Nickel
A 353/A 353M	Pressure Vessel Plates, Alloy Steel, 9 Percent Nickel, Double-Normalized and Tempered
A 387/A 387M	Pressure Vessel Plates, Alloy Steel, Chromium-Molybdenum
A 455/A 455M	Pressure Vessel Plates, Carbon Steel, High Strength Manganese
A 515/A 515M	Pressure Vessel Plates, Carbon Steel, for Intermediate and Higher-Temperature Service
A 516/A 516M	Pressure Vessel Plates, Carbon Steel, for Moderate and Lower-Temperature Service
A 517/A 517M	Pressure Vessel Plates, Alloy Steel, High-Strength, Quenched and Tempered
A 533/A 533M	Pressure Vessel Plates, Alloy Steel, Quenched and Tempered, Manganese-Molybdenum and Manganese-Molybdenum-Nickel
A 537/A 537M	Pressure Vessel Plates, Heat-Treated, Carbon-Manganese-Silicon Steel
A 542/A 542M	Pressure Vessel Plates, Alloy Steel, Quenched-and-Tempered, Chromium-Molybdenum, and Chromium-Molybdenum-Vanadium
A 543/A 543M	Pressure Vessel Plates, Alloy Steel, Quenched and Tempered Nickel-Chromium-Molybdenum
A 553/A 553M	Pressure Vessel Plates, Alloy Steel, Quenched and Tempered, 8 and 9 Percent Nickel
A 562/A 562M	Pressure Vessel Plates, Carbon Steel, Manganese-Titanium for Glass or Diffused Metallic Coatings
A 612/A 612M	Pressure Vessel Plates, Carbon Steel, High Strength, for Moderate and Lower Temperature Service
A 645/A 645M	Pressure Vessel Plates, 5 Percent Nickel Alloy Steel, Specially Heat Treated
A 724/A 724M	Pressure Vessel Plates, Carbon Steel, Quenched and Tempered, for Welded Layered Pressure Vessels
A 734/A 734 M	Pressure Vessel Plates, Alloy Steel and High-Strength Low-Alloy Steel, Quenched and Tempered
A 735/A 735M	Pressure Vessel Plates, Low-Carbon Manganese-Molybdenum-Columbium Alloy Steel, for Moderate and Lower Temperature Service

CARBON AND ALLOY STEEL PLATES FOR BOILERS AND PRESSURE VESSELS (Continued)

ASTM Spec.	Title
A 736/A 736M	Pressure Vessel Plates, Low-Carbon Age-Hardening Nickel-Copper-Chromium-Molybdenum-Columbium and Nickel-Copper-Manganese-Molybdenum-Columbium Alloy Steel
A 737/A 737M	Pressure Vessel Plates, High-Strength, Low-Alloy Steel
A 738/A 738M	Pressure Vessel Plates, Heat-Treated, Carbon-Manganese-Silicon Steel, for Moderate and Lower Temperature Service
A 770/A 770M	Steel Plates for Special Applications, Through-Thickness Tension Testing of
A 782/A 782M	Pressure Vessel Plates, Quenched-and-Tempered, Manganese-Chromium-Molybdenum-Silicon-Zirconium Alloy Steel
A 832/A 832M	Pressure Vessel Plates, Alloy Steel, Chromium-Molybdenum-Vanadium
A 841/A 841M	Pressure Vessel Plates, Produced by Thermo-Mechanical Control Process (TMCP)
A 844/A 844M	Pressure Vessel Plates, 9 % Nickel Alloy, Produced by the Direct-Quenching Process

CHEMICAL COMPOSITIONS OF PRESSURE VESSEL STEELS[a]

ASTM	Grade/Type	C	Mn	P	S	Si	Ni	Cr	Mo	Other
A 202/A 202M	Gr A	0.17	0.97-1.52	0.035	0.035	0.54-0.96	---	0.31-0.64	---	---
	Gr B	0.25	0.97-1.52	0.035	0.035	0.54-0.96	---	0.31-0.64	---	---
A 203/A 203M	Gr A (≤ 2 in.)	0.17	0.78	0.035	0.035	0.13-0.45	2.03-2.57	---	---	---
	Gr B (≤ 2 in.)	0.21	0.78	0.035	0.035	0.13-0.45	2.03-2.57	---	---	---
	Gr D (≤ 2 in.)	0.17	0.78	0.035	0.035	0.13-0.45	3.18-3.82	---	---	---
	Gr E (≤ 2 in.)	0.20	0.78	0.035	0.035	0.13-0.45	3.18-3.82	---	---	---
	Gr F (≤ 2 in.)	0.20	0.78	0.035	0.035	0.13-0.45	3.18-3.82	---	---	---
A 204/A 204M	Gr A (≤ 1 in.)	0.18	0.98	0.035	0.035	0.13-0.45	---	---	0.41-0.64	---
	Gr B (≤ 1 in.)	0.20	0.98	0.035	0.035	0.13-0.45	---	---	0.41-0.64	---
	Gr C (≤ 1 in.)	0.23	0.98	0.035	0.035	0.13-0.45	---	---	0.41-0.64	---
A 225/A 225M	Gr C	0.25	1.72	0.035	0.035	0.13-0.45	0.37-0.73	---	---	0.11-0.20 V
	Gr D	0.20	1.84	0.035	0.035	0.08-0.56	0.37-0.73	---	---	0.08-0.20 V
A 285/A 285M	Gr A	0.17	0.98	0.035	0.035	---	---	---	---	---

CHEMICAL COMPOSITIONS OF PRESSURE VESSEL STEELS[a] (Continued)

ASTM	Grade/Type	C	Mn	P	S	Si	Ni	Cr	Mo	Other
A 285/A 285M	Gr B	0.22	0.98	0.035	0.035	---	---	---	---	---
	Gr C	0.28	0.98	0.035	0.035	---	---	---	---	---
A 299/A 299M	(≤ 1 in.)	0.28	0.84-1.52	0.035	0.035	0.13-0.45	---	---	---	---
A 302/A 302M	Gr A (≤ 1 in.)	0.20	0.87-1.41	0.035	0.035	0.13-0.45	---	---	0.41-0.64	---
	Gr B (≤ 1 in.)	0.20	1.07-1.62	0.035	0.035	0.13-0.45	---	---	0.41-0.64	---
	Gr C (≤ 1 in.)	0.20	1.07-1.62	0.035	0.035	0.13-0.45	0.37-0.73	---	0.41-0.64	---
	Gr D (≤ 1 in.)	0.20	1.07-1.62	0.035	0.035	0.13-0.45	0.67-1.03	---	0.41-0.64	---
A 353/A 353M	---	0.13	0.98	0.035	0.035	0.13-0.45	8.40-9.60	---	---	---
A 387/A 387M	Gr 2	0.04-0.21	0.50-0.88	0.035	0.035	0.13-0.45	---	0.46-0.85	0.40-0.65	---
	Gr 12	0.04-0.17	0.35-0.73	0.035	0.035	0.13-0.45	---	0.74-1.21	0.40-0.65	---
	Gr 11	0.04-0.17	0.35-0.73	0.035	0.035	0.44-0.86	---	0.94-1.56	0.40-0.70	---
	Gr 22	0.04-015[b]	0.25-0.66	0.035	0.035	0.50	---	1.88-2.62	0.85-1.15	---
	Gr 22L	0.12	0.25-0.66	0.035	0.035	0.50	---	1.88-2.62	0.85-1.15	---
A 387/A 387M	Gr 21	0.04-0.15[b]	0.25-0.66	0.035	0.035	0.50	---	2.63-3.37	0.85-1.15	---
	Gr 21L	0.12	0.25-0.66	0.035	0.035	0.50	---	2.63-3.37	0.85-1.15	---
	Gr 5	0.15	0.25-0.66	0.035	0.030	0.55	---	3.90-6.10	0.40-0.70	---
	Gr 9	0.15	0.25-0.66	0.030	0.030	1.05	---	7.90-10.10	0.85-1.15	---
	Gr 91	0.06-0.15	0.25-0.66	0.025	0.012	0.18-0.56	0.43	7.90-9.60	0.80-1.10	0.16-0.27 V, 0.05-0.11 Cb, 0.025-0.080 N, 0.05 Al
A 455/A 455M	---	0.33[c]	0.79-1.30	0.035	0.035	0.13[d]	---	---	---	---
A 515/A 515M	Gr 60 (≤ 1 in.)	0.24	0.98	0.035	0.035	0.13-0.45	---	---	---	---
	Gr 65 (≤ 1 in.)	0.28	0.98	0.035	0.035	0.13-0.45	---	---	---	---
	Gr 70 (≤ 1 in.)	0.31	1.30	0.035	0.035	0.13-0.45	---	---	---	---
A 516/A 516M	Gr 55 (≤ ½ in.)	0.18	0.55-0.98	0.035	0.035	0.13-0.45	---	---	---	---
	Gr 55 (½-2 in.)	0.20	0.55-1.30	0.035	0.035	0.13-0.45	---	---	---	---

CHEMICAL COMPOSITIONS OF PRESSURE VESSEL STEELS[a] (Continued)

ASTM	Grade/Type	C	Mn	P	S	Si	Ni	Cr	Mo	Other
A 516/A 516M	Gr 60 (≤ ½ in.)	0.21	0.55-0.98[e]	0.035	0.035	0.13-0.45	---	---	---	---
	Gr 60 (½-2 in.)	0.23	0.79-1.30[e]	0.035	0.035	0.13-0.45	---	---	---	---
	Gr 65 (≤ ½ in.)	0.24	0.79-1.30	0.035	0.035	0.13-0.45	---	---	---	---
	Gr 65 (½-2 in.)	0.26	0.79-1.30	0.035	0.035	0.13-0.45	---	---	---	---
	Gr 70 (≤ ½ in.)	0.27	0.79-1.30	0.035	0.035	0.13-0.45	---	---	---	---
	Gr 70 (½-2 in.)	0.28	0.79-1.30	0.035	0.035	0.13-0.45	---	---	---	---
A 517/A 517M	Gr A	0.13-0.23	0.74-1.20	0.035	0.035	0.34-0.86	---	0.46-0.84	0.15-0.31	0.0025 B, 0.04-0.16 Zr
	Gr B	0.13-0.23	0.64-1.10	0.035	0.035	0.13-0.37	---	0.36-0.69	0.12-0.28	0.0005-0.005 B, 0.02-0.09 V, 0.01-0.04 Ti
	Gr C	0.08-0.22	1.02-1.62	0.035	0.035	0.13-0.32	---	---	0.17-0.33	0.001-0.005 B
	Gr E	0.10-0.22	0.35-0.78	0.035	0.035	0.08-0.45	---	1.34-2.06	0.36-0.64	0.001-0.005 B, 0.005-0.11 Ti
	Gr F	0.08-0.22	0.55-1.10	0.035	0.035	0.13-0.37	0.67-1.03	0.36-0.69	0.36-0.64	0.0005-0.006 B, 0.02-0.09 V, 0.12-0.53 Cu
	Gr H	0.10-0.23	0.87-1.41	0.035	0.035	0.13-0.37	0.27-0.73	0.36-0.69	0.17-0.33	0.0005 B min, 0.02-0.09 V
	Gr J	0.10-0.23	0.40-0.78	0.035	0.035	0.18-0.37	---	---	0.46-0.69	0.001-0.005 B
	Gr K	0.08-0.22	1.02-1.62	0.035	0.035	0.13-0.32	---	---	0.42-0.50	0.001-0.005 B
	Gr M	0.10-0.23	0.40-0.78	0.035	0.035	0.18-0.37	1.15-1.55	---	0.41-0.64	0.001-0.005 B
	Gr P	0.10-0.23	0.40-0.78	0.035	0.035	0.18-0.37	1.15-1.55	0.79-1.26	0.41-0.64	0.001-0.005 B
	Gr Q	0.12-0.23	0.87-1.41	0.035	0.035	0.13-0.37	1.15-1.55	0.94-1.56	0.36-0.64	0.02-0.09 V
	Gr S	0.10-0.22	1.02-1.62	0.035	0.035	0.13-0.45	---	---	0.10-0.38	0.07 Ti, 0.07 Cb
	Gr T	0.06-0.16	1.12-1.60	0.035	0.010	0.34-0.66	---	---	0.41-0.64	0.001-0.005 B, 0.02-0.09 V

CHEMICAL COMPOSITIONS OF PRESSURE VESSEL STEELS^a (Continued)

ASTM	Grade/Type	C	Mn	P	S	Si	Ni	Cr	Mo	Other
A 533/A 533M	Type A	0.25	1.07-1.62[f]	0.035	0.035	0.13-0.45	---	---	0.41-0.64	---
	Type B	0.25	1.07-1.62[f]	0.035	0.035	0.13-0.45	0.37-0.73	---	0.41-0.64	---
	Type C	0.25	1.07-1.62[f]	0.035	0.035	0.13-0.45	0.67-1.03	---	0.41-0.64	---
	Type D	0.25	1.07-1.62[f]	0.035	0.035	0.13-0.45	0.17-0.43	---	0.41-0.64	---
A 537/A 537M	(≤ 1½ in.)	0.24	0.64-1.46	0.035	0.035	0.13-0.55	0.28	0.29	0.09	0.38 Cu
A 542/A 542M	Type A	0.18[h]	0.25-0.66	0.025	0.025	0.50	0.43	1.88-2.62	0.85-1.15	0.43 Cu, 0.04 V
	Type B	0.09-0.18	0.25-0.66	0.015	0.015	0.50	0.28	1.88-2.62	0.85-1.15	0.28 Cu, 0.03 V
	Type C	0.08-0.18	0.25-0.66	0.025	0.025	0.13	0.28	2.63-3.37	0.85-1.15	0.28 Cu, 0.18-0.33 V, 0.005-0.045 Ti, NA[i] B
	Type D	0.09-0.18	0.25-0.66	0.020	0.015	0.013	0.28	1.08-2.62	0.85-1.15	0.23 Cu, 0.23-0.37 V, 0.035 Ti, NA[i] B, 0.08 Cb, 0.020[j] Ca
A 543/A 543M	Type B	0.20	0.40	0.020	0.020	0.13-0.45	2.18-4.07	0.94-1.96	0.16-0.69	0.03 V
	Type C	0.18	0.40	0.020	0.020	0.13-0.45	1.93-3.57	0.94-1.96	0.16-0.69	0.03 V
A 562/A 562M		0.12	1.30	0.035	0.035	0.15-0.50	---	---	---	0.15 Cu, (4 x C) Ti min
A 612/A 612M	(≤ 0.750 in.)	0.29	0.92-1.46	0.035	0.035	0.13-0.45	0.28[k]	0.29[k]	0.09[k]	0.38 Cu[k], 0.09 V[k]
	(> 0.750-1 in.)	0.29	0.92-1.62	0.035	0.035	0.13-0.55	0.28[k]	0.29[k]	0.09[k]	0.38 Cu[k], 0.09 V[k]
A 645/A 645M		0.15	0.25-0.66	0.035	0.035	0.18-0.45	4.65-5.35	---	0.17-0.38	0.01-0.16 Al total, 0.025 N
A 724/A 724M	Gr A	0.22	0.92-1.72	0.035	0.035	0.60	0.28[k]	0.29[k]	0.09[k]	0.38 Cu[k], 0.09 V[k]
	Gr B	0.24	0.92-1.72	0.035	0.035	0.55	0.28[k]	0.29[k]	0.09[k]	0.38 Cu[k], 0.09 V[k]
	Gr C	0.26	1.02-1.72	0.035	0.035	0.18-0.65	0.28[k]	0.29[k]	0.09[k]	0.38 Cu[k], 0.09 V[k], 0.05[l] B
A 734/A 734M	Type A	0.14	0.40-0.83	0.036	0.016	0.45	0.85-1.25	0.84-1.26	0.22-0.43	---

CHEMICAL COMPOSITIONS OF PRESSURE VESSEL STEELS[a] (Continued)

ASTM	Grade/Type	C	Mn	P	S	Si	Ni	Cr	Mo	Other
A 734/A 734M	Type B	0.19	1.72	0.036	0.016	0.45	---	0.29	---	0.38 Cu[m], 0.13 V, 0.030 N
A 735/A 735M	≤ ⅝ in. (16 mm)	0.08	1.12-2.04	0.035	0.025	0.45	---	---	0.20-0.50	0.02-0.10 Cb, 0.18-0.37 Cu[m]
A 736/A 736M	Gr A	0.09	0.35-0.78	0.025	0.025	0.45	0.67-1.03	0.56-0.94	0.12-0.28	0.95-1.35 Cu, 0.01 Cb min
	Gr C	0.09	1.21-1.77	0.025	0.025	0.45	0.67-1.03	---	0.12-0.28	0.95-1.35 Cu, 0.01 Cb min
A 737/A 737M	Gr B	0.22	1.07-1.62[n]	0.035	0.035	0.10-0.55	---	---	---	0.05 Cb
	Gr C	0.24	1.07-1.62	0.035	0.035	0.10-0.55	---	---	---	0.03-0.12 V, 0.03 N
A 738/A 738M	Gr A (≤ 2½ in.)	0.24	1.62	0.035	0.035	0.13-0.55	0.53	0.29	0.09	0.38 Cu, 0.08 V[o], 0.05 Cb[o], 0.10 Cb+V[o]
	Gr B (≤ 2½ in.)	0.20	0.84-1.62	0.030	0.030	0.13-0.60	0.63	0.34	0.21	0.38 Cu, 0.08 V, 0.05 Cb, 0.10 Cb+V
	Gr C (≤ 2½ in.)	0.20	1.62	0.025	0.025	0.13-0.55	0.53	0.29	0.09	0.38 Cu, 0.05 V
A 782/A 782M	Gr 21V	0.22	0.62-1.30	0.035	0.035	0.34-0.86	---	0.46-1.06	0.17-0.64	0.03-0.16 Zr
A 832/A 832M	Gr 21V	0.08-0.18	0.25-0.66	0.035	---	0.13	---	2.63-3.37	0.85-1.15	0.18-0.33 V, 0.005-0.045 Ti, NA B[i]
	Gr 22V	0.09-0.18	0.25-0.66	0.020	0.015	0.13	0.28	1.88-2.62	0.85-1.15	0.23-0.37 V, 0.035 Ti, NA B[i], 0.23 Cu, 0.08 Cb, 0.20 Ca[j]
	Gr 23V	0.08-0.18	0.25-0.66	---	---	0.13	---	2.63-3.37	0.85-1.15	0.18-0.33, 0.010-0.075 Cb

CHEMICAL COMPOSITIONS OF PRESSURE VESSEL STEELS[a] (Continued)

ASTM	Grade/Type	C	Mn	P	S	Si	Ni	Cr	Mo	Other
A 841/A 841M	Gr A (≤ 1½ in.)	0.20	0.70-1.35[p]	0.030	0.030	0.15-0.50	0.25	0.25	0.08	0.35 Cu, 0.03 Cb, 0.06 V, 0.020 total or 0.015 acid soluble Al min
	Gr B (≤ 1½ in.)	0.15	0.70-1.35	0.030	0.025	0.15-0.50	0.60	0.25	0.30	0.35 Cu, 0.03 Cb, 0.06 V
A 844/A 844M		0.13	0.98	0.020	0.020	0.13-0.45	8.40-9.60	---	---	---

a. When the ASTM specification differs in product and heat analysis, the product analysis is shown in this table. b. The carbon content for plates > 5 in. (125 mm) in thickness is 0.17 maximum on product analysis. c. When the silicon is > 0.10%, the carbon maximum shall be 0.28%. d. At the purchaser's or the producer's option, silicon may be 0.40% maximum on heat analysis, 0.45% maximum on product analysis. e. Grade 60 plates ½ in. (12.5 mm) and under in thickness may have 0.85-1.20% manganese on heat analysis, and 0.79-1.30% manganese on product analysis. f. The maximum manganese content may be increased to 1.60% on heat analysis and 1.65% on product analysis when Class 2 or Class 3 properties are specified and when Supplementary Requirement S3 (see ASTM A 20/A 20M) is specified with a total holding time of more than 1 h/in. (2.4 min/mm) of thickness. g. Manganese may exceed 1.35% on heat analysis, up to a maximum of 1.60%, and nickel may exceed 0.25% on heat analysis, up to a maximum of 0.50%, provided that the heat analysis carbon equivalent does not exceed 0.57% when based upon the following equation: $CE = C + \dfrac{Mn}{6} + \dfrac{Cr + Mo + V}{5} + \dfrac{Ni + Cu}{15\%}$

When this option is exercised, the manganese and nickel contents on product analysis shall not exceed the heat analysis content by more than 0.12% and 0.03%, respectively. h. In A 542/A 542M-82 and earlier editions, for plates ≤ 5 in. (125 mm) in thickness, the carbon was limited to 0.15% maximum. I. NA = Product analysis is not applicable. Footnotes continued on next page. j. Rare earth metals (REM) may be added in place of calcium, subject to agreement between the producer and the purchaser. In that case, the total amount of REM shall be determined and reported. k. When analysis shows that the amount of an element is 0.02% or lower, the value may be reported as ≤ 0.02%. l. If boron is less than 0.001%, the analysis report for the element may be stated as < 0.001%. m. When specified. n. The maximum manganese may be increased to 1.60% on heat analysis and 1.72% on product analysis, provided that the carbon content on heat analysis does not exceed 0.18%. o. Vanadium and columbium may be added only by agreement between the producer and the purchaser. p. Manganese may exceed 1.35% on heat analysis, up to a maximum of 1.60%, provided that the carbon equivalent on heat analysis does not exceed 0.47%, or the value specified in Supplementary Requirement S77 when that requirement is invoked, when based on the following formula:

$CE = C + \dfrac{Mn}{6} + \dfrac{Cr + Mo + V}{5} + \dfrac{Ni + Cu}{15\%}.$

When this option is exercised, the manganese content on product analysis shall not exceed the heat analysis content by more than 0.12%. Single values are maximums, unless otherwise specified.

MECHANICAL PROPERTIES OF PRESSURE VESSEL STEELS

ASTM	Grade/Class	Tensile Strength ksi	Tensile Strength MPa	Yield Strength ksi	Yield Strength MPa	% El	% R A	Charpy Impact[a] ft·lb (J) @ °F(°C)
A 202/A 202M	Gr A	75-95	515-655	45	310	19	---	⊖
	Gr B	85-110	585-760	47	325	18	---	⊖
A 203/A 203M	Gr A, D (≤ 2 in.)	65-85	450-585	37	255	23	---	20 ft·lb (27 J)@optional
	Gr B, E (≤ 2 in.)	70-90	485-620	40	275	21	---	20 ft·lb (27 J)@optional
	Gr F (≤ 2 in.)	80-100	550-690	55	380	20	---	⊖
A 204/A 204M	Gr A	65-85	450-585	37	255	23	---	⊖
	Gr B	70-90	485-620	40	275	21	---	⊖
	Gr C	75-95	515-655	43	295	20	---	⊖
A 225/A 225M	Gr C	105-135	725-930	70	485	20	---	⊖
	Gr D (≤ 3 in.)	80-105	550-725	55	380	19	---	⊖
A 285/A 285M	Gr A	45-65	310-450	24	165	30	---	---
	Gr B	50-70	345-485	27	185	28	---	---
	Gr C	55-75	380-515	30	205	27	---	---
A 299/A 299M	(≤ 1 in.)	75-95	515-655	42	290	19	---	⊖
A 302/A 302M	Gr A	75-95	515-655	45	310	19	---	⊖
	Gr B	80-100	550-690	50	345	18	---	⊖
	Gr C	80-100	550-690	50	345	20	---	⊖
	Gr D	80-100	550-690	50	345	20	---	⊖
A 353/A 353M		100-120	690-825	75	515	20	---	0.015 in. LE@-320 (-195)
A 387/A 387M	Class 1, Gr 2, 12	55-80	380-550	33	---	22	---	⊖
	Class 1, Gr 11	60-85	415-585	35	---	22	---	⊖
	Cl 1, Gr 22, 21, 5, 9, 21L, 22L	60-85	415-585	30	---	18	45[b], 40[c]	⊖
	Class 2[d], Gr 2	70-90	485-620	45	310	22	---	⊖
	Class 2[d], Gr 11	75-100	515-690	45	310	22	---	⊖
	Class 2[d], Gr 12	65-85	450-585	40	275	22	---	⊖

The Metals Black Book – 3rd Edition

MECHANICAL PROPERTIES OF PRESSURE VESSEL STEELS (Continued)

ASTM	Grade/Class	Tensile Strength		Yield Strength		% El	% R A	Charpy Impact[a] ft·lb (J) @ °F(°C)
		ksi	MPa	ksi	MPa			
A 387/A 387M	Class 2[d], Gr 22, 21, 5, 9	75-100	515-690	45	310	18	45[b], 40[c]	e
	Class 2[d], Gr 91	85-110	585-760	60	415	18	---	e
A 455/A 455M	< 0.375 in. (9.5 mm)	75-95	515-655	38	260	22	---	---
	> 0.375-0.580 in. (15 mm)	73-93	505-640	37	255	22	---	---
	> 0.580-0.750 in. (20 mm)	70-90	485-620	35	240	22	---	---
A 515/A 515M	Gr 60	60-80	415-550	32	220	25	---	e
	Gr 65	65-85	450-585	35	240	23	---	e
	Gr 70	70-90	485-620	38	260	21	---	e
A 516/A 516M	Gr 55	55-75	380-515	30	205	27	---	e
	Gr 60	60-80	415-550	32	220	25	---	e
	Gr 65	65-85	450-585	35	240	23	---	e
	Gr 70	70-90	485-620	38	260	21	---	e
A 517/A 517M	≤ 2½ in. (65 mm)	115-135	795-930	100	690	16	35[c], 45[b]	e
	> 2½ in.-6 in. (65-150 mm)	105-135	725-930	90	620	14	45[b]	e
A 533/A 533M	Class 1	80-100	550-690	50	345	18	---	e
	Class 2	90-115	620-795	70	485	16	---	e
	Class 3	100-125	690-860	83	570	16	---	e
A 537/A 537M	Class 1 (≤ 2½ in.)	70-90	485-620	50	345	22	---	e
	Class 2 (≤ 2½ in.)	80-100	550-690	60	415	22	---	e
	Class 3 (≤ 2½ in.)	80-100	550-690	55	380	22	---	e
A 542/A 542M	Class 1	105-125	725-860	85	585	14	---	e
	Class 2	115-135	795-930	100	690	13	---	e
	Class 3	95-115	655-795	75	515	20	---	e
	Class 4	85-110	585-760	55	380	20	---	40 (54)/35 (48)@optional
	Class 4a	85-110	585-760	60	415	18	---	40 (54)/35 (48)@0 (-18)

MECHANICAL PROPERTIES OF PRESSURE VESSEL STEELS (Continued)

ASTM	Grade/Class	Tensile Strength		Yield Strength		% El	% R A	Charpy Impact[a] ft-lb (J) @ °F(°C)
		ksi	MPa	ksi	MPa			
A 543/A 543M	Class 1	105-125	725-860	85	585	14	---	⊖
	Class 2	115-135	795-930	100	690	14	---	⊖
	Class 3	90-115	620-795	70	485	16	---	⊖
A 553/A 553M	Type I	100-120	690-825	85	585	20	---	0.015 in. LE@-320 (-195)
	Type II	100-120	690-825	85	585	20	---	0.015 in. LE@-275 (-170)
A 562/A 562M	≤ ½ in. (12.5 mm)	55-75	380-515	30	205	26	---	⊖
A 612/A 612M	> ½ in.-1 in. (> 12.5-25 mm)	83-105	570-725	50	345	22	---	⊖
		81-101	560-695	50	345	22	---	⊖
A 645/A 645M	Gr A, C	95-115	655-795	65	450	20	---	0.015 in. LE@-275 (-170)
A 724/A 724M	Gr A, C	90-110	620-760	70	485	19	---	⊖
	Gr B	95-115	655-795	75	515	17	---	⊖
A 734/A 734M		77-97	530-670	65	450	20	---	⊖
A 735/A 735M	Class 1[f]	80-100	550-690	65	450	18	---	20 (27) @ -50 (-45)
	Class 2[f]	85-105	585-725	70	485	18	---	20 (27) @ -50 (-45)
	Class 3	90-110	620-760	75	515	18	---	20 (27) @ -50 (-45)
	Class 4	95-115	655-790	80	550	18	---	20 (27) @ -50 (-45)
A 736/A 736M	Gr A, Class 1 (≤ ¾ in.)	90-110	620-760	80	550	20	---	20 (27) @ -50 (-45)
	Gr A, Class 2 (≤ ¾ in.)	72-92	495-635	65	450	20	---	20 (27) @ -50 (-45)
	Gr A, Class 3 (≤ ¾ in.)	85-105	585-725	75	515	20	---	20 (27) @ -50 (-45)
	Gr C, Class 1 (≤ ¾ in.)	100-120	690-825	90	620	20	---	20 (27) @ -50 (-45)
	Gr C, Class 3 (≤ ¾ in.)	95-115	655-795	85	585	20	---	20 (27) @ -50 (-45)
A 737/A 737M	Gr B	70-90	485-620	50	345	23	---	⊖
	Gr C	80-100	550-690	60	415	23	---	⊖
A 738/A 738M	Gr A (≤ 2½ in.) (≤ 65 mm)	75-95	515-655	45	310	20	---	⊖
	Gr B (≤ 2½ in.) (≤ 65 mm)	85-102	585-705	60	415	20	---	⊖

MECHANICAL PROPERTIES OF PRESSURE VESSEL STEELS (Continued)

ASTM	Grade/Class	Tensile Strength		Yield Strength		% El	% R A	Charpy Impact[a] ft·lb (J) @ °F(°C)
		ksi	MPa	ksi	MPa			
A 738/A 738M	Gr C (≤ 2½ in.) (≤ 65 mm)	80-100	550-690	60	415	22	---	e
A 782/A 782M	Class 1	97-119	670-820	80	550	18	---	0.015 in. LE@32 (0)
	Class 2	107-129	740-890	90	620	17	---	0.015 in. LE@32 (0)
	Class 3	115-136	795-940	100	690	16	---	0.015 in. LE@32 (0)
A 832/A 832M		85-110	585-760	60	415	18	45[a], 40[b]	40 (54)/35 (48)@0 (-18)
A 841/A 841M	Class 1 (≤ 2½ in.) (≤ 65 mm)	70-90	485-620	50	345	22	---	15 (20)@-40 (-40)
	Class 1(> 2½ in.) (> 65 mm)	65-85	450-585	45	310	22	---	15 (20)@-40 (-40)
	Class 2 (≤ 2½ in.) (≤ 65 mm)	80-100	550-690	60	415	22	---	15 (20)@-40 (-40)
	Class 2(> 2½ in.) (> 65 mm)	75-95	515-655	55	380	22	---	15 (20)@-40 (-40)
A 844/A 844M		100-120	690-825	85	585	20	---	0.015 in. LE@-320 (-195)

a. Minimum impact energy required for average of each set of three specimens in ft·lb (J), followed by minimum impact energy permitted for one specimen only of a set in ft·lb (J), if applicable, followed by the @ test temperature in °F (°C); all values are for full size specimens 10 mm x 10 mm, for sub-size specimens see the related ASTM Specification. Charpy V-notch impact tests shall be made in accordance with ASTM A 20/A 20M. LE - lateral expansion. b. Measured on round test specimens. c. Measured on flat specimens. d. Not applicable to annealed material. e. Supplemental requirement only. f. Class 1 as-rolled and Class 2 as-rolled are limited to 1 in. (25 mm) maximum thickness. Single values are minimum, unless otherwise specified.

ASME FERROUS METALS - P NUMBERS & S NUMBERS

ASME Spec No.	Type or Grade	Nominal Composition	Product Form	Tensile Strength[a] ksi	P No	Welding Group No.	S No.	Group No.	UNS No.
SA-36	---	C-Mn-Si	Plate	58	1	1	---	---	K02600
SA-53	Type F	C	Furnace welded pipe	48	1	1	---	---	---
SA-53	Type S, Grade A	C	Seamless pipe	48	1	1	---	---	K02504
SA-53	Type E, Grade A	C	ERW pipe	48	1	1	---	---	K02504
SA-53	Type E, Grade B	C-Mn	ERW pipe	60	1	1	---	---	K03005
SA-53	Type S, Grade B	C-Mn	Seamless pipe	60	1	1	---	---	K03005
SA-105	---	C-Si	Pipe flange	70	1	2	---	---	K03504
SA-106	A	C-Si	Seamless pipe	48	1	1	---	---	K02501
SA-106	B	C-Si	Seamless pipe	60	1	1	---	---	K03006
SA-106	C	C-Si	Seamless pipe	70	1	2	---	---	K03501
A 108	1015 CW	C	Bar	60	---	---	1	1	G10150
A 108	1018 CW	C	Bar	60	---	---	1	1	G10180
A 108	1020 CW	C	Bar	60	---	---	1	1	G10200
A 134	A 283 Grade A	C	Welded pipe	45	---	---	1	1	---
A 134	A 285 Grade A	C	Welded pipe	45	---	---	1	1	K01700
A 134	A 283 Grade B	C	Welded pipe	50	---	---	1	1	---
A 134	A 285 Grade B	C	Welded pipe	50	---	---	1	1	K02200
A 134	A 283 Grade C	C	Welded pipe	55	---	---	1	1	---
A 134	A 285 Grade C	C	Welded pipe	55	---	---	1	1	K02801
A 134	A 283 Grade D	C	Welded pipe	60	---	---	1	1	---
SA-134	---	C	Welded pipe	---	1	1	---	---	---
SA-135	A	C	ERW pipe	48	1	1	---	---	---
SA-135	B	C	ERW pipe	60	1	1	---	---	---
A 139	A	C	Welded pipe	48	---	---	1	1	---

ASME FERROUS METALS - P NUMBERS & S NUMBERS (Continued)

ASME Spec No.	Type or Grade	Nominal Composition	Product Form	Tensile Strength[a] ksi	Welding P No	Welding Group No.	Welding S No.	Welding Group No.	UNS No.
A 139	B	C	Welded pipe	60	---	---	1	1	K03003
A 139	C	C	Welded pipe	60	---	---	1	1	K03004
A 139	D	C	Welded pipe	60	---	---	1	1	K03010
A 139	E	C	Welded pipe	66	---	---	1	1	K03012
A 148	90-60	---	Castings	90	---	---	4	3	---
A 167	Type 302	18Cr-8Ni	Plate, sheet, & strip	75	---	---	8	1	S30200
A 167	Type 302B	18Cr-8Ni-2Si	Plate, sheet, & strip	75	---	---	8	1	S30215
A 167	Type 304	18Cr-8Ni	Plate, sheet, & strip	75	---	---	8	1	S30400
A 167	Type 304L	18Cr-8Ni	Plate, sheet, & strip	75	---	---	8	1	S30403
A 167	Type 301	17Cr-7Ni	Plate, sheet, & strip	75	---	---	8	1	S30451
A 167	Type 305	18Cr-11Ni	Plate, sheet, & strip	75	---	---	8	1	S30500
A 167	Type 308	20Cr-10Ni	Plate, sheet, & strip	75	---	---	8	2	S30800
A 167	Type 309	23Cr-12Ni	Plate, sheet, & strip	75	---	---	8	2	S30900
A 167	Type 309S	23Cr-12Ni	Plate, sheet, & strip	75	---	---	8	2	S30908
A 167	Type 310	25Cr-20Ni	Plate, sheet, & strip	75	---	---	8	2	S31000
A 167	Type 310S	25Cr-20Ni	Plate, sheet, & strip	75	---	---	8	2	S31008
A 167	Type 316L	16Cr-12Ni-2Mo	Plate, sheet, & strip	75	---	---	8	1	S31603
A 167	Type 317	18Cr-13Ni-3Mo	Plate, sheet, & strip	75	---	---	8	1	S31700
A 167	Type 317L	18Cr-13Ni-3Mo	Plate, sheet, & strip	75	---	---	8	1	S31703
A 167	Type 321	18Cr-10Ni-Ti	Plate, sheet, & strip	75	---	---	8	1	S32100
A 167	Type 3472	18Cr-10Ni-Cb	Plate, sheet, & strip	75	---	---	8	1	S34700
A 167	Type 348	18Cr-10Ni-Cb	Plate, sheet, & strip	75	---	---	8	1	S34800
SA-178	A	C	ERW tube	47	1	1	---	---	K01200
SA-178	C	C	ERW tube	60	1	1	---	---	K03503

ASME FERROUS METALS - P NUMBERS & S NUMBERS (Continued)									
ASME		**Nominal**		**Tensile**	**Welding**				
Spec No.	**Type or Grade**	**Composition**	**Product Form**	**Strength[a] ksi**	**P No**	**Group No.**	**S No.**	**Group No.**	**UNS No.**
SA-178	D	C-Mn-Si	ERW tube	70	1	2	---	---	K01200
SA-179	---	C	Seamless tube	47	1	1	---	---	K03502
SA-181	Class 60	C-Si	Pipe flange	60	1	1	---	---	K03502
SA-181	Class 70	C-Si	Pipe flange	70	1	2	---	---	K03502
SA-182	F12, Class 1	1Cr-0.5Mo	Forgings	60	4	1	---	---	K11562
SA-182	F12, Class 2	1Cr-0.5Mo	Forgings	70	4	1	---	---	K11564
SA-182	F11, Class 2	1.25Cr-0.5Mo-Si	Forgings	70	4	1	---	---	K11572
SA-182	F11, Class 3	1.25Cr-0.5Mo-Si	Forgings	75	4	1	---	---	K11572
SA-182	F11, Class 1	1.25Cr-0.5Mo-Si	Forgings	60	4	1	---	---	K11597
SA-182	F2	0.5Cr-0.5Mo	Forgings	70	3	2	---	---	K12122
SA-182	F1	C-0.5Mo	Forgings	70	3	2	---	---	K12822
SA-182	F22, Class 1	2.25Cr-1Mo	Forgings	60	5A	1	---	---	K21590
SA-182	F22, Class 3	2.25Cr-1Mo	Forgings	75	5A	1	---	---	K21590
SA-182	FR	2Ni-1Cu	Forgings	63	9A	1	---	---	K22035
SA-182	F21	3Cr-1Mo	Forgings	75	5A	1	---	---	K31545
SA-182	F3V	3Cr-1Mo-Ti-B	Forgings	85	5	1	---	---	K31830
SA-182	F5	5Cr-0.5Mo	Forgings	70	5B	1	---	---	K41545
SA-182	F5a	5Cr-0.5Mo	Forgings	90	5B	1	---	---	K42544
SA-182	F9	9Cr-1Mo	Forgings	85	5B	1	---	---	K90941
SA-182	F91	9Cr-1Mo-V	Forgings	85	5B	2	---	---	---
SA-182	F6a, Class 1	13Cr	Forgings	70	6	1	---	---	K91151
SA-182	F6a, Class 2	13Cr	Forgings	85	6	3	---	---	K91151
SA-182	FXM-19	22Cr-13Ni-5Mn	Forgings	100	8	3	---	---	S20910
SA-182	FXM-11	21Cr-6Ni-9Mn	Forgings	90	8	3	---	---	S21904

ASME FERROUS METALS - P NUMBERS & S NUMBERS (Continued)

ASME Spec No.	Type or Grade	Nominal Composition	Product Form	Tensile Strength[a] ksi	P No	Welding Group No.	Welding S No.	Welding Group No.	UNS No.
SA-182	F304	18Cr-8Ni	Forgings > 5 in.	70	8	1	---	---	S30400
SA-182	F304	18Cr-8Ni	Forgings	75	8	1	---	---	S30400
SA-182	F304L	18Cr-8Ni	Forgings > 5 in.	65	8	1	---	---	S30403
SA-182	F304L	18Cr-8Ni	Forgings	70	8	1	---	---	S30403
SA-182	F304H	18Cr-8Ni	Forgings > 5 in.	70	8	1	---	---	S30409
SA-182	F304H	18Cr-8Ni	Forgings	75	8	1	---	---	S30409
SA-182	F304N	18Cr-8Ni-N	Forgings	80	8	1	---	---	S30451
SA-182	F304LN	18Cr-8Ni-N	Forgings > 5 in.	70	8	1	---	---	S30453
SA-182	F304LN	18Cr-8Ni-N	Forgings	75	8	1	---	---	S30453
SA-182	F46	17Cr-14Ni-4Si	Forgings	78	8	1	---	---	S30600
SA-182	F45	21Cr-11Ni-N	Forgings	87	8	2	---	---	S30815
SA-182	F310	25Cr-20Ni	Forgings > 5 in.	70	8	2	---	---	S31000
SA-182	F310	25Cr-20Ni	Forgings	75	8	2	---	---	S31000
SA-182	F50	25Cr-6Ni-Mo-N	Forgings	100	10H	1	---	---	S31200
SA-182	F44	20Cr-18Ni-6Mo	Forgings	94	8	4	---	---	S31254
SA-182	F316	16Cr-12Ni-2Mo	Forgings > 5 in.	70	8	1	---	---	S31600
SA-182	F316	16Cr-12Ni-2Mo	Forgings	75	8	1	---	---	S31600
SA-182	F316L	16Cr-12Ni-2Mo	Forgings > 5 in.	65	8	1	---	---	S31603
SA-182	F316L	16Cr-12Ni-2Mo	Forgings	70	8	1	---	---	S31603
SA-182	F316H	16Cr-12Ni-2Mo	Forgings > 5 in.	70	8	1	---	---	S31609
SA-182	F316H	16Cr-12Ni-2Mo	Forgings	75	8	1	---	---	S31609
SA-182	F316N	16Cr-12Ni-2Mo-N	Forgings	80	8	1	---	---	S31651
SA-182	F316LN	16Cr-12Ni-2Mo-N	Forgings > 5 in.	70	8	1	---	---	S31653
SA-182	F316LN	16Cr-12Ni-2Mo-N	Forgings	75	8	1	---	---	S31653

ASME FERROUS METALS - P NUMBERS & S NUMBERS (Continued)

ASME Spec No.	Type or Grade	Nominal Composition	Product Form	Tensile Strength[a] ksi	Welding P No	Group No.	S No.	Group No.	UNS No.
SA-182	F317	18Cr-13Ni-3Mo	Forgings > 5 in.	70	8	1	---	---	S31700
SA-182	F317	18Cr-13Ni-3Mo	Forgings	75	8	1	---	---	S31700
SA-182	F317L	18Cr-13Ni-3Mo	Forgings > 5 in.	65	8	1	---	---	S31703
SA-182	F317L	18Cr-13Ni-3Mo	Forgings	70	8	1	---	---	S31703
SA-182	F51	22Cr-5Ni-3Mo-N	Forgings	90	10H	1	---	---	S31803
SA-182	F321	18Cr-10Ni-Ti	Forgings > 5 in.	70	8	1	---	---	S32100
SA-182	F321	18Cr-10Ni-Ti	Forgings	75	8	1	---	---	S32100
SA-182	F321H	18Cr-10Ni-Ti	Forgings > 5 in.	70	8	1	---	---	S32109
SA-182	F321H	18Cr-10Ni-Ti	Forgings	75	8	1	---	---	S32109
SA-182	---	25Cr-7.5Ni-3.5M0-N-Cu-W	Forgings	109	---	---	10H	1	S32760
SA-182	F10	20Ni-8Cr	Forgings	80	8	2	---	---	S33100
SA-182	F347	18Cr-10Ni-Cb	Forgings > 5 in.	70	8	1	---	---	S34700
SA-182	F347	18Cr-10Ni-Cb	Forgings	75	8	1	---	---	S34700
SA-182	F347H	18Cr-10Ni-Cb	Forgings > 5 in.	70	8	1	---	---	S34709
SA-182	F347H	18Cr-10Ni-Cb	Forgings	75	8	1	---	---	S34709
SA-182	F348	18Cr-10Ni-Cb	Forgings > 5 in.	70	8	1	---	---	S34800
SA-182	F348	18Cr-10Ni-Cb	Forgings	75	8	1	---	---	S34800
SA-182	F348H	18Cr-10Ni-Cb	Forgings > 5 in.	70	8	1	---	---	S34809
SA-182	F348H	18Cr-10Ni-Cb	Forgings	75	8	1	---	---	S34809
SA-182	F6b	13Cr-0.5Mo	Forgings	110	6	3	---	---	S41026
SA-182	F6NM	13Cr-4.5Ni-Mo	Forgings	115	6	4	---	---	S41500
SA-182	F429	15Cr	Forgings	60	6	2	---	---	S42900
SA-182	F430	17Cr	Forgings	60	7	2	---	---	S43000
SA-182	FXM-27Cb	27Cr-1Mo	Forgings	60	10I	1	---	---	S44627

ASME FERROUS METALS - P NUMBERS & S NUMBERS (Continued)

ASME Spec No.	Type or Grade	Nominal Composition	Product Form	Tensile Strength[a] ksi	Welding P No	Welding Group No.	Welding S No.	Welding Group No.	UNS No.
SA-182	F6a, Cl. 3	13Cr	Forgings	110	---	---	6	3	S41000
SA-182	F6a, Cl. 4	13Cr-5Mo	Forgings	130	---	---	6	3	S41000
SA-192	---	C-Si	Seamless tube	47	1	1	---	---	K01201
SA-199	T11	1.25Cr-0.5Mo-Si	Seamless tube	60	4	1	---	---	K11597
SA-199	T22	2.25Cr-1Mo	Seamless tube	60	5A	1	---	---	K21590
SA-199	T4	2.25Cr-0.5Mo-0.75Si	Seamless tube	60	5A	1	---	---	K31509
SA-199	T21	3Cr-1Mo	Seamless tube	60	5A	1	---	---	K31545
SA-199	T5	5Cr-0.5Mo	Seamless tube	60	5B	1	---	---	K41545
SA-199	T9	9Cr-1Mo	Seamless tube	60	5B	1	---	---	K81590
SA-199	T91	9Cr-1Mo-V	Seamless tube	85	5B	2	---	---	---
SA-202	A	0.5Cr-1.25Mn-Si	Plate	75	4	1	---	---	K11742
SA-202	B	0.5Cr-1.25Mn-Si	Plate	85	4	1	---	---	K12542
SA-203	A	2.5Ni	Plate	65	9A	1	---	---	K21703
SA-203	B	2.5Ni	Plate	70	9A	1	---	---	K22103
SA-203	D	3.5Ni	Plate	65	9B	1	---	---	K31718
SA-203	E	3.5Ni	Plate	70	9B	1	---	---	K32018
SA-203	F	3.5Ni	Plate, > 2 in.	75	9B	1	---	---	---
SA-203	F	3.5Ni	Plate, ≤ 2 in.	80	9B	1	---	---	---
SA-204	A	C-0.5Mo	Plate	65	3	1	---	---	K11820
SA-204	B	C-0.5Mo	Plate	70	3	2	---	---	K12020
SA-204	C	C-0.5Mo	Plate	75	3	2	---	---	K12320
SA-209	T1b	C-0.5Mo	Seamless tube	53	3	1	---	---	K11422
SA-209	T1a	C-0.5Mo	Seamless tube	60	3	1	---	---	K12023
SA-210	A-1	C-Si	Seamless tube	60	1	1	---	---	K02707

ASME Spec No.	Type or Grade	Nominal Composition	Product Form	Tensile Strength[a] ksi	P No	Welding Group No.	S No.	Group No.	UNS No.
							Welding		
SA-210	C	C-Mn-Si	Seamless tube	70	1	2	---	---	K03501
A 211	A570A	C	Welded pipe	45	---	---	1	1	---
A 211	A570Gr30	C	Welded pipe	49	---	---	1	1	K02502
A 211	A570B	C	Welded pipe	49	---	---	1	1	---
A 211	A570 Gr33	C	Welded pipe	52	---	---	1	1	K02502
A 211	A570C	C	Welded pipe	52	---	---	1	1	---
A 211	A570D	C	Welded pipe	55	---	---	1	1	---
SA-213	T2	0.5Cr-0.5Mo	Seamless tube	60	3	1	---	---	K11547
SA-213	T12	1Cr-0.5Mo	Seamless tube	60	4	1	---	---	K11562
SA-213	T11	1.25Cr-0.5Mo-Si	Seamless tube	60	4	1	---	---	K11597
SA-213	T17	1Cr-V	Seamless tube	60	10B	1	---	---	K12047
SA-213	T22	2.25Cr-1Mo	Seamless tube	60	5A	1	---	---	K21590
SA-213	T21	3Cr-1Mo	Seamless tube	60	5A	1	---	---	K31545
SA-213	T5c	5Cr-0.5Mo-Ti	Seamless tube	60	5B	1	---	---	K41245
SA-213	T5	5Cr-0.5Mo	Seamless tube	60	5B	1	---	---	K41545
SA-213	T5b	5Cr-0.5Mo-Si	Seamless tube	60	5B	1	---	---	K51545
SA-213	T9	9Cr-1Mo	Seamless tube	60	5B	1	---	---	K81590
SA-213	T91	9Cr-1Mo-V	Seamless tube	85	5B	2	---	---	---
SA-213	TP201	17Cr-4Ni-6Mn	Seamless tube	95	8	3	---	---	S20100
SA-213	TP202	18Cr-5Ni-9Mn	Seamless tube	90	8	3	---	---	S20200
SA-213	XM-19	22Cr-13Ni-5Mn	Seamless tube	100	8	3	---	---	S20910
SA-213	TP304	18Cr-8Ni	Seamless tube	75	8	1	---	---	S30400
SA-213	TP304L	18Cr-8Ni	Seamless tube	70	8	1	---	---	S30403
SA-213	TP304H	18Cr-8Ni	Seamless tube	75	8	1	---	---	S30409

ASME FERROUS METALS - P NUMBERS & S NUMBERS (Continued)

ASME FERROUS METALS - P NUMBERS & S NUMBERS (Continued)

ASME Spec No.	Type or Grade	Nominal Composition	Product Form	Tensile Strength[a] ksi	P No	Welding Group No.	Welding S No.	Welding Group No.	UNS No.
SA-213	TP304N	18Cr-8Ni-N	Seamless tube	80	8	1	---	---	S30451
SA-213	TP304LN	18Cr-8Ni-N	Seamless tube	75	8	1	---	---	S30453
SA-213	S30815	21Cr-11Ni-N	Seamless tube	87	8	2	---	---	S30815
SA-213	TP309S	23Cr-12Ni	Seamless tube	75	8	2	---	---	S30908
SA-213	Type 309H	23Cr-12Ni	Seamless tube	75	8	2	---	---	S30909
SA-213	TP309Cb	23Cr-12Ni-Cb	Seamless tube	75	8	2	---	---	S30940
SA-213	Type 309HCb	23Cr-12Ni-Cb	Seamless tube	75	8	2	---	---	S30941
SA-213	TP310S	25Cr-20Ni	Seamless tube	75	8	2	---	---	S31008
SA-213	Type 310H	25Cr-20Ni	Seamless tube	75	8	2	---	---	S31009
SA-213	TP310Cb	25Cr-20Ni-Cb	Seamless tube	75	8	2	---	---	S31040
SA-213	Type 310HCb	25Cr-20Ni-Cb	Seamless tube	75	8	2	---	---	S31041
SA-213	TP310MoLN	25Cr-22Ni-2Mo-N	Seamless tube t >¼ in.	78	8	2	---	---	S31050
SA-213	TP310MoLN	25Cr-22Ni-2Mo-N	Seamless tube t ≤¼ in	84	8	2	---	---	S31050
SA-213	TP316	16Cr-12Ni-2Mo	Seamless tube	75	8	1	---	---	S31600
SA-213	TP316L	16Cr-12Ni-2Mo	Seamless tube	70	8	1	---	---	S31603
SA-213	TP316H	16Cr-12Ni-2Mo	Seamless tube	75	8	1	---	---	S31609
SA-213	TP316N	16Cr-12Ni-2Mo-N	Seamless tube	80	8	1	---	---	S31651
SA-213	TP316LN	16Cr-12Ni-2Mo-N	Seamless tube	75	8	1	---	---	S31653
SA-213	S31725	19Cr-15Ni-4Mo	Seamless tube	75	8	4	---	---	S31725
SA-213	S31726	19Cr-15.5Ni-4Mo	Seamless tube	80	8	4	---	---	S31726
SA-213	TP321	18Cr-10Ni-Ti	Seamless tube	75	8	1	---	---	S32100
SA-213	TP321H	18Cr-10Ni-Ti	Seamless tube	75	8	1	---	---	S32109
SA-213	TP347	18Cr-10Ni-Cb	Seamless tube	75	8	1	---	---	S34700
SA-213	TP347H	18Cr-10Ni-Cb	Seamless tube	75	8	1	---	---	S34709

ASME FERROUS METALS - P NUMBERS & S NUMBERS (Continued)

ASME Spec No.	Type or Grade	Nominal Composition	Product Form	Tensile Strength[a] ksi	P No	Welding Group No.	S No.	Group No.	UNS No.
SA-213	TP348	18Cr-10Ni-Cb	Seamless tube	75	8	1	---	---	S34800
SA-213	TP348H	18Cr-10Ni-Cb	Seamless tube	75	8	1	---	---	S34809
SA-213	XM-15	18Cr-18Ni-2Si	Seamless tube	75	8	1	---	---	S38100
SA-214	---	C	ERW tube	---	1	1	---	---	K01807
SA-216	WCA	C-Si	Castings	60	1	1	---	---	J02502
SA-216	WCC	C-Mn-Si	Castings	70	1	2	---	---	J02503
SA-216	WCB	C-Si	Castings	70	1	2	---	---	J03002
SA-217	WC6	1.25Cr-0.5Mo	Castings	70	4	1	---	---	J12072
SA-217	WC4	1Ni-0.5Cr-0.5Mo	Castings	70	4	1	---	---	J12082
SA-217	WC1	C-0.5Mo	Castings	65	3	1	---	---	J12522
SA-217	WC9	2.25Cr-1Mo	Castings	70	5A	1	---	---	J21890
SA-217	WC5	0.75Ni-1Mo-0.75Cr	Castings	70	4	1	---	---	J22000
SA-217	C5	5Cr-0.5Mo	Castings	90	5B	1	---	---	J42025
SA-217	C12	9Cr-1Mo	Castings	90	5B	1	---	---	J82090
SA-217	CA15	13Cr	Castings	90	6	3	---	---	J91150
SA-225	D	Mn-0.5Ni-V	Plate, > 3 in.	75	10A	1	---	---	---
SA-225	D	Mn-0.5Ni-V	Plate, ≤ 3 in.	80	10A	1	---	---	---
SA-225	C	Mn-0.5Ni-V	Plate	105	10A	1	---	---	K12524
SA-226	---	C-Si	ERW tube	47	1	1	---	---	K01201
SA-234	WPB	C-Si	Piping fitting	60	1	1	---	---	K03006
SA-234	WPC	C-Si	Piping fitting	70	1	2	---	---	K03501
SA-234	WP11, Class 1	1.25Cr-0.5Mo-Si	Piping fitting	60	4	1	---	---	K12524
SA-234	WP12, Class 1	1Cr-0.5Mo	Piping fitting	60	4	1	---	---	K12062
SA-234	WP1	C-0.5Mo	Piping fitting	55	3	1	---	---	K12821

ASME FERROUS METALS - P NUMBERS & S NUMBERS (Continued)

ASME Spec No.	Type or Grade	Nominal Composition	Product Form	Tensile Strength[a] ksi	P No	Welding Group No.	Welding S No.	Welding Group No.	UNS No.
SA-234	WP22, Class 1	2.25Cr-1Mo	Piping fitting	60	5A	1	---	---	K21590
SA-234	WPR	2Ni-1Cu	Piping fitting	63	9A	1	---	---	K22035
SA-234	WP5	5Cr-0.5Mo	Piping fitting	60	5B	1	---	---	K41545
SA-234	WP9	9Cr-1Mo	Piping fitting	60	5B	1	---	---	K90941
SA-234	WP91	9Cr-1Mo-V	Piping fitting	85	5B	2	---	---	---
SA-240	Type 201-1	17Cr-4Ni-6Mn	Plate, Sheet & Strip	95	8	3	---	---	S20100
SA-240	Type 202	18Cr-5Ni-9Mn	Plate, Sheet & Strip	90	8	3	---	---	S20200
SA-240	---	16Cr-9Mn-2Ni-N	Plate	95	8	3	---	---	S20400
SA-240	Type XM-19	22Cr-13Ni-5Mn	Plate	100	8	3	---	---	S20910
SA-240	Type XM-19	22Cr-13Ni-5Mn	Sheet & Strip	105	8	3	---	---	S20910
SA-240	Type XM-17	19Cr-8Mn-6Ni-Mo-N	Plate	90	8	3	---	---	S21600
SA-240	Type XM-17	19Cr-8Mn-6Ni-Mo-N	Sheet & Strip	100	8	3	---	---	S21600
SA-240	Type XM-18	19Cr-8Mn-6Ni-Mo-N	Plate	90	8	3	---	---	S21603
SA-240	Type XM-18	19Cr-8Mn-6Ni-Mo-N	Sheet & Strip	100	8	3	---	---	S21603
SA-240	S21800	18Cr-8Ni-4Si-N	Plate, Sheet & Strip	95	8	3	---	---	S21800
SA-240	Type XM-29	18Cr-3Ni-12Mn	Plate, Sheet & Strip	100	8	3	---	---	S24000
SA-240	Type 302	18Cr-8Ni	Plate, Sheet & Strip	75	8	1	---	---	S30200
SA-240	Type 304	18Cr-8Ni	Plate, Sheet & Strip	75	8	1	---	---	S30400
SA-240	Type 304L	18Cr-8Ni	Plate, Sheet & Strip	70	8	1	---	---	S30403
SA-240	Type 304H	18Cr-8Ni	Plate, Sheet & Strip	75	8	1	---	---	S30409
SA-240	Type 304N	18Cr-8Ni-N	Plate, Sheet & Strip	80	8	1	---	---	S30451
SA-240	Type XM-21	18Cr-8Ni-N	Plate	85	8	1	---	---	S30452
SA-240	Type XM-21	18Cr-8Ni-N	Sheet & Strip	90	8	1	---	---	S30452
SA-240	Type 304LN	18Cr-8Ni-N	Plate, Sheet & Strip	75	8	1	---	---	S30453

ASME FERROUS METALS - P NUMBERS & S NUMBERS (Continued)

ASME Spec No.	Type or Grade	Nominal Composition	Product Form	Tensile Strength[a] ksi	Welding P No	Group No.	S No.	Group No.	UNS No.
SA-240	Type 305	18Cr-11Ni	Plate, Sheet & Strip	70	8	1	---	---	S30500
SA-240	S30600	17Cr-14Ni-4Si	Plate, Sheet & Strip	78	8	1	---	---	S30600
SA-240	S30815	21Cr-11Ni-N	Plate, Sheet & Strip	87	8	2	---	---	S30815
SA-240	Type 309S	23Cr-12Ni	Plate, Sheet & Strip	75	8	2	---	---	S30908
SA-240	Type 309H	23Cr-12Ni	Plate, Sheet & Strip	75	8	2	---	---	S30909
SA-240	Type 309Cb	23Cr-12Ni-Cb	Plate, Sheet & Strip	75	8	2	---	---	S30940
SA-240	Type 309HCb	23Cr-12Ni-Cb	Plate, Sheet & Strip	75	8	2	---	---	S30949
SA-240	Type 310S	25Cr-20Ni	Plate, Sheet & Strip	75	8	2	---	---	S31008
SA-240	Type 310Cb	25Cr-20Ni-Cb	Plate, Sheet & Strip	75	8	2	---	---	S31040
SA-240	Type 310HCb	25Cr-20Ni-Cb	Plate, Sheet & Strip	75	8	2	---	---	S31041
SA-240	Type 310MoLN	25Cr-22Ni-2Mo-N	Plate, Sheet & Strip	80	8	2	---	---	S31050
SA-240	Type S31200	25Cr-6Ni-Mo-N	Plate, Sheet & Strip	100	10H	1	---	---	S31200
SA-240	S31254	20Cr-18Ni-6Mo	Plate, Sheet & Strip	94	8	4	---	---	S31254
SA-240	S31260	25Cr-6.5 Ni-3Mo-N	Plate, Sheet & Strip	100	10H	1	---	---	S31260
SA-240	Type 316	16Cr-12Ni-2Mo	Plate, Sheet & Strip	75	8	1	---	---	S31600
SA-240	Type 316L	16Cr-12Ni-2Mo	Plate, Sheet & Strip	70	8	1	---	---	S31603
SA-240	Type 316H	16Cr-12Ni-2Mo	Plate, Sheet & Strip	75	8	1	---	---	S31609
SA-240	Type 316Ti	16Cr-12Ni-2Mo-Ti	Plate, Sheet & Strip	75	8	1	---	---	S31635
SA-240	Type 316Cb	16Cr-12Ni-2Mo-Cb	Plate, Sheet & Strip	75	8	1	---	---	S31640
SA-240	Type 316N	16Cr-12Ni-2Mo-N	Plate, Sheet & Strip	80	8	1	---	---	S31651
SA-240	Type 316LN	16Cr-12Ni-2Mo-N	Plate, Sheet & Strip	75	8	1	---	---	S31653
SA-240	Type 317	18Cr-13Ni-3Mo	Plate, Sheet & Strip	75	8	1	---	---	S31700
SA-240	Type 317L	18Cr-13Ni-3Mo	Plate, Sheet & Strip	75	8	1	---	---	S31703
SA-240	S31725	19Cr-15Ni-4Mo	Plate, Sheet & Strip	75	8	4	---	---	S31725

ASME FERROUS METALS - P NUMBERS & S NUMBERS (Continued)

ASME Spec No.	Type or Grade	Nominal Composition	Product Form	Tensile Strength[a] ksi	P No	Welding Group No.	Welding S No.	Welding Group No.	UNS No.
SA-240	S31726	19Cr-15.5Ni-4Mo	Plate, Sheet & Strip	80	8	4	---	---	S31726
SA-240	S31753	18Cr-13Ni-3Mo-N	Plate, Sheet & Strip	80	8	1	---	---	S31753
SA-240	S31803	22Cr-5Ni-3Mo-N	Plate, Sheet & Strip	90	10H	1	---	---	S31803
SA-240	Type 321	18Cr-10Ni-Ti	Plate, Sheet & Strip	75	8	1	---	---	S32100
SA-240	Type 321H	18Cr-10Ni-Ti	Plate, Sheet & Strip	75	8	1	---	---	S32109
SA-240	S32550	25Cr-5Ni-3Mo-2Cu	Plate, Sheet & Strip	110	10H	1	---	---	S32550
SA-240	---	25Cr-25Ni-3.5Mo-N-Cu-W	Plate, Sheet & Strip	109	---	---	10H	1	S32760
SA-240	Type 329	26Cr-4Ni-Mo	Plate, Sheet & Strip	90	10H	1	---	---	S32900
SA-240	S32950	26Cr-4Ni-Mo-N	Plate, Sheet & Strip	90	10H	1	---	---	S32950
SA-240	Type 347	18Cr-10Ni-Cb	Plate, Sheet & Strip	75	8	1	---	---	S34700
SA-240	Type 347H	18Cr-10Ni-Cb	Plate, Sheet & Strip	75	8	1	---	---	S34709
SA-240	Type 348	18Cr-10Ni-Cb	Plate, Sheet & Strip	75	8	1	---	---	S34800
SA-240	Type 348H	18Cr-10Ni-Cb	Plate, Sheet & Strip	75	8	1	---	---	S34809
SA-240	Type XM-15	18Cr-18Ni-2Si	Plate, Sheet & Strip	75	8	1	---	---	S38100
SA-240	Type 405	12Cr-1Al	Plate, Sheet & Strip	60	7	1	---	---	S40500
SA-240	Type 409	11Cr-Ti	Plate, Sheet & Strip	55	7	1	---	---	S40900
SA-240	Type 410	13Cr	Plate, Sheet & Strip	65	6	1	---	---	S41000
SA-240	Type 410S	13Cr	Plate, Sheet & Strip	60	7	1	---	---	S41008
SA-240	S41500	13Cr-4.5Ni-Mo	Plate, Sheet & Strip	115	6	4	---	---	S41500
SA-240	Type 429	15Cr	Plate, Sheet & Strip	65	6	2	---	---	S42900
SA-240	Type 430	17Cr	Plate, Sheet & Strip	65	7	2	---	---	S43000
SA-240	Type 439	17Cr-Ti	Plate, Sheet & Strip	65	7	2	---	---	S43035
SA-240	S44400	18Cr-2Mo	Plate, Sheet & Strip	60	7	2	---	---	S44400
SA-240	Type XM-33	27Cr-1Mo-Ti	Plate, Sheet & Strip	68	10I	1	---	---	S44626

ASME FERROUS METALS - P NUMBERS & S NUMBERS (Continued)									
ASME Spec No.	Type or Grade	Nominal Composition	Product Form	Tensile Strength[a] ksi	Welding				UNS No.
					P No	Group No.	S No.	Group No.	
SA-240	Type XM-27	27Cr-1Mo	Plate, Sheet & Strip	65	10I	1	---	---	S44627
SA-240	S44635	25Cr-4Ni-4Mo-Ti	Plate, Sheet & Strip	90	10I	1	---	---	S44635
SA-240	S44660	26Cr-3Ni-3Mo	Plate, Sheet & Strip	85	10K	1	---	---	S44660
SA-240	S44700	29Cr-4Mo	Plate, Sheet & Strip	80	10J	1	---	---	S44700
SA-240	S44800	29Cr-4Mo-2Ni	Plate, Sheet & Strip	80	10K	1	---	---	S44800
SA-240	Type 305	18Cr-11Ni-V	Plate, Sheet & Strip	70	---	---	8	1	S30500
SA-249	TP 201	17Cr-4Ni-6Mn	Welded tube	95	8	3	---	---	S20100
SA-249	TP 202	18Cr-5Ni-9Mn	Welded tube	90	8	3	---	---	S20200
SA-249	TP XM-19	22Cr-13Ni-5Mn	Welded tube	100	8	3	---	---	S20910
SA-249	TP XM-29	18Cr-3Ni-12Mn	Welded tube	100	8	3	---	---	S24000
SA-249	TP304	18Cr-8Ni	Welded tube	75	8	1	---	---	S30400
SA-249	TP304L	18Cr-8Ni	Welded tube	70	8	1	---	---	S30403
SA-249	TP304H	18Cr-8Ni	Welded tube	75	8	1	---	---	S30409
SA-249	TP304N	18Cr-8Ni-N	Welded tube	80	8	1	---	---	S30451
SA-249	TP304LN	18Cr-8Ni-N	Welded tube	75	8	1	---	---	S30453
SA-249	S30815	21Cr-11Ni-N	Welded tube	87	8	2	---	---	S30815
SA-249	TP309S	23Cr-12Ni	Welded tube	75	8	2	---	---	S30908
SA-249	Type 309H	23Cr-12Ni	Welded tube	75	8	2	---	---	S30909
SA-249	TP309Cb	23Cr-12Ni-Cb	Welded tube	75	8	2	---	---	S30940
SA-249	TP309HCb	23Cr-12Ni-Cb	Welded tube	75	8	2	---	---	S30941
SA-249	TP310S	25Cr-20Ni	Welded tube	75	8	2	---	---	S31008
SA-249	TP310H	25Cr-20Ni	Welded tube	75	8	2	---	---	S31009
SA-249	TP310Cb	25Cr-20Ni-Cb	Welded tube	75	8	2	---	---	S31040
SA-249	TP310MoLN	25Cr-22Ni-2Mo-N	Welded tube, t > ¼ in.	78	8	2	---	---	S31050

ASME FERROUS METALS - P NUMBERS & S NUMBERS (Continued)

ASME Spec No.	Type or Grade	Nominal Composition	Product Form	Tensile Strength[a] ksi	P No	Welding Group No.	S No.	Group No.	UNS No.
SA-249	TP310MoLN	25Cr-22Ni-2Mo-N	Welded tube, t ≤ ¼ in.	84	8	2	---	---	S31050
SA-249	S31254	20Cr-18Ni-6Mo	Welded tube	94	8	4	---	---	S31254
SA-249	TP316	16Cr-12Ni-2Mo	Welded tube	75	8	1	---	---	S31600
SA-249	TP316L	16Cr-12Ni-2Mo	Welded tube	70	8	1	---	---	S31603
SA-249	TP316H	16Cr-12Ni-2Mo	Welded tube	75	8	1	---	---	S31609
SA-249	TP316N	16Cr-12Ni-2Mo-N	Welded tube	80	8	1	---	---	S31651
SA-249	TP316LN	16Cr-12Ni-2Mo-N	Welded tube	75	8	1	---	---	S31653
SA-249	TP317	18Cr-13Ni-3Mo	Welded tube	75	8	1	---	---	S31700
SA-249	TP317L	18Cr-13Ni-3Mo	Welded tube	75	8	1	---	---	S31703
SA-249	S31725	19Cr-15Ni-4Mo	Welded tube	75	8	4	---	---	S31725
SA-249	S31726	19Cr-15.5Ni-4Mo	Welded tube	80	8	4	---	---	S31726
SA-249	TP321	18Cr-10Ni-Ti	Welded tube	75	8	1	---	---	S32100
SA-249	TP321H	18Cr-10Ni-Ti	Welded tube	75	8	1	---	---	S32109
SA-249	TP347	18Cr-10Ni-Cb	Welded tube	75	8	1	---	---	S34700
SA-249	TP347H	18Cr-10Ni-Cb	Welded tube	75	8	1	---	---	S34709
SA-249	TP348	18Cr-10Ni-Cb	Welded tube	75	8	1	---	---	S34800
SA-249	TP348H	18Cr-10Ni-Cb	Welded tube	75	8	1	---	---	S34809
SA-249	TP XM-15	18Cr-18Ni-2Si	Welded tube	75	8	1	---	---	S38100
SA-250	T2	0.5Cr-0.5Mo	ERW tube	60	3	1	---	---	K11547
SA-250	T11	1.25Cr-0.5Mo-Si	ERW tube	60	4	1	---	---	K11597
SA-250	T22	2.25Cr-1Mo	ERW tube	60	5A	1	---	---	K21590
SA-250	T1b	C-0.5Mo	ERW tube	53	3	1	---	---	K11422
SA-250	T1	C-0.5Mo	ERW tube	55	3	1	---	---	K11522
SA-250	T1a	C-0.5Mo	ERW tube	60	3	1	---	---	K12023

ASME FERROUS METALS - P NUMBERS & S NUMBERS (Continued)

ASME Spec No.	Type or Grade	Nominal Composition	Product Form	Tensile Strength[a] ksi	Welding P No	Welding Group No.	Welding S No.	Welding Group No.	UNS No.
A 254	Class 1	C	Cu Brazed tube	42			---	---	K01001
A 254	Class 2	C	Cu Brazed tube	42			---	---	K01001
SA-266	4	C-Mn-Si	Forgings	70	1	2	---	---	K03017
SA-266	1	C-Si	Forgings	60	1	1	---	---	K03506
SA-266	2	C-Si	Forgings	70	1	2	---	---	K03506
SA-266	3	C-Si	Forgings	75	1	2	---	---	K05001
SA-268	TP405	12Cr-1Al	Smls & welded tube	60	7	1	---	---	S40500
SA-268	S40800	12Cr-Ti	Smls & welded tube	55	7	1	---	---	S40800
SA-268	TP409	11Cr-Ti	Smls & welded tube	55	7	1	---	---	S40900
SA-268	TP410	13Cr	Smls & welded tube	60	6	1	---	---	S41000
SA-268	S41500	13Cr-4.5Ni-Mo	Smls & welded tube	115	6	4	---	---	S41500
SA-268	TP429	15Cr	Smls & welded tube	60	6	2	---	---	S42900
SA-268	TP430	17Cr	Smls & welded tube	60	7	2	---	---	S43000
SA-268	TP439	17Cr-Ti	Smls & welded tube	60	7	2	---	---	S43035
SA-268	TP430Ti	18Cr-Ti-Cb	Smls & welded tube	60	7	1	---	---	S43036
SA-268	18Cr-2Mo	18Cr-2Mo	Smls & welded tube	60	7	2	---	---	S44400
SA-268	TP446-2	27Cr	Smls & welded tube	65	10I	1	---	---	S44600
SA-268	TP446-1	27Cr	Smls & welded tube	70	10I	1	---	---	S44600
SA-268	TP XM-33	27Cr-1Mo-Ti	Smls & welded tube	68	10I	1	---	---	S44626
SA-268	TP XM-27	27Cr-1Mo	Smls & welded tube	65	10I	1	---	---	S44627
SA-268	25-4-4	25Cr-4Ni-4Mo-Ti	Smls & welded tube	90	10I	1	---	---	S44635
SA-268	26-3-3	26Cr-3Ni-3Mo	Smls & welded tube	85	10K	1	---	---	S44660
SA-268	29-4	29Cr-4Mo	Smls & welded tube	80	10J	1	---	---	S44700
SA-268	S44735	29Cr-4Mo-Ti	Smls & welded tube	75	10J	1	---	---	S44735

ASME FERROUS METALS - P NUMBERS & S NUMBERS (Continued)

ASME Spec No.	Type or Grade	Nominal Composition	Product Form	Tensile Strength[a] ksi	P No	Welding Group No.	S No.	Group No.	UNS No.
SA-268	29-4-2	29Cr-4Mo-2Ni	Smls & welded tube	80	10K	1	---	---	S44800
A 269	TP316L	16Cr-12Ni-2Mo	Smls & welded tube	70	---	---	8	1	---
A 269	TP316	16Cr-12Ni-2Mo	Smls & welded tube	75	---	---	8	1	---
A 269	TP304	18Cr-8Ni	Smls & welded tube	75	---	---	8	1	S30400
A 269	TP304L	18Cr-8Ni	Smls & welded tube	70	---	---	8	1	S30403
A 271	TP304	18Cr-8Ni	Smls tube	75	---	---	8	1	S30400
A 271	TP304L	18Cr-8Ni	Smls tube	70	---	---	8	1	S30403
A 276	TP316L	16Cr-12Ni-2Mo	Bar	70	---	---	8	1	---
A 276	TP316	16Cr-12Ni-2Mo	Bar	75	---	---	8	1	---
A 276	TP304	18Cr-8Ni	Bar	75	---	---	8	1	S30400
A 276	TP304L	18Cr-8Ni	Bar	70	---	---	8	1	S30403
A 276	TP410	13Cr	Bar	65	---	---	6	1	---
SA-283	A	C	Plate	45	1	1	---	---	---
SA-283	B	C	Plate	50	1	1	---	---	---
SA-283	C	C	Plate	55	1	1	---	---	---
SA-283	D	C	Plate	60	1	1	---	---	---
SA-285	A	C	Plate	45	1	1	---	---	K01700
SA-285-	B	C	Plate	50	1	1	---	---	K02200
SA-285	C	C	Plate	55	1	1	---	---	K02801
SA-299	---	C-Mn-Si	Plate	75	1	2	---	---	K02803
SA-302	A	Mn-0.5Mo	Plate	75	3	2	---	---	K12021
SA-302	B	Mo-0.5Mo	Plate	80	3	3	---	---	K12022
SA-302	C	Mn-0.5Mo-0.5Ni	Plate	80	3	3	---	---	K12039
SA-302	D	Mn-0.5Mo-0.75Ni	Plate	80	3	3	---	---	K12054

ASME FERROUS METALS - P NUMBERS & S NUMBERS (Continued)

ASME Spec No.	Type or Grade	Nominal Composition	Product Form	Tensile Strength[a] ksi	P No	Welding Group No.	S No.	Group No.	UNS No.
SA-312	TP XM-19	22Cr-13Ni-5Mn	Smls & welded pipe	100	8	3	---	---	S20910
SA-312	TP XM-11	21Cr-6Ni-9Mn	Smls & welded pipe	90	8	3	---	---	S21904
SA-312	TP XM-29	18Cr-3Ni-12Mn	Smls & welded pipe	100	8	3	---	---	S24000
SA-312	TP304	18Cr-8Ni	Smls & welded pipe	75	8	1	---	---	S30400
SA-312	TP304L	18Cr-8Ni	Smls & welded pipe	70	8	1	---	---	S30403
SA-312	TP304H	18Cr-8Ni	Smls & welded pipe	75	8	1	---	---	S30409
SA-312	TP304N	18Cr-8Ni-N	Smls & welded pipe	80	8	1	---	---	S30451
SA-312	TP304LN	18Cr-8Ni-N	Smls & welded pipe	75	8	1	---	---	S30453
SA-312	S30600	17Cr-14Ni-4Si	Smls & welded pipe	78	8	1	---	---	S30600
SA-312	S30815	21Cr-11Ni-N	Smls & welded pipe	87	8	2	---	---	S30815
SA-312	TP309S	23Cr-12Ni	Smls & welded pipe	75	8	2	---	---	S30908
SA-312	TP309H	23Cr-12Ni	Smls & welded pipe	75	8	2	---	---	S30909
SA-312	TP309Cb	23Cr-12Ni-Cb	Smls & welded pipe	75	8	2	---	---	S30940
SA-312	TP309HCb	23Cr-12Ni-Cb	Smls & welded pipe	75	8	2	---	---	S30941
SA-312	TP310S	25Cr-20Ni	Smls & welded pipe	75	8	2	---	---	S31008
SA-312	TP310H	25Cr-20Ni	Smls & welded pipe	75	8	2	---	---	S31009
SA-312	TP310Cb	25Cr-20Ni-Cb	Smls & welded pipe	75	8	2	---	---	S31040
SA-312	TP310HCb	25Cr-20Ni-Cb	Smls & welded pipe	75	8	2	---	---	S31041
SA-312	TP310MoLN	25Cr-22Ni-2Mo-N	Welded pipe, t > ¼ in.	78	8	2	---	---	S31050
SA-312	TP310MoLN	25Cr-22Ni-2Mo-N	Welded pipe, t ≤ ¼ in.	84	8	2	---	---	S31050
SA-312	S31254	20Cr-18Ni-6Mo	Smls & welded pipe	94	8	4	---	---	S31254
SA-312	TP316	16Cr-12Ni-2Mo	Smls & welded pipe	75	8	1	---	---	S31600
SA-312	TP316L	16Cr-12Ni-2Mo	Smls & welded pipe	70	8	1	---	---	S31603
SA-312	TP316H	16Cr-12Ni-2Mo-	Smls & welded pipe	75	8	1	---	---	S31609

ASME FERROUS METALS - P NUMBERS & S NUMBERS (Continued)

ASME Spec No.	Type or Grade	Nominal Composition	Product Form	Tensile Strength[a] ksi	Welding				UNS No.
					P No	Group No.	S No.	Group No.	
SA-312	TP316N	16Cr-12Ni-2Mo-N	Smls & welded pipe	80	8	1	---	---	S31651
SA-312	TP316LN	16Cr-12Ni-2Mo-N	Smls & welded pipe	75	8	1	---	---	S31653
SA-312	TP317	18Cr-13Ni-3Mo	Smls & welded pipe	75	8	1	---	---	S31700
SA-312	TP317L	18Cr-13Ni-3Mo	Smls & welded pipe	75	8	1	---	---	S31703
SA-312	S31725	19Cr-15Ni-4Mo	Smls & welded pipe	75	8	4	---	---	S31725
SA-312	S31726	19Cr-15.5Ni-4Mo	Smls & welded pipe	80	8	4	---	---	S31726
SA-312	TP321	18Cr-10Ni-Ti	Smls & welded pipe, > $\frac{3}{8}$ in.	70	8	1	---	---	S32100
SA-312	TP321	18Cr-10Ni-Ti	Smls & welded pipe, ≤ $\frac{3}{8}$ in.	75	8	1	---	---	S32100
SA-312	TP321	18Cr-10Ni-Ti	Welded pipe	75	8	1	---	---	S32100
SA-312	TP321H	18Cr-10Ni-Ti	Smls & welded pipe, > $\frac{3}{8}$ in.	70	8	1	---	---	S32109
SA-312	TP321H	18Cr-10Ni-Ti	Smls & welded pipe, ≤ $\frac{3}{8}$ in.	75	8	1	---	---	S32109
SA-312	TP321H	18Cr-10Ni-Ti	Welded pipe	75	8	1	---	---	S32109
SA-312	TP347	18Cr-10Ni-Cb	Smls & welded pipe	75	8	1	---	---	S34700
SA-312	TP347H	18Cr-10Ni-Cb	Smls & welded pipe	75	8	1	---	---	S34709
SA-312	TP348	18Cr-10Ni-Cb	Smls & welded pipe	75	8	1	---	---	S34800
SA-312	TP348H	18Cr-10Ni-Cb	Smls & welded pipe	75	8	1	---	---	S34809
SA-312	TP XM-15	18Cr-18Ni-2Si	Smls & welded pipe	75	8	1	---	---	S38100
A 331	8620 CW	0.5Ni-0.5Cr-Mo	Bar	90	---	---	3	3	G86200
SA-333	6	C-Mn-Si	Smls & welded pipe	60	1	1	---	---	K03006
SA-333	1	C-Mn	Smls & welded pipe	55	1	1	---	---	K03008

ASME Spec No.	Type or Grade	Nominal Composition	Product Form	Tensile Strength[a] ksi	P No	Welding Group No.	S No.	Group No.	UNS No.
SA-333	4	0.75Cr-0.75Ni-Cu-Al	Smls & welded pipe	60	4	2	---	---	K11267
SA-333	7	2.5Ni	Smls & welded pipe	65	9A	1	---	---	K21903
SA-333	9	2Ni-1Cu	Smls & welded pipe	63	9A	1	---	---	K22035
SA-333	3	3.5Ni	Smls & welded pipe	65	9B	1	---	---	K31918
SA-333	10	C-Mn-Si	Smls & welded pipe	80	1	3	---	---	---
SA-333	8	9Ni	Smls & welded pipe	100	11A	1	---	---	K81340
SA-334	6	C-Mn-Si	Welded tube	60	1	1	---	---	K03006
SA-334	1	C-Mn	Welded tube	55	1	1	---	---	K03008
SA-334	7	2.5Ni	Welded tube	65	9A	1	---	---	K21903
SA-334	9	2Ni-1Cu	Welded tube	63	9A	1	---	---	K22035
SA-334	3	3.5Ni	Welded tube	65	9B	1	---	---	K31918
SA-334	8	9Ni	Welded tube	100	11A	1	---	---	K81340
SA-335	P1	C-0.5Mo	Seamless pipe	55	3	1	---	---	K11522
SA-335	P2	0.5Cr-0.5Mo	Seamless pipe	55	3	1	---	---	K11547
SA-335	P12	1Cr-0.5Mo	Seamless pipe	60	4	1	---	---	K11562
SA-335	P15	1.5Si-0.5Mo	Seamless pipe	60	3	1	---	---	K11578
SA-335	P11	1.25Cr-0.5Mo-Si	Seamless pipe	60	4	1	---	---	K11597
SA-335	P22	2.25Cr-1Mo	Seamless pipe	60	5A	1	---	---	K21590
SA-335	P21	3Cr-1Mo	Seamless pipe	60	5A	1	---	---	K31545
SA-335	P5c	5Cr-0.5Mo-Ti	Seamless pipe	60	5B	1	---	---	K41245
SA-335	P5	5Cr-0.5Mo	Seamless pipe	60	5B	1	---	---	K41545
SA-335	P5b	5Cr-0.5Mo-Si	Seamless pipe	60	5B	1	---	---	K51545
SA-335	P9	9Cr-1Mo	Seamless pipe	60	5B	1	---	---	K81590
SA-335	P91	9Cr-1Mo-V	Seamless pipe	85	5B	2	---	---	---

Table title: ASME FERROUS METALS - P NUMBERS & S NUMBERS (Continued)

ASME FERROUS METALS - P NUMBERS & S NUMBERS (Continued)

ASME Spec No.	Type or Grade	Nominal Composition	Product Form	Tensile Strength[a] ksi	P No	Welding Group No.	S No.	Group No.	UNS No.
SA-336	F6	13Cr	Forgings	85	6	3	---	---	---
SA-336	F12	1Cr-0.5Mo	Forgings	70	4	1	---	---	K11564
SA-336	F11, Cl. 1	1.25Cr-0.5Mo-Si	Forgings	60	4	1	---	---	K11597
SA-336	F11, Cl. 2	1.25Cr-0.5Mo-Si	Forgings	70	4	1	---	---	K11572
SA-336	F11, Cl. 3	1.25Cr-0.5Mo-Si	Forgings	75	4	1	---	---	K11572
SA-336	F1	C-0.5Mo	Forgings	70	3	2	---	---	K12520
SA-336	F22, Cl. 1	2.25Cr-1Mo	Forgings	60	5A	1	---	---	K21590
SA-336	F22, Cl. 3	2.25Cr-1Mo	Forgings	75	5A	1	---	---	K21590
SA-336	F21, Cl. 1	3Cr-1Mo	Forgings	60	5A	1	---	---	K31545
SA-336	F21, Cl. 3	3Cr-1Mo	Forgings	75	5A	1	---	---	K31545
SA-336	F3V	3Cr-1Mo-V-Ti-B	Forgings	85	5C	1	---	---	K31830
SA-336	F5	5Cr-0.5Mo	Forgings	60	5B	1	---	---	K41545
SA-336	F5A	5Cr-0.5Mo	Forgings	80	5B	1	---	---	K42544
SA-336	F9	9Cr-1Mo	Forgings	85	5B	1	---	---	K81590
SA-336	F91	9Cr-1Mo-V	Forgings	85	5B	2	---	---	---
SA-336	F46	17Cr-14Ni-4Si	Forgings	78	8	1	---	---	S30600
SA-336	FXM-19	22Cr-13Ni-5Mn	Forgings	100	8	3	---	---	S20910
SA-336	FXM-11	21Cr-6Ni-9Mn	Forgings	90	8	3	---	---	S21904
SA-336	F304	18Cr-8Ni	Forgings	70	8	1	---	---	S30400
SA-336	F304L	18Cr-8Ni	Forgings	65	8	1	---	---	S30403
SA-336	F304H	18Cr-8Ni	Forgings	70	8	1	---	---	S30409
SA-336	F304N	18Cr-8Ni-N	Forgings	80	8	1	---	---	S30451
SA-336	F304LN	18Cr-8Ni-N	Forgings	70	8	1	---	---	S30453
SA-336	F310	25Cr-20Ni	Forgings	75	8	2	---	---	S31000

ASME FERROUS METALS - P NUMBERS & S NUMBERS (Continued)

ASME Spec No.	Type or Grade	Nominal Composition	Product Form	Tensile Strength[a] ksi	P No	Welding Group No.	S No.	Group No.	UNS No.
SA-336	F316	16Cr-12Ni-2Mo	Forgings	70	8	1	---	---	S31600
SA-336	F316L	16Cr-12Ni-2Mo	Forgings	65	8	1	---	---	S31603
SA-336	F316H	16Cr-12Ni-2Mo	Forgings	70	8	1	---	---	S31609
SA-336	F316N	16Cr-12Ni-2Mo-N	Forgings	80	8	1	---	---	S31651
SA-336	F316LN	16Cr-12Ni-2Mo-N	Forgings	70	8	1	---	---	S31653
SA-336	F321	18Cr-10Ni-Ti	Forgings	70	8	1	---	---	S32100
SA-336	F321H	18Cr-10Ni-Ti	Forgings	70	8	1	---	---	S32109
SA-336	F347	18Cr-10Ni-Cb	Forgings	70	8	1	---	---	S34700
SA-336	F347H	18Cr-10Ni-Cb	Forgings	70	8	1	---	---	S34709
SA-336	F348	18Cr-10Ni-Cb	Forgings	70	8	1	---	---	S34800
SA-336	F348H	18Cr-10Ni-Cb	Forgings	65	8	1	---	---	S34809
SA-336	FXM-27Cb	27Cr-1Mo	Forgings	60	10I	1	---	---	S44627
SA-350	LF1, Cl. 1	C-Mn-Si	Forgings	60	1	1	---	---	K03009
SA-350	LF2	C-Mn-Si	Forgings	70	1	2	---	---	K03011
SA-350	LF5 Class 1	1.5 Ni	Forgings	60	9A	1	---	---	K13050
SA-350	LF5 Class 2	1.5 Ni	Forgings	70	9A	1	---	---	K13050
SA-350	LF9	2Ni-1Cu	Forgings	63	9A	1	---	---	K22036
SA-350	LF3, Class 2	3.5Ni	Forgings	70	9B	1	---	---	K32025
SA-351	CF3	18Cr-8Ni	Castings	70	8	1	---	---	J92500
SA-351	CF3A	18Cr-8Ni	Castings	77	8	1	---	---	J92500
SA-351	CF8	18Cr-8Ni	Castings	70	8	1	---	---	J92600
SA-351	CF8A	18Cr-8Ni	Castings	77	8	1	---	---	J92600
SA-351	CF8C	18Cr-10Ni-Cb	Castings	70	8	1	---	---	J92710
SA-351	CF3M	18Cr-12Ni-2Mo	Castings	70	8	1	---	---	J92800

ASME FERROUS METALS - P NUMBERS & S NUMBERS (Continued)

ASME Spec No.	Type or Grade	Nominal Composition	Product Form	Tensile Strength[a] ksi	Welding P No	Group No.	S No.	Group No.	UNS No.
SA-351	CF8M	18Cr-12Ni-2Mo	Castings	70	8	1	---	---	J92900
SA-351	CF10	19Cr-9Ni-0.5Mo	Castings	70	8	1	---	---	---
SA-351	CF10M	19Cr-9Ni-2Mo	Castings	70	8	1	---	---	---
SA-351	CG8M	19Cr-10Ni-3Mo	Castings	75	8	1	---	---	J93000
SA-351	CK3MCuN	20Cr-18Ni-6Mo	Castings	80	8	4	---	---	J93254
SA-351	CE8MN	24Cr-10Ni-Mo-N	Castings	95	10H	1	---	---	J93345
SA-351	CD4MCu	25Cr-5Ni-3Mo-3Cu	Castings	100	10H	1	---	---	J93370
SA-351	CD3MWCuN	25Cr-7.5Ni-3.5Mo-N-Cu-W	Castings	100	---	---	10H	1	J93380
SA-351	CH8	25Cr-12Ni	Castings	65	8	2	---	---	J93400
SA-351	CH20	25Cr-12Ni	Castings	70	8	2	---	---	J93402
SA-351	CG6MMN	22Cr-12Ni-5Mn	Castings	85	8	3	---	---	J93790
SA-351	CK20	25Cr-20Ni	Castings	65	8	2	---	---	J94202
SA-351	CN7M	28Ni-19Cr-3Cu-Mo	Castings	62	45	---	---	---	J95150
SA-351	CT15C	20Cr-32.5Ni-Cb	Castings	63	45	---	---	---	---
A 351	CA15	13Cr	Castings	90	---	---	6	3	---
A 351	CE20N	25Cr-8Ni-N	Castings	80	---	---	8	2	---
A 351	CF10MC	16Cr-14Ni-2Mo	Castings	70	---	---	8	1	J92971
A 351	CH10	25Cr-12Ni	Castings	70	---	---	8	2	J93401
A 351	HK40	25Cr-20Ni-5Mo	Castings	62	---	---	8	2	J94204
A 351	HT30	15Cr-35Ni-5Mo	Castings	65	---	---	45	---	J94603
SA-352	LCA	C-Si	Castings	60	1	1	---	---	J02504
SA-352	LCC	C-Mn-Si	Castings	70	1	2	---	---	J02505
SA-352	LCB	C-Si	Castings	65	1	1	---	---	J03003
SA-352	LC1	C-0.5Mo	Castings	65	3	1	---	---	J12522

ASME FERROUS METALS - P NUMBERS & S NUMBERS (Continued)

ASME Spec No.	Type or Grade	Nominal Composition	Product Form	Tensile Strength[a] ksi	P No	Welding Group No.	S No.	Group No.	UNS No.
SA-352	LC2	2.5Ni	Castings	70	9A	1	---	---	J22500
SA-352	LC3	3.5Ni	Castings	70	9B	1	---	---	J31550
SA-352	LC4	4.5Ni	Castings	70	9C	1	---	---	J41500
SA-352	LC2-1	3Ni-1.5Cr-0.5Mo	Castings	105	11A	5	---	---	J42215
SA-352	CA6NM	13Cr-4Ni	Castings	110	6	4	---	---	J91540
SA-353	---	9Ni	Plate	100	11A	1	---	---	K81340
SA-358	XM-19	22Cr-13Ni-5Mn	Fusion welded pipe	100	8	3	---	---	S20910
SA-358	XM-29	18Cr-3Ni-12Mn	Fusion welded pipe	100	8	3	---	---	S24000
SA-358	304	18Cr-8Ni	Fusion welded pipe	75	8	1	---	---	S30400
SA-358	304L	18Cr-8Ni	Fusion welded pipe	70	8	1	---	---	S30403
SA-358	304H	18Cr-8Ni	Fusion welded pipe	75	8	1	---	---	S30409
SA-358	304N	18Cr-8Ni-N	Fusion welded pipe	80	8	1	---	---	S30451
SA-358	304LN	18Cr-8Ni-N	Fusion welded pipe	75	8	1	---	---	S30453
SA-358	S30815	21Cr-11Ni-N	Fusion welded pipe	87	8	2	---	---	S30815
SA-358	309S	23Cr-12Ni	Fusion welded pipe	75	8	2	---	---	S30908
SA-358	309Cb	23Cr-12Ni-Cb	Fusion welded pipe	75	8	2	---	---	S30940
SA-358	310S	25Cr-20Ni	Fusion welded pipe	75	8	2	---	---	S31008
SA-358	310Cb	25Cr-20Ni-Cb	Fusion welded pipe	75	8	2	---	---	S31040
SA-358	S31254	20Cr-18Ni-6Mo	Fusion welded pipe	94	8	4	---	---	S31254
SA-358	316	16Cr-12Ni-2Mo	Fusion welded pipe	75	8	1	---	---	S31600
SA-358	316L	16Cr-12Ni-2Mo	Fusion welded pipe	70	8	1	---	---	S31603
SA-358	316H	16Cr-12Ni-2Mo	Fusion welded pipe	75	8	1	---	---	S31609
SA-358	316N	16Cr-12Ni-2Mo-N	Fusion welded pipe	80	8	1	---	---	S31651
SA-358	316LN	16Cr-12Ni-2Mo-N	Fusion welded pipe	75	8	1	---	---	S31653

ASME FERROUS METALS - P NUMBERS & S NUMBERS (Continued)

ASME Spec No.	Type or Grade	Nominal Composition	Product Form	Tensile Strength[a] ksi	P No	Welding Group No.	Welding S No.	Welding Group No.	UNS No.
SA-358	S31725	19Cr-15Ni-4Mo	Fusion welded pipe	75	8	4	---	---	S31725
SA-358	S31726	19Cr-15.5Ni-4Mo	Fusion welded pipe	80	8	4	---	---	S31726
SA-358	321	18Cr-10Ni-Ti	Fusion welded pipe	75	8	1	---	---	S32100
SA-358	347	18Cr-10Ni-Cb	Fusion welded pipe	75	8	1	---	---	S34700
SA-358	348	18Cr-10Ni-Cb	Fusion welded pipe	75	8	1	---	---	S34800
SA-369	FPA	C-Si	Forged pipe	48	1	1	---	---	K02501
SA-369	FPB	C-Mn-Si	Forged pipe	60	1	1	---	---	K03006
SA-369	FP1	C-0.5Mo	Forged pipe	55	3	1	---	---	K11522
SA-369	FP2	0.5Cr-0.5Mo	Forged pipe	55	3	1	---	---	K11547
SA-369	FP12	1Cr-0.5Mo	Forged pipe	60	4	1	---	---	K11562
SA-369	FP11	1.25Cr-0.5Mo-Si	Forged pipe	60	4	1	---	---	K11597
SA-369	FP22	2.25Cr-1Mo	Forged pipe	60	5A	1	---	---	K21590
SA-369	FP21	3Cr-1Mo	Forged pipe	60	5A	1	---	---	K31545
SA-369	FP5	5Cr-0.5Mo	Forged pipe	60	5B	1	---	---	S41545
SA-369	FP9	9Cr-1Mo	Forged pipe	60	5B	1	---	---	K90941
SA-369	FP91	9Cr-1Mo-V	Forged pipe	85	5B	2	---	---	---
SA-372	A	C-Si	Forgings	60	1	1	---	---	K03002
SA-372	B	C-Mn-Si	Forgings	75	1	2	---	---	K04001
SA-376	16-8-2H	16Cr-8Ni-2Mo	Seamless pipe	75	8	1	---	---	---
SA-376	TP304	18Cr-8Ni	Seamless pipe ≥ 0.812"	70	8	1	---	---	S30400
SA-376	TP304	18Cr-8Ni	Seamless pipe < 0.812"	75	8	1	---	---	S30400
SA-376	TP304H	18Cr-8Ni	Seamless pipe	75	8	1	---	---	S30409
SA-376	TP304N	18Cr-8Ni-N	Seamless pipe	80	8	1	---	---	S30451
SA-376	TP304LN	18Cr-8Ni-N	Seamless pipe	75	8	1	---	---	S30453

ASME FERROUS METALS - P NUMBERS & S NUMBERS (Continued)

ASME Spec No.	Type or Grade	Nominal Composition	Product Form	Tensile Strength[a] ksi	P No	Welding Group No.	S No.	Group No.	UNS No.
SA-376	TP316	16Cr-12Ni-2Mo	Seamless pipe	75	8	1	---	---	S31600
SA-376	TP316H	16Cr-12Ni-2Mo	Seamless pipe	75	8	1	---	---	S31609
SA-376	TP316N	18Cr-12Ni-2Mo-N	Seamless pipe	80	8	1	---	---	S31651
SA-376	TP316LN	18Cr-12Ni-2Mo-N	Seamless pipe	75	8	1	---	---	S31653
SA-376	S31725	19Cr-15Ni-4Mo	Seamless pipe	75	8	4	---	---	S31725
SA-376	S31726	19Cr-15.5Ni-4Mo	Seamless pipe	80	8	4	---	---	S31726
SA-376	TP321	18Cr-10Ni-Ti	Seamless pipe > ⅜ in.	70	8	1	---	---	S32100
SA-376	TP321	18Cr-10Ni-Ti	Seamless pipe ≤ ⅜ in.	75	8	1	---	---	S32100
SA-376	TP321H	18Cr-10Ni-Ti	Seamless pipe > ⅜ in	70	8	1	---	---	S32109
SA-376	TP321H	18Cr-10Ni-Ti	Seamless pipe ≤ ⅜ in.	75	8	1	---	---	S32109
SA-376	TP347	18Cr-10Ni-Cb	Seamless pipe	75	8	1	---	---	S34700
SA-376	TP347H	18Cr-10Ni-Cb	Seamless pipe	75	8	1	---	---	S34709
SA-376	TP348	18Cr-10Ni-Cb	Seamless pipe	75	8	1	---	---	S34800
A381	Y35	C	Pipe	60	---	---	1	1	K03013
A381	Y42	C	Pipe	60	---	---	1	1	---
A381	Y48	C	Pipe > ⅜ in	62	---	---	1	1	---
A381	Y46	C	Pipe	63	---	---	1	1	---
A381	Y50	C	Pipe > ⅜ in	64	---	---	1	1	---
A381	Y52b	C	Pipe > ⅜ in	66	---	---	1	2	---
A381	Y56b	C	Pipe > ⅜ in	71	---	---	1	2	---
A381	Y52a	C	Pipe to ⅜ in	72	---	---	1	2	---
A381	Y56a	C	Pipe to ⅜ in	75	---	---	1	2	---
A381	Y60b	C	Pipe > ⅜ in	75	---	---	1	2	---

ASME FERROUS METALS - P NUMBERS & S NUMBERS (Continued)									
ASME Spec No.	Type or Grade	Nominal Composition	Product Form	Tensile Strength[a] ksi	P No	Welding			UNS No.
						Group No.	S No.	Group No.	
A381	Y60a	C	Pipe > ⅜ in	78	---	---	1	2	---
SA-387	12, Class 1	1Cr-0.5Mo	Plate	55	4	1	---	---	K11757
SA-387	12, Class 2	1Cr-0.5Mo	Plate	65	4	1	---	---	K11757
SA-387	11, Class 1	1.25Cr-0.5Mo-Si	Plate	60	4	1	---	---	K11789
SA-387	11, Class 2	1.25Cr-0.5Mo-Si	Plate	75	4	1	---	---	K11789
SA-387	Grade 2, Class 1	0.5Cr-0.5Mo	Plate	55	3	1	---	---	K12143
SA-387	Grade 2, Class 2	0.5Cr-0.5Mo	Plate	70	3	2	---	---	K12143
SA-387	22, Class 1	2.25Cr-1Mo	Plate	60	5A	1	---	---	K21590
SA-387	22 Class 2	2.25Cr-1Mo	Plate	75	5A	1	---	---	K21590
SA-387	21, Class1	3Cr-1Mo	Plate	60	5A	1	---	---	K31545
SA-387	21, Class 2	3Cr-1Mo	Plate	75	5A	1	---	---	K31545
SA-387	5, Class 1	5Cr-0.5Mo	Plate	60	5B	1	---	---	K41545
SA-387	5, Class 2	5Cr-0.5Mo	Plate	75	5B	1	---	---	K41545
SA-387	Grade 91, Class 2	9Cr-1Mo-V	Plate	85	5B	2	---	---	---
SA-403	WPXM-19	22Cr-13Ni-5Mn	Wrought pipe fitting	100	8	3	---	---	S20910
SA-403	WP304	18Cr-8Ni	Wrought pipe fitting	75	8	1	---	---	S30400
SA-403	WP304L	18Cr-8Ni	Wrought pipe fitting	70	8	1	---	---	S30403
SA-403	WP304H	18Cr-8Ni	Wrought pipe fitting	75	8	1	---	---	S30409
SA-403	WP304N	18Cr-8Ni-N	Wrought pipe fitting	80	8	1	---	---	S30451
SA-403	WP304LN	18Cr-8Ni-N	Wrought pipe fitting	75	8	1	---	---	S30453
SA-403	WP309	23Cr-12Ni	Wrought pipe fitting	75	8	2	---	---	S30900
SA-403	WP310	25Cr-20Ni	Wrought pipe fitting	75	8	2	---	---	S31000
SA-403	WP316	16Cr-12Ni-2Mo	Wrought pipe fitting	75	8	1	---	---	S31600

ASME FERROUS METALS - P NUMBERS & S NUMBERS (Continued)

ASME Spec No.	Type or Grade	Nominal Composition	Product Form	Tensile Strength[a] ksi	Welding P No	Welding Group No.	Welding S No.	Welding Group No.	UNS No.
SA-403	WP316L	16Cr-12Ni-2Mo	Wrought pipe fitting	70	8	1	---	---	S31603
SA-403	WP316H	16Cr-12Ni-2Mo	Wrought pipe fitting	75	8	1	---	---	S31609
SA-403	WP316N	16Cr-12Ni-2Mo-N	Wrought pipe fitting	80	8	1	---	---	S31651
SA-403	WP316LN	16Cr-12Ni-2Mo-N	Wrought pipe fitting	75	8	1	---	---	S31653
SA-403	WP317	18Cr-13Ni-3Mo	Wrought pipe fitting	75	8	1	---	---	S31700
SA-403	WP317L	18Cr-13Ni-3Mo	Wrought pipe fitting	75	8	1	---	---	S31703
SA-403	WP321	18Cr-10Ni-Ti	Wrought pipe fitting	75	8	1	---	---	S32100
SA-403	WP321H	18Cr-10Ni-Ti	Wrought pipe fitting	75	8	1	---	---	S32109
SA-403	WP347	18Cr-10Ni-Cb	Wrought pipe fitting	75	8	1	---	---	S34700
SA-403	WP347H	18Cr-10Ni-Cb	Wrought pipe fitting	75	8	1	---	---	S34709
SA-403	WP348	18Cr-10Ni-Cb	Wrought pipe fitting	75	8	1	---	---	S34800
SA-403	WP348H	18Cr-10Ni-Cb	Wrought pipe fitting	75	8	1	---	---	S34809
SA-409	TP304	18Cr-8Ni	Welded pipe	75	8	1	---	---	S30400
SA-409	TP304L	18Cr-8Ni	Welded pipe	70	8	1	---	---	S30403
SA-409	S30815	21Cr-11Ni-N	Welded pipe	87	8	2	---	---	S30815
SA-409	TP309S	23Cr-11Ni	Welded pipe	75	8	2	---	---	S30908
SA-409	TP309Cb	23Cr-12Ni-Cb	Welded pipe	75	8	2	---	---	S30940
SA-409	TP310S	25Cr-20Ni	Welded pipe	75	8	2	---	---	S31008
SA-409	TP310Cb	25Cr-20Ni-Cb	Welded pipe	75	8	2	---	---	S31040
SA-409	S31254	20Cr-18Ni-6Mo	Welded pipe	94	8	4	---	---	S31254
SA-409	TP316	16Cr-12Ni-2Mo	Welded pipe	75	8	1	---	---	S31600
SA-409	TP316L	16Cr-12Ni-2Mo	Welded pipe	70	8	1	---	---	S31603
SA-409	TP317	18Cr-13Ni-3Mo	Welded pipe	75	8	1	---	---	S31700
SA-409	S31725	19Cr-15Ni-4Mo	Welded pipe	75	8	4	---	---	S31725

ASME FERROUS METALS - P NUMBERS & S NUMBERS (Continued)

ASME Spec No.	Type or Grade	Nominal Composition	Product Form	Tensile Strength[a] ksi	P No	Welding Group No.	Welding S No.	Welding Group No.	UNS No.
SA-409	S31726	19Cr-15.5Ni-4Mo	Welded pipe	80	8	4	---	---	S31726
SA-409	TP321	18Cr-10Ni-Ti	Welded pipe	75	8	1	---	---	S32100
SA-409	TP347	18Cr-10Ni-Cb	Welded pipe	75	8	1	---	---	S34700
SA-409	TP348	18Cr-10Ni-Cb	Welded pipe	75	8	1	---	---	S34800
SA-414	A	C	Sheet	45	1	1	---	---	K01501
SA-414	B	C	Sheet	50	1	1	---	---	K02201
SA-414	C	C	Sheet	55	1	1	---	---	K02503
SA-414	D	C-Mn	Sheet	60	1	1	---	---	K02505
SA-414	E	C-Mn	Sheet	65	1	1	---	---	K02704
SA-414	F	C-Mn	Sheet	70	1	2	---	---	K03102
SA-414	G	C-Mn	Sheet	75	1	2	---	---	K03103
SA-420	WPL6	C-Mn-Si	Piping fitting	60	1	1	---	---	---
SA-420	WPL9	2Ni-1Cu	Piping fitting	63	9A	1	---	---	K22035
SA-420	WPL3	3.5Ni	Piping fitting	65	9B	1	---	---	---
SA-420	WPL8	9Ni	Piping fitting	100	11A	1	---	---	K81340
SA-423	1	0.75Cr-0.5Ni-Cu	Smls & welded tube	60	4	2	---	---	K11535
SA-423	2	0.75Ni-0.5Cu-Mo	Smls & welded tube	60	4	2	---	---	K11540
SA-426	CP15	C-0.5Mo-Si	Centrifugal Cast pipe	60	3	1	---	---	J11522
SA-426	CP2	0.5Cr-0.5Mo	Centrifugal Cast pipe	60	3	1	---	---	J11547
SA-426	CP12	1Cr-0.5Mo	Centrifugal Cast pipe	60	4	1	---	---	J11562
SA-426	CP11	1.25Cr-0.5Mo	Centrifugal Cast pipe	70	4	1	---	---	J12072
SA-426	CP1	C-0.5Mo	Centrifugal Cast pipe	65	3	1	---	---	J12521
SA-426	CP22	2.25Cr-1Mo	Centrifugal Cast pipe	70	5A	1	---	---	J21890
SA-426	CP21	3Cr-1Mo	Centrifugal Cast pipe	60	5A	1	---	---	J31545

ASME FERROUS METALS - P NUMBERS & S NUMBERS (Continued)

ASME Spec No.	Type or Grade	Nominal Composition	Product Form	Tensile Strength[a] ksi	Welding P No	Welding Group No.	Welding S No.	Welding Group No.	UNS No.
SA-426	CP5	5Cr-0.5Mo	Centrifugal Cast pipe	90	5B	1	---	---	J42045
SA-426	CP5b	5Cr-1.5Si-0.5Mo	Centrifugal Cast pipe	60	5B	1	---	---	J51545
SA-426	CP9	9Cr-1Mo	Centrifugal Cast pipe	90	5B	1	---	---	J82090
SA-426	CPCA15	13Cr	Centrifugal Cast pipe	90	6	3	---	---	J91150
SA-430	FP16-8-2H	16Cr-8Ni-2Mo	Forged pipe	70	8	1	---	---	S16800
SA-430	FP304	18Cr-8Ni	Forged pipe	70	8	1	---	---	S30400
SA-430	FP304H	18Cr-8Ni	Forged pipe	70	8	1	---	---	S30409
SA-430	FP304N	18Cr-8Ni-N	Forged pipe	75	8	1	---	---	S30451
SA-430	FP316	16Cr-12Ni-2Mo	Forged pipe	70	8	1	---	---	S31600
SA-430	FP316H	16Cr-12Ni-2Mo	Forged pipe	70	8	1	---	---	S31609
SA-430	FP316N	16Cr-12Ni-2Mo-N	Forged pipe	75	8	1	---	---	S31651
SA-430	FP321	18Cr-10Ni-Ti	Forged pipe	70	8	1	---	---	S32100
SA-430	FP321H	18Cr-10Ni-Ti	Forged pipe	70	8	1	---	---	S32109
SA-430	FP347	18Cr-10Ni-Cb	Forged pipe	70	8	1	---	---	S34700
SA-430	FP347H	18Cr-10Ni-Cb	Forged pipe	70	8	1	---	---	S34709
A 441	1	Mn-Cu-V	Shapes	70	---	---	---	2	K12211
A 441	2	Mn-Cu-V	Shapes	70	---	---	---	2	K12211
A446	A	C	Sheet	45	---	---	---	1	---
SA-451	CPF8	18Cr-8Ni	Cetrifugal Cast pipe	70	8	1	---	---	J92600
SA-451	CPF8A	18Cr-8Ni	Centrifugal Cast pipe	77	8	1	---	---	J92600
SA-451	CPF8C	18Cr-10Ni-Cb	Centrifugal Cast pipe	70	8	1	---	---	J92710
SA-451	CPF8M	18Cr-12Ni-2Mo	Centrifugal Cast pipe	70	8	1	---	---	J92900
SA-451	CPF3	18Cr-8Ni	Centrifugal Cast pipe	70	8	1	---	---	---

ASME FERROUS METALS - P NUMBERS & S NUMBERS (Continued)

ASME Spec No.	Type or Grade	Nominal Composition	Product Form	Tensile Strength[a] ksi	P No	Welding Group No.	Welding S No.	Welding Group No.	UNS No.
SA-451	CPF3M	16Cr-12Ni-2Mo	Centrifugal Cast pipe	70	8	1	---	---	---
SA-451	CPF3A	18Cr-8Ni	Centrifugal Cast pipe	77	8	1	---	---	---
SA-451	CPH8	25Cr-12Ni	Centrifugal Cast pipe	65	8	2	---	---	J93400
SA-451	CPH20	25Cr-12Ni	Centrifugal Cast pipe	70	8	2	---	---	J93402
SA-451	CPK20	25Cr-20Ni	Centrifugal Cast pipe	65	8	2	---	---	J93402
A-451	CPF10MC	16Cr-14Ni-2Mo	Centrifugal Cast pipe	70	---	---	8	1	J92971
A-451	CPE20N	25Cr-8Ni-N	Centrifugal Cast pipe	80	---	---	8	2	---
SA-452	TP304H	18Cr-8Ni	Centrifugal Cast pipe	75	8	1	---	---	S30409
SA-452	TP316H	16Cr-12Ni-2Mo	Centrifugal Cast pipe	75	8	1	---	---	S31609
SA-452	TP347H	18Cr-10Ni-Cb	Centrifugal Cast pipe	75	8	1	---	---	S34709
SA-455	---	C-Mn-Si	Plate, > 0.580-0.750 in.	70	1	2	---	---	K03300
SA-455	---	C-Mn-Si	Plate, > 0.375-0.580 in.	73	1	2	---	---	K03300
SA-455	---	C-Mn-Si	Plate, up to 0.375 in.	75	1	2	---	---	K03300
SA-479	XM-19	22Cr-13Ni-5Mn	Bar & Shape	100	8	3	---	---	S20910
SA-479	XM-17	19Cr-8Mn-6Ni-Mo-N	Bar & Shape	90	8	3	---	---	S21600
SA-479	XM-18	19Cr-8Mn-6Ni-Mo-N	Bar & Shape	90	8	3	---	---	S21603
SA-479	S21800	18Cr-8Ni-4Si-N	Bar & Shape	95	8	3	---	---	S21800
SA-479	XM-11	21Cr-6Ni-9Mn	Bar & Shape	90	8	3	---	---	S21904
SA-479	XM-29	18Cr-3Ni-12Mn	Bar & Shape	100	8	3	---	---	S24000
SA-479	302	18Cr-8Ni	Bar & Shape	75	8	1	---	---	S30200
SA-479	304	18Cr-8Ni	Bar & Shape	75	8	1	---	---	S30400
SA-479	304L	18Cr-8Ni	Bar & Shape	70	8	1	---	---	S30403
SA-479	304H	18Cr-8Ni	Bar & Shape	75	8	1	---	---	S30409
SA-479	304N	18Cr-8Ni-N	Bar & Shape	80	8	1	---	---	S30451

ASME FERROUS METALS - P NUMBERS & S NUMBERS (Continued)

ASME Spec No.	Type or Grade	Nominal Composition	Product Form	Tensile Strength[a] ksi	Welding P No	Welding Group No.	Welding S No.	Welding Group No.	UNS No.
SA-479	304LN	18Cr-8Ni-N	Bar & Shape	75	8	1	---	---	S30453
SA-479	S30600	18Cr-15Ni-4Si	Bar & Shape	75	8	1	---	---	S30600
SA-479	S30815	21Cr-11Ni-N	Bar & Shape	87	8	2	---	---	S30815
SA-479	309S	23Cr-12Ni	Bar & Shape	75	8	2	---	---	S30908
SA-479	309Cb	23Cr-12Ni-Cb	Bar & Shape	75	8	2	---	---	S30940
SA-479	310S	25Cr-20Ni	Bar & Shape	75	8	2	---	---	S31008
SA-479	310Cb	25Cr-20Ni-Cb	Bar & Shape	75	8	2	---	---	S31040
SA-479	S31254	20Cr-18Ni-6Mo	Bar & Shape	94	8	4	---	---	S31254
SA-479	316	16Cr-12Ni-2Mo	Bar & Shape	75	8	1	---	---	S31600
SA-479	316L	16Cr-12Ni-2Mo	Bar & Shape	70	8	1	---	---	S31603
SA-479	316H	16Cr-12Ni-2Mo	Bar & Shape	75	8	1	---	---	S31609
SA-479	316Ti	16Cr-12Ni-2Mo-Ti	Bar & Shape	75	8	1	---	---	S31635
SA-479	316Cb	16Cr-12Ni-2Mo-Cb	Bar & Shape	75	8	1	---	---	S31640
SA-479	316N	16Cr-12Ni-2Mo-N	Bar & Shape	80	8	1	---	---	S31651
SA-479	316LN	16Cr-12Ni-2Mo-N	Bar & Shape	75	8	1	---	---	S31653
SA-479	S31725	19Cr-15Ni-4Mo	Bar & Shape	75	8	4	---	---	S31725
SA-479	S31726	19Cr-15.5Ni-4Mo	Bar & Shape	80	8	4	---	---	S31726
SA-479	321	18Cr-10Ni-Ti	Bar & Shape	75	8	1	---	---	S32100
SA-479	321H	18Cr-10Ni-Ti	Bar & Shape	75	8	1	---	---	S32109
SA-479	S32550	25Cr-5Ni-3Mo-2Cu	Bar & Shape	110	10H	1	---	---	S32550
SA-479	---	25Cr-7.5Ni-3.5Mo-N-Cu-W	Bar & Shape	109	---	---	10H	1	S32760
SA-479	347	18Cr-10Ni-Cb	Bar & Shape	75	8	1	---	---	S34700
SA-479	347H	18Cr-10Ni-Cb	Bar & Shape	75	8	1	---	---	S34709
SA-479	348	18Cr-10Ni-Cb	Bar & Shape	75	8	1	---	---	S34800

ASME FERROUS METALS - P NUMBERS & S NUMBERS (Continued)

ASME Spec No.	Type or Grade	Nominal Composition	Product Form	Tensile Strength[a] ksi	P No	Welding Group No.	Welding S No.	Welding Group No.	UNS No.
SA-479	348H	18Cr-10Ni-Cb	Bar & Shape	75	8	1	---	---	S34809
SA-479	403	12Cr	Bar & Shape	70	6	1	---	---	S40300
SA-479	405	12Cr-1Al	Bar & Shape	60	7	1	---	---	S40500
SA-479	410	13Cr	Bar & Shape	70	6	1	---	---	S41000
SA-479	414	12.5Cr-2Ni-Si	Bar & Shape	115	6	4	---	---	S41400
SA-479	S41500	13Cr-4.5Ni-Mo	Bar & Shape	115	6	4	---	---	S41500
SA-479	430	17Cr	Bar & Shape	70	7	2	---	---	S43000
SA-479	439	17Cr-Ti	Bar & Shape	70	7	2	---	---	S43035
SA-479	S44400	18Cr-2Mo	Bar & Shape	60	7	2	---	---	S44400
SA-479	XM-27	27Cr-1Mo	Bar & Shape	65	10I	1	---	---	S44627
SA-479	S44700	29Cr-4Mo	Bar & Shape	70	10J	1	---	---	S44700
SA-479	S44800	29Cr-4Mo-2Ni	Bar & Shape	70	10K	1	---	---	S44800
SA-487	Grade 16, Class A	Low C-Mn-Ni	Castings	70	1	2	---	---	---
SA-487	Grade 1, Class A	Mn-V	Castings	85	10A	1	---	---	J03004
SA-487	Grade1, Class B	Mn-V	Castings	90	10A	1	---	---	J03004
SA-487	Grade 2, Class A	Mn-0.25Mo-V	Castings	85	10F	1	---	---	J13005
SA-487	Grade 2, Class B	Mn-0.25Mo-V	Castings	90	10F	1	---	---	J13005
SA-487	Grade 4, Class A	0.5Ni-0.5Cr-0.25Mo-V	Castings	90	10F	1	---	---	J13047
SA-487	Grade 4, Class B	0.5Ni-0.5Cr-0.25Mo-V	Castings	105	11A	3	---	---	J13047
SA-487	Grade 4, Class E	0.5Ni-0.5Cr-0.25Mo-V	Castings	115	11A	3	---	---	J13047
SA-487	Grade 8, Class A	2.25Cr-1Mo	Castings	85	5C	1	---	---	J22091
SA-487	Grade 8, Class C	2.25Cr-1Mo	Castings	100	5C	4	---	---	J22091
SA-487	Grade 8, Class B	2.25Cr-1Mo	Castings	105	5C	4	---	---	J22091
SA-487	CA15M Class A	13Cr	Castings	90	6	3	---	---	J91151

ASME FERROUS METALS - P NUMBERS & S NUMBERS (Continued)

ASME Spec No.	Type or Grade	Nominal Composition	Product Form	Tensile Strength[a] ksi	P No	Welding Group No.	S No.	Group No.	UNS No.
SA-487	CA15 Class C	13Cr	Castings	90	6	3	---	---	---
SA-487	CA15 Class B	13Cr	Castings	90	6	3	---	---	---
SA-487	CA15 Class D	13Cr	Castings	100	6	3	---	---	---
SA-487	CA6NM Class B	13Cr-4Ni	Castings	100	6	4	---	---	J91540
SA-487	CA6NM Class A	13Cr-4Ni	Castings	110	6	4	---	---	J91540
SA-494	CX2MW	59Ni-22Cr-14Mo-4Fe-3W	Castings	80	44	---	---	---	N26022
A 494	CW-6M	59Ni-22Cr-14Mo-4Fe-3W	Castings	72	---	---	44	112	N26022
A 494	---	56Ni-21Mo-13Cr	Coatings	80	---	---	44	112	CX2MW
A 500	C	C	Tube	62	---	---	1	1	K02705
A 500	B	C	Tube	58	---	---	1	1	K03000
A 501	---	C	Tube	58	---	---	1	1	K03000
SA-508	3, Class 1	0.75Ni-0.5Mo-Cr-V	Forgings	80	3	3	---	---	K12042
SA-508	3, Class 2	0.75Ni-0.5Mo-Cr-V	Forgings	90	3	3	---	---	K12042
SA-508	2, Class 1	0.75Ni-0.5Mo-0.3Cr-V	Forgings	80	3	3	---	---	K12766
SA-508	2, Class 2	0.75Ni-0.5Mo-0.3Cr-V	Forgings	90	3	3	---	---	K12766
SA-508	1	C-Si	Forgings	70	1	2	---	---	K13502
SA-508	1a	C-Mn-Si	Forgings	70	1	2	---	---	---
SA-508	4N, Class 3	3.5Ni-1.75Cr-0.5Mo-V	Forgings	90	3	3	---	---	K22375
SA-508	4N, Class 1	3.5Ni-1.75Cr-0.5Mo-V	Forgings	105	11A	5	---	---	K22375
SA-508	4N, Class 2	3.5Ni-1.75Cr-0.5Mo-V	Forgings	115	11A	5	---	---	K22375
SA-508	3V	3Cr-1Mo-V-Ti-B	Forgings	85	5C	1	---	---	K31830
SA-508	5, Class 1	3.5Ni-1.75Cr-0.5Mo-V	Forgings	105	11A	5	---	---	K42365
SA-508	5, Class 2	3.5Ni-1.75Cr-0.5Mo-V	Forgings	115	11A	5	---	---	K42365
A 513	1015 CW	C	Tube	65	---	---	---	1	G10150
A 513	1020 CW	C	Tube	70	---	---	---	2	G10200

ASME FERROUS METALS - P NUMBERS & S NUMBERS (Continued)									
ASME Spec No.	Type or Grade	Nominal Composition	Product Form	Tensile Strength[a] ksi	P No	Welding			UNS No.
						Group No.	S No.	Group No.	
A 513	1025 CW	C	Tube	75	---	---	1	2	G10250
A 513	1026 CW	C	Tube	80	---	---	1	3	G10260
A 514	F	0.75Ni-0.5Cr-0.5Mo-V	Plate, ≤ 2½ in.	110	---	---	11B	3	K11576
A 514	J	C-0.5Mo	Plate, ≤ 1¼ in.	110	---	---	11B	6	K11625
A 514	B	0.5Cr-0.2Mo-V	Plate, ≤ 1¼in.	110	---	---	11B	4	K11630
A 514	D	1Cr-0.25Mo-Si	Plate, ≤1¼in.	110	---	---	11B	5	K11662
A 514	A	0.5Cr-0.25Mo-Si	Plate, ≤ 1¼in.	110	---	---	11B	1	K11856
A 514	E	1.75Cr-0.5Mo-Cu	Plate, >2½in.-6 in. incl.	100	---	---	11B	2	K21604
A 514	E	1.75Cr-0.5Mo-Cu	Plate, ≤ 2½ in	110	---	---	11B	2	K21604
A 514	P	1.25Ni-1Cr-0.5Mo	Plate, >2½in.-6 in. incl.	100	---	---	11B	8	K21650
A 514	P	1.25Ni-1Cr-0.5Mo	Plate, ≤ 2½ in	110	---	---	11B	8	K21650
A 514	Q	1.3Ni-1.3Cr-0.5Mo-V	Plate, >2½in.-6 in. incl.	100	---	---	11B	9	---
A 514	Q	1.3Ni-1.3Cr-0.5Mo-V	Plate, ≤ 2½ in	110	---	---	11B	9	---
SA-515	60	C-Si	Plate	60	1	1	---	---	K02401
SA-515	65	C-Si	Plate	65	1	1	---	---	K02800
SA-515	70	C-Si	Plate	70	1	2	---	---	K03101
SA-516	55	C-Si	Plate	55	1	1	---	---	K01800
SA-516	60	C-Mn-Si	Plate	60	1	1	---	---	K02100
SA-516	65	C-Mn-Si	Plate	65	1	1	---	---	K02403
SA-516	70	C-Mn-Si	Plate	70	1	2	---	---	K02700
SA-517	F	0.75Ni-0.5Cr-0.5Mo-V	Plate > 2½in.-6 in. incl.	105	11B	3	---	---	K11576
SA-517	F	0.75Ni-0.5Cr-0.5Mo-V	Plate ≤ 2½ in	115	11B	3	---	---	K11576
SA-517	J	C-0.5Mo	Plate > 2½in.-6 in. incl.	105	11B	6	---	---	K11625
SA-517	J	C-0.5Mo	Plate ≤ 2½ in	115	11B	6	---	---	K11625

ASME FERROUS METALS - P NUMBERS & S NUMBERS (Continued)

ASME Spec No.	Type or Grade	Nominal Composition	Product Form	Tensile Strength[a] ksi	P No	Welding Group No.	Welding S No.	Welding Group No.	UNS No.
SA-517	B	0.5Cr-0.2Mo-V	Plate > 2½in.-6 in. incl.	105	11B	4	---	---	K11630
SA-517	B	0.5Cr-0.2Mo-V	Plate ≤ 2½ in	115	11B	4	---	---	K11630
SA-517	A	0.5Cr-0.25Mo-Si	Plate > 2½in.-6 in. incl.	105	11B	1	---	---	K11856
SA-517	A	0.5Cr-0.25Mo-Si	Plate ≤ 2½ in	115	11B	1	---	---	K11856
SA-517	E	1.75Cr-0.5 Mo-Cu	Plate > 2½in.-6 in. incl.	105	11B	2	---	---	K21604
SA-517	E	1.75Cr-0.5Mo-Cu	Plate, ≤ 2½ in	115	11B	2	---	---	K21604
SA-517	P	1.25Ni-1Cr-0.5Mo	Plate > 2½in.-6 in. incl.	105	11B	8	---	---	K21650
SA-517	P	1.25Ni-1Cr-0.5Mo	Plate ≤ 2½ in	115	11B	8	---	---	K21650
A 519	1018 HR	C	Tube	50	---	---	1	1	G10180
A 519	1018 CW	C	Tube	70	---	---	1	2	G10180
A 519	1020 HR	C	Tube	50	---	---	1	1	G10200
A 519	1020 CW	C	Tube	70	---	---	1	2	G10200
A 519	1022 HR	C	Tube	50	---	---	1	1	G10220
A 519	1022 CW	C	Tube	70	---	---	1	2	G10220
A 519	1025 HR	C	Tube	55	---	---	1	1	G10250
A 519	1025 CW	C	Tube	75	---	---	1	2	G10250
A 519	1026 HR	C	Tube	55	---	---	1	1	G10260
A 519	1026 CW	C	Tube	75	---	---	1	2	G10260
A 521	Class CC	C	Forgings	60	11A	1	---	---	---
A 521	Class CE	C	Forgings	75	11A	1	---	---	---
SA-522	Type II	8Ni	Forgings	100	1	---	---	---	K71340
SA-522	Type I	9Ni	Forgings	100	1	---	---	---	K81340
SA-524	II	C-Mn-Si	Seamless pipe	55	1	---	---	---	K02104
SA-524	I	C-Mn-Si	Seamless pipe	60	1	---	---	---	K02104

ASME FERROUS METALS - P NUMBERS & S NUMBERS (Continued)									
ASME Spec No.	Type or Grade	Nominal Composition	Product Form	Tensile Strength[a] ksi	P No	Welding			UNS No.
						Group No.	S No.	Group No.	
SA-533	Type A, Class 1	Mn-0.5Mo	Plate	80	3	3	---	---	K12521
SA-533	Type A, Class 2	Mn-0.5Mo	Plate	90	3	3	---	---	K12521
SA-533	Type A, Class 3	Mn-0.5Mo	Plate	100	11A	4	---	---	K12521
SA-533	Type D, Class 1	Mn-0.5Mo-0.25Ni	Plate	80	3	3	---	---	K12529
SA-533	Type D, Class 2	Mn-0.5Mo-0.25Ni	Plate	90	3	3	---	---	K12529
SA-533	Type D, Class 3	Mn-0.5Mo-0.25Ni	Plate	100	11A	4	---	---	K12529
SA-533	Type B, Class 1	Mo-0.5Mo-0.5Ni	Plate	80	3	3	---	---	K12539
SA-533	Type B, Class 2	Mo-0.5Mo-0.5Ni	Plate	90	3	3	---	---	K12539
SA-533	Type B, Class 3	Mo-0.5Mo-0.5Ni	Plate	100	11A	4	---	---	K12539
SA-533	Type C, Class 1	Mn-0.5Mo-0.75Ni	Plate	80	3	3	---	---	K12554
SA-533	Type C, Class 2	Mn-0.5Mo-0.75Ni	Plate	90	3	3	---	---	K12554
SA-533	Type C, Class 3	Mn-0.5Mo-0.75Ni	Plate	100	11A	4	---	---	K12554
SA-537	Class 1	C-Mn-Si	Plate, > 2½-4 in.	65	1	2	---	---	K02400
SA-537	Class 1	C-Mn-Si	Plate, ≤ 2½ in.	70	1	2	---	---	K02400
SA-537	Class 2	C-Mn-Si	Plate, 4-6 in.	70	1	3	---	---	K02400
SA-537	Class 2	C-Mn-Si	Plate, > 2½-4 in.	75	1	3	---	---	K02400
SA-537	Class 2	C-Mn-Si	Plate, ≤ 2½ in.	80	1	3	---	---	K02400
SA-537	Class 3	C-Mn-Si	Plate, > 4 in.	70	1	3	---	---	K12437
SA-537	Class 3	C-Mn-Si	Plate, 2½ in < t < 4 in.	75	1	3	---	---	K12437
SA-537	Class 3	C-Mn-Si	Plate, ≤ 2½ in.	80	1	3	---	---	K12437
SA-541	1	C-Si	Forgings	70	1	2	---	---	K03506
SA-541	1A	C-Mn-Si	Forgings	70	1	2	---	---	---
SA-541	11, Class 4	1.25Cr-0.5Mo-Si	Forgings	80	4	1	---	---	K11572
SA-541	3, Class 1	0.5Ni-0.5Mo-V	Forgings	80	3	3	---	---	K12045

ASME FERROUS METALS - P NUMBERS & S NUMBERS (Continued)

ASME Spec No.	Type or Grade	Nominal Composition	Product Form	Tensile Strength[a] ksi	Welding				UNS No.
					P No	Group No.	S No.	Group No.	
SA-541	3, Class 2	0.5Ni-0.5Mo-V	Forgings	90	3	3	---	---	K12045
SA-541	2, Class 1	0.75Ni-0.5Mo-0.3Cr-V	Forgings	80	3	3	---	---	K12765
SA-541	2, Class 2	0.75Ni-0.5Mo-0.3Cr-V	Forgings	90	3	3	---	---	K12765
SA-541	Class 3V	3Cr-1Mo-V-Ti-B	Forgings	85	5C	1	---	---	K31830
SA-541	22, Class 3	2.25Cr-1Mo	Forgings	85	5C	1	---	---	K21390
SA-541	22, Class 4	2.25Cr-1Mo	Forgings	105	5C	4	---	---	K21390
SA-541	22, Class 5	2.25Cr-1Mo	Forgings	115	5C	5	---	---	K21390
SA-542	Type B, Class 4a	2.25Cr-1Mo	Plate	85	5C	1	---	---	K21590
SA-542	Type B, Class 4	2.25Cr-1Mo	Plate	85	5C	1	---	---	K21590
SA-542	Type A, Class 4	2.25Cr-1Mo	Plate	85	5C	1	---	---	K21590
SA-542	Type A, Class 4a	2.25Cr-1Mo	Plate	85	5C	1	---	---	K21590
SA-542	Type A, Class 3	2.25Cr-1Mo	Plate	95	5C	3	---	---	K21590
SA-542	Type B, Class 3	2.25Cr-1Mo	Plate	95	5C	3	---	---	K21590
SA-542	Type A, Class 1	2.25Cr-1Mo	Plate	105	5C	4	---	---	K21590
SA-542	Type B, Class 1	2.25Cr-1Mo	Plate	105	5C	4	---	---	K21590
SA-542	Type B, Class 2	2.25Cr-1Mo	Plate	115	5C	5	---	---	K21590
SA-542	Type A, Class 2	2.25Cr-1Mo	Plate	115	5C	5	---	---	K21590
SA-542	Type C, Class 4	3Cr-1Mo-V-Ti-B	Plate	85	5C	1	---	---	K31830
SA-542	Type C, Class 4a	3Cr-1Mo-V-Ti-B	Plate	85	5C	1	---	---	K31830
SA-542	Type C, Class 3	3Cr-1Mo-V-Ti-B	Plate	95	5C	3	---	---	K31830
SA-542	Type C, Class 1	3Cr-1Mo-V-Ti-B	Plate	105	5C	4	---	---	K31830
SA-542	Type C, Class 2	3Cr-1Mo-V-Ti-B	Plate	115	5C	5	---	---	K31830
SA-543	B Class 3	3Ni-1.75Cr-0.5Mo	Plate	90	11A	5	---	---	K42339
SA-543	C Class 3	2.75Ni-1.5Cr-0.5Mo	Plate	90	11A	5	---	---	K42338

ASME FERROUS METALS - P NUMBERS & S NUMBERS (Continued)

ASME Spec No.	Type or Grade	Nominal Composition	Product Form	Tensile Strength[a] ksi	Welding P No	Welding Group No.	Welding S No.	Welding Group No.	UNS No.
SA-543	B Class 1	3Ni-1.75Cr-0.5Mo	Plate	105	11A	5	---	---	K42339
SA-543	C Class 1	2.75Ni-1.5Cr-0.5Mo	Plate	105	11A	5	---	---	K42338
SA-543	B Class 2	3Ni-1.75Cr-0.5Mo	Plate	115	11B	10	---	---	K42339
SA-543	C Class 2	2.75Ni-1.5Cr-0.5Mo	Plate	115	11B	10	---	---	K42338
SA-553	Type II	8Ni	Plate	100	11A	1	---	---	K71340
SA-553	Type I	9Ni	Plate	100	11A	1	---	---	K81340
SA-556	A2	C	Seamless tube	47	1	1	---	---	K01807
SA-556	B2	C-Si	Seamless tube	60	1	1	---	---	K02707
SA-556	C2	C-Si	Seamless tube	70	1	2	---	---	K03006
SA-557	A2	C	ERW tube	47	1	1	---	---	K01807
SA-557	B2	C	ERW tube	60	1	1	---	---	K03007
SA-557	C2	C-Mn	ERW tube	70	1	2	---	---	K03505
SA-562	---	C-Mn-Ti	Plate	55	1	1	---	---	K11224
A 570	30	C	Sheet & Strip	49	---	---	1	1	K02502
A 570	33	C	Sheet & Strip	52	---	---	1	1	K02502
A 570	36	C	Sheet & Strip	53	---	---	1	1	K02502
A 570	40	C	Sheet & Strip	55	---	---	1	1	K02502
A 570	45	C	Sheet & Strip	60	---	---	1	1	K02502
A 570	50	C	Sheet & Strip	65	---	---	1	1	K02502
A 572	42	C-Mn-Si	Plate & Shapes	60	---	---	1	1	---
A 572	50	C-Mn-Si	Plate & Shapes	65	---	---	1	1	---
A 572	55	C-Mn-Si	Plate & Shapes	70	---	---	1	2	---
A 573	---	C	Plate	---	---	---	1	1	---
A 575	---	C	Bar	---	---	---	1	1	---

ASME FERROUS METALS - P NUMBERS & S NUMBERS (Continued)

ASME Spec No.	Type or Grade	Nominal Composition	Product Form	Tensile Strength[a] ksi	Welding				UNS No.
					P No	Group No.	S No.	Group No.	
A 576	---	C	Bar	---	---	---	---	---	---
SA-587	---	C	ERW pipe	48	1	1	1	1	K11500
A 588	A, a	Mn-0.5Cr-0.3Cu-Si-V	Plate & Bar	63	---	1	3	1	K11430
A 588	A, b	Mn-0.5Cr-0.3Cu-Si-V	Plate & Bar	67	---	---	3	1	K11430
A 588	A, c	Mn-0.5Cr-0.3Cu-Si-V	Plate & Shapes	70	---	---	3	1	K11430
A 588	B, a	Mn-0.6Cr-0.3Cu-Si-V	Plate & Bar	63	---	---	3	1	K12043
A 588	B, b	Mn-0.6Cr-0.3Cu-Si-V	Plate & Bar	67	---	---	3	1	K12043
A 588	B, c	Mn-0.6Cr-0.3Cu-Si-V	Plate & Shapes	70	---	---	3	1	K12043
SA-592	F	0.75Ni-0.5Cr-0.5Mo-V	Forgings, 2½ - 4 in.	105	11B	3	---	---	K11576
SA-592	F	0.75Ni-0.5Cr-0.5Mo-V	Forgings, ≤ 2½ in.	115	11B	3	---	---	K11576
SA-592	E	1.75Cr-0.5Mo-Cu	Forgings, 2½ - 4 in.	105	11B	2	---	---	K11695
SA-592	E	1.75Cr-0.5Mo-Cu	Forgings, ≤ 2½ in.	115	11B	2	---	---	K11695
SA-592	A	0.5Cr-0.25Mo-Si	Forgings, 2½ - 4 in.	105	11B	1	---	---	K11856
SA-592	A	0.5Cr-0.25Mo-Si	Forgings, ≤ 2½ in.	115	11B	1	---	---	K11856
A 611	A	C	Sheet	42	---	---	1	1	---
A 611	B	C	Sheet	45	---	---	1	1	---
A 611	C	C	Sheet	48	---	---	1	1	---
SA-612	---	C-Mn-Si	Plate, > ½ in. to 1 in.	81	10C	1	---	---	K02900
SA-612	---	C-Mn-Si	Plate, ≤ ½ in.	83	10C	1	---	---	K02900
A 618	II, b	Mn-Cu-V	Tube, > ¾ 1½ in.	67	---	---	1	2	K12609
A 618	II, a	Mn-Cu-V	Tube, ≤ ¾ in.	70	---	---	1	2	K12609
A 618	III	Mn-V	Tube	65	---	---	1	1	K12700
SA-620	---	C	Sheet	40	1	1	1	1	K00040
A 633	A	Mn-Cb	Plate & Shapes	63	---	---	1	1	K01802

ASME FERROUS METALS - P NUMBERS & S NUMBERS (Continued)

ASME Spec No.	Type or Grade	Nominal Composition	Product Form	Tensile Strength[a] ksi	Welding P No	Welding Group No.	Welding S No.	Welding Group No.	UNS No.
A 633	C b	Mn-Cb	Plate, > 2½ - 4 in., Shapes	65	---	---	1	1	K12000
A 633	C a	Mn-Cb	Plate to 2.5 in., Shapes	70	---	---	1	2	K12000
A 633	D b	Mn-Cr-Ni-Cu	Plate, > 2½ - 4 in., Shapes	65	---	---	1	1	K12037
A 633	D a	Mn-Cr-Ni-Cu	Plate to 2.5 in., Shapes	70	---	---	1	2	K12037
A 633	E	C-Mn-Si-V	Plate, Shapes	80	---	---	1	3	K12202
SA-645	---	5Ni-0.25Mo	Plate	95	11A	2	---	---	K41583
SA-660	WCA	C-Si	Centrifugal Cast pipe	60	1	1	---	---	J02504
SA-660	WCC	C-Mn-Si	Centrifugal Cast pipe	70	1	2	---	---	J02505
SA-660	WCB	C-Si	Centrifugal Cast pipe	70	1	2	---	---	J03003
SA-662	A	C-Mn-Si	Plate	58	1	1	---	---	K01701
SA-662	C	C-Mn-Si	Plate	70	1	2	---	---	K02007
SA-662	B	C-Mn-Si	Plate	65	1	1	---	---	K02203
A 663	---	C	Bar	---	---	---	1	1	---
SA-666	201	17Cr-4Ni-6Mn	Plate, Sheet & Strip	95	8	3	---	---	S20100
SA-666	XM-11	21Cr-6Ni-9Mn	Plate, Sheet & Strip	100	8	3	---	---	S21904
SA-666	302	18Cr-8Ni	Plate, Sheet & Strip	75	8	1	---	---	S30200
SA-666	304	18Cr-8Ni	Plate, Sheet & Strip	75	8	1	---	---	S30400
SA-666	304L	18Cr-8Ni	Plate, Sheet & Strip	70	8	1	---	---	S30403
SA-666	304N	18Cr-8Ni-N	Plate, Sheet & Strip	80	8	1	---	---	S30451
SA-666	304LN	18Cr-8Ni-N	Plate, Sheet & Strip	80	8	1	---	---	S30453
SA-666	316	16Cr-12Ni-2Mo	Plate, Sheet & Strip	75	8	1	---	---	S31600
SA-666	316L	16Cr-12Ni-2Mo	Plate, Sheet & Strip	70	8	1	---	---	S31603
SA-666	316N	16Cr-12Ni-2Mo-N	Plate, Sheet & Strip	80	8	1	---	---	S31651

ASME FERROUS METALS - P NUMBERS & S NUMBERS (Continued)

ASME Spec No.	Type or Grade	Nominal Composition	Product Form	Tensile Strength[a] ksi	Welding				UNS No.
					P No	Group No.	S No.	Group No.	
A 668	Class B	C	Forgings	60	---	---	1	1	G10200
A 668	Class C	C	Forgings	66	---	---	1	1	G10250
A 668	Class D	C-Mn	Forgings	75	---	---	1	2	G10300
A 668	Class F b	C-Mn	Forgings, > 4-10 in.	85	---	---	1	3	---
A 668	Class F a	C-Mn	Forgings, to 4 in.	90	---	---	1	3	---
A 668	Class K b	C	Forgings, >7-10 in.	100	---	---	4	3	---
A-668	Class K a	C	Forgings, to 7 in.	105	---	---	4	3	---
A-668	Class L c	C	Forgings, >7-10 in.	110	---	---	4	3	---
A-668	Class L b	C	Forgings, >4-7 in.	115	---	---	4	3	---
A-668	Class L a	C	Forgings, to 4 in.	125	---	---	4	3	---
SA-671	CC60	C-Mn-Si	Fusion welded pipe	60	1	1	---	---	K02100
SA-671	CE55	C-Mn-Si	Fusion welded pipe	55	1	1	---	---	K02202
SA-671	CD70	C-Mn-Si	Fusion welded pipe	70	1	2	---	---	K02400
SA-671	CD80	C-Mn-Si	Fusion welded pipe	80	1	3	---	---	K02400
SA-671	CB60	C-Si	Fusion welded pipe	60	1	1	---	---	K02401
SA-671	CE60	C-Mn-Si	Fusion welded pipe	60	1	1	---	---	K02402
SA-671	CC65	C-Mn-Si	Fusion welded pipe	65	1	1	---	---	K02403
SA-671	CC70	C-Mn-Si	Fusion welded pipe	70	1	2	---	---	K02700
SA-671	CB65	C-Si	Fusion welded pipe	65	1	1	---	---	K02800
SA-671	CA55	C	Fusion welded pipe	55	1	1	---	---	K02801
SA-671	CK75	C-Mn-Si	Fusion welded pipe	75	1	2	---	---	K02803
SA-671	CB70	C-Si	Fusion welded pipe	70	1	2	---	---	K03101
SA-672	A45	C	Fusion welded pipe	45	1	1	---	---	K01700
SA-672	C55	C-Si	Fusion welded pipe	55	1	1	---	---	K01800

ASME Spec No.	Type or Grade	Nominal Composition	Product Form	Tensile Strength[a] ksi	P No	Welding Group No.	Welding S No.	Welding Group No.	UNS No.
SA-672	B55	C-Si	Fusion welded pipe	55	1	1	---	---	K02001
SA-672	C60	C-Mn-Si	Fusion welded pipe	60	1	1	---	---	K02100
SA-672	A50	C	Fusion welded pipe	50	1	1	---	---	K02200
SA-672	E55	C	Fusion welded pipe	55	1	1	---	---	K02202
SA-672	D70	C-Mn-Si	Fusion welded pipe	70	1	2	---	---	K02400
SA-672	D80	C-Mn-Si	Fusion welded pipe	80	1	3	---	---	K02400
SA-672	B60	C-Si	Fusion welded pipe	60	1	1	---	---	K02401
SA-672	E60	C-Mn-Si	Fusion welded pipe	60	1	1	---	---	K02402
SA-672	C65	C-Mn-Si	Fusion welded pipe	65	1	1	---	---	K02403
SA-672	C70	C-Mn-Si	Fusion welded pipe	70	1	2	---	---	K02700
SA-672	B65	C-Si	Fusion welded pipe	65	1	1	---	---	K02800
SA-672	A55	C	Fusion welded pipe	55	1	1	---	---	K02801
SA-672	N75	C-Mn-Si	Fusion welded pipe	75	1	2	---	---	K02803
SA-672	B70	C-Si	Fusion welded pipe	70	1	2	---	---	K03101
SA-672	L65	C-0.5Mo	Fusion welded pipe	65	3	1	---	---	K11820
SA-672	L70	C-0.5Mo	Fusion welded pipe	70	3	2	---	---	K12020
SA-672	H75	Mn-0.5Mo	Fusion welded pipe	75	3	2	---	---	K12021
SA-672	H80	Mn-0.5Mo	Fusion welded pipe	80	3	3	---	---	K12022
SA-672	L75	C-0.5Mo	Fusion welded pipe	75	3	2	---	---	K12320
SA-672	J100	Mn-0.5Mo	Fusion welded pipe	100	11A	4	---	---	K12521
SA-672	J80	Mn-0.5Mo-0.75Ni	Fusion welded pipe	80	3	3	---	---	K12554
SA-672	J90	Mn-0.5Mo-0.75Ni	Fusion welded pipe	90	3	3	---	---	K12554
SA-675	45	C	Bar	45	1	1	---	---	---
SA-675	50	C	Bar	50	1	1	---	---	---

ASME FERROUS METALS - P NUMBERS & S NUMBERS (Continued)

ASME Spec No.	Type or Grade	Nominal Composition	Product Form	Tensile Strength[a] ksi	P No	Welding Group No.	Welding S No.	Welding Group No.	UNS No.
SA-675	55	C	Bar	55	1	1	---	---	---
SA-675	60	C	Bar	60	1	1	---	---	---
SA-675	65	C	Bar	65	1	1	---	---	---
SA-675	70	C	Bar	70	1	2	---	---	---
A 675	75	C	Bar	75	---	---	1	2	---
SA-688	XM-29	18Cr-3Ni-12Mn	Welded tube	100	8	3	---	---	S24000
SA-688	TP304	18Cr-8Ni	Welded tube	75	8	1	---	---	S30400
SA-688	TP304L	18Cr-8Ni	Welded tube	70	8	1	---	---	S30403
SA-688	TP304N	18Cr-8Ni-N	Welded tube	80	8	1	---	---	S30451
SA-688	TP304LN	18Cr-8Ni-N	Welded tube	75	8	1	---	---	S30453
SA-688	TP316	16Cr-12Ni-2Mo	Welded tube	75	8	1	---	---	S31600
SA-688	TP316L	16Cr-12Ni-2Mo	Welded tube	70	8	1	---	---	S31603
SA-688	TP316N	16Cr-12Ni-2Mo-N	Welded tube	80	8	1	---	---	S31651
SA-688	TP316LN	16Cr-12Ni-2Mo-N	Welded tube	75	8	1	---	---	S31653
SA-691	CMSH-70	C-Mn-Si	Fusion welded pipe	70	1	2	---	---	K02400
SA-691	CMSH-80	C-Mn-Si	Fusion welded pipe	80	1	3	---	---	K02400
SA-691	CMS-75	C-Mn-Si	Fusion welded pipe	75	1	2	---	---	K02803
SA-691	1Cr, Class 1	1Cr-0.5Mo	Fusion welded pipe	55	4	1	---	---	K11757
SA-691	1Cr, Class 2	1Cr-0.5Mo	Fusion welded pipe	65	4	1	---	---	K11757
SA-691	1.25Cr, Class 1	1.25Cr-0.5Mo-Si	Fusion welded pipe	60	4	1	---	---	K11789
SA-691	1.25Cr, Class 2	1.25Cr-0.5Mo-Si	Fusion welded pipe	75	4	1	---	---	K11789
SA-691	CM-65	C-0.5Mo	Fusion welded pipe	65	3	1	---	---	K11820
SA-691	CM70	C-0.5Mo	Fusion welded pipe	70	3	2	---	---	K12020
SA-691	0.5Cr, Class 1	0.5Cr-0.5Mo	Fusion welded pipe	55	3	1	---	---	K12143

ASME FERROUS METALS - P NUMBERS & S NUMBERS (Continued)

ASME Spec No.	Type or Grade	Nominal Composition	Product Form	Tensile Strength[a] ksi	P No	Welding Group No.	Welding S No.	Welding Group No.	UNS No.
SA-691	0.5Cr, Class2	0.5Cr-0.5Mo	Fusion welded pipe	70	3	2	---	---	K12143
SA-691	CM75	C-0.5Mo	Fusion welded pipe	75	3	2	---	---	K12320
SA-691	2.25Cr, Class 1	2.25Cr-1Mo	Fusion welded pipe	60	5A	1	---	---	K21590
SA-691	2.25Cr, Class 2	2.25Cr-1Mo	Fusion welded pipe	75	5A	1	---	---	K21590
SA-691	3Cr, Class 1	3Cr-1Mo	Fusion welded pipe	60	5A	1	---	---	K31545
SA-691	3Cr, Class 2	3Cr-1Mo	Fusion welded pipe	75	5A	1	---	---	K31545
SA-691	5Cr, Class 1	5Cr-0.5Mo	Fusion welded pipe	60	5B	1	---	---	K41545
SA-691	5Cr, Class 2	5Cr-0.5Mo	Fusion welded pipe	75	5B	1	---	---	K41545
A 694	---	C	Forgings	---	---	---	1	1	K03014
SA-695	Type B, Grade 35	C-Mn-Si	Bar	60	1	1	---	---	K03504
SA-695	Type B, Grade 40	C-Mn-Si	Bar	70	1	2	---	---	K03504
SA-696	B	C-Mn-Si	Bar	60	1	1	---	---	K03200
SA-696	C	C-Mn-Si	Bar	70	1	2	---	---	K03200
A 714	Grade V, Tp E	2Ni-Cu	Smls & welded pipe	65	---	---	9A	1	K22035
A 714	Grade V	2Ni-Cu	Smls & welded pipe	65	---	---	9A	1	K22035
SA-724	A	C-Mn-Si	Plate	90	1	4	---	---	K11831
SA-724	B	C-Mn-Si	Plate	95	1	4	---	---	K12031
SA-724	C	C-Mn-Si	Plate	90	1	4	---	---	K12037
SA-727	---	C-Mn-Si	Forgings	60	1	1	---	---	K02506
SA-731	S41500	13Cr-4.5Ni-Mo	Smls & welded pipe	115	6	4	---	---	S41500
SA-731	TP439	17Cr-Ti	Smls & welded pipe	60	7	2	---	---	S43035
SA-731	18Cr-2Mo	18Cr-2Mo	Smls & welded pipe	60	7	2	---	---	S44400
SA-731	TPXM-33	27Cr-1Mo-Ti	Smls & welded pipe	65	10I	1	---	---	S44626
SA-731	TPXM-27	27Cr-1Mo	Smls & welded pipe	65	10I	1	---	---	S44627

ASME FERROUS METALS - P NUMBERS & S NUMBERS (Continued)

ASME Spec No.	Type or Grade	Nominal Composition	Product Form	Tensile Strength[a] ksi	P No	Welding Group No.	Welding S No.	Welding Group No.	UNS No.
SA-731	S44660	26Cr-3Ni-3Mo	Smls & welded pipe	85	10K	1	---	---	S44660
SA-731	S44700	29Cr-4Mo	Smls & welded pipe	80	10J	1	---	---	S44700
SA-731	S44800	29Cr-4Mo-2Ni	Smls & welded pipe	80	10K	1	---	---	S44800
SA-737	B	C-Mn-Si-Cb	Plate	70	1	2	---	---	K12001
SA-737	C	C-Mn-Si-V	Plate	80	1	3	---	---	K12202
SA-738	A	C-Mn-Si	Plate, ≤ 2½ in.	75	1	2	---	---	K12447
SA-738	B	C-Mn-Si	Plate, ≤ 2½ in.	85	1	3	---	---	K12447
SA-738	C	C-Mn-Si	Plate, > 4-6 in., incl.	70	1	3	---	---	---
SA-738	C	C-Mn-Si	Plate, > 2½-4 in.	75	1	3	---	---	---
SA-738	C	C-Mn-Si	Plate, ≤ 2½ in.	80	1	3	---	---	---
SA-739	B11	1.25Cr-0.5Mo	Bar	70	4	1	---	---	K11797
SA-739	B22	2.25Cr-1Mo	Bar	75	5A	1	---	---	K21390
SA-765	Type I	C-Mn-Si	Forgings	60	1	1	---	---	K03046
SA-765	Type II	C-Mn-Si	Forgings	70	1	2	---	---	K03047
SA-765	Type III	3.5Ni	Forgings	70	9B	1	---	---	K32026
SA-789	S31200	25Cr-6Ni-Mo-N	Smls & welded tube	100	10H	1	---	---	S31200
SA-789	S31260	25Cr-6.5Ni-3Mo-N	Smls & welded tube	100	10H	1	---	---	S31260
SA-789	S31500	18Cr-5Ni-3Mo-N	Smls & welded tube	92	10H	1	---	---	S31500
SA-789	S31803	22Cr-5Ni-3Mo-N	Smls & welded tube	90	10H	1	---	---	S31803
SA-789	S32304	23Cr-4Ni-Mo-Cu-N	Smls & welded tube	87	10H	1	---	---	S32304
SA-789	S32550	25Cr-5Ni-3Mo-2Cu	Smls & welded tube	110	10H	1	---	---	S32550
SA-789	S32750	25Cr-7Ni-4Mo-N	Smls & welded tube	116	10H	1	---	---	S32750
SA-789	---	25Cr-7.5Ni-3.5Mo-N-Cu-W	Smls & welded tube	109	---	---	10H	1	S32760
SA-789	S32900	26Cr-4Ni-Mo	Smls & welded tube	90	10H	1	---	---	S32900

ASME FERROUS METALS - P NUMBERS & S NUMBERS (Continued)

ASME Spec No.	Type or Grade	Nominal Composition	Product Form	Tensile Strength[a] ksi	P No	Welding Group No.	S No.	Group No.	UNS No.
SA-789	S32950	26Cr-4Ni-Mo-N	Smls & welded tube	90	10H	1	---	---	S32950
SA-790	S31200	25Cr-6Ni-Mo-N	Smls & welded tube	100	10H	1	---	---	S31200
SA-790	S31260	25Cr-6.5Ni-3Mo-N	Smls & welded tube	100	10H	1	---	---	S31260
SA-790	S31500	18Cr-5Ni-3Mo-N	Smls & welded tube	92	10H	1	---	---	S31500
SA-790	S31803	22Cr-5Ni-3Mo-N	Smls & welded tube	90	10H	1	---	---	S31803
SA-790	S32304	23Cr-4Ni-Mo-Cu-N	Smls & welded tube	87	10H	1	---	---	S32304
SA-790	S32550	25Cr-5Ni-3Mo-2Cu	Smls & welded tube	110	10H	1	---	---	S32550
SA-790	S32750	25Cr-7Ni-4Mo-N	Smls & welded tube	116	10H	1	---	---	S32750
SA-790	---	25Cr-7.5Ni-3.5Mo-N-Cu-W	Smls & welded tube	109	---	---	10H	1	S32760
SA-790	S32900	26Cr-4Ni-Mo	Smls & welded tube	90	10H	1	---	---	S32900
SA-790	S32950	26Cr-4Ni-Mo-N	Smls & welded tube	90	10H	1	---	---	S32950
SA-803	TP439	17Cr-Ti	Welded tube	70	7	2	---	---	S43035
SA-803	26-3-3	26Cr-3Ni-3Mo	Welded tube	85	10K	1	---	---	S44660
SA-813	TPXM-19	22Cr-13Ni-Mn	Welded pipe	100	8	3	---	---	S20910
SA-813	TPXM-11	21Cr-6Ni-9Mn	Welded pipe	90	8	3	---	---	S21904
SA-813	TPXM-29	18Cr-3Ni-12Mn	Welded pipe	100	8	3	---	---	S24000
SA-813	TP304	18Cr-8Ni	Welded pipe	75	8	1	---	---	S30400
SA-813	TP304L	18Cr-8Ni	Welded pipe	70	8	1	---	---	S30403
SA-813	TP304H	18Cr-8Ni	Welded pipe	75	8	1	---	---	S30409
SA-813	TP304N	18Cr-8Ni-N	Welded pipe	80	8	1	---	---	S30451
SA-813	TP304LN	18Cr-8Ni-N	Welded pipe	75	8	1	---	---	S30453
SA-813	S30815	21Cr-11Ni-N	Welded pipe	87	8	2	---	---	S30815
SA-813	TP309S	23Cr-12Ni	Welded pipe	75	8	2	---	---	S30908
SA-813	TP309Cb	23Cr-12Ni-Cb	Welded pipe	75	8	2	---	---	S30940

ASME FERROUS METALS - P NUMBERS & S NUMBERS (Continued)									
ASME			Tensile Strength[a]	Welding					
Spec No.	Type or Grade	Nominal Composition	Product Form	ksi	P No	Group No.	S No.	Group No.	UNS No.
SA-813	TP310S	25Cr-20Ni	Welded pipe	75	8	2	---	---	S31008
SA-813	TP310Cb	25Cr-20Ni-Cb	Welded pipe	75	8	2	---	---	S31040
SA-813	S31254	20Cr-18Ni-6Mo	Welded pipe	94	8	4	---	---	S31254
SA-813	TP316	16Cr-12Ni-2Mo	Welded pipe	75	8	1	---	---	S31600
SA-813	TP316L	16Cr-12Ni-2Mo	Welded pipe	70	8	1	---	---	S31603
SA-813	TP316H	16Cr-12Ni-2Mo	Welded pipe	75	8	1	---	---	S31609
SA-813	TP316N	16Cr-12Ni-2Mo-N	Welded pipe	80	8	1	---	---	S31651
SA-813	TP316LN	16Cr-12Ni-2Mo-N	Welded pipe	75	8	1	---	---	S31653
SA-813	TP317	18Cr-13Ni-3Mo	Welded pipe	75	8	1	---	---	S31700
SA-813	TP317L	18Cr-13Ni-3Mo	Welded pipe	75	8	1	---	---	S31703
SA-813	TP321	18Cr-10Ni-Ti	Welded pipe	75	8	1	---	---	S32100
SA-813	TP321H	18Cr-10Ni-Ti	Welded pipe	75	8	1	---	---	S32109
SA-813	TP347	18Cr-10Ni-Cb	Welded pipe	75	8	1	---	---	S34700
SA-813	TP347H	18Cr-10Ni-Cb	Welded pipe	75	8	1	---	---	S34709
SA-813	TP348	18Cr-10Ni-Cb	Welded pipe	75	8	1	---	---	S34800
SA-813	TP348H	18Cr-10Ni-Cb	Welded pipe	75	8	1	---	---	S34809
SA-813	TP XM-15	18Cr-18Ni-2Si	Welded pipe	75	8	1	---	---	S38100
SA-814	TP XM-19	22Cr-13Ni-5Mn	CW welded pipe	100	8	3	---	---	S20910
SA-814	TP XM-11	21Cr-6Ni-9Mn	CW welded pipe	90	8	3	---	---	S21904
SA-814	TP XM-29	18Cr-3Ni-12Mn	CW welded pipe	100	8	3	---	---	S24000
SA-814	TP304	18Cr-8Ni	CW welded pipe	75	8	1	---	---	S30400
SA-814	TP304L	18Cr-8Ni	CW welded pipe	70	8	1	---	---	S30403
SA-814	TP304H	18Cr-8Ni	CW welded pipe	75	8	1	---	---	S30409
SA-814	TP304N	18Cr-8Ni-N	CW welded pipe	80	8	1	---	---	S30451

ASME FERROUS METALS - P NUMBERS & S NUMBERS (Continued)

ASME Spec No.	Type or Grade	Nominal Composition	Product Form	Tensile Strength[a] ksi	P No	Welding Group No.	Welding S No.	Welding Group No.	UNS No.
SA-814	TP304LN	18Cr-8Ni-N	CW welded pipe	75	8	1	---	---	S30453
SA-814	S30815	21Cr-11Ni-N	CW welded pipe	87	8	2	---	---	S30815
SA-814	TP309S	23Cr-12Ni	CW welded pipe	75	8	2	---	---	S30908
SA-814	TP309Cb	23Cr-12Ni-Cb	CW welded pipe	75	8	2	---	---	S30940
SA-814	TP310S	25Cr-20Ni	CW welded pipe	75	8	2	---	---	S31008
SA-814	TP310Cb	25Cr-20Ni-Cb	CW welded pipe	75	8	2	---	---	S31040
SA-814	S31254	20Cr-18Ni-6Mo	CW welded pipe	94	8	4	---	---	S31254
SA-814	TP316	16Cr-12Ni-2Mo	CW welded pipe	75	8	1	---	---	S31600
SA-814	TP316L	16Cr-12Ni-2Mo	CW welded pipe	70	8	1	---	---	S31603
SA-814	TP316H	16Cr-12Ni-2Mo	CW welded pipe	75	8	1	---	---	S31609
SA-814	TP316N	16Cr-12Ni-2Mo-N	CW welded pipe	80	8	1	---	---	S31651
SA-814	TP316LN	16Cr-12Ni-2Mo-N	CW welded pipe	75	8	1	---	---	S31653
SA-814	TP317	18Cr-13Ni-3Mo	CW welded pipe	75	8	1	---	---	S31700
SA-814	TP317L	18Cr-13Ni-3Mo	CW welded pipe	75	8	1	---	---	S31703
SA-814	TP321	18Cr-10Ni-Ti	CW welded pipe	75	8	1	---	---	S32100
SA-814	TP321H	18Cr-10Ni-Ti	CW welded pipe	75	8	1	---	---	S32109
SA-814	TP347	18Cr-10Ni-Cb	CW welded pipe	75	8	1	---	---	S34700
SA-814	TP347H	18Cr-10Ni-Cb	CW welded pipe	75	8	1	---	---	S34709
SA-814	TP348	18Cr-10Ni-Cb	CW welded pipe	75	8	1	---	---	S34800
SA-814	TP348H	18Cr-10Ni-Cb	CW welded pipe	75	8	1	---	---	S34809
SA-814	TP XM-15	18Cr-18Ni-2Si	CW welded pipe	75	8	1	---	---	S38100
SA-815	---	22Cr-5Ni-3Mo-N	Fittings	90	10H	1	---	---	S31803
SA-815	---	25Cr-7.5Ni-3.5Mo-N-Cu-W	Fittings	109	---	---	10H	1	S32760
SA-815	S41500	13Cr-4.5Ni-Mo	Fittings	110	6	4	---	---	S41500

ASME Spec No.	Type or Grade	Nominal Composition	Product Form	Tensile Strength[a] ksi	P No	Welding Group No.	S No.	Group No.	UNS No.
ASME FERROUS METALS - P NUMBERS & S NUMBERS (Continued)									
SA-832	21V	3Cr-1Mo-V-Ti-B	Plate	85	5C	1	---	---	K31830
SA-836	---	C-Si-Ti	Forgings	55	1	1	---	---	---
SA-841	---	C-Mn-Si	Plate, > 2½ in.	65	10C	1	---	---	---
SA-841	---	C-Mn-Si	Plate, ≤ 2½ in.	70	10C	1	---	---	---
A 890	C03MWCuN	25Cr-7.5Ni-3.5Mo-N-Cu-W	Castings	100	---	---	10H	1	J93380
A 928	---	25Cr-7.5Ni-3.5Mo-N-Cu-W	Welded pipe	109	---	---	10H	1	S32760
API 5L	A25, Class I	C-Mn	Smls/welded pipe & tube	45	---	---	1	1	---
API 5L	A25, Class II	C-Mn	Smls/welded pipe & tube	45	---	---	1	1	---
API 5L	A	C-Mn	Smls/welded pipe & tube	48	---	---	1	1	---
API 5L	B	C-Mn	Smls/welded pipe & tube	60	---	---	1	1	---
API 5L	X42	C-Mn	Smls/welded pipe & tube	60	---	---	1	1	---
API 5L	X46	C-Mn	Smls/welded pipe & tube	63	---	---	1	1	---
API 5L	X52	C-Mn	Smls/welded pipe & tube	66	---	---	1	1	---
API 5L	X56	C-Mn	Smls/welded pipe & tube	71	---	---	1	2	---
API 5L	X60	C-Mn	Smls/welded pipe & tube	75	---	---	1	2	---
API 5L	X65	C-Mn	Smls/welded pipe & tube	77	---	---	1	2	---
API 5L	X70	C-Mn	Smls/welded pipe & tube	82	---	---	1	3	---
API 5L	X80	C-Mn	Smls/welded pipe & tube	90	---	---	1	4	---
MSS SP-75	WPHY-42	C-Mn	Smls./welded fittings	60	---	---	1	1	---
MSS SP-75	WPHY-46	C-Mn	Smls./welded fittings	63	---	---	1	1	---
MSS SP-75	WPHY-52	C-Mn	Smls./welded fittings	66	---	---	1	1	---
MSS SP-75	WPHY-56	C-Mn	Smls./welded fittings	71	---	---	1	2	---
MSS SP-75	WPHY-60	C-Mn	Smls./welded fittings	75	---	---	1	2	---
MSS SP-75	WPHY-65	C-Mn	Smls./welded fittings	77	---	---	1	2	---

ASME FERROUS METALS - P NUMBERS & S NUMBERS (Continued)

ASME Spec No.	Type or Grade	Nominal Composition	Product Form	Tensile Strength[a] ksi	Welding P No	Group No.	S No.	Group No.	UNS No.
MSS SP-75	WPHY-70	C-Mn	Smls./welded fittings	82	---	---	---	3	---
SA/CSA-G 40.21	Gr. 38W	C-Mn-Si	Plate, Bar & Shapes	60	1	1	---	---	---

a. Tensile strength values are minimum

FASTENERS CARBON, ALLOY & STAINLESS STEELS:

AMERICAN SPECIFICATION TITLES & DESIGNATIONS, CHEMICAL COMPOSITIONS & MECHANICAL PROPERTIES

CARBON AND ALLOY STEEL FASTENERS

ASTM Spec.	Title
Bolting Materials	
A 193/A 193M	Alloy-Steel and Stainless Steel Bolting Materials for High-Temperature Service
A 194/A 194M	Carbon and Alloy Steel Nuts for Bolts for High-Pressure and High-Temperature Service
A 320/A 320M	Alloy-Steel Bolting Materials for Low-Temperature Service
A 437/A 437M	Alloy-Steel Turbine-Type Bolting Material Specially Heat Treated for High-Temperature Service
A 449	Quenched and Tempered Steel Bolts and Studs
A 453/A 453M	High-Temperature Bolting Materials, with Expansion Coefficients Comparable to Austenitic Stainless Steels
A 489	Carbon Steel Lifting Eyes
A 540/A 540M	Alloy-Steel Bolting Materials for Special Applications
Bolts	
A 307	Carbon Steel Bolts and Studs, 60,000 psi Tensile Strength
A 325	Structural Bolts, Steel, Heat Treated, 120/105 ksi Minimum Tensile Strength
A 325M	High-Strength Bolts for Structural Steel Joints (Metric)
A 354	Quenched and Tempered Alloy Steel Bolts, Studs, and Other Externally Threaded Fasteners
A 394	Steel Transmission Tower Bolts, Zinc-Coated and Bare
A 449	Quenched and Tempered Steel Bolts and Studs
A 489	Carbon Steel Lifting Eyes
A 490	Heat-Treated Steel Structural Bolts, 150 ksi Minimum Tensile Strength
A 490M	High-Strength Steel Bolts, Classes 10.9 and 10.9.3, for Structural Steel Joints (Metric)
A 687	High-Strength Nonheaded Steel Bolts and Studs
F 432	Roof and Rock Bolts and Accessories
F 468	Nonferrous Bolts, Hex Cap Screws, and Studs for General Use
F 468M	Nonferrous Bolts, Hex Cap Screws, and Studs for General Use (Metric)
F 541	Alloy Steel Eyebolts
F 568M	Carbon and Alloy Steel Externally Threaded Metric Fasteners
F 593	Stainless Steel Bolts, Hex Cap Screws, and Studs

CARBON AND ALLOY STEEL FASTENERS (Continued)	
ASTM Spec.	**Title**
Bolting Materials (Continued)	
F 738M	Stainless Steel Metric Bolts, Screws, and Studs
F 788/F 788M	Surface Discontinuities of Bolts, Screws, and Studs, Inch and Metric Series
F 901	Aluminum Transmission Tower Bolts and Nuts
F 1554	Anchor Bolts, Steel, 36, 55, and 105-ksi Yield Strength
Nuts	
A 563	Carbon and Alloy Steel Nuts
A 563M	Carbon and Alloy Steel Nuts (Metric)
F 467	Nonferrous Nuts for General Use
F 467M	Nonferrous Nuts for General Use (Metric)
F 594	Stainless Steel Nuts
F 812/F 812M	Surface Discontinuities of Nuts, Inch and Metric Series
F 836M	Stainless Steel Metric Nuts
F 901	Aluminum Transmission Tower Bolts and Nuts
Rivets	
A 31	Steel Rivets and Bars for Rivets, Pressure Vessel
A 449	Quenched and Tempered Steel Bolts and Studs
A 502	Steel Structural Rivets
A 574	Alloy Steel Socket-Head Cap Screws
A 574M	Alloy Steel Socket-Head Cap Screws (Metric)
F 468	Nonferrous Bolts Hex Cap Screws and Studs for General Use
F 468M	Nonferrous Bolts Hex Cap Screws and Studs for General Use (Metric)
F 568M	Carbon and Alloy Steel Externally Threaded Metric Fasteners
F 593	Stainless Steel Bolts, Hex Cap Screws, and Studs
F 738M	Stainless Steel Metric Bolts, Screws, and Studs
F 788/F 788M	Surface Discontinuities of Bolts, Screws, and Studs, Inch and Metric Series
F 835	Alloy Steel Socket Button and Flat Countersunk Head Cap Screws

CARBON AND ALLOY STEEL FASTENERS (Continued)

ASTM Spec.	Title
Rivets (Continued)	
F 835M	Alloy Steel Socket Button and Flat Countersunk Head Cap Screws (Metric)
F 837	Stainless Steel Socket Head Cap Screws
F 837M	Stainless Steel Socket Head Cap Screws (Metric)
F 879	Stainless Steel Socket Button and Flat Countersunk Head Cap Screws
F 879M	Stainless Steel Socket Button and Flat Countersunk Head Cap Screws (Metric)
F 880	Stainless Steel Socket Set Screws
F 880M	Stainless Steel Socket Set Screws (Metric)
F 912	Alloy Steel Socket Set Screws
F 912M	Alloy Steel Socket Set Screws (Metric)
Studs	
A 307	Carbon Steel Bolts and Studs, 60,000 psi Tensile Strength
A 354	Quenched and Tempered Alloy Steel Bolts, Studs, and Other Externally Threaded Fasteners
A 449	Quenched and Tempered Steel Bolts and Studs
A 687	High-Strength Nonheaded Steel Bolts and Studs
A 738M	Stainless Steel Metric Bolts, Screws, and Studs
F 468	Nonferrous Bolts, Hex Cap Screws, and Studs for General Use
F 468M	Nonferrous Bolts, Hex Cap Screws, and Studs for General Use (Metric)
F 568M	Carbon and Alloy Steel Externally Threaded Metric Fasteners
F 788/F 788M	Surface Discontinuities of Bolts, Screws, and Studs, Inch and Metric Series
Washers	
F 436	Hardened Steel Washers
F 436M	Hardened Steel Washers (Metric)
F 844	Washers, Steel, Plain (Flat), Unhardened, for General Use
SAE Std.	**Title**
J78	Steel Self-Drilling Tapping Screws

CARBON AND ALLOY STEEL FASTENERS (Continued)	
SAE Std.	**Title**
Washers (Continued)	
J81	Threaded Rolling Screws
J82	Mechanical and Quality Requirements for Machine Screws
J121	Decarburization in Hardened and Tempered Threaded Fasteners
J122	Surface Discontinuities on Nuts
J123	Surface Discontinuities on Bolts, Screws and Studs
J429	Mechanical and Material Requirements for Externally Threaded Fasteners
J430	Mechanical and Chemical Requirements for Nonthreaded Fasteners
J933	Mechanical and Quality Requirements foe Tapping Screws
J995	Mechanical and Material Requirements for Steel Nuts
J1061	Surface Discontinuities on General Application Bolts, Screws, and Studs
J1102	Mechanical and Material Requirements for Wheel Bolts
J1199	Mechanical and Material Requirements for Metric Externally Threaded Steel Fasteners
J1216	Test Methods for Metric Threaded Fasteners
J1237	Metric Thread Rolling Screws

CHEMICAL COMPOSITION OF CARBON STEEL BOLTING & OTHER FASTENERS[a]

SAE J429 Grade	Material & Treatment	C	Mn	P	S	B
1	Low or medium carbon steel	0.55	---	0.048	0.058	---
2	Low or medium carbon steel	0.55	---	0.048	0.058b	---
4	Medium carbon cold drawn steel	0.55	---	0.048	0.13	---
5	Medium carbon steel, quenched & tempered	0.28-0.55	---	0.048	0.058	---
5.1	Low or medium carbon steel, quenched or tempered	0.15-0.30	---	0.048	0.058	---
5.2	Low carbon martensite steel, fully killed, fine grained, quenched & tempered	0.15-0.25	0.74 min	0.048	0.058	0.0005 min
7	Medium carbon alloy steel, quenched & tempered	0.28-0.55	---	0.040	0.045	---
8	Medium carbon alloy steel, quenched & tempered	0.28-0.55	---	0.040	0.045	---
8.1	Elevated temperature drawn steel, medium carbon alloy or SAE 1541 or 1541H steel	0.28-0.55	---	0.048	0.058	---
8.2	Low carbon martensite steel, fully killed, fine grained, quenched & tempered.	0.15-0.25	0.74 min	0.048	0.058	0.0005 min

a. All values are product analysis. For heat analysis permissible variations see SAE J409.
All single values are maximums, unless otherwise specified.

CHEMICAL COMPOSITION OF STEEL BOLTING FOR HIGH TEMPERATURE SERVICE[e]								
ASTM A 193	AISI	C	Mn	Si	Cr	Ni	Mo	Others
Ferritic Steels								
B5	501	0.10[c]	1.00	1.00	4.00-6.00	---	0.40-0.65	---
B6, B6X	410	0.15	1.00	1.00	11.50-13.50	---	---	---
B7, B7M	4140[a]	0.37-0.49[d]	0.65-1.10	0.15-0.35	0.75-1.20	---	0.15-0.25	---
B16	---	0.36-0.47	0.45-0.70	0.15-0.35	0.80-1.15	---	0.50-0.65	0.25-0.35 V[f]
Austenitic Steels								
B8, B8A	304	0.08	2.00	1.00	18.00-20.00	8.00-10.50	---	---
B8C, B8CA	347	0.08	2.00	1.00	17.00-19.00	9.00-13.00	---	(Ta + Cb) = 10 X C[c]
B8M, B8MA, B8M2, B8M3	316	0.08	2.00	1.00	16.00-18.00	10.00-14.00	2.00-3.00	---
B8N, B8NA	304N	0.08	2.00	1.00	18.00-20.00	8.00-10.50	---	0.10-0.16 N
B8MN, B8MNA	316N	0.08	2.00	1.00	16.00-18.00	10.00-14.00	2.00-3.00	0.10-0.16 N
B8P, B8PA	305[b]	0.08	2.00	1.00	17.00-19.00	10.50-13.00	---	---
B8MLCuN, B8MLCuNA	---	0.020	1.00	0.80	19.50-20.50	17.50-18.50	6.00-6.50	0.18-0.22 N, 0.50-1.00 Cu
B8T, B8TA	321	0.08	2.00	1.00	17.00-19.00	9.00-12.00	---	Ti = 5 X C[c]
B8R, B8RA	---	0.06	4.00-6.00	1.00	20.50-23.50	11.50-13.50	1.50-3.00	0.20-0.40 N, 0.10-0.30 V, 0.10-0.30 (Cb + Ta)
B8S, B8SA	---	0.10	7.00-9.00	3.50-4.50	16.00-18.00	8.00-9.00	---	0.08-0.18 N
B8LN, B8LNA	304N[b]	0.030	2.00	1.00	18.00-20.00	8.00-10.50	---	0.10-0.16 N
B8MLN, B8MLNA	316N[b]	0.030	2.00	1.00	16.00-18.00	10.00-14.00	2.00-3.00	0.10-0.16 N

a. Also includes AISI types 4142, 4145, 4140H, 4142H, 4145H. b. AISI type as designated, with restricted carbon content. c. Minimum value. d. For bar sizes 3.5" and over, carbon content may be 0.50% max. For B7M grade, a minimum carbon content of 0.28% is permitted, provided that the required tensile properties are met in the section sizes involved; the use of AISI 4130 or 4130H is allowed. e. Phosphorous and sulfur requirements are not listed due to space limitation; see ASTM A193 for details. f. 0.015 Al max. Single values are maximums, unless otherwise specified.

ASTM A 325 - CHEMICAL REQUIREMENTS FOR STRUCTURAL STEEL TYPE 1 BOLTS[a]

Type 1	C	Mn	P	S	Si	B
Carbon Steel	0.30-0.52	0.60	0.040	0.050	---	b
Alloy Steel	0.30-0.52	0.60	0.035	0.040	0.15-0.35	b, c
Carbon Boron Steel	0.30-0.52	0.60	0.040	0.050		0.0005-0.003

a. Heat analysis only; for product analysis see ASTM A 325 Table 1. b. Boron shall not be added. c. Alloying elements permitted.
All single values are maximums unless otherwise stated.

ASTM A 325 - CHEMICAL REQUIREMENTS FOR STRUCTURAL STEEL TYPE 3 BOLTS[a, b]

Type 3	C	Mn	P	S	Si	Cu	Ni	Cr	V	Mo	Ti
Class A	0.33-0.40	0.90-1.20	0.040	0.050	0.15-0.35	0.25-0.45	0.25-0.45	0.45-0.65	---	---	---
Class B	0.38-0.48	0.70-0.90	0.06-0.12	0.050	0.30-0.50	0.20-0.40	0.50-0.80	0.50-0.75	---	0.06	---
Class C	0.15-0.25	0.80-1.35	0.035	0.040	0.15-0.35	0.20-0.50	0.25-0.50	0.30-0.50	0.020 min	---	---
Class D	0.15-0.25	0.40-1.20	0.040	0.050	0.25-0.50	0.30-0.50	0.50-0.80	0.50-1.00	---	0.10	0.05
Class E	0.20-0.25	0.60-1.00	0.040	0.040	0.15-0.35	0.30-0.60	0.30-0.60	0.60-0.90	---	---	---
Class F	0.20-0.25	0.90-1.20	0.040	0.040	0.15-0.35	0.20-0.40	0.20-0.40	0.45-0.65	---	---	---

a. A, B, C, D, E, and F are classes of material used for Type 3 bolts. Selection of a class shall be at the option of the bolt manufacturer.
b. Heat analysis only; for product analysis see ASTM A 325 Table 2.
All single values are maximums unless otherwise stated.

ASTM A 354 - CHEMICAL REQUIREMENTS FOR FASTENERS[a]

All Grades	C		P	S
	For sizes through 1½ in.	For sizes > 1½ in.		
Product Analysis	0.28-0.55	0.33-0.55	0.040	0.045
Heat Analysis	0.30-0.53	0.35-0.53	0.035	0.040

a. Chemical requirements pertain to grades BC and BD. All single values are maximums.

ASTM A 449 - CHEMICAL REQUIREMENTS FOR TYPE 1 AND TYPE 2 BOLTS AND STUDS

Type 1	C	Mn	P	S	B
Product Analysis	0.25-0.58	0.57 min	0.048	0.058	---
Heat Analysis	0.28-0.55	0.60 min	0.040	0.050	---
Type 2					
Product Analysis	0.13-0.14	0.67 min	0.048	0.058	0.0005 min
Heat Analysis	0.15-0.38	0.70 min	0.040	0.050	0.0005 min

Single values are maximums, unless otherwise specified.

ASTM A 489 - CHEMICAL REQUIREMENTS FOR CARBON STEEL LIFTING EYES

Types 1 & 2	C	Mn	P	S	Si
Product Analysis	0.51	1.06	0.048	0.058	0.12-0.38
Heat Analysis	0.48	1.00	0.040	0.050	0.15-0.35

Single values are maximums.

ASTM A 490 - CHEMICAL REQUIREMENTS FOR STRUCTURAL STEEL TYPE 1 BOLTS

For sizes through 1⅜ in.	C	P	S	Alloying Elements
Product Analysis	0.28-0.50	0.045	0.045	a
Heat Analysis	0.30-0.48	0.040	0.040	a
For size 1½ in.				
Product Analysis	0.33-0.55	0.045	0.045	a
Heat Analysis	0.35-0.53	0.040	0.040	a

a. The steel shall contain sufficient alloying elements to qualify it as an alloy steel. See ASTM A 490 paragraph 6.1 for more details.
Single values are maximums.

ASTM A 490 - CHEMICAL REQUIREMENTS FOR STRUCTURAL STEEL TYPE 2 AND TYPE 3 BOLTS

Type 2	C	Mn	P	S	Cu	Ni	Cr	Mo	B
Product Analysis	0.13-0.37	0.67	0.048 max	0.058 max	---	---	---	---	0.0005
Heat Analysis	0.15-0.34	0.70	0.040 max	0.050 max	---	---	---	---	0.0005
Type 3									
Product Analysis	0.19-0.55	0.37	0.045 max	0.055 max	0.63 max	0.17	0.42	0.14	---
Heat Analysis	0.20-0.53	0.40	0.040 max	0.050 max	0.60 max	0.20	0.45	0.15	---

Single values are minimums unless otherwise specified.

ASTM A 687 - CHEMICAL REQUIREMENTS FOR HIGH-STRENGTH NONHEADED STEEL BOLTS AND STUDS[a]

	C	Mn	P	S	Si	Cr	Mo
Product Analysis	0.36-0.45	0.71-1.04	0.040	0.045	0.18-0.37	0.75-1.15	0.13-0.27
Heat Analysis	0.38-0.43	0.75-1.00	0.035	0.040	0.20-0.35	0.80-1.10	0.15-0.25

a. Intentionally added bismuth, selenium, tellurium, or lead shall not be permitted. Single values are maximums.

ASTM F 593[a] – CHEMICAL REQUIREMENTS FOR STAINLESS STEEL BOLTS, HEX CAP SCREWS, AND STUDS

Alloy	UNS No.	Alloy	C	Mn	P	S	Si	Cr	Ni	Cu	Mo	Other
Austenitic Alloys												
Group 1	S30300	303	0.15	2.00	0.20	0.15 min	1.00	17.0-19.0	8.0-10.0	---	0.60[b]	---
Group 1	S30323	303Se	0.15	2.00	0.20	0.060	1.00	17.0-19.0	8.0-10.0	---	---	Se 0.15 min
Group 1	S30400	304	0.08	2.00	0.045	0.030	1.00	18.0-20.0	8.0-10.5	1.00	---	---
---	S30403	304L	0.03	2.00	0.045	0.030	1.00	18.0-20.0	8.0-12.0	1.00	---	---
Group 1	S30500	305	0.12	2.00	0.045	0.030	1.00	17.0-19.0	10.5-13.0	1.00	---	---
Group 1	S38400	384	0.08	2.00	0.045	0.030	1.00	15.0-17.0	17.0-19.0	1.00	0.50[b]	---
Group 1	S20300	XM1	0.08	5.0-6.5	0.040	0.18-0.35	1.00	16.0-18.0	5.0-6.5	1.75-2.25	---	---
Group 1	S30430	XM7	0.10	2.0	0.045	0.030	1.00	17.0-19.0	8.0-10.0	3.00-4.00	---	---

ASTM F 593ᵃ - CHEMICAL REQUIREMENTS FOR STAINLESS STEEL BOLTS, HEX CAP SCREWS, AND STUDS (Continued)

Alloy	UNS No.	Alloy	C	Mn	P	S	Si	Cr	Ni	Cu	Mo	Other
Austenitic Alloys (Continued)												
Group 2	S31600	316	0.08	2.0	0.045	0.030	1.00	16.0-18.0	10.0-14.0	---	2.00-3.00	---
---	S31603	316L	0.03	2.00	0.045	0.030	1.00	16.0-18.0	10.0-14.0	---	---	---
Group 3	S32100	321	0.08	2.0	0.045	0.030	1.00	17.0-19.0	9.0-12.0	---	---	Ti 5 x C min
Group 3	S34700	347	0.08	2.0	0.045	0.030	1.00	17.0-19.0	9.0-13.0	---	---	(Cb + Ta) 10 x C min
Ferritic Alloys												
Group 4	S43000	430	0.12	1.00	0.040	0.030	1.00	16.0-18.0	---	---	---	---
Group 4	S43020	430F	0.12	1.25	0.060	0.15 min	1.00	16.0-18.0	---	---	0.60ᵇ	---
Martensitic Alloys												
Group 5	S41000	410	0.15	1.00	0.040	0.030	1.00	11.5-13.5	---	---	---	---
Group 5	S41600	416	0.15	1.25	0.060	0.15 min	1.00	12.0-14.0	---	---	0.60ᵇ	---
Group 5	S41623	416Se	0.15	1.25	0.060	0.060	1.00	12.0-14.0	---	---	---	Se 0.15 min
Group 6	S43100	431	0.20	1.00	0.040	0.030	1.00	15.0-17.0	1.25-2.50	---	---	---
Precipitation-Hardening Alloy												
Group 7	S17400	630	0.07	1.00	0.040	0.030	1.00	15.0-17.5	3.00-5.00	3.00-5.00	---	(Cb + Ta) 0.15-0.45

a. It is the intent of ASTM F 593 that fasteners shall be ordered by alloy group numbers which include alloys considered to be chemically equivalent for general purpose use. However, ASTM F 593 does allow ordering by a specific alloy number (see ASTM F 593 paragraph 6.2 for more details). The alloy groupings are as follows: Group 1 - alloys 304, 304L, 305, 384, XM7 (when approved by the purchaser, alloys 303, 303Se, or XM1 may be furnished in place of alloy XM7); Group 2 - alloy 316, 316L; Group 3 - alloys 321, 347; Group 4 - alloy 430 (when approved by the purchaser, alloy 430F may be furnished in place of 430); Group 5 - alloy 410 (when approved by the purchaser, alloys 416 or 416SE may be furnished in place of 410); Group 6 - alloy 431; Group 7 - alloy 630.
b. At manufacturer's option, determined only when intentionally added.
Single values are maximums, unless otherwise specified.

ASTM F 1554 - CHEMICAL REQUIREMENTS FOR STEEL ANCHOR BOLTS

Grade 36	Diameter, in. (mm)	C	Mn	P	S	Cu[b]
Product Analysis	≤ ¾ (20)	0.29	a	0.05	0.06	0.18 min
Heat Analysis	≤ ¾ (20)	0.26	a	0.04	0.05	0.20 min
Product Analysis	> ¾ to 1½ (20 - 40), incl.	0.30	0.54-0.98	0.05	0.06	0.18 min
Heat Analysis	> ¾ to 1½ (20 - 40), incl.	0.27	0.60-0.90	0.04	0.05	0.20 min
Product Analysis	> 1½ to 4 (40 - 100), incl.	0.31	0.54-0.98	0.05	0.06	0.18 min
Heat Analysis	> 1½ to 4 (40 - 100), incl.	0.28	0.60-0.90	0.04	0.05	0.20 min
Grades 55 and 105						
Product Analysis	all sizes	---	---	0.048	0.058	0.18 min
Heat Analysis	all sizes	---	---	0.040	0.050	0.20 min

a. Optional with the manufacturer but shall be compatible with weldable steel.
b. When specified.
Single values are maximums unless otherwise specified.

SAE J995 - MECHANICAL AND MATERIAL REQUIREMENTS FOR STEEL NUTS

Nut Grade No.	C	Mn	P	S
2	0.47	---	0.12	0.15
5	0.55	0.30 min	0.05	0.15
8	0.55	0.30 min	0.04	0.05

Single values are maximums, unless otherwise specified.

CHEMICAL COMPOSITION OF STEEL NUTS FOR HIGH-PRESSURE & HIGH-TEMPERATURE SERVICE[e]								
ASTM A 194	AISI	C	Mn	Si	Cr	Ni	Mo	Others
1	carbon	0.15 min	1.00	0.40	---	---	---	---
2, 2HM, 2H	carbon	0.40 min	1.00	0.40	---	---	---	---
4	C-Mo	0.40-0.50	0.70-0.90	0.15-0.35	---	---	0.20-0.30	---
3	501	0.10 min	1.00	1.00	4.00-6.00	---	0.40-0.65	---
6	410	0.15	1.00	1.00	11.50-13.50	---	---	---
6F	416S[a]	0.15	1.25	1.00	12.00-14.00	---	---	---
6F	416Se[b]	0.15	1.25	1.00	12.00-14.00	---	---	---
7, 7M	4140[c]	0.37-0.49	0.65-1.10	0.15-0.35	0.75-1.20	---	0.15-0.25	---
8, 8A	304	0.08	2.00	1.00	18.00-20.00	8.00-10.50	---	---
8C, 8CA	347	0.08	2.00	1.00	17.00-19.00	9.00-13.00	---	Cb = 10 X C min
8M, 8MA	316	0.08	2.00	1.00	16.00-18.00	10.00-14.00	2.00-3.00	---
8T, 8TA	321	0.08	2.00	1.00	17.00-19.00	9.00-12.00	---	Ti = 5 X C min
8F, 8FA	303S[a]	0.15	2.00	1.00	17.00-19.00	8.00-10.00	---	---
8F, 8FA	303Se[b]	0.15	2.00	1.00	17.00-19.00	8.00-10.00	---	---
8P, 8PA	305[d]	0.08	2.00	1.00	17.00-19.00	10.50-13.00	---	---
8N, 8NA	304N	0.08	2.00	1.00	18.00-20.00	8.00-10.50	---	0.10-0.16 N
8LN, 8LNA	304N[d]	0.030	2.00	1.00	18.00-20.00	8.00-10.50	---	0.10-0.16 N
8MN, 8MNA	316N	0.08	2.00	1.00	16.00-18.00	10.00-14.00	2.00-3.00	0.10-0.16 N
8MLN, 8MLNA	316N[d]	0.030	2.00	1.00	16.00-18.00	10.00-14.00	2.00-3.00	0.10-0.16 N
8R, 8RA	XM19	0.06	4.00-6.00	1.00	20.50-23.50	11.50-13.50	1.50-3.00	0.10-0.30 Cb, 0.20-0.40 N, 0.10-0.30 V
8S, 8SA	---	0.10	7.00-9.00	3.50-4.50	16.00-18.00	8.00-9.00	---	0.08-0.18 N
8MLCuN, 8MLCuNA	---	0.020	1.00	0.80	19.50-20.50	17.50-18.50	6.00-6.50	0.18-0.22 N, 0.50-1.00 Cu
16	Cr-Mo-V	0.36-0.47	0.47-0.50	0.15-0.35	0.80-1.15	---	0.50-0.65	0.25-0.35 V, 0.15 Al

CHEMICAL COMPOSITION OF STEEL NUTS FOR HIGH-PRESSURE & HIGH-TEMPERATURE SERVICE[e] (Continued)

a. 0.15% sulfur minimum. b. 0.15% selenium minimum. c. Other similar AISI designations meeting grades 7, 7M are 4142, 4145, 4140H, 4142H, 4145H. d. With restricted carbon. e. Phosphorous and sulfur requirements are not listed due to space limitation; see ASTM A194 for details. Single values are maximums, unless otherwise specified.

ASTM A 563 - CHEMICAL REQUIREMENTS FOR CARBON STEEL NUTS[a]

Grade	C	Mn	P	S
O, A, B, C	0.58	---	0.13[b]	---
D[c]	0.58	0.27 min	0.048	0.058
DH[c]	0.18-0.58	0.57 min	0.048	0.058

a. Product analysis only; for heat analysis see ASTM A 563 Table 1. b. Acid bessemer steel only. c. For D and DH nuts, sulfur content may be 0.05 to 0.15% in which case manganese must be a minimum of 1.35%. Single values are maximums unless otherwise specified.

ASTM A 563 - CHEMICAL REQUIREMENTS FOR ALLOY STEEL NUTS[b]

Grade[a]	C	Mn	P	S	Si	Cu	Ni	Cr	V	Mo
Gr C3 Cl N	---	---	0.07-0.155	0.055	0.15-0.95	0.22-0.58	1.03	0.25-1.30	---	---
Gr C3 Cl A	0.31-0.42	0.86-1.24	0.045	0.055	0.13-0.37	0.22-0.48	0.22-0.48	0.42-0.68	---	---
Gr C3 Cl B	0.36-0.50	0.67-0.93	0.06-0.125	0.055	0.25-0.55	0.17-0.43	0.47-0.83	0.47-0.83	---	0.07
Gr C3 Cl C	0.14-0.26	0.76-1.39	0.040	0.045	0.13-0.37	0.17-0.53	0.22-0.53	0.27-0.53	0.010 min	---
Gr C3 Cl D	0.14-0.26	0.36-1.24	0.045	0.055	0.20-0.55	0.27-0.53	0.47-0.83	0.45-1.05	---	0.11
Gr C3 Cl E	0.18-0.27	0.56-1.04	0.045	0.045	0.13-0.37	0.27-0.63	0.27-0.63	0.55-0.95	---	---
Gr C3 Cl F	0.19-0.26	0.86-1.24	0.045	0.045	0.13-0.37	0.17-0.43	0.17-0.43	0.42-0.68	---	---
Gr DH3	0.19-0.55	0.37 min	0.052	0.055	---	0.17 min	0.17 min	0.42 min	---	0.14 min

a. Grade C3 nuts may be made of any of the listed material classes. Selection of the class shall be at the option of the manufacturer. b. Product analysis only; for heat analysis see ASTM A 563 Table 2. Single values are maximums unless otherwise specified.

F 594 CHEMICAL REQUIREMENTS FOR STAINLESS STEEL NUTS												
Alloy Group	UNS No.	Alloy	C	Mn	P	S	Si	Cr	Ni	Cu	Mo	Other
1	S30300	303	0.15	2.00	0.20	0.15 min	1.00	17.0-19.0	8.0-10.0	---	0.60 max[a]	---
1	S30323	303Se	0.15	2.00	0.20	0.060	1.00	17.0-19.0	8.0-10.0	---	---	Se 0.15 min
1	S30400	304	0.08	2.00	0.045	0.030	1.00	18.0-20.0	8.0-10.5	---	---	---
1	S30500	305	0.12	2.00	0.045	0.030	1.00	17.0-19.0	10.5-13.0	---	---	---
1	S38400	384	0.08	2.00	0.045	0.030	1.00	15.0-17.0	17.0-19.0	---	---	---
1	S20300	XM1	0.08	5.0-6.5	0.040	0.18-0.35	1.00	16.0-18.0	5.0-6.5	1.75-2.25	0.50 max[a]	---
1	S30430	XM7	0.10	2.00	0.045	0.030	1.00	17.0-19.0	8.0-10.0	3.00-4.00	---	---
2	S31600	316	0.08	2.00	0.045	0.030	1.00	16.0-18.0	10.0-14.0	---	2.00-3.00	---
3	S32100	321	0.08	2.00	0.045	0.030	1.00	17.0-19.0	9.0-12.0	---	---	Ti 5 x C min
3	S34700	347	0.08	2.00	0.045	0.030	1.00	17.0-19.0	9.0-13.0	---	---	Cb+Ta 10 x C min
Ferritic Alloys												
4	S43000	430	0.12	1.00	0.040	0.030	1.00	16.0-18.0	---	---	---	---
4	S43020	430F	0.12	1.25	0.060	0.15 min	1.00	16.0-18.0	---	---	0.60 max[a]	---
Martensitic Alloys												
5	S41000	410	0.15	1.00	0.040	0.030	1.00	11.5-13.5	---	---	---	---
5	S41600	416	0.15	1.25	0.060	0.15 min	1.00	12.0-14.0	---	---	0.60[a]	---
5	S41623	416Se	0.15	1.25	0.060	0.060	1.00	12.0-14.0	---	---	---	Se 0.15 min
6	S43100	431	0.20	1.00	0.040	0.030	1.00	15.0-17.0	1.25-2.50	---	---	---
Precipitation-Hardening Alloy												
7	S17400	630	0.07	1.00	0.040	0.030	1.00	15.0-17.5	3.00-5.00	3.00-5.00	---	Cb+Ta 0.15-0.45

a. At manufacturer's option, determined only when intentionally added.
Single values are maximums unless otherwise specified.

ASTM F 436 - CHEMICAL REQUIREMENTS FOR HARDENED STEEL WASHERS						
Type 1	P	S	Si	Cr	Ni	Cu
Product Analysis	0.050	0.060	---	---	---	---
Heat Analysis	0.040	0.050	---	---	---	---
Type 3						
Product Analysis	0.045	0.055	0.13-0.37	0.42-0.68	0.22-0.48	0.22-0.48
Heat Analysis	0.040	0.050	0.15-0.35	0.45-0.65	0.25-0.45	0.25-0.45

Single values are maximums.

MECHANICAL PROPERTIES OF CARBON STEEL BOLTING & OTHER FASTENERS

SAE J429 Grade	Products	Nominal Size Diameter in.	Full Size[a] Proof Load psi	Full Size[a] Tensile Strength ksi	Machine Specimens[b] Yield Strength ksi	Machine Specimens[b] Tensile Strength ksi	Machine Specimens[b] % El	Machine Specimens[b] % R.A.	Surface Hardness HR30N max	Core Hardness max	Grade Indentification Marking
1	B, Sc, St	¼ thru 1½	33	60	36	60	18	35	---	70-100 HRB	None
2	B, Sc, St	¼ thru ¾	55	74	57	74	18	35	---	80-100 HRB	None
2	B, Sc, St	>¾ thru 1½	33	60	36	60	18	35	---	70-100 HRB	None
4	St	¼ thru 1½	65	115	100	115	10	35	---	22-32 HRC	None
5	B, Sc, St	¼ thru 1	85	120	92	120	14	35	54	25-34 HRC	 (line marking)
5	B, Sc, St	>1 thru 1½	74	105	81	105	14	35	50	19-30 HRC	(line marking)
5.1	Se, B, Sc	No. 6 thru ⅝	85	120	---	---	---	---	59.5	25-40 HRC	 (line marking)
5.2	B, Sc	¼ thru 1	85	120	92	120	14	35	56	26-36 HRC	 (line marking)
7	B, Sc	¼ thru 1½	105	133	115	133	12	35	54	28-34 HRC	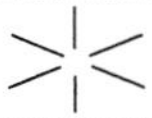 (line marking)
8	B, Sc, St	¼ thru 1½	120	150	130	150	12	35	58.6	33-39 HRC	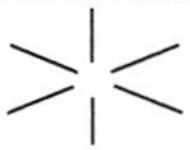 (line marking)
8.1	St	¼ thru 1½	120	150	130	150	10	35	---	32-38 HRC	None
8.2	B, Sc	¼ thru 1	120	150	130	150	10	35	58.6	33-39 HRC	 (line marking)

a. Full size bolts, screws, studs and screw & washer assemblies; for proof load requirements see SAE J429 Table 1.

b. Machine test specimens of bolts, screws and studs.

c. Other restrictions apply, see SAE J429 Table 1 footnotes for details.

B - Bolts; Sc - Screws; St - Studs; Se - Screw & Washer assembly. Single values are minimums, unless otherwise specified.

MECHANICAL PROPERTIES OF FERRITIC STEEL BOLTING FOR HIGH TEMPERATURE SERVICE

ASTM A 193 Grade	Material	Diameter in. (mm)	Tempering °F	Tempering °C	Tensile Strength ksi	Tensile Strength MPa	Yield Strength ksi	Yield Strength MPa	%El (4D)	%RA	Hardness max
B5	4-6 % Cr	up to 4 (100), incl	1100	593	100	690	80	550	16	50	---
B6	13 % Cr	up to 4 (100), incl	1100	593	110	760	85	585	15	50	---
B6X	13 % Cr	up to 4 (100), incl	1100	593	90	620	70	485	16	50	26 HRC
B7	Cr-Mo	2½ (65) and under	1100	593	125	860	105	720	16	50	321 HB or 35 HRC
		over 2½ to 4 (65 to 100)	1100	593	115	795	95	655	16	50	302 HB or 33 HRC
		over 4 to 7 (100 to 180)	1100	593	100	690	75	515	18	50	277 HB or 29 HRC
B7M[a]	Cr-Mo	2½ (65) and under	1150	620	100[a]	690	80[a]	550	18[a]	50[a]	235 HB or 99 HRB[a]
		2½ to 4 (65 to 100)	1150	620	100	690	80	550	18	50	235 HB or 99 HRB
		over 4 to 7 (100 to 180)	1150	620	100	690	75	515	18	50	235 HB or 99 HRB
B16	Cr-Mo-V	2½ (65) and under	1200	650	125	860	105	725	18	50	321 HB or 35 HRC
		over 2½ to 4 (65 to 100)	1200	650	110	760	95	655	17	45	302 HB or 33 HRC
		over 4 to 7 (100 to 180)	1200	650	100	690	85	586	16	45	277 HB or 29 HRC

a. To meet the tensile requirements, the Brinell hardness shall be over 200 HB (93HRB).
Single values are minimums, unless otherwise specified.

MECHANICAL PROPERTIES OF AUSTENITIC STEEL BOLTING FOR HIGH TEMPERATURE SERVICE

ASTM A 193 Class, Grade & Diameter	Tensile Strength ksi (MPa)		Yield Strength ksi (MPa)		%El (4D)	%R.A.	Hardness, max
Classes 1 & 1D: B8, B8C, B8M, B8P, B8T, B8LN, B8MLN, all diameters	75	515	30	205	30	50	223 HB[a] or 96 HRB
Class 1A: B8A, B8CA, B8MA, B8PA, B8TA, B8LNA, B8MLNA, B8NA, B8MNA, B8MLCuNA, all diameters	75	515	30	205	30	50	192 HB or 90 HRB
Classes 1B & 1D: B8N, B8MN, B8MLCuN, all diameters	80	550	35	240	30	40	223 HB[a] or 96 HRB
Classes 1C & 1D: B8R, all diameters	100	690	55	380	35	55	271 HB or 28 HRC
Class 1C: B8RA, all diameters	100	690	55	380	35	55	271 HB or 28 HRC
Classes 1C & 1D: B8S, all diameters	95	655	50	345	35	55	271 HB or 28 HRC
Class 1C: B8SA, all diameters	95	655	50	345	35	55	271 HB or 28 HRC
Class 2: B8, B8C, B8P, B8T, B8N							
¾ in. (20 mm) and under	125	860	100	690	12	35	321 HB or 35 HRC
over ¾ to 1 in. (20 to 25 mm) incl	115	795	80	550	15	35	321 HB or 35 HRC
over 1 to 1¼ in. (25 to 32 mm) incl	105	725	65	450	20	35	321 HB or 35 HRC
over 1¼ to 1½ in. (32 to 40 mm) incl	100	690	50	345	28	45	321 HB or 35 HRC
Class 2: B8M, B8MN, B8MLCuN							
¾ in. (20 mm) and under	110	760	96	665	15	45	321 HB or 35 HRC
over ¾ to 1 in. (20 to 25 mm) incl	100	960	80	550	20	45	321 HB or 35 HRC
over 1 to 1¼ in. (25 to 32 mm) incl	95	655	65	450	25	45	321 HB or 35 HRC
over 1¼ to 1½ in. (32 to 38 mm) incl	90	620	50	345	30	45	321 HB or 35 HRC
Class 2B: B8M2							
2 in. (51 mm) and under	95	655	75	515	25	40	321 HB or 35 HRC

MECHANICAL PROPERTIES OF AUSTENITIC STEEL BOLTING FOR HIGH TEMPERATURE SERVICE (Continued)

ASTM A 193 Class, Grade & Diameter	Tensile Strength ksi (MPa)		Yield Strength ksi (MPa)		%EI (4D)	%R.A.	Hardness, max
Class 2B: Grade B8M2							
over 2 to 2½ (51 to 63)	90	620	65	450	30	40	321 HB or 35 HRC
over 2½ to 3 (63 to 76)	80	550	55	380	30	40	321 HB or 35 HRC
Class 2C: B8M3							
2 (51) and under	85	585	65	450	30	60	321 HB or 35 HRC
over 2 (51)	85	585	60	415	30	60	321 HB or 35 HRC

a. For sizes ¾ in. (20 mm) in diameter and smaller, a maximum hardness of 241 HB (100 HRB) is permitted.
Single values are minimums, unless otherwise specified.

HARDNESS REQUIREMENTS FOR CARBON STEEL BOLTS AND STUDS

ASTM A 307Grade	Length, in. 0	Hardness[a]	
		Brinell (HB)	Rockwell B (HRB)
A	less than 3 x dia[b]	121-241	69-100
	3 x dia and longer	241 max	100 max
B	less than 3 x dia[b]	121-212	69-95
	3 x dia and longer	212 max	95 max
C	All	No hardness required	

a. As measured anywhere on the surface or through the cross section.
b. Also bolts with drilled or undersize heads. These sizes and bolts with modified heads shall meet the minimum and maximum hardness as hardness is the only requirement.

ASTM A 307 - TENSILE REQUIREMENTS FOR MACHINED SPECIMENS OF CARBON STEEL BOLTS AND STUDS[a]

Grade	Tensile Strength		Yield Point		% Elongation
	ksi	MPa	ksi	MPa	
A	60	415	---	---	18
B	60-100	415-690	---	---	18
C	58-80	400-550	36	50	23

a. Full-size tensile testing of these bolts and studs may be required, see ASTM A 307 Table 2 for details.
Single values are minimums.

ASTM A 325 - HARDNESS REQUIREMENTS FOR STRUCTURAL STEEL BOLTS[b]

Bolt Size, in.	Bolt Length, in.	Brinell (HB)	Rockwell C (HRC)
½ to 1, incl.	less than 3D[a]	253-319	25-34
	3D and over	319 max	34 max
1⅛ to 1½, incl.	less than 3D[a]	223-286	19-30
	3D and over	286 max	30 max

a. Bolts having a length less than 3 times the diameter are subject only to minimum/maximum hardness. Such lengths cannot be reasonably tensile tested.
     D = Nominal diameter or thread size.
b. All bolt types and classes.

ASTM A 354 - HARDNESS REQUIREMENTS FOR FULL-SIZE ALLOY STEEL BOLTS, STUDS AND FASTENERS

Grade	Size, in.	Hardness		
		Brinell (HB)	Rockwell C (HRC)	
BC	¼ to 2½	255-331	26-36	
BC	> 2½	235-311	22-33	
BD	¼ to 2½	311-363	33-39	
BD	> 2½	293-363	31-39	

ASTM A 354 - MECHANICAL REQUIREMENTS FOR MACHINED SPECIMENS OF ALLOY STEEL BOLTS, STUDS AND FASTENERS[a]

Grade	Size	Tensile Strength		Yield Strength		% Elongation	% Reduction of Area
		psi	MPa	psi	MPa		
BC	¼ to 2½ incl.	125,000	862	109,000	752	16	50
BC	> 2½	115,000	793	99,000	683	16	45
BD	¼ to 2½ incl.	150,000	1034	130,000	896	14	40
BD	> 2½	140,000	965	115,000	793	14	40

a. Full-size tensile testing of these fasteners may be required, see ASTM A 307 Tables 3a and 3b for details.
Single values are minimums.

ASTM A 449 - HARDNESS REQUIREMENTS FOR BOLTS AND STUDS

Bolt or Stud Diameter, in.	Hardness	
	Brinell (HB)	Rockwell C (HRC)
¼ to 1, incl.	255-321	25-34
> 1 to 1½, incl.	223-285	19-30
> 1½ to 3, incl.	183-235	---

ASTM A 449 - TENSILE REQUIREMENTS FOR SPECIMENS MACHINED OF BOLTS AND STUDS[a]

Bolt or Stud Diameter, in.	Tensile Strength		Yield Strength		% Elongation in 4D	% Reduction of Area
	psi	MPa	psi	MPa		
¼ to 1, incl.	120,000	825	92,000	635	14	35
> 1 to 1½, incl.	105,000	725	81,000	560	14	35
> 1½ to 3, incl.	90,000	620	58,000	400	14	35

a. Full-size tensile testing of these bolts and studs may be required, see ASTM A 449 Tables 3 and 4 for details.
Single values are minimums.

ASTM A 489 - TENSILE PROPERTIES FOR MACHINED SPECIMENS OF CARBON STEEL LIFTING EYES

Types	Tensile Strength		Yield Point		% Elongation 4D	% Reduction of Area
	psi	MPa	psi	MPa		
1 & 2	65,000-90,000	448-620	30,000	207	30	60

Single values are minimums.

ASTM A 489 - IMPACT PROPERTIES FOR CARBON STEEL LIFTING EYES

Types	Impact Strength[a]	Minimum Single Value[b]	Test Temperature
1 & 2	35 ft•lbf (47 J)	23 ft•lbf (31 J)	32°F (0°C)

a. Minimum average value of three specimens tested (Charpy V-notch).
b. Only one test value from the three specimens is allowed to fall below the minimum average value.

ASTM A 490 - HARDNESS REQUIREMENTS FOR STRUCTURAL STEEL BOLTS

Nominal Size ½ to 1½ in.	Length, in.	Brinell (HB)	Rockwell C (HRC)
Types 1, 2, & 3	< 3 x dia.	311-352	33-38
Types 1, 2, & 3	≥ 3 x dia.	352 max	38 max

ASTM A 490 - TENSILE PROPERTIES FOR MACHINED SPECIMENS OF STRUCTURAL STEEL BOLTS[a]

Size, in.	Tensile Strength, psi	Yield Strength psi	%Elongation	%Reduction of Area
½ to 1½, incl.	150,000-170,000	130,000	14	40

a. Full-size tensile testing of these bolts may be required, see ASTM A 490 Table 5 for details.
Single values are minimums.

ASTM A 687 - TENSILE REQUIREMENTS FOR HIGH-STRENGTH NONHEADED STEEL BOLTS AND STUDS

Tensile Strength		Yield Strength		% El	% R A	Impact Strength		Test Temperature
ksi	MPa	ksi	MPa			Min Avg[a]	Min Single Value[b]	
150 max	1034 max	105	724	15	45	15 ft•lbf (20 J)	12 ft•lbf (16 J)	-20°F (-29°C)

a. Minimum average value of three specimens tested (Charpy V-notch).
b. Minimum value for one test specimen.
Single values are minimums unless otherwise specified.

ASTM F 1554 - TENSILE PROPERTIES FOR BARS AND MACHINED SPECIMENS OF STEEL ANCHOR BOLTS

Grade	Tensile Strength		Yield Strength		% Elongation		% Reduction of Area	
	ksi	MPa	ksi	MPa	8 in. (200 mm)[a]	2 in. (50 mm)[a]		
36	58-80	400-552	36	248	20	23	40[b]	40[c]	40[e]
55	75-95	517-655	55	380	18	21	30[b]	22[c]	18[e]
105	125-150	862-1034	105	724	12	15	45[b]	45[c]	---

a. Elongation in 8 in. (200 mm) applies to bars. Elongation in 2 in. (50 mm) applies to tests on machined specimens.
b. ¼ to 2 in. (6.4 - 50 mm), inclusive.
c. > 2 to 2½ in. (50 - 63 mm), inclusive.
d. > 2½ to 3 in. (63 - 76), inclusive.
e. > 3 to 4 in. (76 - 102 mm), inclusive.
Single values are minimums.

ASTM F 593 - MECHANICAL PROPERTY REQUIREMENTS FOR STAINLESS STEEL BOLTS, HEX CAP SCREWS, AND STUDS

Stainless Alloy Group	Cond[a]	Alloy/Mech Property Marking	Nominal Diameter, in.	Full-Size Tests			Machined Specimen Tests		
				Tensile Strength ksi[c]	Yield Strength ksi[b,c]	Rockwell Hardness	Tensile Strength ksi[c]	Yield Strength ksi[b,c]	% Elongation in 4D
Austenitic Alloys									
Group 1: 303, 304, 304L, 305, 384, XM1, XM7, 303Se	AF	F593A	¼ to 1½, incl.	85 max	---	85 HRB max	80 max	50 max	40
	A	F593B	¼ to 1½, incl.	75 to 100	30	65-95 HRB	70	30	30
	CW1	F593C	¼ to ⅝, incl.	100 to 150	65	95 HRB-32 HRC	95	60	20
	CW2	F593D	¾ to 1½, incl.	85 to 140	45	80 HRB-32 HRC	80	40	25
	SH1	F593A	¼ to ⅝, incl.	120 to 160	95	24-36 HRC	115	90	12
	SH2	F593B	¾ to 1, incl.	110 to 150	75	20-32 HRC	105	70	15
	SH3	F593C	1⅛ to 1¼, incl.	100 to 140	60	95 HRB-30 HRC	95	55	20
	SH4	F593D	1⅜ to 1½, incl.	95 to 130	45	90 HRB-28 HRC	90	40	28
Group 2: 316, 316L	AF	F593E	¼ to 1½, incl.	85 max	---	85 HRB max	80 max	50 max	40
	A	F593F	¼ to 1½, incl.	75 to 100	30	65-95 HRB	70	30	30
	CW1	F593G	¼ to ⅝, incl.	100 to 150	65	95 HRB-32 HRC	95	60	20
	CW2	F593H	¾ to 1½, incl.	85 to 140	45	80 HRB-32 HRC	80	40	25
	SH1	F593E	¼ to ⅝, incl.	120 to 160	95	24-36 HRC	115	90	12
	SH2	F593F	¾ to 1, incl.	110 to 150	75	20-32 HRC	105	70	15
	SH3	F593G	1⅛ to 1¼, incl.	100 to 140	60	95 HRB-30 HRC	95	55	20
	SH4	F593H	1⅜ to 1½, incl.	95 to 130	45	90 HRB-28 HRC	90	40	28
Group 3: 321, 347	AF	F593J	¼ to 1½, incl.	85 max	---	85 HRB max	80 max	50 max	40
	A	F593K	¼ to 1½, incl.	75 to 100	30	65-95 HRB	70	30	30
	CW1	F593L	¼ to ⅝, incl.	100 to 150	65	95 HRB-32 HRC	95	60	20
	CW2	F593M	¾ to 1½, incl	85 to 140	45	80 HRB-32 HRC	80	40	25
	SH1	F593J	¼ to ⅝, incl.	120 to 160	95	24-36 HRC	115	90	12
	SH2	F593K	¾ to 1, incl.	110 to 150	75	20-32 HRC	105	70	15

ASTM F 593 - MECHANICAL PROPERTY REQUIREMENTS FOR STAINLESS STEEL BOLTS, HEX CAP SCREWS, AND STUDS (Continued)

Stainless Alloy Group	Cond[a]	Alloy/Mech Property Marking	Nominal Diameter, in.	Full-Size Tests			Machined Specimen Tests		
				Tensile Strength ksi[c]	Yield Strength ksi[b,c]	Rockwell Hardness	Tensile Strength ksi[c]	Yield Strength ksi[b,c]	% Elongation in 4D
Austenitic Alloys (Continued)									
Group 3: 321, 347	SH3	F593L	1⅛ to 1¼, incl.	100 to 140	60	95 HRB-30 HRC	95	55	20
	SH4	F593M	1⅜ to 1½, incl	95 to 130	45	90 HRB-28 HRC	90	40	28
Ferritic Alloys									
Group 4: 430, 430F	A	F593N	¼ to 1½, incl.	70 to 100	35	65-95 HRB	70	35	25
	CW1	F593V	¼ to ⅝, incl.	75 to 105	40	75-98 HRB	70	30	20
	CW2	F593W	¾ to 1½, incl.	70 to 100	35	65-95 HRB	70	30	20
Martensitic Alloys									
Group 5: 410, 416, 416Se	H	F593P	¼ to 1½, incl.	110 to 140	90	20-30 HRC	110	90	18
	HT	F593R	¼ to 1½, incl.	160 to 190	120	34-45 HRC	160	120	12
Group 6: 431	H	F593S	¼ to 1½, incl.	125 to 150	100	25-32 HRC	125	100	15
	HT	F593T	¼ to 1½, incl.	180 to 220	140	40-48 HRC	180	140	10
Precipitation Hardening Alloys									
Group 7: 630	AH	F593U	¼ to 1½, incl.	135 to 170	105	28-38 HRC	135	105	16

a. A - Machined from annealed or solution annealed stock thus retaining the properties of the original material, or hot formed and solution-annealed.
AF - Headed and rolled from annealed stock and then reannealed.
AH - Solution annealed and age-hardened after forming.
CW - Headed and rolled from annealed stock thus acquiring a degree of cold work; sizes 0.75 in. and larger may be hot worked and solution-annealed.
H - Hardened and tempered at 1050°F (565°C) minimum.
HT - Hardened and tempered at 525°F (274°C) minimum.
SH - Machined from strain hardened stock or cold-worked to develop the specified properties.
b. Yield strength is the stress at which an offset of 0.2% gage length occurs.
c. The yield and tensile strength values for full-size products shall be computed by dividing the yield and maximum tensile load values by the stress area for the product size and thread series determined in accordance with Test Methods F 606 Table 4. Single values are minimums unless otherwise specified.

SAE J995 - PROOF LOAD AND HARDNESS REQUIREMENTS FOR NUTS

Nut Grade	Nut Size Diameter, in.	Proof Load Stress[a], ksi		Hardness, HRC
		UNC 8 UN	UNF, 12 UN and Finer	
2	¼ thru 1½	90	90	32 max
5	¼ thru 1	120	109	32 max
	Over 1 thru 1½	105	94	32 max
8	¼ thru ⅝	150	150	24-32
	Over ⅝ thru 1	150	150	26-34
	Over 1 thru 1½	150	150	26-36

a. The proof load in pounds for a nut is computed by multiplying the proof load stress for the nut grade, size, and thread series, as described in SAE J995 Table 2, and the stress area for the applicable size and thread series as described in SAE J995 Table 3 (see SAE J995 Appendix Table 6 for computed values for some products).

ASTM A 194 - HARDNESS REQUIREMENTS OF STEEL NUTS FOR HIGH-PRESSURE & HIGH-TEMPERATURE SERVICE

Grade	Completed Nuts			Sample Nuts[a]	
	HB	HRC	HRB	HB	HRB
1	121	---	70	121	70
2	159-352	---	84	159	84
2H	248-352	24-38	---	179	89
2H to 1½" (38.1 mm) incl.	248-352	24-38	---	179	89
2H over 1½" (38.1 mm)	212-352	38 max	95	147	79
2HM, 7M	159-237	22 max	---	159	84
3, 4, 7, and 16	248-352	24-38	---	201	94
6, 6F	228-271	20-28	---	---	---
8, 8C, 8M, 8T, 8F, 8P, 8N	126-300	---	60-105	---	---
8MN, 8LN, 8MLN, 8MLCuN	126-300	---	60-105	---	---
8A, 8CA, 8MA, 8TA	126-192	---	60-90	---	---
8FA, 8PA, 8NA, 8MNA	126-192	---	60-90	---	---
8LNA, 8MLNA, 8MLCuN	126-192	---	60-90	---	---
8R, 8RA, 8S, 8SA	183-271	88 HRB-25 HRC	---	---	---

a. Sample nuts require a specific heat treatment prior to testing, see ASTM A194. Single values are minimum, unless otherwise specified.

ASTM A 563 - HARDNESS REQUIREMENTS OF CARBON AND ALLOY STEEL NUTS WITH UNC, 8 UN, 6 UN AND COARSER PITCH THREADS

Grade of Nut	Nominal Nut Size, in.	Style of Nut	Hardness	
			Brinell (HB)	Rockwell
O	¼ to 1½	square	103-302	55 HRB-32 HRC
A	¼ to 1½	square	116-302	68 HRB-32 HRC
O	¼ to 1½	hex	103-302	55 HRB-32 HRC
A	¼ to 1½	hex	116-302	68 HRB-32 HRC
B	¼ to 1	hex	121-302	69 HRB-32 HRC

ASTM A 563 - HARDNESS REQUIREMENTS OF CARBON AND ALLOY STEEL NUTS WITH UNC, 8 UN, 6 UN AND COARSER PITCH THREADS (Continued)

Grade of Nut	Nominal Nut Size, in.	Style of Nut	Hardness	
			Brinell (HB)	Rockwell
B	1⅛ to 1½	hex	121-302	69 HRB-32 HRC
Dᵃ	¼ to 1½	hex	159-352	84 HRB-38 HRC
DHᵇ	¼ to 1½	hex	248-352	24 HRC-38 HRC
DH3	½ to 1	hex	248-352	24 HRC-38 HRC
A	¼ to 4	heavy hex	116-302	68 HRB-32 HRC
B	¼ to 1	heavy hex	121-302	69 HRB-32 HRC
B	1⅛ to 1½	heavy hex	121-302	69 HRB-32 HRC
Cᵃ	¼ to 4	heavy hex	143-352	78 HRB-38 HRC
C3	¼ to 4	heavy hex	143-352	78 HRB-38 HRC
Dᵃ	¼ to 4	heavy hex	159-352	84 HRB-38 HRC
DHᵇ	¼ to 4	heavy hex	248-352	24 HRC-38 HRC
DH3	¼ to 4	heavy hex	248-352	24 HRC-38 HRC
A	¼ to 1½	hex thick	116-302	68 HRB-32 HRC
B	¼ to 1	hex thick	121-302	69 HRB-32 HRC
B	1⅛ to 1½	hex thick	121-302	69 HRB-32 HRC
Dᵃ	¼ to 1½	hex thick	159-352	84 HRB-38 HRC
DHᵇ	¼ to 1½	hex thick	248-352	24 HRC-38 HRC

a. Nuts made in accordance to the requirements of Specification A 194/A 194M, Grade 2 or Grade 2H, and marked with their grade symbol are acceptable equivalents for Grades C and D nuts. When A 194 zinc-coated inch series nuts are supplied, the zinc coating, overtapping, lubrication and rotational capacity testing shall be in accordance with Specification A 563.

b. Nuts made in accordance with the requirements of Specification A 194/A 194 M, Grade 2 or Grade 2H, and marked with its grade symbol are an acceptable equivalent for Grade DH nuts. When A 194 zinc-coated inch series nuts are supplied, the zinc coating, overtapping, lubrication and rotational capacity testing shall be in accordance with Specification A 563.

ASTM A 563 - HARDNESS REQUIREMENTS OF CARBON STEEL NUTS WITH UNF, 12 UN, AND FINER PITCH THREADS

Grade of Nut	Nominal Nut Size, in.	Style of Nut	Hardness	
			Brinell (HB)	Rockwell
O	¼ to 1½	hex	103-302	55 HRB-32 HRC
A	¼ to 1½	hex	116-302	68 HRB-32 HRC
B	¼ to 1	hex	121-302	69 HRB-32 HRC
B	1⅛ to 1½	hex	121-302	69 HRB-32 HRC
Dª	¼ to 1½	hex	159-352	84 HRB-38 HRC
DHᵇ	¼ to 1½	hex	248-352	24 HRC-38 HRC
A	¼ to 4	heavy hex	116-302	68 HRB-32 HRC
B	¼ to 1	heavy hex	121-302	69 HRB-32 HRC
B	1⅛ to 1½	heavy hex	121-302	69 HRB-32 HRC
Dª	¼ to 4	heavy hex	159-352	84 HRB-38 HRC
DHᵇ	¼ to 4	heavy hex	248-352	24 HRC-38 HRC
A	¼ to 1½	hex thick	116-302	68 HRB-32 HRC
B	¼ to 1	hex thick	121-302	69 HRB-32 HRC
B	1⅛ to 1½	hex thick	121-302	69 HRB-32 HRC
Dª	¼ to 1½	hex thick	159-352	84 HRB-38 HRC
DHᵇ	¼ to 1½	hex thick	248-352	24 HRC-38 HRC

a. Nuts made in accordance to the requirements of Specification A 194/A 194M, Grade 2 or Grade 2H, and marked with their grade symbol are acceptable equivalents for Grades C and D nuts. When A 194 zinc-coated inch series nuts are supplied, the zinc coating, overtapping, lubrication and rotational capacity testing shall be in accordance with Specification A 563.

b. Nuts made in accordance with the requirements of Specification A 194/A 194 M, Grade 2 or Grade 2H, and marked with its grade symbol are an acceptable equivalent for Grade DH nuts. When A 194 zinc-coated inch series nuts are supplied, the zinc coating, overtapping, lubrication and rotational capacity testing shall be in accordance with Specification A 563.

ASTM F 594 - MECHANICAL PROPERTY REQUIREMENTS FOR STAINLESS STEEL NUTS

Stainless Alloy Group	Condition[a]	Alloy/Mechanical Property Marking	Nominal Diameter, in.	Proof Stress, ksi, min	Rockwell Hardness
Austenitic Alloys					
Group 1: 303, 304, 305, 384, XM1, XM7, 303Se	AF	F594A	¼ to 1½, incl.	70	85 HRB max
	A	F594B	¼ to 1½, incl.	75	65 to 95 HRB, incl.
	CW1	F594C	¼ to ⅝, incl.	100	95 HRB-32 HRC, incl.
	CW2	F594D	¾ to 1½, incl.	85	80 HRB-32 HRC, incl.
	SH1	F594A	¼ to ⅝, incl.	120	24-36 HRC, incl.
	SH2	F594B	¾ to 1, incl.	110	20-32 HRC, incl.
	SH3	F594C	1⅛ to 1¼, incl.	100	95 HRB-30 HRC, incl.
	SH4	F594D	1⅜ to 1½, incl.	85	90 HRB-28 HRC, incl.
Group 2: 316	AF	F594E	¼ to 1½, incl.	70	85 HRB max
	A	F594F	¼ to 1½, incl.	75	65-95 HRB, incl.
	CW1	F594G	¼ to ⅝, incl.	100	95 HRB-32 HRC, incl.
	CW2	F594H	¾ to 1½, incl.	85	80 HRB-32 HRC, incl.
	SH1	F594E	¼ to ⅝, incl.	120	24-36 HRC, incl.
	SH2	F594F	¾ to 1, incl.	110	20-32 HRC, incl.
	SH3	F594G	1⅛ to 1¼, incl.	100	95 HRB-30 HRC, incl.
	SH4	F594H	1⅜ to 1½, incl.	85	90 HRB-28 HRC, incl.
Group 3: 321, 347	AF	F594J	¼ to 1½, incl.	70	85 HRB max
	A	F594K	¼ to 1½, incl.	75	65-95 HRB, incl.
	CW1	F594L	¼ to ⅝, incl.	100	95 HRB-32 HRC, incl.
	CW2	F594M	¾ to 1½, incl.	85	80 HRB-32 HRC, incl.
	SH1	F594J	¼ to ⅝, incl.	120	24-36 HRC, incl.
	SH2	F594K	¾ to 1, incl.	110	20-32, HRC incl.
	SH3	F594L	1⅛ to 1¼, incl.	100	95 HRB-30 HRC, incl.

ASTM F 594 - MECHANICAL PROPERTY REQUIREMENTS FOR STAINLESS STEEL NUTS (Continued)

Stainless Alloy Group	Condition[a]	Alloy/Mechanical Property Marking	Nominal Diameter, in.	Proof Stress, ksi, min	Rockwell Hardness
Austenitic Alloys (continued)					
Group 3: 321, 347	SH4	F594M	1⅜ to 1½, incl.	85	90 HRB-28 HRC, incl.
Ferritic Alloys					
Group 4: 430, 430F	A	F594N	¼ to 1½, incl.	70	65-95 HRB, incl.
Martensitic Alloys					
Group 5: 410, 416,	H	F594P	¼ to 1½, incl.	100	20-30 HRC, incl.
416Se	HT	F594R	¼ to 1½, incl.	160	34-45 HRC, incl.
Group 6: 431	H	F594S	¼ to 1½, incl.	125	25-32 HRC, incl.
	HT	F594T	¼ to 1½, incl.	180	40-48 HRC, incl.
Precipitation Hardening Alloys					
Group 7: 630	AH	F594U	¼ to 1½, incl.	135	28-38 HRC, incl.

a. A - Machined from annealed or solution annealed stock thus retaining the properties of the original material, or hot formed and solution annealed. AF - Annealed after all threading is completed. AH - Solution annealed and age hardened after forming. CW - Annealed and cold worked. Sizes 0.75 in. and larger may be hot worked. H - Hardened and tempered at 1050°F (565°C) minimum. HT - Hardened and tempered at 525°F (274°C) minimum. SH - Machined from strain hardened stock. Single values are minimums, unless otherwise specified.

ASTM F436 - HARDNESS REQUIREMENTS FOR HARDENED STEEL WASHERS

Heat Treatment	Minimum Depth of Hardness	Hardness Rockwell	
		Non-Coated	Zinc-Coated[a]
Through hardened	Through thickness	38-45 HRC	26-45 HRC
Carburized	0.015 in. min	69-73 HRA[b] or 79-83 HR15N[b]	63-73 HRA or 73-83 HR15N
Carburized and hardened	Core	30 HRC min or 65 HRA min	---

a. Zinc-coated by the hot-dip process. b. Surface hardness.

Chapter

17

TOOL STEELS:

AMERICAN SPECIFICATION TITLES & DESIGNATIONS, CHEMICAL COMPOSITIONS, MECHANICAL PROPERTIES & HEAT TREATMENT DATA

Tool Steels

ASTM Spec.	Title
A 561	Macroetch Testing of Tool Steel Bars
A 597	Cast Tool Steel
A 600	Tool Steel High Speed
A 681	Tool Steels Alloy
A 686	Tool Steel, Carbon

CHEMICAL COMPOSITION OF TOOL STEELS[a]

AISI	UNS	C	Mn	Cr	Mo	W	V	Co	Other
Molybdenun High-Speed Tool Steels									
M1	T11301	0.78-0.88	0.15-0.40	3.50-4.00	8.20-9.20	1.40-2.10	1.00-1.35	---	---
M2 (reg. C)	T11302	0.78-0.88	0.15-0.40	3.75-4.50	4.50-5.50	5.50-6.75	1.75-2.20	---	---
M2 (high C)	T11302	0.95-1.05	0.15-0.40	3.75-4.50	4.50-5.50	5.00-6.75	1.75-2.20	---	---
M3 Cl 1	T11313	1.00-1.10	0.15-0.40	3.75-4.50	4.75-6.50	5.00-6.75	2.25-2.75	---	---
M3 Cl 2	T11323	1.15-1.25	0.15-0.40	3.75-4.50	4.75-6.50	5.00-6.75	2.75-3.25	---	---
M4	T11304	1.25-1.40	0.15-0.40	3.75-4.75	4.25-5.50	5.25-6.50	3.75-4.50	---	---
M6	T11306	0.75-0.85	0.15-0.40	3.75-4.50	4.50-5.50	3.75-4.75	1.30-1.70	11.00-13.00	---
M7	T11307	0.97-1.05	0.15-0.40	3.50-4.00	8.20-9.20	1.40-2.10	1.75-2.25	---	---
M10 (reg. C)	T11310	0.84-0.94	0.10-0.40	3.75-4.50	7.75-8.50	---	1.80-2.20	---	---
M10 (high C)	T11310	0.95-1.05	0.10-0.40	3.75-4.50	7.75-8.50	---	1.80-2.20	---	---
M30	T11330	0.75-0.85	0.15-0.40	3.50-4.25	7.75-9.00	1.30-2.30	1.00-1.40	4.50-5.50	---
M33	T11333	0.85-0.92	0.15-0.40	3.50-4.00	9.00-10.00	1.30-2.10	1.00-1.35	7.75-8.75	---
M34	T11334	0.85-0.92	0.15-0.40	3.50-4.00	7.75-9.20	1.40-2.10	1.90-2.30	7.75-8.75	---
M36	T11336	0.80-0.90	0.15-0.40	3.75-4.50	4.50-5.50	5.50-6.50	1.75-2.25	7.75-8.75	---
M41	T11341	1.05-1.15	0.20-0.60	3.75-4.50	3.25-4.25	6.25-7.00	1.75-2.25	4.75-5.75	---

CHEMICAL COMPOSITION OF TOOL STEELS^a (Continued)

AISI	UNS	C	Mn	Cr	Mo	W	V	Co	Other
Molybdenun High-Speed Tool Steels (Continued)									
M42	T11342	1.05-1.15	0.15-0.40	3.50-4.25	9.00-10.00	1.15-1.85	0.95-1.35	7.75-8.75	---
M43	T11343	1.15-1.25	0.20-0.40	3.50-4.25	7.50-8.50	2.25-3.00	1.50-1.75	7.75-8.75	---
M44	T11344	1.10-1.20	0.20-0.40	4.00-4.75	6.00-7.00	5.00-5.75	1.85-2.20	11.00-12.25	---
M46	T11346	1.22-1.30	0.20-0.40	3.70-4.20	8.00-8.50	1.90-2.20	3.00-3.30	7.80-8.80	---
M47	T11347	1.05-1.15	0.15-0.40	3.50-4.00	9.25-10.00	1.30-1.80	1.15-1.35	4.75-5.25	---
M48	T11348	1.42-1.52	0.15-0.40	3.50-4.00	4.75-5.50	9.50-10.50	2.75-3.25	8.00-10.00	---
M62	T11362	1.25-1.35	0.15-0.40	3.50-4.00	10.00-11.00	5.75-6.50	1.80-2.10	---	---
Intermediate High-Speed Tool Steels									
M50	T11350	0.78-0.88	0.15-0.45	3.75-4.50	3.90-4.75	---	0.80-1.25	---	---
M52	T11352	0.85-0.95	0.15-0.45	3.50-4.30	4.00-4.90	0.75-1.50	1.65-2.25	---	---
Tungsten High-Speed Tool Steels									
T1	T12001	0.65-0.80	0.10-0.40	3.75-4.50	---	17.25-18.75	0.90-1.30	---	---
T2	T12002	0.80-0.90	0.20-0.40	3.75-4.50	1.00	17.50-19.00	1.80-2.40	---	---
T4	T12004	0.70-0.80	0.10-0.40	3.75-4.50	0.40-1.00	17.50-19.00	0.80-1.20	4.25-5.75	---
T5	T12005	0.75-0.85	0.20-0.40	3.75-5.00	0.50-1.25	17.50-19.00	1.80-2.40	7.00-9.50	---
T6	T12006	0.75-0.85	0.20-0.40	4.00-4.75	0.40-1.00	18.50-21.00	1.50-2.10	11.00-13.00	---
T8	T12008	0.75-0.85	0.20-0.40	3.75-4.50	0.40-1.00	13.25-14.75	1.80-2.40	4.25-5.75	---
T15	T12015	1.50-1.60	0.15-0.40	3.75-5.00	1.00	11.75-13.00	4.50-5.25	4.75-5.25	---
M50	T11350	0.78-0.88	0.15-0.45	3.75-4.50	3.90-4.75	---	0.80-1.25	---	---
M52	T11352	0.85-0.95	0.15-0.45	3.50-4.30	4.00-4.90	0.75-1.50	1.65-2.25	---	---
Chromium Hot-Work Tool Steels									
H10	T20810	0.35-0.45	0.25-0.70	3.00-3.75	2.00-3.00	---	0.25-0.75	---	---
H11	T20811	0.33-0.43	0.20-0.50	4.75-5.50	1.10-1.60	---	0.30-0.60	---	---
H12	T20812	0.30-0.40	0.20-0.50	4.75-5.50	1.25-1.75	1.00-1.70	0.50	---	---
H13	T20813	0.32-0.45	0.20-0.50^b	4.75-5.50	1.10-1.75	---	0.80-1.20	---	---

CHEMICAL COMPOSITION OF TOOL STEELS[a] (Continued)									
AISI	UNS	C	Mn	Cr	Mo	W	V	Co	Other
Chromium Hot-Work Tool Steels (Continued)									
H14	T20814	0.35-0.45	0.20-0.50	4.75-5.50	---	4.00-5.25	---	---	---
H19	T20819	0.32-0.45	0.20-0.50	4.00-4.75	0.30-0.55	3.75-4.50	1.75-2.20	4.00-4.50	---
Tungsten Hot-Work Tool Steels									
H21	T20821	0.26-0.36	0.15-0.40	3.00-3.75	---	8.50-10.00	0.30-0.60	---	---
H22	T20822	0.30-0.40	0.15-0.40	1.75-3.75	---	10.00-11.75	0.25-0.50	---	---
H23	T20823	0.25-0.35	0.15-0.40	11.00-12.75	---	11.00-12.75	0.75-1.25	---	---
H24	T20824	0.42-0.53	0.15-0.40	2.50-3.50	---	14.00-16.00	0.40-0.60	---	---
H25	T20825	0.22-0.32	0.15-0.40	3.75-4.50	---	14.00-16.00	0.40-0.60	---	---
H26	T20826	0.45-0.55[C]	0.15-0.40	3.75-4.50	---	17.25-19.00	0.75-1.25	---[c]	---
Molybdenum Hot-Work Tool Steels									
H41	T20841	0.60-0.75[C]	0.15-0.40	3.50-4.00	8.20-9.20	1.40-2.10	1.00-1.30	---	---
H42	T20842	0.55-0.70[C]	0.15-0.40	3.75-4.50	4.50-5.50	5.50-6.75	1.75-2.20	---	---
H43	T20843	0.50-0.65[C]	0.15-0.40	3.75-4.50	7.75-8.50	---	1.80-2.20	---	---
Air-Hardening, Medium Alloy, Cold-Work Tool Steels									
A2	T30102	0.95-1.05	1.00	4.75-5.50	0.90-1.40	---	0.15-0.50	---	---
A3	T30103	1.20-1.30	0.40-0.60	4.75-5.50	0.90-1.40	---	0.80-1.40	---	---
A4	T30104	0.95-1.05	1.80-2.20	0.90-2.20	0.90-1.40	---	---	---	---
A6	T30106	0.65-0.75	1.80-2.50	0.90-1.20	0.90-1.40	---	---	---	---
A7	T30107	2.00-2.85	0.80	5.00-5.75	0.90-1.40	0.50-1.50	3.90-5.15	---	---
A8	T30108	0.50-0.60	0.50	4.75-5.50	1.15-1.65	1.00-1.50	---	---	---
A9	T30109	0.45-0.55	0.50	4.75-5.50	1.30-1.80	---	0.80-1.40	---	1.25-1.75 Ni
A10	T30110	1.25-1.50	1.60-2.10	---	1.25-1.75	---	---	---	1.55-2.05 Ni
High-Carbon, High-Chromium, Cold-Work Tool Steels									
D2	T30402	1.40-1.60	0.60	11.00-13.00	0.70-1.20	---	1.10	1.00	---
D3	T30403	2.00-2.35	0.60	11.00-13.50	---	1.00	1.00	---	---

CHEMICAL COMPOSITION OF TOOL STEELS^a (Continued)									
AISI	UNS	C	Mn	Cr	Mo	W	V	Co	Other
High-Carbon, High-Chromium, Cold-Work Tool Steels (Continued)									
D4	T30404	2.05-2.40	0.60	11.00-13.00	0.70-1.20	---	1.00	---	---
D5	T30405	1.40-1.60	0.60	11.00-13.00	0.70-1.20	---	1.00	2.50-3.50	---
D7	T30407	2.15-2.50	0.60	11.50-13.50	0.70-1.20	---	3.80-4.40	---	---
Oil-Hardening Cold-Work Tool Steels									
O1	T31501	0.85-1.00	1.00-1.40	0.40-0.60	---	0.40-0.60	0.30	---	---
O2	T31502	0.85-0.95	1.40-1.80	0.50	0.30	---	0.30	---	---
O6	T31506	1.25-1.55	0.30-1.10	0.30	0.20-0.30	---	---	---	---
O7	T31507	1.10-1.30	1.00	0.35-0.85	0.30	1.00-2.00	0.40	---	---
Shock-Resisting Tool Steels									
S1	T41901	0.40-0.55	0.10-0.40	1.00-1.80	0.50	1.50-3.00	0.15-0.30	---	---
S2	T41902	0.40-0.55	0.30-0.50	---	0.30-0.60	---	0.50	---	---
S4	T41904	0.50-0.65	0.60-0.95	0.10-0.50	---	---	0.15-0.35	---	---
S5	T41905	0.50-0.65	0.60-1.00	0.10-0.50	0.20-1.35	---	0.15-0.35	---	---
S6	T41906	0.40-0.50	1.20-1.50	1.20-1.50	0.30-0.50	---	0.20-0.40	---	---
S7	T41907	0.45-0.55	0.20-0.90	3.00-3.50	1.30-1.80	---	0.20-0.30	---	---
Low-Alloy Special Purpose Tool Steels									
L2	T61202	0.45-1.00^C	0.10-0.90	0.70-1.20	0.25	---	---	0.10-0.30	---
L3	T61203	0.95-1.10	0.25-0.80	1.30-1.70	---	---	0.10-0.30	---	---
L6	T61206	0.65-0.75	0.25-0.80	0.60-1.20	0.50	---	---	---	1.25-2.00 Ni
Cold Work Tool Steels									
F1	T60601	0.95-1.25	0.50	---	---	1.00-1.75	---	---	---
F2	T60602	1.20-1.40	0.10-0.50	0.20-0.40	---	3.00-4.50	---	---	---
Low-Carbon Mold Tool Steels									
P2	T51602	0.10	0.10-0.40	0.75-1.25	0.15-0.40	---	---	---	0.10-0.50 Ni
P3	T51603	0.10	0.20-0.60	0.40-0.75	---	---	---	---	1.00-1.50 Ni

CHEMICAL COMPOSITION OF TOOL STEELS^a (Continued)

AISI	UNS	C	Mn	Cr	Mo	W	V	Co	Other
Low-Carbon Mold Tool Steels (Continued)									
P4	T51604	0.12	0.20-0.60	4.00-5.25	0.40-1.00	---	---	---	---
P5	T51605	0.06-0.10	0.20-0.60	2.00-2.50	---	---	---	---	0.35 Ni
P6	T51606	0.05-0.15	0.35-0.70	1.25-1.75	---	---	---	---	3.25-3.75 Ni
P20	T51620	0.28-0.40	0.60-1.00	1.40-2.00	0.30-0.55	---	---	---	---
P21	T51621	0.18-0.22	0.20-0.40	0.20-0.30	---	---	0.15-0.25	---	4.00-4.25 Ni / 1.05-1.25 Al
Water-Hardening Tool Steels									
W1 Gr A	T72301	d	0.10-0.40	0.15	0.10	0.15	0.10	---	0.20 Cu / 0.20 Ni
W1 Gr C	T72301	d	0.10-0.40	0.30	0.10	0.15	0.10	---	0.20 Cu / 0.20 Ni
W2 Gr A	T72302	e	0.10-0.40	0.15	0.10	0.15	0.15-0.35	---	0.20 Cu / 0.20 Ni
W2 Gr C	T72302	e	0.10-0.40	0.30	0.10	0.15	0.15-0.35	---	0.20 Cu / 0.20 Ni
W5	T72305	1.05-1.15	0.10-0.40	0.40-0.60	0.10	0.15	0.10	---	0.20 Cu / 0.20 Ni

All single values are maximums. a. All tools steels listed, except group W, allow up to 0.75% copper plus nickel. All tool steels listed allow 0.03% (maximum) phosphorus and 0.03% (maximum) sulfur. Where specified, sulfur may be increased to 0.06% to 0.15% to improve machinability for all tool steels listed, except group W. Although the silicon content is not listed due to limited space, it is specified; see appropriate standard grade for details. b. Manganese limit is 1.0% maximum for H13 resulfurized.
c. Available in several carbon ranges. d. The carbon range for W1 and their respective suffix identification, sometimes referred to as tempers, are as follows:

Suffix	8	8½	9	9½	10	10½	11	11½
Carbon Range	0.80-0.90	0.85-0.95	0.90-1.00	0.95-1.05	1.00-1.10	1.05-1.15	1.10-1.20	1.15-1.25

e. The carbon ranges for W2 and their respective suffix identification are as follows:

Suffix	8½	9	9½	13
Carbon Range	0.85-0.95	0.90-1.00	0.95-1.10	1.30-1.50

TYPICAL NORMALIZING AND ANNEALING TEMPERATURES FOR TOOL STEELS

AISI	Normalizing[a] °F	Normalizing[a] °C	Annealing[b] Temperature °F	Annealing[b] Temperature °C	Annealing[b] Rate of Cooling max °F/hr	Annealing[b] Rate of Cooling max °C/hr	Hardness HB
M1, M10	Do not normalize		1500-1600	815-870	40	22	207-235
M2	Do not normalize		1600-1650	870-900	40	22	212-241
M3, M4	Do not normalize		1600-1650	870-900	40	22	223-255
M7	Do not normalize		1500-1600	815-870	40	22	217-255
M30, M33	Do not normalize		1600-1650	870-900	40	22	235-269
M34, M35	Do not normalize		1600-1650	870-900	40	22	235-269
M36, M41	Do not normalize		1600-1650	870-900	40	22	235-269
M42, M46	Do not normalize		1600-1650	870-900	40	22	235-269
M47	Do not normalize		1600-1650	870-900	40	22	235-269
M43	Do not normalize		1600-1650	870-900	40	22	248-269
M44	Do not normalize		1600-1650	870-900	40	22	248-293
M48	Do not normalize		1600-1650	870-900	40	22	285-311
M62	Do not normalize		1600-1650	870-900	40	22	262-285
T1	Do not normalize		1600-1650	870-900	40	22	217-255
T2	Do not normalize		1600-1650	870-900	40	22	223-255
T4	Do not normalize		1600-1650	870-900	40	22	229-269
T5	Do not normalize		1600-1650	870-900	40	22	235-277
T6	Do not normalize		1600-1650	870-900	40	22	248-293
T8	Do not normalize		1600-1650	870-900	40	22	229-255
T15	Do not normalize		1600-1650	870-900	40	22	241-277
M50	Do not normalize		1525-1550	830-845	40	22	197-235
M52	Do not normalize		1525-1550	830-845	40	22	197-235
H10, H11	Do not normalize		1550-1650	845-900	40	22	192-229
H12, H13	Do not normalize		1550-1650	845-900	40	22	192-229

TYPICAL NORMALIZING AND ANNEALING TEMPERATURES FOR TOOL STEELS (Continued)							
	Normalizing[a]		Annealing[b]				
			Temperature		Rate of Cooling max		Hardness
AISI	°F	°C	°F	°C	°F/hr	°C/hr	HB
H14	Do not normalize		1600-1650	870-900	40	22	207-235
H 19	Do not normalize		1600-1650	870-900	40	22	207-241
H21, H22	Do not normalize		1600-1650	870-900	40	22	207-235
H25	Do not normalize		1600-1650	870-900	40	22	207-235
H23	Do not normalize		1600-1650	870-900	40	22	212-255
H24, H26	Do not normalize		1600-1650	870-900	40	22	217-241
H41, H43	Do not normalize		1500-1600	815-870	40	22	207-235
H42	Do not normalize		1550-1650	845-900	40	22	207-235
D2, D3	Do not normalize		1600-1650	870-900	40	22	217-255
D4	Do not normalize		1600-1650	870-900	40	22	217-255
D5	Do not normalize		1600-1650	870-900	40	22	223-255
D7	Do not normalize		1600-1650	870-900	40	22	235-262
A4	Do not normalize		1360-1400	740-760	25	14	200-241
A6	Do not normalize		1350-1375	730-745	25	14	217-248
A7	Do not normalize		1600-1650	870-900	25	14	235-262
A8	Do not normalize		1550-1600	845-870	40	22	192-223
A9	Do not normalize		1550-1600	845-870	25	14	212-248
A10	1450	790	1410-1460	765-795	15	8	235-269
O1	1600	870	1400-1450	760-790	40	22	183-212
O2	1550	845	1375-1425	745-775	40	22	183-212
O6	1600	870	1410-1450	765-790	20	11	183-217
O7	1650	900	1450-1500	790-815	40	22	192-217
S1	Do not normalize		1450-1500	790-815	40	22	183-229[c]
S2	Do not normalize		1400-1450	760-790	40	22	192-217

TYPICAL NORMALIZING AND ANNEALING TEMPERATURES FOR TOOL STEELS (Continued)

AISI	Normalizing[a] °F	Normalizing[a] °C	Annealing[b] Temperature °F	Annealing[b] Temperature °C	Rate of Cooling max °F/hr	Rate of Cooling max °C/hr	Hardness HB
S5	Do not normalize		1425-1475	775-800	25	14	192-229
S7	Do not normalize		1500-1550	815-845	25	14	187-223
P2	Not required		1350-1500	730-815	40	22	103-123
P3	Not required		1350-1500	730-815	40	22	109,-137
P4	Do not normalize		1600-1650	870-900	25	14	116-128
P5	Not required		1550-1600	845-870	40	22	105-116
P6	Not required		1550	845	15	8	183-217
P20	1650	900	1400-1450	760-790	40	22	149-179
P21	1650	900	Do Not Anneal		---	---	---
L2	1600-1650	871-900	1400-1450	760-790	40	22	163-197
L3	1650	900	1450-1500	790-815	40	22	174-201
L6	1600	870	1400-1450	760-790	40	22	183-212
F1	1650	900	1400-1475	760-800	40	22	183-207
F2	1650	900	1450-1500	790-815	40	22	207-235
W1, W2	1450-1700	790-925[d]	1360-1450[e]	740-790[e]	40	22	156-201
W5	1600-1700	870-925	1400-1450	760-790	40	22	163-201

a. Time held at temperature varies from 15 minutes for small sections to 1 hour for large sizes. Cooling is done in still air. Normalizing should not be confused with low-temperature annealing.

b. The upper limit of ranges should be used for large sections, and the lower limit for smaller sections. Time held at temperature varies from 1 hour for light sections to 4 hours for heavy sections and large furnace charges of high-alloy steel.

c. For 0.25 Si type 183 to 207 HB; for 1.00 Si type, 207 to 229 HB.

d. Temperature varies with carbon content: 0.60 to 0.75 C - 1500°F (815°C);   0.75 to 0.90 C - 1450°F (790°C);   0.90 to 1.10 C - 1600°F (970°C); 1.10 to 1.40 C - 1600-1700°F (870-925°C).

e. Temperature varies with carbon content: 0.60 to 0.90 C - 1360-1450°F (740-790°C); 0.90 to 1.40 C - 1400-1450°F (760-790°C).

HEAT TREATING REQUIREMENTS FOR TOOL STEELS

ASTM Spec. A 600 Type	Preheat Temperature °F	Preheat Temperature °C	Austenitizing Temperature[a] Salt Bath °F	Salt Bath °C	Controlled Atmosphere Furnace °F	Controlled Atmosphere Furnace °C	Quenching Medium	Tempering Temperature[b] °F	Tempering Temperature[b] °C	Hardness HRC, min
M1	1350-1550	732-843	2185	1196	2205	1207	Oil or Salt	1025	552	64
M2 (regular C)	1350-1550	732-843	2220	1216	2240	1227	Oil or Salt	1025	552	64
M2 (high C)	1350-1550	732-843	2200	1204	2220	1216	Oil or Salt	1025	552	65
M3, Class 1	1350-1550	732-843	2200	1204	2220	1216	Oil or Salt	1025	552	64
M3, Class 2	1350-1550	732-843	2200	1204	2220	1216	Oil or Salt	1025	552	64
M4	1350-1550	732-843	2200	1204	2220	1216	Oil or Salt	1025	552	64
M6	1350-1550	732-843	2170	1188	2190	1199	Oil or Salt	1025	552	64
M7	1350-1550	732-843	2200	1204	2220	1216	Oil or Salt	1025	552	65
M10 (regular C)	1350-1550	732-843	2185	1196	2205	1207	Oil or Salt	1025	552	63
M10 (high C)	1350-1550	732-843	2185	1196	2205	1207	Oil or Salt	1025	552	64
M30	1350-1550	732-843	2200	1204	2220	1216	Oil or Salt	1025	552	64
M33	1350-1550	732-843	2200	1204	2220	1216	Oil or Salt	1025	552	65
M34	1350-1550	732-843	2200	1204	2220	1216	Oil or Salt	1025	552	64
M36	1350-1550	732-843	2200	1204	2220	1216	Oil or Salt	1025	552	64
M41	1350-1550	732-843	2175	1190	2195	1202	Oil or Salt	1000	538	66
M42	1350-1550	732-843	2150	1177	2170	1188	Oil or Salt	1000	538	66
M43	1350-1550	732-843	2150	1177	2170	1188	Oil or Salt	1000	538	66
M44	1350-1550	732-843	2170	1188	2190	1199	Oil or Salt	1000	538	66
M46	1350-1550	732-843	2200	1204	2220	1216	Oil or Salt	1000	538	66
M47	1350-1550	732-843	2175	1190	2195	1202	Oil or Salt	1000	538	66
M48	1350-1550	732-843	2175	1190	2195	1202	Oil or Salt	1000	538	66
M50	1350-1550	732-843	2020	1104	2040	1116	Oil or Salt	1000	538	61
M52	1350-1550	732-843	2125	1163	2145	1174	Oil or Salt	1000	538	63

HEAT TREATING REQUIREMENTS FOR TOOL STEELS (Continued)

| ASTM Spec. | Preheat Temperature | | Austenitizing Temperature[a] | | | | | | Quenching | Tempering Temperature[b] | | Hardness |
| | | | Salt Bath | | Controlled Atmosphere Furnace | | | | | | | |
	°F	°C	°F	°C	°F	°C			Medium	°F	°C	HRC, min
A 600 Type												
M62	1350-1550	732-843	2175	1190	2195	1202			Oil or Salt	1000	538	66
T1	1500-1600	816-871	2330	1277	2350	1288			Oil or Salt	1025	552	63
T2	1500-1600	816-871	2330	1277	2350	1288			Oil or Salt	1025	552	63
T4	1500-1600	816-871	2330	1277	2350	1288			Oil or Salt	1025	552	63
T5	1500-1600	816-871	2330	1277	2350	1288			Oil or Salt	1025	552	63
T6	1500-1600	816-871	2330	1277	2350	1288			Oil or Salt	1025	552	63
T8	1500-1600	816-871	2330	1277	2350	1288			Oil or Salt	1025	552	63
T15	1500-1600	816-871	2240	1227	2260	1238			Oil or Salt	1000	538	65
A 681 Type												
H10	1450	788	1850	1010	1875	1024			Air	1025	552	55
H11	1450	788	1825	996	1850	1010			Air	1025	552	53
H12	1450	788	1825	996	1850	1010			Air	1025	552	53
H13	1450	788	1825	996	1850	1010			Air	1025	552	52
H14	1450	788	1900	1038	1925	1052			Air	1025	552	55
H19	1450	788	2150	1177	2175	1191			Air	1025	552	55
H21	1450	788	2150	1177	2175	1191			Air	1025	552	52
H22	1450	788	2150	1177	2175	1191			Air	1025	552	53
H23	1500	816	2275	1246	2300	1260			Oil	1200	649	42
H24	1450	788	2200	1204	2225	1218			Air	1025	552	55
H25	1450	788	2250	1232	2275	1246			Air	1025	552	44
H26	1550	843	2275	1246	2300	1260			Air	1025	552	58
H41	1450	788	2125	1163	2150	1177			Air	1025	552	60

HEAT TREATING REQUIREMENTS FOR TOOL STEELS (Continued)

ASTM Spec. A 681 Type	Preheat Temperature		Austenitizing Temperature[a]				Quenching Medium	Tempering Temperature[b]		Hardness HRC, min
			Salt Bath		Controlled Atmosphere Furnace					
	°F	°C	°F	°C	°F	°C		°F	°C	
H42	1450	788	2175	1191	2200	1204	Air	1025	552	60
H43	1450	788	2150	1177	2175	1191	Air	1025	552	58
A2	1450	788	1725	941	1750	954	Air	400	204	60
A3	1450	788	1775	968	1800	982	Air	400	204	63
A4	1250	677	1550	843	1575	857	Air	400	204	61
A6	1200	649	1525	829	1550	843	Air	400	204	58
A7	1500	816	1750	954	1775	968	Air	400	204	63
A8	1450	788	1825	996	1850	1010	Air	950	510	56
A9	1450	788	1825	996	1850	1010	Air	950	510	56
A10	1200	649	1475	802	1500	816	Air	400	204	59
D2	1500	816	1825	996	1850	1010	Air	400	204	59
D3	1500	816	1750	954	1775	968	Oil	400	204	61
D4	1500	816	1800	982	1825	996	Air	400	204	62
D5	1500	816	1825	996	1850	1010	Air	400	204	61
D7	1500	816	1925	1052	1950	1066	Air	400	204	63
O1	1200	649	1450	788	1475	802	Oil	400	204	59
O2	1200	649	1450	788	1475	802	Oil	400	204	59
O6	---	---	1450	788	1475	802	Oil	400	204	59
O7	1200	649	1575	857	1600	871	Oil	400	204	62
S1	1250	677	1725	941	1750	954	Oil	400	204	56
S2	1250	677	1625	885	1650	899	Brine	400	204	58

HEAT TREATING REQUIREMENTS FOR TOOL STEELS (Continued)

ASTM Spec. A 681 Type	Preheat Temperature		Austenitizing Temperature[a]				Quenching Medium	Tempering Temperature[b]		Hardness HRC, min
	°F	°C	Salt Bath		Controlled Atmosphere Furnace			°F	°C	
			°F	°C	°F	°C				
S4	1250	677	1625	885	1650	899	Oil	400	204	58
S5	1250	677	1625	885	1650	899	Oil	400	204	58
S6	1450	788	1700	927	1725	941	Oil	400	204	56
S7	1250	677	1725	941	1750	954	Air	400	204	56
L2[c]	1200	649	1575	857	1600	871	Oil	400	204	53[a]
L3	1200	649	1525	829	1550	843	Oil	400	204	62
L6	1200	649	1500	816	1525	829	Oil	400	204	58
F1	1200	649	1525	829	1550	843	Brine	400	204	64
F2	1200	649	1525	829	1550	843	Brine	400	204	64

a. The austenitizing and tempering temperatures shall be ± 10°F (± 5°C). If samples are austenized in salt, the sample shall be immersed in the austenitizing salt bath for 5 minutes minimum. If austenized in a controlled atmosphere furnace, the sample shall be at the austenizing temperature for 5 to 15 minutes (10 to 20 minutes for D types). The time at temperature is the time after the sample reaches the austenizing temperature. This range in time is given because of the difficulty in determining when the sample reaches the austenizing temperature in some types of controlled atmosphere furnaces.

b. Those steels tempered at 400°F (204°C) shall have a single 2 hour temper, while those tempered at 950 °F (510 °C), 1025 °F (552 °C), 1200°F (649°C) shall be double-tempered for 2 hours each cycle, and those tempered at 1000°F (538°C) shall be tripled-tempered for 2 hours each cycle.

Note: The austenitizing temperatures are stipulated for the response to hardening test only according to ASTM A 600 and ASTM A 681. Other combinations of austenizing and tempering temperatures may be used for particular applications.

c. 0.45-0.55% carbon type.

HEAT TREAT REQUIREMENTS FOR TYPE W TOOL STEELS

ASTM A 686	Carbon	Austenitizing Temperature		Quenching Medium	Hardened HRC, min	Annealed, HB max	Cold Drawn, HB max
		°F	°C				
Type W1	0.70-0.85	1475	802	brine	64	202	241
	0.85-0.95	1475	802	brine	65		
	0.95-1.50	1450	788	brine	65		
Type W2	0.85-0.95	1475	802	brine	65	202	241
	0.95-1.50	1450	788	brine	65		
Type W5	1.05-1.15	1475	802	brine	65	202	241

HARDNESS REQUIREMENTS FOR TOOL STEELS

ASTM A 600 Type	Annealed, HB	Cold Drawn, HB	Cold Drawn Annealed, HB
M1	248	262	255
M2 (regular C)	248	262	255
M2 (high C)	255	269	262
M3, Class 1 and Class 2	255	269	262
M4	255	269	262
M6	277	293	285
M7	255	269	262
M10 (regular C)	248	262	255
M10 (high C)	255	269	262
M30	269	285	277
M33	269	285	277
M34	269	285	277
M36	269	285	277
M41	269	285	277
M42	269	285	277

HARDNESS REQUIREMENTS FOR TOOL STEELS (Continued)

ASTM A 600 Type	Annealed, HB	Cold Drawn, HB	Cold Drawn Annealed, HB
M43	269	285	297
M44	285	302	293
M46	269	285	277
M47	269	285	277
M48	311	331	321
M50	248	262	255
M52	248	262	255
M62	285	302	293
T1	255	269	262
T2	255	269	262
T4	269	285	277
T5	285	302	293
T6	302	321	311
T8	255	269	262
T15	277	293	285
ASTM A 681 Type			
H10	229	---	255
H11	235	---	262
H12	235	---	262
H13	235	---	262
H14	235	---	262
H19	241	---	262
H21	235	---	262
H22	235	---	262
H23	255	---	269
H24	241	---	262
H25	235	---	262

HARDNESS REQUIREMENTS FOR TOOL STEELS (Continued)			
ASTM A 681 Type	Annealed, HB	Cold Drawn, HB	Cold Drawn Annealed, HB
H26	241	---	262
H41	235	---	262
H42	235	---	262
H43	235	---	262
A2	248	---	262
A3	229	---	255
A4	241	---	262
A6	248	---	262
A7	269	---	285
A8	241	---	262
A9	248	---	262
A10	269	---	285
D2	255	---	269
D3	255	---	269
D4	255	---	269
D5	255	---	269
D7	262	---	277
O1	212	---	241
O2	217	---	241
O6	229	---	241
O7	241	---	255
S1	229	---	255
S2	217	---	241
S4	229	---	255
S5	229	---	255
S6	229	---	255
S7	229	---	255

HARDNESS REQUIREMENTS FOR TOOL STEELS (Continued)

ASTM A 681 Type	Annealed, HB	Cold Drawn, HB	Cold Drawn Annealed, HB
L2	197	---	241
L3	201	---	241
L6	235	---	262
F1	207	---	241
F2	235	---	262
P2	100	---	---
P3	143	---	---
P4	131	---	---
P5	131	---	---
P6	212	---	---
P20	a	---	---
P21	a	---	---

a. Normally furnished in prehardened condition.

SELECTED PROPERTIES OF TOOL STEELS

AISI	Major Factors[a] Wear Resistance	Toughness	Hot Hardness	Working[f] Hardness HRC	Depth[b] of Hardening	Finest Grain Size at Full Hardness	As-Quenched Surface Hardness HRC	Core Hardness HRC[g]
Molybdenun High-Speed Steels								
M1	7	3	8	63-65	D	9½	64-66	64-66
M2	7	3	8	63-65	D	9½	64-66	64-66
M3[d]	8	3	8	63-66	D	9½	64-66	64-66
M3[e]	8	3	8	63-66	D	9½	64-66	64-66
M4	9	3	8	63-66	D	9½	6547	65-67
M7	8	3	8	63-66	D	9½	64-66	64-66
M10	7	3	8	63-65	D	9½	64-66	64-66
M30	7	2	8	63-65	D	9½	64-66	64-66
M33	8	1	9	63-65	D	9½	64-66	64-66
M34	8	1	9	63-65	D	9½	64-66	64-66
M36	7	1	9	63-65	D	9½	64-66	64-66
M41	8	1	9	66-70	D	9½	63-65	63-65
M42	8	1	9	66-70	D	9½	63-65	63-65
M43	8	1	9	66-70	D	9½	63-65	63-65
M44	8	1	9	66-70	D	9½	63-65	63-65
M46	8	1	9	66-69	D	9½	63-65	63-65
M47	8	1	9	66-70	D	9½	63-65	63-65
Intermediate High-Speed Steels								
M50	6	3	6	61-63	D	8½	63-65	63-65
M52	6	3	6	62-64	D	8½	63-65	63-65
T1	7	3	8	63-65	D	9½	64-66	64-66
T2	8	3	8	63-66	D	9½	65-67	65-67

SELECTED PROPERTIES OF TOOL STEELS (Continued)

AISI	Major Factors[a]			Working[f] Hardness HRC	Depth[b] of Hardening	Finest Grain Size at Full Hardness	As-Quenched Surface Hardness HRC	Core Hardness HRC[g]
	Wear Resistance	Toughness	Hot Hardness					
Intermediate High-Speed Steels (Continued)								
T4	7	2	8	63-65	D	9½	63-66	63-66
T5	7	1	9	63-65	D	9½	64-66	64-66
T6	8	1	9	63-65	D	9½	64-66	64-66
T8	8	2	8	63-65	D	9½	64-66	64-66
T15	9	1	9	64-68	D	9½	65-68	65-68
Chromium Hot-Work Steels								
H10	3	9	6	39-56	D	8	52-59	52-59
H11	3	9	6	38-55	D	8	53-55	53-55
H12	3	9	6	38-55	D	8	53-55	53-55
H13	3	9	6	40-53	D	8	51-54	51-54
H14	4	6	7	40-54	D	8	53-57	53-56
H19	5	6	7	40-55	D	8½	48-57	48-57
Tungsten Hot-Work Steels								
H21	4	6	8	40-55	D	9	45-63	45-63
H22	5	5	8	36-54	D	9	48-56	48-56
H23	5	5	8	38-48	D	7	34-40	34-40
H24	5	5	8	40-55	D	9	52-56	52-56
H25	4	6	8	35-45	D	9	33-46	33-46
H26	6	4	8	50-58	D	9	51-59	51-59
Molybdenum Hot-Work Steels								
H42	6	4	7	45-62	D	8½	54-62	54-62
Air-Hardening, Medium-Alloy, Cold-Work Steels								
A2	6	4	5	57-62	D	8½	63-65	63-65

SELECTED PROPERTIES OF TOOL STEELS (Continued)

AISI	Major Factors[a]			Working[f] Hardness HRC	Depth[b] of Hardening	Finest Grain Size at Full Hardness	As-Quenched Surface Hardness HRC	Core Hardness HRC[g]
	Wear Resistance	Toughness	Hot Hardness					
Air-Hardening, Medium-Alloy, Cold-Work Steels (Continued)								
A3	7	3	5	58-63	D	8½	63-65	63-65
A4	5	4	4	54-62	D	8½	61-63	61-63
A5	5	4	4	54-60	D	8½	60-62	60-62
A6	4	5	4	54-60	D	8½	60-62	60-62
A7	9	1	6	58-66	D	8½	64-66	64-66
A8	4	8	6	48-57	D	8	60-62	60-62
A9	4	8	6	40-56	D	8	55-57	55-57
A10	3	3	3	55-62	D	8	60-63	60-63
High-Carbon, High-Chromium, Cold-Work Steels								
D2	9	2	6	58-64	D	7½	61-64	61-64
D3	8	1	6	58-64	D	7½	64-66	64-66
D4	8	1	6	58-64	D	7½	64-66	64-66
D5	8	2	7	58-63	D	7½	61-64	61-64
D7	9	1	6	58-66	D	7½	64-68	64-68
Oil-Hardening Cold-Work Steels								
O1	4	3	3	57-62	M	9	61-64	59-61
O2	4	3	3	57-62	M	9	61-64	59-61
O6	3	3	2	59-63	M	9	65-67	50-55
O7	5	3	3	58-64	M	9	61-64	59-61
Shock-Resisting Steels								
S1	4	8	5	50-58	M	8	55-58	55-58
S2	2	8	2	50-60	M	8	61-63	56-60
S5	2	8	3	50-60	M	9	61-63	58-62

SELECTED PROPERTIES OF TOOL STEELS (Continued)

AISI	Major Factors[a] Wear Resistance	Toughness	Hot Hardness	Working[f] Hardness HRC	Depth[b] of Hardening	Finest Grain Size at Full Hardness	As-Quenched Surface Hardness HRC	Core Hardness HRC[g]
Shock-Resisting Steels								
S6	2	8	3	50-56	M	8	56-58	56-58
S7	3	8	5	47-57	D	8	59-61	59-61
Low-Alloy Special-Purpose Steels								
L2	1	7	2	45-62	M	8½	56-62	54-58
L6	3	6	2	45-62	M	8	58-63	58-60
Low-Carbon Mold Steels								
For hubbed and/or carburized cavities								
P2	1c	9	2c	58-64c	S	---	62-65c	15-21
P3	1c	9	2c	58-64c	S	---	62-64c	15-21
P4	1c	9	4c	58-64c	M	---	62-65c	33-35
P5	1c	9	2c	50-64c	S	---	62-65c	20-25
P6	1c	9	3c	58-61c	M	---	60-62c	35-37
For machined cavities								
P20	1c	8	2c	30-50	M	7½	52-54	45-50
P21	1	8	4	36-39	D	---	22-26	22-26
Water-Hardening Tool Steels								
W1	2-4	3-7	1	58-65	S	9	65-67	38-43
W2	2-4	3-7	1	58-65	S	9	65-67	38-43
W5	3-4	3-7	1	58-65	S	9	65-67	38-43

a. Rating range from 1 (low) to 9 (high). b. S - shallow; M - medium; D - deep. c. After carburizing. d. AISI M3 Class 1. e. AISI M3 Class 2. f Usual working hardness. g Core hardness related to 1 in (25 mm) diameter round.

Note: Wear resistance increases with increasing carbon content. Toughness decreases with increasing carbon content and depth of hardening.

WROUGHT STAINLESS STEELS:

AMERICAN SPECIFICATION TITLES & DESIGNATIONS, CHEMICAL COMPOSITIONS & MECHANICAL PROPERTIES

WROUGHT STAINLESS STEELS

ASTM Spec.	Title
Stainless and Heat Resisting Steel Bars	
A 276	Stainless Steel Bars and Shapes
A 314	Stainless Steel Billets and Bars for Forging
A 479/A 479M	Stainless Steel Bars and Shapes for Use in Boilers and Other Pressure Vessels
A 484/A 484M	Stainless Steel Bars, Billets, and Forgings, General Requirements for
A 564/A 564M	Hot-Rolled and Cold-Finished Age-Hardening Stainless Steel Bars and Shapes
A 565	Martensitic Stainless Steel Bars, Forgings, and Forging Stock for High-Temperature Service
A 582/A 582M	Free-Machining Stainless Steel Bars
A 831/A 831M	Austenitic and Martensitic Stainless Steel Bars, Billets, and Forgings for Liquid Metal Cooled Reactor Core Components
Stainless Steel Plate, Sheet, Strip	
A 167	Stainless and Heat-Resisting Chromium-Nickel Steel Plate, Sheet, and Strip
A 176	Stainless and Heat-Resisting Chromium Steel Plate, Sheet, and Strip
A 240/A 240M	Heat-Resisting Chromium and Chromium-Nickel Stainless Steel Plate, Sheet, and Strip for Pressure Vessels
A 263	Corrosion-Resisting Chromium Steel-Clad Plate, Sheet, and Strip
A 264	Stainless Chromium-Nickel Steel-Clad Plate, Sheet, and Strip
A 480/A 480M	Flat-Rolled Stainless and Heat-Resisting Steel Plate, Sheet, and Strip, General Requirements for
A 666	Austenitic Stainless Steel, Sheet, Strip, Plate, and Flat Bar
A 693	Precipitation-Hardening Stainless and Heat-Resisting Steel Plate, Sheet, and Strip
A 793	Steel, Stainless, for Rolled Floor Plate
A 887	Borated Stainless Steel Plate, Sheet, and Strip for Nuclear Application
A 895	Stainless Steel Plate, Sheet, and Strip, Free Machining
Stainless Steel Wire	
A 313/A313M	Stainless Steel Spring Wire
A 368	Stainless Steel Wire Strand
A 478	Chromium-Nickel Stainless Steel Weaving and Knitting Wire
A 492	Stainless Steel Rope Wire
A 493	Stainless Steel Wire and Wire Rods for Cold Heading and Cold Forging

WROUGHT STAINLESS STEELS (Continued)	
ASTM Spec.	**Title**
Stainless Steel Wire (Continued)	
A 555/A 555M	Stainless Steel Wire and Wire Rods, General Requirements for
A 580/A 580M	Stainless Steel Wire
A 581/A 581M	Free-Machining Stainless Steel Wire
Stainless Steel Bearing	
A 756	Stainless Anti-Friction Bearing Steel
Stainless Steel Fittings	
A 403	Wrought Austenitic Stainless Steel Piping Fittings
A 774/A 774M	As-Welded Wrought Austenitic Stainless Steel Fittings for General Corrosion Service at Low and Moderate Temperatures
A 815/A 815M	Wrought Ferritic, Ferritic/Austenitic, and Martensitic Stainless Steel Piping Fittings
Stainless Steel Surgical Implants	
F 138	Stainless Steel Bars and Wire for Surgical Implants (Special Quality)
F 139	Stainless Steel, Sheet and Strip for Surgical Implants (Special Quality)

CHEMICAL COMPOSITION OF WROUGHT MARTENSITIC STAINLESS STEELS[a]

AISI Type	UNS	C	Mn	Si	Cr	Ni	Mo	Other
403	S40300	0.15	1.00	0.50	11.50-13.00	---	---	---
410	S41000	0.15	1.00	1.00	11.50-13.50	---	---	---
414	S41400	0.15	1.00	1.00	11.50-13.50	1.25-2.50	---	---
416	S41600	0.15	1.25	1.00	12.00-14.00	---	0.60[b]	0.15 S min
416 Se	S41623	0.15	1.25	1.00	12.00-14.00	---	---	0.15 min Se
420	S42000	0.15 min	1.00	1.00	12.00-14.00	---	---	---
420 F	S42020	0.15 min	1.25	1.00	12.00-14.00	---	0.60[b]	0.15 S min
422	S42200	0.20-0.25	1.00	0.75	11.00-13.50	0.50-1.00	0.75-1.25	0.75-1.25 W, 0.15-0.30 V
431	S43100	0.20	1.00	1.00	15.00-17.00	1.25-2.50	---	---
440 A	S44002	0.60-0.75	1.00	1.00	16.00-18.00	---	0.75	---
440 B	S44003	0.75-0.95	1.00	1.00	16.00-18.00	---	0.75	---
440 C	S44004	0.95-1.20	1.00	1.00	16.00-18.00	---	0.75	---
Non-AISI/Common Name								
410 Cb (XM-30)	S41040	0.15	0.15	1.00	11.50-13.50	---	---	0.05-0.20 Cb
416 Plus X (XM-6)	S41610	0.15	1.50-2.50	1.00	12.00-14.00	---	0.6	---

a. Although sulfur and phosphorous contents are not listed in this table due to limited space, they are specified, see appropriate material standard for more details.
b. Optional.
Single values are maximums, unless otherwise specified.

CHEMICAL COMPOSITION WROUGHT FERRITIC STAINLESS STEELS[a]

AISI Type	UNS	C	Mn	Si	Cr	Ni	Mo	Other
405	S40500	0.08	1.00	1.00	11.50-14.50	---	---	0.10-0.30 Al
409	S40900	0.08	1.00	1.00	10.50-11.75	0.50	---	(6 x C) min to 0.75 max Ti
429	S42900	0.12	1.00	1.00	14.00-16.00	---	---	---
430	S43000	0.12	1.00	1.00	16.00-18.00	---	---	---

CHEMICAL COMPOSITION WROUGHT FERRITIC STAINLESS STEELS[a] (Continued)

AISI Type	UNS	C	Mn	Si	Cr	Ni	Mo	Other
430 F	S43020	0.12	1.25	1.00	16.00-18.00	---	0.6[b]	0.15 S min
430 F Se	S43023	0.12	1.25	1.00	16.00-18.00	---	---	0.15 min Se
434	S43400	0.12	1.00	1.00	16.00-18.00	---	0.75-1.25	---
436	S43600	0.12	1.00	1.00	16.00-18.00	---	0.75-1.25	(5 x C) min to 0.70 (Cb+Ta) max
439	S43035	0.07	1.00	1.00	17.00-19.00	0.50	---	(12 x C) min to 1.10 Ti max, 0.15 Al
442	S44200	0.20	1.00	1.00	18.00-23.00	---	---	---
446	S44600	0.20	1.50	1.00	23.00-27.00	---	---	0.25 N
Non-AISI/Common Name								
18-2	S44400	0.025	1.00	1.00	17.50-19.50	1.00	1.75-2.50	0.20 + 4 (C + N) min to 0.80 (Ti + Cb) max, 0.025 N
E-Brite, 26-1	S44627	0.01	0.40	0.40	25.00-27.00	0.50	0.75-1.50	0.05-0.20 Cb, 0.015 N, 0.20 Cu
25-4-4	S44635	0.025	1.00	0.75	24.5-26.0	3.50-4.50	3.50-4.50	0.20 + 4 (C + N) min to 0.80 (Ti + Cb) max, 0.035 N
SC-1	S44660	0.030	1.00	1.00	25.00-28.00	1.00-3.50	3.00-4.00	0.20 + 6 (C + N) min to 1.0 (Ti + Cb) max, 0.040 N
AL 29-4	S44700	0.010	0.30	0.20	28.0-30.0	0.15	3.50-4.20	0.025 (C + N) max, 0.15 Cu, 0.020 N
AL 29-4C	S44735	0.030	1.00	1.00	28.0-30.0	1.00	3.60-4.20	0.20-1.00 (Ti + Cb) and 6 (C + N) min (Ti + Cb), 0.45 N
AL 29-4-2	S44800	0.010	0.30	0.20	28.0-30.0	2.00-2.50	3.50-4.20	0.025 (C + N) max, 0.15 Cu, 0.020 N
AL 29-4-2C	S44736	0.030	1.00	1.00	28.0-30.0	2.00-4.50	3.60-4.20	0.20 to 1.00 (Cb + Ti) and 6 (C + N) min, 0.045 N

a. Although sulfur and phosphorous contents are not listed in this table due to limited space, they are specified, see appropriate material standard for more details.

b. Optional.

Single values are maximums, unless otherwise specified.

CHEMICAL COMPOSITION WROUGHT AUSTENITIC STAINLESS STEELS[a]

AISI Type	UNS	C	Mn	Si	Cr	Ni	Mo	Other
201	S20100	0.15	5.50-7.50	1.00	16.00-18.00	3.50-5.50	---	0.25 N
202	S20200	0.15	7.50-10.00	1.00	17.00-19.00	4.00-6.00	---	0.25 N
205	S20500	0.12-0.25	14.00-15.50	1.00	16.00-18.00	1.00-1.75	---	0.32-0.40 N
301	S30100	0.15	2.00	1.00	16.00-18.00	6.00-8.00	---	---
302	S30200	0.15	2.00	1.00	17.00-19.00	8.00-10.00	---	---
302 B	S30215	0.15	2.00	2.0-3.0	17.00-19.00	8.00-10.00	---	---
303	S30300	0.15	2.00	1.00	17.00-19.00	8.00-10.00	---	0.6 Mo optional, 0.15 S min
303 Se	S30323	0.15	2.00	1.00	17.00-19.00	8.00-10.00	---	0.15 min Se
304	S30400	0.08	2.00	1.00	18.00-20.00	8.00-10.50	---	---
304 L	S30403	0.03	2.00	1.00	18.00-20.00	8.00-12.00	---	---
304 N	S30451	0.08	2.00	1.00	18.00-20.00	8.00-10.50	---	0.10-016 N
305	S30500	0.12	2.00	1.00	17.00-19.00	10.00-13.00	---	---
308	S30800	0.08	2.00	1.00	19.00-21.00	10.00-12.00	---	---
309	S30900	0.20	2.00	1.00	22.00-24.00	12.00-15.00	---	---
309 S	S30908	0.08	2.00	1.00	22.00-24.00	12.00-15.00	---	---
310	S31000	0.25	2.00	1.50	24.00-26.00	19.00-22.00	---	---
310 S	S31008	0.08	2.00	1.50	24.00-26.00	19.00-22.00	---	---
314	S31400	0.25	2.00	1.50-3.00	23.00-26.00	19.00-22.00	---	---
316	S31600	0.08	2.00	1.00	16.00-18.00	10.00-14.00	2.00-3.00	---
316 F	S31620	0.08	2.00	1.00	17.00-19.00	12.00-14.00	1.75-2.50	0.120 S min
316 L	S31603	0.030	2.00	1.00	16.00-18.00	10.00-14.00	2.00-3.00	---
316 N	S31651	0.08	2.00	1.00	16.00-18.00	10.00-14.00	2.00-3.00	0.10-0.16 N
317	S31700	0.08	2.00	1.00	18.00-20.00	11.00-15.00	3.00-4.00	---
317 L	S31703	0.03	2.00	1.00	18.00-20.00	11.00-15.00	3.00-4.00	---
321	S32100	0.08	2.00	1.00	17.00-19.00	9.00-12.00	---	(5 x C) Ti min
321H	S32109	0.04-0.10	2.00	1.00	17.00-20.00	9.00-12.00	---	(4 x C) min to 0.60 Ti max

CHEMICAL COMPOSITION WROUGHT AUSTENITIC STAINLESS STEELSᵃ (Continued)

AISI Type	UNS	C	Mn	Si	Cr	Ni	Mo	Other
347	S34700	0.08	2.00	1.00	17.00-19.00	9.00-13.00	---	(10 x C) min Cb
348	S34800	0.08	2.00	1.00	17.00-19.00	9.00-13.00	---	(10 x C) min Cb, 0.10 Ta, 0.20 Co
384	S38400	0.08	2.00	1.00	15.00-17.00	17.00-19.00	---	---
Non-AISI/Common Name								
ASTM F 138 Gr 1	---	0.08	2.00	0.75	17.00-19.00	13.00-15.50	2.00-3.00	0.10 N, 0.50 Cu
ASTM F 138 Gr 2	---	0.030	2.00	0.75	17.00-19.00	13.00-15.50	2.00-3.00	0.10 N, 0.50 Cu
ASTM F 139 Gr 1	---	0.08	2.00	0.75	17.00-19.00	13.00-15.50	2.00-3.00	0.10 N, 0.50 Cu
ASTM F 139 Gr 2	---	0.030	2.00	0.75	17.00-19.00	13.00-15.50	2.00-3.00	0.10 N, 0.50 Cu
18-18 Plus	S28200	0.15	17.00-19.00	1.00	17.00-19.00	---	0.50-1.50	0.5-1.50 Cu, 0.40-0.60 N
18-15	S30600	0.18	2.00	3.70-4.30	17.0-18.5	14.0-15.5	0.20	0.50 Cu
20Cb-3	N08020	0.07	2.00	1.00	19.00-21.00	32.00-38.00	2.00-3.00	(8 x C) min to 1.00 Cb max, 3.00-4.00 Cu
20Mo-4	N08024	0.03	1.00	0.50	22.50-25.00	35.00-40.00	3.50-5.00	0.50-1.50 Cu, 0.15-0.35 Cb
20Mo-6	N08026	0.03	1.00	0.50	22.00-26.00	33.00-37.20	5.00-6.70	2.00-4.00 Cu, 0.10-0.16 N
254 SMO	S31254	0.020	1.00	0.80	19.50-20.50	17.50-18.50	6.00-6.50	0.50-1.00 Cu, 0.180-0.220 N
304 H	S30409	0.04-0.10	2.00	1.00	18.00-20.00	8.00-11.00	---	---
304 HN, (XM-21)	S30452	0.08	2.00	1.00	18.00-20.00	8.00-10.50	---	0.16-0.30 N
304 LN	S30453	0.03	2.00	1.00	18.00-20.00	8.00-12.00	---	0.10-0.16 N
310 Cb	S31040	0.08	2.00	1.50	24.00-26.00	19.00-22.00	---	(10 x C) min to 1.10 (Cb + Ta) max, 0.10 N
316 H	S31609	0.04-0.10	2.00	1.00	16.00-18.00	10.00-14.00	2.00-3.00	---
316 Ti	S31635	0.08	2.00	1.00	16.00-18.00	10.00-14.00	2.00-3.00	5 x (C + N) min to 0.70 Ti max, 0.10 N
316 Cb	S31640	0.08	2.00	1.50	16.00-18.00	10.00-14.00	2.00-3.00	(10 x C) min to 1.10 (Cb + Ta) max, 0.10 N
316 LN	S31653	0.03	2.00	1.00	16.00-18.00	10.00-14.00	2.00-3.00	0.10-0.16 N
317 LN	S31753	0.03	2.00	1.00	18.00-20.00	11.00-15.00	3.00-4.00	0.10-0.22 N
347H	S34709	0.04-0.10	2.00	1.00	17.00-20.00	9.00-13.00	---	(8 x C) min to 1.0 Cb max
348H	S34809	0.04-0.10	2.00	1.00	17.00-20.00	9.00-13.00	---	(8 x C) min to 1.0 Cb max, 0.10 Ta, 0.2 Co
Nitronic 60	S21800	0.10	7.00-9.00	3.50-4.50	16.00-18.00	8.00-9.00	---	0.08-0.18 N

CHEMICAL COMPOSITION WROUGHT AUSTENITIC STAINLESS STEELS[a] (Continued)

Non-AISI Type - Common Name	UNS	C	Mn	Si	Cr	Ni	Mo	Other
XM-10	S21900	0.08	8.00-10.00	1.00	19.00-21.50	5.50-7.50	---	0.15-0.40 N
XM-19	S20910	0.06	4.00-6.00	1.00	20.50-23.50	11.50-13.50	1.50-3.00	0.20-0.40 N, 0.10-0.30 Cb, 0.10-0.30 V
XM-28	S24100	0.15	11.00-14.00	1.00	16.50-19.50	0.50-2.50	---	0.20-0.45 N
XM-29	S24000	0.08	11.50-14.50	1.00	17.00-19.00	2.50-3.75	---	0.20-0.40 N
20Cb-3	N08020	0.07	2.00	1.00	19.00-21.00	32.00-38.00	2.00-3.00	3.00-4.00 Cu, (8 x C) min to 1.00 (Cb) max
---	N08024	0.03	1.00	0.50	22.50-25.00	35.00-40.00	3.50-5.00	0.50-1.50 Cu, 0.15-0.35 (Cb)
20Mo6	N08026	0.03	1.00	0.50	22.00-26.00	33.00-37.20	5.00-6.70	2.00-4.00 Cu, 0.10-0.16 N
Sanicro 28	N08028	0.03	2.50	1.00	26.0-28.0	30.0-34.0	3.0-4.0	0.6-1.4 Cu
330 (RA-330)	N08330	0.08	2.00	0.75-1.50	17.0-20.0	34.0-37.0	---	1.00 Cu, 0.005 Pb, 0.025 Sn
AL-6X	N08366	0.035	2.00	1.00	20.0-22.0	23.5-25.5	6.0-7.0	---
AL-6XN	N08367	0.030	2.00	1.00	20.0-22.0	23.50-25.50	6.0-7.0	0.18-0.25 N
904L	N08904	0.020	2.00	1.00	19.0-23.0	23.0-28.0	4.00-5.00	1.00-2.00 Cu
---	N08925	0.020	1.00	0.50	19.00-21.00	24.00-26.00	6.0-7.0	0.8-1.5 Cu, 0.10-0.20 N
1925 hMo, 26-3 Mo	N08926	0.020	2.00	0.50	19.00-21.00	24.00-26.00	6.0-7.0	0.5-1.5 Cu, 0.15-0.25 N
---	N08931	0.015	2.0	0.3	26.0-28.0	30.0-32.0	6.0-7.0	1.0-1.4 Cu, 0.15-0.25 N
---	N08932	0.020	2.00	0.50	24.0-26.0	24.0-26.0	4.7-5.7	1.0-2.0 Cu, 0.17-0.25 N

a. Although sulfur and phosphorous contents are not listed in this table due to limited space, they are specified, see appropriate material standard for more details.   Single values are maximums, unless otherwise specified.

CHEMICAL COMPOSITION WROUGHT PRECIPITATION HARDENING STAINLESS STEELS[a]

Common Name	UNS	C	Mn	Si	Cr	Ni	Mo	Other
PH 13-8 Mo	S13800	0.05	0.20	0.10	12.25-13.25	7.50-8.50	2.00-2.50	0.90-1.35 Al, 0.01 N
PH 14-8 Mo	S14800	0.05	1.00	1.00	13.75-15.00	7.75-8.75	2.00-3.00	0.75-1.50 Al
15-5 PH	S15500	0.07	1.00	1.00	14.00-15.50	3.50-5.50	---	2.50-4.50 Cu, 0.15-0.45 Cb

CHEMICAL COMPOSITION WROUGHT PRECIPITATION HARDENING STAINLESS STEELS[a] (Continued)

Common Name	UNS	C	Mn	Si	Cr	Ni	Mo	Other
PH 15-7 Mo	S15700	0.09	1.00	1.00	14.00-16.00	6.50-7.75	2.00-3.00	0.75-1.50 Al
17-4PH	S17400	0.07	1.00	1.00	15.00-17.50	3.00-5.00	---	3.00-5.00 Cu, 0.15-0.45 Cb
17-7PH	S17700	0.09	1.00	1.00	16.00-18.00	6.50-7.75	---	0.75-1.50 Al
AM-350 (633)	S35000	0.07-0.11	0.5-1.25	0.50	16.00-17.00	4.00-5.00	2.50-3.25	0.07-0.13 N
AM-355 (634)	S35500	0.10-0.15	0.5-1.25	0.50	15.00-16.00	4.00-5.00	2.50-3.25	0.07-0.13 N
Custom 450 (XM-25)	S45000	0.05	1.00	1.00	14.00-16.00	5.00-7.00	0.50-1.00	(8 x C) Cb min, 1.25-1.75 Cu
Custom 455 (XM-16)	S45500	0.05	0.50	0.50	11.00-12.50	7.50-9.50	0.50 Mo	1.50-2.50 Cu, 0.80-1.40 Ti, 0.10-0.50 Cb

a. Although sulfur and phosphorous contents are not listed in this table due to limited space, they are specified, see appropriate material standard for more details. Single values are maximums, unless otherwise specified.

CHEMICAL COMPOSITION WROUGHT DUPLEX STAINLESS STEELS[a]

Common Name	UNS	C	Mn	Si	Cr	Ni	Mo	Other
3RE60	S31500	0.030	1.20-2.00	1.40-2.00	18.0-19.0	4.25-5.25	2.50-3.00	---
7-Mo Plus	S32950	0.030	2.00	0.60	26.0-29.0	3.50-5.20	1.00-2.50	0.15-0.35 N
44LN	S31200	0.030	2.00	1.00	24.0-26.0	5.50-6.50	1.20-2.00	0.14-0.20 N
329	S32900	0.08	1.00	0.75	23.0-28.0	2.50-5.00	1.00-2.00	---
2205	S31803	0.030	2.00	1.00	21.0-23.0	4.50-6.50	2.50-3.50	0.08-0.20 N
2304	S32304	0.030	2.50	1.0	21.5-24.5	3.0-5.5	---	0.05-0.60 Cu, 0.05-0.20 N
2507	S32750	0.030	1.20	0.8	24.0-26.0	6.0-8.0	3.0-5.0	0.24-0.32 N
AF918	S39277	0.025	---	0.80	24.0-26.0	6.5-8.0	3.0-4.0	1.2-2.0 Cu, 0.23-0.33 N, 0.80-1.20 W
DP3	S31260	0.03	1.00	0.75	24.0-26.0	5.50-7.50	2.50-3.50	0.20-0.80 Cu, 0.10-0.30 N, 0.10-0.50 W
DP3W	S39274	0.030	1.0	0.80	24.0-26.0	6.0-8.0	2.50-3.50	0.20-0.80 Cu, 0.24-0.32 N, 1.50-2.50 W
Ferralium 255	S32550	0.04	1.50	1.00	24.0-27.0	4.50-6.50	2.9-3.9	1.50-2.50 Cu, 0.10-0.25 N
Uranus 50	S32404	0.04	2.00	1.0	20.5-22.5	5.5-8.5	2.0-3.0	1.0-2.0 Cu, 0.20 N
Zeron 100	S32760	0.03	1.0	1.0	24.0-26.0	6.0-8.0	3.0-4.0	0.5-1.0 Cu, 0.2-0.3 N, 0.5-1.0 W, (Cr + 3.3 Mo)+16 x N = 40 min

CHEMICAL COMPOSITION WROUGHT DUPLEX STAINLESS STEELS[a] (Continued)

a. Although sulfur and phosphorous contents are not listed in this table due to limited space, they are specified, see appropriate material standard for more details. Single values are maximums, unless otherwise specified.

MECHANICAL PROPERTIES OF WROUGHT MARTENSITIC STAINLESS STEELS

AISI Type (UNS No.) Product Form	ASTM Specification	Heat Treat Condition[a]	Tensile Strength		Yield Strength		% El	% RA	Hardness, max
			ksi	MPa	ksi	MPa			
UNS S41000 (Type 410)									
Plate, sheet, strip	A 240	---	65	450	30	205	20	---	217 HB or 96 HRB
Bar, Shape	A 276	A, HF	70	480	40	275	20	45	---
	A 479	A	70	485	40	275	20	45	223 HB
		1	70	485	40	275	20	45	223 HB
		2	110	760	85	585	15	45	269 HB
		3	130	895	100	690	12	35	331 HB
Bar, Wire	A 493	A	82	565	---	---	---	---	---
		LD	85	585	---	---	---	---	---
Wire	A 580	A	70	485	40	275	20	45	---
		A, CF	70	485	40	275	16	45	---
		T, CF	100	690	80	550	12	40	---
		H, CF	120	830	90	620	12	40	---
Type 416 (UNS S41600)									
Wire	A 581	A	85-125	585-860	---	---	---	---	---
		T	115-145	790-1000	---	---	---	---	---
		H	140-175	965-1210	---	---	---	---	---

MECHANICAL PROPERTIES OF WROUGHT MARTENSITIC STAINLESS STEELS (Continued)

AISI Type (UNS No.) Product Form	ASTM Specification	Heat Treat Condition[a]	Tensile Strength		Yield Strength		% El	% RA	Hardness, max
			Ksi	MPa	ksi	MPa			
Type 416 (UNS S41600)									
Bar, Plate, Sheet, Strip	A 582, A 895	A	---	---	---	---	---	---	262 HB
		T	---	---	---	---	---	---	248-302 HB
		H	---	---	---	---	---	---	293-352 HB
Type 440A (UNS S44002)									
Bar, Shape	A 276	A, HF	---	---	---	---	---	---	269 HB
		A, CF	---	---	---	---	---	---	285 HB
Wire	A 580	A, CF	140 max	965 max	---	---	---	---	---
Type 440B (UNS S44003)									
Bar, Shape	A 276	A, HF	---	---	---	---	---	---	269 HB
		A, CF	---	---	---	---	---	---	285 HB
Wire	A 580	A, CF	140 max	965 max	---	---	---	---	---
Type 440C (UNS S44004)									
Bar, Shape	A 276	A, HF	---	---	---	---	---	---	269 HB
		A, CF	---	---	---	---	---	---	285 HB
Wire	A 580	A, CF	140 max	965 max	---	---	---	---	---

a. A - annealed, HF - hot finished, CF - cold finished, T - hardened & tempered at relatively high temperature, H - hardened and tempered at relatively low temperature, NT - normalized and tempered, LD - lightly drafted.

b. Typical values.

c. Tempered at 600°F (315°C).

Single values are minimums, except for hardness values that are maximums, unless otherwise specified.

MECHANICAL PROPERTIES OF WROUGHT FERRITIC STAINLESS STEELS

Common Name (UNS No.) Product Form	ASTM Specification	Heat Treat Condition[a]	Tensile Strength ksi	Tensile Strength MPa	Yield Strength ksi	Yield Strength MPa	% El	% RA	Hardness, max
AISI Type 409 (UNS S40900)									
Plate, sheet, strip	A 240	---	55	380	25	170	20.0	---	180 HB or 88 HRB
AISI Type 430 (UNS S43000)									
Plate, sheet, strip	A 240	---	65	450	30	205	22[b]	---	183 HB or 89 HRB
Bar, Shape	A 276	A, HF or CF	60	415	30	207	20	45	---
	A 479	A	70	485	40	275	20	45	192 HB
Wire	A 493	A	75	520	---	---	---	---	---
		LD	86	595	---	---	---	---	---
	A 580	A	70	485	40	275	20	45	---
		A, CF	70	485	40	275	16	45	---
Fitting	A 815 Gr WP430	---	65-90	450-620	35	240	20	---	190 HB
AISI Type 439 (UNS S43900)									
Plate, sheet, strip	A 240	A	60	415	30	205	22.0	---	183 HB or 89 HRB
Bar, Shapes	A 479	A	70	485	40	275	20[c]	45[c]	192 HB
E-Brite, Type XM-27, 26-1 (UNS S44627)									
Bar, Shape	A 276	A, HF	65	450	40	275	20	45	219 HB
		A, CF	65	450	40	275	16	45	219 HB
	A 479	A	65	450	40	275		45[c]	217 HB
Plate, sheet, strip	A 240	---	65	450	40	275	22	---	187 HB or 90 HRB
SC-1, 26-3-3 (UNS S44660)									
Plate, sheet, strip	A 240	---	85	585	65	450	18	---	241 HB or 100 HRB
25-4-4 (UNS S44635)									
Plate, sheet, strip	A 240	---	90	620	75	515	20	---	269 HB or 28 HRC
29-4C (UNS S44735)									
Plate, sheet, strip	A 240	---	80	550	60	415	18	---	255 HB or 25 HRC

MECHANICAL PROPERTIES OF WROUGHT FERRITIC STAINLESS STEELS (Continued)

Common Name (UNS No.) Product Form	ASTM Specification	Heat Treat Condition[a]	Tensile Strength		Yield Strength		% El	% RA	Hardness, max
			ksi	MPa	ksi	MPa			
29-4-2 (UNS S44800)									
Bar, Shape	A 276	HF	70	480	55	380	20	40	---
		CF	75	520	60	415	15	30	---
	A 479	A	70	485	55	380	20	40	---
Plate, sheet, strip	A 240	A	80	550	60	415	20	---	223 HB or 20 HRC
Wire	A 493	A	100	690	---	---	---	---	---
		LD	105	725	---	---	---	---	---

a. A - annealed, HF - hot finished, CF - cold finished, LD - lightly drafted.
b. 20% elongation minimum for 0.050 in (1.3mm) and under in thickness.
c. Elongation in 2 in. (50 mm) of 12% minimum and reduction of area of 35% minimum permitted for cold-finished bars.
Single values are minimums, except for hardness values that are maximums, unless otherwise specified.

MECHANICAL PROPERTIES OF WROUGHT AUSTENITIC STAINLESS STEELS[a]

Common Name (UNS No.) Product Form	ASTM Specification	Heat Treat Condition[b], (Size)	Tensile Strength ksi	MPa	Yield Strength ksi	MPa	% El	% RA	Hardness, max
Type 304 (UNS S30400)									
Plate, Sheet, Strip	A 240	A	75	515	30	205	40.0	---	201 HB or 92 HRB
Bar, Shape	A 276	A, HF	75	515	30	205	40[c]	50	---
		A, CF (≤ ½ in.)	90	620	45	310	30	40	---
		A, CF (> ½ in.)	75	515	30	205	30	40	---
Wire	A 479	A	75	515	30	205	30	40	---
	A 580	A	75	520	30	210	35[d]	50[d]	---
		CF	90	620	45	310	30[d]	40[d]	---
Plate, Sheet, Strip, Flat Bar	A 666	A	75	515	30	205	40	---	201 HB or 92 HRB
		⅛ H	100	690	55	380	35	---	---
		¼ H	125	860	75	515	10[d]	---	---
		½ H	150	1035	110	760	7[f]	---	---
Type 304L (UNS S30403)									
Plate, Sheet, Strip	A 240	A	70	485	25	170	40.0	---	201HB or 92 HRB
Bar, Shape	A 276	A, HF	70	485	25	170	40[c]	50	---
		A, CF[b] (≤ ½ in.)	90	620	45	310	30	40	---
		A, CF (> ½ in.)	70	485	25	170	30	40	---
Wire	A 479	A	70	485	25	170	30	40	---
	A 580	A, CF	70	485	25	170	35[d]	50[d]	---
			90	620	45	310	30[d]	40[d]	---
Plate, Sheet, Strip, Flat Bar	A 666	A	70	485	25	170	40	---	201 HB or 92 HRB
		⅛ H	100	690	55	380	30	---	---
		¼ H	125	860	75	515	8[e]	---	---
		½ H	150	1035	110	760	6[f]	---	---

MECHANICAL PROPERTIES OF WROUGHT AUSTENITIC STAINLESS STEELS^a (Continued)

Common Name (UNS No.) / Product Form	ASTM Specification	Heat Treat Condition^b, (Size)	Tensile Strength		Yield Strength		% El	% RA	Hardness, max
			ksi	MPa	ksi	MPa			
Type 308 (UNS S30800)									
Bar, Shape	A 276	A, HF	75	515	30	205	40[c]	50	---
		A, CF (≤ ½ in.)	90	620	45	310	30	40	---
		A, CF (> ½ in.)	75	515	30	205	30	40	---
Wire	A 580	A	75	520	30	210	35[d]	50[d]	---
		CF	90	620	45	310	30[d]	40[d]	---
Type 309 (UNS S30900)									
Bar, Shape	A 276	A, HF	75	515	30	205	40[c]	50	---
		A, CF (≤ ½ in.)	90	620	45	310	30	40	---
		A, CF (> ½ in.)	75	515	30	205	30	40	---
Wire	A 580	A	75	520	30	210	35[d]	50[d]	---
		CF	90	620	45	310	30[d]	40[d]	---
Type 310 (UNS S31000)									
Bar, Shape	A 276	A, HF	75	515	30	205	40[c]	50	---
		A, CF (≤ ½ in.)	90	620	45	310	30	40	---
		A, CF (> ½ in.)	75	515	30	205	30	40	---
Wire	A 580	A	75	520	30	210	35[d]	50[d]	---
		CF	90	620	45	310	30[d]	40[d]	---
Type 316 (UNS S31600)									
Plate, Sheet, Strip	A 240	---	75	515	30	205	40.0	---	---
Bar, Shape	A 276	A, HF	75	515	30	205	40[c]	50	217 HB or 95 HRB
		A, CF (≤ ½ in.)	90	620	45	310	30	40	---
		A, CF (> ½ in.)	75	515	30	205	30	40	---
	A 479	A	75	515	30	205	30	40	---

MECHANICAL PROPERTIES OF WROUGHT AUSTENITIC STAINLESS STEELS[a] (Continued)									
Common Name (UNS No.) Product Form	ASTM Specification	Heat Treat Condition[b], (Size)	Tensile Strength		Yield Strength		% El	% RA	Hardness, max
			ksi	MPa	ksi	MPa			
Type 316 (UNS S31600) (Continued)									
Bar, Shape	A 479	SH, L-1	85	585	65	450	30	60	---
		SH, level 2, ≤ 2 in.	95	655	75	515	25	40	---
		SH level 2, > 2 to 2½ in., incl.	90	620	65	450	30	40	---
		SH, level 2, > 2½ to 3 in., incl.	80	550	55	380	30	40	---
Wire	A 580	A	75	520	30	210	35[d]	50[d]	---
		CF	90	620	45	310	30[d]	40[d]	---
Plate, Sheet, Strip, Flat Bar	A 666	A	75	515	30	205	40	---	217 HB or 95 HRB
		⅛ H	100	690	55	380	30	---	---
		¼ H	125	860	75	515	10	---	---
		½ H	150	1035	110	760	6[f]	---	---
Type 316L (UNS S31603)									
Plate, Sheet, Strip	A 240	A	70	485	25	170	40.0	---	217 HB or 95 HRB
Bar, Shape	A 276	A, HF	70	485	25	170	40	50	---
		A, CF[b] (≤ ½ in.)	90	620	45	310	30	40	---
		A, CF[c] (> ½ in.)	70	480	25	170	30	40	---
Wire	A 479	A	70	485	25	170	30	40	---
	A 580	A	70	485	25	170	35[d]	50[d]	---
		A, CF	90	620	45	310	30[d]	40[d]	---
Plate, Sheet, Strip, Flat Bar	A 666	A	70	485	25	170	40	---	217 HB or 95 HRB
		⅛ H	100	690	55	380	25	---	---
		¼ H	125	860	75	515	8	---	---

MECHANICAL PROPERTIES OF WROUGHT AUSTENITIC STAINLESS STEELS[a] (Continued)									
Common Name (UNS No.) Product Form	ASTM Specification	Heat Treat Condition[b], (Size)	Tensile Strength		Yield Strength		% El	% RA	Hardness, max
			ksi	MPa	ksi	MPa			
Type 316L (UNS S31603) (Continued)									
Plate, Sheet, Strip, Flat Bar	A 666	½ H	150	1035	110	760	6[f]	---	---
Type 316Cb (UNS S31640)									
Plate, Sheet, Strip	A 240	A	75	515	30	205	30.0	---	217 HB or 95 HRB
Bar, Shape	A 276	A, HF	75	515	30	205	40[c]	50	---
		A, CF (≤ ½ in.)	90	620	45	310	30	40	---
		A, CF (> ½ in.)	75	515	30	205	30	40	---
	A 479	A	75	515	30	205	30	40	---
Type 317 (UNS S31700)									
Plate, Sheet, Strip	A 240	A	75	515	30	205	35.0	---	217 HB or 95 HRB
Bar, Shape	A 276	A, HF	75	515	30	205	40[c]	50	---
		A, CF (≤ ½ in.)	90	620	45	310	30	40	---
		A, CF (> ½ in.)	75	515	30	205	30	40	---
	A 479	A	75	515	30	205	30	40	---
Wire	A 580	A	75	520	30	210	35[d]	50[d]	---
		CF	90	620	45	310	30[d]	40[d]	---
Type 317L (UNS S31703)									
Plate, Sheet, Strip	A 240	A	75	515	30	205	40.0	---	217 HB or 95 HRB
Type 321 (UNS S32100)									
Plate, Sheet, Strip	A 240	A	75	515	30	205	40.0	---	217 HB or 95 HRB
Bar, Shape	A 276	A, HF	75	515	30	205	40[c]	50	---
		A, CF (≤ ½ in.)	90	620	45	310	30		

Common Name (UNS No.) Product Form	ASTM Specification	Heat Treat Condition^b, (Size)	Tensile Strength		Yield Strength		% El	% RA	Hardness, max
			ksi	MPa	ksi	MPa			
Type 321 (UNS S32100) (Continued)									
Bar, Shape (Continued)	A 479	A, CF (> ½ in.)	75	515	30	205	30	40	---
		A	75	515	30	205	30	40	---
Wire	A 580	A	75	520	30	210	35^d	50^d	---
		CF	90	620	45	310	30^d	40^d	---
Type 347 (UNS S34700)									
Plate, Sheet, Strip	A 240	A	75	515	30	205	40.0	---	201 HB or 92 HRB
Bar, Shape	A 276	A, HF	75	515	30	205	40^c	50	---
		A, CF (≤ ½ in.)	90	620	45	310	30	40	---
		A, CF (> ½ in.)	75	515	30	205	30	40	---
Wire	A 479	A	75	515	30	205	30	40	---
	A 580	A	75	520	30	210	35^d	50^d	---
		CF	90	620	45	310	30^d	40^d	---
Type 348 (UNS S34800)									
Plate, Sheet, Strip	A 240	A	75	515	30	205	40.0	---	201 HB or 92 HRB
Bar, Shape	A 276	A, HF	75	515	30	205	40^c	50	---
		A, CF (≤ ½ in.)	90	620	45	310	30	40	---
		A, CF (> ½ in.)	75	515	30	205	30	40	---
Wire	A 479	A	75	515	30	205	30	40	---
	A 580	A	75	520	30	210	35^d	50^d	---
		CF	90	620	45	310	30^d	40^d	---
20Cb-3 (UNS N08020), 20Mo-4 (UNS N08024), 20Mo-6 (UNS N08026)									
Pipe Flanges, Forged Fittings, Valves & Parts	B 462	A	80	551	35	241	30.0	50.0	---

MECHANICAL PROPERTIES OF WROUGHT AUSTENITIC STAINLESS STEELS[a] (Continued)

Common Name (UNS No.) Product Form	ASTM Specification	Heat Treat Condition[b] (Size)	Tensile Strength		Yield Strength		% El	% RA	Hardness, max
			ksi	MPa	ksi	MPa			
20Cb-3 (UNS N08020), 20Mo-4 (UNS N08024), 20Mo-6 (UNS N08026) (Continued)									
Plate, Sheet, Strip	B 463	A	80	551	35	241	30.0	---	217 HB or 95 HRC
Bar, Wire[g]	B 473	A, HF or CF	80	551	35	241	30.0	50.0	---
		A, SH	90	620	60	415	15.0	40.0	---
Sanicro 28 (UNS N08028)									
Plate, Sheet, Strip	B 709	A	73	500	31	214	40	---	70-90 HRB[h]
Type 330 (RA-330) (UNS N08330)									
Bar	B 511	A	70	483	30	207	30	---	---
Plate, Sheet, Strip	B 536	A	70	483	30	207	30	---	70-90 HRB[h]
AL-6X (UNS N08366)									
Plate, Sheet, Strip	B 688	A	75	515	35	240	30	---	95 HRB[h] (< $^3/_{16}$ in.)
Rod, Bar, Wire	B 691	A, HF or CF	75	517	30	206	30	---	---
Type 904L (UNS N08904)									
Plate, Sheet, Strip	B 625	A	71	490	31	215	35	---	70-90 HRB[h]
Bar	B 649	A, HF or CF	71	490	31	220	35	---	---
Wire	B 649	A, CF	90-120	620-830	---	---	---	---	---

a. Fe-Ni-Cr-Mo alloys and their variations have been included in this table of "austenitic stainless steels".
b. A - annealed, HF - hot finished, CF - cold finished, H - hard, SH - strain hardened, CW - cold worked.
c. For shapes having section thickness of ¼ in. (6.5 mm) or less, 30% minimum elongation is acceptable.
d. For wire $^9/_{32}$ in. (3.96 mm) and under, elongation and reduction in area are 25% and 40%, respectively.
e. For thickness up to 0.030 in.
f. For thickness 0.015 and greater.
g. For wire only, tensile strength 90 to 120.0 ksi (620 to 830 MPa); no requirements on yield strength, elongation, and reduction of area.
h. Hardness values shown for information only, not a specified property for the ASTM standard.
Single values are minimums, except for hardness values that are maximums, unless otherwise specified.

MECHANICAL PROPERTIES OF AUSTENITIC STAINLESS STEELS FOR SURGICAL IMPLANTS

Product Form	ASTM Specification	Heat Treat Condition	Tensile Strength		Yield Strength		% El	Hardness, max
			ksi	MPa	ksi	MPa		
Bar and Wire	F 138 Gr 1	Annealed	75	515	30	205	40	---
		Hot Worked	---	---	---	---	---	275 HB
		Cold Worked (1/16 - 3/4 in.)	125	860	100	690	12	---
	F 138 Gr 2	Hot Worked	---	---	---	---	---	250 HB
		Annealed	70	480	25	170	40	---
		Cold Worked (1/16 - 3/4 in.)	125	860	100	690	12	---
Sheet and Strip	F 139 Gr 1	Cold Worked	125	860	100	690	10	95 HRB
		Annealed	75	515	30	205	40	---
	F 139 Gr 2	Cold Worked	125	860	100	690	10	---
		Annealed	70	480	25	170	40	95 HRB

Single values are minimums, except for hardness values that are maximums. See page 465 for chemical compositions of F 138 and F 139 grades.

MECHANICAL PROPERTIES OF WROUGHT HIGH NITROGEN AUSTENITIC STAINLESS STEELS[a]

Common Name (UNS No.) / Product Form	ASTM Specification	Heat Treat Condition[b]	Tensile Strength		Yield Strength		% El	% RA	Hardness, max
			ksi	MPa	ksi	MPa			
Type 304N (UNS S30451)									
Plate, Sheet, Strip	A 240	A	80	550	35	240	30	---	201 HB or 92 HRB
Bar, Shape	A 276	A, HF or CF	80	550	35	240	30	---	---
	A 479	A	80	550	35	240	30	40	---
Type 304LN (UNS 30453)									
Plate, Sheet, Strip	A 240	A	75	515	30	205	40.0	---	201 HB or 92 HRB
Bar, Shape	A 276	A, HF	75	515	30	205	40	50	---
		CF to 1/2 in., incl.	90	620	45	310	30	40	---
		CF > 1/2 in.	75	515	30	205	30	40	---
	A 479	A	75	515	30	205	30	40	---

MECHANICAL PROPERTIES OF WROUGHT HIGH NITROGEN AUSTENITIC STAINLESS STEELS[a] (Continued)

Common Name (UNS No.) Product Form	ASTM Specification	Heat Treat Condition[b]	Tensile Strength ksi	MPa	Yield Strength ksi	MPa	% El	% RA	Hardness, max
Type 316N (UNSS31651)									
Plate, Sheet, Strip	A 240	A	80	550	35	240	35.0	---	217 HB or 95 HRB
Bar, Shape	A 276	A, HF or CF	80	550	35	240	30	---	---
	A 479	A	80	550	35	240	30	40	---
316LN (UNS S31653)									
Plate, Sheet, Strip	A 240	A	75	515	30	205	40.0	---	217 HB or 95 HRB
Bar, Shape	A 276	A, HF	75	515	30	205	40	50	---
		CF to ½ in., incl.	90	620	45	310	30	40	---
		CF > ½ in.	75	515	30	205	30	40	---
	A 479	A	75	515	30	205	30	40	---
317LN (UNS S31753)									
Plate, Sheet, Strip	A 240	A	80	550	35	240	40.0	---	217 HB or 95 HRB
Nitronic 50 (XM-19) (UNS S20910)									
Sheet, Strip	A 240	A	105	725	60	415	30.0	---	241 HB or 100 HRB
Plate	A 240	A	100	690	55	380	35.0	---	241 HB or 100 HRB
Bar, Shape	A 276	A, HF or CF	100	690	55	380	35	55	---
		HR, HF or CF, to 2 in., incl.	135	930	105	725	20	50	---
		HR, HF or CF, > 2 to 3 in., incl.	115	795	75	515	25	50	---
		HR, HF or CF, > 3 to 8 in., incl.	100	690	60	415	30	50	---
	A 479	A	100	690	55	380	35	55	293 HB
		HR, to 2 in., incl.	135	930	105	725	20	50	---

MECHANICAL PROPERTIES OF WROUGHT HIGH NITROGEN AUSTENITIC STAINLESS STEELS[a] (Continued)

Common Name (UNS No.) / Product Form	ASTM Specification	Heat Treat Condition[b]	Tensile Strength		Yield Strength		% El	% RA	Hardness, max
			ksi	MPa	ksi	MPa			
Nitronic 50 (XM-19) (UNS S20910) (Continued)									
Bar, Shape	A 479	HR, > 2 to 3 in., incl.	115	795	75	515	25	50	---
		HR, > 3 to 8 in., incl.	100	690	60	415	30	50	---
		SH, to 1½ in., incl	145	1000	125	860	12	40	---
		SH, > 1½ to 2¼ in., incl	120	825	105	725	15	45	---
Wire	A 580	A, CF	100	690	55	380	35	55	---
Nitronic 40 (XM-10) (UNS S21900)									
Bar, Shape	A 276	A, HF or CF	90	620	50	345	45	60	---
Wire	A 580	A, HF or CF	90	620	50	345	45	60	---
Nitronic 33 (XM-29) (UNS S24000)									
Sheet, Strip	A 240	A	100	690	60	415	40.0	---	241 HB or 100 HRB
Plate	A 240	A	100	690	55	380	40.0	---	241 HB or 100 HRB
Bar, Shape	A 276	A, HF or CF	100	690	55	380	30	50	---
	A 479	A	100	690	55	380	30	50	293 HB
Wire	A 580	A, CF	100	690	55	380	30	50	---
Nitronic 32 (XM-28) (UNS S24100)									
Bar, Shape	A 276	A, HF or CF	100	690	55	380	30	50	---
Wire	A 580	A, CF	100	690	55	380	30	50	---
254 SMO (UNS S31254)									
Plate, Sheet, Strip	A 240	A	94	650	44	300	35.0	---	223 HB or 96 HRB
Bar, Shape	A 276	A, HF or CF	95	650	44	300	35	50	---

MECHANICAL PROPERTIES OF WROUGHT HIGH NITROGEN AUSTENITIC STAINLESS STEELS^a (Continued)

Common Name (UNS No.) Product Form	ASTM Specification	Heat Treat Condition^b	Tensile Strength		Yield Strength		% El	% RA	Hardness, max
			ksi	MPa	ksi	MPa			
254 SMO (UNS S31254) (Continued)									
Bar, Shape	A 479	A	95	655	44	305	35	50	---
AL-6XN (UNS N08367)									
Pipe Flanges, Forged Fittings, Valves & Parts	B 462	A	95	655	45	310	30.0	50.0	---
Plate, Sheet, Strip	B 688	A (< 3/16 in.)	104	717	46	317	30	---	100 HRB^h
		A (3/16 - 3/4 in.)	100	690	45	310	30	---	240 HB^h
		A (> 3/4 in.)	95	655	45	310	30	---	233 HB^h
Rod, Bar, Wire	B 691	A, HF or CF	95	655	45	310	30	---	---
UNS N08925									
Plate, Sheet, Strip	B 625	A	87	600	43	295	40	---	---
Bar	B 649	A, HF or CF	87	600	43	300	40	---	---
Wire	B 649	A, CF	90-120	620-830	---	---	---	---	---
UNS N08926									
Plate, Sheet, Strip	B 625	A	94	650	43	295	35	---	---
Bar	B 649	A, HF or CF	94	650	40	270	40	---	---
Wire	B 649	A, CF	90-120	620-830	---	---	---	---	---
UNS N08931									
Plate, Sheet, Strip	B 625	A	94	650	40	276	40	---	---
Bar	B 649	A, HF or CF	94	650	40	270	40	---	---
Wire	B 649	A, CF	90-120	620-830	---	---	---	---	---
UNS N08932									
Plate	B 625	A	87	600	44	305	40	---	---

MECHANICAL PROPERTIES OF WROUGHT HIGH NITROGEN AUSTENITIC STAINLESS STEELSᵃ (Continued)

a. Fe-Ni-Cr-Mo alloys and their variations have been included in this table of "austenitic stainless steels".
b. A - annealed, HF - hot finished, CF - cold finished, SH – strain-hardened.
c. Hardness values shown for information only, not a specified property for the ASTM standard.
Single values are minimums, except for hardness values that are maximums, unless otherwise specified.

MECHANICAL PROPERTIES OF WROUGHT DUPLEX STAINLESS STEELS

Common Name (UNS No.) / Product Form	ASTM Specification	Heat Treat Conditionᵃ	Tensile Strength ksi	MPa	Yield Strength ksi	MPa	% El	% RA	Hardness, max
44LN (UNS S31200)									
Sheet, Plate, Strip	A 240	---	100	690	65	450	25.0	---	293 HB or 31 HRC
DP-3 (UNS S31260)									
Sheet, Plate, Strip	A 240	---	100	690	70	485	20.0	---	290 HB
2205 (UNS S31803)									
Plate, Sheet, Strip	A 240	---	90	620	65	450	25.0	---	293 HB or 31 HRC
Bar, Shape	A 276	A, HF or CF	90	620	65	450	25	---	290 HB
	A 479	A	90	620	65	450	25	---	290 HB
Piping Fitting	A 815	---	90	620	65	450	20.0	---	290 HB
2304 (UNS S32304)									
Sheet, Plate, Strip	A 240	---	87	600	58	400	25.0	---	290 HB or 32 HRC
Ferralium 255 (UNS S32550)									
Plate, Sheet, Strip	A 240	---	110	760	80	550	15.0	---	302 HB or 32 HRC
Bar, Shapes	A 479	A	110	760	80	550	15	---	297 HB
2507 (UNS S32750)									
Plate, Sheet, Strip	A 240	A	116	795	80	550	15.0	---	310 HB or 32 HRC
Bar, Shape	A 479	A (≤ 2 in.)	116	800	80	550	15	---	310 HB
		A (> 2 in.)	110	760	75	515	15	---	310 HB

MECHANICAL PROPERTIES OF WROUGHT DUPLEX STAINLESS STEELS (Continued)									
Common Name (UNS No.)	ASTM	Heat Treat	Tensile Strength		Yield Strength				Hardness, max
Product Form	Specification	Condition[a]	ksi	MPa	ksi	MPa	% El	% RA	
Zeron 100 (UNS S32760)									
Plate, Sheet, Strip	A 240	---	108	750	80	550	25.0	---	270 HB
AISI Type 329 (UNS S32900)									
Plate, Sheet, Strip	A 240	---	90	620	70	485	15.0	---	269 HB or 28 HRC
Tube	A 789	---	90	620	70	485	20	---	271 HB or 28 HRC
Pipe	A 790	---	90	620	70	485	20	---	271 HB or 28 HRC
7-Mo PLUS (UNS S32950)									
Plate, Sheet, Strip	A 240	---	100	690	70	485	15.0	---	293 HB or 32 HRC
Bar, Shapes	A 479	A	100	690	70	485	15	---	297 HB
Tube	A 789	---	100	690	70	480	20	---	290 HB or 30.5 HRC
Pipe	A 790	---	100	690	70	480	20	---	290 HB or 30.5 HRC
Piping Fitting	A 815	---	100	690	70	485	15.0	---	290 HB

a. A - annealed, HF - hot finished, CF - cold finished.
Single values are minimums, except for hardness values that are maximums, unless otherwise specified.

MECHANICAL PROPERTIES OF WROUGHT PRECIPITATION-HARDENING STAINLESS STEELS

Common Names (UNS No.)	ASTM Spec.	Heat Treat Condition[a]	Tensile Strength ksi	Tensile Strength MPa	Yield Strength ksi	Yield Strength MPa	% El[b] L, T	% RA[b] L, T	Hardness, max	Charpy ft·lb/J
PH 13-8 Mo, XM-13 (UNS S13800)										
Bar, Shape	A 564	A	---	---	---	---	---	---	38 HRC/363 HB	---
		H950	220	1520	205	1410	10	45, 35	45 HRC/430 HB	---
		H1000	205	1410	190	1310	10	50, 40	43 HRC/400 HB	---
		H1025	185	1275	175	1210	11	50, 45	41 HRC/380 HB	---
		H1050	175	1210	165	1140	12	50, 45	40 HRC/372 HB	---
		H1100	150	1030	135	930	14	50	34 HRC/313 HB	---
		H1150	135	930	90	620	14	50	30 HRC/283 HB	---
		H1150M	125	860	85	585	16	55	26 HRC/259 HB	---
Plate, Sheet, Strip (0.1875-0.625 in.)	A 693	A	---	---	---	---	---	---	38 HRC/363 HB	---
		H950	220	1515	205	1410	10	50	45 HRC	---
		H1000	200	1380	190	1310	10	55	43 HRC	---
15-5 PH, XM-12 (UNS S15500)										
Bar, Shape	A 564	A	---	---	---	---	---	---	38 HRC/363 HB	---
		H900	190	1310	170	1170	10, 6	35, 15	40 HRC/388 HB	---
		H925	170	1170	155	1070	10, 7	38, 20	38 HRC/375 HB	5/6.8 L, ---
		H1025	155	1070	145	1000	12, 8	45, 27	35 HRC/331 HB	15/20 L, 10/14 T
		H1075	145	1000	125	860	13, 9	45, 28	32 HRC/311 HB	20/27 L, 15/20 T
		H1100	140	965	115	795	14, 10	45, 29	31 HRC/302 HB	25/34 L, 15/20 T
		H1150	135	930	105	725	16, 11	50, 30	28 HRC/277 HB	30/41 L, 20/27 T
		H1150M	115	795	75	515	18, 14	55, 35	24 HRC/255 HB	55/75 L, 35/47 T
Plate, Sheet, Strip (0.1875-0.625 in.)	A 693	A	---	---	---	---	---	---	38 HRC/363 HB	---
		H900	190	1310	170	1170	8	25	40-48 HRC/388-477 HB	---
		H925	170	1170	155	1070	8	25	38-47 HRC/375-477 HB	---
		H1025	155	1070	145	1000	8	30	33-42 HRC/321-415 HB	10/14 optional

MECHANICAL PROPERTIES OF WROUGHT PRECIPITATION-HARDENING STAINLESS STEELS (Continued)

Common Names (UNS No.)	ASTM Spec.	Heat Treat Condition[a]	Tensile Strength ksi	Tensile Strength MPa	Yield Strength ksi	Yield Strength MPa	% El[b] L, T	% RA[b] L, T	Hardness, max	Charpy ft-lb/J
15-5 PH, XM-12 (UNS S15500) (Continued)										
Plate, Sheet, Strip	A 693	H1075	145	1000	125	860	9	30	29-38 HRC/293-375 HB	15/20 optional
		H1100	140	965	115	790	10	30	29-38 HRC/293-375 HB	15/20 optional
		H1150	135	930	105	725	10	35	26-36 HRC/269-352 HB	25/34 optional
		H1150M	115	790	75	515	11	40	24-34 HRC/248-321 HB	55/75 optional
PH 15-7 Mo, Type 632 (UNS S15700)										
Bar, Shape	A 564	A	---	---	---	---	---	---	100 HRB/269 HB	---
		RH950	200	1380	175	1210	7	25	415 HB	---
		TH1050	180	1240	160	1100	8	25	375 HB	---
Plate, Sheet, Strip	A 693	A	150 max	1035 max	65 max	450 max	25	---	100 HRB	---
		CR	200	1380	175	1205	1	---	41 HRC	---
		CR+H900	240	1655	230	1585	1	---	46 HRC	---
17-4 PH, Type 630 (UNS S17400)										
Bar, Shape (≤ 3 in. - 75 mm)	A 564	A	---	---	---	---	---	---	38 HRC/363 HB	---
		H900	190	1310	170	1170	10	40	40 HRC/388 HB	---
		H925	170	1170	155	1070	10	44	38 HRC/375 HB	5/6.8 L
(≤ 8 in. - 200 mm)		H1025	155	1070	145	1000	12	45	35 HRC/331 HB	15/20 L
		H1075	145	1000	125	860	13	45	32 HRC/311 HB	20/27 L
		H1100	140	965	115	795	14	45	31 HRC/302 HB	25/34 L
		H1150	135	930	105	725	16	50	28 HRC/277 HB	30/41 L
		H1150M	115	795	75	515	18	55	24 HRC/255 HB	55/75 L
Plate, Sheet, Strip (0.1875-0.625 in.)	A 693	A	185	1255	160	1105	3	---	38 HRC/363 HB	---
		H900	190	1310	170	1170	8	25	40-48 HRC/388-477 HB	---
		H925	170	1170	155	1070	8	25	38-47 HRC/375-477 HB	---

MECHANICAL PROPERTIES OF WROUGHT PRECIPITATION-HARDENING STAINLESS STEELS (Continued)

Common Names (UNS No.)	ASTM Spec.	Heat Treat Condition[a]	Tensile Strength		Yield Strength		% El[b]	% RA[b]	Hardness, max	Charpy
			ksi	MPa	ksi	MPa	L, T	L, T		ft·lb/J
17-4 PH, Type 630 (UNS S17400) (Continued)										
Plate, Sheet, Strip (0.1875-0.625 in.)	A 693	H1025	155	1070	145	1000	8	30	33-42 HRC/321-415 HB	10/14 optional
		H1075	145	1000	125	860	9	30	29-38 HRC/293-375 HB	15/20 optional
		H1100	140	965	115	790	10	30	29-38 HRC/293-375 HB	15/20 optional
		H1150	135	930	105	725	10	35	26-36 HRC/269-352 HB	25/34 optional
		H1150M	115	790	75	515	11	40	24-34 HRC/248-321 HB	55/75 optional
Stainless W, Type 635 (UNS S17600)										
Bar, Shape	A 564	A	120	825	75	515	10	45	32 HRC or 302 HB	--
		H950	190	1310	170	1170	8	25	39 HRC/363 HB	--
		H1000	180	1240	160	1100	8	30	37 HRC/352 HB	--
		H1050	170	1170	150	1030	10	40	35 HRC/331 HB	--
Plate (only)	A 693	A	120 max	825 max	75 max	515 max	5	--	32 HRC	--
		H950	190	1310	170	1170	8	25	39 HRC/363 HB	--
		H1000	180	1240	160	1105	8	30	38 HRC/352 HB	--
		H1050	170	1170	150	1035	8	30	36 HRC/331 HB	--
17-7 PH, Type 631 (UNS S17700)										
Bar, Shape	A 564	A	--	--	--	--	--	--	98 HRB/229 HB	--
		RH950	185	1275	150	1035	6	10	41 HRC/388 HB	--
		TH1050	170	1170	140	965	6	25	38 HRC/352 HB	--
Plate, Sheet, Strip	A 693	A	150 max	1035 max	55 max	380 max	20	--	92 HRB	--
		CR	200	1380	175	1205	1	--	41 HRC	--
		CR+H900	240	1655	230	1580	1	--	46 HRC	--

MECHANICAL PROPERTIES OF WROUGHT PRECIPITATION-HARDENING STAINLESS STEELS (Continued)

Common Names (UNS No.)	ASTM Spec.	Heat Treat Condition[a]	Tensile Strength ksi	Tensile Strength MPa	Yield Strength ksi	Yield Strength MPa	% El[b] L, T	% RA[b] L, T	Hardness, max	Charpy ft-lb/J
AM-350, Type 633 (UNS S35000)										
Plate, Sheet, Strip (> 0.010 in.)	A 693	A	200 max	1380 max	85 max	585 max	12	---	30 HRC	---
(0.0100-0.1875 in.)		H850	185	1275	150	1035	8	---	42 HRC	---
(0.0015-0.0020 in.)		H1000	165	1140	145	1000	4	---	36 HRC	---
AM-355, Type 634 (UNS S35500)										
Bar, Shape	A 564	A	---	---	---	---	---	---	363 HB	---
		H1000	170	1170	155	1070	12	25	37 HRC/341 HB	---
Plate	A 693	A	---	---	---	---	---	---	40 HRC	---
Plate, Sheet, Strip	A 693	H850	190	1310	165	1140	10	---	---	---
		H1000	170	1170	150	1035	12	---	37 HRC	---
Custom 450, XM-25 (UNS S45000)										
Bar, Shape (≤ 8 in. - 200 mm)	A 564	A	130[c]	895[c]	95	655	10	40	32 HRC/321 HB	---
		H900	180	1240	170	1170	10	40	39 HRC/363 HB	---
		H950	170	1170	160	1100	10	40	37 HRC/341 HB	---
		H1000	160	1100	150	1030	12	45	36 HRC/331 HB	---
		H1025	150	1030	140	965	12	45	34 HRC/321 HB	---
		H1050	145	1000	135	930	12	45	34 HRC/321 HB	---
		H1100	130	895	105	725	16	50	30 HRC/285 HB	---
		H1150	125	860	75	520	15	50	26 HRC/262 HB	---
Plate, Sheet, Strip (> 0.062 in.)	A 693	A	165 max	1205 max	150 max	1035 max	4	---	33 HRC/311 HB	---
		H900	180	1240	170	1170	5	---	40 HRC	---
		H1000	160	1105	150	1035	7	---	36 HRC	---
		H1150	125	860	75	515	10	---	26 HRC	---

MECHANICAL PROPERTIES OF WROUGHT PRECIPITATION-HARDENING STAINLESS STEELS (Continued)

Common Names (UNS No.)	ASTM Spec.	Heat Treat Condition[a]	Tensile Strength ksi	MPa	Yield Strength ksi	MPa	% El[b] L, T	% RA[b] L, T	Hardness, max	Charpy ft·lb/J
Custom 455, XM-16 (UNS S45500)										
Bar, Shape (≤ 6 in. - 150 mm)	A 564	A	---	---	---	---	---	---	36 HRC/331 HB	---
		H900	235	1620	220	1520	8	30	47 HRC/444 HB	---
		H950	220	1520	205	1410	10	40	44 HRC/415 HB	---
		H1000	205	1410	185	1280	10	40	40 HRC/363 HB	---
Plate, Sheet, Strip (> 0.062 in.)	A 693	A	175 max	1205 max	160 max	1105 max	3	---	36 HRC/331 HB	---
		H950	222	1525	205	1410	4	---	44 HRC	---

a Heat treatment condition A is solution treated, CR - cold rolled.
b First value is longitudinal, second value is transverse, if single value is listed it represents both longitudinal and transverse tests.
c Tensile strengths of 130 to 165 ksi (895 to 1140 MPa) for sizes up to ½ in. (13 mm).
Single values are minimums, unless otherwise specified.

Chapter

19

CAST STAINLESS STEELS:

AMERICAN SPECIFICATION TITLES & DESIGNATIONS, CHEMICAL COMPOSITIONS & MECHANICAL PROPERTIES

STAINLESS STEEL CASTINGS

ASTM Spec.	Title
A 217/A 217M	Steel Castings, Martensitic Stainless and Alloy, for Pressure-Containing Parts, Suitable for High-Temperature Service
A 297/A 297M	Steel Castings, Iron-Chromium and Iron Chromium-Nickel, Heat Resistant, for General Application
A 351/A 351M	Castings, Austenitic, Austenitic-Ferritic (Duplex), for Pressure-Containing Parts
A 352/A 352M	Steel Castings, Ferritic and Martensitic, for Pressure-Containing Parts, Suitable for Low-Temperature Service
A 356/A 356M	Steel Castings, Carbon, Low Alloy, and Stainless Steel, Heavy-Walled for Steam Turbines
A 447/A 447M	Steel Castings, Chromium-Nickel-Iron Alloy (25-12 Class), for High-Temperature Service
A 451	Centrifugally Cast Austenitic Steel Pipe for High-Temperature Service
A 452	Centrifugally Cast Austenitic Steel Cold-Wrought Pipe for High-Temperature Service
A 487/A 487M	Steel Castings, Suitable for Pressure Service
A 703/A 703M	Steel Castings, General Requirements, for Pressure-Containing Parts
A 743/A 743M	Castings, Iron-Chromium, Iron Chromium-Nickel, Corrosion Resistant, for General Application
A 744/A 744M	Castings, Iron-Chromium-Nickel, Corrosion Resistant, for Severe Service
A 747/A 747M	Steel Castings, Stainless, Precipitation Hardening
A 757/A 757M	Steel Castings, Ferritic and Martensitic, for Pressure-Containing and Other Applications, for Low-Temperature Service
A 799/A 799M	Steel Castings, Stainless, Instrument Calibration, for Estimating Ferrite Content
A 800/A 800M	Steel Castings, Austenitic Alloy, Estimating Ferrite Content Thereof
A 890/A 890M	Castings, Iron-Chromium-Nickel-Molybdenum Corrosion-Resistant, Duplex (Austenitic/Ferritic) for General Application

CHEMICAL COMPOSITION OF CORROSION-RESISTANT CAST STAINLESS STEELS[a]

Grade	UNS	ASTM Spec.	C	Si	Cr	Ni	Other	Similar[b] Wrought
CA-15	J91540	A 217, A 487, A 743	0.15	1.50	11.5-14.0	1.0	0.50 Mo	410
CA-15M	J91151	A 487, A743	0.15	0.65	11.5-14.0	1.0	0.15-1.00 Mo	410Mo
CA-28MWV	---	A 743	0.20-0.28	1.00	11.0-12.5	0.50-1.00	0.9-1.25 Mo, 0.9-1.25 W, 0.2-0.3 V	---
CA-40	J91153	A 743	0.20-0.40	1.50	11.5-14.0	1.0	0.5 Mo	420
CA-40F	---	A 743	0.20-0.40	1.50	11.5-14.0	1.0	0.5 Mo, 0.20-0.40 S	---
CA-6N	---	A 743	0.06	1.00	10.5-12.5	6.0-8.0	---	---
CA-6NM	J91540	A 352, A 356, A 487, A 743	0.06	1.00	11.5-14.0	3.5-4.5	0.4-1.0 Mo	---
CA-28MWV	---	A 743	0.20-0.28	1.0	11.0-12.5	0.50-1.00	0.90-1.25 Mo, 0.90-1.25 W, 0.20-0.30V	---
CB-6	---	A 743	0.06	1.00	15.5-17.5	3.5-5.5	0.5 Mo	---
CB-7Cu-1	---	A 747	0.07	1.00	15.50-17.70	3.60-4.60	2.50-3.20 Cu, 0.15-0.35 Cb, 0.05 N	---
CB-7Cu-2	---	A 747	0.07	1.00	14.0-15.50	4.50-5.50	2.50-3.20 Cu, 0.15-0.35 Cb, 0.05 N	---
CB-30	J91803	A 743	0.30	1.50	18.0-21.0	2.0	0.90-1.20 Cu optional	431, 442
CC-50	J92615	A 743	0.50	1.50	26.0-30.0	4.0	---	446
CD-3MWCuN	J93380	A 351, A 890 Gr 6A	0.03	1.00	24.0-26.0	6.5-8.5	3.0-4.0 Mo, 0.5-1.0 Cu, 0.5-1.0 W, 0.20-0.30 N	---
CD-4MCu	J93370	A 351, A 744, A 890 Gr 1A	0.04	1.00	24.5-26.5	4.75-6.00	1.75-2.25 Mo, 2.75-3.25 Cu	---
CE-8N	J92805	---	0.08	1.50	23.0-26.0	8.0-11.0	0.50 Mo, 0.20-0.30 N	---
CE-8MN	J93345	A 890 Gr 2A	0.08	1.50	22.5-25.5	8.00-11.00	3.00-4.50 Mo, 0.10-0.30 N	---
CE-30	J93423	A 743	0.30	2.00	26.0-30.0	8.0-11.0	---	312
CF-3	J92500	A 351, A 743, A 744	0.03	2.00	17.0-21.0	8.0-12.0	---	304L
CF-3M	J92800	A 351, A 743, A 744	0.03	1.50	17.0-21.0	9.0-13.0	2.0-3.0 Mo	316L
CF-3MN	---	A 743	0.03	1.50	17.0-22.0	9.0-13.0	2.0-3.0 Mo, 0. 10-0.20 N	---

CHEMICAL COMPOSITION OF CORROSION-RESISTANT CAST STAINLESS STEELS[a] (Continued)

Grade	UNS	ASTM Spec.	C	Si	Cr	Ni	Other	Similar[b] Wrought
CF-8	J92600	A 351, A 743, A 744	0.08	2.00	18.0-21.0	8.0-11.0	---	304
CF-8C	J92710	A 351, A 743, A 744	0.08	2.00	18.0-21.0	9.0-12.0	(8 x C) Cb min to 1.00 or (9 x C) Cb+Ta min to 1.1	347
CF-8M	J92900	A 351, A 743, A 744	0.08	2.00	18.0-21.0	9.0-12.0	2.0-3.0 Mo	316
CF-10	---	A 351	0.04-0.10	2.00	18.0-21.0	8.0-11.0	0.50 Mo	---
CF-10M	---	A 351	0.04-0.10	1.50	18.0-21.0	9.0-12.0	2.0-3.0 Mo	---
CF-10MC	J92971	A 351	0.10	1.50	15.0-18.0	13.0-16.0	1.75-2.25 Mo, (10 x C) Cb min to 1.20	---
CF10SMnN	---	A 351, A 743	0.10	3.50-4.50	16.0-18.0	8.0-9.0	7.00-9.00 Mn, 0.08-0.18 N	---
CF-16F	J92701	A 743	0.16	2.00	18.0-21.0	9.0-12.0	1.50 Mo, 0.20-0.35 Se	303Se
CF-16Fa	---	A 743	0.16	2.00	18.0-21.0	9.0-12.0	0.40-0.80 Mo, 0.20-0.40 S	---
CF-20	J92602	A 743	0.20	2.00	18.0-21.0	8.0-11.0	---	302
CG-3M	---	A 351, A 743, A 744	0.03	1.50	18.0-21.0	9.0-13.0	3.0-4.0 Mo	---
CG-6MMN	---	A 743	0.06	1.00	20.5-23.5	11.5-13.5	1.50-3.00 Mo, 4.00-6.00Mn, 0.10-0.30 Cb, 0.10-0.30 V, 0.20-0.40 N	---
CG-8M	---	A 351, A 743, A 744	0.08	1.50	18.0-21.0	9.0-13.0	3.0-4.0 Mo	317
CG-12	J93001	A 743	0.12	2.00	20.0-23.0	10.0-13.0	---	---
CH-8	J93400	A 351	0.08	1.50	22.0-26.0	12.0-15.0	0.50 Mo	---
CH-10	J93401	A 351	0.04-0.10	2.00	22.0-26.0	12.0-15.0	0.50 Mo	---
CH-20	J93402	A 351, A 743	0.20	2.00	22.0-26.0	12.0-15.0	---	309
CK-3MCuN	---	A 351, A 743, A 744	0.025	1.00	19.5-20.5	17.5-19.5	6.0-7.0 Mo, 0.50-1.00 Cu, 0.180-0.240 N	---
CK-20	J94202	A 743	0.20	2.00	23.0-27.0	19.0-22.0	---	310
CN-3M	---	A 743	0.03	1.00	20.0-22.0	23.0-27.0	4.5-5.5 Mo	---

CHEMICAL COMPOSITION OF CORROSION-RESISTANT CAST STAINLESS STEELS[a] (Continued)

Grade	UNS	ASTM Spec.	C	Si	Cr	Ni	Other	Similar[b] Wrought
CN-3MN	---	A 743, A 744	0.03	1.00	20.0-22.0	23.5-25.5	6.00-7.00 Mo, 0.75 Cu, 0.18-0.26 N	AL-6XN
CN-7M	---	A 351, A 743, A 744	0.07	1.50	19.0-22.0	27.5-30.5	2.0-3.0 Mo, 3.0-4.0 Cu	---
CN-7MS	---	A 743, A744	0.07	2.50-3.50	18.0-20.0	22.0-25.0	2.5-3.0 Mo, 1.5-2.0 Cu	---
---	J91550	A 757 Gr E3N	0.06	1.00	11.5-14.0	3.5-4.5	0.40-1.0 Mo	---
---	---	A 890 Gr 3A	0.06	1.00	24.0-27.0	4.00-6.00	1.75-2.50 Mo, 0.15-0.25 N	---
---	J92205	A 890 Gr 4A	0.03	1.00	21.0-23.5	4.5-6.5	2.5-3.5 Mo, 1.00 Cu, 0.10-0.30 N	2205
---	J93404	A 890 Gr 5A	0.03	1.00	24.0-26.0	6.0-8.0	4.0-5.0 Mo, 0.10-0.30 N	2507

a. Although manganese, sulphur and phosphorous contents are not listed in this table due to limited space, they are specified; see appropriate ASTM standard for more details.

b. Similar wrought designations are listed only as a guide for comparison to cast grades; they are not equivalent.

All values are maximums, unless otherwise specified.

CHEMICAL COMPOSITION OF HEAT-RESISTANT CAST STAINLESS STEELS[a]

Grade	UNS	ASTM Specification	C	Mn	Si	P	S	Cr	Ni
HC	J92605	A 297	0.50	1.00	2.00	0.04	0.04	26.0-30.0	4.00
HC30	J92613	A 608	0.25-0.35	0.5-1.0	0.50-2.00	0.04	0.04	26-30	4.0
HD	J93005	A 297	0.50	1.50	2.00	0.04	0.04	26.-30.0	4.0-7.0
HD50	J93015	A 608	0.45-0.55	1.50	0.50-2.00	0.04	0.04	26-30	4-7
HE	J93403	A 297	0.20-0.50	2.00	2.00	0.04	0.04	26.0-30.0	8.0-11.0
HE35	J93413	A 608	0.30-0.40	1.50	0.50-2.00	0.04	0.04	26-30	8-11
HF	J92603	A 297	0.20-0.40	2.00	2.00	0.04	0.04	18.0-23.0	8.0-12.0
HF30	J92803	A 608	0.25-0.35	1.50	0.50-2.00	0.04	0.04	19-23	9-12
HH	J93503	A 297, A 447	0.20-0.50	2.00	2.00	0.04	0.04	24.0-28.0	11.0-14.0

CHEMICAL COMPOSITION OF HEAT-RESISTANT CAST STAINLESS STEELS[a] (Continued)

Grade	UNS	ASTM Specification	C	Mn	P	S	Si	Cr	Ni
HH30	J93513	A 608	0.25-0.35	1.50	0.04	0.04	0.50-2.00	24-28	11-14
HH33	J93633	A 608	0.28-0.38	1.50	0.04	0.04	0.50-2.00	24-26	12-14
HI	J94003	A 297	0.20-0.50	2.00	0.04	0.04	2.00	26.0-30.0	14.0-18.0
HI35	J94013	A 608	0.30-0.40	1.50	0.04	0.04	0.50-2.00	26-30	14-18
HK	J94224	A 297	0.20-0.60	2.00	0.04	0.04	2.00	24.0-28.0	18.0-22.0
HK30	J94203	A 351, A 608	0.25-0.35	1.50	0.04	0.04	0.50-2.00	23.0-27.0	19.0-22.0
HK40	J94204	A 351, A 608	0.35-0.45	1.50	0.04	0.04	0.50-2.00	23.0-27.0	19.0-22.0
HL	N08604	A297	0.20-0.60	2.00	0.04	0.04	2.00	28.0-32.0	18.0-22.0
HL30	N08613	A 608	0.25-0.35	1.50	0.04	0.04	0.50-2.00	28-32	18-22
HL40	N08614	A 608	0.35-0.45	1.50	0.04	0.04	0.50-2.00	28-32	18-22
HN	J94213	A 297	0.20-0.50	2.00	0.04	0.04	2.00	19.0-23.0	23.0-27.0
HN40	J94214	A 608	0.35-0.45	1.50	0.04	0.04	0.50-2.00	19-23	23-27
HP	N08705	A 297	0.35-0.75	2.00	0.04	0.04	2.50	24-28	33-37
HT	N08605	A 297	0.35-0.75	2.00	0.04	0.04	2.50	15.0-19.0	33.0-37.0
HT30	N08603	---	0.25-0.35	2.00	0.04	0.04	2.50	13.0-17.0	33.0-37.0
HT50	N08050	A 608	0.40-0.60	1.50	0.04	0.04	0.50-2.00	15-19	33-37
HU	N08004	A 297	0.35-0.75	2.00	0.04	0.04	2.50	17.0-21.0	37.0-41.0
HU50	N08005	A 608	0.40-0.60	1.50	0.04	0.04	0.50-2.00	17-21	37-41
HW	N08001	A 297	0.35-0.75	2.00	0.04	0.04	2.50	10.0-14.0	58.0-62.0
HW50	N08006	A 608	0.40-0.60	1.50	0.04	0.04	0.50-2.00	10-14	58-62
HX	N06006	A 297	0.35-0.75	2.00	0.04	0.04	2.50	15.0-19.0	64.0-68.0
HX50	N06050	A 608	0.40-0.60	1.50	0.04	0.04	0.50-2.00	15-19	64-68

a. Molybdenum is for all alloys 0.5% max.
All values are maximums, unless otherwise specified.

MECHANICAL PROPERTIES OF CORROSION-RESISTANT CAST STAINLESS STEELS

ASTM Spec. Grade	Heat Treatment[a]	Tensile Strength		Yield Strength		% El	% RA	Hardness HB max[c]
		ksi	MPa	Ksi	MPa			
ASTM A 743								
CA-6N	---	140	965	135	930	15	50	---
CA-6NM	---	110	755	80	550	15	35	285
CA-15	---	90	620	65	450	18	30	241
CA-15M	---	90	620	65	450	18	30	241
CA-28MWV	---	140	965	110	760	10	24	302-352[d]
CA-40	---	100	690	70	485	15	25	269
CA-40F	---	100	690	70	485	12	---	269
CB-30	---	65	450	30	205	---	---	241
CC-50	---	55	380	---	---	---	---	241
CD-4MCu	---	100	690	70	485	16	---	---
CE-30	---	80	550	40	275	10	---	---
CF-3	---	70	485	30	205	35	---	---
CF-3M	---	70	485	30	205	30	---	---
CF-3MN	---	75	515	37	255	35	---	---
CF-8	---	70[b]	485[b]	30[b]	205[b]	35	---	---
CF-8C	---	70	485	30	205	30	---	---
CF-8M	---	70	485	30	205	30	---	---
CF10SMnN	---	85	585	42	290	30	---	---
CF-16F	---	70	485	30	205	25	---	---
CF-16Fa	---	70	485	30	205	25	---	---
CF-20	---	70	485	30	205	30	---	---
CG-3M	---	75	515	35	240	25	---	---
CG-6MMN	---	85	585	42	290	30	---	---
CG-8M	---	75	520	35	240	25	---	---

MECHANICAL PROPERTIES OF CORROSION-RESISTANT CAST STAINLESS STEELS (Continued)

ASTM Spec.		Tensile Strength		Yield Strength				Hardness
Grade	Heat Treatment[a]	Ksi	MPa	Ksi	MPa	% El	% RA	HB max[c]
ASTM A 743 (Continued)								
CG-12	---	70	485	28	195	35	---	---
CH-10	---	70	485	30	205	30	---	---
CK-3MCuN	---	80	550	38	260	35	---	---
CK-20	---	65	450	28	195	30	---	---
CN-3M	---	63	435	25	170	30	---	---
CN-3MN	---	80	550	38	260	35	---	---
CN-7M	---	62	425	25	170	35	---	---
CN-7MS	---	70	485	30	205	35	---	---
ASTM A 890								
1A[e]	---	100	690	70	485	16	---	---
2A[e]	---	95	655	65	450	25	---	---
3A[e]	---	95	655	65	450	25	---	---
4A[e]	---	90	620	60	415	25	---	---
5A[e]	---	100	690	75	515	18	---	---
6A[e]	---	100	700	65	450	25	---	---
ASTM A 747								
CB7Cu-1	H900	170	1170	145	1000	5	---	375 min
	H925	175	1205	150	1035	5	---	375 min
	H1025	150	1035	140	965	9	---	311 min
	H1075	145	1000	115	795	9	---	277 min
	H1100	135	930	110	760	9	---	269 min
	H1150	125	860	97	670	10	---	269 min
	H1150M	---	---	---	---	---	---	310
	H1150 DBL	---	---	---	---	---	---	310

MECHANICAL PROPERTIES OF CORROSION-RESISTANT CAST STAINLESS STEELS (Continued)

ASTM Spec.		Tensile Strength		Yield Strength				Hardness
Grade	Heat Treatment[a]	ksi	MPa	Ksi	MPa	% El	% RA	HB max[c]
ASTM A 747 (Continued)								
CB7Cu-2	H900	170	1170	145	1000	5	---	375 min
	H925	175	1205	150	1035	5	---	375 min
	H1025	150	1035	140	965	9	---	311 min
	H1075	145	1000	115	795	9	---	277 min
	H1100	135	930	110	760	9	---	269 min
	H1150	125	860	97	670	10	---	269 min
	H1150M	---	---	---	---	---	---	310
	H1150 DBL	---	---	---	---	---	---	310

a. ST - solution treated; NT - normalized and tempered; N - normalized; See ASTM A 743 or ASTM A 890, Table 1, for detailed heat treatment requirements.

b. For low ferrite or nonmagnetic castings of this grade, the following values apply: tensile strength 65 ksi (450 MPa) min., yield strength 28 ksi (195 MPa) min.

c. Hardness values are supplementary requirements in ASTM A 743 S14 and are maximum values, unless otherwise specified.

d. The hardness requirement for this grade in the annealed condition is 269 HB max.

e. Supplementary requirement S32 of ASTM A 890. All values are maximums, unless otherwise specified.

TYPICAL ROOM-TEMPERATURE MECHANICAL PROPERTIES OF HEAT-RESISTANT CAST ALLOYS

Grade	Condition	Tensile Strength		Yield Strength		% Elongation	Hardness, HB
		ksi	MPa	ksi	MPa		
HC	As-cast	110	760	75	515	19	223
	Aged[a]	115	790	80	550	18	---
HD	As-cast	85	585	48	330	16	90
HE	As-cast	95	655	45	310	20	200
	Aged[a]	90	620	55	380	10	270
HF	As-cast	92	635	45	310	38	165
	Aged[a]	100	690	50	345	25	190

TYPICAL ROOM-TEMPERATURE MECHANICAL PROPERTIES OF HEAT-RESISTANT CAST ALLOYS (Continued)

Grade	Condition	Tensile Strength		Yield Strength		% Elongation	Hardness, HB
		ksi	MPa	ksi	MPa		
HH, type 1	As-cast	85	585	50	345	25	185
	Aged[a]	86	595	55	380	11	200
HH, type 2	As-cast	80	550	40	275	15	180
	Aged[a]	92	635	45	310	8	200
HI	As-cast	80	550	45	310	12	180
	Aged[a]	90	620	65	450	6	200
HK	As-cast	75	515	50	345	17	170
	Aged[b]	85	585	50	345	10	190
HL	As-cast	82	565	52	360	19	192
HN	As-cast	68	470	38	260	13	160
HP	As-cast	71	490	40	275	11	170
HT	As-cast	70	485	40	275	10	180
	Aged[b]	75	515	45	310	5	200
HU	As-cast	70	485	40	275	9	170
	Aged[c]	73	505	43	295	5	190
HW	As-cast	68	470	36	250	4	185
HW	Aged[d]	84	580	52	360	4	205
HX	As-cast	65	450	36	250	9	176
	Aged[c]	73	505	44	305	9	185

a. Aging treatment: 24 hours at 1400°F (760°C), furnace cool.
b. Aging treatment: 24 hours at 1400°F (760°C), air cool.
c. Aging treatment: 48 hours at 1800°F (980°C), air cool.
d. Aging treatment: 48 hours at 1800°F (980°C), furnace cool.
All values are maximums, unless otherwise specified.

Chapter
20

MACHINING DATA:

CARBON, ALLOY & STAINLESS STEELS

TURNING DATA FOR FREE-MACHINING CARBON STEELS

Material AISI	Hardness HB / Heat Treatment[a]	Cut Depth, in.	High Speed Steel Tool			Coated Carbide Tool		
			Speed fpm	Feed ipr	Tool Material	Speed fpm	Feed ipr	Tool Material[b]
1116, 1117, 1118, 1119, 1211, 1212	100-150 HR or A	0.040	200	0.007	M2, M3	1200	0.007	CC-7
		0.150	150	0.015	M2, M3	775	0.015	CC-6
		0.300	120	0.020	M2, M3	625	0.020	CC-6
		0.625	90	0.030	M2, M3	---	---	---
	150-200 CD	0.040	210	0.007	M2, M3	1225	0.007	CC-7
		0.150	160	0.015	M2, M3	800	0.015	CC-6
		0.300	125	0.020	M2, M3	650	0.020	CC-6
		0.625	100	0.030	M2, M3	---	---	---
1213, 1215	100-150 HR or A	0.040	295	0.008	M2, M3	1300	0.007	CC-7
		0.150	225	0.015	M2, M3	850	0.015	CC-6
		0.300	175	0.020	M2, M3	675	0.020	CC-6
		0.625	140	0.030	M2, M3	---	---	---
	150-200 CD	0.040	300	0.008	M2. M3	1350	0.007	CC-7
		0.150	230	0.015	M2, M3	900	0.015	CC-6
		0.300	180	0.020	M2. M3	725	0.020	CC-6
		0.625	140	0.030	M2, M3	---	---	---
1108, 1109, 1110, 1115	100-150 HR or A	0.040	180	0.008	M2, M3	1050	0.007	CC-7
		0.150	135	0.015	M2, M3	700	0.015	CC-6
		0.300	110	0.020	M2, M3	550	0.020	CC-6
		0.625	85	0.030	M2, M3	---	---	---
1108, 1109, 1110, 1115	150-200 CD	0.040	190	0.008	M2, M3	1100	0.007	CC-7
		0.150	145	0.015	M2, M3	700	0.015	CC-6
		0.300	110	0.020	M2. M3	575	0.020	CC-6
		0.625	90	0.030	M2, M3	---	---	---

TURNING DATA FOR FREE-MACHINING CARBON STEELS (Continued)								
Material AISI	Hardness HB Heat Treatment[a]	Cut Depth, in.	High Speed Steel Tool			Coated Carbide Tool		
			Speed fpm	Feed ipr	Tool Material	Speed fpm	Feed ipr	Tool Material[b]
1132, 1137, 1139, 1140, 1141, 1144, 1145, 1146, 1151	175-225 HR, N, A, CD	0.040	170	0.008	M2. M3	1000	0.007	CC-7
		0.150	130	0.015	M2. M3	650	0.015	CC-6
		0.300	100	0.020	M2, M3	525	0.020	CC-6
		0.625	80	0.030	M2, M3	---	---	---
	275-325 QT	0.040	115	0.007	T15, M42	750	0.007	CC-7
		0.150	90	0.015	T15, M42	500	0.015	CC-6
		0.300	70	0.020	T15, M42	400	0.020	CC-6
		0.625	---	---	---	---	---	---
	325-375 QT	0.040	800	0.007	T15, M42	650	0.007	CC-7
		0.150	605	0.015	T15, M42	425	0.015	CC-6
		0.300	500	0.020	T15, M42	350	0.020	CC-6
		0.625	---	---	---	---	---	---
	375-425 QT	0.040	65	0.007	T15, M42	525	0.007	CC-7
		0.150	50	0.015	T15, M42	350	0.015	CC-6
		0.300	45	0.020	T15, M42	300	0.020	CC-6
		0.625	---	---	---	---	---	---

a. Heat treatment abbreviations are as follows: HR - hot rolled; A - annealed; CD - cold drawn; N - normalized; QT - quenched & tempered.
b. Coated carbide tool material grades are designated by an additional prefix letter C, to distinguish it from the Industry Code system for uncoated carbide tools.

TURNING DATA FOR LOW CARBON STEELS

Material AISI	Hardness HB Heat Treatment[a]	Cut Depth, in.	High Speed Steel Tool			Coated Carbide Tool		
			Speed fpm	Feed ipr	Tool Material	Speed fpm	Feed ipr	Tool Material[b]
1009, 1010, 1012, 1015, 1017, 1020, 1023, 1025	125-175 HR, N, A, CD	0.040	150	0.007	M2, M3	950	0.007	CC-7
		0.150	125	0.015	M2, M3	625	0.015	CC-6
		0.300	100	0.020	M2, M3	500	0.020	CC-6
		0.625	80	0.030	M2, M3	---	---	---
	175-225 HR, N, A, CD	0.040	145	0.007	M2, M3	850	0.007	CC-7
		0.150	115	0.015	M2, M3	550	0.015	CC-6
		0.300	95	0.020	M2, M3	450	0.020	CC-6
		0.625	75	0.030	M2, M3	---	---	---
	225-275 A, CD	0.040	125	0.007	M2, M3	750	0.007	CC-7
		0.150	95	0.015	M2, M3	500	0.015	CC-6
		0.300	75	0.020	M2, M3	400	0.020	CC-6
		0.625	60	0.030	M2, M3	---	---	---

a. Heat treatment abbreviations are as follows: HR - hot rolled; A - annealed; CD - cold drawn; N - normalized.
b. Coated carbide tool material grades are designated by an additional prefix letter C, to distinguish it from the Industry Code system for uncoated carbide tools.

TURNING DATA FOR MEDIUM CARBON STEELS

Material AISI	Hardness HB Heat Treatment[a]	Cut Depth, in.	High Speed Steel Tool			Coated Carbide Tool		
			Speed fpm	Feed ipr	Tool Material	Speed fpm	Feed ipr	Tool Material[b]
1030, 1033, 1035, 1037, 1038, 1039, 1040, 1042, 1043, 1044, 1045, 1046, 1049, 1050, 1053, 1055	125-175 HR, N, A, CD	0.040	140	0.007	M2, M3	925	0.007	CC-7
		0.150	115	0.015	M2, M3	600	0.015	CC-6
		0.300	90	0.020	M2, M3	475	0.020	CC-6
		0.625	70	0.030	M2, M3	---	---	---
	175-225 HR, N, A, CD	0.040	130	0.007	M2, M3	785	0.007	CC-7
		0.150	100	0.015	M2, M3	525	0.015	CC-6
		0.300	85	0.020	M2, M3	415	0.020	CC-6
		0.625	65	0.030	M2, M3	---	---	---
	225-275 HR, N, A, CD, QT	0.040	115	0.007	M2, M3	750	0.007	CC-7
		0.150	90	0.015	M2, M3	500	0.015	CC-6
		0.300	70	0.020	M2, M3	400	0.020	CC-6
		0.625	55	0.030	M2, M3	---	---	---
	275-325 HR, N, A, QT	0.040	100	0.007	T15, M42	700	0.007	CC-7
		0.150	75	0.015	T15, M42	450	0.015	CC-6
		0.300	60	0.020	T15, M42	375	0.020	CC-6
		0.625	--0	---	---	---	---	---
	325-375 QT	0.040	70	0.007	T15, M42	575	0.007	CC-7
		0.150	55	0.015	T15, M42	375	0.015	CC-6
		0.300	45	0.020	T15, M42	315	0.020	CC-6
		0.625	---	---	---	---	---	---
	375-425 QT	0.040	60	0.007	T15, M42	475	0.007	CC-7
		0.150	45	0.015	T15, M42	300	0.015	CC-6
		0.300	35	0.020	T15, M42	250	0.020	CC-6

a. Heat treatment abbreviations are as follows: HR - hot rolled; A - annealed; CD - cold drawn; N - normalized; QT - quenched & tempered.
b. Coated carbide tool material grades are designated by an additional prefix letter C, to distinguish it from the Industry Code system for uncoated carbide tools.

TURNING DATA FOR HIGH CARBON STEELS

Material AISI	Hardness HB Heat Treatment[a]	Cut Depth, in.	High Speed Steel Tool			Coated Carbide Tool		
			Speed fpm	Feed ipr	Tool Material	Speed fpm	Feed ipr	Tool Material[b]
1060, 1064, 1065, 1069, 1070, 1074, 1075, 1078	175-225 HR, N, A, CD	0.040	120	0.007	M2, M3	750	0.007	CC-7
		0.150	90	0.015	M2, M3	500	0.015	CC-6
		0.300	70	0.020	M2, M3	400	0.020	CC-6
		0.625	50	0.030	M2, M3	---	---	---
1080, 1084, 1085, 1086, 1090, 1095	225-275 HR, N, A, CD, QT	0.040	100	0.007	M2, M3	700	0.007	CC-7
		0.150	80	0.015	M2, M3	475	0.015	CC-6
		0.300	60	0.020	M2, M3	375	0.020	CC-6
		0.625		0.030	M2, M3	---	---	---
	275-325 HR, N, A, CD, QT	0.040	80	0.007	T15, M42	675	0.007	CC-7
		0.150	60	0.015	T15, M42	450	0.015	CC-6
		0.300	50	0.020	T15, M42	350	0.020	CC-6
		0.625	---	---	---	---	---	---
	325-375 QT	0.040	60	0.007	T15, M42	550	0.007	CC-7
		0.150	45	0.015	T15, M42	375	0.015	CC-6
		0.300	35	0.020	T15, M42	300	0.020	CC-6
		0.625	---	---	---	---	---	---
	375-425 QT	0.040	45	0.005	T15, M42	425	0.007	CC-7
		0.150	35	0.010	T15, M42	275	0.015	CC-6
		0.300	25	0.020	T15, M42	225	0.020	CC-6
		0.625	---	---	---	---	---	---

a. Heat treatment abbreviations are as follows: HR - hot rolled; A - annealed; CD - cold drawn; N - normalized; QT - quenched & tempered.

b. Coated carbide tool material grades are designated by an additional prefix letter C, to distinguish it from the Industry Code system for uncoated carbide tools.

TURNING DATA FOR LOW CARBON ALLOY STEELS

Material AISI	Hardness HB Heat Treatment[a]	Cut Depth, in.	High Speed Steel Tool			Coated Carbide Tool		
			Speed fpm	Feed ipr	Tool Material	Speed fpm	Feed ipr	Tool Material[b]
4012, 4024, 4118, 4320, 4419, 4422, 4615, 4620,	175-225 HR, A, CD	0.040	135	0.007	M2, M3	725	0.007	CC-7
		0.150	105	0.015	M2, M3	575	0.015	CC-6
		0.300	80	0.020	M2, M3	450	0.020	CC-6
		0.625	65	0.030	M2, M3	---	---	---
4720, 4815, 4820, 5015, 5115, 5120, 6118, 8115,	225-275 HR, N, A, CD	0.040	110	0.007	M2, M3	650	0.007	CC-7
		0.150	85	0.015	M2, M3	500	0.015	CC-6
		0.300	65	0.020	M2, M3	400	0.020	CC-6
		0.625	50	0.030	M2, M3	---	---	---
8617, 8620, 9310, 94B17	275-325 N, QT	0.040	95	0.007	T15, M42	575	0.007	CC-7
		0.150	75	0.015	T15, M42	450	0.015	CC-6
		0.300	60	0.020	T15, M42	350	0.020	CC-6
		0.625	---	---	---	---	---	---
	325-375 N, QT	0.040	80	0.005	T15, M42	475	0.007	CC-7
		0.150	60	0.010	T15, M42	375	0.015	CC-6
		0.300	45	0.015	T15, M42t	300	0.020	CC-6
		0.625	---	---	---	---	---	---
	375-425 QT	0.040	65	0.005	T15, M42	375	0.007	CC-7
		0.150	50	0.010	T15, M42	300	0.015	CC-6
		0.300	40	0.015	T15, M42	225	0.020	CC-6
		0.625	---	---	---	---	---	---

a. Heat treatment abbreviations are as follows: HR - hot rolled; A - annealed; CD - cold drawn; N - normalized; QT - quenched & tempered.
b. Coated carbide tool material grades are designated by an additional prefix letter C, to distinguish it from the Industry Code system for uncoated carbide tools.

TURNING DATA FOR MEDIUM CARBON ALLOY STEELS

Material AISI	Hardness HB Heat Treatment[a]	Cut Depth, in.	High Speed Steel Tool			Coated Carbide Tool		
			Speed fpm	Feed ipr	Tool Material	Speed fpm	Feed ipr	Tool Material[b]
1330, 1335,	175-225	0.040	135	0.007	M2, M3	650	0.007	CC-7
4028, 4037,	HR, A, CD	0.150	105	0.015	M2. M3	525	0.015	CC-6
4130, 4137,		0.300	80	0.020	M2, M3	400	0.020	CC-6
4427, 4626,		0.625	65	0.030	M2, M3	---	---	---
5130, 5135,	225-275	0.040	115	00.007	M2, M3	600	0.007	CC-7
8625, 8630,	A, N, CD, QT	0.0150	90	0.015	M2, M3	450	0.015	CC-6
8637, 94B30		0.300	70	0.020	M2, M3	275	0.020	CC-6
		0.625	55	0.030	M2, M3	---	---	---
	275-325 N, QT	0.040	95	0.007	T15, M42	550	0.007	CC-7
		0.150	75	0.015	T15, M42	425	0.015	CC-6
		0.300	60	0.020	T15, M42	350	0.020	CC-6
	325-375 N, QT	0.040	70	0.005	T15, M42	500	0.007	CC-7
		0.150	55	0.010	T15, M42	350	0.015	CC-6
		0.300	40	0.015	T15, M42	300	0.020	CC-6
	375-425 QT	0.040	60	0.005	T15, M42	375	0.007	CC-7
		0.150	45	0.010	T15, M42	300	0.015	CC-6
		0.300	35	0.015	T15, M42	225	0.020	CC-6

a. Heat treatment abbreviations are as follows: HR - hot rolled; A - annealed; CD - cold drawn; N - normalized; QT - quenched & tempered.

b. Coated carbide tool material grades are designated by an additional prefix letter C, to distinguish it from the Industry Code system for uncoated carbide tools.

TURNING DATA FOR MEDIUM CARBON ALLOY STEELS

Material AISI	Hardness HB Heat Treatment[a]	Cut Depth, in.	High Speed Steel Tool			Coated Carbide Tool		
			Speed fpm	Feed ipr	Tool Material	Speed fpm	Feed ipr	Tool Material[b]
1340, 1345, 4042, 4047, 4140, 4147, 4340, 50B46, 5140, 5147, 81B45, 8640, 8645, 86B45, 8740, 8742	175-225 HR, A, CD	0.040	135	0.007	M2, M3	650	0.007	CC-7
		0.150	105	0.015	M2, M3	525	0.015	CC-6
		0.300	80	0.020	M2, M3	400	00020	CC-6
		0.625	65	0.030	M2, M3	---	---	---
	225-275 A, N, CD, QT	0.040	115	0.007	M2, M3	600	0.007	CC-7
		0.150	90	0.015	M2, M3	475	0.015	CC-6
		0.300	70	0.020	M2, M3	375	0.020	CC-6
		0.625	55	0.030	M2, M3	---	---	---
	275-325 N, QT	0.040	90	0.007	T15, M42	575	0.007	CC-7
		0.150	70	0.015	T15, M42	450	0.015	CC-6
		0.300	55	0.020	TI5, M42	350	0.020	CC-6
		0.625	---	---	---	---	---	---
	325-375 N, QT	0.040	70	0.005	T15, M42	500	0.007	CC-7
		0.150	55	0.010	T15, M42	400	0.015	CC-6
		0.300	40	0.015	T15, M42	300	0.020	CC-6
		0.625						
	375-425 QT	0.040	60	0.005	T15, M42	400	0.007	CC-7
		0.150	45	0.010	T15, M42	300	0.015	CC-6
		0.300	35	0.015	T15, M42	250	0.020	CC-6
		0.625				---		---

a. Heat treatment abbreviations are as follows: HR - hot rolled; A - annealed; CD - cold drawn; N - normalized; QT - quenched & tempered.

b. Coated carbide tool material grades are designated by an additional prefix letter C, to distinguish it from the Industry Code system for uncoated carbide tools.

FACE MILLING DATA FOR FREE MACHINING CARBON STEELS

Material AISI	Hardness HB Heat Treatment[a]	Cut Depth, in.	High Speed Steel Tool			Coated Carbide Tool		
			Speed fpm	Feed per Tooth, in.	Tool Material	Speed fpm	Feed per Tooth, in.	Tool Material[b]
1116, 1117, 1118, 1119, 1211, 1212	100-150 HR or A	0.040	260	0.008	M2, M7	1200	0.008	CC-6
		0.150	200	0.012	M2, M7	785	0.012	CC-6
		0.300	155	0.016	M2, M7	610	0.016	CC-5
	150-200 CD	0.040	250	0.008	M2, M7	1100	0.008	CC-6
		0.150	190	0.012	M2, M7	715	0.012	CC-6
		0.300	150	0.016	M2, M7	560	0.016	CC-5
1213, 1215	100-150 HR or A	0.040	375	0.008	M2, M7	1325	.00081	CC-6
		0.150	290	0.012	M2, M7	865	0.012	CC-6
		0.300	225	0.016	M2, M7	670	0.016	CC-5
	150-200 CD	0.040	360	0.008	M2, M7	1200	0.008	CC-6
		0.150	275	0.012	M2, M7	800	0.012	CC-6
		0.300	215	0.016	M2, M7	625	0.016	CC-5
1108, 1109, 1110, 1115	100-150 HR or A	0.040	235	0.008	M2, M7	1050	0.008	CC-6
		0.150	180	0.012	M2, M7	710	0.012	CC-6
		0.300	140	0.016	M2, M7	550	0.016	CC-5
	150-200 CD	0.040	225	0.008	M2, M7	950	0.008	CC-6
		0.150	170	0.012	M2, M7	645	0.012	CC-6
		0.300	130	0.016	M2. M7	500	0.016	CC-5
1132, 1137, 1139, 1140, 1141, 1144	175-225 HR, N, A, CD	0.040	225	0.008	M2, M7	925	0.008	CC-6
		0.150	165	0.012	M2, M7	730	0.012	CC-6
		0.300	125	0.016	M2, M7	565	0.016	CC-5
1145, 1146, 1151	275-325 QT	0.040	155	0.006	T15, M42	715	0.005	CC-6
		0.150	115	0.009	T15, M42	560	0.007	CC-6
		0.300	90	0.012	T15, M42	435	0.009	CC-5

FACE MILLING DATA FOR FREE MACHINING CARBON STEELS (Continued)

Material AISI	Hardness HB Heat Treatment[a]	Cut Depth, in.	High Speed Steel Tool			Coated Carbide Tool		
			Speed fpm	Feed per Tooth, in.	Tool Material	Speed fpm	Feed per Tooth, in.	Tool Material[b]
1132, 1137,	325-375 QT	0.040	90	0.005	T15, M42	600	0.004	CC-6
1139, 1140,		0.150	70	0.008	T15, M42	455	0.006	CC-6
1141, 1144		0.300	55	0.010	T15, M42	350	0.008	CC-5
1145, 1146,	375-425 QT	0.040	75	0.004	T15, M42	490	0.003	CC-6
1151		0.150	55	0.006	T15, M42	355	0.005	CC-6
		0.300	40	0.008	T15, M42	280	0.007	CC-5

a. Heat treatment abbreviations are as follows: HR - hot rolled; A - annealed; CD - cold drawn; N - normalized; QT - quenched & tempered.
b. Coated carbide tool material grades are designated by an additional prefix letter C, to distinguish it from the Industry Code system for uncoated carbide tools.

FACE MILLING DATA FOR LOW CARBON STEELS

Material AISI	Hardness HB Heat Treatment[a]	Cut Depth, in.	High Speed Steel Tool			Coated Carbide Tool		
			Speed fpm	Feed per Tooth, in.	Tool Material	Speed fpm	Feed per Tooth, in.	Tool Material[b]
1009, 1010,	125-175 HR, N, A, CD	0.040	210	0.008	M2, M7	1075	0.008	CC-6
1012, 1015,		0.150	160	0.012	M2, M7	730	0.012	CC-6
1017, 1020,		0.300	125	0.016	M2, M7	565	0.016	CC-5
1023, 1025	175-225 HR, N, A, CD	0.040	190	0.008	M2, M7	885	0.008	CC-6
		0.150	140	0.012	M2, M7	635	0.012	CC-6
		0.300	110	0.016	M2, M7	495	0.016	CC-5
	225-275 A, CD	0.040	160	0.006	M2, M7	790	0.007	CC-6
		0.150	115	0.010	M2, M7	570	0.010	CC-6
		0.300	90	0.014	M2, M7	450	0.014	CC-5

a. Heat treatment abbreviations are as follows: HR - hot rolled; A - annealed; CD - cold drawn; N - normalized.
b. Coated carbide tool material grades are designated by an additional prefix letter C, to distinguish it from the Industry Code system for uncoated carbide tools.

FACE MILLING DATA FOR MEDIUM CARBON STEELS

Material AISI	Hardness HB Heat Treatment[a]	Cut Depth, in.	High Speed Steel Tool			Coated Carbide Tool		
			Speed fpm	Feed per Tooth, in.	Tool Material	Speed fpm	Feed per Tooth, in.	Tool Material[b]
1030, 1033,	125-175	0.040	180	0.008	M2, M7	1050	0.008	CC-6
1035, 1037,	HR, N, A, CD	0.150	150	0.012	M2, M7	700	0.012	CC-6
1038, 1039,		0.300	115	0.016	M2, M7	545	0.016	CC-5
1040, 1042,	175-225	0.040	160	0.008	M2, M7	840	0.008	CC-6
1043, 1044,	HR, N, A, CD	0.150	125	0.012	M2, M7	625	0.012	CC-6
1045, 1046,		0.300	100	0.016	M2, M7	485	0.016	CC-5
1049, 1050,	225-275	0.040	125	0.006	T15, M42	750	0.007	CC-6
1053, 1055	HR, N, A, CD, QT	0.150	100	0.010	T15, M42	540	0.010	CC-6
		0.300	80	0.014	T15, M42	420	0.014	CC-5
	275-325	0.040	110	0.006	T15, M42	700	0.005	CC-6
	HR, N, A, QT	0.150	90	0.009	T15, M42	470	0.007	CC-6
		0.300	70	0.012	T15, M42	355	0.009	CC-5
	325-375	0.040	80	0.005	T15, M42	540	0.004	CC-6
	QT	0.150	60	0.008	T15, M42	390	0.006	CC-6
		0.300	50	0.010	T15, M42	305	0.008	CC-5
	375-425	0.040	70	0.003	T15, M42	435	0.003	CC-6
	QT	0.150	50	0.004	T15, M42	310	0.005	CC-6
		0.300	35	0.005	T15, M42	240	0.007	CC-6

a. Heat treatment abbreviations are as follows: HR - hot rolled; A - annealed; CD - cold drawn; N - normalized; QT - quenched & tempered.

b. Coated carbide tool material grades are designated by an additional prefix letter C, to distinguish it from the Industry Code system for uncoated carbide tools.

FACE MILLING DATA FOR HIGH CARBON STEELS

Material AISI	Hardness HB Heat Treatment[a]	Cut Depth in.	High Speed Steel Tool				Coated Carbide Tool		
			Speed fpm	Feed per Tooth, in.	Tool Material	Speed fpm	Feed per Tooth, in.	Tool Material[b]	
1060, 1064,	175-225	0.040	145	0.008	M2, M7	800	0.008	CC-6
1065, 1069,	HR, N, A, CD	0.150	110	0.012	M2, M7	585	0.012	CC-6
1070, 1074,		0.300	85	0.016	M2, M7	455	0.016	CC-5
1075, 1078,	225-275	0.040	115	0.006	M2, M7	725	0.007	CC-6
1080, 1084,	HR, N, A, CD, QT	0.150	90	0.010	M2, M7	520	0.010	CC-6
1085, 1086,		0.300	70	0.014	M2, M7	405	0.014	CC-5
1090, 1095	275-325	0.040	100	0.006	T15, M42	625	0.005	CC-6
	HR, N, A, CD, QT	0.150	80	0.009	T15, M42	430	0.007	CC-6
		0.300	60	0.012	T15, M42	330	0.009	CC-5
	325-375	0.040	70	0.005	T15, M42	500	0.004	CC-6
	QT	0.150	50	0.008	T15, M42	375	0.006	CC-6
		0.300	40	0.010	T15, M42	290	0.008	CC-5
	375-425	0.040	65	0.004	T15, M42	390	0.003	CC-6
	QT	0.150	45	0.006	T15, M42	280	0.005	CC-6
		0.300	30	0.008	T15, M42	215	0.007	CC-6

a. Heat treatment abbreviations are as follows: HR - hot rolled; A - annealed; CD - cold drawn; N - normalized; QT - quenched & tempered.

b. Coated carbide tool material grades are designated by an additional prefix letter C, to distinguish it from the Industry Code system for uncoated carbide tools.

FACE MILLING DATA FOR LOW CARBON ALLOY STEELS

Material AISI	Hardness HB Heat Treatment[a]	Cut Depth, in.	High Speed Steel Tool			Coated Carbide Tool		
			Speed fpm	Feed per Tooth, in.	Tool Material	Speed fpm	Feed per Tooth, in.	Tool Material[b]
4012, 4024, 4118, 4320, 4419, 4422,	175-225 HR, A, CD	0.040	180	0.008	M2, M7	870	0.008	CC-6
		0.150	140	0.012	M2, M7	665	0.012	CC-6
		0.300	110	0.016	M2, M7	515	0.016	CC-5
4615, 4620, 4720, 4815, 4820, 5015,	225-275 HR, N, A, CD	0.040	150	0.006	M2, M7	765	0.007	CC-6
		0.150	110	0.010	M2, M7	585	0.010	CC-6
		0.300	85	0.014	M2, M7	455	0.014	CC-5
5115, 5120, 6118, 8115, 8617, 8620,	275-325 N, QT	0.040	105	0.006	T15, M42	630	0.005	CC-6
		0.150	85	0.009	T15, M42	485	0.007	CC-6
		0.300	65	0.012	T15, M42	375	0.009	CC-5
9310, 94B17	325-375 N, QT	0.040	85	0.005	T15, M42	500	0.004	CC-6
		0.150	65	0.008	T15, M42	350	0.006	CC-6
		0.300	50	0.010	T15, M42	275	0.008	CC-5
	375-425 QT	0.040	65	0.004	T15, M42	425	0.003	CC-6
		0.150	50	0.006	T15, M42	275	0.005	CC-6
		0.300	40	0.008	T15, M42	200	0.007	CC-6

a. Heat treatment abbreviations are as follows: HR - hot rolled; A - annealed; CD - cold drawn; N - normalized; QT - quenched & tempered.
b. Coated carbide tool material grades are designated by an additional prefix letter C, to distinguish it from the Industry Code system for uncoated carbide tools.

FACE MILLING DATA FOR MEDIUM CARBON ALLOY STEELS

Material AISI	Hardness HB / Heat Treatment[a]	Cut Depth, in.	High Speed Steel Tool			Coated Carbide Tool		
			Speed fpm	Feed per Tooth, in.	Tool Material	Speed fpm	Feed per Tooth, in.	Tool Material[b]
1330, 1335,	175-225 / HR, A, CD	0.040	170	0.008	M2, M7	825	0.008	CC-6
4028, 4037,		0.150	130	0.012	M2, M7	550	0.012	CC-6
4130, 4137,		0.300	110	0.016	M2, M7	400	0.016	CC-5
4427, 4626,	225-275 / A, N, CD, QT	0.040	145	0.006	M2, M7	750	0.007	CC-6
5130, 5135,		0.150	105	0.010	M2, M7	500	0.010	CC-6
8625, 8630,		0.300	80	0.014	M2, M7	400	0.014	CC-5
8637, 94B30	275-325 / N, QT	0.040	100	0.006	T15, M42	700	0.005	CC-6
		0.150	80	0.009	T15, M42	475	0.007	CC-6
		0.300	60	0.012	T15, M42	375	0.009	CC-5
	325-375 / N, QT	0.040	80	0.005	T15, M42	625	0.004	CC-6
		0.150	60	0.008	T15, M42	400	0.006	CC-6
		0.300	50	0.010	T15, M42	300	0.008	CC-5
	375-425 / QT	0.040	60	0.004	T15, M42	475	0.003	CC-6
		0.150	50	0.006	T15, M42	325	0.005	CC-6
		0.300	40	0.008	T15, M42	250	0.007	CC-6

a. Heat treatment abbreviations are as follows: HR - hot rolled; A - annealed; CD - cold drawn; N - normalized; QT - quenched & tempered.

b. Coated carbide tool material grades are designated by an additional prefix letter C, to distinguish it from the Industry Code system for uncoated carbide tools.

FACE MILLING DATA FOR MEDIUM CARBON ALLOY STEELS

Material AISI	Hardness HB Heat Treatment[a]	Cut Depth, in.	High Speed Steel Tool			Coated Carbide Tool		
			Speed fpm	Feed per Tooth, in.	Tool Material	Speed fpm	Feed per Tooth, in.	Tool Material[b]
1340, 1345,	175-225	0.040	175	0.008	M2, M7	825	0.008	CC-6
4042, 4047,	HR, A, CD	0.150	135	0.012	M2, M7	570	0.012	CC-6
4140, 4147,		0.300	115	0.016	M2. M7	450	0.016	CC-5
4340, 50B46,	225-275	0.040	150	0.006	M2, M7	765	0.007	CC-6
5140, 5147,	A, N, CD, QT	0.150	110	0.010	M2, M7	520	0.010	CC-6
81B45, 8640,		0.300	85	0.014	M2, M7	410	0.014	CC-5
8645, 86B45,	275-325	0.040	105	0.006	T15, M42	725	0.005	CC-6
8740, 8742	N, QT	0.150	85	0.009	T15, M42	485	0.007	CC-6
		0.300	65	0.012	T15, M42	375	0.009	CC-5
	325-375	0.040	80	0.005	T15, M42	630	0.004	CC-6
	N, QT	0.150	60	0.008	T15, M42	420	0.006	CC-6
		0.300	50	0.010	T15, M42	325	0.008	CC-5
	375-425	0.040	60	0.004	T15, M42	485	0.003	CC-6
	QT	0.150	50	0.006	T15, M42	345	0.005	CC-6
		0.300	40	0.008	T15, M42	265	0.007	CC-6

a. Heat treatment abbreviations are as follows: HR - hot rolled; A - annealed; CD - cold drawn; N - normalized; QT - quenched & tempered.

b. Coated carbide tool material grades are designated by an additional prefix letter C, to distinguish it from the Industry Code system for uncoated carbide tools.

TURNING DATA FOR FREE MACHINING MARTENSITIC STAINLESS STEELS

Material AISI	Hardness HB Heat Treatment[a]	Cut Depth, in.	High Speed Steel Tool			Coated Carbide Tool		
			Speed fpm	Feed ipr	Tool Material	Speed fpm	Feed ipr	Tool Material[b]
416, 416Se, 420F, 420F Se, 440F, 440F Se	185-240 A, CD	0.040	175	0.007	M2, M3	775	0.007	CC-7
		0.150	150	0.015	M2, M3	675	0.015	CC-6
		0.300	120	0.020	M2, M3	400	0.020	CC-6
		0.625	90	0.030	M2, M3	---	---	---
	275-325 QT	0.040	100	0.007	T15, M42	575	0.007	CC-7
		0.150	80	0.015	T15, M42	525	0.015	CC-6
		0.300	60	0.020	T15, M42	400	0.020	CC-6
		0.625	---	---	T15, M42	---	---	---
	375-425 QT	0.040	60	0.005	T15, M42	300	0.005	CC-7
		0.150	50	0.010	T15, M42	250	0.010	CC-6
		0.300	40	0.015	T15, M42	200	0.015	CC-6
		0.625			T15, M42	---	---	---

a. Heat treatment abbreviations are as follows: A - annealed; CD - cold drawn; QT - quenched & tempered.
b. Coated carbide tool material grades are designated by an additional prefix letter C, to distinguish it from the Industry Code system for uncoated carbide tools.

TURNING DATA FOR MARTENSITIC STAINLESS STEELS

Material AISI	Hardness HB Heat Treatment[a]	Cut Depth, in.	High Speed Steel Tool			Coated Carbide Tool		
			Speed fpm	Feed ipr	Tool Material	Speed fpm	Feed ipr	Tool Material[b]
403, 410, 420, 422, 501, 502	175-225 A	0.040	145	0.007	M2, M3	850	0.007	CC-7
		0.150	115	0.015	M2, M3	550	0.015	CC-6
		0.300	90	0.020	M2, M3	450	0.020	CC-6
		0.625	70	0.030	M2. M3	--	--	--
	275-325 QT	0.040	95	0.007	T15, M42	700	0.007	CC-7
		0.150	75	0.015	T15, M42	450	0.015	CC-6
		0.300	60	0.020	T15, M42	375	0.020	CC-6
		0.625	--	--	T15, M42	--	--	--
	375-425 QT	0.040	65	0.007	T15, M42	475	0.007	CC-7
		0.150	50	0.015	T15, M42	300	0.015	CC-6
		0.300	40	0.020	T15, M42	250	0.020	CC-6
		0.625	--	--	T15, M42	--	--	--
440A, 440B, 440C	225-275 A	0.040	80	0.007	T15, M42	525	0.007	CC-7
		0.150	65	0.015	T15, M42	400	0.015	CC-6
		0.300	50	0.020	T15, M42	325	0.020	CC-6
		0.625	35	0.030	T15, M42	--	--	--
	275-325 QT	0.040	65	0.007	T15, M42	400	0.007	CC-7
		0.150	50	0.015	T15, M42	325	0.015	CC-6
		0.300	40	0.020	T15, M42	250	0.020	CC-6
		0.625	--	--	T15, M42	--	--	--

a. Heat treatment abbreviations are as follows: A - annealed; QT - quenched & tempered.
b. Coated carbide tool material grades are designated by an additional prefix letter C, to distinguish it from the Industry Code system for uncoated carbide tools.

TURNING DATA FOR FERRITIC STAINLESS STEELS

Material AISI	Hardness HB Heat Treatment[a]	Cut Depth, in.	High Speed Steel Tool			Coated Carbide Tool		
			Speed fpm	Feed ipr	Tool Material	Speed fpm	Feed ipr	Tool Material[b]
405, 409,	135-185	0.040	150	0.007	M2, M3	850	0.007	CC-7
429, 430,	A	0.150	120	0.015	M2, M3	650	0.015	CC-6
434, 436,		0.300	95	0.020	M2, M3	525	0.020	CC-6
442, 446		0.625	75	0.030	M2. M3	---	---	---

a. Heat treatment abbreviations are as follows: A - annealed.
b. Coated carbide tool material grades are designated by an additional prefix letter C, to distinguish it from the Industry Code system for uncoated carbide tools.

TURNING DATA FOR AUSTENTIC STAINLESS STEELS

Material AISI	Hardness HB Heat Treatment[a]	Cut Depth, in.	High Speed Steel Tool			Coated Carbide Tool		
			Speed fpm	Feed ipr	Tool Material	Speed fpm	Feed ipr	Tool Material[b]
201, 202,	135-185	0.040	110	0.007	M2, M3	525	0.007	CC-3
301, 302,	A	0.150	90	0.015	M2, M3	450	0.015	CC-3
304, 304L,		0.300	70	0.020	M2, M3	350	0.020	CC-2
308, 321,		0.625	55	0.030	M2. M3	---	---	---
347, 348	225-275	0.040	100	0.007	T15, M42	450	0.007	CC-3
	CD	0.150	80	0.015	T15, M42	400	0.015	CC-3
		0.300	65	0.020	T15, M42	300	0.020	CC-2
		0.625	50	0.030	T15, M42	---	---	---
302B, 309,	135-185	0.040	95	0.007	M2, M3	500	0.007	CC-3
309S, 310,	A	0.150	75	0.015	M2, M3	425	0.015	CC-3
316, 316L,		0.300	60	0.020	M2, M3	325	0.020	CC-2
317, 330		0.625	45	0.030	M2. M3	---	---	---

TURNING DATA FOR AUSTENITIC STAINLESS STEELS (Continued)

Material AISI	Hardness HB Heat Treatment[a]	Cut Depth, in.	High Speed Steel Tool			Coated Carbide Tool		
			Speed fpm	Feed ipr	Tool Material	Speed fpm	Feed ipr	Tool Material[b]
302B, 309,	225-275	0.040	80	0.007	T15, M42	425	0.007	CC-3
309S, 310,	CD	0.150	65	0.015	T15, M42	350	0.015	CC-3
316, 316L,		0.300	50	0.020	T15, M42	275	0.020	CC-2
317, 330		0.625	40	0.030	T15, M42	---	---	---
Nitronic 32,	210-250	0.040	60	0.007	M2, M3	275	0.007	CC-3
Nitronic 40,	A	0.150	50	0.015	M2, M3	250	0.015	CC-3
Nitronic 50,		0.300	40	0.020	M2, M3	200	0.020	CC-2
Nitronic 60		0.625	25	0.030	M2. M3	---	---	---

a. Heat treatment abbreviations are as follows: A - annealed; CD - cold drawn.

b. Coated carbide tool material grades are designated by an additional prefix letter C, to distinguish it from the Industry Code system for uncoated carbide tools.

TURNING DATA FOR PRECIPITATION HARDENING STAINLESS STEELS

Material AISI	Hardness HB	Heat Treatment[a]	Cut Depth, in.	High Speed Steel Tool			Coated Carbide Tool		
				Speed fpm	Feed ipr	Tool Material	Speed fpm	Feed ipr	Tool Material[b]
15-5 PH, 16-6 PH, 17-4 PH, 17-7 PH,	150-200	ST	0.040	95	0.007	M2, M3	525	0.007	CC-7
			0.150	80	0.015	M2, M3	450	0.015	CC-6
			0.300	60	0.020	M2, M3	350	0.020	CC-6
			0.625	45	0.030	M2, M3	---	---	---
AM-350, AM-355, AM-363,	275-325	ST, H	0.040	75	0.007	T15, M42	450	0.007	CC-7
			0.150	60	0.015	T15, M42	400	0.015	CC-6
			0.300	45	0.020	T15, M42	300	0.020	CC-6
			0.625	---	---	T15, M42	---	---	---
Custom 450, Custom 455, PH 13-8 Mo, PH 14-8 Mo, PH 15-7 Mo,	325-375	ST, H	0.040	70	0.007	T15, M42	---	---	---
			0.150	55	0.015	T15, M42	---	---	---
			0.300	40	0.020	T15, M42	---	---	---
			0.625	---	---	T15, M42	---	---	---
Stainless W	375-440	H	0.040	45	0.005	T15, M42	---	---	---
			0.150	30	0.010	T15, M42	---	---	---
			0.300	25	0.015	T15, M42	---	---	---
			0.625	---	---	T15, M42	---	---	---

a. Heat treatment abbreviations are as follows: ST - solution treated; H - hardened.

b. Coated carbide tool material grades are designated by an additional prefix letter C, to distinguish it from the Industry Code system for uncoated carbide tools.

TURNING DATA FOR CAST MARTENSITIC STAINLESS STEELS

Material AISI	Hardness HB Heat Treatment[a]	Cut Depth, in.	High Speed Steel Tool			Coated Carbide Tool		
			Speed fpm	Feed ipr	Tool Material	Speed fpm	Feed ipr	Tool Material[b]
CA-15,	175-225	0.040	115	0.007	M2, M3	475	0.007	CC-7
CA-15a,	A, N, NT	0.150	95	0.015	M2, M3	425	0.015	CC-6
CA-15M,		0.300	75	0.020	M2, M3	325	0.020	CC-6
CA-40		0.625	60	0.030	M2. M3	---	---	---
	275-325	0.040	75	0.007	T15, M42	400	0.007	CC-7
	QT	0.150	60	0.015	T15, M42	325	0.015	CC-6
		0.300	45	0.020	T15, M42	250	0.020	CC-6
		0.625	---	---	T15, M42	---	---	---
	375-425	0.040	50	0.005	T15, M42	225	0.005	CC-7
	QT	0.150	40	0.010	T15, M42	200	0.010	CC-6
		0.300	30	0.015	T15, M42	150	0.015	CC-6
		0.625	---	---	T15, M42	---	---	---

a. Heat treatment abbreviations are as follows: A - annealed; N - normalized; NT - normalized & tempered; QT - quenched & tempered.
b. Coated carbide tool material grades are designated by an additional prefix letter C, to distinguish it from the Industry Code system for uncoated carbide tools.

TURNING DATA FOR CAST AUSTENITIC STAINLESS STEELS

Material AISI	Hardness HB Heat Treatment[a]	Cut Depth, in.	High Speed Steel Tool			Coated Carbide Tool		
			Speed fpm	Feed ipr	Tool Material	Speed fpm	Feed ipr	Tool Material[b]
CF-3, CF-8,	135-185	0.040	80	0.007	M2, M3	450	0.007	CC-3
CF-20	A	0.150	65	0.015	M2, M3	400	0.015	CC-3
		0.300	50	0.020	M2, M3	300	0.020	CC-2
		0.625	40	0.030	M2. M3	---	---	---
CF-3M, CF-8M,	135-185	0.040	80	0.007	M2, M3	375	0.007	CC-3
CG-8M, CG-12,	A	0.150	65	0.015	M2, M3	300	0.015	CC-3
CH-20, HK-40		0.300	50	0.020	M2, M3	250	0.020	CC-2
		0.625	40	0.030	M2. M3	---	---	---
CN-7M,	140-170	0.040	110	0.007	M2, M3	450	0.007	CC-3
CF-16F	A	0.150	90	0.015	M2, M3	400	0.015	CC-3
		0.300	70	0.020	M2, M3	300	0.020	CC-2
		0.625	55	0.030	M2. M3	---	---	---

a. Heat treatment abbreviations are as follows: A - annealed.
b. Coated carbide tool material grades are designated by an additional prefix letter C, to distinguish it from the Industry Code system for uncoated carbide tools.

TURNING DATA FOR CAST PRECIPITATION HARDENING STAINLESS STEELS

Material AISI	Hardness HB Heat Treatment[a]	Cut Depth, in.	High Speed Steel Tool			Coated Carbide Tool		
			Speed fpm	Feed ipr	Tool Material	Speed fpm	Feed ipr	Tool Material[b]
CD-4MCu,	325-375	0.040	55	0.007	T15, M42	400	0.007	CC-7
CB-7Cu	ST	0.150	45	0.015	T15, M42	350	0.015	CC-6
		0.300	35	0.020	T15, M42	250	0.020	CC-6
		0.625	---	---	T15, M42	---	---	---

TURNING DATA FOR CAST PRECIPITATION HARDENING STAINLESS STEELS (Continued)

Material AISI	Hardness HB Heat Treatment[a]	Cut Depth, in.	High Speed Steel Tool			Coated Carbide Tool		
			Speed fpm	Feed ipr	Tool Material	Speed fpm	Feed ipr	Tool Material[b]
CD-4MCu,	400-450	0.040	40	0.007	T15, M42	250	0.005	CC-7
CB-7Cu	STH	0.150	35	0.015	T15, M42	225	0.010	CC-6
		0.300	25	0.020	T15, M42	175	0.015	CC-6
		0.625	---	---	T15, M42	---	---	---

a. Heat treatment abbreviations are as follows: ST - solution treated; STH - solution treated and hardened.
b. Coated carbide tool material grades are designated by an additional prefix letter C, to distinguish it from the Industry Code system for uncoated carbide tools.

FACE MILLING DATA FOR FREE MACHINING MARTENSITIC STAINLESS STEELS

Material AISI	Hardness HB Heat Treatment[a]	Cut Depth, in.	High Speed Steel Tool			Coated Carbide Tool		
			Speed fpm	Feed per Tooth, in.	Tool Material	Speed fpm	Feed per Tooth, in.	Tool Material[b]
416, 416Se,	185-240	0.040	175	0.008	M2, M7	915	0.007	CC-6
420F, 420F Se,	A, CD	0.150	135	0.012	M2, M7	600	0.010	CC-6
440F, 440F Se		0.300	105	0.016	M2, M7	480	0.014	CC-5
	275-325	0.040	105	0.006	T15, M42	650	0.005	CC-6
	QT	0.150	90	0.009	T15, M42	455	0.007	CC-6
		0.300	70	0.012	T15, M42	355	0.009	CC-5
	375-425	0.040	60	0.004	T15, M42	375	0.003	CC-6
	QT	0.150	45	0.006	T15, M42	300	0.005	CC-6
		0.300	35	0.008	T15, M42	225	0.007	CC-6

a. Heat treatment abbreviations are as follows: A - annealed; CD - cold drawn; QT - quenched & tempered.
b. Coated carbide tool material grades are designated by an additional prefix letter C, to distinguish it from the Industry Code system for uncoated carbide tools.

FACE MILLING DATA FOR MARTENSITIC STAINLESS STEELS

Material AISI	Hardness HB Heat Treatment[a]	Cut Depth, in.	High Speed Steel Tool			Coated Carbide Tool		
			Speed fpm	Feed per Tooth, in.	Tool Material	Speed fpm	Feed per Tooth, in.	Tool Material[b]
403, 410, 420, 422, 501, 502	175-225 A	0.040	145	0.008	M2, M7	765	0.008	CC-6
		0.150	115	0.012	M2, M7	500	0.012	CC-6
		0.300	90	0.016	M2, M7	375	0.016	CC-5
	275-325 QT	0.040	100	0.006	T15, M42	625	0.005	CC-6
		0.150	75	0.009	T15, M42	400	0.007	CC-6
		0.300	60	0.012	T15, M42	315	0.009	CC-5
	375-425 QT	0.040	60	0.004	T15, M42	300	0.003	CC-6
		0.150	45	0.006	T15, M42	200	0.005	CC-6
		0.300	35	0.008	T15, M42	150	0.007	CC-6
440A, 440B, 440C	225-275 A	0.040	105	0.006	M2, M7	700	0.007	CC-6
		0.150	80	0.010	M2, M7	465	0.010	CC-6
		0.300	60	0.014	M2, M7	365	0.014	CC-5
	275-325 QT	0.040	80	0.006	T15, M42	625	0.005	CC-6
		0.150	60	0.009	T15, M42	400	0.007	CC-6
		0.300	45	0.012	T15, M42	300	0.009	CC-5

a. Heat treatment abbreviations are as follows: A - annealed; QT - quenched & tempered.
b. Coated carbide tool material grades are designated by an additional prefix letter C, to distinguish it from the Industry Code system for uncoated carbide tools.

FACE MILLING DATA FOR FERRITIC STAINLESS STEELS

Material AISI	Hardness HB Heat Treatment[a]	Cut Depth, in.	High Speed Steel Tool Speed fpm	Feed per Tooth, in.	Tool Material	Coated Carbide Tool Speed fpm	Feed per Tooth, in.	Tool Material[b]
405, 409,	135-185	0.040	190	0.008	M2, M7	925	0.008	CC-6
429, 430,	A	0.150	145	0.012	M2, M7	600	0.012	CC-6
434, 436, 442, 446		0.300	115	0.016	M2, M7	475	0.016	CC-5

a. Heat treatment abbreviations are as follows: A - annealed.
b. Coated carbide tool material grades are designated by an additional prefix letter C, to distinguish it from the Industry Code system for uncoated carbide tools.

FACE MILLING DATA FOR AUSTENITIC STAINLESS STEELS

Material AISI	Hardness HB Heat Treatment[a]	Cut Depth, in.	High Speed Steel Tool Speed fpm	Feed per Tooth, in.	Tool Material	Coated Carbide Tool Speed fpm	Feed per Tooth, in.	Tool Material[b]
201, 202,	135-185	0.040	130	0.008	M2, M7	700	0.008	CC-2
301, 302,	A	0.150	100	0.012	M2, M7	475	0.012	CC-2
304, 304L,		0.300	80	0.016	M2, M7	350	0.016	CC-2
308, 321,	225-275	0.040	115	0.006	M2, M7	650	0.007	CC-2
347, 348	CD	0.150	90	0.010	M2, M7	425	0.010	CC-2
		0.300	70	0.014	M2, M7	325	0.014	CC-2
302B, 309,	135-185	0.040	115	0.008	M2, M7	650	0.008	CC-2
309S, 310,	A	0.150	90	0.012	M2, M7	425	0.012	CC-2
316, 316L,		0.300	70	0.016	M2, M7	325	0.016	CC-2
317, 330	225-275	0.040	100	0.006	M2, M7	525	0.007	CC-2
	CD	0.150	75	0.010	M2, M7	400	0.010	CC-2
		0.300	55	0.014	M2, M7	300	0.014	CC-2

FACE MILLING DATA FOR AUSTENITIC STAINLESS STEELS (Continued)

Material AISI	Hardness HB Heat Treatment[a]	Cut Depth, in.	High Speed Steel Tool			Coated Carbide Tool		
			Speed fpm	Feed per Tooth, in.	Tool Material	Speed fpm	Feed per Tooth, in.	Tool Material[b]
Nitronic 40,	210-250	0.040	80	0.006	M2, M7	350	0.007	CC-2
Nitronic 50,	A	0.150	60	0.010	M2, M7	275	0.010	CC-2
Nitronic 60		0.300	45	0.014	M2, M7	225	0.014	CC-2

a. Heat treatment abbreviations are as follows: A - annealed; CD - cold drawn.
b. Coated carbide tool material grades are designated by an additional prefix letter C, to distinguish it from the Industry Code system for uncoated carbide tools.

FACE MILLING DATA FOR PRECIPITATION HARDENING STAINLESS STEELS

Material AISI	Hardness HB Heat Treatment[a]	Cut Depth, in.	High Speed Steel Tool			Coated Carbide Tool		
			Speed fpm	Feed per Tooth, in.	Tool Material	Speed fpm	Feed per Tooth, in.	Tool Material[b]
15-5 PH,	150-200	0.040	110	0.008	M2, M7	575	0.008	CC-6
16-6 PH,	ST	0.150	85	0.012	M2, M7	475	0.012	CC-6
17-4 PH,		0.300	65	0.016	M2, M7	375	0.016	CC-5
17-7 PH,	275-325	0.040	80	0.006	T15, M42	425	0.005	CC-6
AM-350,	ST, H	0.150	60	0.009	T15, M42	350	0.007	CC-6
AM-355,		0.300	45	0.012	T15, M42	275	0.009	CC-5
AM-363,	325-375	0.040	70	0.005	T15, M42	---	---	---
Custom 450,	ST, H	0.150	55	0.008	T15, M42	---	---	---
Custom 455		0.300	40	0.010	T15, M42	---	---	---
PH 13-8 Mo,	375-440	0.040	60	0.004	T15, M42	---	---	---
PH 14-8 Mo,	H	0.150	45	0.006	T15, M42	---	---	---
PH 15-7 Mo,		0.300	35	0.008	T15, M42	---	---	---
Stainless W								

FACE MILLING DATA FOR PRECIPITATION HARDENING STAINLESS STEELS (Continued)

a. Heat treatment abbreviations are as follows: ST - solution treated; H - hardened.

b. Coated carbide tool material grades are designated by an additional prefix letter C, to distinguish it from the Industry Code system for uncoated carbide tools.

FACE MILLING DATA FOR CAST MARTENSITIC STAINLESS STEELS

Material AISI	Hardness HB Heat Treatment[a]		Cut Depth, in.	High Speed Steel Tool			Coated Carbide Tool		
				Speed fpm	Feed per Tooth, in.	Tool Material	Speed fpm	Feed per Tooth, in.	Tool Material[b]
CA-15, CA-15a, CA-15M,	175-225	A, N, NT	0.040	135	0.008	M2, M7	625	0.008	CC-6
			0.150	110	0.012	M2, M7	460	0.012	CC-6
			0.300	85	0.016	M2, M7	355	0.016	CC-5
CA-40	275-325	QT	0.040	90	0.006	T15, M42	475	0.005	CC-6
			0.150	70	0.009	T15, M42	355	0.007	CC-6
			0.300	50	0.012	T15, M42	280	0.009	CC-5
	375-425	QT	0.040	70	0.004	T15, M42	300	0.003	CC-6
			0.150	50	0.006	T15, M42	215	0.005	CC-6
			0.300	35	0.008	T15, M42	160	0.007	CC-6

a. Heat treatment abbreviations are as follows: A - annealed; N - normalized; NT - normalized & tempered; QT - quenched & tempered.

b. Coated carbide tool material grades are designated by an additional prefix letter C, to distinguish it from the Industry Code system for uncoated carbide tools.

FACE MILLING DATA FOR CAST AUSTENITIC STAINLESS STEELS

Material AISI	Hardness HB Heat Treatment[a]	Cut Depth, in.	High Speed Steel Tool			Coated Carbide Tool		
			Speed fpm	Feed per Tooth, in.	Tool Material	Speed fpm	Feed per Tooth, in.	Tool Material[b]
CF-3, CF-8,	135-185	0.040	100	0.008	M2, M7	575	0.008	CC-2
CF-20	A	0.150	75	0.012	M2, M7	425	0.012	CC-2
		0.300	55	0.016	M2, M7	325	0.016	CC-2
CF-3M, CF-8M,	135-185	0.040	100	0.008	M2, M7	475	0.008	CC-2
CG-8M, CG-12,	A	0.150	80	0.012	M2, M7	355	0.012	CC-2
CH-20, HK-40		0.300	60	0.016	M2, M7	280	0.016	CC-2
CF-16F,	140-170	0.040	130	0.008	M2, M7	575	0.008	CC-2
CN-7M	A	0.150	100	0.012	M2, M7	425	0.012	CC-2
		0.300	75	0.016	M2, M7	325	0.016	CC-2

a. Heat treatment abbreviations are as follows: A - annealed.
b. Coated carbide tool material grades are designated by an additional prefix letter C, to distinguish it from the Industry Code system for uncoated carbide tools.

FACE MILLING DATA FOR CAST PRECIPITATION HARDENING STAINLESS STEELS

Material AISI	Hardness HB Heat Treatment[a]	Cut Depth, in.	High Speed Steel Tool			Coated Carbide Tool		
			Speed fpm	Feed per Tooth, in.	Tool Material	Speed fpm	Feed per Tooth, in.	Tool Material[b]
CD-4MCu,	325-375	0.040	75	0.005	T15, M42	475	0.004	CC-6
CB-7Cu	ST	0.150	55	0.008	T15, M42	350	0.006	CC-6
		0.300	40	0.010	T15, M42	280	0.008	CC-5
	400-450	0.040	55	0.003	T15, M42	325	0.003	CC-6
	STH	0.150	45	0.005	T15, M42	250	0.005	CC-6
		0.300	30	0.007	T15, M42	195	0.007	CC-6

a. Heat treatment abbreviations are as follows: ST - solution treated; STH - solution treated and hardened.
b. Coated carbide tool material grades are designated by an additional prefix letter C, to distinguish it from the Industry Code system for uncoated carbide tools.

Chapter

21

PHYSICAL PROPERTIES:

THE ELEMENTS, CARBON, ALLOY & STAINLESS STEELS

PHYSICAL PROPERTIES OF THE ELEMENTS

Element	Sym.	Atomic No.	Atomic Wt.	Electrons In Shell							Melting Pt. °C	Boiling Pt. °C	Density[a]	Val.[b]
				K	L	M	N	O	P	Q				
Actinium	Ac	89	227	2	8	18	32	18	9	2	1600	---	---	---
Aluminum	Al	13	26.98	2	8	3	---	---	---	---	660.2	2060	s 2.699	3+
Americium	Am	95	241	2	8	18	32	24	9	---	---	---	---	---
Antimony	Sb	51	121.8	2	8	18	18	5	---	---	630.5	1440	s 6.62	5+
Argon	Ar	18	39.95	2	8	8	---	---	---	---	-189.4	-185.8	g 1.784 L 1.40 s 1.65	Inert
Arsenic	As	33	74.92	2	8	18	5	---	---	---	814 (36 atm.)	610	s 5.73	3+ 5+
Astatine	At	85	210	2	8	18	32	18	7	---	---	---	---	---
Barium	Ba	56	137.3	2	8	18	18	8	2	---	704	1640	s 3.5	2+
Berkelium	Bk	97	249	2	8	18	32	26	9	2	---	---	---	---
Beryllium	Be	4	9.01	2	2	---	---	---	---	---	1350	1530	s 1.85	2+
Bismuth	Bi	93	209.0	2	8	18	32	18	5	---	271.3	1420	s 9.80	---
Boron	B	5	10.81	2	3	---	---	---	---	---	2300	2550	s 2.3	3+
Bromine	Br	35	79.91	2	8	15	7	---	---	---	-7.2	19.0	s 3.12	---
Cadmium	Cd	48	112.4	2	8	18	18	2	---	---	320.9	765	s 8.65	2+
Calcium	Ca	20	40.08	2	8	8	2	---	---	---	850	1440	s 1.55	2+
Californium	Cf	98	252	2	8	18	32	27	9	2	~3500	4200(?)	---	---
Carbon	C	6	12.01	2	4	---	---	---	---	---	~3500	4200(?)	s 3. 51	4+
Cesium	Cs	55	132.9	2	8	18	18	8	1	---	28	690	s 1.9	1+
Chlorine	Cl	17	35.45	2	8	7	---	---	---	---	-101	-34.7	g 3.214 L 1.557 s 1.9	1-
Chromium	Cr	24	52.00	2	8	13	1	---	---	---	1890	2500	s 7.19	3+
Cobalt	Co	27	58.93	2	8	15	2	---	---	---	1495	2900	s 8.9	2+

PHYSICAL PROPERTIES OF THE ELEMENTS (Continued)

Element	Sym.	Atomic No.	Atomic Wt.	Electrons In Shell							Melting Pt. °C	Boiling Pt. °C	Density[a]	Val.[b]
				K	L	M	N	O	P	Q				
Copper	Cu	29	63.54	2	8	18	1	---	---	---	1083	2600	s 8.96	1+
Curium	Cm	96	242	2	8	18	32	25	9	2	---	---	---	---
Einsteinium	E	99	254	2	8	18	32	28	9	2	---	---	---	---
Fluorine	F	9	19.00	2	7	---	---	---	---	---	-223	-188.2	g 1.69 L 1.108	1-
Francium	Fa	87	223	2	8	18	32	18	8	1	---	---	---	---
Gallium	Ga	31	69.72	2	8	18	3	---	---	---	29.78	2070	s 5.91	3+
Germanium	Ge	32	72.59	2	8	18	4	---	---	---	958	2970	s 5.36	4+
Gold	Au	79	197.0	2	8	18	32	18	1	---	1063	2970	g 19.32	1+
Hafnium	Hf	72	178.5	2	8	18	32	10	2	---	1700	---	s 11.4	4+
Helium	He	2	4.003	2	---	---	---	---	---	---	-272.2 (26 atm.)	-268.9	g 0.1785 L 0.147	Inert
Hydrogen	H	1	1.008	1	---	---	---	---	---	---	-259.18	-252.8	g 0.0899 L 0.070	1+
Indium	In	49	114.8	2	8	18	18	3	---	---	156.4	---	s 7.31	3+
Iodine	I	53	126.9	2	8	18	18	7	---	---	114	183	s 4.93	1-
Iridium	Ir	77	192.2	2	8	18	32	17		---	2454	5300	s 22.5	4+
Iron	Fe	26	55.85	2	8	14	2	---		---	1539	2740	s 7.87	2+ 3+
Krypton	Kr	36	83.80	2	8	18	8	---	---	---	-157	-152	g 3.708 L 2.155	Inert
Lead	Pb	82	207.2	2	8	18	32	18	4	---	327.4	1740	s 11.34	2+ 4+
Lithium	Li	3	6.94	2	1	---	---	---	---	---	186	1609	s 0.534	1+
Magnesium	Mg	12	24.31	2	8	2	---	---	---	---	650	1110	s 1.74	2+
Manganese	Mn	25	54.94	2	8	13	2	---	---	---	1245	2150	s 7.43	2+

PHYSICAL PROPERTIES OF THE ELEMENTS (Continued)

Element	Sym.	Atomic No.	Atomic Wt.	Electrons In Shell							Melting Pt. °C	Boiling Pt. °C	Density[a]	Val.[b]
				K	L	M	N	O	P	Q				
Mercury	Hg	80	200.6	2	8	18	32	18	2	--	-38.87	357	s 13.55	2+
Molybdenum	Mo	42	95.94	2	8	18	13	1	--	--	2625	4800	s 10.2	4+
Neon	Ne	10	20.18	2	8	--	--	--	--	--	-248.67	-245.9	g 0.9002 L 1.204	Inert
Neptunium	Np	93	237	2	8	18	32	22	9	2	--	--	--	--
Nickel	Ni	28	58.71	2	8	16	2	--	--	--	1455	2730	s 8.90	2+
Niobium (Columbium)	Nb (Cb)	41	92.91	2	8	18	12	1	--	--	2415	--	s 8.57	5+
Nitrogen	N	7	14.01	2	5	--	--	--	--	--	-209.86	-195.8	g 1.2506 L 0.808 s 1.026	3-
Osmium	Os	76	190.2	2	8	18	32	14	2	--	2700	5500	s 22.5	4+
Oxygen	O	8	16.00	2	6	--	--	--	--	--	-218.4	-183.0	g 1.429 L 1.14 s 1.426	2-
Palladium	Pd	46	106.4	2	8	18	18	--	--	--	1554	4000	s 12.0	--
Phosphorus	P	15	30.97	2	8	5	--	--	--	--	44.1	280	s 1.82	5+
Platinum	Pt	78	195.1	2	8	18	32	17	1	--	1773	4410	s 21.45	--
Plutonium	Pu	94	239	2	8	18	32	23	9	2	--	--	--	--
Polonium	Po	84	210	2	8	18	32	18	6	--	600	--	--	--
Potassium	K	19	39.10	2	8	8	1	--	--	--	63	770	s 0.86	1+
Protactinium	Pa	91	231	2	8	18	32	20	9	2	3000	--	--	--
Radium	Ra	88	226	2	8	18	32	18	8	2	700	s 5.0	--	--
Radon	Rn	86	222	2	8	18	32	18	8	--	-71	-61.8	--	Inert
Rare earths	La	57	138.9	2	8	18	18	9	2	--	--	--	--	3+
Rare earths	Lu	71	175.0	2	8	18	32	9	2	--	--	--	--	--

PHYSICAL PROPERTIES OF THE ELEMENTS (Continued)

Element	Sym.	Atomic No.	Atomic Wt.	Electrons In Shell							Melting Pt. °C	Boiling Pt. °C	Density[a]	Val.[b]
				K	L	M	N	O	P	Q				
Rhenium	Re	75	186.2	2	8	18	32	13	2	---	3170	---	s 20	---
Rhodium	Rh	45	102.9	2	8	18	16	1	---	---	1966	4500	s 12.44	3+
Rubidium	Rb	37	85.47	2	8	18	8	1	---	---	39	680	s 1.53	1+
Ruthenium	Ru	44	101.1	2	8	18	15	1	---	---	2500	4900	s 12.2	4+
Scandium	Sc	21	44.96	2	8	9	2	---	---	---	1200	---	s 2.5	3+
Selenium	Se	34	78.96	2	8	18	6	---	---	---	220	680	s 4.81	2-
Silicon	Si	14	28.09	2	8	4	---	---	---	---	1430	2300	s 2.4	4+
Silver	Gg	47	107.9	2	8	18	18	1	---	---	960.5	2210	s 10.49	1+
Sodium	Ng	11	22.99	2	8	1	---	---	---	---	97.5	880	s 0.97	1+
Strontium	Sr	38	87.62	2	8	18	8	2	---	---	770	1380	s 2.6	2+
Sulfur	S	16	32.06	2	8	6	---	---	---	---	119.0	246.2	L 1.803 / s 2.07	2- 6+
Tantalum	Ta	73	181.0	2	8	18	32	11	2	---	2996	---	s 16.6	5+
Tellurium	Te	52	127.6	2	8	18	18	6	---	---	450	1390	s 6.24	2-
Thallium	Tl	81	204.4	2	8	18	32	18	3	---	300	1460	g 11.85	3+
Thorium	Th	90	232	2	8	18	32	18	10	2	1800	s 11.5	---	4+
Tin	Sn	50	118.7	2	8	18	18	4	---	---	231.9	2270	s 7.298	4+
Titanium	Ti	22	47.90	2	8	10	2	---	---	---	1820	---	s 4.54	4+
Tungsten	W	74	183.9	2	8	18	32	12	2	---	3410	5930	s 19.3	4+
Uranium	U	92	238	2	8	18	32	21	9	2	1130	s 18.7	---	4+
Vanadium	V	23	50.94	2	8	11	2	---	---	---	1735	3400	s 6.0	3+ 5+
Xenon	Xe	54	131.3	2	8	18	18	8	---	---	-112	-108	g 5.851 / L 3.52 / s 2.7	Inert
Yttrium	Y	39	88.91	2	8	18	9	2	---	---	1490	---	s 5.51	3+

The Metals Black Book – 3rd Edition

PHYSICAL PROPERTIES OF THE ELEMENTS (Continued)

Element	Sym.	Atomic No.	Atomic Wt.	Electrons In Shell							Melting Pt. °C	Boiling Pt. °C	Density[a]	Val.[b]
				K	L	M	N	O	P	Q				
Zinc	Zn	30	65.37	2	8	18	2	---	---	---	419.46	906	s 7.133	2+
Zirconium	Zr	40	91.22	2	8	18	10	2	---	---	1750	---	s 6.5	4+

a. Density measured in g - g/l; L g/cm³; s g/cm³. g - gas; L - liquid; s - solid.

b. Most common valence.

Densities of Selected Carbon & Alloy Steels

Iron & Carbon Steel	g/cm³	lb/in.³
Pure iron	7.874	0.2845
Ingot iron	7.866	0.2842
Wrought iron	7.7	0.28
Gray cast iron	7.15[a]	0.258[a]
Malleable iron	7.27[b]	0.262[b]
0.06% C steel	7.871	0.2844
0.23% C steel	7.859	0.2839
0.435% C steel	7.844	0.2834
1.22% C steel	7.830	0.2829
Low-Carbon Chromium-Molybdenum Steels		
0.5% Mo steel	7.86	0.283
1Cr-0.5% Mo steel	7.86	0.283
1.25Cr-0.5% Mo steel	7.86	0.283
2.25Cr-1.0% Mo steel	7.86	0.283
5Cr-0.5Mo% Steel	7.78	0.278
7Cr-0.5Mo% steel	7.78	0.278
9Cr-1Mo% steel	7.67	0.276

Densities of Selected Carbon & Alloy Steels (Continued)

Iron & Carbon Steel (Continued)	g/cm³	lb/in.³
Medium-Carbon Alloy Steels		
1Cr-0.35Mo-0.25V steel	7.86	0.283
H11 die steel (5Cr-1.5Mo-0.4V)	7.79	0.281

a. 6.95-7.35 g/cm³ (0.251-0.265 lb/in.³)
b. 7.20-7.34 g/cm³ (0.260-0.265 lb/in.³)

COEFFICIENTS OF LINEAR EXPANSION OF CARBON & ALLOY STEELS[a,j]

AISI	Condition	20-100°C 68-212°F	20-200°C 68-392°F	20-300°C 68-572°F	20-400°C 68-752°F	20-500°C 68-932°F	20-600°C 68-1112°F	20-700°C 68-1292°F
1008	Annealed	12.6[b]	13.1[b]	13.5[b]	13.8[b]	14.2[b]	14.6[b]	15.0[b]
1008	Annealed	11.6	12.5	13.0	13.6	14.2	14.6	15.0
1010	Annealed	12.2[b]	13.0[b]	13.5[b]	13.9[b]	14.3[b]	14.7[b]	15.0[b]
1010	Unknown	11.9	12.6	13.3	13.8	14.3	14.7	14.9
1010	Unknown	---	---	---	15.1	---	---	---
1015	Rolled	11.9[b]	12.5[b]	13.0[b]	13.6[b]	14.2[b]	---	---
1015	Annealed	12.2[b]	---	---	13.4[b]	---	14.2[b]	---
1016	Annealed	12.0[b]	---	---	13.5[b]	---	14.4[b]	---
1017	Unknown	12.2[b]	---	---	13.5[b]	---	14.5[b]	---
1018	Annealed	12.0[b]	---	---	13.5[b]	---	14.4[b]	---
1019	Unknown	12.2[b]	---	---	13.5[b]	---	14.7[b]	---
1020	Annealed	11.7	12.1	12.8	13.4	13.9	14.4	14.8
1020	Unknown	12.2[b]	---	---	13.5[b]	---	14.2[b]	---
1021	Unknown	12.0[b]	---	---	13.5[b]	---	14.3[b]	---
1022	Annealed	12.2[b]	12.7[b]	13.1[b]	13.5[b]	13.9[b]	14.4[b]	14.9[b]

COEFFICIENTS OF LINEAR EXPANSION OF CARBON & ALLOY STEELS[aj] (Continued)

AISI	Condition	20-100°C 68-212°F	20-200°C 68-392°F	20-300°C 68-572°F	20-400°C 68-752°F	20-500°C 68-932°F	20-600°C 68-1112°F	20-700°C 68-1292°F
1023	Unknown	12.2[b]	---	---	13.5[b]	---	14.4[b]	---
1025	Annealed	12.0[b]	---	---	13.5[b]	---	14.4[b]	---
1026	Annealed	12.0[b]	---	---	13.5[b]	---	14.4[b]	---
1029	Annealed	12.0[b]	---	---	13.5[b]	---	14.4[b]	---
1030	Annealed	11.7[b]	---	---	13.5[b]	---	14.4[b]	---
1035	Annealed	11.1	11.9	12.7	13.4	14.0	14.4	14.8
1037	Annealed	11.1[b]	---	---	13.5[b]	---	14.6[b]	---
1039	Annealed	11.1[b]	---	---	13.5[b]	---	14.6[b]	---
1040	Annealed	11.3	12.0	12.5	13.3	13.9	14.4	14.8
1043	Annealed	11.3[b]	---	---	13.5[b]	---	14.6[b]	---
1044	Annealed	11.1[b]	12.0[b]	---	13.3[b]	---	---	---
1045	Annealed	11.6[b]	12.3[b]	13.1[b]	13.7[b]	14.2[b]	14.7[b]	15.1[b]
1045	Annealed	11.2	11.9[c]	12.7[c]	13.5	14.1[c]	14.5[c]	14.8[c]
1046	Unknown	11.1[b]	---	---	13.5[b]	---	---	---
1050	Annealed	11.1[b]	12.0[b]	---	13.5[b]	---	---	---
1052	Annealed	11.3[c]	11.8[c]	12.7[c]	13.7[c]	14.5[c]	14.7[c]	15.0[c]
1053	Unknown	11.1[b]	---	---	13.5[b]	---	---	---
1055	Annealed	11.0	11.8	12.6	13.4	14.0	14.5	14.8
1060	Annealed	11.1[c]	11.9[c]	12.9[c]	13.5[c]	14.1[c]	14.6[c]	14.9[c]
1064	Unknown	11.1[b]	---	---	13.5[b]	---	---	---
1065	Unknown	11.1[b]	---	---	13.5[b]	---	---	---
1070	Unknown	11.5[b]	---	---	13.3[b]	---	---	---
1078	Unknown	11.3[b]	---	---	13.3	---	---	---

COEFFICIENTS OF LINEAR EXPANSION OF CARBON & ALLOY STEELS[a-j] (Continued)

AISI	Condition	20-100°C 68-212°F	20-200°C 68-392°F	20-300°C 68-572°F	20-400°C 68-752°F	20-500°C 68-932°F	20-600°C 68-1112°F	20-700°C 68-1292°F
1090	Annealed	11.0	11.6	12.4	13.2	13.8	14.2	14.7
1080	Unknown	11.7[b]	12.2[b]	---	---	---	---	---
1085	Annealed	11.1[b]	11.7[b]	12.5[b]	13.2[b]	13.6[b]	14.2[b]	14.7
1086	Unknown	11.1[b]	---	---	13.1[b]	---	---	---
1095	Unknown	---	---	---	---	---	---	14.6
1095	Annealed	11.4	---	---	---	---	---	---
1095	Hardened	13.0[b]	---	---	---	---	---	---
1117	Unknown	12.2[d]	---	---	13.1[e]	---	---	---
1118	Unknown	12.2[d]	---	---	13.3[e]	---	---	---
1132	Unknown	12.6[d]	---	---	---	---	---	---
1137	Unknown	12.8	---	---	---	---	---	---
1139	Unknown	12.6[d]	---	---	---	---	---	---
1140	Unknown	12.6[d]	---	---	---	---	---	---
1141	Unknown	---	12.6[b]	---	---	---	---	---
1144	Unknown	13.3	---	---	---	---	---	---
1145	Annealed	11.2[b]	12.1[b]	13.0[b]	13.6[b]	14.0[b]	14.6[b]	14.8[b]
1145	Annealed	11.6[b]	12.3[b]	13.1[b]	13.7[b]	14.2[b]	14.7[b]	15.1[b]
1146	Unknown	12.8	---	---	---	---	---	---
1151	Unknown	---	12.6[b]	---	---	---	---	---
1330	Unknown	12.0	12.8	13.3	---	---	---	---
1335	Unknown	12.2	12.8	13.3	---	---	---	---
1345	Unknown	12.0	12.6	13.3	---	---	---	---
1522	Annealed	12.0[b]	---	---	13.5[b]	---	14.4[b]	---
1524	Unknown	11.9	12.7	---	13.9	---	14.7	---

COEFFICIENTS OF LINEAR EXPANSION OF CARBON & ALLOY STEELS[a,j] (Continued)

AISI	Condition	20-100°C 68-212°F	20-200°C 68-392°F	20-300°C 68-572°F	20-400°C 68-752°F	20-500°C 68-932°F	20-600°C 68-1112°F	20-700°C 68-1292°F
1524	Annealed	12.0[b]	---	---	13.5[b]	---	14.4[b]	---
1526	Annealed	12.0[b]	---	---	13.5[b]	---	14.4[b]	---
1541	Annealed	12.0[b]	---	---	13.5[b]	---	14.4[b]	---
1548	Unknown	11.0[b]	---	---	13.3[b]	---	14.6[b]	---
1551	Annealed	11.7[b]	---	---	13.5[b]	---	14.6[b]	---
1552	Unknown	11.1[b]	---	---	13.5[b]	---	---	---
1561	Annealed	11.1[b]	---	---	13.5[b]	---	14.6[b]	---
1566	Annealed	11.5[b]	---	---	13.5[b]	---	14.7[b]	---
2330	Annealed	10.9[c]	11.2[c]	12.1[c]	12.9[c]	13.4[c]	13.8[c]	---
2515	Unknown	10.9[d]	---	12.6[f]	---	13.50[g]	---	---
3120	Unknown	11.3[h]	---	---	---	---	---	14.6[i]
3130	Unknown	11.3[h]	---	---	---	---	---	14.6[i]
3140	Unknown	11.3[h]	---	---	---	---	---	14.6[i]
3150	Unknown	11.3[h]	---	---	---	---	---	14.6[i]
4023	Unknown	11.7[h]	---	---	---	---	---	---
4027	Unknown	11.7[h]	---	---	---	---	---	---
4028	Unknown	11.9	12.4	12.9	---	---	---	---
4032	Unknown	11.9	12.4	12.9	---	---	---	---
4042	Unknown	11.9	12.4	12.9	---	---	---	---
4047	Unknown	11.9	12.4	12.9	---	---	---	---
4130	Unknown	12.2	---	---	13.7	---	14.6	---
4135	Unknown	11.7	12.2	12.8	---	---	---	---
4137	Unknown	11.7	12.2	12.8	---	---	---	---

COEFFICIENTS OF LINEAR EXPANSION OF CARBON & ALLOY STEELS[a,j] (Continued)								
AISI	Condition	20-100°C 68-212°F	20-200°C 68-392°F	20-300°C 68-572°F	20-400°C 68-752°F	20-500°C 68-932°F	20-600°C 68-1112°F	20-700°C 68-1292°F
4140	OH & T	12.3	12.7	---	13.7	---	14.5	---
4142	Unknown	11.7	12.2	12.8	---	---	---	---
4145	OH & T	11.7	12.2	12.8	---	---	---	---
4147	Unknown	11.7	12.2	12.8	---	---	---	---
4161	Unknown	11.5	12.2	12.9	---	---	---	---
4320	Unknown	11.3[h]	---	---	---	---	---	14.6[i]
4337	Unknown	11.3[h]	---	---	---	---	---	14.6[i]
4340	OH & T 630°C	12.3	12.7	---	13.7	---	14.5	---
4340	OH & T 630°C	---	12.4	---	13.6	---	14.3	---
4422	Unknown	11.7[h]	---	---	---	---	---	---
4427	Unknown	12.6	---	13.8	---	---	15.1	---
4615	Unknown	11.5	12.1	12.7	13.2	13.7	14.1	---
4617	C & T	12.5	13.1	---	---	---	---	---
4626	N & T	11.7[d]	---	12.6[f]	---	---	13.8[g]	---
4718	Unknown	11.3	12.2	13.1	---	---	---	---
4815	Unknown	11.5[d]	12.2	13.1	---	---	---	---
4820	Unknown	11.3[d]	12.2	12.9	---	---	---	---
5046	Unknown	11.9	12.4	12.9	---	---	---	---
50B60	Unknown	11.9	12.4	12.9	---	---	---	---
5117	Unknown	12.0	12.8	13.5	---	---	---	---
5120	Unknown	12.0	12.8	13.5	---	---	---	---
5130	Unknown	12.2	12.9	13.5	---	---	---	---
5132	Unknown	12.2	12.9	13.5	---	---	---	---
5135	Unknown	12.0	12.8	13.5	---	---	---	---
5140	Annealed	---	12.6	13.4	13.9	14.3	14.6	15.0

COEFFICIENTS OF LINEAR EXPANSION OF CARBON & ALLOY STEELS[a-j] (Continued)

AISI	Condition	20-100°C 68-212°F	20-200°C 68-392°F	20-300°C 68-572°F	20-400°C 68-752°F	20-500°C 68-932°F	20-600°C 68-1112°F	20-700°C 68-1292°F
5145	Unknown	12.2	---	---	---	---	---	---
5150	Unknown	12.8	---	---	---	---	---	---
5155	Unknown	12.2	---	---	---	---	---	---
52100	Annealed	11.9[b]	---	---	---	---	---	---
52100	Hardened	12.6[b]	---	---	---	---	---	---
6150	Annealed	12.2	12.7	13.3	13.7	14.1	14.4	---
6150	H & T 205°C	12.0	12.5	12.9	13.0	13.3	13.7	---
6150	Annealed	12.4[c]	12.6[c]	13.3[c]	13.8[c]	14.2[c]	14.5[c]	14.7[c]
8115	Unknown	11.9	12.6	13.3	---	---	---	---
81B45	Unknown	11.9	12.6	13.3	---	---	---	---
8617	Unknown	11.9[h]	---	---	---	---	---	---
8622	Unknown	11.1	12.2	12.9	---	---	---	---
8625	Unknown	11.1	12.2	12.9	---	---	---	---
8627	Unknown	11.3	12.2	12.9	---	---	---	---
8630	Unknown	11.3[h]	---	---	---	---	---	---
8637	Unknown	11.3	12.2	12.8	---	---	---	---
8645	OH & T	11.7	12.2	12.8	---	---	---	---
8650	OH & T	11.7	12.2	12.8	---	---	---	---
8655	OH & T	11.7	12.2	12.8	---	---	---	---

a. To obtain coefficients in μin./in. °F, multiply values by 0.556. b. Value represents average coefficient between 0°C (32°F) and indicated temperature. c. Value represents average coefficient between 25°C (75°F) and indicated temperature. d. Value represents average coefficient between 20°C (68°F) and 95°C (200°F). e. Value represents average coefficient between 20°C (68°F) and 370°C (700°F). f. Value represents average coefficient between 20°C (68°F) and 260°C (500°F). g. Value represents average coefficient between 20°C (68°F) and 540°C (1000°F). h. Value represents average coefficient between -18°C (0°F) and 95°C (200°F). i. Value represents average coefficient between -18°C (0°F) and 650°C (1200°F). j Average coefficient of linear expansion, μm/m-°K, at °C. H & T - Hardened & Tempered; OH & T - Oil hardened & tempered; C & T - Carburized & tempered; N & T - Normalized & Tempered.

THERMAL CONDUCTIVITIES OF CARBON & ALLOY STEELS[a,d]

AISI	Condition	0°C 32°F	100°C 212°F	200°C 392°F	300°C 572°F	400°C 752°F	500°C 932°F	600°C 1112°F	700°C 1292°F	800°C 1472°F	1000°C 1832°F	1200°C 2192°F
1008	Unknown	59.5	57.8	53.2	49.4	45.6	41.0	36.8	33.1	28.5	27.6	29.7
1008	Annealed	65.3[b]	60.3	54.9	---	45.2	---	36.4	---	28.5	27.6	---
1010	Unknown	---	46.7	---	---	---	---	---	---	---	---	---
1015	Annealed	51.9	51.0	48.9	---	---	---	---	---	---	---	---
1016	Annealed	51.9	50.2	47.6	---	---	---	---	---	---	---	---
1018	Annealed	51.9	50.8	48.9	---	---	---	---	---	---	---	---
1020	Unknown	51.9	51.0	48.9	---	---	---	---	---	---	---	---
1022	Annealed	51.9	50.8	48.8	---	---	---	---	---	---	---	---
1025	Annealed	51.9	51.1	49.0	46.1	42.7	39.4	35.6	31.8	26.0	27.2	29.7
1026	Annealed	51.9	50.1	48.4	---	---	---	---	---	---	---	---
1029	Annealed	51.9	50.1	48.4	---	---	---	---	---	---	---	---
1030	Annealed	---	51.0	---	---	---	---	---	---	---	---	---
1035	Annealed	---	50.8	---	---	---	---	---	---	---	---	---
1037	Annealed	---	51.0	---	---	---	---	---	---	---	---	---
1039	Annealed	---	50.7	---	---	---	---	---	---	---	---	---
1040	Annealed	---	50.7	---	---	---	---	---	---	---	---	---
1042	Annealed	51.9	50.7	48.2	45.6	41.9	38.1	33.9	30.1	24.7	26.8	29.7
1043	Annealed	---	50.8	---	---	---	---	---	---	---	---	---
1044	Annealed	---	50.8	---	---	---	---	---	---	---	---	---
1045	Annealed	---	50.8	---	---	---	---	---	---	---	---	---
1046	Unknown	51.2	49.7	---	---	---	---	---	---	---	---	---
1050	Annealed	51.2	49.7	46.8	---	---	---	---	---	---	---	---
1055	Unknown	51.2	49.7	---	---	---	---	---	---	---	---	---
1060	Unknown	50.5	---	46.8	---	---	---	---	---	---	---	---
1064	Unknown	51.2	49.7	---	---	---	---	---	---	---	---	---

THERMAL CONDUCTIVITIES OF CARBON & ALLOY STEELS[a,d] (Continued)

AISI	Condition	0°C 32°F	100°C 212°F	200°C 392°F	300°C 572°F	400°C 752°F	500°C 932°F	600°C 1112°F	700°C 1292°F	800°C 1472°F	1000°C 1832°F	1200°C 2192°F
1070	Unknown	49.9	48.4	--	--	--	--	--	--	--	--	--
1078	Annealed	47.8	48.2	45.2	41.4	38.1	35.2	32.7	30.1	24.3	26.8	30.1
1078	Unknown	49.6	48.1	--	--	--	--	--	--	--	--	--
1080	Unknown	50.5	--	46.8	--	--	--	--	--	--	--	--
1086	Unknown	49.9	48.4	--	--	--	--	--	--	--	--	--
1095	Unknown	50.5	46.7	--	--	--	--	--	--	--	--	--
1117	Unknown	51.9[b]	--	--	--	--	--	--	--	--	--	--
1118	Unknown	51.5[b]	--	--	--	--	--	--	--	--	--	--
1141	Unknown	--	50.5	47.6	--	--	--	--	--	--	--	--
1151	Unknown	--	50.5	47.6	--	--	--	--	--	--	--	--
1522	Annealed	51.9	50.1	48.4	--	--	--	--	--	--	--	--
1524	Annealed	51.9	50.1	48.4	--	--	--	--	--	--	--	--
1526	Annealed	51.9	50.1	48.4	--	--	--	--	--	--	--	--
1541	Annealed	51.9	50.1	48.4	--	--	--	--	--	--	--	--
1548	Unknown	50.5	49.0	48.3	--	--	--	--	--	--	--	--
1551	Annealed	50.7	49.3	48.4	--	--	--	--	--	--	--	--
1561	Annealed	51.2	49.7	--	--	--	--	--	--	--	--	--
1566	Annealed	51.2	49.7	--	--	--	--	--	--	--	--	--
2515	Unknown	34.3[b]	--	--	--	--	--	--	--	--	--	--
4037	H & T	48.2	45.6	--	39.4	--	33.9	--	--	--	--	--
4130	H & T	42.7	--	40.6	--	37.3	--	31.0	--	28.1	30.1	--
4140	H & T	42.7	42.3	--	37.7	--	33.1	--	--	--	--	--
4145	H & T	41.8[b]	--	--	--	--	--	--	--	--	--	--
4161	H & T	42.7[b]	--	--	--	--	--	--	--	--	--	--
4427	Unknown	36.8[b]	--	--	--	--	--	--	--	--	--	--

THERMAL CONDUCTIVITIES OF CARBON & ALLOY STEELS[a,d] (Continued)

AISI	Condition	0°C 32°F	100°C 212°F	200°C 392°F	300°C 572°F	400°C 752°F	500°C 932°F	600°C 1112°F	700°C 1292°F	800°C 1472°F	1000°C 1832°F	1200°C 2192°F
4626	Unknown	---	44.1	---	---	---	---	---	---	---	---	---
5132	Unknown	48.6	46.5	44.4	42.3	38.5	35.6	31.8	28.9	26.0	28.1	30.1
5140	H & T	44.8	43.5	---	37.7	---	31.4	---	---	---	---	---
8617	Unknown	---	43.3	---	---	---	---	---	---	---	---	---
8622	Unknown	---	---	37.5[c]	---	---	---	---	---	---	---	---
8627	Unknown	---	---	37.5[c]	---	---	---	---	---	---	---	---
8637	Unknown	---	---	37.5[c]	---	---	---	---	---	---	---	---
8822	Unknown	37.5[c]	---	---	---	---	---	---	---	---	---	---

a. To obtain conductivities in Btu/ft-h-°F, multiply value by 0.5777893; to obtain conductivities in Btu-in./ft²-h-°F, multiply value by 6.933472; to obtain conductivities in cal/cm-s-°C, multiply value by 0.0023884.

b. Value represents thermal conductivity at 20°C (68°F).

c. Value represents thermal conductivity at 50°C (120°F). H & T - Hardened & tempered.

d. Thermal conductivity values in W/m-°K, at °C.

SPECIFIC HEATS OF CARBON & ALLOY STEELS[e]

AISI	Condition	50-100°C	150-200°C	200-250°C	250-300°C	300-350°C	350-400°C	450-500°C	550-600°C	650-700°C
1008	Annealed	481	519	536	553	574	595	662	754	867
1010	Unknown	450	500	520	535	565	590	650	730	825
1015	Annealed	486	519	---	---	---	599	---	---	---
1016	Annealed	481	515	---	---	---	595	---	---	---
1017	Unknown	481[a]	---	---	---	---	---	---	---	---
1018	Annealed	486	519	---	---	---	599	---	---	---
1020	Unknown	486	519	---	---	---	599	---	---	---
1025	Annealed	486	519	532	557	574	599	662	749	846
1030	Annealed	486	519	---	---	---	599	---	---	---
1035	Annealed	486	519	---	---	---	586	---	---	---
1040	Annealed	486	519	---	---	---	586	---	---	---
1042	Annealed	486	515	528	548	569	586	649	708	770
1045	Annealed	486	519	---	---	---	586	---	---	---
1050	Annealed	486	519	---	---	---	590	---	---	---
1060	Unknown	502	544	---	---	---	---	---	---	---
1070	Unknown	490	532	---	---	---	---	---	---	---
1078	Annealed	490	532	548	565	586	607	670	712	770
1086	Unknown	500	532	---	---	---	---	---	---	---
1095	Unknown	461[b]	---	---	---	---	---	---	---	---
1117	Unknown	481	---	---	---	---	---	---	---	---
1140	Unknown	461[b]	---	---	---	---	---	---	---	---
1151	Unknown	502[b]	---	---	---	---	---	---	---	---
1522	Annealed	486	519	---	---	---	599	---	---	---
1524	Annealed	477	511	528	544	565	590	649	741	837
1561	Annealed	486	519	---	---	---	---	---	---	---
4032	Unknown	---	461[c]	---	---	---	---	---	---	---

SPECIFIC HEATS OF CARBON & ALLOY STEELS[e] (Continued)										
AISI	Condition	50-100°C	150-200°C	200-250°C	250-300°C	300-350°C	350-400°C	450-500°C	550-600°C	650-700°C
4130	H & T	477	515	---	544	---	595	657	737	825
4140	H & T	---	473[d]	---	---	---	519[d]	---	561[d]	---
4142	Unknown	---	502[c]	---	---	---	---	---	---	---
4626	N & T	335[g]		---	---	---	---	---	---	615[f]
4815	Unknown	481[b]	---	---	---	---	---	---	---	---
5132	Unknown	494	523	536	553	574	595	657	741	837
5140	H & T	452[d]	473[d]	---	---	---	519[d]	---	561 [d]	---
8115	Unknown	461[b]	---	---	---	---	---	---	---	---
8617	Unknown	481[a]	---	---	---	---	---	---	---	---
8637	Unknown	---	502[c]	---	---	---	---	---	---	---

a. Value represents specific heat at 25 - 95°C.
b. Value represents specific heat at 20 - 100°C.
c. Value represents specific heat at 20 - 200°C.
d. Value represented is a mean value of temperatures between 20°C and the higher of the given temperatures.
e. Mean apparent specific heat, J/Kg.°K, at°C.
f. Value represents specific heat at 25 - 540°C.
g. Value represents specific heat at 10 - 25°C.
H & T - Hardened & tempered; N & T - Normalized & tempered.

ELECTRICAL RESISTIVITIES OF CARBON & ALLOY STEELS[e]

AISI	Condition	20°C	100°C	200°C	400°C	600°C	700°C	800°C	900°C	1000°C	1100°C	1200°C
1008	Annealed	0.142	0.190	0.263	0.458	0.734	0.905	1.081	1.130	1.165	1.193	1.216
1010	Unknown	0.143	---	---	---	---	---	---	---	---	---	---
1015	Annealed	0.159[a]	0.219	0.292	---	---	---	---	---	---	---	---
1016	Annealed	0.160[a]	0.220	0.290	---	---	---	---	---	---	---	---
1018	Annealed	0.159[a]	0.219	0.293	---	---	---	---	---	---	---	---
1020	Unknown	0.159[a]	0.219	0.292	---	---	---	---	---	---	---	---
1022	Annealed	0.159[a]	0.219	0.293	---	---	---	---	---	---	---	---
1025	Annealed	0.159[a]	0.219	0.292	0.487	0.758	0.925	1.094	1.136	1.167	1.194	1.219
1029	Annealed	0.160[a]	0.220	0.290	---	---	---	---	---	---	---	---
1030	Annealed	0.166	---	---	---	---	---	---	---	---	---	---
1035	Annealed	0.163[a]	0.217	---	---	---	---	---	---	---	---	---
1040	Annealed	0.160[a]	0.221	---	---	---	---	---	---	---	---	---
1042	Annealed	0.171	0.221	0.296	0.493	0.766	0.932	1.111	1.149	1.179	1.207	1.230
1043	Annealed	0.163[a]	0.219	---	---	---	---	---	---	---	---	---
1045	Annealed	0.162[a]	0.223	---	---	---	---	---	---	---	---	---
1046	Unknown	0.163[a]	0.224	---	---	---	---	---	---	---	---	---
1050	Annealed	0.163[a]	0.224	0.300	---	---	---	---	---	---	---	---
1055	Unknown	0.163[a]	0.224	---	---	---	---	---	---	---	---	---
1060	Unknown	0.180	---	---	---	---	---	---	---	---	---	---
1065	Unknown	0.163[a]	0.224	---	---	---	---	---	---	---	---	---
1070	Unknown	0.168[a]	0.230	---	---	---	---	---	---	---	---	---
1078	Annealed	0.180	0.232	0.308	0.505	0.772	0.935	1.129	1.164	1.191	1.214	1.231
1080	Unknown	0.180	---	---	---	---	---	---	---	---	---	---
1095	Unknown	0.180	---	---	---	---	---	---	---	---	---	---

ELECTRICAL RESISTIVITIES OF CARBON & ALLOY STEELS[e] (Continued)

AISI	Condition	20°C	100°C	200°C	400°C	600°C	700°C	800°C	900°C	1000°C	1100°C	1200°C
1137	Unknown	0.170	---	---	---	---	---	---	---	---	---	---
1141	Unknown	0.170	---	---	---	---	---	---	---	---	---	---
1151	Unknown	0.170	---	---	---	---	---	---	---	---	---	---
1524	Unknown	0.208	0.259	0.333	0.523	0.786	0.946	1.103	1.143	1.174	1.202	1.227
1524	Annealed	0.160[a]	0.220	0.290	---	---	---	---	---	---	---	---
1552	Unknown	0.163[a]	0.224	---	---	---	---	---	---	---	---	---
4130	H & T	0.223	0.271	0.342	0.529	0.786	---	1.103	---	1.171	---	1.222
4140	H & T	0.220	0.260	0.330	0.480	0.650	---	---	---	---	---	---
4626	N & T	0.200[b]	---	---	---	---	---	---	---	---	---	---
4815	Unknown	0.260[a]	0.310	---	---	---	---	---	---	---	---	---
5132	Unknown	0.210	0.259	0.330	0.517	0.778	0.934	1.106	1.145	1.177	1.205	1.230
5140	H & T	0.228	0.281	0.352	0.530	0.785	---	---	---	---	---	---
8615	Unknown	---	0.300[c]	---	---	---	---	---	---	---	---	---
8625	Unknown	---	0.300[c]	---	---	---	---	---	---	---	---	---
8720	Unknown	---	0.300[c]	---	---	---	---	---	---	---	---	---

a. Value represents resistivity at 0°C.
b. Value represents resistivity at -18°C.
c. Value represents resistivity at 50°C.
d. Value represents resistivity at 300°C.
e. Resistivity values in μΩ-m, at °C.　H & T - Hardened & tempered;　N & T - Normalized & tempered.

PHYSICAL PROPERTIES OF WROUGHT STAINLESS STEELS[a]

AISI Type	Density g/cm³ (lb/in.³)	Elastic Modulus GPa (10⁶ psi)	Mean CTE[b] from 0°C (32°F) to: 315°C (600°F) μm/m°C (μin./in.°F)	538°C (1000°F) μm/m°C (μin./in.°F)	Thermal Conductivity at 100°C (212°F) W/m°K (Btu/ft h°F)	at 500°C (932°F) W/m°K (Btu/ft h°F)	Specific Heat at 0-100°C (32-212°F) J/kg°K (Btu/lb°F)	Electrical Resistivity ηΩm	Magnetic Permeability
201	7.8 (0.28)	197 (28.6)	17.5 (9.7)	18.4 (10.2)	16.2 (9.4)	21.5 (12.4)	500 (0.12)	690	1.02
202	7.8 (0.28)	197 (28.6)	18.4 (10.2)	19.2 (10.7)	16.2 (9.4)	21.6 (12.5)	500 (0.12)	690	1.02
205	7.8 (0.28)	197 (28.6)	17.9 (9.9)	19.1 (10.6)	---	---	500 (0.12)	---	---
301	8.0 (0.29)	193 (28.0)	17.2 (9.6)	18.2 (10.1)	16.2 (9.4)	21.5 (12.4)	500 (0.12)	720	1.02
302	8.0 (0.29)	193 (28.0)	17.8 (9.9)	18.4 (10.2)	16.2 (9.4)	21.5 (12.4)	500 (0.12)	71-0	1.02
302B	8.0 (0.29)	193 (28.0)	18.0 (10.0)	19.4 (10.8)	15.9 (9.2)	21.6 (12.5)	500 (0.12)	720	1.02
303	8.0 (0.29)	193 (28.0)	17.8 (9.9)	18.4 (10.2)	16.2 (9.4)	21.5 (12.4)	500 (0.12)	720	1.02
304	8.0 (0.29)	193 (28.0)	17.8 (9.9)	18.4 (10.2)	16.2 (9.4)	21.5 (12.4)	500 (0.12)	720	1.02
304L	8.0 (0.29)	---	---	---	---	---	---	---	1.02
302Cu	8.0 (0.29)	193 (28.0)	17.8 (9-9)	---	11.2 (6.5)	21.5 (12.4)	500 (0.12)	720	1.02
304N	8.0 (0.29)	196 (28.5)	---	---	---	---	500 (0.12)	720	1.02
305	8.0 (0.29)	193 (28.0)	17.8 (9.9)	18.4 (10.2)	16.2 (9.4)	21.5 (12.4)	500 (0.12)	720	1.02
308	8.0 (0.29)	193 (28.0)	17.8 (9.9)	18.4 (10.2)	15.2 (8.8)	21.6 (12.5)	500 (0.12)	720	---
309	8.0 (0.29)	200 (29.0)	16.6 (9.2)	17.2 (9.6)	15.6 (9.0)	18.7 (10.8)	500 (0.12)	780	1.02
310	8.0 (0.29)	200 (29.0)	16.2 (9.0)	17.0 (9.4)	14.2 (8.2)	18.7 (10.8)	500 (0.12)	780	1.02
314	7.8 (0.28)	200 (29.0)	15.1 (8.4)	17.5 (9.7)	17.5 (10.1)	20.9 (12.1)	500 (0.12)	770	1.02
316	8.0 (0.29)	193 (28.0)	16.2 (9.0)	17.5 (9.7)	16.2 (9.4)	21.5 (12.4)	500 (0.12)	740	1.02
316L	8.0 (0.29)	---	---	---	---	---	---	---	1.02
316N	8.0 (0.29)	196 (28.5)	---	---	---	---	500 (0.12)	740	1.02
317	8.0 (0.29)	193 (28.0)	16.2 (9.0)	17.5 (9.7)	16.2 (9.4)	21.5 (12.4)	500 (0.12)	740	1.02
317L	8.0 (0.29)	200 (29.0)	---	18.1 (10.1)	14.4 (8.3)	---	500 (0.12)	790	---

PHYSICAL PROPERTIES OF WROUGHT STAINLESS STEELS[a] (Continued)

AISI Type	Density g/cm³ (lb/in.³)	Elastic Modulus GPa (10⁶ psi)	Mean CTE[b] from 0°C (32°F) to: 315°C (600°F) µm/m°C (µin./in.°F)	538°C (1000°F) µm/m°C (µin./in.°F)	Thermal Conductivity at 100°C (212°F) W/m°K (Btu/ft h°F)	at 500°C (932°F) W/m°K (Btu/ft h°F)	Specific Heat at 0-100°C (32-212°F) J/kg°K (Btu/lb°F)	Electrical Resistivity ηΩm	Magnetic Permeability
321	8.0 (0.29)	193 (28.0)	17.2 (9.6)	18.6 (10.3)	16.1 (9.3)	22.2 (12.8)	500 (0.12)	720	1.02
329	7.8 (0.28)	---	---	---	---	---	460 (0.11)	750	---
330	8.0 (0.29)	196 (28.5)	16.0 (8.9)	16.7 (9.3)	---	---	460 (0.11)	1020	1.02
347	8.0 (0.29)	193 (28.0)	17.2 (9.6)	18.6 (10.3)	16.1 (9.3)	22.2 (12.8)	500 (0.12)	730	1.02
384	8.0 (0.29)	193 (28.0)	17.8 (9.9)	18.4 (10.2)	16.2 (9.4)	21.5 (12.4)	500 (0.12)	790	1.02
405	7.8 (0.28)	2(X) (29.0)	11.6 (6.4)	12.1 (6.7)	27.0 (15.6)	---	460 (0.11)	600	---
409	7.8 (0.28)	---	---	---	---	---	---	---	---
410	7.8 (0.28)	200 (29.0)	11.4 (6.3)	11.6 (6.4)	24.9 (14.4)	28.7 (16.6)	460 (0.11)	570	700-1000
414	7.8 (0.28)	200 (29.0)	11.0 (6.1)	12.1 (6.7)	24.9 (14.4)	28.7 (16.6)	460 (0.11)	700	---
416	7.8 (0.28)	200 (29.0)	11.0 (6.1)	11.6 (6.4)	24.9 (14.4)	28.7 (16.6)	460 (0.11)	570	700-1000
420	7.8 (0.28)	200 (29.0)	10.8 (6.0)	11.7 (6.5)	24.9 (14.4)	---	460 (0.11)	550	---
422	7.8 (0.28)	---	11.4 (6.3)	11.9 (6.6)	23.9 (13.8)	27.3 (15.8)	460 (0.11)	---	---
429	7.8 (0.28)	200 (29.0)	---	---	25.6 (14.8)	---	460 (0.11)	590	---
430	7.8 (0.28)	200 (29.0)	11.0 (6.1)	11.4 (6.3)	26.1 (15.1)	26.3 (15.2)	460 (0.11)	600	600-1100
430F	7.8 (0.28)	200 (29.0)	11.0 (6.1)	11.4 (6.3)	26.1 (15.1)	26.3 (15.2)	460 (0.11)	600	---
431	7.8 (0.28)	200 (29.0)	12.1 (6.7)	---	20.2 (11.7)	---	460 (0.11)	720	---
434	7.8 (0.28)	200 (29.0)	11.0 (6.1)	11.4 (6.3)	---	26.3 (15.2)	460 (0.11)	600	600-1100
436	7.8 (0.28)	200 (29.0)	---	---	23.9 (13.8)	26.0 (15.0)	460 (0.11)	600	600-1100
439	7.7 (0.28)	200 (29.0)	11.0 (6.1)	11.4 (6.3)	24.2 (14.0)	---	460 (0.11)	630	---
440A	7.8 (0.28)	200 (29.0)	---	---	24.2 (14.0)	---	460 (0.11)	600	---
440C	7.8 (0.28)	200 (29.0)	---	---	24.2 (14.0)	---	460 (0.11)	600	---

PHYSICAL PROPERTIES OF WROUGHT STAINLESS STEELS[a] (Continued)								
			Mean CTE[b] from 0°C (32°F) to:		Thermal Conductivity		Specific Heat at 0-100°C (32-212°F) J/kg°K (Btu/lb°F)	Electrical Resistivity ηΩm	Magnetic Permeability
AISI Type	Density g/cm³ (lb/in.³)	Elastic Modulus GPa (10⁶ psi)	315°C (600°F) µm/m°C (µin./in.°F)	538°C (1000°F) µm/m°C (µin./in.°F)	at 100°C (212°F) W/m°K (Btu/ft h°F)	at 500°C (932°F) W/m°K (Btu/ft h°F)			
444	7.8 (0.28)	200 (29.0)	10.6 (5.9)	11.4 (6.3)	26.8 (15.5)	---	420 (0.10)	620	---
446	7.5 (0.27)	200 (29-0)	10.8 (6.0)	11.2 (6.2)	20.9 (12.1)	24.4 (14.1)	500 (0.12)	670	400-700
13-8Mo	7.8 (0.28)	203 (29.4)	11.2 (6.2)	11.9 (6.6)	14.0 (8.1)	22.0 (12.7)	460 (0.11)	1020	---
15-5	7.8 (0.28)	196 (28.5)	11.4 (6.3)	---	17.8 (10.3)	23.0 (13.1)	420 (0.10)	770	95[c]
17-4	7.8 (0.28)	196 (28.5)	11.6 (6.4)	---	18.3 (10.6)	23.0 (13.1)	460 (0.11)	800	95[c]
17-7	7.8 (0.28)	204 (29.5)	11.6 (6.4)	---	16.4 (9.5)	21.8 (12.6)	460 (0.11)	830	---

a. Wrought stainless steels in the annealed condition.
b. Coefficient of thermal expansion.
c. Magentic permeability approximate value.

Chapter

22

AMERICAN STANDARDS CROSS REFERENCES CARBON, ALLOY & STAINLESS STEELS

AMERICAN STANDARDS CROSS REFERENCES - AISI CARBON STEELS

AISI	UNS No.	ASTM	SAE	AMS	MIL	FED
1005	G10050	A 29 (1005), A 510 (1005)	J403 (1005), J412 (1005)	---	MIL-S-11310 (CS1005)	---
1006	G10060	A 29 (1006), A 510 (1006), A 545 (1006), A 635 (1006), A 830	J403 (1006), J412 (1006), J1397 (1006)	---	MIL-S-11310 (CS1006)	---
1008	G10080	A 29 (1008), A 108 (1008), A 510 (1008), A 512 (1008), A 513 (1008), A 519 (1008), A 545 (1008), A 549 (1008), A 575 (1008), A 576 (1008), A 635 (1008), A 830	J403 (1008), J412 (1008), J1397 (1008)	---	MIL-S-11310 (CS1008)	---
1009	G10090	A 635 (1009), A 827, A 830	J1249 (1009)	---	---	---
1010	G10100	A 29 (1010), A 108, A 510, A 512 (1010, MT 1010), A 513, (1010, MT 1010), A519 (1010, MT 1010), A 545, A 549 (1010), A 576 (1010), A 635 (1010), A 787 (MT 1010), A 830	J403 (1010), J412 (1010), J1397 (1010)	5040, 5042, 5044, 5047, 5050, 5053, 7225	MIL-S-11310 (CS1010)	---
1011	G10110	A 29 (1011), A 510 (1011)	J412 (1011), J1249 (1011)	---	---	---
1012	G10120	A 29 (1012), A 510 (1012), A 512 (1012), A 519 (1012), A 545 (1012), A 549 (1012), A 575 (M1012), A 576 (1012), A 611 (Grade C&D), A 635 (1012), A 830	J403 (1012), J412 (1012), J1397 (1012)	---	MIL-S-11310 (CS1012)	---
---	G10130	A 29 (1013), A 510 (1013), A 548 (1013)	J403 (1013), J412 (1013)	---	---	---
1015	G10150	A 29 (1015), A 510 (1015), A 512 (1015, MT 1015), A 513 (1015, MT 1015), A 519 (1015, MT 1015), A 545 (1015), A 549 (1015), A 575 (M1015), A 576 (1015), A 635 (1015), A 659 (1015), A 787 (MT 1015), A 794 (1015), A 830	J403 (1015), J412 (1015), J1397 (1015)	5060	---	---

AMERICAN STANDARDS CROSS REFERENCES - AISI CARBON STEELS (Continued)						
AISI	UNS No.	ASTM	SAE	AMS	MIL	FED
1016	G10160	A 29 (1016), A 108 (1016), A 510 (1016), A 512 (1016), A 513 (1016), A 519 (1016), A 545 (1016), A 548 (1016), A 549 (1016), A 576 (1016), A 635 (1016), A 659 (1016), A 794 (1016), A 830	J403 (1016), J412 (1016), J1397 (1016)	---	---	---
1017	G10170	A 29 (1017), A 510 (1017), A 513 (1017), A 519 (1017), A 544 (1017), A 549 (1017), A 575 (M1017), A 576 (1017), A 611 (A, B, C, E), A 635 (1017), A 659 (1017), A 794 (1017), A 830	J403 (1017), J412 (1017), J1397 (1017)	---	MIL-S-11310 (CS1017)	---
1018	G10180	A 29 (1018), A 510 (1018), A 512 (1018), A 513 (1018), A 519 (1018), A 544 (1018), A 545 (1018), A 548 (1018), A 549 (1018), A 576 (1018), A 611 (D-1), A 635 (1018), A 659 (1018), A 794 (1018), A 830	J403 (1018), J412 (1018), J1397 (1018)	5069	MIL-S-11310 (CS1018)	---
1019	G10190	A 29 (1019), A 510 (1019), A 512 (1019), A 513 (1019), A 519 (1019), A 545 (1019), A 548 (1019), A 576 (1019), A 635 (1019), A 830	J403 (1019), J412 (1019), J1397 (1019)	---	---	---
1020	G10200	A 29 (1020), A 510 (1020), A 512 (1020, MT 1020), A 513 (1020, MT 1020), A 519 (1020, MT 1020), A 544 (1020), A 575 (M1020), A 576 (1020), A 635 (1020), A 659 (1020), A 787 (MT 1020), A 794 (1020), A 827, A 830	J403 (1020), J412 (1020), J1397 (1020)	5032, 5045, 5046	MIL-T-3520, MIL-S-7952, MIL-S-11310 (CS1020), MIL-S-16788	---
1021	G10210	A 29 (1021), A 510 (1021), A 512 (1021), A 513 (1021), A 519 (1021), A 545 (1021), A 548 (1021), A 576 (1021), A 635 (1021), A 659 (1021), A 794 (1021), A 830	J403 (1021), J412 (1021), J1397 (1021)	---	---	---

AMERICAN STANDARDS CROSS REFERENCES - AISI CARBON STEELS (Continued)						
AISI	UNS No.	ASTM	SAE	AMS	MIL	FED
1022	G10220	A 29 (1022), A 510 (1022), A 513 (1022), A 519 (1022), A 544 (1022), A 545 (1022), A 548 (1022), A 576 (1022), A 635 (1022), A 830	J403 (1022), J412 (1022), J1397 (1022)	5070	MIL-S-11310 (CS1022)	---
1023	G10230	A 29 (1023), A 510 (1023), A 513 (1023), A 576 (1023), A 635 (1023), A 659 (1023), A 794 (1023), A 830	J403 (1023), J412 (1023), J1397 (1023)	---	---	---
1025	G10250	A 29 (1025), A 510 (1025), A 512 (1025), A 513 (1025), A 519 (1025), A 575 (M1025), A 576 (1025), A 830	J403 (1025), J412 (1025), J1397 (0125)	5046, 5075, 5077	MIL-T-3520, MIL-T-5066, MIL-S-7952, MIL-S-11310 (CS1025)	QQ-S-700 (C1025)
1026	G10260	A 29 (1026), A 510 (1026), A 512 (1026), A 513 (1026), A 519 (1026), A 545 (1026), A 576 (1026), A 830	J403 (1026), J412 (1026), J1397 (1026)	---	MIL-F-20670, MIL-S-22698, MIL-S-24093	---
1029	G10290	A 29 (1029), A 510 (1029), A 576 (1029)	J403 (1029), J412 (1029)	---	---	---
1030	G10300	A 29 (1030), A 510 (1030), A 512 (1030), A 513 (1030), A 519 (1030), A 544 (1030), A 545 (1030), A 546 (1030), A 576 (1030), A 682 (1030), A 830	J403 (1030), J412 (1030), J1397 (1030)	---	MIL-S-11310 (CS1030), MIL-S-46070	QQ-S-700 (C1030)
1033	G10330	A 513 (1033), A 830	J1249 (1033)	---	---	---
1034	G10340	A 29 (1034), A 510 (1034)	J412 (1034), J1249 (1034)	---	---	---
1035	G10350	A 29 (1035), A 510 (1035), A512 (1035), A 513 (1035), A 519 (1035), A 544 (1035), A 545 (1035), A 546 (1035), A 576 (1035), A 682 (1035), A 827, A 830	J403 (1035), J412 (1035), J1397 (1035)	5080, 5082	MIL-S-3289, MIL-S-19434, MIL-S-46070	QQ-S-700 (C1035)
1037	G10370	A 29 (1037), A 510 (1037), A 576 (1037), A 830	J403 (1037), J412 (1037), J1397 (1037)	---	---	---
1038	G10380	A 29 (1038), A 510 (1038), A 544 (1038), A 545 (1038), A 546 (1038), A 576 (1038), A 830	J403 (1038), J412 (1038), J1397 (1038)	---	---	---
1038H	H10380	A 304 (1038 H)	J1268 (1038H)	---	---	---

AMERICAN STANDARDS CROSS REFERENCES - AISI CARBON STEELS (Continued)

AISI	UNS No.	ASTM	SAE	AMS	MIL	FED
1039	G10390	A 29 (1039), A 510 (1039), A 546 (1039), A 576 (1039), A 830	J403 (1039), J412 (1039), J1397 (1039)	---	---	---
1040	G10400	A 29 (1040), A 510 (1040), A 513 (1040), A 519 (1040), A 546 (1040), A 576 (1040), A 682 (1040), A 827, A 830	J403 (1040), J412 (1040), J1397 (1040)	---	MIL-S-11310 (CS1040), MIL-S-16788, MIL-S-46070	---
1042	G10420	A 29 (1042), A 183 (2-Nuts), A 510 (1042), A 576 (1042), A 830	J403 (1042), J412 (1042), J1397 (1042)	---	---	---
1043	G10430	A 29 (1043), A 183 (2-Nuts), A 510 (1043), A 576 (1043), A 830	J403 (1043), J412 (1043), J1397 (1043)	---	---	---
1044	G10440	A 183 (2-Nuts), A 510 (1044), A 575 (M1044), A 576 (1044)	J403 (1044), J412 (1044), J1397 (1044)	---	---	---
1045	G10450	A 29 (1045), A 183 (2-Nuts), A 236 (1045), A 266 (1045), A 510 (1045), A 519 (1045), A 576 (1045), A 682 (1045), A 827, A 830	J403 (1045), J412 (1045), J1397 (1045)	---	---	QQ-S-700 (C1045)
1045H	H10450	A 304 (1045 H)	J1268 (1045H)	---	---	---
1046	G10460	A 29 (1046), A 510 (1046), A 576 (1046), A 830	J403 (1046), J412 (1046), J1397 (1046)	---	---	---
1049	G10490	A 29 (1049), A 510 (1049), A 576 (1049), A 830	J403 (1049), J412 (1049), J1397 (1049)	---	---	---
1050	G10500	A 29 (1050), A 510 (1050), A 513 (1050), A 519 (1050), A 576 (1090), A 682 (1050), A 827, A 830	J403 (1050), J412 (1050), J1397 (1050)	5085	---	QQ-S-700 (C1050)
1053	G10530	A 510 (1053), A 576 (1053)	J403 (1053), J412 (1053)	---	---	---
1055	G10550	A 510 (1055), A 576 (1055), A 682 (1055), A 713 (1055), A 830	J403 (1055), J412 (1055), J1397 (1055)	---	MIL-S-10520	QQ-S-700 (C1055)
1059	G10590	A 29 (1059), A 510 (1059), A 713 (1059)	J412 (1059), J1249 (1059)	---	---	---
1060	G10600	A 29 (1060), A 510 (1060), A 513 (1060), A 576 (1060), A 682 (1060), A 713 (1060)	J403 (1060), J412 (1060), J1397 (1060)	7240	---	---

AMERICAN STANDARDS CROSS REFERENCES - AISI CARBON STEELS (Continued)

AISI	UNS No.	ASTM	SAE	AMS	MIL	FED
1064	G10640	A 29 (1064), A 510 (1064), A 682 (1064), A 713 (1064)	J412 (1064), J1249 (1064), J1397 (1064)	---	---	---
1065	G10650	A 29 (1065), A 510 (1065), A 682 (1065), A 713 (1065)	J403 (1065), J412 (1065), J1397 (1065)	---	MIL-S-46049 (1065), MIL-S46409 (1065)	QQ-S-700 (C1065)
1069	G10690	A 29 (1069), A 510 (1069), A 713 (1069)	J403 (1069), J412 (1069)	---	MIL-S-11713	---
1070	G10700	A 29 (1070), A 510 (1070), A 576 (1070), A 682 (1070), A 713 (1070), A 830	J403 (1070), J412 (1070), J1397 (1070)	5115	MIL-S-11713 (2), MIL-S-12504	---
1074	G10740	A 29 (1074), A 510 (1074), A 682 (1074), A 713 (1074), A 830	J403 (1074), J412 (1074), J1397 (1074)	5120	MIL-S-46049 (1074)	QQ-S-700 (C1074)
1075	G10750	A 29 (1075), A 510 (1075), A 713 (1075)	J403 (1075), J412 (1075)	---	---	---
1078	G10780	A 29 (1078), A 510 (1078), A 576 (1078), A 713 (1078), A 830	J403 (1078), J412 (1078), J1397 (1078)	---	---	---
1080	G10800	A 29 (1080), A 510 (1080), A 576 (1080), A 682 (1080), A 713 (1080), A 830	J403 (1080), J412 (1080), J1397 (1080)	5110	---	QQ-S-700 (C1080)
1084	G10840	A 29 (1084), A 510 (1084), A 576 (1084), A 713 (1084), A 830	J403 (1084), J412 (1084), J1397 (1084)	---	---	QQ-S-700 (C1084)
1085	G10850	A 510 (1085), A 682 (1085), A 830	J403 (1085), J412 (1085), J1397 (1085)	---	---	QQ-S-700 (C1085)
1086	G10860	A 29 (1086), A 510 (1086), A 682 (1086), A 713 (1086), A 830	J403 (1086), J412 (1086), J1397 (1086)	---	---	QQ-S-700 (C1086)
1090	G10900	A 29 (1090), A 510 (1090), A 576 (1090), A 713 (1090), A 830	J403 (1090), J412 (1090), J1397 (1090)	5112	---	QQ-S-700 (C1090)
1095	G10950	A 29 (1095), A 510 (1095), A 576 (1095), A 682 (1095), A 713 (1095), A 830	J403 (1095), J412 (1095), J1397 (1095)	5121, 5122, 5132, 7304	MIL-S-7947, MIL-S-8559, MIL-S-16788 (C10)	QQ-S-700 (C1095)

AMERICAN STANDARDS CROSS REFERENCES - AISI RESULFURIZED CARBON STEELS

AISI	UNS No.	ASTM	SAE	AMS
1108	G11080	A 29 (1108), A 510 (1108), A 544 (1108)	J403 (1108), J412 (1108)	---
1109	G11090	A 29 (1109), A 510 (1109), A 544 (1109), A 576 (1109)	J412 (1109), J1249 (1109), J1397 (1109)	---
1110	G11100	A 29 (1110), A 510 (1110), A 512 (1110), A 544 (1110), A 549 (1110), A 576 (1110)	J403 (1110), J412 (1110)	---
1116	G11160	A 29 (1116), A 510 (1116)	J412 (1116)	---
1117	G11170	A 29 (1117), A 108 (1117), A 510 (1117), A 512 (1117), A 576 (1117)	J403 (1117), J412 (1117), J1397 (1117)	5022
1118	G11180	A 29 (1118), A 510 (1118), A 576 (1118)	J403 (1118), J412 (1118), J1397 (1118)	---
1119	G11190	A 29 (1119), A 183 (1-Bolts, 1-Nuts), A 510 (1119), A 576 (1119)	J412 (1119), J1249 (1119), J1397 (1119)	---
1123	G11230	---	J403 (1123)	---
1132	G11320	A 29 (1132), A 510 (1132)	J412 (1132), J1249 (1132), J1397 (1132)	---
1137	G11370	A 29 (1137), A 311 (1137), A 510 (1137), A 576 (1137)	J403 (1137), J412 (1137), J1397 (1137)	5024
1139	G11390	A 29 (1139), A 510 (1139), A 576 (1139)	J403 (1139)	---
1140	G11400	A 29 (1140), A 510 (1140), A 576 (1140)	J403 (1140), J412 (1140), J1397 (1140)	---
1141	G11410	A 29 (1141), A 108 (1141), A 311 (1141), A 510 (1141), A 576 (1141)	J403 (1141), J412 (1141), J1397 (1141)	---
1144	G11440	A 29 (1144), A 311 (1144), A 510 (1144)	J403 (1144), J412 (1144), J1397 (1144)	---
1145	G11450	A 29 (1145), A 510 (1145)	J412 (1145), J1249 (1145), J1397 (1145)	---
1146	G11460	A 29 (1146), A 510 (1146), A 576 (1146)	J403 (1146), J412 (1146), J1397 (1146)	---
1151	G11510	A 29 (1151), A 108 (1151), A 510 (1151), A 576 (1151)	J403 (1151), J412 (1151), J1397 (1151)	---

AMERICAN STANDARDS CROSS REFERENCES - AISI REPHOSPHORIZED & RESULFURIZED CARBON STEELS

AISI	UNS No.	ASTM	SAE	AMS
1211	G12110	A 29 (1211), A 108 (1211), A 576 (1211)	J403 (1211)	---
1212	G12120	A 29 (1212), A 108 (1212), A 576 (1212)	J403 (1212)	5010
1213	G12130	A 29 (1213), A 108 (1213), A 576 (1213)	J403 (1213)	---
12L13	G12134	A 29 (12L13)	J403 (12L13)	---
12L14	G12144	A 576 (12L14)	A 576 (12L14), J412 (12L14), J1397 (12L14)	---
1215	G12150	A 29 (1215), A 108 (1215), A 576 (1215)	J403 (1215), J412 (1215)	5010

AMERICAN STANDARDS CROSS REFERENCES - AISI CARBON-MANGANESE STEELS

AISI	UNS No.	ASTM	SAE
1513	G15130	A 29 (1513), A 510 (1513)	J403 (1513), J412 (1513)
1518	G15180	A 29 (1518), A 510 (1518), A 519 (1518)	J412 (1518), J1249 (1518)
1522	G15220	A 29 (1522), A 510 (1522)	J403 (1522), J412 (1522)
15B21H	H15211	---	J1268 (15B21H)
1522H	H15220	A 304 (1522 H)	J1268 (1522H)
1024, 1524	G15240	A 29 (1524), A 510 (1524), A 513 (1524), A 519 (1524), A 545 (1524), A 635 (1524), A 830	J403 (1524), J412 (1524), J1397 (1524)
1524H	H15240	A 304 (1524 H)	J1268 (1524H)
1525	G15250	A 29 (1525), A 510 (1525)	J412 (1525), J1249 (1525)
1526	G15260	A 29 (1526), A 510 (1526)	J403 (1526), J412 (1526)
1526H	H15260	A 304 (1526 H)	J1268 (1526H)
1027, 1527	G15270	A 29 (1527), A 510 (1527), A 513 (1527), A 830	J403 (1527), J412 (1527)
1533	G15330	---	J403 (1533)
1534	G15340	---	J403 (1534)
15B35H	H15351	A 304 (15B35 H)	J1268 (15B35H)
1036, 1536	G15360	A 29 (1536), A 510 (1536), A 830	J403 (1536), J412 (1536), J1397 (1536)
15B37H	H15371	A 304 (15B37 H)	J1268 (15B37H)
1041, 1541	G15410	A 29 (1541), A 510 (1541), A 519 (1541), A 545 (1541), A 546 (1541), A 830	J403 (1541), J412 (1541), J1397 (1541)
1541H	H15410	A 304 (1541 H)	J775 (NV 1), J1268 (1541H)
15B41H	H15411	A 304 (15B41 H)	J1268 (15B41H)
1547	G15470	A 29 (1547), A 510 (1547)	J403 (1547), J412 (1547), J775 (NV-2), J1249 (1547), J1397 (1547)
1048, 1548	G15480	A 29 (1548), A 510 (1548), A 830	J403 (1548), J412 (1548), J1397 (1548)
15B48H	H15481	A 304 (15B48 H)	J1268 (15B48H)
1051, 1551	G15510	A 29 (1551), A 510 (1551)	J403 (1551), J412 (1551)
1052, 1552	G15520	A 29 (1552), A 510 (1552), A 830	J403 (1552), J412 (1552), J1397 (1552)
1553	G15530	---	J403 (1553)
1061, 1561	G15610	A 29 (1561), A 510 (1561), A 713 (1561)	J403 (1561), J412 (1561)

AMERICAN STANDARDS CROSS REFERENCES - AISI CARBON-MANGANESE STEELS (Continued)

AISI	UNS No.	ASTM	SAE
15B62H	H15621	A 304 (15B62 H)	J1268 (15B62H)
1066, 1566	G15660	A 29 (1566), A510 (1566), A 713 (1566)	J403 (1566), J412 (1566)
1570	G15700	---	J403 (1570)
1072, 1572	G15720	A 29 (1572), A 510 (1572), A 713 (1572)	J403 (1572), J412 (1572)
1580	G15800	---	J403 (1580)
1590	G15900	---	J403 (1590)

AMERICAN STANDARDS CROSS REFERENCES - AISI MANGANESE STEELS

AISI	UNS No.	ASTM	SAE
1330	G13300	A 322 (1330), A 331 (1330), A 519 (1330), A 711, A 752 (1330), A 829	J404 (1330), J412 (1330) J1397 (1330)
1330H	H13300	A 304 (1330 H)	J1268 (1330H)
1335	G13350	A 331 (1335), A 519 (1335), A 547 (1335), A 711 (1335), A 752 (1335), A 829	J404 (1335), J412 (1335), J1397 (1335)
1335H	H13350	A 304 (1335 H), A 547 (1335 H)	J1268 (1335H)
1340	G13400	A 322 (1340), A 331 (1340), A 513 (1340), A 519 (1340), A 547 (1340), A 752 (1340), A 829	J404 (1340), J412 (1340), J1397 (1340)
1340H	H13400	A 304 (1340 H), A 547 (1340 H)	J1268 (1340H)
1345	G13450	A 322 (1345), A 331 (1345), A 519 (1345), A 752 (1345), A 829	J404 (1345), J412 (1345), J1397 (1345)
1345H	H13450	A 304 (1345 H)	J1268 (1345H)

AMERICAN STANDARDS CROSS REFERENCES - AISI NICKEL-CHROMIUM STEELS

AISI	UNS No.	ASTM	SAE	MIL
3140	G31400	A 331 (3140), A 519 (3140)	J775 (NV 4), J1249(3140)	
E3310	G33106	A 331 (E3310), A 519 (E3310), A 646 (3310-1)	J1249 (3310)	MIL-S-7393, MIL-S-8503

AMERICAN STANDARDS CROSS REFERENCES - AISI MOLYBDENUM STEELS

AISI	UNS No.	ASTM	SAE	AMS
4012	G40120	A 331 (4012), A 519 (4012), A 752 (4012)	J412 (4012), J1249 (4012), J1397 (4012)	---
4023	G40230	A 29 (4023), A 322 (4023), A 331 (4023), A 519 (4023), A 752 (4023)	J404 (4023), J412 (4023), J1397 (4023)	---
4024	G40240	A 29 (4024), A 322 (4024), A 331 (4024), A 519 (4024), A 752 (4024)	J404 (4024), J412 (4024), J1397 (4024)	---
4027	G40270	A 29 (4027), A 322 (4027), A 331 (4027), A 519 (4027), A 752 (4027)	J404 (4027), J412 (4027), J1397 (4027)	---
4027H	H40270	A 304 (4027 H)	J1268 (4027H)	---
4028	G40280	A 29 (4028), A 322 (4028), A 331 (4028), A 519 (4028), A 752 (4028)	J404 (4028), J412 (4028), J1397 (4028)	---
4028H	H40280	A 304 (4028 H)	J1268 (4028H)	---
4032	G40320	---	J404 (4032), J412 (4032), J1397 (4032)	---
4032H	H40320	A 304 (4032 H)	J1268 (4032H)	---
4037	G40370	A 29 (4037), A 320 (L7A, L71), A 322 (4037), A 331 (4037), A 519 (4037), A 752 (4037)	J404 (4037), J412 (4037), J1397 (4037)	6300
4037H	H40370	A 304 (4037 H), A 547 (4037 H)	J1268 (4037H)	---
4042	G40420	A 29 (4042), A 194 (4042-7), A 331 (4042), A519 (4042)	J404 (4042), J412 (4042), J1397 (4042)	---
4042H	H40420	A 304 (4042 H)	J1268 (4042H)	---
4047	G40470	A 29 (4047), A 322 (4047), A 331 (4047), A 519 (4047), A 752 (4047)	J404 (4047), J412 (4047), J1397 (4047)	---
4047H	H40470	A 304 (4047 H)	J1268 (4047H)	---
4063	G40630	A 331 (4063), A 519 (4063)	J1249 (4063)	---
4419	G44190	A 29 (4419), A 331 (4419), A 752 (4419)	J404 (4419), J412 (4419), J1397 (4419)	---
4419H	H44190	A 304 (4419 H)	J1268 (4419H)	---
4422	G44220	A 29 (4422), A 331 (4422), A 519 (4422)	J404 (4422), J412 (4422), J1397 (4422)	---
4427	G44270	A 29 (4427), A 331 (4427), A 519 (4427)	J404 (4427), J412 (4427), J1397 (4427)	---

AMERICAN STANDARDS CROSS REFERENCES - AISI CHROMIUM-MOLYBDENUM STEELS

AISI	UNS No.	ASTM	SAE	AMS	MIL
4118	G41180	A 29 (4118), A 322 (4118), A 331 (4118), A 506 (4118), A 507 (4118), A 513 (4118), A 519 (4118), A 752 (4118), A 829	J404 (4118), J412 (4118), J1397 (4118)	---	---
4118H	H41180	A 304 (4118 H)	J1268 (4118H)	---	---
4120	G41200	---	J403 (EX15)	---	---
4121	G41210	---	J403 (EX24)	---	---
4130	G41300	A 29 (4130), A 322 (4130), A 331 (4130), A 506 (4130), A 507 (4130), A 513 (4130), A 519 (4130), A 646 (4130), A 752 (4130), A 829	J404 (4130), J412 (4130), J1397 (4130)	6348, 6350, 6351, 6360, 6361, 6362, 6370, 6371, 6373, 6374, 6528, 7496	MIL-S-6758, MIL-S-18729
4130H	H41300	A 304 (4130 H)	J1268 (4130H)	---	---
4135	G41350	A 29 (4135), A 331 (4135), A 372 (V-2), A 519 (4135), A 829	J404 (4135), J412 (4135), J1397 (4135)	6352, 6365, 6372	---
4135H	H41350	A 304 (4135 H)	J1268 (4135H)		---
4137	G41370	A 29 (4137), A 320 (L7B, L72), A 322 (4137), A 331 (4137), A 372 (VIII), A 519 (4137), A 752 (4137), A 829	J404 (4137), J412 (4137), J1397 (4137)	---	---
4137H	H41370	A 304 (4137 H)	J1268 (4137H)		
4140	G41400	A 29 (4140), A 193 (B7, B7M), A 194 (7, 7M), A 320 (L7, L7M, L7D), A 322 (4140), A 331 (4140), A 506 (4140), A 513 (4140), A 519 (4140), A 646 (4140), A 711, A 752 (4140), A 829	J404 (4140), J412 (4140), J1397 (4140)	6349, 6381, 6382, 6390, 6395, 6529	MIL-S-5626
4140H	H41400	A 304 (4140 H)	J775 (NV 7), J1268 (4140H)	---	---
4142	G41420	A 29 (4142), A 322 (4142), A 331 (4142), A 372 (VIII), A 519 (4142), A 547 (4142), A 711, A 752 (4142), A 829	J404 (4142), J412 (4142), J1397 (4142)	---	---
4142H	H41420	A 304 (4142 H), A 540 (B22)	J1268 (4142H)	---	---

AMERICAN STANDARDS CROSS REFERENCES - AISI CHROMIUM-MOLYBDENUM STEELS (Continued)

AISI	UNS No.	ASTM	SAE	AMS	MIL
4145	G41450	A 29 (4145), A 322 (4145), A 331 (4145), A 519 (4145), A 711, A 752 (4145), A 829	J404 (4145), J412 (4145), J1397 (4145)	---	---
4145H	H41450	A 304 (4145 H)	J1268 (4145H)	---	---
4147	G41470	A 29 (4147), A 322 (4147), A 331 (4147), A 372 (VIII), A 519 (4147), A 752 (4147)	J404 (4147), J412 (4147), J1397 (4142)	---	---
4147H	H41470	A 304 (4147 H)	J1268 (4147H)	---	---
4150	G41500	A 29 (4150), A 322 (4150), A 331 (4150), A 519 (4150), A 711, A 752 (4150)	J404 (4150), J412 (4150), J1397 (4150)	---	MIL-S-11595 (ORD4150)
4150H	H41500	A 304 (4150 H)	J1268 (4150H)	---	---
4161	G41610	A 29 (4161), A 322 (4161), A 331 (4161), A 752 (4161)	J404 (4161), J412 (4161), J1397 (4161)	---	---
4161H	H41610	A 304 (4161 H)	J1268 (4161H)		

AMERICAN STANDARDS CROSS REFERENCES - AISI NICKEL-CHROMIUM-MOLYDENUM STEELS

AISI	UNS No.	ASTM	SAE	AMS	MIL
4320	G43200	A 29 (4320), A 322 (4320), A 331 (4320), A 519 (4320), A 535 (4320), A 752 (4320)	J404 (4320), J412 (4320), J1397 (4320)	---	---
4320H	H43200	A 304 (4320 H)	J1268 (4320H)	6299	---
4337	G43370	A 519 (4337)	J1249 (4337)	6412, 6413, 7496	---
E4337	G43376	A 519 (E4337)	---	---	
4340	G43400	A 29 (4340), A 320 (L43), A 322 (4340), A 331 (4340), A 506 (4340), A 519 (4340), A 646 (4340-7), A 711, A 752 (4340), A 829	J404 (4340), J412 (4340), J1397 (4340)	6359, 6409, 6414, 6415, 6454	MIL-S-5000
4340H	H43400	A 304 (4340 H), A 540 (B23), A 547 (4340 H)	J1268 (4340H)	---	---
E4340	G43406	A 331 (E4340), A 519 (E4340), A 752 (E4340), A 829	J 404 (E4340), J1397 (E4340)	6415	MIL-S-5000, MIL-S-83135

AMERICAN STANDARDS CROSS REFERENCES - AISI NICKEL-CHROMIUM-MOLYDENUM STEELS (Continued)

AISI	UNS No.	ASTM	SAE	AMS	MIL
E4340H	H43406	A 304 (E4340 H)	J1268 (E4340H)	---	---
4715	G47150	---	J403 (EX30)	---	---
4718	G47180	A 29 (4718), A 331 (4718), A 519 (4718), A 752 (4718)	J404 (4718), J412 (4718), J1397 (4718)	---	---
4718H	H47180	A 304 (4718 H)	J1268 (4718H)	---	---
4720	G47200	A 29 (4720), A 322 (4720), A 331 (4720), A 519 (4720), A 535 (4720), A 711, A 752 (4720)	J404 (4720), J412 (4720), J1397 (4720)	---	---
4720H	H47200	A 304 (4720 H)	J1268 (4720H)	---	---
8115	G81150	A 29 (8115), A 519 (8115)	J404 (8115), J1397 (8115)	---	---
81B45	G81451	A 322, A 331 (81B45), A 519 (81B45), A 752 (81B45)	J404 (81B45), J412 (81B45), J1397 (81B45)	---	---
81B45H	H81451	A 304 (81B45 H)	J1268 (81B45H)	---	---
8615	G86150	A 29 (8615), A 322 (8615), A 331 (8615), A 506 (8615), A 507 (8615), A 519 (8615), A 752 (8615), A 829	J404 (8615), J1397 (8615)	6270	---
8617	G86170	A 29 (8617), A 322 (8617), A 331 (8617), A 519 (8617), A 752 (8617), A 829	J404 (8617), J1397 (8617)	6272	---
8617H	H86170	A 304 (8617 H)	J1268 (8617H)	---	---
8620	G86200	A 29 (8620), A 322 (8620), A 331 (8620), A 506 (8620), A 507 (8620), A 513 (8620), A 519 (8620), A 646 (8620-4), A 752 (8620), A 829	J404 (8620), J1397 (8620)	6274, 6276, 6277, 6375	MIL-S-8690
8620H	H86200	A 304 (8620 H)	J1268 (8620H)	---	---
8622	G86220	A 331 (8622), A 519 (8622), A 752 (8622), A 829	J404 (8622), J1397 (8622)	---	---
8622H	H86220	A 304 (8622 H)	J1268 (8622H)	---	---
8625	G86250	A 29 (8625), A 322 (8625), A 331 (8625), A 519 (8625), A 752 (8625), A 829	J404 (8625), J1397 (8625)	---	---
8625H	H86250	A 304 (8625 H)	J1268 (8625H)	---	---

AMERICAN STANDARDS CROSS REFERENCES - AISI NICKEL-CHROMIUM-MOLYDENUM STEELS (Continued)

AISI	UNS No.	ASTM	SAE	AMS	MIL
8627	G86270	A 29 (8627), A 322 (8627), A 331 (8627), A 519 (8627), A 752 (8627), A 829	J404 (8627), J1397 (8627)	---	---
8627H	H86270	A 304 (8627 H)	J1268 (8627H)	---	---
8630	G86300	A 29 (8630), A 322 (8630), A 331 (8630), A 513 (8630), A 519 (8630), A 752 (8630), A 829	J404 (8630), J412 (8630), J1397 (8630)	6280, 6281, 6530, 6535, 7496	MIL-S-6050, MIL-S-18728
8630H	H86300	A 304 (8630 H)	J1268 (8630H)	---	---
86B30H	H86301	A 304 (86B30 H)	J1268 (86B30H)	---	---
8637	G86370	A 29 (8637), A 322 (8637), A 331 (8637), A 519 (8637), A 689 (8637), A 752 (8637), A 829	J404 (8637), J412 (8637), J1397 (8637)	---	---
8637H	H86370	A 304 (8637 H), A 547 (8637 H)	J1268 (8637H)	---	---
8640	G86400	A 29 (8640), A 322 (8640), A 331 (8640), A 519 (8640), A 752 (8640), A 829	J404 (8640), J412 (8640), J1397 (8640)	---	---
8640H	H86400	A 304 (8640 H), A 547 (8640 H)	J1268 (8640H)	---	---
8642	G86420	A 29 (8642), A 322 (8642), A 331 (8642), A 519 (8642), A 752 (8642)	J404 (8642), J412 (8642), J1397 (8642)	---	---
8642H	H86420	A 304 (8642 H), A 547 (8642 H)	J1268 (8642H)	---	---
8645	G86450	A 29 (8645), A 322 (8645), A 331 (8645), A 519 (8645), A 752 (8645)	J404 (8645), J412 (8645), J775 (NV 5), J1397 (8645)	---	---
8645H	H86450	A 304 (8645 H)	J1268 (8645H)	---	---
86B45	G86451	A 519 (86B45)	J404 (86B45), J412 (86B45), J1397 (86B45)	---	---
86B45H	H86451	A 304 (86B45 H)	J1268 (86B45H)	---	---
8650	G86500	A 29 (8650), A 331 (8650), A 519 (8650), A 689	J404 (8650), J412 (8650), J1397 (8650)	---	---
8650H	H86500	A 304 (8650 H)	J1268 (8650H)	---	---
8655	G86550	A 29 (8655), A 322 (8655), A 331 (8655), A 519 (8655), A 689, A 752 (8655), A 829	J404 (8655), J412 (8655), J1397 (8655)	---	---

AMERICAN STANDARDS CROSS REFERENCES - AISI NICKEL-CHROMIUM-MOLYDENUM STEELS (Continued)					
AISI	UNS No.	ASTM	SAE	AMS	MIL
8655H	H86550	A 304 (8655 H)	J1268 (8655H)	---	---
8660	G86600	A 322 (8660), A 519 (8660), A 689, A 711	J404 (8660), J412 (8660), J1397 (8660)	---	---
8660H	H86600	A 304 (8660 H)	J1268 (8660H)	---	---
8720	G87200	A 29 (8720), A 322 (8720), A 331 (8720), A 519 (8720), A 752 (8720)	J404 (8720), J1397 (8720)	---	---
8720H	H87200	A 304 (8720 H)	J1268 (8720H)		
8735	G87350	A 519 (8735)	J1249 (8735)	6282, 6320, 6331, 6357, 7496	MIL-S-6098
8740	G87400	A 29 (8740), A 320 (L7C, L73), A 322 (8740), A 331 (8740), A 519 (8740), A 752 (8740)	J404 (8740), J412 (8740), J1397 (8740)	6322, 6323, 6325, 6327, 6358, 7496	MIL-S-6049
8740H	H87400	A 304 (8740 H)	J1268 (8740H)		
8742	G87420	A 331 (8742), A 519 (8742), A 829	J1249 (8742)		
8822	G88220	A 29 (8822), A 322 (8822), A 331 (8822), A 519 (8822)	J404 (8822), J1397 (8822)	---	---
8822H	H88220	A 304 (8822 H)	J1268 (8822H)	---	---
9310H	H93100	A 304 (9310 H)	J1268 (9310H)	---	---
E9310	G93106	A 29 (E9310), A 331 (E9310), A 646 (E9310-2), A 519 (E9310)	J404 (9310), J1397 (9310)	6260, 6265, 6267	---
94B15	G94151	A 519 (94B15)	J404 (94B15), J1397 (94B15)	---	---
94B15H	H94151	A 304 (94B15 H)	J1268 (94B15H)	---	---
94B17	G94171	A 519 (94B17), A 752 (94B17)	J404 (94B17), J1397 (94B17)	6275	---
94B17H	H94171	A 304 (94B17 H)	J1268 (94B17H)	---	---
94B30	G94301	A 322 (94B30), A 331 (94B30), A 519 (94B30), A 752 (94B30)	J404 (94B30), J412 (94B30), J1397 (94B30)	---	---
94B30H	H94301	A 304 (94B30 H)	J1268 (94B30H)	---	---
94B40	G94401	A 519 (94B40)	J1249 (94B40)	---	---
9840	G98400	A 519 (9840), A 711	J1249 (9840)	6342	---

AMERICAN STANDARDS CROSS REFERENCES - AISI NICKEL-CHROMIUM-MOLYDENUM STEELS (Continued)

AISI	UNS No.	ASTM	SAE	AMS	MIL
9850	G98500	A 519 (9850)	J1249 (9850)	---	MIL-S-19434

AMERICAN STANDARDS CROSS REFERENCES - AISI NICKEL-MOLYDENUM STEELS

AISI	UNS No.	ASTM	SAE	AMS
4615	G46150	A 29 (4615), A 322 (4615), A 331 (4615), A 752 (4615), A 829	J404 (4615), J412 (4615), J1397 (4615)	6290
4617	G46170	A 519 (4617), A 829	J404 (4617), J412 (4617), J1397 (4617)	6292
4620	G46200	A 29 (4620), A 322 (4620), A 331 (4620), A 519 (4620), A 535 (4620), A 646 (4620-3), A 752 (4620), A 829	J404 (4620), J412 (4620), J1397 (4620)	6294
4620H	H46200	A 304 (4620 H)	J1268 (4620H)	---
4621	G46210	A 322 (4621), A 331 (4621), A 519 (4621), A 752 (4621)	J404 (4621), J412 (4621), J1397 (4621)	---
4621H	H46210	A 304 (4621 H)	J1268 (4621H)	---
4626	G46260	A 29 (4626), A 322 (4626), A 331 (4626), A 752 (4626)	J404 (4626), J412 (4626), J1397 (4626)	---
4626H	H46260	A 304 (4626 H)	---	---
4815	G48150	A 29 (4815), A 322 (4815), A 331 (4815), A 519 (4815), A 752 (4815)	J404 (4815), J412 (4815), J1397 (4815)	---
4815H	H48150	A 304 (4815 H)	J1268 (4815H)	---
4817	G48170	A 29 (4817), A 322 (4817), A 331 (4817), A 519 (4817), A 752 (4817)	J404 (4817), J412 (4817), J1397 (4817)	---
4817H	H48170	A 304 (4817 H)	J1268 (4817H)	---
4820	G48200	A 29 (4820), A 322 (4820), A 331 (4820), A 519 (4820), A 535 (4820), A 752 (4820)	J404 (4820), J412 (4820), J1397 (4820)	---
4820H	H48200	A 304 (4820 H)	J1268 (4820H)	---

AMERICAN STANDARDS CROSS REFERENCES - AISI CHROMIUM STEELS

AISI	UNS No.	ASTM	SAE	AMS	MIL
5015	G50150	A 29 (5015), A 331 (5015), A 519 (5015), A 752 (5015)	J404 (5015), J412 (5015), J1397 (5015)	---	---
50B40	G50401	A 519 (50B40)	J404 (50B40), J412 (50B40), J1397 (50B40)	---	---
50B40H	H50401	A 304 (50B40 H)	J1268 (50B40H)	---	---
50B44	G50441	A 322 (50B44), A 331 (50B44), A 519 (50B44), A 752 (50B44)	J404 (50B44), J412 (50B44), J1397 (50B44)	---	---
50B44H	H50441	A 304 (50B44 H)	J1268 (50B44H)	---	---
5046	G50460	A 29 (5046), A 519 (5046)	J404 (5046), J412 (5046), J1397 (5046)	---	---
5046H	H50460	A 304 (5046 H)	J1268 (5046H)	---	---
50B46	G50461	A 322 (50B46), A 331 (50B46), A 519 (50B46), A 752 (50B46)	J404 (50B46), J412 (50B46), J1397 (50B46)	---	---
50B46H	H50461	A 304 (50B46 H)	J1268 (50B46H)	---	---
50B50	G50501	A 322 (50B50), A 331 (50B50), A 519 (50B50), A 752 (50B50)	J404 (50B50), J412 (50B50), J1397 (50B50)	---	---
50B50H	H50501	A 304 (50B50 H)	J1268 (50B50H)	---	---
5060	G50600	---	J404 (5060), J1397 (5060)	---	---
50B60	G50601	A 322 (50B60), A 331 (50B60), A 519 (50B60), A 752 (50B60)	J404 (50B60), J412 (50B60), J1397 (50B60)	---	---
50B60H	H50601	A 304 (50B60 H)	J1268 (50B60H)	---	---
E50100	G50986	A 29 (E50100), A 295 (50100), A 519 (E50100)	J404 (50100), J412 (50100), J1397 (50100)	6442	---
5115	G51150	A 29 (5115), A 519 (5115)	J404 (5115), J1397 (5115)	---	---
5117	G51170	A 322 (5117), A 331 (5117)	J404 (5117)	---	---
---	G51190		J1081 (PS 64)	---	---
5120	G51200	A 29 (5120), A 322 (5120), A 331 (5120), A 519 (5120)	J404 (5120), J1397 (5120)	---	---
5120H	H51200	A 304 (5120 H)	J1268 (5120H)	---	---
---	G51210	---	J1081 (PS 59)	---	---

AMERICAN STANDARDS CROSS REFERENCES - AISI CHROMIUM STEELS (Continued)					
AISI	UNS No.	ASTM	SAE	AMS	MIL
---	G51240	---	J1081 (PS 65)	---	---
5130	G51300	A 29 (5130), A 322 (5130), A 331 (5130), A 513 (5130), A 752 (5130)	J404 (5130), J412 (5130), J1397 (5130)	---	---
5130H	H51300	A 304 (5130 H)	J1268 (5130H)	---	---
5132	G51320	A 29 (5132), A 322 (5132), A 331 (5132), A 519 (5132), A 752 (5132)	J404 (5132), J412 (5132), J1397 (5132)	---	---
5132H	H51320	A 304 (5132 H)	J1268 (5132H)	---	---
5135	G51350	A 29 (5135), A 322 (5135), A 331 (5135), A 519 (5135), A 752 (5135)	J404 (5135), J412 (5135), J1397 (5135)	---	---
5135H	H51350	A 304 (5135 H)	J1268 (5135H)	---	---
5140	G51400	A 29 (5140), A 322 (5140), A 331 (5140), A 506 (5140), A 519 (5140), A 752 (5140)	J404 (5140), J412 (5140), J1397 (5140)	---	---
5140 H	H51400	A 304 (5140 H)	J1268 (5140H)	---	---
5145	G51450	A 29 (5145), A 331 (5145), A 519 (5145), A 752 (5145)	J404 (5145), J412 (5145), J1397 (5145)	---	---
5145H	H51450	A 304 (5145 H)	---	---	---
5147	G51470	A 29 (5147), A 331 (5147), A 519 (5147), A 752 (5147)	J404 (5147), J412 (5147), J1397 (5147)	---	---
5147H	H51470	A 304 (5147 H)	J1268 (5147H)	---	---
5150	G51500	A 29 (5150), A 322 (5150), A 331 (5150), A 506 (5150), A 519 (5150), A 752 (5150)	J404 (5150), J412 (5150), 1397 (5150)	---	---
5150H	H51500	A 304 (5150 H)	J775 (NV 6), J1268 (5150H)	---	---
5155	G51550	A 29 (5155), A 322 (5155), A 331 (5155), A 519 (5155), A 752 (5155)	J404 (5155), J412 (5155), J1397 (5155)	---	---
5155H	H51550	A 304 (5155 H)	J1268 (5155H)	---	---

AMERICAN STANDARDS CROSS REFERENCES - AISI CHROMIUM STEELS (Continued)

AISI	UNS No.	ASTM	SAE	AMS	MIL
5160	G51600	A 29 (5160), A 322 (5160), A 331 (5160), A 506 (5160), A 519 (5160), A 752 (5160), A 829	J404 (5160), J412 (5160), J1397 (5160)	---	---
5160H	H51600	A 304 (5160 H)	J1268 (5160H)	---	---
51B60	G51601	A 322 (51B60), A 331 (51B60), A 519 (51B60), A 752 (51B60)	J404 (51B60), J412 (51B60), J1397 (51B60)	---	---
51B60H	H51601	A 304 (51B60 H)	J1268 (51B60H)	---	---
E51100	G51986	A 29 (E51100), A 295 (E51100), A 322 (E51100), A 519 (E51100), A 711, A 752 (E51100)	J404 (51100), J412 (51100), J1397 (51100)	6443, 6446, 6449	---
E52100	G52986	A 29 (E52100), A 322 (E52100), A 331 (E52100), A 519 (E52100), A 535 (E52100), A 646 (E52100), A 752 (E52100)	J404 (52100), J412 (52100), J1397 (52100)	6440, 6444, 6447	MIL-S-980 (52100), MIL-S-7420, MIL-S-22141 (52100)

AMERICAN STANDARDS CROSS REFERENCES - AISI CHROMIUM-VANADIUM STEELS

AISI	UNS No.	ASTM	SAE	AMS	MIL
6118	G61180	A 29 (6118), A 322 (6118), A 331 (6118), A 519 (6118), A 752 (6118)	J404 (6118), J1397 (6118)	---	---
6118H	H61180	A 304 (6118 H)	J1268 (6118H)		
6120	G61200	A 331 (6120), A 519 (6120), A 711	J1249 (6120)	---	---
6150	G61500	A 29 (6150), A 322 (6150), A 331 (6150), A 519 (6150), A 752 (6150), A 829	J404 (6150), J412 (6150), J1397 (6150)	6448, 6450, 6445, 7301	MIL-S-8503
6150H	H61500	A 304 (6150 H)	J1268 (6150H)	---	---

AMERICAN STANDARDS CROSS REFERENCES - AISI SILICON-MANGANESE STEELS

AISI	UNS No.	ASTM	SAE	AMS
9254	G92540	A 29 (9254), A 401, A 752 (9254), A 877	J157, J404 (9254), J412 (9254), J1397 (9254)	6451
9255	G92550	A 29 (9255), A 519 (9255), A 752 (9255)	J404 (9255), J412 (9255), J1397 (9255)	---

AMERICAN STANDARDS CROSS REFERENCES - AISI SILICON-MANGANESE STEELS (Continued)

AISI	UNS No.	ASTM	SAE	AMS
9260	G92600	A 29 (9260), A 322 (9260), A 331 (9260), A 519 (9260), A 689, A 752 (9260)	J404 (9260), J412 (9260), J1397 (9260)	---
9260H	H92600	A 304 (9260 H)	J1268 (9260H)	---
9262	G92620	A 519 (9262)	J1249 (9262)	---

AMERICAN STANDARDS CROSS REFERENCES - NON-AISI CARBON STEELS

ASTM	UNS No.	SAE	AMS	MIL
A 105, A 695 (B)	K03504	---	---	---
A 106 (A), A 369 (FPA)	K02501	---	---	---
A 106 (B), A 234 (WPB), A 333 (6), A 334 (6), A 369 (FPB), A 556 (C2)	K03006	---	---	---
A 106 (C), A 210 (C), A 234 (WPC)	K03501	---	---	---
A 131 (B)	K02102	---	---	MIL-S-22698
A 178 (A), A 179	K01200	---	---	---
A 179, A 192, A 226	K01201	---	---	---
A 21 (F, G, H), A 730 (F)	K05200	---	---	---
A 21 (U), A 383, A 730 (C, D, E)	K04700	---	---	---
A 210 (A-1), A 556 (B2)	K02707	---	---	---
A 214, A 556 (A2), A 557 (A2)	K01807	---	---	---
A 228	K08500	J178	---	---
A 229	K07001	J316 (A)	---	---
A 230	K06701	J172, J316 (B)	---	---
A 25 (A), A 504 (A, L), A 631 (A)	K05700	---	---	---
A 25 (B), A 504 (B), A 631 (B)	K06200	---	---	---
A 25 (C), A 504 (C), A 631 (C)	K07201	---	---	---
A 25 (U), A 504 (U), A 631 (U)	K07200	---	---	---

AMERICAN STANDARDS CROSS REFERENCES - NON-AISI CARBON STEELS (Continued)

ASTM	UNS No.	SAE	AMS	MIL
A 266 (1, 2), A 541 (1)	K03506	---	---	---
A 266 (3), A 649 (2)	K05001	---	---	---
A 283 (C), A 284 (C), A 515 (60)	K02401	---	---	---
A 283 (D), A 284 (C, D)	K02702	---	---	---
A 284, A 515 (55)	K02001	---	---	---
A 333 (1), A 334 (1)	K03008	---	---	---
A 350 (LF2), A706	K03011	---	---	---
A 500 (A, B), A 501	K03000	---	---	---
A 53 (E-A, S-A), A 523 (A)	K02504	---	---	---
A 53 (E-B, S-B), A 523 (B)	K03005	---	---	---
A 570 (45, 50, 55), A 857	K02507	---	---	---
A 648 (II), A 821	K06700	---	---	---
---	K00095	---	7706, 7707	---
---	K02508	---	5062, 7496	---

AMERICAN STANDARDS CROSS REFERENCES - NON-AISI ALLOY STEELS

ASTM	Nominal Composition	UNS No.	AMS	SAE	MIL
A 131 (AH36, DH36, EH36), A 808	1¼Mn	K11852	---	---	---
A 161 (T1), A 209 (T1), A 250 (T1), A 335 (P1), A 369 (FP1)	½Mo	K11522	---	---	---
A 182 (F 21b), A 508 (3V), A 541 (3V), A 542 (C), A 832	3Cr-1Mo-¼V	K31830	---	---	---
A 182 (F11-1), A 199 (T11), A 200 (T11) A 213 (T11), A 250 (T11), A 335 (P11), A 336 (F11B), A 369 (FP11)	1¼Cr-½Mo-Si	K11597	---	---	---
A 182 (F11-2), A 336 (FN, F11A), A 541 (11C)	1¼Cr-½Mo-Si	K11572	---	---	---

AMERICAN STANDARDS CROSS REFERENCES - NON-AISI ALLOY STEELS (Continued)

ASTM	Nominal Composition	UNS No.	AMS	SAE	MIL
A 182 (F12-1), A 213 (T12), A 250 (T12), A 335 (P12), A 369 (FP12)	1Cr-½Mo	K11562	---	---	---
A 182 (F12-2), A 336 (F12)	1Cr-½Mo	K11564	---	---	---
A 182 (F21), A 199 (T21), A 200 (T21), A 213 (T21), A 335 (P21), A 336 (F21, F21a), A 369 (FP21), A 387 (21)	3Cr-1Mo	K31545	---	---	---
A 182 (F22-1), A 199 (T22), A 200 (T22), A 213 (T22), A 234 (WP22), A 250 (T22), A 335 (P22), A 336 (F22, F22a), A 369 (FP22), A 387 (22, 22L), A 508 (22B), A 542 (A, B)	2¼Cr-1Mo	K21590	---	---	---
A 182 (F5), A 199 (T5), A 200 (T5), A 213 (T5), A 234 (WP5), A 335 (P5), A 336 (F5), A 369 (FP5), A 387 (5)	5Cr-½Mo	K41545	---	---	---
A 182 (F5a), A 336 (F5a)	5Cr-½Mo	K42544	---	---	---
A 182 (F9), A 234 (WP9), A 369 (FP9)	9Cr-1Mo	K90941	---	---	---
A 182 (FR), A 234 (WPR), A 333 (9), A 334 (9), A 714 (V)	2Ni-1Cu	K22035	---	---	---
A 193 (B16), A 437 (B4D)	1Cr-½Mo-V	K14072	---	J775 (K14072)	---
A 199 (T4), A 200 (T4)	2½Cr-½Mo-¾Si	K31509	---	---	---
A 204 (A)	½Mo	K11820	---	---	---
A 204 (B)	½Mo	K12020	---	---	---
A 204 (C)	½Mo	K12320	---	---	---
A 209 (T1a), A 250 (T1a)	½Mo	K12023	---	---	---
A 209 (T1b), A 250 (T1b)	½Mo	K11422	---	---	---
A 213 (T2), A 250 (T2), A 335 (P2), A 369 (FP2)	½Cr-½Mo	K11547	---	---	---
A 213 (T5b), A 335 (P5b)	5Cr-½Mo-Si	K51545	---	---	---

AMERICAN STANDARDS CROSS REFERENCES - NON-AISI ALLOY STEELS (Continued)

ASTM	Nominal Composition	UNS No.	AMS	SAE	MIL
A 213 (T5c), A 335 (P5c)	5Cr-½Mo-Ti	K41245	---	---	---
A 231, A 372 (IV)	1Cr-V	K15048	---	---	---
A 232	1Cr-V	K15047	---	J132	---
A 302 (A)	Mn-½Mo	K12021	---	---	---
A 302 (B)	Mn-½Mo	K12022	---	---	---
A 302 (C)	Mn-½Mo-½Ni	K12039	---	---	---
A 302 (D)	Mn-½Mo-¾Ni	K12054	---	---	---
A 325 (C), A 563 (C3C)	Cr-Ni-Cu-V	K12033	---	---	---
A 325 (D), A 563 (C3D)	Cr-Ni-Cu	K12059	---	---	---
A 325 (A), A 563 (C3-A)	Cr-Ni-Cu	K13643	---	---	---
A 325 (B), A 563 (C3-B)	Cr-Ni-Cu	K14358	---	---	---
A 325 (E), A 563 (C3-E)	Cr-Ni-Cu	K12254	---	---	---
A 325 (F), A 563 (C3-F)	Cr-Ni-Cu	K12238	---	---	---
A 333 (7), A 334 (7)	2½Ni	K21903	---	---	---
A 333 (8), A 334 (8), A 353/A 353M, A 522 (I), A553/A 553M (I), A 844	9Ni	K81340	---	---	---
A 350 (LF6), A 633 (E), A 678 (D), A 737 (C)	C-Mn-Si(V)	K12202	---	---	---
A 355 (A), A 519 (E7140)	1½Al-1½Cr-⅓Mo	K24065	6470, 6471, 6472	---	MIL-S-6709
A 372 (V-1), A 649 (3)	1Cr-¼Mo	K13047	---	---	---
A 372 (VI)	2½Ni-1½Cr-Mo	K31820	---	---	MIL-S-16216, MIL-S-21952, MIL-S-22664, MIL-S-23009, MIL-S-24451
A 387 (11)	1¼Cr-½Mo-Si	K11789	---	---	---
A 387 (12)	1Cr-½Mo	K11757	---	---	---
A 387 (2)	½Cr-½Mo	K12143	---	---	---

AMERICAN STANDARDS CROSS REFERENCES - NON-AISI ALLOY STEELS (Continued)

ASTM	Nominal Composition	UNS No.	AMS	SAE	MIL
A 469 (1), A 470 (1)	V	K14501	---	---	---
A 469 (6, 7, 8), A 470 (5, 6, 7)	3½Ni-1½Cr-½Mo-V	K42885	---	---	---
A 502 (3A), A 588 (A)	1Mn-½Cr-⅓Cu-V	K11430	---	---	---
A 508 (3, 3a)	¾Ni-½Mo-Cr-V	K12042	---	---	MIL-S-24238 (A)
A 514 (A), A 517 (A), A 592 (A)	½Cr-¼Mo-Si	K11856	6386 (1)	---	---
A 514 (B), A 517 (B)	½Cr-⅕Mo-V	K11630	6386 (2)	---	---
A 514 (C), A 517 (C)	1⅓Mn-¼Mo-B	K11511	6386 (3)	---	---
A 514 (E), A 517 (E)	1¾Cr-½Mo-Ti	K21604	---	---	---
A 514 (F), A 517 (F), A 592 (F)	¾Ni-½Cr-½Mo-V	K11576	---	---	---
A 514 (H), A 517 (H)	1⅕Mn-½Cr-¼Mo-V-B	K11646	---	---	---
A 514 (J), A 517 (J)	½Mo	K11625	6386 (5)	---	---
A 514 (M), A 517 (M)	1¼Ni-½Mo-B	K11683	---	---	---
A 514 (P), A 517 (P)	1¼Ni-1Cr-½Mo	K21650	---	---	---
A 522 (II), A 553 (II)	8Ni	K71340	---	---	---
A 533 (A)	Mn-½Mo	K12521	---	---	---
A 533 (B)	Mn-½Mo-½Ni	K12539	---	---	---
A 533 (D)	Mn-½Mo-¼Ni	K12529	---	---	---
A 538 (A, B)	18Ni-7¾Co-4¼Mo-Ti-Al	K92810	---	---	MIL-S-46850
A 538 (C), A 579 (73)	18½Ni-8¾Co-5Mo-⅔Ti-Al	K93120	6514, 6521	J1099 (A538C)	MIL-S-46850
A 538/A 538M (B)	18Ni-7¾Co-4⅞Mo-⅖Ti-Al	K92890	6501, 6512, 6520	J1099 (A538B)	MIL-S-46850
A 540 (B24V), A 579 (22)	2Ni-¾Cr-⅓Mo-V	K24070	---	---	---
A 541 (22B, 22C, 22D), A 739 (B 22)	2¼Cr-1Mo	K21390	---	---	---
A 579 (12)	5Ni-½Cr-½Mo-V	K51255	---	---	MIL-S-24371, MIL-S-24512
A 579 (21), A 646 (4335 Mod.-6)	1¾Ni-¾Cr-½Mo-⅕V	K23477	6428	---	---

AMERICAN STANDARDS CROSS REFERENCES - NON-AISI ALLOY STEELS (Continued)

ASTM	Nominal Composition	UNS No.	AMS	SAE	MIL
A 579 (23)	1Cr-1Mo-½Ni-V	K24728	6431, 6432, 6438, 6439	---	MIL-S-8949
A 579 (31)	1¾Ni-1½Si-1¼Mn-⅓Cr-⅖Mo	K32550	6418, 7496	---	MIL-S-7108
A 579 (32), A 646 (300M-8)	1¾Ni-¾Cr-½Mo-V	K44220	6257, 6417, 6419	---	MIL-S-8844 (3)
A 579 (71)	18Ni-8½Co-3¼Mo-⅕Ti-Al	K92820	---	J1099 (A538A)	MIL-S-46850
A 579 (81)	8Ni-4Co-½Cr-½Mo-V	K91122	6546	---	---
A 579 (82)	7¾Ni-4½Co-1Cr-1Mo-V	K91283	6524, 6526	---	---
A 618 (II), A 714 (II)	Mn-Cu-V	K12609	---	---	---
A 633 (D), A 724 (C)	C-Mn-Si	K12037	---	---	---
A 646 (4330 Mod.-5)	1¾Ni-⅞Cr-⅖Mo-V	K23080	6411, 6427	---	MIL-S-8699, MIL-S-46128
A 710 (A), A 736	Ni-Cu-Cr-Mo-Cb	K20747	---	---	---
A 737 (B), A 738 (B), A 812 (65, 80)	C-Mn-Si(Cb)	K12001	---	---	---
---	1Cr-½Mo-V	K14675	6304, 6305	---	MIL-S-24502
---	1¼Cr-½Mo-¼V	K23015	6302, 6385, 6458, 7496	---	---
---	1¼Cr-⅞V-½Mo	K22770	6303, 6436, 7496	---	---
---	1¾Ni-¼Mo	K22440	6312, 6317	---	---
---	1¾Ni-¾Cr-¼Mo	K22950	6396, 6424	---	---
---	1¾Ni-¾Cr-⅜Mo-⅕V	K33517	6429, 6430, 6433, 6434, 6435	---	---
---	3Ni-1½Cr-½Mo	K32045	---	---	MIL-S-16216, MIL-S-21952, MIL-S-22664, MIL-S-23009, MIL-S-24451

AMERICAN STANDARDS CROSS REFERENCES - NON-AISI ALLOY STEELS (Continued)

ASTM	Nominal Composition	UNS No.	AMS	SAE	MIL
---	3¼Ni-1¼Cr-⅛Mo	K44414	6263, 6264	---	---
---	4½Mo-3Ni-1Cr-⅜V-Al	K71350	6256	---	---
---	9½Ni-5½Mn	K91456	5623, 5625	---	---
---	9Ni-4½Co-¾Cr-1Mo-V	K91472	6523, 6525	---	---
---	12½Ni-4½Mn-4Cr	K91505	5624	---	---
---	10Ni-14Co-2Cr-1Mo	K92571	6522, 6527, 6543, 6544	---	---

AMERICAN STANDARDS CROSS REFERENCES - AISI STAINLESS STEELS

AISI	UNS	ASTM	SAE	AMS	MIL	FED
201	S20100	A 213 (TP201), A 240 (201), A 249 (TP201), A 412 (201), A 666 (201)	J405 (30201)	---	---	QQ-S-766 (201)
202	S20200	A 213 (TP202), A 240 (202), A 249 (TP202), A 314 (202), A 412 (202), A 473 (202), A 666 (202)	J405 (30202)	---	---	QQ-S-763 (202), QQ-S-766 (202)
205	S20500	A 666 (205)	---	---	---	---
301	S30100	A 167 (301), A 177 (301), A 554 (301), A 666 (301)	J405 (30301)	5517, 5518, 5519	MIL-S-5059 (301)	QQ-S-766 (301)
302	S30200	A 167 (302), A 240 (302), A 276 (302), A 313 (302), A 314 (302), A 368 (302), A 473 (302), A 478 (302), A 479 (302), A 492 (302), A 493 (302), A 511 (302), A 554 (302), A 580 (302), A 666 (302)	J230, J405 (30302)	5515, 5516, 5600, 5636, 5637, 5688, 5693, 7210, 7241	MIL-S-5059, MIL-S-7720	QQ-S-763 (302), QQ-S-766 (302)
302B	S30215	A 167 (302 B), A 276 (302 B), A 314 (302 B), A 473 (302 B), A 580 (302 B)	J405 (30302 B)	---	---	---
303	S30300	A 194 (303, 8F, 8FA), A 314 (303), A 320 (303, B8F, B8FA), A 473 (303), A 581 (303), A 582 (303), A 895 (303)	J405 (30303)	5640 (Type 1)	---	---

AMERICAN STANDARDS CROSS REFERENCES - AISI STAINLESS STEELS (Continued)						
AISI	UNS	ASTM	SAE	AMS	MIL	FED
303 Se	S30323	A 194 (303 Se, 8F, 8FA), A 314 (303 Se), A 320 (303 Se, B8F, B8FA), A 473 (303 Se), A 581 (303 Se), A 582 (303 Se), A 895 (303 Se)	J405 (30303 Se)	5640 (Type 2), 5641, 5738	---	---
304	S30400	A 167 (304), A 182 (304), A 193 (304, B8, B8A), A 194 (304, 8, 8A), A 213 (304), A 240 (304), A 249 (304), A 269 (304), A 270 (304), A 271 (304), A 276 (304), A 312 (304), A 313 (304), A 314 (304), A 320 (304, B8, B8A), A 336 (F304), A 358 (304), A 368 (304), A 376 (304), A 409 (304), A 430 (304), A 473 (304), A 478 (304), A 479 (304), A 492 (304), A 493 (304), A 511 (304), A 554 (304), A 580 (304), A 632 (304), A 666 (304), A 688 (304), A 793 (304), A 813 (TP 304), A 814 (TP 304), A 851 (TP 304)	J405 (30304)	5501, 5513, 5560, 5563, 5564, 5565, 5566, 5567, 5639, 5697, 7228, 7245	MIL-S-5059 (304), MIL-T-5695, MIL-T-6845 (304), MIL-T-8504 (304), MIL-T-8506 (304), MIL-S-23195 (304), MIL-S-23196 (304), MIL-S-27419	QQ-S-763 (304), QQ-S-766 (304)
304 L	S30403	A 167 (304 L), A 182 (304 L), A 213 (304 L), A 240 (304 L), A 249 (304 L), A 269 (304 L), A 270 (TP 304 L), A 276 (304 L), A 312 (304 L), A 314 (304 L), A 336 (F 304 L), A 403 (304 L), A 473 (304 L), A 478 (304 L), A 479 (304 L), A 511 (304 L), A 554 (304 L), A 580 (304 L), A 632 (304 L), A 666 (304 L), A 688 (304 L), A 774 (TP 304 L), A 778 (TP 304 L), A 813 (TP 304 L), A 814 (TP 304 L), A 851 (TP 304 L)	J405 (30304 L)	5511, 5647	MIL-S-4043, MIL-S-23195 (304 L), MIL-S-23196 (304 L)	QQ-S-763 (304 L), QQ-S-766 (304 L)

AISI	UNS	ASTM	SAE	AMS	MIL	FED
		AMERICAN STANDARDS CROSS REFERENCES - AISI STAINLESS STEELS (Continued)				
304 N	S30451	A 182 (304 N), A 193 (B8N, B8NA), A 194 (8N, 8NA), A 213 (304 N), A 240 (304 N), A 249 (304 N), A 312 (304 N), A 358 (304 N), A 376 (304 N), A 403 (304 N), A 430 (304 N), A 479 (304 N), A 666 (304 N), A 688 (TP 304 N), A 813 (TP 304 N), A 814 (TP 304 N)	---	---	---	---
305	S30500	A 167 (305), A 193 (B8P, B8PA), A 194 (8P, 8PA), A 240 (305), A 249 (305), A 276 (305), A 313 (305), A 314 (305), A 320 (B8P, B8PA), A 368 (305), A 473 (305), A 478 (305), A 492 (305), A 493 (305), A 511 (305), A 554 (305), A 580 (305)	J405 (30305)	5514, 5685, 5686	---	QQ-S-763 (305), QQ-S-766 (305)
308	S30800	A 167 (308), A 276 (308), A 314 (308), A 473 (308), A 580 (308)	J405 (30308)	---	---	---
309	S30900	A 167 (309), A 249 (309), A 276 (309), A 314 (309), A 358 (309), A 403 (309), A 409 (309), A 473 (309), A 580 (309)	J405 (30309)	---	---	QQ-S-763 (309), QQ-S-766 (309)
309 S	S30908	A 167 (309 S), A 213 (TP 309 S), A 240 (309 S), A 249 (TP 309 S), A 276 (309 S), A 312 (TP 309 S), A 314 (309 S), A 473 (309 S), A 479 (309 S), A 511 (309 S), A 554 (309 S), A 580 (309 S), A 813 (TP 309 S), A 814 (TP 309 S)	J405 (30309 S)	5523, 5574, 5650, 7490	---	---
310	S31000	A 167 (310), A 182 (310), A 213 (310), A 249 (310), A 276 (310), A 314 (310), A 336 (F 310), A 358 (310), A 403 (310), A 409 (310), A 473 (310), A 580 (310), A 632 (310)	J405 (30310)	---	---	QQ-S-763 (310), QQ-S-766 (310)

AMERICAN STANDARDS CROSS REFERENCES - AISI STAINLESS STEELS (Continued)

AISI	UNS	ASTM	SAE	AMS	MIL	FED
310 S	S31008	A 167 (310 S), A 213 (TP 310 S), A 240 (310 S), A 249 (TP 310 S), A 276 (310 S), A 312 (TP 310 S), A 314 (310 S), A 473 (310 S), A 479 (310 S), A 511 (310 S), A 554 (310 S), A 580 (310 S), A 813 (TP 310 S), A 814 (TP 310 S)	J405 (30310 S)	5521, 5572, 5577, 5651, 7490	---	QQ-S-763
314	S31400	A 276 (314), A 314 (314), A 473 (314), A 580 (314)	J405 (30314)	5652, 7490	---	---
316	S31600	A 167 (316), A 182 (316), A 193 (316, B8M, B8MA, B8M2, B8M3), A 194 (316, 8M, 8MA), A 213 (316), A 240 (316), A 249 (316), A 269 (316), A 270 (TP 316), A 271 (TP 316), A 276 (316), A 312 (316), A 313 (316), A 314 (316), A 320 (316, B8M, B8MA), A 336 (F 316), A 358 (316), A 368 (316), A 376 (316), A 403 (316), A 409 (316), A 430 (316), A 473 (316), A 478 (316), A 479 (316), A 492 (316), A 493 (316), A 511 (316), A 554 (316), A 580 (316), A 632 (TP 316), A 666 (316), A 688 (TP 316), A 771 (TP 316), A 813 (TP 316), A 814 (TP 316), A 826 (TP 316)	J405 (30316)	5524, 5573, 5648, 5690, 5696, 7490	MIL-S-5059 (316), MIL-S-7720 (316)	QQ-S-763 (316), QQ-S-766 (316)

AMERICAN STANDARDS CROSS REFERENCES - AISI STAINLESS STEELS (Continued)

AISI	UNS	ASTM	SAE	AMS	MIL	FED
316 L	S31603	A 167 (316 L), A 182 (316 L), A 213 (316 L), A 240 (316 L), A 249 (316 L), A 269 (316 L), A 270 (TP 316 L), A 276 (316 L), A 312 (316 L), A 314 (316 L), A 336 (F 316 L), A 403 (316 L), A 473 (316 L), A 478 (316 L), A 479 (316 L), A 511 (316 L), A 554 (316 L), A 580 (316 L), A 632 (316 L), A 666 (316 L), A 688 (316 L), A 774 (TP 316 L), A 778 (TP 316 L), A 813 (TP 316 L), A 814 (316 L)	J405 (30316 L)	5507, 5653	---	QQ-S-763 (316 L), QQ-S-766 (316 L)
316 F	S31620	---	---	5649	---	---
316 N	S31651	A 182 (316 N), A 193 (B8MN, B8MNA), A 194 (8MN, 8MNA), A 213 (316 N), A 240 (316 N), A 249 (316 N), A 276 (316 N), A 312 (316 N), A 336 (F 316 N), A 358 (316 N), A 376 (316 N), A 403 (316 N), A 430 (316 N), A 479 (316 N), A 666 (316 N), A 688 (TP 316 N), A 813 (TP 316 N), A 814 (TP 316 N)	---	---	---	---
317	S31700	A 167 (317), A 182 (F 317), A 213 (TP 317), A 240 (317), A 249 (317), A 269 (317), A 276 (317), A 312 (317), A 314 (317), A 403 (317), A 409 (317), A 473 (317), A 478 (317), A 511 (317), A 554 (317), A 580 (317), A 632 (317), A 813 (TP 317), A 814 (TP 317)	J405 (30317)	---	---	QQ-S-763 (317)
317 L	S31703	A 167 (317 L), A 182 (F 317 L), A 213 (TP 317 L), A 240 (317 L), A 249 (TP 317 L), A 312 TP 317 L), A 774 (TP 317 L), A 778 (TP 317 L), A 813 (TP 317 L), A 814 (TP 317 L)	---	---	---	---

AMERICAN STANDARDS CROSS REFERENCES - AISI STAINLESS STEELS (Continued)						
AISI	UNS	ASTM	SAE	AMS	MIL	FED
321	S32100	A 167 (321), A 182 (321), A 193 (321, B8T, B8TA), A 194 (321, 8T, 8TA), A 213 (321), A 240 (321), A 249 (321), A 269 (321), A 271 (321), A 276 (321), A 312 (321), A 314 (321), A 320 (321, B8T, B8TA), A 336 (F321), A 358 (321), A 376 (321), A 403 (321), A 409 (321), A 430 (321), A 473 (321), A 479 (321), A 511 (321), A 554 (321), A 580 (321), A 632 (321), A 774 (TP 321), A 778 (TP 321), A 813 (TP 321), A 814 (TP 321)	J405 (30321)	5510, 5557, 5559, 5570, 5576, 5645, 5689, 7211, 7490	MIL-S-27419	QQ-S-763 (321), QQ-S-766 (321)
329	S32900	A 240 (329), A 268 (329), A 789 (329), A 790	---	---	---	---
347	S34700	A 167 (347), A 182 (347), A 193 (347, B8CA), A 194 (347, 8C, 8CA), A 213 (347), A 240 (347), A 249 (347), A 269 (347), A 271 (347), A 276 (347), A 312 (347), A 314 (347), A 320 (347, B8C, B8CA), A 336 (F347), A 358 (347), A 376 (347), A 403 (347), A 409 (347), A 430 (347), A 473 (347), A 479 (347), A 511 (347), A 554 (347), A 580 (347), A 632 (347), A 774 (TP 347), A 778 (TP 347), A 813 (TP 347), A 814 (TP 347)	J405 (30347)	5512, 5556, 5558, 5571, 5575, 5646, 5654, 5674, 7229, 7490	MIL-S-23195 (347), MIL-S-23196 (347)	QQ-S-763 (347), QQ-S-766 (347)
348	S34800	A 167 (348), A 182 (348), A 213 (348), A 240 (348), A 249 (348), A 269 (348), A 276 (348), A 312 (348), A 314 (348), A 336 (348), A 358 (348), A 376 (348), A 403 (348), A 409 (348), A 479 (348), A 580 (348), A 632 (348), A 813 (TP 348), A 814 (TP 348)	J405 (30348)	---	MIL-S-23195 (348), MIL-S-23196 (348)	QQ-S-766 (348)

AMERICAN STANDARDS CROSS REFERENCES - AISI STAINLESS STEELS (Continued)

AISI	UNS	ASTM	SAE	AMS	MIL	FED
384	S38400	A 493 (384)	J405 (30384)	---	---	---
403	S40300	A 176 (403), A 276 (403), A 314 (403), A 473 (403), A 479 (403), A 511 (403), A 580 (403)	J405 (51403)	---	---	QQ-S-763 (403)
405	S40500	A 176 (405), A 240 (405), A 268 (405), A 276 (405), A 473 (405), A 479 (405), A 511 (405), A 580 (405)	J405 (51405)	---	---	QQ-S-763 (405)
409	S40900	A 176 (409), A 240 (409), A 268 (409), A 791 (TP 409), A 803 (TP 409)	J405 (51409)	---	---	---
410	S41000	A 176 (410), A 182 (F62, F8a), A 193 (410, B6, B6X), A 194 (410, 6), A 240 (410), A 268 (410), A 276 (410), A 314 (410), A 336 (F6), A 473 (410), A 479 (410), A 493 (410), A 511 (410), A 580 (410)	J405 (51410), J412 (51410)	5504, 5505, 5591, 5613, 5776, 7493	---	QQ-S-763 (410), QQ-S-766 (410)
414	S41400	A 276 (414), A 314 (414), A 473 (414), A 479 (414), A 511 (414), A 580 (414)	J405 (51414)	---	---	QQ-S-763 (414)
416	S41600	A 194 (416, 6F), A 314 (416), A 473 (416), A 581 (416), A 582 (416), A 895 (416)	J405 (51416)	5610	---	---
416 Se	S41623	A 194 (416 Se, 6F), A 314 (416 Se), A 473 (314 Se), A 511 (416 Se), A 581 (416 Se), A 582 (416 Se), A 895 (416 Se)	J405 (51416 (Se)	5610	---	QQ-S-763
420	S42000	A 276 (420), A 314 (420), A 473 (420), A 580 (420)	J405 (51420)	5506, 5621, 7207	---	QQ-S-763 (420), QQ-S-766 (420)
420 F	S42020	A 895 (420 F)	J405 (51420F)	5620	---	---
422	S42200	A 565 (616)	J467 (422), J775 (HNV-8)	5655	---	---

AMERICAN STANDARDS CROSS REFERENCES - AISI STAINLESS STEELS (Continued)						
AISI	UNS	ASTM	SAE	AMS	MIL	FED
429	S42900	A 176 (429), A 182 (429), A 240 (429), A 268 (429), A 276 (429), A 314 (429), A 473 (429), A 493 (429), A 511 (429), A 554 (429), A 815 (WP 429)	J405 (51429)	---	---	QQ-S-763 (429), QQ-S-766 (429)
430	S43000	A 176 (430), A 182 (430), A 240 (430), A 268 (430), A 276 (430), A 314 (430), A 473 (430), A 479 (430), A 493 (430), A 511 (430), A 554 (430), A 580 (430), A 815 (WP 430)	J405 (51430)	5503, 5627	---	QQ-S-764 (430), QQ-S-766 (430)
430 F	S43020	A 314 (430 F), A 473 (430 F), A 581 (430 F), A 582 (430 F), A 895 (430 F)	J405 (51430 F)	---	---	---
430FSe	S43023	A 314 (430 F Se), A 473 (430 F Se), A 581 (430 F Se), A 582 (430 F Se), A 895 (430 F Se)	J405 (51430 F Se)	---	---	---
439	S43035	A 240 (439), A 268 (439), A 479 (439), A 791 (TP 439), A 803 (TP 439)	---	---	---	---
431	S43100	A 276 (431), A 473 (431), A 493 (431), A 511 (MT 431), A 579 (63), A 580 (431)	J405 (51431)	5628	MIL-S-8967 MIL-S-18732	---
434	S43400	---	J405 (51434)	---		---
436	S43600	---	J405 (51436)	---		---
440 A	S44002	A 276 (440 A), A 314 (440 A), A 473 (440 A), A 511 (440 A), A 580 (440 A)	J405 (51440 A)	5631, 5632, 7445	---	QQ-S-763 (440 A)
440 B	S44003	A 276 (440 B), A 314 (440 B), A 473 (440 B), A 580 (440 B)	J405 (51440 B)	7445	---	QQ-S-763 (440 B)
440 C	S44004	A 276 (440 C), A 314 (440 C), A 473 (440 C), A 493 (440 C), A 580 (440 C)	J405 (51440 C)	5618, 5630, 5880, 7445	---	QQ-S-763 (440 C)
442	S44200	A 176 (442)	J405 (51442)	---	---	---

AMERICAN STANDARDS CROSS REFERENCES - AISI STAINLESS STEELS (Continued)

AISI	UNS	ASTM	SAE	AMS	MIL	FED
446	S44600	A 176 (446), A 268 (446-1, 446-2), A 276 (446), A 314 (446), A 473 (446), A 511 (446), A 580 (446), A 815 (WP 446)	J405 (51446)	---	---	QQ-S-763 (446), QQ-S-766 (446)
501	S50100	A 182 (B5, F7), A 193 (501, B5), A 194 (501, 3), A 314 (501), A 387 (5), A 473 (501)	J405 (51501)	---	---	---
502	S50200	A 314 (502), A 387 (5), A 473 (502)	J405 (51502)	---	MIL-T-16286	---

AMERICAN STANDARDS CROSS REFERENCES - NON-AISI STAINLESS STEELS

UNS	ASTM	SAE	AMS	MIL
S13800	A 564 (XM-13), A 693 (XM-13), A 705 (XM-13)	---	5629, 5864	---
S15500	A 564 (XM-12), A 693 (XM-12), A 705 (XM-12)	---	5658, 5659, 5826, 5862	---
S15700	A 564 (632), A 579 (632), A 693 (632), A 705 (632)	J467 (PH15-7-Mo)	5520	---
S16800	A 376 (16-8-2-H), A 430 (16-8-2-H)		---	---
S17400	A 564 (630), A 693 (630), A 705 (630)	J467 (17-4PH)	5604, 5622, 5643	MIL-C-24111, MIL-S-81591
S17600	A 564 (635), A 693 (635), A 705 (635)	---		---
S17700	A 313 (631), A 564 (631), A 579 (62), A 693 (631), A 705 (631)	J217, J467 (17-7PH)	5528, 5529, 5568, 5644, 5673, 5678	MIL-S-25043
S18200	A 581 (XM-34), A 582 (XM-34)	---	---	---
S18235	A 581, A 582	---	---	---
S20103	A 240, A 480	---	---	---
S20153	A 240, A 480	---	---	---
S20161	A 167, A 314, A 412	---	---	---
S20300	A 581 (XM-1), A 582 (XM-1)	J405 (203 EZ)	5762	---
S20400	A 240, A 666	---	---	---

AMERICAN STANDARDS CROSS REFERENCES - NON-AISI STAINLESS STEELS (Continued)

UNS	ASTM	SAE	AMS	MIL
S20910	A 182 (XM-19), A 184 (TP XM-19), A 193 (B8R, B8RA), A 194 (8R, 8RA), A 240 (XM-19), A 249 (XM-19), A 269 (XM-19), A 312 (XM-19), A 336 (FXM-19), A 403 (XM-19), A 412 (XM-19), A 479 (XM-19), A 580 (XM-19), A 813 (TP XM-19)	---	5764	---
S21400	A 240 (XM-31), A 580 (XM-31)	---	---	---
S21460	A 412 (XM-14), A 666 (XM-14)	---	---	---
S21500	A 213	---	---	---
S21600	A 240 (XM-17), A 479 (XM-17), A 492 (XM-17)	---	---	---
S21603	A 240 (XM-18), A 479 (XM-18), A 492 (XM-18)	---	---	---
S21800	A 193 (B8S, B8SA), A 194 (8S, 8SA), A240, A 276, A 479, A 555, A 580	---	5848	---
S21900	A 269 (TP XM-10), A 276 (XM-10), A 312 (TP XM-10), A 314 (XM-10), A 412 (XM-10), A 473 (XM-10), A 580 (XM-10), A 813 (TP XM-10), A 814 (TP XM-10)	---	5561	---
S21904	A 182 (F XM-11), A 276 (XM-11), A 314 (XM-11), A 412 (XM-11), A 473 (XM-11), A 479 (XM-11), A 580 (XM-11), A 666 (XM-11)	---	5562, 5595, 5656	---
S24000	A 240 (XM-29), A 249 (XM-29), A 269 (XM-29), A 312 (XM-29), A 412 (XM-29), A 479 (XM-29), A 580 (XM-29), A 688 (XM-29), A 813 (TP XM-29), A 814 (TP XM-29)	---	---	---
S24100	A 580 (XM-28)	---	---	---
S28200	A 276, A 314, A 473, A 493, A 580	---	---	---
S30310	A 581 (XM-5), A 582 (XM-5)	J405 (303 plus X), J412 (303 plus X)	5640 (Type 3)	---
S30345	A 581 (XM-2), A 582 (XM-2)	---	5638	---
S30360	A 581 (XM-3), A 582 (XM-3)	---	5635	---

AMERICAN STANDARDS CROSS REFERENCES - NON-AISI STAINLESS STEELS (Continued)

UNS	ASTM	SAE	AMS	MIL
S30409	A 182 (304 H), A 213 (304 H), A 240 (304 H), A 249 (304 H), A 271 (304 H), A 312 (304 H), A 336 (F 304 H), A 358 (304 H), A 376 (304 H), A 403 (304 H), A 430 (304 H), A 479 (304H), A 813 (TP 304 H), A 814 (TP 304 H)	---	---	---
S30452	A 240 (XM-21), A 276 (XM-21)	---	---	---
S30453	A 182 (F 304 LN), A 193 (B8LN, B8LNA), A 194 (8LN, 8LNA), A 213 (TP 304 LN), A 240 (304 LN), A 249 (TP 304 LN), A 269 (TP 304 LN), A 276, A 312 (TP 304 LN), A 320 (B8LN, B8LNA), A 336 (F 304 LN), A 376 (TP 304 LN), A 403 (WP 304 LN), A 479 (304 LN), A 666 (304 LN), A 688 (TP 304 LN), A 813 (TP 304 LN), A 814 (TP 304 LN)	---	---	---
S30600	A 182 (F 46), A 240, A 269, A 312, A 358, A 479	---	---	---
S30601	A 240, A 312	---	---	---
S30815	A 167, A 182 (F 45), A 213, A 240, A 249, A 276, A 312, A 358, A 473, A 479, A 813, A 814	---	---	---
S30880	A 479 (ER308)	---	---	MIL-E-19933 (MIL-308, MIL-308Co), MIL-I-23413 (MIL-308, MIL-308Co)
S30883	---	---	---	MIL-E-19933 (MIL-308L, MIL-308CoL), MIL-I-23413 (MIL-308L, MIL-308CoL)
S30884	---	---	---	MIL-E-19933D (MIL-308HC)
S30909	A 213 (TP 309 H), A 240 (309 H), A 249 (TP 309 H), A 312 (TP 309 H), A 312M, A 336 (F 309 H), A 479 (309 H)	---	---	---

AMERICAN STANDARDS CROSS REFERENCES - NON-AISI STAINLESS STEELS (Continued)

UNS	ASTM	SAE	AMS	MIL
S30940	A 213 (TP 309 Cb), A 240 (309 Cb), A 249 (TP 309 Cb), A 312 (TP 309 Cb), A 478 (309 Cb), A 479 (309 Cb), A 554 (309-S-Cb), A 580 (309 Cb), A 813 (TP 309 Cb), A 814 (TP 309 Cb)	---	---	---
S30941	A 213 (TP 309 HCb), A 240 (309 HCb), A 249 (TP 309 HCb), A 312 (TP 309 HCb)	---	---	---
S31009	A 213 (TP 310 H), A 240 (310 H), A 249 (310 H), A 312 (TP 310 H), A 336 (F 310 H), A 479 (310 H)	---	---	---
S31040	A 213 (TP 310 Cb), A 240 (310 Cb), A 249 (TP 310 Cb), A 312 (TP 310 Cb), A 478 (310 Cb), A 479 (310 Cb), A 813 (TP 310 Cb), A 814 (TP 310 Cb)	---	---	---
S31041	A 213 (TP 310 HCb), A 249 (TP 310 HCb), A 312 (TP 310 HCb), A 312H	---	---	---
S31050	A 213, A 240, A 249, A 312, A 479	---	---	---
S31080	---	---	5694	MIL-E-19933 (MIL-310), MIL-I-23413 (MIL-310)
S31200	A 182 (F50), A 240, A 789, A 790	---	---	---
S31254	A 182 (F44), A 193, A 194, A 240, A 249, A 269, A 276, A 312, A358, A 403, A 409, A 479, A 813, A 814	---	---	---
S31272	A 213, A 312	---	---	---
S31609	A 182 (316 H), A 213 (316 H), A 240 (316 H), A 249 (316 H), A 271 (TP 316 H), A 312 (316 H), A 336 (F 316 H), A 358 (316 H), A 376 (316 H), A 403 (316 H), A 430 (316 H), A 479 (316 H), A 813 (TP 316 H), A 814 (TP 316 H)	---	---	---
S31635	A 240 (316 Ti), A 368 (316 Ti), A 478 (316 Ti), A 479 (316 Ti)	---	---	---
S31640	A 240 (316 Cb), A 368 (316 Cb), A 478 (316 Cb), A 479 (316 Cb)	---	---	---

AMERICAN STANDARDS CROSS REFERENCES - NON-AISI STAINLESS STEELS (Continued)

UNS	ASTM	SAE	AMS	MIL
S31653	A 182 (F 316 LN), A 193 (B8MLN, B8MLNA), A 194 (8MLN, 8MLNA), A 213 (TP 316 LN), A 240 (316 LN), A 249 (TP 316 LN), A 269 (TP 316 LN), A 276 (316 LN), A 312 (TP 316 LN), A 320 (B8MLN, B8MLNA), A 336 (F 316 LN), A 376 (TP 316 LN), A 403 (WP 316 LN), A 479 (316 LN), A 688 (TP 316 LN), A 813 (TP 316 LN), A 814 (TP 316 LN)	---	---	
S31654	A 276	---	---	---
S31680	---	---	5692	MIL-E-19933 (MIL-316), MIL-I-23413 (MIL-316)
S31683	---	---	---	MIL-E-19933 (MIL-316L), MIL-I-23413 (MIL-316L)
S31725	A 167, A 182 (F47), A 213, A 240, A 249, A 269, A 276, A 312, A 358, A 376, A 409, A 479	---	---	---
S31726	A 167, A 182 (F48), A 213, A 240, A 249, A 269, A 276, A 312, A 358, A 376, A 409, A 479	---	---	---
S31753	A 167, A 240	---	---	---
S32109	A 182 (321 H), A 213 (321 H), A 240 (321 H), A 249 (321 H), A 271 (321 H), A 312 (321 H), A 336 (F321 H), A 376 (321 H), A 403 (321 H), A 430 (321 H), A 479 (321 H), A 813 TP 321 H), A 814 (TP 321 H)	---	---	---
S32615	A 213, A 240, A 312, A 351	---	---	---
S32654	A 167, A 240, A 480	---	---	---
S32803	A 167, A 240, A 268	---	---	---
S34565	A 167, A 182, A 240, A 249, A 358, A 403, A 479, A 480	---	---	---
S34709	A 182 (347 H), A 213 (347 H), A 240 (347 H), A 249 (347 H), A 271 (347 H), A 312 (347 H), A 336 (F347 H), A 376 (347 H), A 403 (347 H), A 430 (347 H), A 479 (347 H), A 813 (TP 347 H), A 814 (TP 347 H)	---	---	---

UNS	ASTM	SAE	AMS	MIL
AMERICAN STANDARDS CROSS REFERENCES - NON-AISI STAINLESS STEELS (Continued)				
S34780	---	---	5790	MIL-E-19933 (MIL-347, MIL-347Co), MIL I-23413 (MIL-348, MIL-348Co)
S34809	A 182 (348 H), A213 (348 H), A 240 (348 H), A 249 (348 H), A 312 (348 H), A 336 (F348 H), A 479 (348 H), A 813 (TP 348 H), A 814 (TP 348 H)	---	---	---
S35000	A 579 (61), A 693 (633)	J467 (AM-350)	5546, 5548, 5554, 5745	MIL-S-8840
S35315	A 167, A 240, A 480	---	---	---
S35500	A 564 (634), A 579 (634), A 693 (634), A 705 (634)	J467 (AM-355)	5547, 5549, 5743, 5744	MIL-S-8840
S36200	A 564 (XM-9), A 693 (XM-9), A 705 (XM-9)	---	5739, 5740	---
S38100	A 167 (XM-15), A 213 (XM-15), A 240 (XM-15), A 249 (XM-15), A 269 (XM-15), A 312 (XM-15), A 813 (TP XM-15), A 814 (TP XM-15)	---	---	---
S38660	A 771, A 826	---	---	---
S39274	A 789, A 790, A 815	---	---	---
S40945	A 176, A 240, A 480	---	---	---
S41001	A 579 (51)	---	5611, 5612	---
S41008	A 176 (410 S), A 240 (410 S), A 473 (410 S)	---	---	---
S41040	A 479 (XM-30)	---	5609	---
S41045	A 176, A 240, A 480	---	---	---
S41050	A 176: A 240	---	---	---
S41080	---	---	5778	MIL-E-19933 (MIL-410)
S41500	A 176, A 182 (F6NM), A 240, A 268, A 479, A 815	---	---	---
S41610	A 581 (XM-6), A 582 (XM-6)	J405 (416 plus X)	---	---

AMERICAN STANDARDS CROSS REFERENCES - NON-AISI STAINLESS STEELS (Continued)				
UNS	ASTM	SAE	AMS	MIL
S41780	---	---	5822, 5823	---
S41800	A 565 (615)	J467	5508, 5616	---
S42010	A 276, A 314, A 493	---	---	---
S42023	A 582 (420 F Se), A 895 (420 F Se)	J405 (51420F Se)	5620	---
S43036	A 268 (430 Ti), A 554 (430 Ti), A 815 (430 Ti)	---	---	---
S44020	---	J405 (51440 F)	5632 (1)	---
S44023	---	J405 (51440 F Se)	5632 (2)	---
S44300	A 268 (443), A 511 (443)	---	---	---
S44400	A 176, A 240, A 268 (18Cr-2Mo), A 276, A 479, A 791 (18-2), A 803 (18-2)	---	---	---
S44626	A 176 (XM-33), A 240 (XM-33), A 268 (XM-33), A 791 (TP XM-33), A 803 (TP XM-33)	---	---	---
S44627	A 176, A 182 (FXM-27 Cb), A 240, A 268, A 276 (XM-27), A 314, A 336 (FXM-27 Cb), A 479, A 731, A 791 (TP XM-27), A 803 (TP XM-27)	---	---	---
S44635	A 176, A 240, A 268 (25-4-4), A 791 (25-4-4), A 803 (25-4-4)	---	---	---
S44660	A 176, A 240, A 268 (26-3-3), A 791 (26-3-3), A 803 (26-3-3)	---	---	---
S44700	A 176, A 240, A 268 (29-4), A 276, A 479, A 493, A 511 (29-4), A 580, A 791 (29-4), A 803 (29-4)	---	---	---
S44735	A 176, A 240, A 268, A 511, A 791 (29-4C), A 803 (29-4C)	---	---	---
S44800	A 176, A 240, A 268 (29-4-2), A 276, A 479, A 493, A 511 (29-4-2), A 791 (29-4-2), A 803 (29-4-2)	---	---	---
S45000	A 564 (XM-25), A 693 (XM-25), A 705 (XM-25)	---	5763, 5773, 5859, 5863	---
S45500	A 313 (XM-16), A 564 (XM-16), A 693 (XM-16), A 705 (XM-16)	---	5578, 5617, 5672, 5860	MIL-S-83311
S50280	---	---	6466	MIL-I-23431 (MIL-505)

AMERICAN STANDARDS CROSS REFERENCES - NON-AISI STAINLESS STEELS (Continued)

UNS	ASTM	SAE	AMS	MIL
S50300	A 182 (F7), A 199 (T7), A 200 (T7), A 369 (FP7), A 387 (7)	---	---	---
S50400	A 199 (T9), A 200 (T9), A 213 (T9), A 276 (9), A 335 (P9), A 387 (9)	---	---	---
S50460	A 182, A 213, A 234, A 335, A 336, A 369, A 387			---
S63198	A 453 (651), A 457 (651), A 458 (651), A 477 (651)	J467 (19-9DL)	5526, 5579	---
S63199	---	J467 (19-9DX)	5782	---
S64152	A 565 (XM-32)		5718, 5719	---
S65006	---	J775 (HNV-6)	5710	---
S66009	---	J775 (EV-9)	5700	---
S66220	A 453 (662), A 638 (662)	J467	5733	---
S66286	A 453 (660), A 638 (660)	J467 (A286), J775 (HEV-7)	5525, 5726, 5731, 5732, 5734, 5737, 5804, 5805, 5853, 5858, 5895, 7235	---
S66545	A 453 (665)	J467 (W545)	---	---

AMERICAN STANDARDS CROSS REFERENCES - AISI TOOL STEELS

AISI	UNS No.	ASTM	SAE	AMS	MIL SPEC
M-1	T11301	A 600 (M-1)	J437 (M1), J438 (M1)	---	---
M-2	T11302	A 597 (CM-2), A 600 (M-2)	J437 (M2), J438 (M2)	---	---
M-4	T11304	A 600 (M-4)	J437 (M4), J438 (M4)	---	---
M-6	T11306	A 600 (M-6)	---	---	---
M-7	T11307	A 600 (M-7)	---	---	---
M-10	T11310	A 600 (M-10)			---
M-3 (Cl 1)	T11313	A 600 (M-3, Cl.1)	J437 (M3), J438 (M3)	---	---
M-3 (Cl 2)	T11323	A 600 (M-3, Cl.2)	J437 (M3), J438 (M3)	---	---
M-30	T11330	A 600 (M-30)	---	---	---

AMERICAN STANDARDS CROSS REFERENCES - AISI TOOL STEELS (Continued)

AISI	UNS No.	ASTM	SAE	AMS	MIL SPEC
M-33	T11333	A 600 (M-33)	---	---	---
M-34	T11334	A 600 (M-34)	---	---	---
M-36	T11336	A 600 (M-36)	---	---	---
M-41	T11341	A 600 (M-41)	---	---	---
M-42	T11342	A 600 (M-42)	---	---	---
M-46	T11346	A 600 (M-46)	---	---	---
M-48	T11348	A 600 (M-48)	---	---	---
M-52	T11352	A 600 (M-52)	---	---	---
M-62	T11362	A 600 (M-62)	---	---	---
T-1	T12001	A 600 (T-1)	J437 (T1), J438 (T1)	5626	---
T-4	T12004	A 600 (T-4)	J437 (T4), J438 (T4)	---	---
T-5	T12005	A 600 (T-5)	J437 (T5), J438 (T5)	---	---
T-6	T12006	A 600 (T-6)	---	---	---
T-8	T12008	A 600 (T-8)	J437 (T8), J438 (T8)	---	---
T-15	T12015	A 600 (T-15)	---	---	---
H-10	T20810	A 681 (H-10)	---	---	---
H-11	T20811	A 579 (H-11), A 681 (H-11)	J437 (H11), J438 (H11), J467 (H11)	6437, 6485, 6487, 6488	MIL-S-47262
H-12	T20812	A 681 (H-12)	J437 (H12), J438 (H12), J467 (H12)	---	---
H-13	T20813	A 681 (H-13)	J437 (H13), J438 (H13), J467 (H13)	6408	---
H-14	T20814	A 681 (H-14)	---	---	---
H-19	T20819	A 681 (H-19)	---	---	---
H-21	T20821	A 681 (H-21)	J437 (H21), J438 (H21)	---	---
H-22	T20822	A 681 (H-22)	---	---	---
H-23	T20823	A 681 (H-23)	---	---	---
H-24	T20824	A 681 (H-24)	---	---	---
H-26	T20826	A 681 (H-26)	---	---	---
H-42	T20842	A 681 (H-42)	---	---	---

AMERICAN STANDARDS CROSS REFERENCES - AISI TOOL STEELS (Continued)

AISI	UNS No.	ASTM	SAE	AMS	MIL SPEC
A-2	T30102	A 681 (A-2)	J437 (A2), J438 (A2)	---	---
A-4	T30104	A 681 (A-4)	---	---	---
A-6	T30106	A 681 (A-6)	---	---	---
A-7	T30107	A 681 (A-7)	---	---	---
A-8	T30108	A 681 (A-8)	---	---	---
A-9	T30109	A 681 (A-9)	---	---	---
A-10	T30110	A 681 (A-10)	---	---	---
D-2	T30402	A 681 (D-2)	J437 (D2), J438 (D2)	---	---
D-3	T30403	A 681 (D-3)	J437 (D3), J438 (D3)	---	---
D-4	T30404	A 681 (D-4)	---	---	---
D-5	T30405	A 681 (D-5)	J437 (D5), J438 (D5)	---	---
D-7	T30407	A 681 (D-7)	J437 (D7), J438 (D7)	---	---
O-1	T31501	A 681 (O-1)	J437 (O1), J438 (O1)	---	---
O-2	T31502	A 681 (O-2)	J437 (O2), J438 (O2)	---	---
O-6	T31506	A 681 (O-6)	J437 (O6), J438 (O6)	---	---
O-7	T31507	A 681 (O-7)	---	---	---
S-1	T41901	A 681 (S-1)	J437 (S1), J438 (S1)	---	---
S-2	T41902	A 681 (S-2)	J437 (S2), J438 (S2)	---	---
S-4	T41904	A 681 (S-4)	---	---	---
S-5	T41905	A 681 (S-5)	J437 (S5), J438 (S5)	---	---
S-6	T41906	A 681 (S-6)	---	---	---
S-7	T41907	A 597 (CS-7), A 681 (S-7)	---	---	---
P-6	T51606	A 681 (P-6)	---	---	---
P-20	T51620	A 681 (P-20)	---	---	---
P-21	T51621	A 681 (P-21)	---	---	---
L-2	T61202	A 681 (L-2)	---	---	---

AMERICAN STANDARDS CROSS REFERENCES - AISI TOOL STEELS (Continued)

AISI	UNS No.	ASTM	SAE	AMS	MIL SPEC
L-6	T61206	A 681 (L-6)	J437 (L6), J438 (L6)	---	---
W-1	T72301	A 686 (W-1)	J437 (W108, W109, W110, W112), J438 (W108, W109, W110, W112)	---	---
W-2	T72302	A 686 (W-2)	J437 (W209, W210), J438 (W209, W210)	---	---
W-5	T72305	A 686 (W-5)	---	---	---

AMERICAN STANDARDS CROSS REFERENCES - CAST STEELS

ACI	UNS No.	ASTM	AMS	MIL
CA-6NM	J91540	A 352 (CA-6NM), A 356 (CA-6NM), A 487 (CA-15, CA-6NM)		---
CA-15	J91150	A 217 (CA-15), A 426 (CPCA-15), A 743 (CA-15)	5351	---
CA-15M	J91151	A 487 (CA-15M), A 743 (CA-15M)	---	---
CA28MWV	J91422	A 743	---	---
CA-40	J91153	A 743 (CA-40)	---	---
CA40F	J91154	A 743	---	---
CB-30	J91803	A 743 (CB-30)	---	---
CC-50	J92615	A 743 (CC-50)	---	---
CD-3MN	J92205	A 890 (4A)	---	---
CE-30	J93423	A 743 (CE-30)	---	---
CF-3	J92500	A 351 (CF-3, CF-3A), A 451 (CPF3, CPF3A), A 743 (CF-3), A 744 (CF-3)	---	---
CF-3M	J92800	A 351 (CF-3M, CF-3MA), A 451 (CPF-3M), A 744 (CF-3M)	---	---
CF-8	J92600	A 351 (CF-8, CF-8A), A 451 (CPF8, CPF8A), A 743 (CF-8), A 744 (CF-8)	---	---

AMERICAN STANDARDS CROSS REFERENCES - CAST STEELS (Continued)

ACI	UNS No.	ASTM	AMS	MIL
CF-8C	J92710	A 351 (CF-8C), A 451 (CPF8C), A 743 (CF-8C), A 744 (CF-8C)	---	---
CF-8M	J92900	A 351 (CF-8M), A 451 (CPF8M), A 743 (CF-8M), A 744 (CF-8M)	---	---
CF3MN	J92804	A 743 (CF3MN)	---	---
CF-10MC	J92971	A 351 (CF-10MC)	---	---
CF10SMnN	J92972	A 743	---	---
CF-16F	J92701	A 743 (CF-16F)	---	---
CF-20	J92602	A 743 (CF-20)	5358	---
CG-12	J93001	A 743 (CG-12)	---	---
CH-8	J93400	A 351 (CH-8), A 451 (CPH-8)	---	---
CH-10	J93401	A 351 (CH-10)	---	---
CH-20	J93402	A 351 (CH-20), A 451 (CH-2, CPH-10, CPH-20), A 743 (CH-20)	---	---
CK-3MCuN	J93254	A 351, A 743, A 744	---	---
CK-20	J94202	A 351 (CK-20), A 451 (CPK20), A 743 (CK-20)	---	---
HC	J92605	A 297 (HC)	---	---
HC-30	J92613	A 608 (HC-30)	---	---
HD	J93005	A 297 (HD)	---	---
HE	J93403	A 297 (HE)	---	---
HE-35	J93413	A 608 (HE-35)	---	---
HF	J92603	A 297 (HF)	---	---
HH	J93503	A 297 (HH)	---	---
HH-30	J93513	A 608 (HH-30)	---	---
HI	J94003	A 297 (HI)	---	---
HK	J94224	A 297 (HK)	---	---
HK-30	J94203	A 351 (HK-30), A 608 (HK-30)	---	---

AMERICAN STANDARDS CROSS REFERENCES - CAST STEELS (Continued)

ACI	UNS No.	ASTM	AMS	MIL
HK-40	J94204	A 351 (HK-40), A 608 (HK-40)	---	---
HN	J94213	A 297 (HN)	---	---
---	J02503	A 216 (WCC), A 757 (A2Q)	---	---
---	J02504	A 352 (LCA), A 660 (WCA)	---	---
---	J02505	A 352 (LCC), A 660 (WCC)	---	---
---	J03002	A 216 (WCB), A 757 (A1Q)	---	---
---	J03003	A 352 (LCB), A 660 (WCB)	---	---
---	J12072	A 217 (WC6), A 426 (CP11)	---	---
---	J12082	A 217 (WC4), A 487 (11)	---	---
---	J21890	A 217 (WC9), A 426 (CP22)	---	---
---	J22000	A 217 (WC5), A 487 (12)	---	---
---	J22092	A 643 (C), A 757 (D1N, D1Q, D1N1, D1Q1, D1N2, D1Q2, D1N3, D1Q3)	---	---
---	J23260	---	5328, 5329	---
---	J24060	---	5330, 5331	---
---	J42015	---	---	MIL-S-16216 (HY80), MIL-S-23008 (HY80), MIL-S-23009 (HY80)
---	J42045	A 217 (C5), A 426 (CP5)	---	---
---	J42240	---	---	MIL-S-16216 (HY100), MIL-S-23008 (HY100), MIL-S-23009 (HY100)
---	J82090	A 217 (C12), A 426 (CP9)	---	---
---	J91152	---	5350	---
---	J91161	---	5349	---
---	J91171	A 426 (CPCA-15), A 487 (CA-15)	---	---
---	J91459	A 128 (D)	---	---

AMERICAN STANDARDS CROSS REFERENCES - CAST STEELS (Continued)				
ACI	UNS No.	ASTM	AMS	MIL
---	J92001	---	5359, 5368	---
---	J92110	A 747 (CB7Cu-2)	5346, 5347, 5348, 5356, 5357, 5400	---
---	J92200	---	5342, 5343, 5344, 5355, 5398	---
---	J92240	---	5340	---
---	J92590	A 351 (CF-10), A 452 (TP304H)	---	---
---	J92620	---	5370, 5371	MIL-S-81591 (304L)
---	J92641	---	5363, 5364	---
---	J93000	A 351 (CG-8M), A 743 (CG-8M), A 744 (CG-8M)	---	---
---	J93183	A 872	---	---
---	J93370	A 744 (CD-4MCu), A 890 (1A)	---	---
---	J93380	A 351, A 743	---	---
---	J94211	---	5365, 5366	---

Chapter

23

INTERNATIONAL SPECIFICATION TITLES & DESIGNATIONS

INTERNATIONAL SPECIFICATION DESIGNATIONS & TITLES OF CAST IRONS

Specification	Title
ISO 13	Grey iron pipes, special castings and grey iron parts for pressure main lines
ISO 49	Malleable cast iron fittings threaded to ISO 7/1
ISO 185	Grey cast iron - classification
ISO 945	Cast iron - designation of microstructure of graphite
ISO 946	Grey cast iron - beam unnotched impact test
ISO 1083	Spheroidal graphite cast iron - classification
ISO 2892	Austenitic cast iron
ISO 5922	Malleable cast iron
ISO 5996	Cast iron gate valves
ISO 6594	Cast iron drainage pipes and fittings - spigot series
ISO 7005 Part 2	Metallic flanges - part 2: cast iron flanges first edition; (supersedes ISO 2084, 2229 and 2441)
ISO 7259	Predominantly key-operated cast iron gate valves for underground use
ISO 8512 Part 1	Surface plates - part 1: cast iron
ISO 9147	Pig-irons - definition and classification
CEN prEN 598	Ductile cast iron pipes, fittings, accessories and their joints for sewerage application - requirements and tests methods
CEN prEN 877 Part 1	Cast iron pipes and fittings, their joints and accessories for the evacuation of water from buildings - part 1: technical specifications
CEN prEN 877 Part 2	Cast iron pipes and fittings, their joints and accessories for the evacuation of water from buildings - part 2: testing and quality control
CEN EN 945	Cast iron - designation of microstructure of graphite
CEN prEN 1092 Part 2	Flanges and their joints - part 2: cast iron flanges
CEN prEN 1171	Cast iron gate valves
CEN prEN 1503 Part 3	Valves - shell materials - part 3:cast irons
CEN prEN 1560	Founding - designation system for cast iron - material symbols and material numbers
CEN prEN 1561	Founding - grey cast irons
CEN prEN 1562	Founding - malleable cast irons
CEN prEN 1563	Founding - spheroidal graphite cast irons
CEN prEN 1564	Founding - austempered ductile cast irons
BS 143 & 1256	Malleable cast iron and cast copper alloy threaded pipe fittings

INTERNATIONAL SPECIFICATION DESIGNATIONS & TITLES OF CAST IRONS (Continued)

Specification	Title
BS 1452	Flake graphite cast iron
BS 1560 Sec 3.1	Steel pipe flanges and flanged fittings (nominal sizes half an inch to 24 inch) for the petroleum industry part 3: steel, cast iron and copper alloy flanges section 3.1: steel flanges
BS 1560 Sec 3.2	Steel pipe flanges and flanged fittings (nominal sizes half an inch to 24 inch) for the petroleum industrypart 3: steel, cast iron and copper alloy flanges section 3.2: cast iron flanges
BS 1560 Sec 3.3	Steel pipe flanges and flanged fittings (nominal sizes half an inch to 24 inch) for the petroleum industry part 3: steel cast iron and copper alloy flanges section 3.3: copper alloy and composite flanges
BS 2035	Cast iron flanged pipes and flanged fittings
BS 2789	Spheroidal graphite or nodular graphite cast iron
BS 3468	Austenitic cast iron
BS 4844	Abrasion resisting white cast iron
BS 5150	Cast iron gate valves
BS 5153	Cast iron check valves for general purposes
BS 5158	Cast iron plug valves
BS 5159	Cast iron and carbon steel ball valves for general purposes
BS 6681	Malleable cast iron
DIN 1685 Part 1	Raw castings made from nodular graphite cast iron; general tolerances, machining allowances
DIN 1691	Flake graphite cast iron (grey cast iron); properties
DIN 1691 Supp 1	Flake graphite cast iron (grey cast iron); general information on the selection of material and design; guide values of mechanical and physical properties
DIN 1692	Malleable cast iron; concept, properties
DIN 1693 Part 1	Cast iron with nodular graphite; unalloyed and low alloy grades
DIN 1693 Part 2	Cast iron with nodular graphite; unalloyed and low alloy grades; properties in cast-on test piece
DIN 1694	Austenitic cast iron
DIN 1694 Supp 1	Austenitic cast iron; reference data on mechanical and physical properties
DIN 1695	Abrasion resisting alloy cast iron
DIN 1695 Supp 1	Wear resisting alloyed cast iron; information on heat treatment, mechanical and physical properties and microstructure

INTERNATIONAL SPECIFICATION DESIGNATIONS & TITLES OF CAST IRONS (Continued)

Specification	Title
DIN 2410 Part 2	Pipes; survey of standards for tubes of ductile cast iron
DIN 2950	Malleable cast iron fittings
DIN 17007 Part 3	Material numbers; material group 0: pig iron, master alloys and cast iron
DIN 50109	Tensile testing of lamellar graphite cast iron
DIN 50149	Testing of malleable cast iron; tensile test
JIS G 5501	Grey iron castings
JIS G 5502	Spheroidal graphite iron castings
JIS G 5503	Austempered spheroidal graphite iron castings
JIS G 5504	Heavy-wllwed ferritic speroidal graphite iron castings for low temperature service
JIS G 5510	Austenitic iron castings
JIS G 5511	Low thermal expansion fe-alloy castings
JIS G 5525	Cast-iron soil pipes and fittings
JIS G 5526	Ductile iron pipes
JIS G 5527	Ductile iron fittings
JIS G 5702	Blackheart malleable iron castings
JIS G 5703	Whiteheart malleable iron castings
JIS G 5704	Pearlitic malleable iron castings

INTERNATIONAL SPECIFICATION DESIGNATIONS & TITLES OF STEEL BARS

Specification	Titles
ISO 722	Rock drilling equipment - hollow drill steels in bar form, hexagonal and round
ISO 1035 Part 1	Hot-rolled steel bars - part 1: dimensions of round bars
ISO 1035 Part 2	Hot-rolled steel bars - part 2: dimensions of square bars
ISO 1035 Part 3	Hot-rolled steel bars - part 3: dimensions of flat bars
ISO 1035 Part 4	Hot-rolled steel bars - part 4: tolerances
ISO 2938	Hollow steel bars for machining
ISO 4951	High yield strength steel bars and sections
ISO 6934 Part 5	Steel for the prestressing of concrete - part 5: hot-rolled steel bars with or without subsequent processing
ISO 6935 Part 1	Steel for the reinforcement of concrete - part 1: plain bars
ISO 6935 Part 2	Steel for the reinforcement of concrete - part 2: ribbed bars
ISO 9443	Heat-treatable and alloy steels - surface quality classes for hot-rolled round bars and wire rods - technical delivery conditions
ISO 10065	Steel bars for reinforcement of concrete - bend and rebend tests
ISO 10144	Certification scheme for steel bars and wires for the reinforcement of concrete structures
CEN prEN 3145	Round bars - normal tolerance hot formed in steel 8 mm < D less than or equal to 250 mm dimensions
CEN prEN 3146	Round bars - close tolerances in steel hot formed 8 mm < D less than or equal to 250 mm dimensions
CEN prEN 3517	Aerospace series steel FE-PL2105 air melted hardened and tempered bar for machining De ≤150 mm 1080 MPa ≤ Rm ≤ 1280 MPa
CEN prEN 3519	Aerospace series steel FE-PL2105 air melted hardened and tempered bar for machining De ≤ 150 mm 880 MPa ≤ Rm≤ 1080 MPa
CEN prEN 3906	Steel FE-PM 3801 air melted solution treated bar D less than or equal to 150 mm for the manufacture of fasteners 1100 MPa less than or equal to Rm less than or equal to 1300 MPa
CEN prEN 3906	Aerospace series martensitic corrosion resisting steel FE-PM3801 air melted solution treated bar D is less than or equal to 50 mm for the manufacture of fasteners 1100 MPa is less than or equal to Rm is less than or equal to 1300 MPa
CEN prEN 4235 Part 3	Aerospace series steel wrought products technical specification part 3 : bar and section (replaces prEN 2069-3 and prEN 2069-6)
CEN prEN 10080	Steel for the reinforcement of concrete weldable ribbed reinforcing steel B 500 - technical delivery condition for bars, coils and welded fabric
CEN prEN 10138 Part 4	Prestressing steel - part 4: hot rolled and processed bars

INTERNATIONAL SPECIFICATION DESIGNATIONS & TITLES OF STEEL BARS (Continued)

Specification	Titles
CEN prEN 10163 Part 4	Delivery requirements for surface quality of hot rolled steel products; part 4: round bars and wire rod
CEN EN 10207	Steels for simple pressure vessels - technical delivery requirements for plates, strips and bars
BS 970 Part 3	Wrought steel for mechanical and allied engineering purposes part 3: bright bars for general engineering purposes
BS 1502	Steels for fired and unfired pressure vessels: sections and bars
BS 1506	Carbon, low alloy and stainless steel bars and billets for bolting material to be used in pressure retaining applications
BS 4449	Carbon steel bars for the reinforcement of concrete
BS 4486	Hot rolled and hot rolled and processed high tensile alloy steel bars for the prestressing of concrete
BS 5892 Part 5	Railway rolling stock material (metric) part 5: steel bars for retaining rings for tyred wheels
BS 6258	Hollow steel bars for machining
BS 6744	Austenitic stainless steel bars for the reinforcement of concrete
BS 7295 Part 1	Fusion bonded epoxy coated carbon steel bars for the reinforcement of concrete part 1: specification for coated bars
BS 7295 Part 2	Fusion bonded epoxy coated carbon steel bars for the reinforcement of concrete part 2: coatings
BS HR 650	High expansion heat- resisting steel bar and wire for the manufacture of bolts, studs, set screws and nuts (Ni 25.5, Cr 15, Ti 2, Mn 1.5, Mo 1.25, Si 0.7, V 0.3) (limiting ruling section 20 mm)
BS S 150	Chromium-molybdenum-vanadium-niobium heat- resisting steel billets, bars, forgings and parts (930-1080 MPa) (Cr 10.5, Mo 0.6, V 0.2, Nb 0.3)
BS S 151	Chromium-nickel-molybdenum-vanadium heat-resisting steel billets, bars, forgings and parts (930-1130 MPa: limiting ruling section 150 mm) (Cr 12, Ni 2.5, Mo 1.7, V 0.3)
BS S 152	Chromium-cobalt-molybdenum-vanadium- niobium heat-resisting steel billets, bars, forgings and parts (consumable electrode remelted) (1000-1140 MPa) (Cr 10.5, Co 6, Mo 0.8, V 0.2, Nb 0.3)
BS S 153	2½ percent nickel-chromium-molybdenum steel billets, bars, forgings and parts (1230-1430 MPa: limiting ruling section 40 mm)
BS S 154	2½ per cent nickel-chromium- molybdenum steel billets, bars, forging and parts (880-1080 MPa: limiting ruling section 150 mm)
BS S 155	Nickel-silicon-chromium-molybdenum-vanadium steel (vacuum arc remelted) billets, bars, forgings and parts (1900-2100 MPa: limiting ruling section 75mm)
BS S 156	4 per cent nickel-chromium-molybdenum case-hardening steel (vacuum arc remelted) billets, bars, forgings and parts (1320-1520 MPa)
BS S 157	3 per cent nickel-chromium-molybdenum case-hardening steel billets, bars, forgings and parts (1180-1380 MPa)

INTERNATIONAL SPECIFICATION DESIGNATIONS & TITLES OF STEEL BARS (Continued)	
Specification	**Titles**
BS S 158	1 per cent chromium-molybdenum steel bars for the manufacture of forged bolts and forged nuts
BS S 159	12% chromium-nickel-molybdenum-vanadium heat-resisting steel bars for the manufacture of fasteners (1100-1300 MPa: limiting ruling section 50 mm)
BS S 160	Chromium-nickel corrosion-resisting steel - bright bars and parts (540 MPa, limiting ruling section 160mm) (Cr 18, Ni 10.5, controlled nitrogen content)
BS S 161	Chromium-nickel-molybdenum corrosion- resisting steel bright bars and parts (540 MPa, limiting ruling section 160mm) (Cr 17.5, Ni 12.5, Mo 2.25, controlled nitrogen content)
BS S 162	18 percent nickel-cobalt-molybdenum maraging steel for forging stock, forgings, bars and parts (1800 MPa: double vacuum melted)
BS S 163	Nickel-chromium-molybdenum-vanadium steel forging stock, forgings bars and parts (1900 MPa: limiting ruling section 100 mm)
BS 2S 130	18/9 chromium-nickel corrosion-resisting steel (niobium stabilized) billets, bars, forgings and parts (540 MPa; limiting ruling section 150 mm)
BS 2S 131	High thermal expansion steel billets, bars, forgings and parts (620 MPa: limiting ruling section 150 mm)
BS 2S 135	1 per cent carbon- chromium steel billets, bars, forgings and parts (limiting ruling section 25 mm)
BS 2S 136	1 per cent carbon-chromium steel (vacuum arc remelted) billets, bars, forgings and parts (limiting ruling section 25 mm)
BS 2S 137	High chromium-nickel corrosion-resisting steel bright bars (free machining) (880-1080 MPa; limiting ruling section 70 mm)
BS 2S 140	2½ per cent nickel-chromium-molybdenum steel billets, bars, forgings and parts (1080-1280 MPa: limiting ruling section 150 mm)
BS 2S 142	1 per cent chromium-molybdenum steel billets, bars, forgings and parts (900-1100 MPa: limiting ruling section 40 mm)
BS 2S 143	Chromium-nickel-copper-molybdenum corrosion- resisting steel (precipitation hardening) billets, bars, forgings and parts (930-1080 MPa)
BS 2S 144	Chromium-nickel-copper-molybdenum corrosion- resisting steel (precipitation hardening) billets, bars, forgings and parts (1130-1330 MPa)
BS 2S 145	Chromium-nickel-copper-molybdenum corrosion- resisting steel (precipitation hardening) billets, bars, forgings and parts (1270-1470 MPa)
BS 2S 147	0.5 per cent nickel- chromium-molybdenum steel bars for the manufacture of forged bolts and forged nuts
BS 2S 149	1.75 per cent nickel-chromium-molybdenum steel bars for the manufacture of forged bolts and forged nuts
BS 3S 91	Mild steel billets, bars, forgings and parts (suitable for bearing shells)
BS 3S 132	3 per cent chromium-molybdenum-vanadium steel billets, bars, forgings and parts (1320-1470 MPa: limiting ruling section 70 mm) (suitable for nitriding)

INTERNATIONAL SPECIFICATION DESIGNATIONS & TITLES OF STEEL BARS (Continued)

Specification	Titles
BS 3S 144	Chromium-nickel-copper-molybdenum corrosion resisting steel (precipitation hardening) billets, bars, forgings and parts (1130-1330 MPa)
BS 4S 106	3 per cent chromium-molybdenum steel billets, bars, forgings and parts (930-1080 MPa: limiting ruling section 150 mm) (suitable for nitriding)
BS 5S 82	4 per cent nickel- chromium-molybdenum case-hardening steel billets, bars, forgings and parts (1320-1520 MPa)
BS 5S 99	2.5 percent nickel- chromium-molybdenum (high carbon) steel billets, bars, forgings and parts (1230-1420 MPa: limiting ruling section 150 mm)
BS 6S 80	High chromium-nickel corrosion-resisting steel forging stock, bars, forgings and parts (880-1080 MPa, limiting ruling section 100 mm)
DIN 1017 Part 2	Steel bars; hot rolled flat steel for special purpose (in bar drawing mills, bolt and screw factories, etc.) dimensions, weights, permissible variations
DIN 1478	Turnbuckles made from steel tube or round steel bar
DIN 2090	Helical compression springs made of flat bar steel; calculation
JIS G 3104	Steel bars for rivet
JIS G 3105	Steel bars for chains
JIS G 3108	Rolled carbon steel for cold-finished steel bars
JIS G 3109	Steel bars for prestressed concrete
JIS G 3112	Steel bars for concrete reinforcement
JIS G 3117	Rerolled steel bars for concrete reinforcement
JIS G 3123	Cold finished carbon and alloy steel bars
JIS G 3137	Small size-deformed steel bars for prestressed concrete
JIS G 3191	Shape, dimension, weight and tolerance for hot rolled steel bar and bar-in-coil
JIS G 4108	Alloy steel bars for special application bolting materials
JIS G 4303	Stainless steel bars
JIS G 4311	Heat-resisting steel bars
JIS G 4318	Cold finished stainless steel bars
JIS G 4901	Corrosion-resisting and heat-resisting superalloy bars

INTERNATIONAL SPECIFICATION DESIGNATIONS & TITLES OF FORGINGS

Specification	Title
ISO 2604 Part 1	Steel products for pressure purposes - quality requirements - part 1: forgings
ISO 4779	Forged steel lifting hooks with point and eye for use with steel chains of grade M(4)
ISO 7597	Forged steel lifting hooks with point and eye for use with steel chains of grade T(8)
ISO 8539	Forged steel lifting components for use with grade T(8) chain
CEN EN 2157 Part 1	Aerospace series steel forging stock and forgings - technical specifications - part 1: general requirements
CEN EN 2157 Part 2	Aerospace series steel forging stock and forgings - technical specification - part 2: forging stock
CEN EN 2157 Part 3	Aerospace series steel forging stock and forgings - technical specification - part 3: pre-production and production forgings
CEN prEN 2451	Steel FE-PL2105 air melted hardened and tempered forgings De ≤ 40 mm 1230 MPa ≤ Rm ≤ 1420 MPa
CEN prEN 2453	Steel FE-PL73 1080 MPa ≤ Rm ≤ 1280 MPa forgings De ≤ 150 mm
CEN prEN 2455	Steel FE-PL2105 air melted hardened and tempered forgings De ≤ 150 mm 880 MPa ≤ Rm ≤ 1080 MPa
CEN prEN 2507	Steel FE-PM66 1270 ≤ Rm ≤ 1470 MPa forgings De ≤ 100 mm
CEN prEN 3518	Steel FE-PL2105 air melted hardened and tempered forgings De ≤ 150 mm 1080 ≤ Rm ≤ 1280 MPa
CEN prEN 3518	Aerospace series steel FE-PL2105 air melted hardened and tempered forgings De ≤ 150 mm 1080 MPa ≤ Rm ≤ 1280 MPa
CEN prEN 3520	Aerospace series steel FE-PL2105 air melted softened forging stock a or d ≤ 300 mm
CEN prEN 10222 Part 1	Steel forgings for pressure purposes - part 1: general requirements for open die forgings
CEN prEN 10222 Part 3	Steel forgings for pressure purposes - part 3: ferritic and martensitic steels with elevated temperature properties
CEN prEN 10222 Part 4	Steel forgings for pressure purposes - part 4: nickel steels with specified low temperature properties
CEN prEN 10222 Part 5	Steel forgings for pressure purposes - part 5: fine grain steels with high proof stress
CEN prEN 10222 Part 6	Steel forgings for pressure purposes - part 6: austenitic, martensitic and austenitic-ferritic stainless steels
CEN prEN 10228 Part 1	Non-destructive testing of steel forgings - part 1: magnetic particle inspection
CEN prEN 10228 Part 2	Non-destructive testing of steel forgings part 2: penetrant testing
CEN prEN 10228 Part 3	Non-destructive testing of steel forgings - part 3: ultrasonic testing of ferritic or martensitic steel forgings
BS 29	Carbon steel forgings above 150 mm ruling section
BS 1503	Steel forgings for pressure purposes
BS 1503	Steel forgings for pressure purposes
BS 3111 Part 1	Steel wire for cold forged fasteners for similar components part 1: carbon and low alloy steel wire
BS 3111 Part 2	Steel wire for cold forged fasteners for similar components part 2: stainless steel

INTERNATIONAL SPECIFICATION DESIGNATIONS & TITLES OF FORGINGS (Continued)

Specification	Title
BS 4124	Method for ultrasonic detection of imperfections in steel forgings
BS 4670	Alloy steel forgings
BS 7254 Sec 6.1	Orthopaedic implants part 6: forgings - section 6.1: production of forgings
BS 7254 Sec 6.2	Orthopaedic implants part 6: forgings - section 6.2: method for specifying forgings
BS S 150	Chromium-molybdenum-vanadium-niobium heat-resisting steel billets, bars, forgings and parts (930-1080 MPa) (Cr 10.5, Mo 0.6, V 0.2, Nb 0.3)
BS S 151	Chromium-nickel-molybdenum-vanadium heat-resisting steel billets, bars, forgings and parts (930-1130 MPa: limiting ruling section 150 mm) (Cr 12, Ni 2.5, Mo 1.7, V 0.3)
BS S 152	Chromium-cobalt-molybdenum-vanadium-niobium heat-resisting steel billets, bars, forgings and parts (consumable electrode remelted) (1000-1140 MPa) (Cr 10.5, Co 6, Mo 0.8, V 0.2, Nb 0.3)
BS S 153	2½ percent nickel-chromium-molybdenum steel billets, bars, forgings and parts (1230-1430 MPa: limiting ruling section 40 mm)
BS S 154	2½ per cent nickel-chromium-molybdenum steel billets, bars, bars, forging and parts (880- 1080 MPa: limiting ruling section 150 mm)
BS S 155	Nickel-silicon-chromium-molybdenum-vanadium steel (vacuum arc remelted) billets, bars, forgings and parts (1900-2100 MPa: limiting ruling section 75 mm)
BS S 156	4 per cent nickel-chromium-molybdenum case-hardening steel (vacuum arc remelted) billets, bars, forgings and parts (1320-1520 MPa)
BS S 157	3 per cent nickel-chromium-molybdenum case-hardening steel billets, bars, forgings and parts (1180-1380 MPa)
BS S 158	1 per cent chromium-molybdenum steel bars for the manufacture of forged bolts and forged nuts
BS S 162	18 percent nickel-cobalt-molybdenum maraging steel for forging stock, forgings, bars and parts (1800 MPa: double vacuum melted)
BS S 163	Nickel-chromium-molybdenum-vanadium steel forging stock, forgings bars and parts (1900 MPa: limiting ruling section 100mm)
BS 2S 130	18/9 chromium-nickel corrosion-resisting steel (niobium stabilized) billets, bars, forgings and parts (540 MPa; limiting ruling section 150 mm)
BS 2S 131	High thermal expansion steel billets, bars, forgings and parts (620 MPa: limiting ruling section 150 mm)
BS 2S 135	1 per cent carbon-chromium steel billets, bars, forgings and parts (limiting ruling section 25 mm)
BS 2S 136	1 per cent carbon-chromium steel (vacuum arc remelted) billets, bars, forgings and parts (limiting ruling section 25 mm)
BS 2S 140	2½ per cent nickel-chromium-molybdenum steel billets, bars, forgings and parts (1080-1280 MPa: limiting ruling section 150 mm)
BS 2S 142	1 per cent chromium-molybdenum steel billets, bars, forgings and parts (900-1100 MPa: limiting ruling section 40 mm)

INTERNATIONAL SPECIFICATION DESIGNATIONS & TITLES OF FORGINGS (Continued)

Specification	Title
BS 2S 143	Chromium-nickel-copper-molybdenum corrosion-resisting steel (precipitation hardening) billets, bars, forgings and parts (930-1080 MPa)
BS 2S 143	Chromium-nickel-copper-molybdenum corrosion-resisting steel (precipitation hardening) billets, bars, forgings and parts (930-1080 MPa)
BS 2S 143	Chromium-nickel-copper-molybdenum corrosion-resisting steel (precipitation hardening) billets, bars, forgings and parts (930-1080 MPa)
BS 2S 144	Chromium-nickel-copper-molybdenum corrosion-resisting steel (precipitation hardening) billets, bars, forgings and parts (1130-1330 MPa)
BS 2S 144	Chromium-nickel-copper-molybdenum corrosion-resisting steel (precipitation hardening) billets, bars, forgings and parts (1130-1330 MPa)
BS 2S 145	Chromium-nickel-copper-molybdenum corrosion-resisting steel (precipitation hardening) billets, bars, forgings and parts (1270-1470 MPa)
BS 2S 145	Chromium-nickel-copper-molybdenum corrosion-resisting steel (precipitation hardening) billets, bars, forgings and parts (1270-1470 MPa)
BS 2S 146	4 per cent nickel-chromium-molybdenum air-hardening steel (vacuum arc remelted) billets, bars, forgings and parts (1760-1960 MPa: limiting ruling section 100 mm)
BS 2S 147	0.5 per cent nickel-chromium-molybdenum steel bars for the manufacture of forged bolts and forged nuts
BS 2S 147	0.5 per cent nickel-chromium-molybdenum steel bars for the manufacture of forged bolts and forged nuts
BS 2S 149	1.75 per cent nickel-chromium-molybdenum steel bars for the manufacture of forged bolts and forged nuts
BS 2S 149	1.75 per cent nickel-chromium-molybdenum steel bars for the manufacture of forged bolts and forged nuts
BS 3S 132	3 per cent chromium-molybdenum-vanadium steel billets, bars, forgings and parts (1320-1470 MPa: limiting ruling section 70 mm) (suitable for nitriding)
BS 3S 144	Chromium-nickel-copper-molybdenum corrosion-resisting steel (precipitation hardening) billets, bars, forgings and parts (1130-1330 MPa)
BS 3S 91	Mild steel billets, bars, forgings and parts (suitable for bearing shells)
BS 4S 106	3 per cent chromium-molybdenum steel billets, bars, forgings and parts (930-1080 MPa: limiting ruling section 150 mm) (suitable for nitriding)

INTERNATIONAL SPECIFICATION DESIGNATIONS & TITLES OF FORGINGS (Continued)

Specification	Title
BS 5S 82	4 per cent nickel-chromium-molybdenum case-hardening steel billets, bars, forgings and parts (1320-1520 MPa)
BS 5S 99	2.5 percent nickel-chromium-molybdenum (high carbon) steel billets, bars, forgings and parts (1230-1420 MPa: limiting ruling section 150 mm)
BS 6S 80	High chromium-nickel corrosion-resisting steel forging stock, bars, forgings and parts (880-1080 MPa, limiting ruling section 100 mm)
DIN 7521	Steel forgings: technical conditions of delivery
DIN 7526	Steel forgings: tolerances and permissible variations for drop forgings
DIN 7526 Supp 1	Steel forgings: tolerances and permissible variations for drop forgings; examples for application
DIN 17103	Weldable fine grain structural steel forgings; technical delivery conditions
DIN 17243	Weldable heat resisting steel forgings and rolled or forged steel bars; technical delivery conditions
DIN 17440	Stainless steels; technical delivery conditions for plate and sheet, cold and hot rolled strip, wire rod, drawn wire, steel bars, forgings and semi-finished products
DIN 17460	High-temperature austenitic steel plate and sheet, cold and hot rolled strip, bars and forgings; technical delivery conditions
DIN 17480	Valve materials; technical delivery conditions
JIS F 7329	Forged steel 40 k globe valves for marine use
JIS F 7330	Forged steel 40 k angle valves for marine use
JIS F 7336	Forged steel globe air valves for marine use
JIS F 7337	Forged steel angle air valves for marine use
JIS F 7341	Forged steel 100 k pressure gauge valves for marine use
JIS F 7421	Forged steel 20 k globe valves for marine use
JIS F 7422	Forged steel 20 k angle valves for marine use
JIS G 0306	Steel forgings - general technical requirements
JIS G 0587	Methods for ultrasonic examination for carbon and low alloy steel forgings
JIS G 3201	Carbon steel forgings for general use
JIS G 3202	Carbon steel forgings for pressure vessels
JIS G 3203	Alloy steel forgings for pressure vessels for high-temperature service
JIS G 3204	Quenched and tempered alloy steel forgings for pressure vessels
JIS G 3205	Carbon and alloy steel forgings for pressure vessels for low-temperature service

INTERNATIONAL SPECIFICATION DESIGNATIONS & TITLES OF FORGINGS (Continued)

Specification	Title
JIS G 3206	High strength chromium-molybdenum alloy steel forgings for pressure vessels under high-temperature service
JIS G 3214	Stainless steel forgings for pressure vessels
JIS G 3221	Chromium molybdenum steel forgings for general use
JIS G 3222	Nickel chromium molybdenum steel forgings for general use
JIS G 3223	High tensile strength steel forgings for tower flanges
JIS G 3251	Carbon steel blooms and billets for forgings
JIS G 3507	Carbon steel wire rods for cold heading and cold forging
JIS G 3508	Boron steel wire rods for cold headingand cold forging
JIS G 3539	Carbon steel wires for cold heading and cold forging
JIS G 3545	Boron steel wires for cold heading and cold forging
JIS G 4315	Stainless steel wires for cold heading and cold forging
JIS G 4319	Stainless steel blooms and billets for forgings

INTERNATIONAL SPECIFICATION DESIGNATIONS & TITLES OF STEEL FASTENERS

Specification	Title
ISO 3506	Corrosion-resistant stainless steel fasteners - specifications
ISO 4775	Hexagon nuts for high- strength structural bolting with large width across flats - product grade B - property classes 8 and 10
ISO 6305 Part 4	Railway components - technical delivery requirements - part 4: untreated steel nuts and bolts and high-strength nuts and bolts for fish-plates and fastenings
ISO 7411	Hexagon bolts for high-strength structural bolting with large width across flats (thread lengths according to ISO 888) - product grade C - property classes 8.8 and 10.9
ISO 7412	Hexagon bolts for high-strength structural bolting with large width across flats (short thread length) - product grade C - property classes 8.8 and 10.9
CEN prEN 3903	Washers, laminated, in corrosion resisting steel FE-PA13
CEN prEN 3906	Aerospace series martensitic corrosion resisting steel FE-PM3801 air melted solution treated bar D is less than or equal to 50 mm for the manufacture of fasteners 1100 MPa is less than or equal to Rm is less than or equal to 1300 MPa
CEN prEN 4200	Aerospace series washers, 100 degrees dimpled, in steel, cadmium plated
BS 3111 Part 1	Steel wire for cold forged fasteners for similar components part 1: carbon and low alloy steel wire
BS 3111 Part 2	Steel wire for cold forged fasteners for similar components part 2: stainless steel
BS 4395 Part 1	High strength friction grip bolts and associated nuts and washers for structural engineering part 1: general grade
BS 4395 Part 2	High strength friction grip bolts and associated nuts and washers for structural engineering part 2: higher grade bolts and nuts and general grade washers
BS 4604 Part 2	The use of high strength friction grip bolts in structural steelwork. metric series part 2: higher grade (parallel shank)
BS 6105	Corrosion-resistant stainless steel fasteners
BS HR 650	High expansion heat-resisting steel bar and wire for the manufacture of bolts, studs, set screws and nuts (Ni 25.5, Cr 15, Ti 2, Mn 1.5, Mo 1.25, Si 0.7, V 0.3) (limiting ruling section 20 mm)
BS S 159	12% chromium-nickel-molybdenum-vanadium heat-resisting steel bars for the manufacture of fasteners (1100-1300 MPa: limiting ruling section 50 mm)
DIN 267 Part 3	Fasteners; technical delivery conditions; property classes for carbon steel and alloy steel bolts and screws; conversion of property classes
DIN 267 Part 11	Fasteners; technical delivery conditions (with additions to ISO 3506); corrosion-resistant stainless steel fasteners
DIN 267 Part 26	Fasteners; technical delivery conditions; steel spring washers for bolt/nut assemblies

INTERNATIONAL SPECIFICATION DESIGNATIONS & TITLES OF STEEL FASTENERS (Continued)

Specification	Title
DIN 267 Part 27	Fasteners; adhesive-coated steel screws, bolts and studs; technical delivery conditions
DIN 267 Part 28	Fasteners; steel-screws, bolts and studs with locking coating; technical delivery conditions
DIN 6914	High-strength hexagon head bolts with large widths across flats for structural steel bolting
DIN 6915	High-strength hexagon nuts with large widths across flats for structural steel bolting
DIN 6916	Round washers for high-strength structural steel bolting
DIN 6917	Square taper washers for high-strengthstructural bolting of steel I sections
DIN 7969	Slotted countersunk head screws for structural steel bolting for supply with or without nut
DIN 7989	Plain washers for steel construction
DIN 7990	Hexagon head bolts for structural steel bolting for supply with nut
DIN 17240	Heat resisting and highly heat resisting materials for bolts and nuts; quality specifications
DIN LN 65013	Specification for steel fasteners with minimum tensile strengths of 1550 N/mm^2 and 1800 N/mm^2 for temperatures up to 260°C
JIS B 1052	Mechanical properties of steel nuts
JIS B 1054	Specification for corrosion-resistant stainless steel fasteners
JIS B 1055	Mechanical properties for heat-treated steel and stainless steel tapping screws
JIS G 4107	Alloy steel bolting materials for high temperature service
JIS G 4108	Alloy steel bars for special application bolting materials

INTERNATIONAL SPECIFICATION DESIGNATIONS & TITLES OF STEEL PIPING, TUBING, FITTINGS & TESTING

Specification	Title
ISO 1106 Part 3	Recommended practice for radiographic examination of fusion welded joints - part 3: fusion welded circumferential joints in steel pipes of up to 50 mm wall thickness
ISO 1127	Stainless steel tubes - dimensions, tolerances and conventional masses per unit length
ISO 2037	Stainless steel tubes for the food industry
ISO 2604 Part 2	Steel products for pressure purposes - quality requirements - part 2: wrought seamless tubes
ISO 2604 Part 3	Steel products for pressure purposes - quality requirements - part 3: electric resistance and induction-welded tubes
ISO 2851	Metal pipes and fittings - stainless steel bends and tees for the food industry
ISO 2853	Stainless steel threaded couplings for the food industry
ISO 2937	Plain end seamless steel tubes for mechanical application
ISO 3183	Oil and natural gas industries - steel line pipe first edition
ISO 3304	Plain end seamless precision steel tubes - technical conditions for delivery
ISO 3305	Plain end welded precision steel tubes - technical conditions for delivery
ISO 3306	Plain end as-welded and sized precision steel tubes - technical conditions for delivery
ISO 3419	Non-alloy and alloy steel butt-welding fittings second edition
ISO 3545 Part 3	Steel tubes and fittings - symbols for use in specifications - part 3: tubular fittings with circular cross-section
ISO 4144	Stainless steel fittings threaded to ISO 7/1
ISO 4145	Non-alloy steel fittings threaded to ISO 7/1
ISO 4200	Plain end steel tubes, welded and seamless - general tables of dimensions and masses per unit length
ISO 5251	Stainless steel butt- welding fittings
ISO 5252	Steel tubes - tolerance systems
ISO 5256	Steel pipes and fittings for buried or submerged pipelines - external and internal coating by bitumen or coal tar derived materials
ISO 9303	Seamless and welded (except submerged arc- welded) steel tubes for pressure purposes - full peripheral ultrasonic testing for the detection of longitudinal imperfections
ISO 9304	Seamless and welded (except submerged arc- welded) steel tubes for pressure purposes - eddy current testing for the detection of imperfections
ISO 9305	Seamless steel tubes for pressure purposes - full peripheral ultrasonic testing for the detection of transverse imperfections

INTERNATIONAL SPECIFICATION DESIGNATIONS & TITLES OF STEEL PIPING, TUBING, FITTINGS & TESTING (Continued)	
Specification	**Title**
ISO 9329 Part 1	Seamless steel tubes for pressure purposes - technical delivery conditions - part 1: unalloyed steels with specified room temperature properties
ISO 9330 Part 1	Welded steel tubes for pressure purposes - technical delivery conditions - part 1: unalloyed steel tubes with specified room temperature properties
ISO 9402	Seamless and welded (except submerged arc-welded) steel tubes for pressure purposes - full peripheral magnetic transducer/flux leakage testing of ferromagnetic steel tubes for the detection of longitudinal imperfections
ISO 9598	Seamless steel tubes for pressure purposes - full peripheral magnetic transducer/flux leakage testing of ferromagnetic steel tubes for the detection of transverse imperfections
ISO 9626	Stainless steel needle tubing for manufacture of medical devices
ISO 9764	Electric resistance and induction welded steel tubes for pressure purposes - ultrasonic testing of the weld seam for the detection of longitudinal imperfections
ISO 9765	Submerged arc-welded steel tubes for pressure purposes - ultrasonic testing of the weld seam for the detection of longitudinal and/or transverse imperfections
ISO 10409	Petroleum and natural gas industries - application of cement lining to steel tubular goods, handling, installation and joining
CEN prEN 1515 Part 2	Flanges and their joints - bolting - part 2: combination of flange and bolting materials for steel flanges - pn designated
CEN prEN 10208 Part 1	Steel pipes for pipe lines for combustible fluids - technical delivery conditions - part 1: pipes of requirement class a
CEN prEN 10208 Part 2	Oil and natural gas industries - technical delivery conditions for steel pipes for pipe lines for combustible fluids part 2: pipes of quality level B
CEN prEN 10216 Part 1	Seamless steel tubes for pressure purposes - technical delivery conditions - part 1: non-alloy steel with specified room temperature properties
CEN prENV 10220	Seamless and welded steel tubes - dimensions and masses per unit length
CEN prEN 10224	Steel pipes, joints and fittings for the conveyance of aqueous liquids including potable water
CEN prEN 10241	Threaded steel fittings
CEN prEN 10253 Part 1	Butt - welding pipe fittings wrought carbon steel without specific inspection requirements
CEN prEN ISO 11961	Petroleum and natural gas industries - steel pipes for use as drill pipe - specification (based on API Spec 7)
CEN prEN 29303	Seamless and welded (except submerged arc- welded) steel tubes for pressure purposes - full peripheral ultrasonic testing for the detection of longitudinal imperfections

The Metals Black Book – 3rd Edition

INTERNATIONAL SPECIFICATION DESIGNATIONS & TITLES OF STEEL PIPING, TUBING, FITTINGS & TESTING (Continued)

Specification	Title
BS 1560 Sec 3.1	Steel pipe flanges and flanged fittings (nominal sizes half an inch to 24 inch) for the petroleum industry part 3: steel, cast iron and copper alloy flanges section 3.1: steel flanges
BS 1560 Sec 3.1	Circular flanges for pipes, valves and fittings (class designated) part 3: steel, cast iron and copper alloy flanges section 3.1: specification for steel flanges
BS 1560 Sec 3.2	Steel pipe flanges and flanged fittings (nominal sizes half an inch to 24 inch) for the petroleum industry part 3: steel, cast iron and copper alloy flanges section 3.2: cast iron flanges
BS 1560 Sec 3.3	Steel pipe flanges and flanged fittings (nominal sizes half an inch to 24 inch) for the petroleum industry part 3: steel cast iron and copper alloy flanges section 3.3: copper alloy and composite flanges
BS 1600	Dimensions of steel pipe for the petroleum industry
BS 1640 Part 1	Steel butt-welding pipe fittings for the petroleum industry part 1: wrought carbon and ferritic alloy steel fittings
BS 1640 Part 2	Steel butt-welding pipe fittings for the petroleum industry part 2: wrought and cast austenitic chromium-nickel steel fittings
BS 1640 Part 3	Steel butt-welding pipe fittings for the petroleum industry part 3: wrought carbon and ferritic alloy steel fittings
BS 1640 Part 4	Steel butt-welding pipe fittings for the petroleum industry part 4: wrought and cast austenitic chromium-nickel steel fittings
BS 1740 Part 1	Wrought steel pipe fittings (screwedbs p thread) part 1: metric units
BS 1965 Part 1	Butt-welding pipe fittings for pressure purposes part 1: carbon steel
BS 2910	Radiographic examination of fusion welded circumferential butt-joints in steel pipes
BS 3293	Carbon steel pipe flanges (over 24 in nominal size) for the petroleum industry
BS 3600	Specification for dimensions and masses per unit length of welded and seamless steel pipes and tubes for pressure purposes
BS 3601	Carbon steel pipes and tubes with specified room temperature properties for pressure purposes
BS 3602 Part 1	Steel pipes and tubes for pressure purposes carbon and carbon manganese steel with specified elevated temperature properties: seamless, electric resistance welded and induction welded tubes
BS 3602 Part 2	Steel pipes and tubes for pressure purposes: carbon and carbon manganese steel with specified elevated temperature properties part 2: specification for longitudinally arc welded tubes
BS 3603	Carbon and alloy steel pipes and tubes with specified low temperature properties for pressure purpose
BS 3604 Part 1	Steel pipes and tubes for pressure purposes: ferric alloy steel with specified elevated temperature properties part 1: seamless and electric resistance welded tubes

INTERNATIONAL SPECIFICATION TITLES &DESIGNATIONS

INTERNATIONAL SPECIFICATION DESIGNATIONS & TITLES OF STEEL PIPING, TUBING, FITTINGS & TESTING (Continued)

Specification	Title
BS 3604 Part 2	Steel pipes and tubes for pressure purposes: ferritic alloy steel with specified elevated temperature properties part 2: specification for longitudinally arc welded tubes
BS 3605 Part 1	Austenitic stainless steel pipes and tubes for pressure purposes part 1: specification for seamless tubes
BS 3605 Part 2	Austenitic stainless steel pipes and tubes for pressure purposes part 2: specification for longitudinally welded tubes
BS 3799	Steel pipe fittings, screwed and socket-welding for the petroleum industry
BS 4515	Process of welding of steel pipelines on land and offshore
BS 4825 Part 1	Stainless steel tubes and fittings for the food industry and other hygienic applications part 1: specification for tubes
BS 4825 Part 2	Stainless steel tubes and fittings for the food industry and other hygienic applications part 2: specification for bends and tees
BS 7416	Precision finished seamless cold-drawn low carbon steel tubes for use in hydraulic fluid power systems
DIN 1615	Welded circular unalloyed steel tubes not subject to special requirements; technical delivery conditions
DIN 1626	Welded circular unalloyed steel tubes subject to special requirements; technical delivery conditions
DIN 1628	High performance welded circular unalloyed steel tubes; technical delivery conditions
DIN 1629	Seamless circular unalloyed steel tubes subject to special requirements; technical delivery conditions
DIN 1630	High performance seamless circular unalloyed steel tubes; technical delivery conditions
DIN 2410 Part 1	Tubes; general review of standards for steel tubes
DIN 2413	Steel pipes; calculation of wall thickness subjected to internal pressure
DIN 2413 Part 1	Design of steel pressure pipes
DIN 2413 Part 2	Design of steel bends used in pressure pipelines
DIN 2445 Part 1	Seamless steel tubes for dynamic loads; hot finished tubes
DIN 2445 Supp 1	Seamless steel tubes for dynamic loads; basis for calculation of straight tubes
DIN 2448	Seamless steel pipes and tubes; dimensions, conventional masses per unit length
DIN 2449	Seamless steel tubes in st 00; dimensions and range of application
DIN 2450	Seamless steel tubes in st 35; dimensions and range of application
DIN 2458	Welded steel pipes and tubes; dimensions, conventional masses per unit length
DIN 2460	Steel water pipes
DIN 2462 Part 1	Seamless stainless steel tubes; dimensions; masses per unit length
DIN 2463 Part 1	Welded austenitic stainless steel; pipes and tubes; dimensions, conventional masses per unit length

INTERNATIONAL SPECIFICATION DESIGNATIONS & TITLES OF STEEL PIPING, TUBING, FITTINGS & TESTING (Continued)

Specification	Title
DIN 2470 Part 1	Steel gas pipelines for permissible working pressures up to 16 bar; pipes and fittings
DIN 2470 Part 2	Steel gas pipelines for permissible working pressures exceeding 16 bar; requirements for pipeline components
DIN 2519	Steel flanges; technical conditions of delivery
DIN 2528	Steel flanges; technical delivery conditions
DIN 2543	Cast steel flanges
DIN 2614	Cement mortar linings for ductile iron and steel pipes and fittings; application, requirements and testing
DIN 2615 Part 1	Steel butt-welding pipes fittings; tees with reduced pressure factor
DIN 2615 Part 2	Steel butt-welding pipe fittings; tees for use at full service pressure
DIN 2616 Part 1	Steel butt-welding pipe fittings; eccentric reducers with reduced pressure factor
DIN 2616 Part 2	Steel butt-welding pipe fittings; reducers for use at full service pressure
DIN 2617	Steel butt-welding pipe fittings; caps
DIN 2618	Butt welding steel fittings; welding saddles
DIN 2619	Butt welding steel fittings; bends for welding
DIN 2848 Part 12	Flanged lined steel fittings
DIN 2916	Bending radii for seamless and welded steel tubes; design sheet
DIN 2917	Seamless steel tubes for superheated steam pipelines and headers
DIN 2980	Screwed steel pipe fittings
DIN 2981	Threaded steel pipe fittings; fittings with long screw thread
DIN 8564 Part 1	Welding of pipelines; steel pipelines; manufacturing; testing of welds
DIN 17172	Steel pipes for long-distance pipelines for combustible liquids and gases; technical conditions of delivery
DIN 17173	Seamless circular tubes made from steels with low temperature toughness; technical delivery conditions
DIN 17175	Seamless tubes of heat-resistant steels; technical conditions of delivery
DIN 17177	Electric pressure-welded steel tubes for elevated temperatures; technical conditions of delivery
DIN 17178	Welded circular fine grain steel tubes subject to special requirements; technical delivery conditions
DIN 17179	Seamless circular fine grain steel tubes subject to special requirements; technical delivery conditions
DIN 17455	General purpose welded circular stainless steel tubes; technical delivery conditions
DIN 17456	General purpose seamless circular stainless steel tubes; technical delivery conditions

INTERNATIONAL SPECIFICATION DESIGNATIONS & TITLES OF STEEL PIPING, TUBING, FITTINGS & TESTING (Continued)

Specification	Title
DIN 17457	Welded circular austenitic stainless steel tubes subject to special requirements; technical delivery conditions
DIN 17458	Seamless circular austenitic stainless steel tubes subject to special requirements; technical delivery conditions
DIN 28038	Flanges for welding, with cylindrical hub, for use on stainless steel pressure vessels
JIS B 0151	Glossary of terms for iron and steel pipe fittings
JIS B 2220	Steel welding pipe flanges
JIS B 2302	Screwed type steel pipe fittings
JIS B 2311	Steel butt-welding pipe fittings for ordinary use
JIS B 2312	Steel butt-welding pipe fittings
JIS B 2313	Steel plate butt-welding pipe fittings
JIS B 2316	Steel socket-welding pipe fittings
JIS G 0582	Ultrasonic examination for steel pipes and tubes
JIS G 0583	Eddy current examination of steel pipes and tubes
JIS G 0584	Ultrasonic examination for arc welded steel pipes
JIS G 3132	Hot-rolled carbon steel strip for pipes and tubes
JIS G 3429	Seamless steel tubes for high pressure gas cylinder
JIS G 3439	Seamless steel oil well casing, tubing and drill pipe
JIS G 3441	Alloy steel tubes for machine purposes
JIS G 3442	Galvanized steel pipes for water service
JIS G 3443	Coated steel pipes for water service
JIS G 3444	Carbon steel tubes for general structural purposes
JIS G 3445	Carbon steel tubes for machine structural purposes
JIS G 3446	Stainless steel pipes for machine and structural purposes
JIS G 3447	Stainless steel sanitary pipes
JIS G 3448	Light gauge stainless steel pipes for ordinary piping
JIS G 3452	Carbon steel pipes for ordinary piping
JIS G 3454	Carbon steel pipes for pressure service
JIS G 3455	Carbon steel pipes for high pressure service

INTERNATIONAL SPECIFICATION DESIGNATIONS & TITLES OF STEEL PIPING, TUBING, FITTINGS & TESTING (Continued)

Specification	Title
JIS G 3456	Carbon steel pipes for high temperature service
JIS G 3457	Arc welded carbon steel pipes
JIS G 3458	Alloy steel pipes
JIS G 3459	Stainless steel pipes
JIS G 3460	Steel pipes for low temperature service
JIS G 6463	Stainless steel boiler & heat exchanger tubes
JIS G 3465	Seamless steel tubes for drilling
JIS G 3466	Carbon steel square pipes for general structural purposes
JIS G 3468	Large diameter welded stainless steel pipes
JIS G 3469	Polyethylene coated steel pipes
JIS G 3471	Corrugated steel pipes and sections
JIS G 3474	High tensile strength steel tubes for tower structural purposes
JIS G 3491	Asphalt protective coatings for steel water pipe
JIS G 3491	Asphalt protective coatings for steel water pipe
JIS G 3492	Coal-tar enamel protective coatings for steel water pipe
JIS G 3492	Coal-tar enamel protective coatings for steel water pipe
JIS G 5201	Centrifugally cast steel pipes for welded structure
JIS G 5202	Centrifugally cast steel pipes for high temperature and high pressure service
JIS Z 2315	Test methods for performance characteristics of eddy current flaw detecting system

INTERNATIONAL SPECIFICATION DESIGNATIONS & TITLES OF STAINLESS STEEL PRODUCTS

Specification	Title
ISO 683 Part 13	Heat-treatable steels, alloy steels and free- cutting steels - part 13: wrought stainless steels
ISO 683 Part 16	Heat-treated steels, alloy steels and free- cutting steels - part 16: precipitation hardening stainless steels
ISO 3651 Part 1	Austenitic stainless steels - determination of resistance to intergranular corrosion - part 1: corrosion test in nitric acid medium by measurement of loss in mass (huey test)
ISO 3651 Part 2	Austenitic stainless steels - determination of resistance to intergranular corrosion - part 2: corrosion test in a sulphuric acid/copper sulphate medium in the presence of copper turnings (monypenny strauss test)
ISO 5755 Part 3	Sintered metal materials - specifications - part 3: sintered alloyed and sintered stainless steels used for structural parts
ISO 5832 Part 1	Implants for surgery - metallic materials - part 1: wrought stainless steel
ISO 5832 Part 9	Implants for surgery - metallic materials - part 9: wrought high nitrogen stainless steel
ISO 6931 Part 1	Stainless steels for springs - part 1:wire
ISO 6931 Part 2	Stainless steels for springs - part 2:strip
ISO 7153 Part 1	Instruments for surgery - metallic materials - part 1: stainless steel
ISO 8442	Stainless steel and silver-plated table cutlery - requirements
ISO 9444	Hot-rolled stainless steel wide strip and sheet - tolerances on dimensions and form
ISO 9445	Cold-rolled stainless steel wide strip and sheet - tolerances on dimensions and form
ISO 9446	Hot-rolled stainless steel narrow strip - tolerances on dimensions and form
ISO 9447	Cold-rolled stainless steel narrow strip - tolerances on dimensions and form
CEN prEN 10222 Part 6	Steel forgings for pressure purposes - part 6 : austenitic, martensitic and austenitic-ferritic stainless steels
CEN prEN 10258	Cold-rolled stainless steel narrow strip - tolerances on dimensions and shape
CEN prEN 10259	Cold-rolled stainless steel wide strip and plate/sheet - tolerances on dimensions and shape
CEN prEN 28442 Part 2	Materials and articles in contact with foodstuffs - cutlery and table holloware - part 2: specification for stainless steel and silver-plated cutlery
BS 1247 Part 1	Manhole steps part 1: galvanized ferrous or stainless steel manhole steps
BS 1449 Part 2	Steel plate, sheet and strip part 2: specification for stainless and heat-resisting steel plate, sheet and strip
BS 1449 Part 2	Steel plate, sheet and strip part 2: stainless and heat resisting steel plate, sheet and strip
BS 1501 Part 3	Steels for pressure purposes: plates, sheet and strip part 3: specification for corrosion-and heat-resisting steels

INTERNATIONAL SPECIFICATION DESIGNATIONS & TITLES OF STAINLESS STEEL PRODUCTS (Continued)

Specification	Title
BS 1501 Part 3	Steels for pressure purposes: plates, sheet and strip part 3: specification for corrosion - and heat-resisting steels
BS 1501 Part 3	Steels for pressure purposes part 3: specification for corrosion - and heat-resisting steels:plates, sheet and strip
BS 1506	Carbon, low alloy and stainless steel bars and billets for bolting material to be used in pressure retaining applications
BS 1554	Stainless and heat-resisting steel round wire
BS 1554	Stainless and heat-resisting steel round wire
BS 1823	Stainless steel hollow-ware for use in hospital operating-theatres and wards
BS 2056	Stainless steel wire for mechanical springs
BS 3059 Part 2	Steel boiler and superheater tubes part 2: carbon, alloy and austenitic stainless steel tubes with specified elevated temperature properties
BS 3146 Part 2	Investment castings in metal part 2: corrosion and heat resisting steels, nickel and cobalt base alloys
BS 4880 Part 1	Urinals part 1: stainless steel slab urinals
BS 5194 Part 1	Surgical instruments part 1: specification for stainless steel (ISO 7153-1: 1991)
BS 5194 Part 1	Surgical instruments with bow handles (excluding scissors) part 1: specification for stainless steels
BS 5194 Part 2	Surgical instruments with bow handles (excluding scissors) part 2: instruments with pivot joints (excluding cutting instruments)
BS 5600 Sec 5.3	Powder metallurgical materials and products part 5: material specifications for sintered metal products, excluding hardmetals section 5.3: sintered alloyed and sintered stainless steels used for structural parts
BS 5770 Part 4	Steel strip intended for the manufacture of springs part 4: martensitic and austenitic stainless steel
BS 5903	Determination of resistance to intergranular corrosion of austenitic stainless steels: copper sulphate-sulphuric acid method (moneypenny strauss test)
BS 6744	Austenitic stainless steel bars for the reinforcement of concrete
BS 7252 Part 1	Metallic materials for surgical implants part 1: wrought stainless steel
BS 7252 Part 9	Metallic materials for surgical implants part 9: specification for high-nitrogen stainless steel
BS 7252 Part 9	Metallic materials for surgical implants part 9: high-nitrogen stainless steel
BS 7252 Part 9	Metallic materials for surgical implants part 9: specification for high-nitrogen stainless steel
BS 7547	Stainless steel needle tubing for the manufacture of medical devices (ISO 9626: 1991)
BS HC 10	Nickel-chromium- molybdenum steel castings (1150-1300 MPa)
BS 2HC 101	Precipitation hardening chromium-nickel-copper- molybdenum steel castings (950 MPa) (Cr 14.0, Ni 4.5, Cu 2.5, Mo 1.5)

INTERNATIONAL SPECIFICATION DESIGNATIONS & TITLES OF STAINLESS STEEL PRODUCTS (Continued)

Specification	Title
BS HC 103	23% chromium-nickel-tungsten corrosion- resisting steel castings (Cr 23, Ni 11, W 3)
BS HC 104	19% chromium-10% nickel niobium-stabilized corrosion-resisting steel castings (460 MPa)
BS HC 105	18% chromium- 11% nickel-2.5% molybdenum niobium-stabilized corrosion-resisting steel castings
BS S 150	Chromium-molybdenum-vanadium-niobium heat- resisting steel billets, bars, forgings and parts (930-1080 MPa) (Cr 10.5, Mo 0.6, V 0.2, Nb 0.3)
BS S 151	Chromium-nickel-molybdenum-vanadium heat-resisting steel billets, bars, forgings and parts (930-1130 MPa: Limiting Ruling Section 150 mm) (Cr 12, Ni 2.5, Mo 1.7, V 0.3)
BS S 152	Chromium-cobalt-molybdenum-vanadium- niobium heat-resisting steel billets, bars, forgings and parts (consumable electrode remelted) (1000-1140 MPa) (Cr10.5, Co 6, Mo 0.8, V 0.2, Nb 0.3)
BS S 530	24/17 chromium- nickel heat-resisting steel sheet and strip (titanium stabilized: 54 Hbar) (elevated temperature properties not verified)
BS S 538	Chromium-cobalt- molybdenum-vanadium heat-resisting steel sheet and strip (930- 1130 MPa: maximum thickness 6 mm) (Cr 12, Co 2.5, Mo 1.7, V 0.3)
DIN 3352 Part 9	Gate valves of heat-resistant steel
DIN 3352 Part 10	Gate valves of stainless steel
DIN 3356 Part 4	Globe valves; high temperature steel stop valves
DIN 3356 Part 5	Globe valves; stainless steel stop valves
DIN 4189 Part 1	Screening surfaces; woven wire cloth made of steel, stainless steel and non-ferrous metals; dimensions
DIN 5512 Part 3	Stainless steel flats for use in rail vehicle construction
DIN 17224	Stainless steel wire and strip for springs; technical delivery conditions
DIN 17440	Stainless steels; technical delivery conditions for plate and sheet, hot rolled strip, wire rod, drawn wire, steel bars, forgings and semi-finished products
DIN 17441	Stainless steels; technical delivery conditions for cold rolled strip and slit strip and for plate and sheet cuttherefrom
DIN 17442	Rolled, wrought or cast stainless steel products for medical instruments
DIN 17443	Rolled and wrought stainless steel products for surgical implants; technical delivery conditions
DIN 17445	Stainless steel castings; technical delivery conditions
DIN 17465	Heat resisting steel castings; technical delivery conditions
DIN 25512	Corrugated sheet sections for rail vehicles; dimensions; weights; static values

INTERNATIONAL SPECIFICATION DESIGNATIONS & TITLES OF STAINLESS STEEL PRODUCTS (Continued)

Specification	Title
DIN 28025 Part 1	Nozzles of stainless steel; PN 10 and PN 16
DIN 28025 Part 2	Nozzles of stainless steel; PN 25 and PN 40
DIN 28124 Part 1	Manhole closures for unalloyed and stainless steel unpressurized vessels
DIN 28124 Part 3	Manhole closures for stainless steel pressure vessels
DIN 28136 Part 2	Unalloyed and stainless steel mixing vessels; arrangement and size of cover nozzles and manholes
DIN 28136 Part 4	Mixing vessels for use in process engineering; wall thickness of unalloyed and stainless steel vessels
DIN 28141	Connecting flanges for unalloyed and stainless steel mixing vessels for use in process engineering
DIN 50914	Testing the resistance of stainless steels to intercrystalline corrosion; copper sulfate/sulfuric acid method; straub test
DIN 50921	Corrosion of metals; testing of austenitic stainless steels for resistance to local corrosion in highly oxidizing acids; corrosion test in nitric acid medium by measurement of loss in mass (huey test)
DIN 59381	Flat products of steel; cold rolled strip of stainless steel and of heat resisting steels; dimensions, permissible dimensional deviations, deviations of form and weight
DIN 59382	Flat steel products: cold rolled wide strip and sheet of stainless steels: dimensions, permissible variations on dimensions and form
JIS E 4049	Welded joints of stainless steel for railway rolling stock - design methods
JIS G 0571	Method of 10 per cent oxalic acid etch test for stainless steels
JIS G 0572	Method of ferric sulfate-sulfuric acid test for stainless steels
JIS G 0573	Method of 65 per cent nitric acid test for stainless steels
JIS G 0574	Method of nitric- hydrofluoric acid test for stainless steels
JIS G 0575	Method of copper sulfate- sulfuric acid test for stainless steels
JIS G 0576	Method of 42 percent magnesium chloride test for stainless steels
JIS G 0577	Method of pitting potential measurement for stainless steels
JIS G 0578	Method of ferric chloride test for stainless steels
JIS G 0579	Method for making anodic polarization measurement for stainless steels
JIS G 0580	Method of electrochemical potentiokinetic reactivation ratio measurement for stainless steels
JIS G 0591	Method of 5 per cent sulfuric acid test for stainless steels
JIS G 3320	Coated stainless steel sheets
JIS G 3554	Hexagonal wire netting

INTERNATIONAL SPECIFICATION DESIGNATIONS & TITLES OF STAINLESS STEEL PRODUCTS (Continued)

Specification	Title
JIS G 3601	Stainless-clad steels
JIS G 4303	Stainless steel bars
JIS G 4304	Hot rolled stainless steel plates, sheets and strip
JIS G 4305	Cold rolled stainless steel plates, sheets and strip
JIS G 4308	Stainless steel wire rods
JIS G 4309	Stainless steel wires
JIS G 4310	Method of mass calculation for stainless steel plates and sheets, and heat-resisting steel plates and sheets
JIS G 4311	Heat-resisting steel bars
JIS G 4312	Heat-resisting steel plates and sheets
JIS G 4313	Cold rolled stainless steel strip for springs
JIS G 4314	Stainless steel wires for springs
JIS G 4315	Stainless steel wires for cold heading and cold forging
JIS G 4317	Hot rolled stainless steel equal leg angles
JIS G 4318	Cold finished stainless steel bars
JIS G 4319	Stainless steel blooms and billets for forgings
JIS G 4320	Cold formed stainless steel equal leg angles
JIS G 5121	Stainless steel castings
JIS G 5122	Heat resisting steel castings
JIS S 2053	Stainless steel vacuum bottles
JIS T 6103	Stainless steel wires for dental use
JIS T 6103	Stainless steel wires for dental use

Chapter

24

EUROPEAN STANDARDS:

CHEMICAL COMPOSITIONS & MECHANICAL PROPERTIES

EN 10025 - CHEMICAL COMPOSITION OF HOT ROLLED PRODUCTS OF NON-ALLOY STRUCTURAL STEELS								
Name	Number	Size	C	Si	Mn	P	S	Others^a, b
S185	1.0035	---	---	---	---	---	---	---
S235JR	1.0037	≤ 16 mm	0.17	---	1.40	0.045	0.045	N 0.009
		> 16 ≤ 25 mm	0.20	---	1.40	0.045	0.045	N 0.009
S235JRG1	1.0036	≤ 16 mm	0.17	---	1.40	0.045	0.045	N 0.007
		> 16 ≤ 25 mm	0.20	---	1.40	0.045	0.045	N 0.007
S235JRG2	1.0038	≤ 40 mm	0.17	---	1.40	0.045	0.045	N 0.009
		> 40 mm	0.20	---	1.40	0.045	0.045	N 0.009
S235J0	1.0114	---	0.17	---	1.40	0.040	0.040	N 0.009
S235J2G3	1.0116	---	0.17	---	1.40	0.035	0.035	---
S235J2G4	1.0117	---	0.17	---	1.40	0.035	0.035	---
S275JR	1.0044	≤ 40 mm	0.21	---	1.50	0.045	0.045	N 0.009
		> 40 mm	0.22	---	1.50	0.045	0.045	N 0.009
S275J0	1. 0143	≤ 150 mm	0.18	---	1.50	0.040	0.040	N 0.009
		> 150 mm	0.20	---	1.50	0.040	0.040	N 0.009
S275J2G3	1.0144	≤ 150 mm	0.18	---	1.50	0.035	0.035	---
		> 150 mm	0.20	---	1.50	0.035	0.035	---
S275J2G4	1.0145	≤ 150 mm	0.18	---	1.50	0.035	0.035	---
		> 150 mm	0.20	---	1.50	0.035	0.035	---
E295	1.0050	---	---	---	---	0.045	0.045	N 0.009

a. For each increase of 0.001% N (max. 0.012%) the P-content will be reduced by 0.005%.

b. The value for N does not apply if a minimum Al-content of 0.020% or sufficient other N binding elements are present.

Single values are maximums, unless otherwise specified.

EN 10025 - MECHANICAL PROPERTIES OF HOT ROLLED PRODUCTS OF NON-ALLOY STRUCTURAL STEELS									
		Tensile Strength, N/mm²		Upper Yield Stress, N/mm²			% Elongation[a]		Impact Value[b], J
Name	Number	< 3 mm	≥ 3 mm to ≤ 100 mm	≤ 16 mm	> 16 mm to ≤ 40 mm	> 40 mm to ≤ 63 mm	≥ 3 mm to ≤ 40 mm	> 40 mm to ≤ 63 mm	> 10 mm to ≤ 150 mm
S185[c]	1.0035	310-540	290-510	185	175	---	18	---	---
S235JR[c]	1.0037	---	---	235	235	---	26	25	27 @ 20
S235JRG1[c]	1.0036	360-510	340-470	235	225	---	26	25	27 @ 20
S235JRG2	1.0038	360-510	340-470	235	225	215	26	25	27 @ 20
S235J0	1.0114	360-510	340-470	235	225	215	26	25	27 @ 0
S235 J2G3	1.0116	360-510	340-470	235	225	215	26	25	27 @ -20
S235J2G4	1.0117	360-510	340-470	235	225	215	26	25	27 @ -20
S275JR	1.0044	430-580	410-560	275	265	255	22	21	27 @ 20
S275J0	1.0143	430-580	410-560	275	265	255	22	21	27 @ 0
S275 J2G3	1.0144	430-580	410-560	275	265	255	22	21	27 @ -20
S275 J2G4	1.0145	430-580	410-560	275	265	255	22	21	27 @ -20
S355JR	1.0045	510-680	490-630	355	345	335	22	21	27 @ 20
S355J0	1.0553	510-680	490-630	355	345	335	22	21	27 @ 0
S355J2G3	1.0570	510-680	490-630	355	345	335	22	21	27 @ -20
S355J2G4	1.0577	510-680	490-630	355	345	335	22	21	27 @ -20
S355K2G3	1.0595	510-680	490-630	355	345	335	22	21	40 @ -20
S355K2G4	1.0596	510-680	490-630	355	345	335	22	21	40 @ -20
E295	1.0050	490-660	470-610	295	285	275	20	19	---
E335	1.0060	590-770	570-710	335	325	315	16	15	---
E360	1.0070	690-900	670-830	360	355	345	11	10	---

a. Elongation values based on longitudinal test specimen.
b. Charpy V notched specimens, average from 3 specimens.
c. For thicknesses up to 25 mm.
Single values are minimums, unless otherwise specified.

EN 10028-2 - CHEMICAL COMPOSITION OF FLAT PRODUCTS MADE FROM STEEL FOR PRESSURE PURPOSES PART 2 - NON-ALLOY AND ALLOY STEELS WITH SPECIFIED ELEVATED TEMPERATURE PROPERTIES

Name	Number	C	Si	Mn	P	S	Cr	Mo	Ni	Others
P235GH	1.0345	0.16	0.35	0.40-1.20	0.030	0.025	0.30	0.08	0.30	Al 0.20[a], Cu 0.03, Nb 010, Ti 0.03, V 0.02[b]
P265GH	1.0425	0.20	0.40	0.50-1.40	0.030	0.025	0.30	0.08	0.30	Al 0.20[a], Cu 0.03, Nb 010, Ti 0.03, V 0.02[b]
P295GH	1.0481	0.08-0.20	0.40	0.90-1.50	0.030	0.025	0.30	0.08	0.30	Al 0.20[a], Cu 0.03, Nb 010, Ti 0.03, V 0.02[b]
P355GH	1.0473	0.10-0.22	0.60	1.00-1.70	0.030	0.025	0.30	0.08	0.30	Al 0.20[a], Cu 0.03, Nb 010, Ti 0.03, V 0.02[b]
16Mo3	1.5415	0.12-0.20	0.35	0.40-0.90	0.030	0.025	0.30	0.25-0.35	0.30	Cu 0.30
13CrMo4-5	1.7335	0.08-0.18	0.35	0.40-1.00	0.030	0.025	0.70-1.15	0.40-0.60	---	Cu 0.30
10CrMo9-10	1.7380	0.08-0.14	0.50	0.40-0.80	0.030	0.025	2.00-2.50	0.90-1.10	---	Cu 0.30
11CrMo9-10	1.7383	0.08-0.15	0.50	0.40-0.80	0.030	0.025	2.00-2.50	0.90-1.10	---	Cu 0.30

a. Total aluminum, minimum.   b. Cr + Cu + Mo + Ni $\leq$ 0.70.   Single values are maximums, unless otherwise specified.

EN 10028-2 - GUIDELINES TEMPERATURES FOR HEAT TREATMENT OF FLAT PRODUCTS MADE FROM STEEL FOR PRESSURE PURPOSES PART 2 - NON-ALLOY AND ALLOY STEELS WITH SPECIFIED ELEVATED TEMPERATURE PROPERTIES

Name	Number	Normalizing[a] °C	Quenching		Tempering[b] °C
			Austenitizing °C		
P235GH	1.0345	890 to 950	---		---
P265GH	1.0425	890 to 950	---		---
P295GH	1.0481	890 to 950	---		---
P355GH	1.0473	890 to 950	---		---
16Mo3	1.5415	890 to 950	---		c
13CrMo4-5	1.7335	---	890 to 950		630 to 730
10CrMo9-10	1.7380	---	920 to 980		680 to 760
11CrMo9-10	1.7383	920 to 980			670 to 750

a. When normalizing, after the specified temperatures have been reached over the whole cross section, no further holding is necessary and should usually be avoided.   b. When tempering, after the specified temperatures have been reached over the whole cross section, they shall be maintained for at least 30 minutes.
c. In certain cases, tempering at 590 to 650°C may be necessary.   This table contains reference data for information only, not mandatory data of EN 10028-2.

EN 10028-2 - MECHANICAL PROPERTIES OF FLAT PRODUCTS MADE FROM STEEL FOR PRESSURE PURPOSES
PART 2 - NON-ALLOY AND ALLOY STEELS WITH SPECIFIED ELEVATED TEMPERATURE PROPERTIES

Name	Number	Usual Delivery Condition[a]	Product Thickness, mm		Yield Strength[b] R_{eH} N/mm²	Tensile Strength, R_m N/mm²	% El	Impact Energy[m], KV J@°C
			Over[d]	Up to				
P235GH	1.0345	N[c]	---	16	235	360 to 480	25[e]	27@0
			16	40	225	360 to 480	25[e]	27@0
			40	60	215	360 to 480	25[e]	27@0
			60	100	200	360 to 480	24	27@0
			100	150	185	350 to 480	24	27@0
P265GH	1.0425	N[c]	---	16	265	410 to 530	23[f]	27@0
			16	40	255	410 to 530	23[f]	27@0
			40	60	245	410 to 530	23[f]	27@0
			60	100	215	410 to 530	22	27@0
			100	150	200	400 to 530	22	27@0
P295GH	1.0481	N[c]	---	16	295	460 to 580	22	27@0
			16	40	290	460 to 580	22	27@0
			40	60	285	460 to 580	22	27@0
			60	100	260	460 to 580	21	27@0
			100	150	235	440 to 570	21	27@0
P355GH	1.0473	N[c]	---	16	355	510 to 650	21	27@0
			16	40	345	510 to 650	21	27@0
			40	60	335	510 to 650	21	27@0
			60	100	315	490 to 630	20	27@0
			100	150	295	480 to 630	20	27@0

EN 10028-2 - MECHANICAL PROPERTIES OF FLAT PRODUCTS MADE FROM STEEL FOR PRESSURE PURPOSES PART 2 - NON-ALLOY AND ALLOY STEELS WITH SPECIFIED ELEVATED TEMPERATURE PROPERTIES (Continued)

Name	Number	Usual Delivery Condition[a]	Product Thickness, mm		Yield Strength[b] R_{eH} N/mm^2	Tensile Strength, R_m N/mm^2	% El	Impact Energy[m], KV J@°C
			Over[d]	Up to				
16Mo3	1.5415	N[g]	---	16	275	440 to 590	24	31@20[h]
			16	40	270	440 to 590	24	31@20[h]
			40	60	260	440 to 590	23	31@20[h]
			60	100	240	430 to 580	22	27@20[h]
			100	150	220	420 to 570	19	27@20[h]
13CrMo4-5	1.7335	N+T	---	16	300	450 to 600	20	31@20[l]
			16	60	295	450 to 600	20	31@20[l]
		N+T or QA or QL	60	100	275	440 to 590	19	27@20[h]
		QL	100	150	255	430 to 580	19	27@20[h]
10CrMo9-10	1.7380	N+T	---	16	310	480 to 630	18	31@20
			16	40	300	480 to 630	18	31@20
			40	60	290	480 to 630	18	31@20
		N+T or QA or QL	60	100	270	470 to 620	17	27@20
		QL	100	150	250	460 to 610	17	27@20
11CrMo9-10	1.7383	N + T or QA or QL	---	60	310	520 to 670	18	31@20[l]
		QL	60	100	310	520 to 670	17	27@20[h]

a. N - normalized, QA - quenched in air, QL - quenched in liquid, T - tempered.

b. Until the yield point criteria are harmonized in the various national codes, determination of R_{eH} may be replaced by determination of $R_{p0.2}$.

In this case, 10 N/mm^2 lower minimum values apply for $R_{p0.2}$.

c. Normalizing rolling may be carried out instead of normalizing for steel grades P235GH, P265GH, P295GH and P355GH. This means that the requirements are still to be met even after subsequent normalizing.

d. Product thickness over 150 mm is subject to agreement.

EN 10028-2 - MECHANICAL PROPERTIES OF FLAT PRODUCTS MADE FROM STEEL FOR PRESSURE PURPOSES PART 2 - NON-ALLOY AND ALLOY STEELS WITH SPECIFIED ELEVATED TEMPERATURE PROPERTIES (Continued)

---

e. Where the elongation at fracture has been determined on tensile test pieces having an original gauge length, L_0, equal to 80 mm and a width of 20 mm, a minimum value of 19% shall apply for product thicknesses over 2 mm up to 2.5 mm, or 20% for product thicknesses over 2.5 mm to less than 3 mm.

f. Where the elongation at fracture has been determined on tensile test pieces having an original gauge length, L_0, equal to 80 mm and a width of 20 mm, a minimum value of 17% shall apply for product thicknesses over 2 mm up to 2.5 mm, or 20% for product thicknesses over 2.5 mm to less than 3 mm.

g. This steel may also be supplied in the N + T condition, at the manufacturer's discretion.

h. If a test at 0°C has been agreed, a minimum value of 24 J shall apply.

l. If a test at 0°C has been agreed, a minimum value of 27 J shall apply.

m. Mean value from three test specimens.

All values are minimums, unless otherwise specified.

EN 10028-3 - CHEMICAL COMPOSITION OF FLAT PRODUCTS MADE FROM STEEL FOR PRESSURE PURPOSES
PART 3 - WELDABLE FINE GRAIN STEELS, NORMALIZED

Name	Number	C	Si	Mn	P	S	Cr	Mo	Ni	Others[d]
P275N	1.0486	0.18	0.40	0.50-1.40	0.030	0.025	0.30[a]	0.08	0.50	Al 0.020 min, Cu 0.30, Nb 0.05, N 0.020, Ti 0.03, V 0.05, a
P275NH	1.0487	0.18	0.40	0.50-1.40	0.030	0.025	0.30[a]	0.08	0.50	Al 0.020 min, Cu 0.30, Nb 0.05, N 0.020, Ti 0.03, V 0.05, a
P275NL1	1.0488	0.16	0.40	0.50-1.50	0.030	0.020	0.30[a]	0.08	0.50	Al 0.020 min, Cu 0.30, Nb 0.05, N 0.020, Ti 0.03, V 0.05, a
P275NL2	1.1104	0.16	0.40	0.50-1.50	0.025	0.015	0.30[a]	0.08	0.50	Al 0.020 min, Cu 0.30, Nb 0.05, N 0.020, Ti 0.03, V 0.05, a
P355N	1.0562	0.20	0.50	0.90-1.70	0.030	0.025	0.30[a]	0.08	0.50	Al 0.020 min, Cu 0.30, Nb 0.05, N 0.020, Ti 0.03, V 0.10, b
P355NH	1.0565	0.20	0.50	0.90-1.70	0.030	0.025	0.30[a]	0.08	0.50	Al 0.020 min, Cu 0.30, Nb 0.05, N 0.020, Ti 0.03, V 0.10, b
P355NL1	1.0566	0.18	0.50	0.90-1.70	0.030	0.020	0.30[a]	0.08	0.50	Al 0.020 min, Cu 0.30, Nb 0.05, N 0.020, Ti 0.03, V 0.10, b
P355NL2	1.1106	0.18	0.50	0.90-1.70	0.025	0.015	0.30[a]	0.08	0.50	Al 0.020 min, Cu 0.30, Nb 0.05, N 0.020, Ti 0.03, V 0.10, b
P460N	1.8905	0.20	0.60	1.00-1.70	0.030	0.025	0.30	0.10	0.80	Al 0.020 min, Cu 0.70[f], N 0.025, Nb 0.05, Ti 0.03, V 0.20, c
P460NH	1.8935	0.20	0.60	1.00-1.70	0.030	0.025	0.30	0.10	0.80	Al 0.020 min, Cu 0.70[f], N 0.025, Nb 0.05, Ti 0.03, V 0.20, c
P460NL1	1.8915	0.20	0.60	1.00-1.70	0.030	0.020	0.30	0.10	0.80	Al 0.020 min, Cu 0.70[f], N 0.025, Nb 0.05, Ti 0.03, V 0.20, c
P460NL2	1.8918	0.20	0.60	1.00-1.70	0.025	0.015	0.30	0.10	0.80	Al 0.020 min, Cu 0.70[f], N 0.025, Nb 0.05, Ti 0.03, V 0.20, c

a. Nb+Ti+V = 0.05
b. Nb+Ti+V = 0.12
c. Nb+T+V = 0.22
d. Where nitrogen is additionally fixed by niobium, titanium or vanadium, the minimum aluminum content specified does not apply.
e. Cr+Cu+Mo shall not exceed 0.45% by mass.
f. If the content of copper exceeds 0.30% by mass, that of nickel shall be at least half that value, by mass.
Single values are maximums, unless otherwise specified.

EN 10028-3 - MECHANICAL PROPERTIES OF FLAT PRODUCTS MADE FROM STEEL FOR PRESSURE PURPOSES
PART 3 - WELDABLE FINE GRAIN STEELS, NORMALIZED

Name	Number	Usual Condition[b]	Yield Strength[a], R_{eH} N/mm²				Tensile Strength, R_m N/mm², ≤ 70 mm	% Elongation ≤ 70 mm
			≤ 16 mm	> 16 to ≤ 35 mm	> 35 to ≤ 50 mm	> 50 to ≤ 70 mm		
P275N	1.0486	Normalized	275	275	265	255	390 to 510	24
P275NH	1.0487							
P275L1	1.0488							
P275L2	1.1104							
P355N	1.0562	Normalized	355	355	345	325	490 to 630	22
P355NH	1.0565							
P355NL1	1.0566							
P355NL2	1.1106							
P460N	1.8905	Normalized	460	460	440	420	570 to 720[c]	17
P460NH	1.8935							
P460NL1	1.8915							
P460NL2	1.8918							

a. Until the yield point criteria are harmonized in the various national codes, determination of R_{eH} may be replaced by determination of $R_{p0.2}$. In this case, the $R_{p0.2}$ values are 10 N/mm² for R_{eH} values up to 355 N/mm² and 15 N/mm² lower for R_{eH} values greater than 355 N/mm².

b. Normalizing rolling may be carried out instead of normalizing. This means that the requirements are still to be met even after subsequent normalizing.

c. For thicknesses up to 16 mm, a maximum value of 730 N/mm² shall be permitted.

All values are minimums, unless otherwise specified.

EN 10028-3 - IMPACT PROPERTIES OF FLAT PRODUCTS MADE FROM STEEL FOR PRESSURE PURPOSES
PART 3 - WELDABLE FINE GRAIN STEELS, NORMALIZED

Name		Product Thickness	Impact Energy Values[a], KV, in J at Test Temperatures in °C											
			Longitudinal Test Pieces						Transverse Test Pieces					
Series	Heat Treatment[b]	mm[c]	-50	-40	-20	0	+20	-50	-40	-20	0	+20
P ... N	Normalized	5 to 150	---	---	40	47	55	---	---	20	27	31
P ... NH												
P ... NL1	Normalized	5 to 150	27	34	47	55	63	16	20	27	34	40
P ... NL2	Normalized	5 to 150	30	40	65	90	100	27	30	40	60	70

a. Charpy V-notch requirements are for minimum average value of a set of three specimens, however, one value may be below the specified average, provided that it is not less than 70% of that value.
b. Normalizing rolling may be carried out instead of normalizing. This means that the requirements are still to be met even after subsequent normalizing.
c. For product thicknesses up to 10 mm, see EN 10028-1 paragraph 9.5.3.
All values are minimums, unless otherwise specified.

EN 10028-4 - MECHANICAL PROPERTIES OF FLAT PRODUCTS MADE FROM STEEL FOR PRESSURE PURPOSES
PART 4 - NICKEL ALLOY STEELS WITH SPECIFIED LOW TEMPERATURE PROPERTIES

Name	Number	C	Si	Mn	P	S	Mo	Ni	Others
11MnNi5-3	1.6212	0.14	0.50	0.70-1.50	0.025	0.015	---	0.30[a]-0.80	Al 0.020 min, Nb 0.05, V 0.05
13MnNi6-3	1.6217	0.16	0.50	0.85-1.70	0.025	0.015	---	0.30[a]-0.85	Al 0.020 min, Nb 0.05, V 0.05
15NiMn6	1.6228	0.18	0.35	0.80-1.50	0.025	0.015	---	1.30-1.70	V 0.05
12Ni14	1.5637	0.15	0.35	0.30-0.80	0.020	0.010	---	3.25-3.75	V 0.05
12Ni19	1.5680	0.15	0.35	0.30-0.80	0.020	0.010	---	4.75-5.25	V 0.05
X8Ni9	1.5662	0.10	0.35	0.30-01.80	0.020	0.010	0.10	8.50-10.0	V 0.05
X7Ni9	1.5663	0.10	0.35	0.30-01.80	0.015	0.005	0.10	8.50-10.0	V 0.01

a. For thicknesses ≤ 25 mm, a minimum nickel content of 0.15% is permitted.
Single values are maximums, unless otherwise specified.

EN 10028-4 - MECHANICAL PROPERTIES OF FLAT PRODUCTS MADE FROM STEEL FOR PRESSURE PURPOSES PART 4 - NICKEL ALLOY STEELS WITH SPECIFIED LOW TEMPERATURE PROPERTIES

Name	Number	Usual Condition[a, b]	Product Thickness, mm Over	Product Thickness, mm Up to	Yield Point[c] R_{eH} - N/mm²	Tensile Strength R_m - N/mm²	%Elongation
11 MnNi5-3	1.6212	N (+ T)[d]	---	30	285	420 to 530	24
			30	50	275	420 to 530	24
13 MnNi6-3	1.6217	N (+ T)[d]	---	30	355	490 to 610	22
			30	50	345	490 to 610	22
15 NiMn6	1.6228	N or N + T	---	30	355	490 to 640	22
		or Q + T	30	50	345	490 to 640	22
12 Ni 4	1.5637	N or N + T	---	30	355	490 to 640	22
		or Q + T	30	50	345	490 to 640	22
12 Ni19	1.5680	N or N + T	---	30	390	530 to 710	20
		or Q + T	30	50	380	530 to 710	20
X8Ni9	1.5662	HT 640	N + N + T	---	30	490	640 to 840	18
			or Q + T	30	50	480	640 to 840	18
X8Ni9	1.5662	HT 680	Q + T[e]	---	30	585	680 to 820	18
			30	50	575	680 to 820	18
X7Ni9	1.5663	Q + T[e]	---	30	585	680 to 820	18
			30	50	575	680 to 820	18

a. N - normalized, QL - liquid quenched, T - tempered, HT 640/680 - heat treatment variant with minimum tensile strength of 640 or 680 N/mm².

b. For temperature and cooling conditions, EN 10028-3 see Table 1.

c. Until the yield point criteria are harmonized in the various national codes, determination of R_{eH} may be replaced by determination of $R_{p0.2}$.

d. Normalizing rolling may be carried out instead of normalizing for grades 11 MnNi 5-3 and 13 MnNi 6-3. This means that the requirements are still to be met even after subsequent normalizing.

e. For thicknesses < 15 mm, delivery conditions N + N + T are also applicable.

All values are minimums, unless otherwise specified.

EN 10028-4 - IMPACT PROPERTIES OF FLAT PRODUCTS MADE FROM STEEL FOR PRESSURE PURPOSES
PART 4 - NICKEL ALLOY STEELS WITH SPECIFIED LOW TEMPERATURE PROPERTIES

Name	Heat Treatment[a,b]	Test Direction	Impact Energy[c] KV in J at Test Temperatures in °C											
			20	0	-20	-40	-50	-60	-80	-100	-120	-150	-170	-196
11MnNi5-3	N (+T)[d]	longitudinal	70	60	55	50	45	40	-	-	-	-	-	-
13MnNi6-3	N (+T)[d]	transverse	45	40	40	35	30	27	-	-	-	-	-	-
15NiMn6	N or N + T or Q + T	longitudinal	65	65	65	60	50	50	40	40	-	-	-	-
		transverse	45	45	45	40	35	35	27	27	-	-	-	-
12Ni14	N or N + T or Q + T	longitudinal	65	60	55	55	50	50	45	40	-	-	-	-
		transverse	45	40	40	35	35	35	30	27	-	-	-	-
12Ni19	N or N + T or Q + T	longitudinal	70	70	70	65	65	65	60	50	40[e]	-	-	-
		transverse	50	50	50	45	45	45	40	30	27[e]	-	-	-
X8 Ni9	HT 640	longitudinal	70	70	70	70	70	70	70	60	50	50	45	40
		transverse	50	50	50	50	50	50	50	40	35	35	30	27
X8 Ni9	HT 680	longitudinal	120	120	120	120	120	120	120	110	100	90	80	70
		transverse	100	100	100	100	100	100	100	90	80	70	60	50
X7 Ni9	Q + T	longitudinal	120	120	120	120	120	120	120	120	120	120	110	100
		transverse	100	100	100	100	100	100	100	100	100	100	90	80

a. N - normalized, QL - liquid quenched, T - tempered.
b. For temperatures and cooling conditions, see EN 10028-4 Table A.1.
c. Charpy V-notch requirements are for minimum average value of a set of three specimens, however, one value may be below the specified average, provided that it is not less than 70% of that value.
d. Normalizing rolling may be carried out instead of normalizing for grades 11 MnNi 5-3 and 13 MnNi 6-3. This means that the requirements are still to be met even after subsequent normalizing.
e. The values are applicable for thicknesses ≤ 25 mm at -110°C and for thicknesses of 25 mm to 30 mm at -115°C.

EN 10028-4 - GUIDELINES FOR HEAT TREATMENT OF FLAT PRODUCTS MADE FROM STEEL FOR PRESSURE PURPOSES
PART 4 - NICKEL ALLOY STEELS WITH SPECIFIED LOW TEMPERATURE PROPERTIES

Name	Heat Treatment Symbol[a]	Austenitizing °C	Cooling[b]	Tempering °C	Cooling[b]
11MnNi5-3	N (+T)	880 to 940	air	580 to 640	a
13MnNi6-3	N (+T)	880 to 940	air	580 to 640	a
15NiMn6	N	850 to 900	air	---	---
	N + T	850 to 900	air	600 to 660	a or w
	Q + T	850 to 900	water or oil	600 to 660	a or w
12Ni14	N	830 to 880	air	---	---
	N + T	830 to 880	air	580 to 640	a or w
	Q + T	820 to 870	water or oil	580 to 640	a or w
12Ni19	N	800 to 850	air	---	---
	N + T	800 to 850	air	580 to 660	a or w
	Q + T	800 to 850	water or oil	580 to 660	a or w
X8Ni9,	N + N + T	880 to 930	air	540 to 600	a or w
X7Ni9	N + N + T	770 to 830	air	540 to 600	a or w
	Q + T	770 to 830	water or oil	540 to 600	a or w

a. N - normalized, Q - liquid quenched, T - tempered.
b. a - air, o - oil, w - water.
Reference data for information only, not mandatory data in EN 10028-4.

EN 10028-5 - CHEMICAL COMPOSITION OF FLAT PRODUCTS MADE FROM STEEL FOR PRESSURE PURPOSES

PART 5 - WELDABLE FINE GRAIN STEELS THERMOMECHANICALLY ROLLED[a]

Name	Number	C	Si	Mn	P	S	Al[b] min	N	Mo	Nb[c]	Ni	Ti[c]	V[c]	Other
P355M	1.8821	0.14	0.50	1.60	0.025	0.020	0.020	0.015	0.20	0.20	0.50	0.05	0.10	Cr + Cu + Mo ≤ 0.60%
P355ML	1.8833	0.14	0.50	1.60	0.020	0.015	0.020	0.015	0.20	0.20	0.50	0.05	0.10	Cr + Cu + Mo ≤ 0.60%
P420M	1.8824	0.16	0.50	1.70	0.025	0.020	0.020	0.020	0.20	0.20	0.50	0.05	0.10	Cr + Cu + Mo ≤ 0.60%
P420M	1.8835	0.16	0.50	1.70	0.020	0.015	0.020	0.020	0.20	0.20	0.50	0.05	0.10	Cr + Cu + Mo ≤ 0.60%
P460ML	1.8826	0.16	0.60	1.70	0.025	0.020	0.020	0.020	0.20	0.20	0.50	0.05	0.10	Cr + Cu + Mo ≤ 0.60%
P460ML	1.8837	0.16	0.60	1.70	0.020	0.015	0.020	0.020	0.20	0.20	0.50	0.05	0.10	Cr + Cu + Mo ≤ 0.60%
P550M	1.8830	0.16	0.60	1.80	0.025	0.020	0.020	0.020	0.30	0.06	0.80	0.05	0.12	Cr + Cu + Mo ≤ 0.60%

a. Elements not listed in this table shall not be intentionally added to the steel without the agreement of the purchaser except for finishing the cast. All appropriate measures shall be taken to prevent the addition from scrap and other materials used in steelmaking of these elements which may adversely affect the mechanical properties and usability.

b. The minimum value for Al$_{total}$ does not apply if adequate contents of nitrogen-fixing elements are present.

c. The total of V + Nb + Ti shall not exceed a value of 0.15% (P 550 M: 0.18%). Single values are maximums, unless otherwise specified.

EN 10028-5 - MECHANICAL PROPERTIES OF FLAT PRODUCTS MADE FROM STEEL FOR PRESSURE PURPOSES

PART 5 - WELDABLE FINE GRAIN STEELS THERMOMECHANICALLY ROLLED

Name	Number	Yield Strength, R_{eH} N/mm²			Tensile Strength, R_m N/mm²	% Elongation
		≤ 16 mm	> 16 to ≤ 40 mm	> 40 to ≤ 70 mm		
P355M	1.8821	355	345	335	450 to 610	22
P355ML	1.8833	355	345	335	450 to 610	22
P420	1.8824	420	400	390	500 to 660	19
P420ML	1.8835	420	400	390	500 to 660	19
P460M	1.8826	460	440	430	530 to 720	17
P460ML	1.8837	460	440	430	530 to 720	17
P550M	1.8830	550	530[a]	520	600 to 790	16

a. Only for thicknesses ≤ 25 mm.

EN 10028-5 - IMPACT PROPERTIES OF FLAT PRODUCTS MADE FROM STEEL FOR PRESSURE PURPOSES
PART 5 - WELDABLE FINE GRAIN STEELS THERMOMECHNICALLY ROLLED

Steel Name Series	Impact Energy Values[a], KV, in J at Test Temperatures in °C									
	Longitudinal Test Pieces					Transverse Test Pieces				
	-50	-40	-20	0	+20	-50	-40	-20	0	+20
P ...M	---	---	40	47	55	---	---	27	27	40
P ...ML	30	40	65	90	100	27	30	40	60	70

a. For undersize test pieces, see EN 10028-1 paragraph 8.4.3.   Single values are minimums, unless otherwise specified.

EN 10028-6 - CHEMICAL COMPOSITION OF FLAT PRODUCTS MADE FROM STEEL FOR PRESSURE PURPOSES
PART 6 - WELDABLE FINE GRAIN STEELS QUENCHED AND TEMPERED[a, b]

Name	Number	C	Si	Mn	P	S	Cr	Mo	Ni	Nb[c]	Ti[c]	V[c]	Zr[c]
P460Q	1.8870	0.18	0.80	1.70	0.025	0.015	1.50	0.70	2.00	0.06	0.05	0.12	0.15
P460QH	1.8871	0.18	0.80	1.70	0.025	0.015	1.50	0.70	2.00	0.06	0.05	0.12	0.15
P460QL	1.8872	0.18	0.80	1.70	0.020	0.020	1.50	0.70	2.00	0.06	0.05	0.12	0.15
P500Q	1.8873	0.18	0.80	1.70	0.025	0.015	1.50	0.70	2.00	0.06	0.05	0.12	0.15
P500QH	1.8874	0.18	0.80	1.70	0.025	0.015	1.50	0.70	2.00	0.06	0.05	0.12	0.15
P500QL	1.8875	0.18	0.80	1.70	0.020	0.020	1.50	0.70	2.00	0.06	0.05	0.12	0.15
P550Q	1.8876	0.18	0.80	1.70	0.025	0.015	1.50	0.70	2.00	0.06	0.05	0.12	0.15
P550QH	1.8877	0.18	0.80	1.70	0.025	0.015	1.50	0.70	2.00	0.06	0.05	0.12	0.15
P550QL	1.8878	0.18	0.80	1.70	0.020	0.020	1.50	0.70	2.00	0.06	0.05	0.12	0.15
P690Q	1.8879	0.20	0.80	1.70	0.025	0.015	1.50	0.70	2.00	0.06	0.05	0.12	0.15
P690QH	1.8880	0.20	0.80	1.70	0.025	0.015	1.50	0.70	2.00	0.06	0.05	0.12	0.15
P690QL	1.8881	0.20	0.80	1.70	0.020	0.020	1.50	0.70	2.00	0.06	0.05	0.12	0.15

a. Elements not listed in this table shall not be intentionally added to the steel without the agreement of the purchaser except for finishing the cast. All appropriate measures shall be taken to prevent the addition from scrap and other materials used in steelmaking of these elements which may adversely affect the mechanical properties and usability. b. Other limits: N 0.015, B 0.0050, Cu 0.50. c. The percentage of grain refining elements shall be at least 0.015%. Single values are maximums, unless otherwise specified.

EN 10028-6 - MECHANICAL PROPERTIES OF FLAT PRODUCTS MADE FROM STEEL FOR PRESSURE PURPOSES
PART 6 - WELDABLE FINE GRAIN STEELS QUENCHED AND TEMPERED

Name	Number	Yield Point, R_eH			Tensile Strength, R_m N/mm²			%Elongation
		≤ 50 mm	< 50 ≤ 100 mm	< 100 to ≤ 150 mm	≤ 100 mm	< 100 to ≤ 150 mm		
P460Q	1.8870							
P460QH	1.8871	460	440	400	550 to 720	500 to 670		17
P460QL	1.8872							
P500Q	1.8873							
P500QH	1.8874	500	480	440	590 to 770	540 to 720		17
P500QL	1.8875							
P550Q	1.8876							
P550QH	1.8877	550	530	490	640 to 820	590 to 770		16
P550QL	1.8878							
P690Q	1.8879							
P690QH	1.8880	690	670	630	770 to 940	720 to 900		14
P690QL	1.8881							

Single values are minimums, unless otherwise specified.

EN 10028-6 - IMPACT PROPERTIES OF FLAT PRODUCTS MADE FROM STEEL FOR PRESSURE PURPOSES
PART 6 - WELDABLE FINE GRAIN STEELS QUENCHED AND TEMPERED

Steel Name Series	Direction of Test Piece	Impact Energy Values, KV, in J at Test Temperatures in °C Product Thicknesses 10 ≤ S ≤ 70 mm				
		-60	-40	-20	0	
P ... Q	longitudinal	---	---	30	40	
P ... QH	transverse	---	---	27	30	
P ... QL	longitudinal	30	40	50	60	
	transverse	27	30	35	40	

Single values are minimums, unless otherwise specified.

EN 10083-1 - CHEMICAL COMPOSITIONS OF QUENCHED AND TEMPERED STEELS - PART 1 - FOR SPECIAL STEELS										
Name	Number	C	Si	Mn	P	S	Cr	Mo	Ni	Others
25CrMo4	1.7218	0.22-0.29	0.40	0.60-0.90	0.035	0.035	0.90-1.20	0.15-0.30	---	---
25CrMoS4	1.7213	0.22-0.29	0.40	0.60-0.90	0.035	0.020-0.040	0.90-1.20	0.15-0.30	---	---
28Mn6	1.1170	0.25-0.32	0.40	1.30-1.65	0.035	0.035	0.40	0.10	0.40	Cr+Mo+Ni 0.63
30CrNiMo8	1.6580	0.26-0.34	0.40	0.30-0.60	0.035	0.035	1.80-2.20	0.30-0.50	1.80-2.20	-
34Cr4	1.7033	0.30-0.37	0.40	0.60-0.90	0.035	0.035	0.90-1.20	-	-	-
34CrS4	1.7037	0.30-0.37	0.40	0.60-0.90	0.035	0.020-0.040	0.90-1.20	-	-	-
34CrMo4	1.7220	0.30-0.37	0.40	0.60-0.90	0.035	0.035	0.90-1.20	0.15-0.30	-	-
34CrMoS4	1.7226	0.30-0.37	0.40	0.60-0.90	0.035	0.020-0.040	0.90-1.20	0.15-0.30	-	-
34CrNiMo6	1.6582	0.30-0.38	0.40	0.50-0.80	0.035	0.035	1.30-1.70	0.15-0.30	1.30-1.70	-
36CrNiMo4	1.6511	0.32-0.40	0.40	0.50-0.80	0.035	0.035	0.90-1.20	0.15-0.30	0.90-1.20	-
36NiCrMo16	---	0.32-0.39	0.40	0.30-0.60	0.035	0.025	1.60-2.00	0.25-0.45	3.60-4.10	-
37Cr4	1.7034	0.34-0.41	0.40	0.60-0.90	0.035	0.035	0.90-1.20	-	-	-
37CrS4	1.7038	0.34-0.41	0.40	0.60-0.90	0.035	0.020-0.040	0.90-1.20	-	-	-
38Cr2	1.7003	0.35-0.42	0.40	0.50-0.80	0.035	0.035	0.40-0.60	-	-	-
38CrS2	1.7023	0.35-0.42	0.40	0.50-0.80	0.035	0.020-0.040	0.40-0.60	-	-	-
41Cr4	1.7035	0.38-0.45	0.40	0.60-0.90	0.035	0.035	0.90-1.20	-	-	-
41CrS4	1.7039	0.38-0.45	0.40	0.60-0.90	0.035	0.020-0.040	0.90-1.20	-	-	-
42CrMo4	1.7225	0.38-0.45	0.40	0.60-0.90	0.035	0.035	0.90-1.20	0.15-0.30	-	-
42CrMoS4	1.7227	0.38-0.45	0.40	0.60-0.90	0.035	0.020-0.040	0.90-1.20	0.15-0.30	-	-
46Cr2	1.7006	0.42-0.50	0.40	0.50-0.80	0.035	0.035	0.40-0.60	-	-	-
46CrS2	1.7025	0.42-0.50	0.40	0.50-0.80	0.035	0.020-0.040	0.40-0.60	-	-	-
50CrMo4	1.7228	0.46-0.54	0.40	0.50-0.80	0.035	0.035	0.90-1.20	0.15-0.30	-	-
51CrV4	1.8159	0.47-0.55	0.40	0.70-1.10	0.035	0.035	0.90-1.20	-	-	V 0.10-0.25
C22E	1.1151	0.17-0.24	0.40	0.40-0.70	0.035	0.035	0.40	0.10	0.40	Cr+Mo+Ni 0.63
C22R	1.1149	0.17-0.24	0.40	0.40-0.70	0.035	0.020-0.040	0.40	0.10	0.40	Cr+Mo+Ni 0.63
C25E	1.1158	0.22-0.29	0.40	0.40-0.70	0.035	0.035	0.40	0.10	0.40	Cr+Mo+Ni 0.63
C25R	1.1163	0.22-0.29	0.40	0.40-0.70	0.035	0.020-0.040	0.40	0.10	0.40	Cr+Mo+Ni 0.63

EN 10083-1 - CHEMICAL COMPOSITIONS OF QUENCHED AND TEMPERED STEELS - PART 1 - FOR SPECIAL STEELS (Continued)

Name	Number	C	Si	Mn	P	S	Cr	Mo	Ni	Others
C30E	1.1178	0.27-0.34	0.40	0.50-0.80	0.035	0.035	0.40	0.10	0.40	Cr+Mo+Ni 0.63
C30R	1.1179	0.27-0.34	0.40	0.50-0.80	0.035	0.020-0.040	0.40	0.10	0.40	Cr+Mo+Ni 0.63
C35E	1.1181	0.32-0.39	0.40	0.50-0.80	0.035	0.035	0.40	0.10	0.40	Cr+Mo+Ni 0.63
C35R	1.1180	0.32-0.39	0.40	0.50-0.80	0.035	0.020-0.040	0.40	0.10	0.40	Cr+Mo+Ni 0.63
C40E	1.1186	0.37-0.44	0.40	0.50-0.80	0.035	0.035	0.40	0.10	0.40	Cr+Mo+Ni 0.63
C40R	1.1189	0.37-0.44	0.40	0.50-0.80	0.035	0.020-0.040	0.40	0.10	0.40	Cr+Mo+Ni 0.63
C45E	1.1191	0.42-0.50	0.40	0.50-0.80	0.035	0.035	0.40	0.10	0.40	Cr+Mo+Ni 0.63
C45R	1.1201	0.42-0.50	0.40	0.50-0.80	0.035	0.020-0.040	0.40	0.10	0.40	Cr+Mo+Ni 0.63
C50E	1.1206	0.47-0.55	0.40	0.60-0.90	0.035	0.035	0.40	0.10	0.40	Cr+Mo+Ni 0.63
C50R	1.1241	0.47-0.55	0.40	0.60-0.90	0.035	0.020-0.040	0.40	0.10	0.40	Cr+Mo+Ni 0.63
C55E	1.1203	0.52-0.60	0.40	0.60-0.90	0.035	0.035	0.40	0.10	0.40	Cr+Mo+Ni 0.63
C55R	1.1209	0.52-0.60	0.40	0.60-0.90	0.035	0.020-0.040	0.40	0.10	0.40	Cr+Mo+Ni 0.63
C60E	1.1221	0.57-0.65	0.40	0.60-0.90	0.035	0.035	0.40	0.10	0.40	Cr+Mo+Ni 0.63
C60R	1.1223	0.57-0.65	0.40	0.60-0.90	0.035	0.020-0.040	0.40	0.10	0.40	Cr+Mo+Ni 0.63

Single values are maximums, unless otherwise specified.

EN 10083-2 - CHEMICAL COMPOSITIONS OF QUENCHED AND TEMPERED STEELS - PART 2 - FOR UNALLOYED QUALITY STEELS

Name	Number	C	Si	Mn	P	S	Cr	Mo	Ni	Others
C25	1.0406	0.22-0.29	0.40	0.40-0.70	0.045	0.045	0.40	0.10	0.40	Cr+Mo+Ni 0.63
C30	1.0528	0.27-0.34	0.40	0.50-0.80	0.045	0.045	0.40	0.10	0.40	Cr+Mo+Ni 0.63
C35	1.0501	0.32-0.39	0.40	0.50-0.80	0.045	0.045	0.40	0.10	0.40	Cr+Mo+Ni 0.63
C40	1.0511	0.37-0.44	0.40	0.50-0.80	0.045	0.045	0.40	0.10	0.40	Cr+Mo+Ni 0.63
C45	1.0503	0.42-0.50	0.40	0.50-0.80	0.045	0.045	0.40	0.10	0.40	Cr+Mo+Ni 0.63
C50	1.0540	0.47-0.55	0.40	0.60-0.90	0.045	0.045	0.40	0.10	0.40	Cr+Mo+Ni 0.63
C55	1.0535	0.52-0.60	0.40	0.60-0.90	0.045	0.045	0.40	0.10	0.40	Cr+Mo+Ni 0.63
C60	1.0601	0.57-0.65	0.40	0.60-0.90	0.045	0.045	0.40	0.10	0.40	Cr+Mo+Ni 0.63

EN 10083-2 - CHEMICAL COMPOSITIONS OF QUENCHED AND TEMPERED STEELS - PART 2 - FOR UNALLOYED QUALITY STEELS (Continued)

Single values are maximums, unless otherwise specified.

EN 10083-3 - CHEMICAL COMPOSITIONS OF QUENCHED AND TEMPERED STEELS - PART 3 - BORON STEELS

Name	Number	C	Si	Mn	P	S	Cr	B
20MnB5	1.5530	0.17-0.23	0.40	1.10-1.40	0.035	0.040	---	B 0.0008-0.0050
27MnCrB5-2	1.7182	0.24-0.30	0.40	1.10-1.40	0.035	0.040	0.30-0.60	B 0.0008-0.0050
30MnB5	1.5531	0.27-0.33	0.40	1.15-1.45	0.035	0.040	---	B 0.0008-0.0050
33MnCrB5-2	1.7185	0.30-0.36	0.40	1.20-1.50	0.035	0.040	0.30-0.60	B 0.0008-0.0050
38MnB5	1.5532	0.36-0.42	0.40	1.15-1.45	0.035	0.040	---	B 0.0008-0.0050
39MnCrB6-2	1.7189	0.36-0.42	0.40	1.40-1.70	0.035	0.040	0.30-0.60	B 0.0008-0.0050

Single values are maximums, unless otherwise specified.

EN 10113-2 - CHEMICAL COMPOSITIONS OF HOT-ROLLED PRODUCTS IN WELDED FINE GRAINED STRUCTURAL STEELS PART 2 - NORMALIZED/NORMALIZED ROLLED STEEL

Name	Number	C	Si	Mn	P	S	Cr	Mo	Ni	Others
S275N	1.0490	0.18	0.40	0.50-1.40	0.035	0.030	0.30	0.10	0.30	Al 0.02 min, Cu 0.35, N 0.015, Nb 0.05, Ti 0.03, V 0.05
S275NL	1.0491	0.16	0.40	0.50-1.40	0.030	0.025	0.30	0.10	0.30	Al 0.02 min, Cu 0.35, N 0.015, Nb 0.05, Ti 0.03, V 0.05
S355N	1.0545	0.20	0.50	0.90-1.65	0.035	0.030	0.30	0.10	0.50	Al 0.02 min, Cu 0.35, N 0.015, Nb 0.05, Ti 0.03, V 0.12
S355NL	1.0546	0.18	0.50	0.90-1.65	0.030	0.025	0.30	0.10	0.50	Al 0.02 min, Cu 0.35, N 0.015, Nb 0.05, Ti 0.03, V 0.12
S420N	1.8902	0.20	0.60	1.00-1.70	0.035	0.030	0.30	0.10	0.80	Al 0.02 min, Cu 0.70, N 0.025, Nb 0.05, Ti 0.03, V 0.20
S420NL	1.8912	0.20	0.60	1.00-1.70	0.030	0.025	0.30	0.10	0.80	Al 0.02 min, Cu 0.70, N 0.025, Nb 0.05, Ti 0.03, V 0.20
S460N	1.8901	0.20	0.60	1.00-1.70	0.035	0.030	0.30	0.10	0.80	Al 0.02 min, Cu 0.70, N 0.025, Nb 0.05, Ti 0.03, V 0.20
S460NL	1.8903	0.20	0.60	1.00-1.70	0.030	0.025	0.30	0.10	0.80	Al 0.02 min, Cu 0.70, N 0.025, Nb 0.05, Ti 0.03, V 0.20

Single values are maximums, unless otherwise specified.

EN 10113-3 - CHEMICAL COMPOSITIONS OF HOT-ROLLED PRODUCTS IN WELDED FINE GRAINED STRUCTURAL STEELS PART 3 - THERMOMECHANICAL ROLLED STEEL

Name	Number	C	Si	Mn	P	S	Mo	Ni	Others
S275M	1.8818	0.13	0.50	1.50	0.035	0.030	0.20	0.30	Al 0.02 min, N 0.015, Nb 0.05, Ti 0.05, V 0.08
S275ML	1.8819	0.13	0.50	1.50	0.030	0.025	0.20	0.30	Al 0.02 min, N 0.015, Nb 0.05, Ti 0.05, V 0.08
S355M	1.8823	0.14	0.50	1.60	0.035	0.030	0.20	0.30	Al 0.02 min, N 0.015, Nb 0.05, Ti 0.05, V 0.10
S355ML	1.8834	0.14	0.50	1.60	0.030	0.025	0.20	0.30	Al 0.02 min, N 0.015, Nb 0.05, Ti 0.05, V 0.10
S420M	1.8825	0.16	0.50	1.70	0.035	0.030	0.20	0.30	Al 0.02 min, N 0.020, Nb 0.05, Ti 0.05, V 0.12
S420ML	1.8836	0.16	0.50	1.70	0.030	0.025	0.20	0.30	Al 0.02 min, N 0.020, Nb 0.05, Ti 0.05, V 0.12
S460M	1.8827	0.16	0.60	1.70	0.035	0.030	0.20	0.45	Al 0.02 min, N 0.025, Nb 0.05, Ti 0.05, V 0.12
S460ML	1.8838	0.16	0.60	1.70	0.030	0.025	0.20	0.45	Al 0.02 min, N 0.025, Nb 0.05, Ti 0.05, V 0.12

Single values are maximums, unless otherwise specified.

EN 10130 - CHEMICAL COMPOSITIONS OF COLD ROLLED LOW CARBON STEEL FLAT PRODUCTS FOR COLD FORMING

Name	Number	C	Si	Mn	P	S	Others
FeP01/DC01	1.0330	0.12	---	0.60	0.045	0.045	---
FeP03/DC03	1.0347	0.10	---	0.45	0.035	0.035	---
FeP04/DC04	1.0338	0.08	---	0.40	0.030	0.030	---
FeP05/DC05	(1.0312)	0.06	---	0.35	0.025	0.025	---
FeP06/DC06	(1.0873)	0.02	---	0.25	0.020	0.020	Ti 0.30 or Nb[a]

a. C + N must be fixed completely.
Single values are maximums, unless otherwise specified.

EN 10149-2 - CHEMICAL COMPOSITIONS OF STEELS WITH HIGH YIELD STRESS FOR COLD FORMING
PART 2 THERMOMECHANICAL ROLLING

Name	Number	C	Si	Mn	P	S	Mo	Others
S315MC	1.0972	0.12	0.50	1.30	0.025	0.020	---	Al 0.015 min, Nb 0.09, V 0.20, Ti 0.15, (Nb+V+Ti 0.22)
S355MC	1.0976	0.12	0.50	1.50	0.025	0.020	---	Al 0.015 min, Nb 0.09, V 0.20, Ti 0.15, (Nb+V+Ti 0.22)
S420MC	1.0980	0.12	0.50	1.60	0.025	0.015	---	Al 0.015 min, Nb 0.09, V 0.20, Ti 0.15, (Nb+V+Ti 0.22)
S460MC	1.0982	0.12	0.50	1.60	0.025	0.015	---	Al 0.015 min, Nb 0.09, V 0.20, Ti 0.15, (Nb+V+Ti 0.22)
S500MC	1.0984	0.12	0.50	1.70	0.025	0.015	---	Al 0.015 min, Nb 0.09, V 0.20, Ti 0.15, (Nb+V+Ti 0.22)
S550MC	1.0986	0.12	0.50	1.80	0.025	0.015	---	Al 0.015 min, Nb 0.09, V 0.20, Ti 0.15, (Nb+V+Ti 0.22)
S600MC	1.8969	0.12	0.50	1.90	0.015	0.015	0.50	Al 0.015 min, Nb 0.09, V 0.20, Ti 0.22, B 0.005
S650MC	1.8976	0.12	0.60	2.00	0.025	0.015	0.50	Al 0.015 min, Nb 0.09, V 0.20, Ti 0.22, B 0.005
S700MC	1.8974	0.12	0.60	2.10	0.025	0.015	0.50	Al 0.015 min, Nb 0.09, V 0.20, Ti 0.22, B 0.005

Single values are maximums, unless otherwise specified.

EN 10149-3 - CHEMICAL COMPOSITIONS OF STEELS WITH HIGH YIELD STRESS FOR COLD FORMING
PART 3 - ANNEALED/NORMALIZED

Name	Number	C	Si	Mn	P	S	Others
S260NC	1.0971	0.16	0.50	1.20	0.025	0.020	Al 0.015 min, Nb 0.09, V 0.10, Ti 0.15, (Nb+V+Ti 0.22)
S315NC	1.0973	0.16	0.50	1.40	0.025	0.020	Al 0.015 min, Nb 0.09, V 0.10, Ti 0.15, (Nb+V+Ti 0.22)
S355NC	1.0977	0.18	0.50	1.60	0.025	0.015	Al 0.015 min, Nb 0.09, V 0.10, Ti 0.15, (Nb+V+Ti 0.22)
S420NC	1.0981	0.20	0.50	1.60	0.025	0.015	Al 0.015 min, Nb 0.09, V 0.10, Ti 0.15, (Nb+V+Ti 0.22)

Single values are maximums, unless otherwise specified.

EN 10155 - CHEMICAL COMPOSITIONS OF STRUCTURAL STEELS WITH IMPROVED ATMOSPHERIC RESISTANCE

Name	Number	C	Si	Mn	P	S	Cr	Mo	Ni	Others
S235J0W	1.8958	0.13	0.40	0.20-0.60	0.040	0.040	0.40-0.80	-	(0.65)	N 0.009[a] , Cu 0.25-0.55
S235J2W	1.8961	0.13	0.40	0.20-0.60	0.040	0.035	0.40-0.80	-	(0.65)	Cu 0.25-0.55
S355J0WP	1.8945	0.12	0.75	1.00	0.06-0.15	0.040	0.30-1.25	-	(0.65)	N 0.009[a] , Cu 0.25-0.55
S355J2WP	1.8946	0.12	0.75	1.00	0.06-0.15	0.035	0.30-1.25	-	(0.65)	Cu 0.25-0.55
S355J0W	1.8959	0.16	0.50	0.50-1.50	0.040	0.040	0.40-0.80	(0.030)	(0.65)	N 0.009[a] , Cu 0.25-0.55
S355J2G1W	1.8963	0.16	0.50	0.50-1.50	0.035	0.035	0.40-0.80	(0.030)	(0.65)	Cu 0.25-0.55, (Zr 0.15)
S355J2G2W	1.8965	0.16	0.50	0.50-1.50	0.035	0.035	0.40-0.80	(0.030)	(0.65)	Cu 0.25-0.55, (Zr 0.15)
S355K2G1W	1.8966	0.16	0.50	0.50-1.50	0.035	0.035	0.40-0.80	(0.030)	(0.65)	Cu 0.25-0.55, (Zr 0.15)
S355K2G2W	1.8967	0.16	0.50	0.50-1.50	0.035	0.035	0.40-0.80	(0.030)	(0.65)	Cu 0.25-0.55, (Zr 0.15)

a. The maximum values do not apply, if the steel contains at least 0.020% Al or other sufficient fixation elements.
Single values are maximums, unless otherwise specified.

EN 10207 - CHEMICAL COMPOSITION OF STEELS FOR SIMPLE PRESSURE VESSELS

Name[a]	Number	C	Si	Mn	P	S	Others[b]
SPH235 (UQ)	1.0112	0.16	0.35	0.40-1.20	0.035	0.030	Al 0.020 min
SPH265 (UQ)	1.0130	0.20	0.40	0.50-1.50	0.035	0.030	Al 0.020 min
SPHL275 (US)	1.1100	0.16	0.40	0.50-1.50	0.030	0.025	Al 0.020 min

a. UQ - non-alloy quality steel, US - non-alloy special steel.
b. If sufficient other nitrogen binding elements are present, the minimum total aluminum content does not apply. The content of other nitrogen binding elemets shall be reported.
Single values are maximums, unless otherwise specified.

EN 10207 - MECHANICAL PROPERTIES OF STEELS FOR SIMPLE PRESSURE VESSELS

Name	Yield Strength, N/mm² For Nominal Thicknesses, mm			Tensile Strength, N/mm²	% Elongation (Longitudinal) For Nominal Thicknesses, mm				Charpy V-Notch[a] J@°C
	≤ 16	> 16 to ≤ 40	> 40 to ≤ 60		> 2 to ≤ 2.5	> 2.5 to < 3	≥ 3 to ≤ 40	> 40 to ≤ 60	
SPH235	235	225	215	360 to 480	20	21	26	25	28@-20
SPH 65	265	255	245	410 to 530	17	18	22	22	28@-20
SPHL275	275	265	255	390 to 510	19	20	24	24	28@-20

a. Charpy V-notch requirements are for longitudinal tests. Minimum average value of a set of three specimens, however, one value may be below the specified average, provided that it is not less than 70% of that value.
Single values are minimums, unless otherwise specified.

EN 10210 - CHEMICAL COMPOSITIONS OF HOT FINISHED STRUCTURAL HOLLOW SECTIONS OF NON-ALLOY FINE GRAIN STRUCTURAL STEELS

Name	Number	C	Si	Mn	P	S	Cr	Mo	Ni	Others
S235JRH[a]	1.0039	0.17	---	1.40	0.045	0.045	---	---	---	N 0.009
S275J0H[a]	1.0149	0.20	---	1.50	0.040	0.040	---	---	---	N 0.009
S275J2H[a]	1.0138	0.20	---	1.50	0.035	0.035	---	---	---	---
S355J0H	1.0547	0.22	0.55	1.60	0.040	0.040	---	---	---	N 0.009
S355J2H	1.0576	0.22	0.55	1.60	0.035	0.035	---	---	---	---
S275NH	1.0493	0.20	0.40	0.50-1.40	0.035	0.030	0.30	0.10	0.30	Nb 0.050, V 0.05, Ti 0.03, Cu 0.35, N 0.015, Al 0.020 min[b]
S275NLH	1.0497	0.20	0.40	0.50-1.40	0.030	0.025	0.30	0.10	0.30	Nb 0.050, V 0.05, Ti 0.03, Cu 0.35, N 0.015, Al 0.020 min[b]
S355NH	1.0539	0.20	0.50	0.90-1.65	0.035	0.030	0.30	0.10	0.50	Nb 0.050, V 0.12, Ti 0.03, Cu 0.35, N 0.015, Al 0.020 min[b]
S355NLH	1.0549	0.18	0.50	0.90-1.65	0.030	0.025	0.30	0.10	0.50	Nb 0.050, V 0.12, Ti 0.03, Cu 0.35, N 0.015, Al 0.020 min[b]
S460NH	1.8953	0.20	0.60	1.00-1.70	0.035	0.030	0.30	0.10	0.80	Nb 0.050, V 0.20, Ti 0.03, Cu 0.70, N 0.025, Al 0.025 min[b]
S460NLH	1.8956	0.20	0.60	1.00-1.70	0.030	0.025	0.30	0.10	0.80	Nb 0.050, V 0.20, Ti 0.03, Cu 0.70, N 0.025, Al 0.025 min[b]

a. Section size limit ≤ 40 mm. b. The maximum values do not apply, if the steel contains at least 0.020% Al or other sufficient fixation elements.
Single values are maximums, unless otherwise specified.

EN 10088- PART 1 - CHEMICAL COMPOSITION OF FERRITIC STAINLESS STEELS[a]

Name	Number	C	Si	Mn	Cr	Mo	Ni	Others
X2CrNi12	1.4003	0.030	1.00	1.50	10.50-12.50	---	0.30-1.00	N ≤ 0.030
X2CrTi12	1.4512	0.030	1.00	1.00	10.50-12.50	---	---	Ti 6x(C+N) ≤ 0.60
X6CrNiTi12	1.4516	0.080	0.70	1.50	10.50-12.50	---	0.50-1.50	Ti 0.05-0.35
X6Cr13	1.4000	0.080	1.00	1.00	12.00-14.00	---	---	---
X6CrAl13	1.4002	0.080	1.00	1.00	12.00-14.00	---	---	Al 0.10-0.30
X2CrTi17	1.4520	0.025	0.50	0.50	16.00-18.00	---	---	N ≤ 0.015 , Ti 0.30-0.60
X6Cr17	1.4016	0.080	1.00	1.00	16.00-18.00	---	---	---
X3CrTi17	1.4510	0.050	1.00	1.00	16.00-18.00	---	---	Ti 4x(C+N)+0.15 ≤ 0.80[b]
X3CrNb17	1.4511	0.050	1.00	1.00	16.00-18.00	---	---	Nb 12xC ≤ 1.00
X6CrMo17-1	1.4113	0.080	1.00	1.00	16.00-18.00	0.90-1.40	---	---
X6CrMoS17	1.4105	0.080	1.00	1.50	16.00-18.00	0.20-0.60	---	S 0.15-0.35
X2CrMoTi17-1	1.4513	0.025	1.00	1.00	16.00-18.00	1.00-1.50	---	N ≤ 0.015, Ti 0.30-0.60
X2CrMoTi18-2	1.4521	0.025	1.00	1.00	17.00-20.00	1.80-2.50	---	N ≤ 0.030, Ti 4x(C+N)+0.15 ≤ 0.80[b]
X2CrMoTiS18-2	1.4523	0.030	1.00	0.50	17.50-19.00	2.00-2.50	---	S 0.15-0.35, (C+N) ≤ 0.040, Ti 0.30-0.85
X6CrNi17-1	1.4017	0.080	1.00	1.00	16.00-18.00	---	1.20-1.60	---
X6CrMoNb17-1	1.4526	0.080	1.00	1.00	16.00-18.00	0.90-1.40	---	N ≤ 0.040, Nb 7x(C+N)+0.10 ≤ 1.00
X2CrNbZr17	1.4590	0.030	1.00	1.00	16.00-17.50	---	---	Zr ≥ 7x(C+N)+0.15, Nb 0.35-0.55
X2CrAlTi18-2	1.4605	0.030	1.00	1.00	17.00-18.00	---	---	Al 1.70-2.10, Ti 4x(C+N)+0.15 ≤ 0.80[b]
X2CrTiNb18	1.4509	0.030	1.00	1.00	17.50-18.50	---	---	Nb 9xC+0.30 ≤ 1.00, Ti 0.10-0.60
X2CrMoTi29-4	1.4592	0.025	1.00	1.00	28.00-30.00	3.50-4.50	---	N ≤ 0.045, 4x(C+N)+0.15 ≤ 0.80[b]

a. Although sulfur and phosphorus are not listed in this table due to space limitation, they are specified in prEN 10088 Part 1.
b. The stabilization may be made by use of titanium or niobium or zirconium. See prEN 10088 Part 1 table 1 for more details.
Single values are maximums, unless otherwise specified.

EN 10088- PART 1 - CHEMICAL COMPOSITION OF MARTENSITIC AND PRECIPITATION HARDENING STAINLESS STEELS[a]

Name	Number	C	Si	Mn	Cr	Mo	Ni	Others
X12Cr13	1.4006	0.08-0.15	1.00	1.00	12.00-14.00	---	---	---
X12CrS13	1.4005	0.08-0.15	1.00	1.50	12.00-14.00	0.60	---	0.15-0.35 S
X20Cr13	1.4021	0.16-0.25	1.00	1.00	12.00-14.00	---	---	---
X30Cr13	1.4028	0.26-0.35	1.00	1.00	12.00-14.00	---	---	---
X29CrS 13	1.4029	0.25-0.32	1.00	1.50	12.00-13.50	0.60	---	---
X39Cr13	1.4031	0.36-0.42	1.00	1.00	12.50-14.50	---	---	---
X46Cr13	1.4034	0.43-0.50	1.00	1.00	12.50-14.50	---	---	---
X50CrMoV15	1.4116	0.45-0.55	1.00	1.00	14.00-15.00	0.50-0.80	---	V 0.10-0.20
X70CrMo15	1.4109	0.65-0.75	0.70	1.00	14.00-16.00	0.40-0.80	---	---
X14CrMoS17	1.4104	0.10-0.17	1.00	1.50	15.50-17.50	0.20-0.60	---	---
X39CrMo17-1	1.4122	0.33-0.45	1.00	1.00	15.50-17.50	0.80-1.30	---	---
X105CrMo17	1.4125	0.95-1.20	1.00	1.00	16.00-18.00	0.40-0.80	---	---
X90CrMoV18	1.4112	0.85-0.95	1.00	1.00	17.00-19.00	0.90-1.30	---	V 0.07-0.12
X17CrNi16-2	1.4057	0.12-0.22	1.00	1.50	15.00-17.00	---	1.50-2.50	---
X3CrNiMo13-4	1.4313	0.05	0.70	1.00	12.00-14.00	0.30-0.70	3.50-4.50	---
X4CrNiMo16-5-1	1.4418	0.06	0.70	1.50	15.00-17.00	0.80-1.50	4.00-6.00	---
X5CrNiCuNb16-4	1.4542	0.07	0.70	1.00	15.00-17.00	0.60	3.00-5.00	(Nb+Ta) 0.15-0.45, Cu 3.00-5.00
X7CrNiAl17-7	1.4568	0.09	0.70	1.00	16.00-18.00	---	6.50-7.80	Al 0.70-1.50
X8CrNiMoAl15-7-2	1.4532	0.10	0.70	1.20	14.00-16.00	2.00-3.00	6.50-7.80	Al 0.70-1.50
X5CrNiMoCuNb14-5	1.4594	0.07	0.70	1.00	13.00-15.00	1.20-2.00	5.00-6.00	(Nb+Ta) 0.15-0.60,Cu 1.20-2.00

a. Although sulfur and phosphorus are not listed in this table due to space limitation, they are specified in prEN 10088 Part 1.
Single values are maximums, unless otherwise specified.

EN 10088- PART 1 - CHEMICAL COMPOSITION OF AUSTENITIC STAINLESS STEELS[a]								
Name	Number	C	Si	Mn	Cr	Mo	Ni	Others
X10CrNi18-8	1.4310	0.05-0.15	2.00	2.00	18.00-19.00	0.08	6.00-9.50	N 0.11
X2CrNiN18-7	1.4318	0.030	1.00	2.00	16.50-18.50	---	6.00-8.00	N 0.10-0.20
X2CrNi18-9	1.4307	0.030	1.00	2.00	17.50-19.50	---	8.00-10.0	N 0.11
X2CrNi19-11	1.4306	0.030	1.00	2.00	18.00-20.00	---	10.0-12.0	N 0.11
X2CrNiN18-10	1.4311	0.030	1.00	2.00	17.00-19.00	---	8.50-11.5	N 0.12-0.22
X5CrNi18-10	1.4301	0.07	1.00	2.00	17.00-19.50	---	8.00-10.50	N0.11
X8CrNiS18-9	1.4305	0.10	1.00	2.00	17.00-19.00	---	8.00-10.0	N 0.11, S 0.15-0.35, Cu ≤ 1.00
X6CrNiTi18-10	1.4541	0.08	1.00	2.00	17.00-19.00	---	9.00-12.0	Ti 5xC ≤ 0.70
X6CrNiNb18-10	1.4550	0.08	1.00	2.00	17.00-19.00	---	9.00-12.0	Nb 10xC ≤ 1.00
X4CrNi18-12	1.4303	0.08	1.00	2.00	17.00-19.00	---	11.0-13.0	N 0.11
X1CrNi25-21	1.4335	0.020	0.25	2.00	24.00-26.00	0.20	20.0-22.0	N 0.11
X2CrNiMo17-12-2	1.4404	0.030	1.00	2.00	16.50-18.50	2.00-2.50	10.0-13.0	N 0.11
X2CrNiMoN17-11-2	1.4406	0.030	1.00	2.00	16.50-18.50	2.00-2.50	10.0-12.0	N 0.12-0.22
X5CrNiMo17-12-2	1.4401	0.07	1.00	2.00	16.50-18.50	2.00-2.50	10.00-13.00	N 0.11
X1CrNiMoN25-22-2	1.4466	0.020	0.70	2.00	24.00-26.00	2.00-2.50	21.0-23.0	N 0.10-0.18
X6CrNiMoTi17-12-2	1.4571	0.08	1.00	2.00	16.50-18.50	2.00-2.50	10.5-13.5	Ti 5xC ≤ 0.70
X6CrNiMoNb17-12-2	1.4580	0.08	1.00	2.00	16.50-18.50	2.00-2.50	10.5-13.5	Nb 10xC ≤ 1.00
X2CrNiMo17-12-3	1.4432	0.030	1.00	2.00	16.50-18.50	2.50-3.00	11.0-12.5	N 0.11
X2CrNiMoN17-13-3	1.4429	0.030	1.00	2.00	16.50-18.50	2.50-3.00	11.0-14.0	N 0.12-0.22
X3CrNiMo17-13-3	1.4438	0.05	1.00	2.00	16.50-18.50	2.50-3.00	10.50-13.00	N 0.11
X2CrNiMo18-14-3	1.4435	0.030	1.00	2.00	17.00-19.00	2.50-3.00	12.5-15.0	N 0.11
X2CrNiMoN18-12-4	1.4434	0.030	1.00	2.00	18.50-19.50	3.00-4.00	10.50-14.00	N 0.10-0.20
X2CrNiMo18-15-4	1.4438	0.030	1.00	2.00	17.50-19.50	3.00-4.00	13.0-16.0	N 0.11
X2CrNiMoN 17-13-5	1.4439	0.030	1.00	2.00	16.50-18.50	4.00-5.00	12.5-14.5	N 0.12-0.22
X1CrNiSi18-15-4	1.4361	0.015	3.70-4.50	2.00	16.50-18.50	0.20	14.0-16.0	N 0.11
X12CrMnNiN17-7-5	1.4372	0.15	1.00	5.50-7.50	16.00-18.00	---	3.50-5.50	N 0.05-0.25

EN 10088- PART 1 - CHEMICAL COMPOSITION OF AUSTENITIC STAINLESS STEELS[a] (Continued)

Name	Number	C	Si	Mn	Cr	Mo	Ni	Others
X2CrMnNiN17-7-5	1.4371	0.030	1.00	6.00-8.00	16.00-17.00	---	3.50-5.50	N 0.15-0.20
X12CrMnNiN18-9-5	1.4373	0.15	1.00	7.50-10.5	17.00-19.00	---	4.00-6.00	N 0.05-0.25
X3CrNiCu19-9-2	1.4580	0.035	1.00	1.50-2.00	18.00-19.00	---	8.00-9.00	N 0.11, Cu 1.50-2.00
X8CrNiCuS18-9-2	1.4570	0.08	1.00	1.00	17.00-19.00	0.60	8.00-10.0	N 0.11, S 0.15-0.35, Cu 1.40-1.80
X3CrNiCu18-9-4	1.4567	0.04	1.00	2.00	17.00-19.00	---	8.50-10.5	N 0.11, Cu 3.00-4.00
X3CrNiCuMo17-11-3-2	1.4578	0.04	1.00	1.00	10.00-11.00	2.00-2.50	10.0-11.0	N 0.11, Cu 3.00-3.50
X1NiCrMoCu31-27-4	1.4563	0.020	0.70	2.00	26.00-28.00	3.00-4.00	30.0-32.0	N 0.11, Cu 0.70-1.50
X1NiCrMoCu25-20-5	1.4539	0.020	0.70	2.00	19.00-21.00	4.00-5.00	24.0-26.0	N 0.15, Cu 1.20-2.00
X1CrNiMoCuN25-25-5	1.4537	0.020	0.70	2.00	24.00-26.00	470-5.70	24.0-27.0	N 0.17-0.25, Cu 1.00-2.00
X1CrNiMoCuN20-18-7	1.4547	0.020	0.70	1.00	19.50-20.50	6.00-7.00	17.5-18.5	N 0.18-0.25, Cu 0.50-1.00
X1NiCrMoCu25-20-7	1.4529	0.020	1.00	2.00	19.00-21.00	6.00-7.00	24.0-26.0	N 0.10-0.25, Cu 0.50-1.50

a. Although sulfur and phosphorus are not listed in this table due to space limitation, they are specified in prEN 10088 Part 1.
Single values are maximums, unless otherwise specified.

EN 10088- PART 1 - CHEMICAL COMPOSITION OF AUSTENITIC-FERRITIC STAINLESS STEELS[a]

Name	Number	C	Si	Mn	Cr	Mo	Ni	Others
X2CrNiN23-4	1.4362	0.030	1.00	2.00	22.00-24.00	0.10-0.60	3.50-5.50	N 0.05-0.20, Cu 0.10-0.60
X3CrNiMoN27-5-2	1.4460	0.050	1.00	2.00	25.00-28.00	1.30-2.00	4.50-6.50	N 0.05-0.20
X2CrNiMoN22-5-3	1.4462	0.030	1.00	2.00	21.00-23.00	2.50-3.50	4.50-6.50	N 0.10-0.22
X2CrNiMoCuN25-6-3	1.4507	0.030	0.70	2.00	24.00-26.00	2.70-4.00	5.50-7.50	N 0.15-0.30, Cu 1.00-2.00
X2CrNiMoN25-7-4	1.4410	0.030	1.00	2.00	24.00-26.00	3.00-4.50	6.00-8.00	N 020-0.35
X2CrNiMoCuWN25-7-4	1.4501	0.030	1.00	1.00	24.00-26.00	3.00-4.00	6.00-8.00	N 0.20-0.30, Cu 0.50-1.00, W 0.50-1.00

a. Although sulfur and phosphorus are not listed in this table due to space limitation, they are specified in prEN 10088 Part 1.
Single values are maximums, unless otherwise specified.

EN 10088-2 - MECHANICAL PROPERTIES OF FERRITIC STAINLESS STEELS PART 2 - SHEET, PLATE AND STRIP FOR GENERAL PURPOSES

Name	Number	Product Form[a]	Max Size, mm	Proof Stress, $R_{p0.2}$ N/mm^2		Tensile Strength R_m - N/mm^2	% Elongation[a]
				Longitudinal	Transverse		
Standard Grades							
X2CrNi12	1.4003	Cold Rolled Strip	6	280	320	450 to 650	20
		Hot Rolled Strip	12	280	320		20
		Hot Rolled Plate	25^d	250	280		18
X2CrTi12	1.4512	Cold Rolled Strip	6	210	220	380 to 560	25
		Hot Rolled Strip	12	210	220		
X6CrNiTi12	1.4516	Cold Rolled Strip	6	280	320	450 to 650	23
		Hot Rolled Strip	12	280	320		23
		Hot Rolled Plate	25^d	250	280		20
X6Cr13	1.4000	Cold Rolled Strip	6	240	250	400 to 600	19
		Hot Rolled Strip	12	220	230		
		Hot Rolled Plate	25^d	220	230		
X6CrAl13	1.4002	Cold Rolled Strip	6	230	250	400 to 600	17
		Hot Rolled Strip	12	210	230		
		Hot Rolled Plate	25^d	210	230		
X6Cr17	1.4016	Cold Rolled Strip	6	260	280	450 to 600	20
		Hot Rolled Strip	12	240	260	450 to 600	18
		Hot Rolled Plate	25^d	240	260	430 to 630	20
X3CrTi17	1.4510	Cold Rolled Strip	6	230	240	420 to 600	23
		Hot Rolled Strip	12	230	240		
X3CrNb17	1.4511	Cold Rolled Strip	6	230	240	420 to 600	23
X6CrMo17-1	1.4113	Cold Rolled Strip	6	260	280	450 to 630	18
		Hot Rolled Strip	12	260	280		

EN 10088-2 - MECHANICAL PROPERTIES OF FERRITIC STAINLESS STEELS
PART 2 - SHEET, PLATE AND STRIP FOR GENERAL PURPOSES (Continued)

Name	Number	Product Form[a]	Max Size, mm	Proof Stress, $R_{p0.2}$ N/mm²		Tensile Strength R_m - N/mm²	% Elongation[a]
				Longitudinal	Transverse		
Standard Grades (Continued)							
X2CrMoTi18-2	1.4521	Cold Rolled Strip	6	300	320	420 to 640	20
		Hot Rolled Strip	12	280	300	400 to 600	
		Hot Rolled Plate	12	280	300	420 to 620	
Special Grades							
X2CrTi17	1.4520	Cold Rolled Strip	6	180	200	380 to 530	24
X2CrMoTi17-1	1.4513	Cold Rolled Strip	6	200	220	400 to 550	23
X6CrNi17-1	1.4017	Cold Rolled Strip	6	480	500	650 to 750	12
X6CrMoNb17-1	1.4526	Cold Rolled Strip	6	280	300	480 to 560	25
X2CrNbZr17	1.4590	Cold Rolled Strip	6	230	250	400 to 550	23
X2CrAlTi18-2	1.4605	Cold Rolled Strip	6	280	300	500 to 650	25
X2CrTiNb18	1.4509	Cold Rolled Strip	6	230	250	430 to 630	18
X2CrMoTi29-4	1.4592	Cold Rolled Strip	6	430	450	550 to 700	20

a. The values apply for test pieces with a gauge length of 80 mm and a width of 20 mm, or for test pieces with a gauge length of 50 mm and a width of 12.5 mm or for test pieces with a gauge length of 5.65 √So. Values apply to longitudinal and transverse test pieces.

b. For thicknesses above 25 mm, the mechanical properties can be agreed.
Single values are minimums, unless otherwise specified.

EN 10088-2 - MECHANICAL PROPERTIES OF MARTENSITIC STAINLESS STEELS
PART 2 - SHEET, PLATE AND STRIP FOR GENERAL PURPOSES

Name	Number	Product Form	Max Size mm	HTC^a	Max Hardness^b		Proof Stress $R_{p0.2}$ - N/mm²	Tensile Strength R_m - N/mm²	%El^c
					Rockwell	HV or HB			
X12Cr13	1.4006	Cold Rolled Strip	6	A	90 HRB	200	---	600 max	20
		Hot Rolled Strip	12	A	90 HRB	200	---	600 max	20
		Hot Rolled Plate^d	75	QT550	---		400	550-750	15
				QT650			450	650-850	12
X20Cr13	1.4021	Cold Rolled Strip	3	QT	44-50 HRC	440-530	---	---	---
		Cold Rolled Strip	6	A	95 HRB	225	---	700 max	15
		Hot Rolled Strip	12	A	95 HRB	225	---	700 max	15
		Hot Rolled Plate^d	75	QT650	---		450	650-850	12
				QT750			550	750-950	10
X30Cr13	1.4028	Cold Rolled Strip	3	QT	45-51 HRC	450-550	---	---	---
		Cold Rolled Strip	6	A	97 HRB	235	---	740 max	15
		Hot Rolled Strip	12	A	97 HRB	235 HV	---	740 max	15
		Hot Rolled Plate^d	75	QT800	---		600	800-1000	10
X39Cr13	1.4031	Cold Rolled Strip	3	QT	47-53 HRC	480-580	---	---	---
		Cold Rolled Strip	6	A	98 HRB	240	---	760 max	12
		Hot Rolled Strip	12	A	98 HRB	240	---	760 max	12
X46Cr13	1.4034	Cold Rolled Strip	6	A	99 HRB	245	---	780 max	12
		Hot Rolled Strip	12	A	99 HRB	245	---	780 max	12
X50CrMoV15	1.4116	Cold Rolled Strip	6	A	100 HRB	280	---	850 max	12
		Hot Rolled Strip	12	A	100 HRB	280	---	850 max	12
X39CrMo17-1	1.4122	Cold Rolled Strip	3	QT	47-53 HRC	480-580	---	---	---
		Cold Rolled Strip	6	A	100 HRB	280	---	900 max	12
		Hot Rolled Strip	12	A	100 HRB	280	---	900 max	12
X3CrNiMo13-4	1.4313	Hot Rolled Plate	75	QT780	---		650	780-980	14
		Hot Rolled Plate	75	QT900	---		800	900-1100	11

EN 10088-2 - MECHANICAL PROPERTIES OF MARTENSITIC STAINLESS STEELS
PART 2 - SHEET, PLATE AND STRIP FOR GENERAL PURPOSES (Continued)

Name	Number	Product Form	Max Size mm	HTC[a]	Max Hardness[b]		Proof Stress $R_{p0.2}$ - N/mm²	Tensile Strength R_m - N/mm²	%El[c]
					Rockwell	HV or HB			
X4CrNiMo16-5-1	1.4418	Hot Rolled Plate	75	QT840	---		680	840-980	14

a. HTC - Heat treatment condition, A - annealed; QT - quenched and tempered.

b. The Brinell or Vickers or Rockwell hardness is normally determined for cold and hot rolled strip in heat treatment condition A.   The tensile test shall be carried out in referee testing.

c. The values apply for test pieces with a gauge length of 80 mm and a width of 20 mm, or for test pieces with a gauge length of 50 mm and a width of 12.5 mm or for test pieces with a gauge length of 5.65 $\sqrt{S_o}$. Values apply to longitudinal and transverse test pieces.

d. Plates may also be delivered in the annealed condition; in such cases, the mechanical properties are to be agreed at the time of enquiry and order.

Single values are minimums, unless otherwise specified.

EN 10088-2 - MECHANICAL PROPERTIES OF AUSTENITIC STAINLESS STEELS
PART 2 - SHEET, PLATE AND STRIP FOR GENERAL PURPOSES[a]

Name	Number	Product Form	Max Size mm	Proof Stress[b,c] $R_{p0.2}$ - N/mm²	Tensile Strength R_m - N/mm²	% Elongation		Impact Energy, KV > 10 mm, J	
						A_{80}^d < 3 mm	A^e ≥ 3 mm	Long.	Trans.
Standard Grades									
X10CrNi18-8	1.4310	Cold Rolled Strip	6	250	600-950	40	40	---	---
X2CrNiN18-7	1.4318	Cold Rolled Strip	6	350	650-850	35	40	---	---
		Hot Rolled Strip	12	330	650-850	35	40	90	60
		Hot Rolled Plate	75	330	630-830	45	45	90	60
X2CrNi18-9	1.4307	Cold Rolled Strip	6	220	520-670	45	45	---	---
		Hot Rolled Strip	12	200	520-670	45	45	90	60
		Hot Rolled Plate	75	200	500-650	45	45	90	60

EN 10088-2 - MECHANICAL PROPERTIES OF AUSTENITIC STAINLESS STEELS
PART 2 - SHEET, PLATE AND STRIP FOR GENERAL PURPOSES[a] (Continued)

Name	Number	Product Form	Max Size mm	Proof Stress[b,c] $R_{p0.2}$ - N/mm²	Tensile Strength R_m - N/mm²	% Elongation A_{80}^d < 3 mm	% Elongation A^e ≥ 3 mm	Impact Energy, KV > 10 mm, J Long.	Impact Energy, KV > 10 mm, J Trans.
Standard Grades (Continued)									
X2CrNi19-11	1.4306	Cold Rolled Strip	6	220	520-670	45	45	---	---
		Hot Rolled Strip	12	200	520-670	45	45	90	60
		Hot Rolled Plate	75	200	500-650	45	45	90	60
X2CrNiN18-10	1.4311	Cold Rolled Strip	6	290	550-750	40	40	---	---
		Hot Rolled Strip	12	270	550-750	40	40	90	60
		Hot Rolled Plate	75	270	550-750	40	40	90	60
X5CrNi18-10	1.4301	Cold Rolled Strip	6	230	540-750	45[f]	45[f]	---	---
		Hot Rolled Strip	12	210	520-720	45[f]	45[f]	90	60
		Hot Rolled Plate	75	210	520-720	45	45	90	60
X8CrNiS18-9	1.4305	Hot Rolled Plate	75	190	500-700	35	35	---	---
X6CrNiTi18-10	1.4541	Cold Rolled Strip	6	220	520-720	40	40	---	---
		Hot Rolled Strip	12	200	520-720	40	40	90	60
		Hot Rolled Plate	75	200	500-700	40	40	90	60
X4CrNi18-12	1.4303	Cold Rolled Strip	6	220	500-650	45	45	---	---
X2CrNiMo17-12-2	1.4404	Cold Rolled Strip	6	240	530-680	40	40	---	---
		Hot Rolled Strip	12	220	530-680	40	40	90	60
		Hot Rolled Plate	75	220	520-670	45	45	90	60
X2CrNiMoN17-11-2	1.4406	Cold Rolled Strip	6	300	580-780	40	40	---	---
		Hot Rolled Strip	12	280	580-780	40	40	90	60
		Hot Rolled Plate	75	280	580-780	40	40	90	60
X5CrNiMo17-12-2	1.4401	Cold Rolled Strip	6	240	530-680	40	40	---	---
		Hot Rolled Strip	12	220	530-680	40	40	90	60
		Hot Rolled Plate	75	220	520-670	45	45	90	60

EN 10088-2 - MECHANICAL PROPERTIES OF AUSTENITIC STAINLESS STEELS

PART 2 - SHEET, PLATE AND STRIP FOR GENERAL PURPOSES[a] (Continued)

Name	Number	Product Form	Max Size mm	Proof Stress[b,c] $R_{p0.2}$ - N/mm²	Tensile Strength R_m - N/mm²	% Elongation A_{80}^d < 3 mm	% Elongation A^e ≥ 3 mm	Impact Energy, KV > 10 mm, J Long.	Impact Energy, KV > 10 mm, J Trans.
Standard Grades									
X5CrNiMoTi17-12-2	1.4571	Cold Rolled Strip	6	240	540-690	40	40	---	---
		Hot Rolled Strip	12	220	540-690	40	40	90	60
		Hot Rolled Plate	75	220	520-670	40	40	90	60
X2CrNiMo17-12-3	1.4432	Cold Rolled Strip	6	240	550-700	40	40	---	---
		Hot Rolled Strip	12	220	550-700	45	45	90	60
		Hot Rolled Plate	75	220	520-670	45	45	90	60
X2CrNiMo18-14-3	1.4435	Cold Rolled Strip	6	240	550-700	40	40	---	---
		Hot Rolled Strip	12	220	550-700	45	45	90	60
		Hot Rolled Plate	75	220	520-670	45	45	90	60
X2CrNiMoN17-13-5	1.4439	Cold Rolled Strip	6	290	580-780	35	35	---	---
		Hot Rolled Strip	12	270	580-780	35	35	90	60
		Hot Rolled Plate	75	270	580-780	40	40	90	60
X1NiCrMoCu25-20-5	1.4539	Cold Rolled Strip	6	240	530-730	35	35	---	---
		Hot Rolled Strip	12	220	530-730	35	35	90	60
		Hot Rolled Plate	75	220	520-720	35	35	90	60
Special Grades									
X1CrNi25-21	1.4335	Hot Rolled Plate	75	200	470-670	40	40	90	60
X6CrNiNb18-10	1.4550	Cold Rolled Strip	6	220	520-720	40	40	---	---
		Hot Rolled Strip	12	200	520-720	40	40	90	60
		Hot Rolled Plate	75	200	500-700	40	40	90	60
X1CrNiMoN25-22-2	1.4466	Hot Rolled Plate	75	250	540-740	40	40	90	60
X6CrNiMoNb17-12-2	1.4580	Hot Rolled Plate	75	220	520-720	40	40	90	60

EN 10088-2 - MECHANICAL PROPERTIES OF AUSTENITIC STAINLESS STEELS
PART 2 - SHEET, PLATE AND STRIP FOR GENERAL PURPOSES[a] (Continued)

Name	Number	Product Form	Max Size mm	Proof Stress[b,c] $R_{p0.2}$ - N/mm²	Tensile Strength R_m - N/mm²	% Elongation A_{80}[d] < 3 mm	% Elongation A[e] ≥ 3 mm	Impact Energy, KV > 10 mm, J Long.	Impact Energy, KV > 10 mm, J Trans.
Special Grades (Continued)									
X2CrNiMoN17-13-3	1.4429	Cold Rolled Strip	6	300	580-780	35	35	--	--
		Hot Rolled Strip	12	280	580-780	35	35	90	60
		Hot Rolled Plate	75	280	580-780	40	40	90	60
X3CrNiMo17-13-3	1.4436	Cold Rolled Strip	6	240	550-700	40	40	--	--
		Hot Rolled Strip	12	220	550-700	40	40	90	60
		Hot Rolled Plate	75	220	530-730	40	40	90	60
X2CrNiMoN18-12-4	1.4434	Cold Rolled Strip	6	290	570-770	35	35	--	--
		Hot Rolled Strip	12	270	570-770	35	35	90	60
		Hot Rolled Plate	75	270	540-740	40	40	90	60
X2CrNiMo18-15-4	1.4438	Cold Rolled Strip	6	240	550-700	35	35	--	--
		Hot Rolled Strip	12	220	550-700	35	35	90	60
		Hot Rolled Plate	75	220	520-720	40	40	90	60
X2CrNiSi18-15-4	1.4361	Hot Rolled Plate	75	220	530-730	40	40	90	60
X12CrMnNiN 17-7-5	1.4372	Cold Rolled Strip	6	350	750-950	45	45	--	--
		Hot Rolled Strip	12	330	750-950	45	45	90	60
		Hot Rolled Plate	75	330	750-950	40	40	90	60
X2CrMnNiN17-7-5	1.4371	Cold Rolled Strip	6	300	650-850	45	45	--	--
		Hot Rolled Strip	12	280	650-850	45	45	90	60
		Hot Rolled Plate	75	280	630-830	35	35	90	60
X12CrMnNiN18-9-5	1.4373	Cold Rolled Strip	6	340	680-880	45	45	--	--
		Hot Rolled Strip	12	320	680-880	45	45	90	60
		Hot Rolled Plate	75	320	600-800	35	35	90	60
X1NiCrMoCu31-27-4	1.4563	Hot Rolled Plate	75	220	500-700	40	40	90	60

EN 10088-2 - MECHANICAL PROPERTIES OF AUSTENITIC STAINLESS STEELS

PART 2 - SHEET, PLATE AND STRIP FOR GENERAL PURPOSES[a] (Continued)

Name	Number	Product Form	Max Size mm	Proof Stress[b,c] $R_{p0.2}$ - N/mm²	Tensile Strength R_m - N/mm²	% Elongation A_{80}[d] < 3 mm	% Elongation A[e] ≥ 3 mm	Impact Energy, KV > 10 mm, J Long.	Impact Energy, KV > 10 mm, J Trans.
Special Grades (Continued)									
X1CrNiMoCuN25-25-5	1.4537	Hot Rolled Plate	75	290	600-800	40	40	90	60
X1CrNiMoCuN20-18-7	1.4547	Cold Rolled Strip	6	320	650-850	35	35	---	---
		Hot Rolled Strip	12	300	650-850	35	35	90	60
		Hot Rolled Plate	75	300	650-850	40	40	90	60
X1NiCrMoCuN25-20-7	1.4529	Hot Rolled Plate	75	300	650-850	40	40	90	60

a. The solution treatment may be omitted if the conditions for the working and subsequent cooling are such that the requirements for the mechanical properties of the product and the resistance-intergranular corrosion as defined in EU 114 are obtained.

b. Transverse test pieces. If, in the case of strip in rolling widths < 300 mm, longitudinal test pieces are taken, the minimum values are reduced as follows:

proof stress: N/mm²

elongation for constant gauge length: minus 5%

elongation for proportional gauge length: minus 2%

c. For continuously hot rolled products, 20 N/mm² higher minimum values of $R_{p0.2}$ and 10 N/mm² higher minimum values of $R_{p1.0}$ may be agreed at the time of enquiry and order.

d. The values apply for transverse test pieces with a gauge length of 80 mm and a width of 20 mm; transverse test pieces with a gauge length of 50 mm and a width of 12.5 mm can also be used.

e. The values apply for transverse test pieces with a gauge length of √5.65 S_o.

f. For stretcher levelled material, the minimum value is 5% lower.

Single values are minimums, unless otherwise specified.

EN 10088-2 - MECHANICAL PROPERTIES OF PRECIPITATION HARDENING STAINLESS STEELS
PART 2 - SHEET, PLATE AND STRIP FOR GENERAL PURPOSES

Name	Number	Product Form	Max Size mm	HTC[a]	Proof Stress $R_{p0.2}$ - N/mm²	Tensile Strength R_m - N/mm²	% Elongation	
							A_{80}^b < 3 mm	$A^c \geq 3$ mm
Special Grade (Martensitic Steel)								
X5CrNiCuNb16-4	1.4542	Cold Rolled Strip	6	SA[d]	---	≤ 1275	5	
				P1300[e]	1150	≥ 1300	3	
				P900[e]	700	≥ 900	6	
		Hot Rolled Plate	50	P1070[f]	1000	1070-1270	8	10
				P950[f]	800	950-1150	10	12
				P850[f]	600	850-1050	12	14
		Hot Rolled Plate	50	SR630[g]	---	≤ 1050	---	
Special Grades (Semi-Austenitic Steels)								
X7CrNiAl17-7	1.4568	Cold Rolled Strip	6	Sa[d, h]	---	≤ 1030	19	
				P1450	1310	≥ 1450	2	
X8CrNiMoAl15-7-2	1.4532	Cold Rolled Strip	6	SA[d]	---	≤ 1100	20	
				P1550[e]	1380	≥ 1550	2	

a. SA - solution annealed; P - precipitation hardened; SR - stress relieved.
b. The values apply for test pieces with a gauge length of 80 mm and a width of 20 mm; test pieces with a gauge length of 50 mm and a width of 12.5 mm can also be used.
c. The values apply for test pieces with a gauge length of 5.65 √So. Values apply-longitudinal and transverse test pieces.
d. Delivery condition.
e. Condition of application; other precipitation hardening temperatures may be agreed.
f. If ordered in the finally treated condition.
g. Delivery condition for further processing; final treatment according-EN 10088 Part 2 Table A.3.
h. For spring-hard rolled condition, see EURONORM 151-2.
Single values are minimums, unless otherwise specified.

EN 10088-2 - MECHANICAL PROPERTIES OF AUSTENITIC-FERRITIC STAINLESS STEELS PART 2 - SHEET, PLATE AND STRIP FOR GENERAL PURPOSES

Name	Number	Product Form	Max Size mm	Proof Stress[a,b] $R_{p0.2}$ N/mm²	Tensile Strength R_m - N/mm²	% Elongation $A_{80}{}^{c}$ < 3 mm	% Elongation A^d ≥ 3 mm	Impact Energy, KV > 10 mm, J Long.	Impact Energy, KV > 10 mm, J Trans.
Standard Grades									
X2CrNiN23-4	1.4362	Cold Rolled Strip	6	420	600-850	20	20	---	---
		Hot Rolled Strip	12	400	600-850	20	20	90	60
		Hot Rolled Plate	75	400	630-800	25	25	90	60
X2CrNiMoN22-5-3	1.4462	Cold Rolled Strip	6	480	660-950	20	20	---	---
		Hot Rolled Strip	12	460	660-950	25	25	90	60
		Hot Rolled Plate	75	460	640-840	25	25	90	60
Special Grades									
X2CrNiMoCuN25-6-3	1.4507	Cold Rolled Strip	6	510	690-940	17	17	---	---
		Hot Rolled Strip	12	590	690-940	17	17	90	60
		Hot Rolled Plate	75	590	690-890	25	25	90	60
X2CrNiMoN25-7-4	1.4410	Cold Rolled Strip	6	550	750-1000	15	15	---	---
		Hot Rolled Strip	12	530	750-1000	15	15	90	60
		Hot Rolled Plate	75	530	730-930	20	20	90	60
X2CrNiMoCuWN25-7-4	1.4501	Hot Rolled Plate	75	530	730-930	25	25	90	60

a. Transverse test pieces. If, in the case of strip in rolling widths < 300 mm, longitudinal test pieces are taken, the minimum proof stress values are reduced by 15 N/mm².

b. For continuously hot rolled products, 20 N/mm² higher minimum values of $R_{p0.2}$ may be agreed at the time of enquiry and order.

c. The values apply for transverse test pieces with a gauge length of 80 mm and a width of 20 mm; transverse test pieces with a gauge length of 50 mm and a width of 12.5 mm can also be used.

d. The values apply for transverse test pieces with a gauge length of √5.65 S_o

Single values are minimums, unless otherwise specified.

Chapter

25

INTERNATIONAL CROSS REFERENCES:

CARBON & ALLOY STEELS

INTERNATIONAL CROSS REFERENCES - CARBON & ALLOY STEELS[a]

Number	Germany DIN Name	USA[b] ASTM/AISI	United Kingdom BS	Japan JIS	France NF	Sweden SS	Russia GOST
1.0028	USt34-2 (S250G1T)	---	---	SS 330	A 34-2	---	---
1.0034	RSt34-2 (S250G2T)	---	1449 34/20 HR, JS, CR, CS		A 34-2 NE	---	St2sp
1.0035	S185 (Fe 310-0) St 33	---	Fe 310-0; 1449 15 HR, HS	---	A 33	1300	St0
1.0036	S235JRG1 (Fe 360 B) USt37-2	A 570 Gr 33, 36	Fe 360 B; 4360-40 B	---	---	1311; 1312	16D; 18kp; St3kp
1.0037	S235JR (Fe 360 B) St 37-2	---	Fe 360 B; 1449 37/23 HR	STKM 12A; C	E 24-2	1311	---
1.0038	S235JRG2 (Fe 360 B) RSt 37-2	A 570 Gr 36	Fe 360 B FU; 1449 27/23 CR; 4360-40 B	---	E 24-2 NE	1312	St3ps; sp
1.0044	S275JR (Fe 430 B) St 44-2	A 570 Gr 40	Fe 430 B FN; 1449 43/25 HR, HS; 4360-43 B	SM 400 A;B;C	E 28-2	1412	St4ps; sp
1.0045	S355JR	---	4360-50 B	---	E 36-2	2172	---
1.0050	E295 (Fe 490-2) St 50-2	A 570 Gr 50; A 572 Gr 50	Fe 490-2 FN; 4360-50 B	SS 490	A 50-2	1550; 2172	St5ps; sp
1.0060	E335 (Fe 590-2) St 60-2	A 572 Gr 65	Fe 590-2 FN; 4360-55 E; 55 C	SM 570	A 60-2	1650	St6ps; sp
1.0070	E360 (Fe 690-2) St 70-2	---	Fe 690-2 FN	---	A 70-2	1655	---
1.0112	P235S	---	1501-164-360B LT20	---	A 37 AP	---	---
1.0114	S235J0; St 37-3 U	---	4360-40 C	---	E 24-3	---	---
1.0116	S235J2G3 (Fe 360 D 1) St 37-3	A 284 Gr D; A 573 Gr 58; A 570 Gr 36, C; A 611 Gr C	Fe 360 D1 FF; 1449 37/23 CR; 4360-40 D	---	E 24-3; E 24-4	1312; 1313	St3kp; ps; sp; 16D
1.0130	P265S	---	1501-164-400B LT 20	---	A 42 AP	---	---
1.0143	S275J0; St 44-3 U	---	4360-43 C	---	E 28-3	1414-01	---

Number	Germany DIN Name	USA[b] ASTM/AISI	United Kingdom BS	Japan JIS	France NF	Sweden SS	Russia GOST
	INTERNATIONAL CROSS REFERENCES - CARBON & ALLOY STEELS[a] (Continued)						
1.0144	S275J2G3 (Fe 430 D 1) St 44-3	A 573 Gr 70; A 611 Gr D	Fe 430 D1 FF; 4360-43 C; 43 D	SM 400 A; B; C	E 28-3; E 28-4	1411; 1412; 1414	St4kp; ps; sp
1.0149	S275J0H; RoSt 44-2	---	4360-43 C	---	---	1412-04	---
1.0226	DX51D; St 02 Z	---	Z2	---	GC	1151-10	---
1.0301	C10	M1010	040 A 10; 045 M 10; 1449 10 CS	S 10 C	AF 34 C 10; XC 10	---	10
1.0330	DC01 St 2; St 12	A 366; 1008	1449 4 CR; 1449 4 CS	SPCC	TC	1142	---
1.0332	DD11; StW 22	A 621; 1008	1449 4 HR; 14 HR	SPHD	1 C	---	15kp
1.0333	USt 3 (DC03G1) USt 13	A 619; 1008	1449 2 CR; 3 CR	SPCD	E	---	---
1.0334	UStW 23 (DD12G1)	A 621; 1008	---	SPHE	2 C	---	10kp
1.0335	DD13; StW 24	A 622; 1008	1449 1 HR	SPHE	3 C	---	08kp
1.0338	DC04 St 4; St 14	A 620; 1008	1449 1 CR; 2 CR	SPCE	ES	1147	08Ju; JuA
1.0345	P235GH HI	A 516 Gr 65, 55; A 515 Gr 65, 55; A 414 Gr C; A 442 Gr 55	1501 Gr. 141-360; 1501 Gr. 161-360 151-360; 1501 Gr. 161-400, 154-360; 1501 Gr. 164-360, 161-360	SGV 410; SGV 450; SGV 480; SPV 450; SPV 480	A 37 CP; AP	1331; 1330	---
1.0347	DC03; RRSt 3; RRSt 13	A 619	1449 3 CR; 1449 2 CR	---	E	1146	08Ju
1.0401	C15	M1015; M1016; M1017	080 A 15; 080 M 15; 1449 17 CS	S 15 C	AF 37 C 12; XC 18	1350	---
1.0402	C22	1020; M1020; M1023	055 M 15; 070 M 20; 1449 22 HS, CS	S 20 C; S 22 C	AF 42 C 20; XC 25; 1 C 22	1450	20
1.0406	C25	1025; M1025	070 M 26	---	1 C 25	---	---

INTERNATIONAL CROSS REFERENCES - CARBON & ALLOY STEELS[a] (Continued)

Number	Germany DIN Name	USA[b] ASTM/AISI	United Kingdom BS	Japan JIS	France NF	Sweden SS	Russia GOST
1.0425	P265GH; H II	---	1501 Gr.161-400, 151-400; 1501 Gr. 164-360, 161-400; 1501 Gr. 164-400; 154-400	SPV 315; SPV 355; SG 295; SGV 410; SGV 450; SGV 480	A 42 CP; AP	1431; 1430; 1432	16K; 20K
1.0473	P355GH; 19Mn6	A 537 Cl1; A 414 Gr G; A612	---	SGV 410; SGV 450; SGV 480	A 52 CP; AP	2101; 2102	---
1.0481	P295GH; 17Mn4	A 516 Gr 70; A 515 Gr 70; A 414 Gr F, G	1501 Gr. 224	SG 365; SGV 410; SGV 450; SGV 480	A 48 CP; AP	---	14G2
1.0501	C35	1035	080 A 32; 080 A 35; 080 M 36; 1449 40 CS	S 35 C	1 C 35; AF 55 C 35; XC 38	1572; 1550	35
1.0503	C45	1045	060 A 47; 080 M 46; 1449 50 HS, CS	S 45 C	1 C 45; AF 65 C 45	1672; 1650	45
1.0511	C40	1040	080 M 40	---	1 C 40; AF 60 C 40	---	---
1.0535	C55	1055	070 M 55	S 55 C	1 C 55; AF 70 C 55	1655	55
1.0539	S355NH; StE 335	---	---	---	TSE 355-4	2134-04	---
1.0540	C50	---	---	---	---	1674	---
1.0545	S355N; StE 355	---	4360-50E	---	E 355 R	2334-01	---
1.0546	S355NL; TStE 355	---	4360-50EE	---	E 355 FP	2135-01	---
1.0547	S355J0H	---	4360-50C	---	TSE 355-3	2172-04	---
1.0549	S355NLH; TStE 355	---	---	---	---	2135	---
1.0553	S355J0; St 52-3U	---	4360-50C	---	E 36-3	---	---

INTERNATIONAL CROSS REFERENCES - CARBON & ALLOY STEELS^a (Continued)

Germany DIN Number	Germany DIN Name	USA^b ASTM/AISI	United Kingdom BS	Japan JIS	France NF	Sweden SS	Russia GOST
1.0562	P355N; StE 355	A 633 Gr C; A588	1501 Gr. 225-490A LT 20	SM 490 A; B; C; YA; YB	FeE 355 KG N; E 355 R/FP; A 510 AP	2106	15GF
1.0565	P355NH; WStE 355	---	1501-225-490B LT 20	---	A 510 AP	2106	---
1.0566	P355NL1; TStE 355	---	1501-225-490A LT 50	---	A 510 FP	2107-01	---
1.0570	S355J2G3; St 52-3	---	Fe 510 D1 FF; 1449 50/35 HR, HS; 4360-50D	SM 490 A; B, C; YA; YB	E 36-3; E 36-4	2132; 2133; 2134; 2174	17GS; 17G1S
1.0577	S355H2G4 (Fe 510 D 2)	A 738	Fe 510 D2 FF; 1501 Gr. 224-460; 1501 Gr. 224-490		A 52 FP	2174	---
1.0601	C60	1060	060 A 62; 1449 HS, CS	S 58 C	1 C 60; AF 70 C 55	---	60(G)
1.0603	C67	1070	080 A 67; 1449 70 HS	---	XC 65	---	---
1.0605	C75	1074; 1075	1449 80 HS	---	---	---	75
1.0614	C76 D; D 75-2	1074	---	---	XC 75	---	---
1.0616	C86 D; D 85-2	1086	---	---	XC 80	---	---
1.0618	C92 D; D 95-2	1095	---	---	XC 90	---	---
1.0715	9SMn28 (11SMn30)	1213	230 M 07	SUM 22	S 250	1912	---
1.0718	9SMnPb28 (11SMnPb30)	12L13	---	SUM 22 L; SUM 23 L; SUM 24 L	S 250 Pb	1914	---
1.0721	10 S 20	1108; 1109	(210 M 15)	---	10 F 1	---	---
1.0722	10 SPb 20	11L08	---	---	10 PbF 2	---	---
1.0723	15S22; 15S20	---	210 A 15; 210 M 15	SUM 32	---	1922	---
1.0726	35 S 20	1140	212 M 36	---	35 MF 6	1957	---
1.0727	45 S 20 (46S20)	1146	---	---	45 MF 4	---	---

INTERNATIONAL CROSS REFERENCES - CARBON & ALLOY STEELS[a] (Continued)

Number	Germany DIN Name	USA[b] ASTM/AISI	United Kingdom BS	Japan JIS	France NF	Sweden SS	Russia GOST
1.0736	9SMn36 (11SMn37)	1215	---	SUM 25	S 300	---	---
1.0737	9SMnPb36 (11SMnPb37)	12L14	---	---	S 300 Pb	1926	---
1.0972	S315MC; QStE 300 TM	---	1501-40F30	---	E 315 D	2642	---
1.0976	S355MC; QStE 360 TM	---	1501-43F35	---	E 355 D	2642	---
1.0982	S460MC; QStE 460 TM	---	1501-50F45	---	---	---	---
1.0984	S500MC; QStE 500 TM	---	---	---	E 490 D	2662	---
1.0986	S550MC; QStE 550 TM	---	1501-60F55	---	E 560 D	---	---
1.1121	Ck 10 (C10E)	1010	040 A 10	S 9 CK; S 10 C	XC 10	1265	08; 10
1.1133	20Mn5	1022; 1518	120 M 19	SMnC 420	20 M 5	2132	20GSL
1.1141	Ck 15 (C15E)	1015	040 A 15; 080 M 15	S 15; S 15 CK	XC 12; XC 15; XC 18	1370	15
1.1151	C22E; Ck 22	1020; 1023	055 M 15; (070 M 20)	S 20 C; S 20 CK; S 22 C	2 C 22; XC 18; XC 25	1450	20
1.1157	40Mn4	1035; 1041	150 M 36	---	35 M5; 40 M 5	---	40G
1.1158	C25E; Ck 25	1025	(070 M 26)	S 25 C; S 28 C	2 C 25; XC 25	---	25
1.1165	30Mn5	1036; 1330	120 M 36 (150 M 28)	SMn 433 H; SCMn 2	35 M 5	---	27ChGSNM DTL; 30GSL
1.1166	34Mn5	1536	---	SMn 433 H	---	---	---
1.1167	36Mn5	1335	150 M 36	SMn 438 (H); SCMn 3	35 M 5; 40 M 5	2120	35G2; 35GL
1.1170	28Mn6	1330	(150 M28); (150 M 19)	SCMn 1	20 M 5; 28 M 6	---	30G
1.1178	C30E; Ck 30	---	080 M 30	---	XC 32	---	---
1.1180	C35R; Cm 35	1035	080 A 35	---	3 C 35; XC 32	1572	---
1.1181	C35E; Ck 35	1035; 1038	080 A 35 (080 M 36)	S 35 C	2 C 35; XC 32; XC 38 H 1	1550; 1572	35
1.1183	Cf 35 (C35G)	1035	080 A 35	S 35 C	XC 38 H 1 TS	1572	35

INTERNATIONAL CROSS REFERENCES - CARBON & ALLOY STEELS[a] (Continued)

Germany DIN Number	Germany DIN Name	USA[b] ASTM/AISI	United Kingdom BS	Japan JIS	France NF	Sweden SS	Russia GOST
1.1186	C40E; Ck 40	1040	060 A 40; 080 A 40; 080 M 40	S 40 C	2 C 40; XC 42 H 1	---	40
1.1191	C45E; Ck 45	1045	080 M 46; 060 A 47	S 45 C; S 48 C	2 C 45; XC 42 H 1; XC 45; XC 48 H 1	1672	45
1.1193	Cf 45 (C45G)	1045	060 A 47; 080 M 46	S 45 C	XC 42 H 1 TS	1672	45
1.1201	C45R; Cm 45	1049	080 M 46	S 50 C	3 C 45; XC 42 H 1; XC 48 H 1	1660	-
1.1203	C55E; Ck 55	1055	060 A 57	S 55 C	2 C 55; XC 55 H 1	1655	55
1.1206	C50E; Ck 50	1049; 1050	080 M 50	---	2 C 50; XC 48 H 1; XC 50 H 1	1674	50
1.1209	C55R; Cm 55	1055	070 M 55	---	3 C 55; XC 55 H 1	---	---
1.1213	Cf 53 (C53G)	1050, 1055	070 M 55	S 50 C	XC 48 H 1 TS	1674	50
1.1221	C60E Ck 60	1060; 1064	060 A 62	S 58 C	2 C 60; XC 60 H 1	1665; 1678	60; 60G; 60GA
1.1231	Ck 67 (C67E)	1070	060 A 67	---	XC 68	1770	65GA; 68GA; 70
1.1248	Ck 75 (C75E)	1074; 1075; 1078	060 A 78	---	XC 75	1774	75(A)
1.1269	Ck 85 (C85E)	1086	---	---	XC 90	---	85(A)
1.1274	Ck 101 (C101E)	1095	---	SUP 4	XC 100	1870	---
1.3401	X120Mn12	---	---	SCMn H 1; SCMn H 11	Z 120 M 12	2183	110G13L
1.3505	100Cr6	52100	2 S 135; 535 A 99	SUJ 2	100 C 6	2258	SchCh 15
1.5024	46Si7	---	---	---	45 S 7; Y 46 S 7; 46 Si 7	---	---
1.5025	51Si7	9255	---	---	51 S 7; 51 Si 7	2090	---
1.5026	55Si7	9255	251 A 58	---	55 S 7	2085; 2090	55S2

INTERNATIONAL CROSS REFERENCES - CARBON & ALLOY STEELS[a] (Continued)								
	Germany DIN	USA[b]	United Kingdom	Japan	France	Sweden	Russia
Number	Name	ASTM/AISI	BS	JIS	NF	SS	GOST
1.5027	60Si7	9260	251 A 60; 251 H 60	---	60 S 7	---	60S2
1.5028	65Si7	9260 H	---	50 P 7; SUP 6	60 S 7	---	---
1.5415	16Mo3; 15Mo3	A 204 Gr A; 4017	1503-243 B	---	15 D 3	2912	---
1.5419	22Mo4	4419	1503-243-430	SCPH 11	---	(2512)	---
1.5423	16Mo5	4520	---	SB 450 M; SB 480 M	---	---	---
1.5622	14Ni6	A 350 Gr LF 5	---	---	16 N 6	---	---
1.5637	12Ni14; 10Ni14	A 350 Gr LF 3	1501-503; 5 S 15	SL 3 N 26; 45	12 N 14; 3.5 Ni 355	---	---
1.5662	X8Ni9	A 353	1501-510; 1502-502-650; 1503-509-690	SL 9 N 53; 60	9 Ni 490	---	---
1.5680	X12Ni5; 12Ni19	2515; 2517	---	---	Z 18 N 5; 5 Ni 390	---	---
1.5711	40NiCr6	3140	---	---	---	---	40ChN
1.5713	13NiCr6	3115	---	---	10 NC 6	---	---
1.5732	14NiCr10	3415	---	SNC 415 (H)	14 NC 11	---	---
1.5736	36NiCr10	3435	---	SNC 631 (H)	30 NC 11	---	---
1.5752	14NiCr14	3310; 3415; 9314	655 H 13	SNC 815 (H)	12 NC 15; 14 NC 12	---	---
1.5919	15CrNi6	3115	---	---	16 NC 6	---	---
1.6511	36CrNiMo4	4340; 9840	817 M 37	---	36 CrNiMo 4; 35 NCD 5; 40 NCD 3	---	40ChN2MA
1.6523	21NiCrMo2	8620	805 H 20; 805 M 20; 806 M 20	SNCM 220 (H)	20 NCD 2	2506	40ChN2MA
1.6546	40NiCrMo2-2	8740	3111-Type 7	SNCM 240	40 NCD 2	---	38ChGNM
1.6562	40NiCrMo8-4	4340	---	SNB 24-1-5	---	---	---
1.6565	40NiCrMo6	4340; 9850	817 A 37; 818 M 40	SNCM 439	---	---	40Ch2N2MA
1.6580	30CrNiMo8	---	823 M 30	SNCM 431	30 CrNiMo 8; 30 NCD 8	---	---

INTERNATIONAL CROSS REFERENCES - CARBON & ALLOY STEELS[a] (Continued)

Germany DIN		USA[b]	United Kingdom	Japan	France	Sweden	Russia
Number	Name	ASTM/AISI	BS	JIS	NF	SS	GOST
1.6582	34CrNiMo6	4337; 4340	816 M 40; 817 M 40	SNCM 447	34 CrNiMo 8; 35 NCD 6	2541	38Ch2N2MA
1.6587	17CrNiMo6	---	---	---	18 NCD 6	---	---
1.6657	14NiCrMo13-4	9310	832 H 13; 832 M 13; S 157	---	16 NCD 13	---	---
1.6746	32NiCrMo14-5	---	---	---	35 NCD 14	---	---
1.6747	30NiCrMo16-6	---	835 M 30	---	35 NCD 16	---	---
1.7003	38Cr2	---	120 M 36	---	38 C 2; 38 Cr 2	---	---
1.7006	46Cr2	5045; 5046	---	---	42 C 2; 42 Cr 2	---	---
1.7015	15Cr3	5015; 5115	523 M 15	SCr 415 (H)	12 C 3; 18 C 3	---	15Ch
1.7030	28Cr4	5130	530 A 30	---	---	---	30Ch
1.7033	34Cr4	5132	530 A 32; 530 H 32; 530 M 32	SCr 430 (H)	32 C 4; 34 Cr 4	---	35Ch
1.7034	37Cr4	5135	31111-3/1; 530 A 36; 530 H 36; 530 M 36	SCr 435 H	37 Cr 4; 38 C 4	---	SchCh10; 40Ch
1.7035	41Cr4	5140	530 A 40; 530 H 40; 530 M 40	SCr 440 (H)	41 Cr 4; 42 C 4	---	40Ch
1.7045	42Cr4	5140	530 A 40	SCr 440	42 C 4 TS	2245	40Ch
1.7108	60SiCr7	9262	---	---	60 SC 7	---	-
1.7131	16MnCr5	5115	527 M 17; 590 H 17; 590 M 17	---	16 MC 4	2173	18ChG
1.7147	20MnCr5	5120	---	SMnC 420 H	20 MC 5	---	18ChG
1.7176	55Cr3	5155; 5160	525 A 58; 525 A 60; 525 H 60	SUP 9 (A)	55 C 3	2253	50ChGA
1.7218	25CrMo4	4130	708 A 25	SCM 420; SCM 430; SCCrM 1	25 CD 4; 25 CrMo 4	2225	20ChM; 30ChM

INTERNATIONAL CROSS REFERENCES - CARBON & ALLOY STEELS[a] (Continued)

Number	Germany DIN Name	USA[b] ASTM/AISI	United Kingdom BS	Japan JIS	France NF	Sweden SS	Russia GOST
1.7220	34CrMo4	4135; 4137	708 A 37	SCM 432; SCCrM 3; SCM 435 H	34 CrMo 4; 35 CD 4	2234	AS38ChGM; 35ChM; 35ChML
1.7223	41CrMo4	4140; 4142	708 M 40; 3111-5/1	SCM 440	42 CD 4 TS	2244	40ChFA
1.7225	42CrMo4	4140; 4142	708 A 42; 708 M 40; 709 M 40	SCM 440 (H); SNB 7	42 CD 4; 42 CrMo 4	2244	---
1.7228	50CrMo4	4150	708 A 47	SCM 445 (H)	50 CrMo 4	---	---
1.7242	16CrMo4	---	---	SCM 418 H	---	---	---
1.7262	15CrMo5	---	---	SCM 415 (H)	12 CD 4	---	---
1.7264	20CrMo5	---	---	SCM 420 H; SCM 421	18 CD 4	---	---
1.7335	13CrMo4-5; 13CrMo4-4	A 182 Gr F11,F12; A 387 Gr 12 Cl. 2	620-440; 620-470; 620-540; 1501-620, 621	SFVA F 12	15 CD 3.5; 15 CD 4.5	2216	12ChM; 15ChM
1.7337	16CrMo4-4	A 387 Gr 12 Cl. 2	---	---	15 CD 4.5	2216	15ChM
1.7361	32CrMo12	---	722 M 24	---	30 CD 12	2240	---
1.7380	10CrMo9-10	A 182 Gr F22; A 387 Gr. 22 Cl. 2	1501-622/515; 1501-622/690; 1502-622; 3604-622	SFVA F 22A, B; SCMV 4; SCPH 32-CF	12 CD 9.10; 10 CD 9.10	2218	12Ch8
1.7715	14MoV6-3	---	1503-660-460	---	---	---	---
1.8159	51CrV4; 50CrV4	6145; 6150	735 A 51; 735 H 51	SUP 10	50 CV 4; 51 CrV 4	2230	50ChGFA; 50ChFA
1.8507	34CrAlMo5	A 355 Cl D	---	---	30 CAD 6.12	---	---
1.8509	41CrAlMo7	A 355 Cl A	905 M 39	SACM 645	40 CAD 6.12	2940	38ChMJuA
1.8515	31CrMo12	---	722 M 24	---	30 CD 12	2240	---
1.8523	39CrMoV13-9	---	897 M 39; 3 S. 132	---	---	---	---

INTERNATIONAL CROSS REFERENCES - CARBON & ALLOY STEELS^a (Continued)

	Germany DIN		United Kingdom	Japan	France	Sweden	Russia
Number		Name	BS	JIS	NF	SS	GOST
			USA^b ASTM/AISI				
1.8902	S420N; StE 420		---	SM 490 A, B, C; YA, YB	FeE 420 KG N; E 420 RIFP	2143	16G2AF
1.8903	S460NL; TStE 460		4360-55 EE	---	E 460 FP	---	---
1.8905	P460N; StE 460		4360-55 F	SM 520 B	FeE 460 KG N; E 460 RIFP	2143	18G2AFps
1.8906	S460QL; TStE 460V		4360-55 F	---	S 460 Q	---	---

a. It is not practical to directly correlate the various metal designations from country to country, let alone comparing several countries and their metal designations; from the view that chemical composition and test methods may be similar, but not identical, and that manufacturing technologies may differ greatly. Consequently, the cross references made in this table are, at best, only listed as a practical guide to assist in finding comparable metal designations, and not equivalent metal designations.

b. Those USA designations beginning with the letter A are ASTM Standards, while those designations beginning with a number or the letter M are AISI Standards.

Chapter

26

INTERNATIONAL CROSS REFERENCES:

CAST IRONS

INTERNATIONAL CROSS REFERENCES - CAST IRONS[a]

Number	Germany DIN Name	USA ASTM	UK BS	Japan JIS	France NF	Sweden SS	Russia GOST
0.6010	GG 10	A 48 Class 20 B	---	G 5501 FC 100	Ft 10 D	01 10-00	Sc 10
0.6015	GG 15	A 48 Class 25 B	1452 Grade 150	G 5501 FC 150	Ft 15 D	01 15-00	Sc 15
0.6020	GG 20	A 48 Class 30 B	1452 Grade 220	G 5501 FC 200	Ft 20 D	01 20-00	Sc 20
0.6025	GG 25	A 48 Class 40 B	1452 Grade 260	G 5501 FC 250	Ft 25 D	01 25-00	Sc 25
0.6030	GG 30	A 48 Class 45 B	1452 Grade 300	G 5501 FC 300	Ft 30 D	01 30-00	Sc 30
0.6035	GG 35	A 48 Class 50 B	1452 Grade 350	G 5501 FC 350	Ft 35 D	01 35-00	Sc 35
0.6040	GG 40	A 48 Class 60 B	1452 Grade 400	---	Ft 40 D	01 40-00	Sc 40
0.6652	GGL-NiMn 137	---	L-NiMn 13 7	---	L-NM 13 7	---	---
0.6655	GGL-NiCuCr 15 6 2	A 436 Type 1	L-NiCuCr 15 6 2	---	L-NUC 15 6 2	---	---
0.6656	GGL-NiCuCr 15 6 3	A 436 Type 1b	L-NiCuCr 15 6 3	---	L-NUC 15 6 3	---	---
0.6660	GGL-NiCr 20 2	A 436 Type 2	L-NiCr 20 2	---	L-NC 20 2	05 23-00	---
0.6661	GGL-NiCr 20 3	A 436 Type 2b	L-NiCr 20 3	---	L-NC 20 3	---	---
0.6667	GGL-NiSiCr 20 5 3	---	L-NiSiCr 20 5 3	---	L-NSC 20 5 3	---	---
0.6676	GGL-NiCr 30 3	A 436 Type 3	L-NiCr 30 3	---	L-NC 30 3	---	---
0.6680	GGL-NiSiCr 30 5 5	A 436 Type 4	L-NiSiCr 30 5 5	---	L-NSC 30 5 5	---	---
0.7040	GGG-40	A 536 Grade 60-40-18	2789 Grade 420/12	G 5502-FCD 400	FGS 400-12	0717-02	VC 42-12
0.7043	GGG-40.3	---	370/17	---	FGS 370-17	0717-15	VC 42-12
---	---	A 536Grade 65-45-12	---	G 5502 FCD 450	---	---	---
0.7050	1693 GGG-50	---	2789 Grade 500/7	G 5502 FCD 500	A32-201 FGS 500-7	0727-02	VC 50-2
0.7060	1693 GGG-60	A 536 Grade 80-55-06	2789 Grade 600/3	G 5502 FCD 600	A32-201 FGS 600-3	0732-03	VC 60-2
0.7070	1693 GGG-70	A 536 Grade 100-70-03	2789 Grade 700/2	G 5502 FCD 700	A31-201 FGS 700-2	0737-01	VC 70-2
0.7080	GGG-80	A 536 Grade 120-90-02	800/2	---	FGS 800-2	---	VC 80-2
0.7652	GGG-NiMn 13 7	---	S-NiMn 13 7	---	S-NM 13 7	---	---
0.7660	GGG-NiCr 20 2	A 439 Type D-2	S-NiCr 20 2	---	S-NC 20 2	---	---
0.7661	GGG-NiCr 20 3	A 439 Type D-2 B	S-NiCr 20 3	---	S-NC 20 3	---	---

INTERNATIONAL CROSS REFERENCES - CAST IRONS[a] (Continued)

Germany DIN		USA	UK	Japan	France	Sweden	Russia
Number	Name	ASTM	BS	JIS	NF	SS	GOST
0.7665	GGG-NiSiCr 20 5 2	---	S-NiSiCr 20 5 2	---	S-NSC 20 5 2	---	---
0.7670	GGG-Ni 22	A 439 Type D-2 C	S-Ni 22	---	S-N 22	---	---
0.7673	GGG-NiMn 23 4	A 571 Type D-2 M	S-NiMn 23 4	---	S-NM 23 4	---	---
0.7676	GGG-NiCr 30 3	A 439 Type D-3	S-NiCr 30 3	---	S-NC 30 3	---	---
0.7677	GGG-NiCr 30 1	A 439 Type D-3 A	S-NiCr 30 1	---	S-NC 30 1	---	---
0.7680	GGG-NiSiCr 30 5 5	A 439 Type D-4	S-NiSiCr 30 5 5	---	S-NSC 30 5 5	---	---
0.7683	GGG-Ni 35	A 439 Type D-5	S-Ni 35	---	S-N 35	---	---
0.7685	GGG-NiCr 35 3	A 439 Type D-5 B	S-NiCr 35 3	---	S-NC 35 3	---	---
0.9620	G-X 260 NiCr 4 2	A 532 I B NiCr-LC	Grade 2 A	---	---	0512-00	---
0.9625	G-X 330 NiCr 4 2	A 532 I A NiCr-HC	Grade 2 B	---	---	0513-00	---
0.9630	G-X 300 CrNiSi 9 5 2	A 532 I D Ni-HiCr	Grade 2 C, D, E	---	---	0457-00	---
0.9635	G-X 300 CrMo 15 3	A 532 II C 15% CrMo-HC	Grade 3 A, B	---	---	---	---
0.9640	G-X 300 CrMoNi 15 2 1	---	Grade 3 A, B	---	---	---	---
0.9645	G-X 260 CrMoNi 20 2 1	A 532 II D 20% CrMo-LC	Grade 3 C	---	---	---	---
0.9650	G-X 260 Cr 27	A 532 III A 25% Cr	Grade 3 D	---	---	0466-00	---
0.9655	G-X 300 CrMo 27 1	A 532 III A 25% Cr	Grade 3 E	---	---	---	---
---	---	---	6681 B 310/10	G 5702 FCMB 310	---	---	---
---	1692 GTS-35-10	A 47M-22010	310 B 35-12	G 5702 FCMB 340	---	---	---
---	1692 GTW-35-04	---	309 W 35-04	G 5703 FCMW 330	---	---	---
---	1692 GTW-40-05	---	309 W 40-05	G 5703 FCMW 370	A 32-701 MB 380-12	---	---
---	1692 GTW-45-07	---	---	G 5703 FCMWP 440	A 32-701 MB 450-7	---	---
---	---	A 220M Grade 310M8	---	G 5704 FCMP 440	---	---	---
---	---	A 220M Grade 340M5	6681 P 50-5	G 5704 FCMP 490	---	---	---
---	---	A 220M Grade 410M4	6681 P 55-04	G 5704 FCMP 540	---	---	---
---	---	A 220 Grade 70003	6681 P 60-3	G 5704 FCMP 590	---	---	---

INTERNATIONAL CROSS REFERENCES - CAST IRONS^a (Continued)

Germany DIN		USA	UK	Japan	France	Sweden	Russia
Number	Name	ASTM	BS	JIS	NF	SS	GOST
---	---	A 220M Grade 620M1	6681 P 70-02	G 5704 FCMP 690	---	---	---
---	---	A 536 Grade 120-90-02	---	G 5502 FCD 800	---	---	---

a. It is not practical to directly correlate the various metal designations from country to country, let alone comparing several countries and their metal designations; from the view that chemical composition and test methods may be similar, but not identical, and that manufacturing technologies may differ greatly. Consequently, the cross references made in this table are, at best, only listed as a practical guide to assist in finding comparable metal designations, and not equivalent metal designations.

Chapter
27

INTERNATIONAL CROSS REFERENCES:

TOOL STEELS

INTERNATIONAL CROSS REFERENCES - TOOL STEELS[a]

Number	Germany DIN Name	USA[b] SAE/ASTM	UK BS	Japan JIS	France NF	International ISO	Russia GOST
1.1525	17350 C 80 W 1	J 438 W108	---	G 4401 SK 5; SK 6	C90E2U; A 35-590 Y$_1$ 80	4957 TC 80	U8A-1; 2
1.1545	17350 C 105 W 1	J 438 W110	---	G 4401 SK 3	C105E2U; A 35-590 Y$_1$ 105	4957 TC 105	U10A1; 2
---	---	---	---	G 4401 SK 4	A 35-590 Y$_1$ 90	4957 TC 90	---
1.1620	17350 C 70 W2	---	---	G 4401 SK 7	A 35-590 Y$_1$ 70	---	---
1.1625	C 80 W 2	A 686 W1	4659 BW 1B	---	---	---	U8-1
1.1645	C 105 W 2	J 438 W110	---	---	(C105E2U) (A 35-590 Y$_2$ 105)	---	U10-1
1.1663	C 125 W	J 438 W112	---	G 4401 SK 2	C120E3U; A 35-590 Y$_2$ 120	4957 TC 120	U13-1
1.1673	C 135 W	---	---	G 4401 SK 1	C140E3U; A 35-590 Y$_2$ 140	4957 TC 140	---
1.1750	C 75 W	A 686 W1	4659 BW 1A	-	---	---	---
1.2067	17350 102Cr6	A 681 L3	4659 BL 3	SUJ 2	100Cr6; A 35-590 Y$_2$ 100 C 6	4957 100 Cr 2	Ch
1.2080	17350 X210Cr12	A 686 D3	4659 BD 3	G 4404 SKD 1	X200Cr12; A 35-590 Z 200 C 12	4957 210 Cr 12	Ch 12
1.2083	X42Cr13	---	---	SUS 420 J 2	X40Cr14; A 35-590 Z 40 C 14	---	---
1.2210	17350 115CrV3	A 686 L2	---	---	---	---	---
1.2330	35CrMo4	AISI 4135; A 686 P20	708 A 37; 4659 BP 20	---	A 35-590 34 CD 4	---	---
1.2332	47CrMo4	AISI 4142	708 M 40	---	---	---	---
1.2343	17350 X38CrMoV5-1	A 686 H11	4659 BH 11	G 4404 SKD 6	X38CrMoV5; A 35-590 Z 38 CDV 5	4957 35 CrMoV 5	4Ch5MFS
1.2344	17350 X40CrMoV5-1	A 686 H13	4659 BH 13	G 4404 SKD 61	X40CrMoV5; A 35-590 Z 40 CDV 5	4957 40 CrMoV 5	4Ch5MF1S
1.2363	X100CrMoV5-1	A 686 A2	4659 BA 2	G 4404 SKD 12	X100CrMoV5; A 35-590 Z 100 CDV 5	4957 100 CrMoV 5	---
1.2365	17350 X32CrMoV3-3	A 686 H10	4659 BH 10	G 4404 SKD 7	32CrMoV12-28; A 35-590 32 CDV 12-28	4957 30 CrMoV 3	3Ch3M3F
---	---	A 686 H19	4659 BH 19	G 4404 SKD 8	---	---	---
1.2379	17350 X155CrVMo12-1	A 686 D2	4659 BD 2	G 4404 SKD 11	X160CrMoV12; A 35-590 Z 160 CDV 12	4957 160 CrMoV 12	---

INTERNATIONAL CROSS REFERENCES- TOOL STEELS[a] (Continued)							
Germany DIN		USA[b]	UK	Japan	France	International	Russia
Number	Name	SAE/ASTM	BS	JIS	NF	ISO	GOST
1.2419	17350 105wcr6	---	---	G 4404 SKS 2; SKS 3; SKS 31	A 35-590 105WCr 5; 105WC 13	4957 105 WCr 1	ChWG
1.2436	17350 X210CrW12	---	---		X210CrVW12-1; A 35-590 Z 210 CW 12-01	4957 210 CrW 12	---
1.2510	100MnCrW4	A 686 O1	4659 BO 1	---	90MnWCrV5; A 35-590 90 MWCV 5	4957 95 MnCrW 1	---
1.2542	45WCrV7	A 686 S1	4659	---	45WCrV8; A 35-590 45 WCV 20	4957 45 WCrV 2	5ChW2SF
1.2550	17350 60WCrV7	A 686 S1	4659 BS 1	---	A 35-590 55 WC 20	4957 60 WCrV 2	---
1.2567	X30WCrV5 3; 30WCrV17-1	---	---	G 4404 SKD 4	X32WCrV5; A 35-590 Z 32 WCV 5	4957 30 WCrV 5	---
1.2581	X30WCrV9-3	A 686 H21	4659 BH 21	G 4404 SKD 5	X30WCrV9; A 35-590 Z 30 WCV 9	4957 30 WCrV 9	3Ch2W8F
1.2601	17350 X165CrMoV12	---	-	---	---	4957 160 CrMoV 12	---
1.2606	X37CrMoW5-1	A 686 H12	4659 BH 12	G 4404 SKD 62	X35CrWMoV5; A 35-590 Z 35 CWDV 5	---	---
---	---	---	---	G 4404 SKT 3	A 35-590 55 CNDV 4	---	---
1.2713	17350 55NiCrMoV6	A 686 L6	4659 BH 224/5	G 4404 SKT 4, SKS 51	55NiCrMoV7; A 35-590 55 NCDV 7	4957 55 NiCrMoV 2	5ChNM
1.2833	100V1	J 438 W210	4659 BW 2	G 4404SKS 43	C105E2UV1; A 35-590 Y_1 105V	4957 TCV 105	---
1.2842	17350 90MnCrV8	A 686 O2	4659 BO 2	---	90MnV8	4957 90 MnV 2	---
1.2885	X32CrMoCoV3-3-3	---	4659 BH 10A	---	---	---	---
1.3202	17350 S 12-1-4-5	A 600 T15	4659 BT 15	G 4403 SKH 10	A 35-590 Z 160WKVC 12-05-05-04	4957 HS 12-1-5-5	---
1.3207	17350 S 10-4-3-10	---	4659 BT 42	G 4403 SKH 57	A 35-590 Z 130WKCDV10-10-04-04-03	4957 HS 10-4-3-10	---
1.3243	17350 S 6-5-2-5	---	4659 BM 35	G 4403 SKH 55	A 35-590 Z 85WDKCV06-05-05-04-02; A 35-590 Z 90WDKCV06-05-04-02	4957 HS 6-5-2-5	R6M5K5
---	---	A 600 M36	---	G 4403 SKH 56	---	---	---
1.3246	17350 S 7-4-2-5	A 600 M41	---	---	A 35-590 Z 110WKCDV07-05-04-04-02	4957 HS 7-4-2-5	---
1.3247	17350 S 2-10-1-8	A 600 M42	4659 BM 42	G 4403 SKH 59	A 35-590 Z 110DKCWV09-08-04-02-01	4957 HS 2-9-1-8	---

INTERNATIONAL CROSS REFERENCES- TOOL STEELS^a (Continued)

Germany DIN Number	Germany DIN Name	USA^b SAE/ASTM	UK BS	Japan JIS	France NF	International ISO	Russia GOST
1.3249	S 2-9-2-8	A 600 M33; M34	4659 BM 34	---	---	---	---
1.3255	17350 S 18-1-2-5	A 600 T4	4659 BT 4	G 4403 SKH 3	A 35-590 Z 80WKCV18-05-04-01	4957 HS 18-1-1-5	---
1.3265	S 18-1-2-10	A 600 T5	4659 BT 5	G 4403 SKH 4	A 35-590 Z 80 WKCV 18-10-04-02	4957 HS 18-0-1-10	---
1.3342	17350 SC 6-5-2	A 600 M3	---	---	A 35-590 Z 90WDCV06-05-04-02	---	---
1.3343	17350 S 6-5-2	A 600 M2	4659 BM 2	G 4403 SKH 51	A 35-590 Z 85WDCV06-05-04-02	4957 HS 6-5-2	(R6AM5) R6M5
---	---	A 600 M3 Class 1	---	G 4403 SKH 52	---	---	---
1.3344	17350 S 6-5-3	A 600 M3 Class 2	---	G 4403 SKH 53	A 35-590 Z 120WDCV06-05-04-03	4957 HS 6-5-3	---
---	---	A 600 M4	4659 BM 4	G 4403 SKH 54	A 35-590 Z 130WDCV06-05-04-04	---	---
1.3346	S 2-9-1	A 686 H41; A 600 M1	4659 BM 1	---	A 35-590 Z 85DCWV08-04-02-01	4957 HS 1-8-1	---
1.3348	17350 S 2-9-2	A 600 M7	---	G 4403 SKH 58	A 35-590 Z 100DCWV09-04-02-02	4957 HS 2-9-2	---
1.3355	S 18-0-1	A 600 T1	4659 BT 1	G 4403 SKH 2	A 35-590 Z 80WCV18-04-01	4957 HS 18-0-1	R18

a. It is not practical to directly correlate the various metal designations from country to country, let alone comparing several countries and their metal designations; from the view that chemical composition and test methods may be similar, but not identical, and that manufacturing technologies may differ greatly. Consequently, the cross references made in this table are, at best, only listed as a practical guide to assist in finding comparable metal designations, and not equivalent metal designations.

b. Those USA designations beginning with the letter A are ASTM Standards, while those beginning with the letter J are SAE Standards. AISI Standards are identified in the table.

Chapter
28

INTERNATIONAL CROSS REFERENCES:

STAINLESS STEELS

INTERNATIONAL CROSS REFERENCES - STAINLESS STEELS

Germany DIN		USA[b]	United Kingdom	Japan	France	International	Russia
Number	Name	UNS/Other	BS	JIS	FN	ISO	GOST
1.4000	X6Cr13	S40300; 403; S41008; 410S; S42900; 429	403 S 17	SUS 403; SUS 410 S; SUS 429	Z 8 C 12	683-13/1	08Ch13
1.4001	X7Cr14	S40300; 403; S41008; 410S; S42900; 429	403 S 17	SUS 403; SUS 410 S; SUS 429	Z 8 C 13 FF	---	08Ch13
1.4002	X6CrAl13	S40500; 405	405 S 17	SUS 405	Z 8 CA 12	683-13/2	-
1.4005	X12CrS13	S41600; 416	416 S 21	SUS 416	Z 11 CF 13	683-13/7	-
1.4006	X12Cr13; X10Cr13	S41000; 410	410 S 21; ANC 1A	SUS 410	Z 10 C 13	683-13/3	12Ch13; 15Ch13L
1.4006	GX12Cr13	J91540; CA-15	410 C 21	---	---	---	---
1.4008	GX8CrNi13	---	410 C 21	SCS 1	Z 12 CN 13 M	---	---
1.4016	X6Cr17	S43000; 430	430 S 17; 430 S 18	SUS 430	Z 8 C 17	683-13/8	12Ch17
1.4021	X20Cr13	S42000; 420	420 S 37	SUS 420 J 1	Z 20 C 13	683-13/4	20Ch13
1.4024	X15Cr13	---	420 S 29	SUS 410 J 1	Z 13 C 13	---	
1.4027	GX20Cr14	---	ANC 1 B, C; 420 C 24; 420 C 29	SCS 2	Z 20 C 13 M	---	20Ch13L
1.4028	X30Cr13	S42020; 420F	420 S 45	SUS 420 J 2	Z 30 C13; Z 33 C 13	683-13/5	30Ch13
1.4031	X38Cr13; X39Cr13	---	---	SUS 420 J 2	Z 40 C 14	---	40Ch13
1.4034	X46Cr13	---	420 S 45	---	Z 44 C 14; Z 38 C 13 M	---	40Ch13
1.4057	X20CrNi17 2; X19CrNi17-2	S43100; 431	431 S 29; 6 S. 80	SUS 431	Z 15 CN 16-02	683-13/9 B	20Ch17N2
1.4104	X12CrMoS17; X14CrMoS17	S43020; 430F	---	SUS 430 F	Z 13 CF 17	683-13/8 C	
1.4113	X6CrMo17-1	S43400; 434	434 S 17	SUS 434	---	683-13/9 C	---
---	---	S44002; 440 A	---	SUS 440 A	---	---	---

INTERNATIONAL CROSS REFERENCES - STAINLESS STEELS (Continued)							
Germany DIN		USA[b]	United Kingdom	Japan	France	International	Russia
Number	Name	UNS/Other	BS	JIS	FN	ISO	GOST
---	---	S44003; 440B	---	SUS 440 B	---	---	---
1.4125	X105CrMo17	S44004; 440C	---	SUS 440 C	Z 100 CD 17	---	95Ch18
---	---	S20100; 201	---	SUS 201	Z 12 CMN 17-07 AZ	683-13/A-2	---
---	---	S20200; 202	284 S 16	SUS 202	---	683-13/A-3	---
1.4301	X5CrNi18 10; X4CrNi18-10	S30400; 304	304 S 11; 304 S 15; 304 S 16; 304 S 17; LW 21; LWCF 21; 304 S 31	SUS 304	Z 4 CN 19-10 FF; Z 5 CN 17-08; Z 6 CN 18-09; Z 7 CN 18-09	683-13/11	08Ch18N10
1.4303	X5CrNi18 12; X4CrNi18-12	S30500; 305 S30800; 308	305 S 17; 305 S 19	SUS 305 J 1; SUS 305	Z 5 CN 18-11 FF	---	06Ch18N11
---	---	S30900; 309	309 S 24	SUH 309	Z 12 CN 24-13	---	---
1.4305	X10CrNiS18 9; X8CrNiS18-9	30300; 303	303 S 22; 303 S 31	SUS 303	Z 8 CNF 18-09	683-13/17	---
---	---	S30323 303 Se	303 S 41	SUS 303 Se	---	683-13/17a	---
1.4306	X2CrNi19-11	S30403; 304L	304 S 11; LW 20; LWCF 20; S. 536; T. 74; 304 C 12 (LT 196); 305 S 11	SCS 19; SUS 304 L	Z 1 CN 18-12; Z 2 CN 18-10; Z 3 CN 19.10 M; Z 3 CN 18-10; Z 3 CN 19-11; Z 3 CN 19-11 FF	683-13/10	03Ch18N11
1.4308	GX5CrNi9-10 G-X 6 CrNi 18 9	J92600 CF-8	304 C 15 (LT 196)	SCS 13	Z 6 CN 18.10 M	---	07Ch18N9L
1.4310	X12CrNi 17 7; X9CrNi18-8	S30100; 301	301 S 21; 301 S 22	SUS 301	Z 11 CN 17-08; Z 11 CN 18-08; Z 12 CN 18-09	683-13/14	---
1.4311	X2CrNiN18-10	S30453; 304LN	304 S 61	SUS 304 LN	Z 3 CN 18-07 Az; Z 3 CN 18-10 Az	683-13/10N	---

INTERNATIONAL CROSS REFERENCES - STAINLESS STEELS (Continued)

Germany DIN		USA[b]	United Kingdom	Japan	France	International	Russia
Number	Name	UNS/Other	BS	JIS	FN	ISO	GOST
1.4312	GX10CrNi18-8	---	302 C 25; ANC 3 A	SCS 12; SCS 13 A	Z 10 CN 18.9 M	---	10Ch18N9L
1.4313	GX5CrNi13-4	J91540; CA6-NM	425 C 11; 425 C12	SCS 5; SCS 6	Z 4 CND 13.4 M; Z 6 CN 13-4; Z 8 CD 17-01	---	---
1.4319	X3CrNiN17-8	S30200; 302	301 S 26; 302 S 26	SUS 302	Z 12 CN 18.09	683-13/12	---
1.4401	X5CrNiMo17 12 2; X4CrNiMo17-12-2	S31600; 316	316 S 13; 316 S 17; 316 S 19; 316 S 31; 316 S 33	SUS 316	Z 3 CND 17-11-01; Z 6 CND 17-11; Z 6 CND 17-11-02 FF; Z 7 CND 17-11-02; Z 7 CND 17-12-02	---	---
1.4404	X2CrNiMo17 13 2; X2CrNiMo17-12-2	S31603; 316L	316 S 11; 316 S 13; 316 S 14; 316 S 31; 316 S 42; S. 537; S.161	SUS 316 L	Z 2 CND 17-12; Z 2 CND 18-13; Z 3 CND 17-11-02; Z CND 17-12-02 FF; Z 3 CND 18-12-02; Z 3 CND 18-12-03; Z 3 CND 19.10 M	---	---
1.4406	X2CrNiMoN 17 12 2; X2CrNiMoN17-11-2	S31653; 316LN	316 S 61; 316 S 63	SUS 316 LN	Z 3 CND 17-11 Az	683-13/19N	---
1.4408	GX5CrNiMo19-11; G-X6CrNiMo18 10	J92900; CF-8M	316 C 16 (LT 196); ANC 4 B	SCS 14	---	---	07Ch18N10G2S2M2L
1.4429	X2CrNiMoN17-13-3	S31653; 316LN	316 S 63	SUS 316 LN	Z 3 CND 17-12 Az	683-13/19N	---
1.4435	X2CrNiMo18-14-3	S31603; 316L	316 S 11; 316 S 13; 316 S 14; 316 S 31; LW 22; LWCF 22	SUS 316 L	Z 3 CND 17-12-03; Z 3 CND 18-14-03	683-13/19	03Ch17N14M3

INTERNATIONAL CROSS REFERENCES - STAINLESS STEELS (Continued)

Germany DIN		USA[b]	United Kingdom	Japan	France	International	Russia
Number	Name	UNS/Other	BS	JIS	FN	ISO	GOST
1.4436	X5CrNiMo17 13 3; X4CrNiMo17-13-3	S31600; 316	316 S 19; 316 S 31; 316 S 33; LW 23; LWCF 23	SUS 316	Z 6 CND 18-12-03; Z 7 CND 18-12-03	683-13/20	---
1.4438	X2CrNiMo 18 16 4; X2CrNiMo18-15-4	S31703; 317L	317 S 12	SUS 317 L	Z 2 CND 19-15-04; Z 3 CND 19-15-04	683-13/24	---
1.4449	X5CrNiMo17 13	S31700; 317	317 S 16	SUS 317	---	---	---
1.4460	X4CrNiMoN 27 5 2; X3CrNiMoN27-5-2	S32900; 329	---	SUS 329 J 1	Z 3 CND 25-07 Az; Z 5 CND 27-05 Az	---	---
1.4462	X2CrNiMoN22-5-3	---	318 S 13	SUS 329 J3L	Z 3 CND 22-05 Az; Z 2 CND 24-08 Az; Z 3 CND 25-06-03 Az	---	---
---	---	---	331 S 42	SUH 31	---	---	---
---	---	---	349 S 52	SUH 35	Z 52 CMN 21-09	---	---
1.4510	X6CrTi17; X3CrTi17	S43036; 430Ti	---	SUS 430 LX	Z 4 CT 17	---	08Ch17T
1.4511	X6CrNb17; X3CrNb17	---	---	SUS 430 LX	Z 4 CNb 17	---	---
1.4512	X6CrTi12; X2CrTi12	S40900; 409	LW 19; 409 S 19	SUH 409	Z 3 CT 12	683-13/1 Ti	---
1.4521	X2CrMoTi18-2	S44300; 443	---	SUS 444	---	683-13/E 1	---
1.4539	X1NiCrMoCuN25-20-5	N08904; 904L	---	---	Z 2 NCDU 25-20	---	---
1.4541	X6CrNiTi18-10	S32100; 321	321 S 31; 321 S 51 (1010); 321 S 51 (1105); LW 24; LWCF 24	SUS 321	Z 6 CNT 18-10	683-13/15	06Ch18N10T; 08Ch18N10T; 09Ch18N10T; 12Ch18N10T
1.4542	X5CrNiCuNb17 4; X5CrNiCuNb16-4	S17400; 630	---	SCS 24; SUS 630	Z 7 CNU 15-05; Z 7 CNU 17-04	683-16/1	---

INTERNATIONAL CROSS REFERENCES - STAINLESS STEELS (Continued)

Germany DIN Number	Germany DIN Name	USA[b] UNS/Other	United Kingdom BS	Japan JIS	France FN	International ISO	Russia GOST
1.4544	---	---	S.524; S.526	---	---	---	08Ch18N12T
1.4546	X5CrNiNb18-10	S34800; 348	347 S 31; 2 S 130;	---	---	---	---
1.4550	X6CrNiNb18-10	S34700; 347	347 S 20; 347 S 31; 347 S 51; ANC 3 B	SUS 347	Z 6 CNNb 18-10	683-13/16	08Ch18N12B
1.4552	GX5CrNiNb19-10 G-X 5 CrNiNb18 9	J92710; CF-8C	347 C 17	SCS 21	Z 6 CNNb 18.10 M	---	---
1.4568	X7CrNiAl17-7	S17700; 631	301 S 81	SUS 631	Z 9 CNA 17-07	683-16/2	09Ch17N7Ju1
1.4571	X6CrNiMoTi17-12-2	S31635; 316Ti	320 S 18; 320 S 31	SUS 316 Ti	Z 6 CNDT 17-12	683-13/21	10Ch17N13M2T
1.4573	X10CrNiMoTi18 12	S31635; 316Ti	320 S 33	SUS 316 Ti	---	683-13/21	10Ch17N13M3T; 08Ch17N13M2T
1.4580	X6CrNiMoNb17-12-2	S31640; 316Cb	318 S 17	---	Z 6 CNDNb 17-12	---	08Ch16N13M2B
1.4581	GX5CrNiMoNb19-11; G-X5CrNiMoNb18 10	---	318 C 17; ANC 4 C	SCS 22	Z 4 CNDNb 18.12 M	---	---
1.4583	X10CrNiMoNb18-12	---	---	---	---	---	---
1.4718	X45CrSi9-3	S65007; J775 HNV 3	401 S 45	SUH 1	Z 45 CS 9	---	40Ch9S2
1.4724	X10CrAl13	---	---	---	Z 13 C 13	---	10Ch13SJu
1.4731	X40CrSiMo10-2	---	---	SUH 3	Z 40 CSD 10	---	40Ch10S2M
1.4742	X10CrAl18	---	---	SUH 21	Z 12 CAS 18	---	15Ch18SJu
1.4747	X80CrNiSi20	S65006; J775 HNV 6	443 S 65	SUH 4	Z 80 CNS 20-02	---	---
---	---	S42200; J775 HNV-8	---	SUH 616	---	---	---

INTERNATIONAL CROSS REFERENCES - STAINLESS STEELS (Continued)

Germany DIN		USA[b]	United Kingdom	Japan	France	International	Russia
Number	Name	UNS/Other	BS	JIS	FN	ISO	GOST
1.4762	X10CrAl24	S44600; 446	---	(SUH 446)	Z 12 CAS 25	---	---
1.4828	X15CrNiSi20-12	S30900; 309	309 S 24	SUH 309	Z 9 CN 24-13; Z 17 CNS 20-12	---	20Ch20N14S2
1.4833	X12CrNi24-12; X7CrNi23 14	S30908; 309S	---	SUS 309 S	Z 15 CN 23-13; Z 15 CN 24-13; Z 20 CN 24-13	---	---
1.4837	GX40CrNiSi25-12	---	309 C 30	SCH 13 A; SCH 17; SCS 17	---	---	40Ch24N12SL
1.4841	X15CrNiSi25-20	S31400; S31000	314 S 25	SUH 310	Z 15 CNS 25-20	---	20Ch25N20S2
1.4842	X12CrNi25-20	S31008; 310S	---	SUS 310 S	Z 12 CN 26-12	---	---
1.4845	X12CrNi25-12	S31008; 310S	310 S 16; 310 S 24; 310 S 25; 310 S 31	SUH 310; SUS 310 S	Z 8 CN 25-20; Z 12 CN 25-20; Z 12 CN 26-21	683-13/H15	20Ch23N18
1.4848	GX40CrNiSi25-20	HK40; J94204	310 C 40; 310 C 45	SCH 21; SCH 22	---	---	---
1.4864	X12NiCrSi36-16	---	NA 17	SUH 330	Z 20 NCS 33-16	---	---
1.4865	GX40NiCrSi38-18	---	330 C 11; 330 C 40; 331 C 40	SCH 15; SCH 16	---	---	---
1.4871	X53CrMnNiN21-9	S63008; J775 EV-8	349 S 54	SUH 35; SUH 36	Z 53 CMNS 21-09 Az; Z 53 CMN 21-09 Az	---	55Ch20G9AN4
---	---	S63017; J775 EV-4	381 S 34	SUH 37	---	---	---
1.4873	X45CrNiW18-9	---	---	SUH 31	Z 35 CNWS 14-14; Z 45 CNW 18-09	---	---
1.4876	X10NiCrAlTi32-20	---	NA 15 (H)	NCF 800 (TP)	Z 8 NC 33-21; Z 10 NC 32-21	---	---
1.4878	X12CrNiTi18-9	S32100; 321	321 S 51	SUS 321	Z 6 CNT 18-10	683-13/15	---
1.4922	X20CrMoV12-1	---	---	---	---	---	---

INTERNATIONAL CROSS REFERENCES - STAINLESS STEELS (Continued)							
Germany DIN		USA[b]	United Kingdom	Japan	France	International	Russia
Number	Name	UNS/Other	BS	JIS	FN	ISO	GOST
1.4944	---	S66286; 660	HR 51	---	Z 6 NCTDV 25-15 B	---	---

a. It is not practical to directly correlate the various metal designations from country to country, let alone comparing several countries and their metal designations; from the view that chemical composition and test methods may be similar, but not identical, and that manufacturing technologies may differ greatly. Consequently, the cross references made in this table are, at best, only listed as a practical guide to assist in finding comparable metal designations, and not equivalent metal designations.

b. Those USA designations beginning with the letters S, J or N and followed by five numbers are UNS Numbers, while those beginning with J775 are SAE Standards. Other designations not beginning with a letter are AISI Standards or common names.

Appendix

1

HARDNESS CONVERSION TABLES

APPROXIMATE HARDNESS CONVERSION NUMBERS FOR NONAUSTENITIC STEELS[a, b]

Rockwell C 150 kgf Diamond HRC	Vickers HV	Brinell 3000 kgf 10mm ball HB	Knoop 500 gf HK	Rockwell A 60 kgf Diamond HRA	Rockwell Superficial Hardness			Approximate Tensile Strength ksi (MPa)
					15 kgf Diamond HR15N	30 kgf Diamond HR30N	45 kgf Diamond HR45N	
68	940	---	920	85.6	93.2	84.4	75.4	---
67	900	---	895	85.0	92.9	83.6	74.2	---
66	865	---	870	84.5	92.5	82.8	73.3	---
65	832	739[d]	846	83.9	92.2	81.9	72.0	---
64	800	722[d]	822	83.4	91.8	81.1	71.0	---
63	772	706[d]	799	82.8	91.4	80.1	69.9	---
62	746	688[d]	776	82.3	91.1	79.3	68.8	---
61	720	670[d]	754	81.8	90.7	78.4	67.7	---
60	697	654[d]	732	81.2	90.2	77.5	66.6	---
59	674	634[d]	710	80.7	89.8	76.6	65.5	351 (2420)
58	653	615	690	80.1	89.3	75.7	64.3	338 (2330)
57	633	595	670	79.6	88.9	74.8	63.2	325 (2240)
56	613	577	650	79.0	88.3	73.9	62.0	313 (2160)
55	595	560	630	78.5	87.9	73.0	60.9	301 (2070)
54	577	543	612	78.0	87.4	72.0	59.8	292 (2010)
53	560	525	594	77.4	86.9	71.2	58.6	283 (1950)
52	544	512	576	76.8	86.4	70.2	57.4	273 (1880)
51	528	496	558	76.3	85.9	69.4	56.1	264 (1820)
50	513	482	542	75.9	85.5	68.5	55.0	255 (1760)
49	498	468	526	75.2	85.0	67.6	53.8	246 (1700)
48	484	455	510	74.7	84.5	66.7	52.5	238 (1640)
47	471	442	495	74.1	83.9	65.8	51.4	229 (1580)
46	458	432	480	73.6	83.5	64.8	50.3	221 (1520)

APPROXIMATE HARDNESS CONVERSION NUMBERS FOR NONAUSTENITIC STEELS[a, b] (Continued)

Rockwell C 150 kgf Diamond HRC	Vickers HV	Brinell 3000 kgf 10mm ball HB	Knoop 500 gf HK	Rockwell A 60 kgf Diamond HRA	Rockwell Superficial Hardness			Approximate Tensile Strength ksi (MPa)
					15 kgf Diamond HR15N	30 kgf Diamond HR30N	45 kgf Diamond HR45N	
45	446	421	466	73.1	83.0	64.0	49.0	215 (1480)
44	434	409	452	72.5	82.5	63.1	47.8	208 (1430)
43	423	400	438	72.0	82.0	62.2	46.7	201 (1390)
42	412	390	426	71.5	81.5	61.3	45.5	194 (1340)
41	402	381	414	70.9	80.9	60.4	44.3	188 (1300)
40	392	371	402	70.4	80.4	59.5	43.1	182 (1250)
39	382	362	391	69.9	79.9	58.6	41.9	177 (1220)
38	372	353	380	69.4	79.4	57.7	40.8	171 (1180)
37	363	344	370	68.9	78.8	56.8	39.6	166 (1140)
36	354	336	360	68.4	78.3	55.9	38.4	161 (1110)
35	345	327	351	67.9	77.7	55.0	37.2	156 (1080)
34	336	319	342	67.4	77.2	54.2	36.1	152 (1050)
33	327	311	334	66.8	76.6	53.3	34.9	149 (1030)
32	318	301	326	66.3	76.1	52.1	33.7	146 (1010)
31	310	294	318	65.8	75.6	51.3	32.5	141 (970)
30	302	286	311	65.3	75.0	50.4	31.3	138 (950)
29	294	279	304	64.6	74.5	49.5	30.1	135 (930)
28	286	271	297	64.3	73.9	48.6	28.9	131 (900)
27	279	264	290	63.8	73.3	47.7	27.8	128 (880)
26	272	258	284	63.3	72.8	46.8	26.7	125 (860)
25	266	253	278	62.8	72.2	45.9	25.5	123 (850)
24	260	247	272	62.4	71.6	45.0	24.3	119 (820)
23	254	243	266	62.0	71.0	44.0	23.1	117 (810)
22	248	237	261	61.5	70.5	43.2	22.0	115 (790)

APPROXIMATE HARDNESS CONVERSION NUMBERS FOR NONAUSTENITIC STEELS[a, b] (Continued)

Rockwell C 150 kgf Diamond HRC	Vickers HV	Brinell 3000 kgf 10mm ball HB	Knoop 500 gf HK	Rockwell A 60 kgf Diamond HRA	Rockwell Superficial Hardness			Approximate Tensile Strength ksi (MPa)
					15 kgf Diamond HR15N	30 kgf Diamond HR30N	45 kgf Diamond HR45N	
21	243	231	256	61.0	69.9	42.3	20.7	112 (770)
20	238	226	251	60.5	69.4	41.5	19.6	110 (760)

a. This table gives the approximate interrelationships of hardness values and approximate tensile strength of steels. It is possible that steels of various compositions and processing histories will deviate in hardness-tensile strength relationship from the data presented in this table. The data in this table should not be used for austenitic stainless steels, but have been shown to be applicable for ferritic and martensitic stainless steels. Where more precise conversions are required, they should be developed specially for each steel composition, heat treatment, and part.

b. All relative hardness values in this table are averages of tests on various metals whose different properties prevent establishment of exact mathematical conversions. These values are consistent with ASTM A 370-91 for nonaustenitic steels. It is recommended that ASTM standards A 370, E 140, E 10, E 18, E 92, E 110 and E 384, involving hardness tests on metals, be reviewed prior to interpreting hardness conversion values.

c. Carbide ball, 10mm.

d. This Brinell hardness value is outside the recommended range for hardness testing in accordance with ASTM E 10.

APPROXIMATE HARDNESS CONVERSION NUMBERS FOR NONAUSTENITIC STEELS [a, b]

Rockwell B 100 kgf 1/16" ball HRB	Vickers HV	Brinell 3000 kgf 10 mm HB	Knoop 500 gf HK	Rockwell A 60 kgf Diamond HRA	Rockwell Superficial hardness			Approximate Tensile Strength ksi (MPa)
					15 kgf 1/16" ball HR15T	30 kgf 1/16" ball HR30T	45 kgf 1/16" ball HR45T	
100	240	240	251	61.5	93.1	83.1	72.9	116 (800)
99	234	234	246	60.9	92.8	82.5	71.9	114 (785)
98	228	228	241	60.2	92.5	81.8	70.9	109 (750)
97	222	222	236	59.5	92.1	81.1	69.9	104 (715)
96	216	216	231	58.9	91.8	80.4	68.9	102 (705)
95	210	210	226	58.3	91.5	79.8	67.9	100 (690)
94	205	205	221	57.6	91.2	79.1	66.9	98 (675)
93	200	200	216	57.0	90.8	78.4	65.9	94 (650)
92	195	195	211	56.4	90.5	77.8	64.8	92 (635)
91	190	190	206	55.8	90.2	77.1	63.8	90 (620)
90	185	185	201	55.2	89.9	76.4	62.8	89 (615)
89	180	180	196	54.6	89.5	75.8	61.8	88 (605)
88	176	176	192	54.0	89.2	75.1	60.8	86 (590)
87	172	172	188	53.4	88.9	74.4	59.8	84 (580)
86	169	169	184	52.8	88.6	73.8	58.8	83 (570)
85	165	165	180	52.3	88.2	73.1	57.8	82 (565)
84	162	162	176	51.7	87.9	72.4	56.8	81 (560)
83	159	159	173	51.1	87.6	71.8	55.8	80 (550)
82	156	156	170	50.6	87.3	71.1	54.8	77 (530)
81	153	153	167	50.0	86.9	70.4	53.8	73 (505)
80	150	150	164	49.5	86.6	69.7	52.8	72 (495)
79	147	147	161	48.9	86.3	69.1	51.8	70 (485)
78	144	144	158	48.4	86.0	68.4	50.8	69 (475)
77	141	141	155	47.9	85.6	67.7	49.8	68 (470)

APPROXIMATE HARDNESS CONVERSION NUMBERS FOR NONAUSTENTIC STEELS [a, b] (Continued)

Rockwell B 100 kgf 1/16" ball HRB	Vickers HV	Brinell 3000 kgf 10 mm HB	Knoop 500 gf HK	Rockwell A 60 kgf Diamond HRA	Rockwell Superficial hardness			Approximate Tensile Strength ksi (MPa)
					15 kgf 1/16" ball HR15T	30 kgf 1/16" ball HR30T	45 kgf 1/16" ball HR45T	
76	139	139	152	47.3	85.3	67.1	48.8	67 (460)
75	137	137	150	46.8	85.0	66.4	47.8	66 (455)
74	135	135	147	46.3	84.7	65.7	46.8	65 (450)
73	132	132	145	45.8	84.3	65.1	45.8	64 (440)
72	130	130	143	45.3	84.0	64.4	44.8	63 (435)
71	127	127	141	44.8	83.7	63.7	43.8	62 (425)
70	125	125	139	44.3	83.4	63.1	42.8	61 (420)
69	123	123	137	43.8	83.0	62.4	41.8	60 (415)
68	121	121	135	43.3	82.7	61.7	40.8	59 (405)
67	119	119	133	42.8	82.4	61.0	39.8	58 (400)
66	117	117	131	42.3	82.1	60.4	38.7	57 (395)
65	116	116	129	41.8	81.8	59.7	37.7	56 (385)
64	114	114	127	41.4	81.4	59.0	36.7	---
63	112	112	125	40.9	81.1	58.4	35.7	---
62	110	110	124	40.4	80.8	57.7	34.7	---
61	108	108	122	40.0	80.5	57.0	33.7	---
60	107	107	120	39.5	80.1	56.4	32.7	---
59	106	106	118	39.0	79.8	55.7	31.7	---
58	104	104	117	38.6	79.5	55.0	30.7	---
57	103	103	115	38.1	79.2	54.4	29.7	---
56	101	101	114	37.7	78.8	53.7	28.7	---
55	100	100	112	37.2	78.5	53.0	27.7	---
54	---	---	111	36.8	78.2	52.4	26.7	---
53	---	---	110	36.3	77.9	51.7	25.7	---

APPROXIMATE HARDNESS CONVERSION NUMBERS FOR NONAUSTENITIC STEELS [a, b] (Continued)

Rockwell B 100 kgf 1/16" ball HRB	Vickers HV	Brinell 3000 kgf 10 mm HB	Knoop 500 gf HK	Rockwell A 60 kgf Diamond HRA	Rockwell Superficial hardness			Approximate Tensile Strength ksi (MPa)
					15 kgf 1/16" ball HR15T	30 kgf 1/16" ball HR30T	45 kgf 1/16" ball HR45T	
52	--	--	109	35.9	77.5	51.0	24.7	--
51	--	--	108	35.5	77.2	50.3	23.7	--
50	--	--	107	35.0	76.9	49.7	22.7	--
49	--	--	106	34.6	76.6	49.0	21.7	--
48	--	--	105	34.1	76.2	48.3	20.7	--
47	--	--	104	33.7	75.9	47.7	19.7	--
46	--	--	103	33.3	75.6	47.0	18.7	--
45	--	--	102	32.9	75.3	46.3	17.7	--
44	--	--	101	32.4	74.9	45.7	16.7	--
43	--	--	100	32.0	74.6	45.0	15.7	--
42	--	--	99	31.6	74.3	44.3	14.7	--
41	--	--	98	31.2	74.0	43.7	13.6	--
40	--	--	97	30.7	73.6	43.0	12.6	--
39	--	--	96	30.3	73.3	42.3	11.6	--
38	--	--	95	29.9	73.0	41.6	10.6	--
37	--	94	29.5	78.0	41.0	9.6		--
36	--	--	93	29.1	72.3	40.3	8.6	--
35	--	--	92	28.7	72.0	39.6	7.6	--
34	--	--	91	28.2	71.7	39.0	6.6	--
33	--	--	90	27.8	71.4	38.3	5.6	--
32	--	--	89	27.4	71.0	37.6	4.6	--
31	--	--	88	27.0	70.7	37.0	3.6	--
30	--	--	87	26.6	70.4	36.3	2.6	--

APPROXIMATE HARDNESS CONVERSION NUMBERS FOR NONAUSTENITIC STEELS [a, b] (Continued)

a.   This table gives the approximate interrelationships of hardness values and approximate tensile strength of steels.   It is possible that steels of various compositions and processing histories will deviate in hardness-tensile strength relationship from the data presented in this table.   The data in this table should not be used for austenitic stainless steels, but have been shown to be applicable for ferritic and martensitic stainless steels.   Where more precise conversions are required, they should be developed specially for each steel composition, heat treatment, and part. b. All relative hardness values in this table are averages of tests on various metals whose different properties prevent establishment of exact mathematical conversions. These values are consistent with ASTM A 370-91 for nonaustenitic steels. It is recommended that ASTM standards A 370, E 140, E 10, E 18, E 92, E 110 and E 384, involving hardness tests on metals, be reviewed prior to interpreting hardness conversion values.

APPROXIMATE HARDNESS CONVERSION NUMBERS FOR AUSTENITIC STEELS[a]

| Rockwell C | Rockwell A | Rockwell Superficial Hardness | | |
150 kgf, Diamond HRC	60 kgf, Diamond HRA	15 kgf, Diamond HR15N	30 kgf, Diamond HR30N	45 kgf, Diamond HR45N
48	74.4	84.1	66.2	52.1
47	73.9	83.6	65.3	50.9
46	73.4	83.1	64.5	49.8
45	72.9	82.6	63.6	48.7
44	72.4	82.1	62.7	47.5
43	71.9	81.6	61.8	46.4
42	71.4	81.0	61.0	45.2
41	70.9	80.5	60.1	44.1
40	70.4	80.0	59.2	43.0
39	69.9	79.5	58.4	41.8
38	69.3	79.0	57.5	40.7
37	68.8	78.5	56.6	39.6
36	68.3	78.0	55.7	38.4
35	67.8	77.5	54.9	37.3
34	67.3	77.0	54.0	36.1

APPROXIMATE HARDNESS NUMBERS FOR AUSTENITIC STEELS[a] (Continued)

Rockwell C	Rockwell A	Rockwell Superficial Hardness		
150 kgf, Diamond HRC	60 kgf, Diamond HRA	15 kgf, Diamond HR15N	30 kgf, Diamond HR30N	45 kgf, Diamond HR45N
33	66.8	76.5	53.1	35.0
32	66.3	75.9	52.3	33.9
31	65.8	75.4	51.4	32.7
30	65.3	74.9	50.5	31.6
29	64.8	74.4	49.6	30.4
28	64.3	73.9	48.8	29.3
27	63.8	73.4	47.9	28.2
26	63.3	72.9	47.0	27.0
25	62.8	72.4	46.2	25.9
24	62.3	71.9	45.3	24.8
23	61.8	71.3	44.4	23.6
22	61.3	70.8	43.5	22.5
21	60.8	70.3	42.7	21.3
20	60.3	69.8	41.8	20.2

a. All relative hardness values in this table are averages of tests on various metals whose different properties prevent establishment of exact mathematical conversions. These values are consistent with ASTM A 370-91 for austenitic steels. It is recommended that ASTM standards A 370, E 140, E 10, E 18, E 92, E 110 and E 384, involving hardness tests on metals, be reviewed prior to interpreting hardness conversion values.

APPROXIMATE HARDNESS CONVERSION VALUES FOR AUSTENITIC STEELS[a]

Rockwell B 100 kgf 1/16" ball HRB	Brinell Indentation Diameter, mm	Brinell 3000 kgf 10 mm Ball HB	Rockwell A 60 kgf Diamond HRA	Rockwell Superficial Hardness		
				15 kgf 1/16" ball HR15T	30 kgf 1/16" ball HR30T	45 kgf 1/16" ball HR45T
100	3.79	256	61.5	91.5	80.4	70.2
99	3.85	248	60.9	91.2	79.7	69.2
98	3.91	240	60.3	90.8	79.0	68.2
97	3.96	233	59.7	90.4	78.3	67.2
96	4.02	226	59.1	90.1	77.7	66.1
95	4.08	219	58.5	89.7	77.0	65.1
94	4.14	213	58.0	89.3	76.3	64.1
93	4.20	207	57.4	88.9	75.6	63.1
92	4.24	202	56.8	88.6	74.9	62.1
91	4.30	197	56.2	88.2	74.2	61.1
90	4.35	192	55.6	87.8	73.5	60.1
89	4.40	187	55.0	87.5	72.8	59.0
88	4.45	183	54.5	87.1	72.1	58.0
87	4.51	178	53.9	86.7	71.4	57.0
86	4.55	174	53.3	86.4	70.7	56.0
85	4.60	170	52.7	86.0	70.0	55.0
84	4.65	167	52.1	85.6	69.3	54.0
83	4.70	163	51.5	85.2	68.6	52.9
82	4.74	160	50.9	84.9	67.9	51.9
81	4.79	156	50.4	84.5	67.2	50.9
80	4.84	153	49.8	84.1	66.5	49.9

a. All relative hardness values in this table are averages of tests on various metals whose different properties prevent establishment of exact mathematical conversions. These values are consistent with ASTM A 370-91 for austenitic steels. It is recommended that ASTM standards A 370, E 140, E 10, E 18, E 92, E 110 and E 384, involving hardness tests on metals, be reviewed prior to interpreting hardness conversion values.

Appendix

2

SI UNIT CONVERSIONS

METRIC CONVERSION FACTORS

To Convert From	To	Multiply By
Angle		
degree	rad	1.745 329 E -02
Area		
in.²	mm²	6.451 600 E + 02
in.²	cm²	6.451 600 E + 00
in.²	m²	6.451 600 E - 04
ft²	m²	9.290 304 E - 02
Bending moment or torque		
lbf - in.	N - m	1.129 848 E - 01
lbf - ft	N - m	1.355 818 E + 00
kgf - m	N - m	9.806 650 E + 00
ozf - in.	N-m	7.061 552 E - 03
Bending moment or torque per unit length		
lbf - in./in.	N - m/m	4.448 222 E + 00
lbf - ft/in.	N - m/m	5.337 866 E + 01
Corrosion rate		
mils/yr	mm/yr	2.540 000 E - 02
mils/yr	µ/yr	2.540 000 E + 01
Current density		
A/in.²	A/cm²	1.550 003 E - 01
A/in.²	A/mm²	1.550 003 E - 03
A/ft²	A/m²	1.076 400 E + 01
Electricity and magnetism		
gauss	T	1.000 000 E - 04

To Convert From	To	Multiply By
Mass per unit length		
lb/ft	kg/m	1.488 164 E + 00
lb/ft	kg/m	1.785 797 E + 01.
Mass per unit time		
lb/h	kg/s	1.259 979 E - 04
lb/min	kg/s	7.559 873 E - 03
lb/s	kg/s	4.535 924 E - 01
Mass per unit volume (includes density)		
g/cm³	kg/m³	1.000 000 E + 03
lb/ft³	g/cm³	1.601 846 E - 02
lb/ft³	kg/m³	1.601 846 E + 01
lb/in.³	g/cm³	2.767 990 E + 01
lb/in.³	kg/m³	2.767 990 E + 04
Power		
Btu/s	kW	1.055 056 E + 00
Btu/min	kW	1.758 426 E - 02
Btu/h	W	2.928 751 E - 01
erg/s	W	1.000 000 E - 07
ft - lbf/s	W	1.355 818 E + 00
ft - lbf/min	W	2.259 697 E - 02
ft - lbf/h	W	3.766 161 E - 04
hp (550 ft - lbf/s)	kW	7.456 999 E - 01
hp (electric)	kW	7.460 000 E - 01

METRIC CONVERSION FACTORS (Continued)

To Convert From	To	Multiply By	To Convert From	To	Multiply By
Electricity and magnetism (Continued)			**Power density**		
maxwell	μWb	1.000 000 E - 02	W/in.^{2}	W/m^2	1.550 003 E + 03
mho	S	1.000 000 E + 00	**Pressure (fluid)**		
Oersted	A/m	7.957 700 E + 01	atm (standard)	Pa	1.013 250 E + 05
Ω - cm	Ω - m	1.000 000 E - 02	bar	Pa	1.000 000 E + 05
Ω circular - mil/ft	μΩ - m	1.662 426 E - 03	in. Hg (32°F)	Pa	3.386 380 E + 03
Energy (impact other)			in. Hg (60°F)	Pa	3.376 850 E + 03
ft - lbf	J	1.355 818 E + 00	lbf/in.^{2} (psi)	Pa	6.894 757 E + 03
Btu (thermochemical)	J	1.054 350 E + 03	torr (mm Hg, 0°C)	Pa	1.333 220 E + 02
cal (thermochemical)	J	4.184 000 E + 00	**Specific heat**		
kW - h	J	3.600 000 E + 06	Btu/lb - °F	J/kg - K	4.186 800 E + 03
W - h	J	3.600 000 E + 03	cal/g - °C	J/kg - K	4.186 800 E + 03
Flow rate			**Stress (force per unit area)**		
ft^3/h	L/min	4.719 475 E - 01	tonf/in.^{2} (tsi)	MPa	1.378 951 E + 01
ft^3/min	L/min	2.831 000 E + 01	kgf/mm^2	MPa	9.806 650 E + 00
gal/h	L/min	6.309 020 E - 02	ksi	MPa	6.894 757 E + 00
gal/min	L/min	3.785 412 E + 00	lbf/in.^{2} (psi)	MPa	6.894 757 E - 03
Force			MN/m^2	MPa	1.000 000 E + 00
lbf	N	4.448 222 E + 00	**Temperature**		
kip (1000 lbf)	N	4.448 222 E + 03	°F	°C	5/9 (°F - 32)
tonf	kN	8.896 443 E + 00	R	K	5/9
kgf	N	9.806 650 E + 00	**Temperature interval**		
			°F	°C	5/9
Force per unit length			**Thermal conductivity**		
lbf/ft	N/m	1.459 390 E + 01	Btu - in./s - ft^2 - °F	W/m - K	5.192 204 E + 02
lbf/in.	N/m	1.751 268 E + 02	Btu/ft - h - °F	W/m - K	1.730 735 E + 00

METRIC CONVERSION FACTORS (Continued)

To Convert From	To	Multiply By	To Convert From	To	Multiply By
Fracture toughness			**Thermal conductivity (Continued)**		
ksi √in.	MPa √m	1.098 800 E + 00	Btu - in./h . ft² - °F	W/m - K	1.442 279 E - 01
Heat content			cal/cm - s - °C	W/m - K	4.184 000 E + 02
Btu/lb	kJ/kg	2.326 000 E + 00	**Thermal expansion**		
cal/g	kJ/kg	4.186 800 E + 00	in./in. - °C	m/m - K	1.000 000 E + 00
Heat input			in./in. - °F	m/m - K	1.800 000 E + 00
J/in.	J/m	3.937 008 E + 01	**Velocity**		
kJ/in.	kJ/m	3.937 008 E + 01	ft/h	m/s	8.466 667 E - 05
Length			ft/min	m/s	5.080 000 E - 03
A	nm	1.000 000 E - 01	ft/s	m/s	3.048 000 E - 01
μin.	μm	2.540 000 E - 02	in./s	m/s	2.540 000 E - 02
mil	μm	2.540 000 E + 01	km/h	m/s	2.777 778 E - 01
in.	mm	2.540 000 E + 01	mph	km/h	1.609 344 E + 00
in.	cm	2.540 000 E + 00	**Velocity of rotation**		
ft	m	3.048 000 E - 01	rev/min (rpm)	rad/s	1.047 164 E - 01
yd	m	9.144 000 E -01	rev/s	rad/s	6.283 185 E + 00
mile	km	1.609 300 E + 00	**Viscosity**		
Mass			poise	Pa - s	1.000 000 E - 01
oz	kg	2.834 952 E - 02	stokes	m²/s	1.000 000 E - 04
lb	kg	4.535 924 E - 01	ft²/s	m²/s	9.290 304 E - 02
ton (short 2000 lb)	kg	9.071 847 E + 02	in.²/s	mm²/s	6.451 600 E + 02
ton (short 2000 lb)	kg x 10³	9.071 847 E - 01	**Volume**		
ton (long 2240 lb)	kg	1.016 047 E + 03	in.³	m³	1.638 706 E - 05
kg x 10³ = 1 metric ton			ft³	m³	2.831 685 E - 02
			fluid oz	m³	2.957 353 E - 05

METRIC CONVERSION FACTORS (Continued)

To Convert From	To	Multiply By	To Convert From	To	Multiply By
Mass per unit area			**Volume (Continued)**		
oz/in.^{2}	kg/m^2	4.395 000 E + 01	gal (U.S. liquid)	m^3	3.785 412 E - 03
oz/ft^2	kg/m^2	3.051 517 E - 01	**Volume per unit time**		
oz/yd^2	kg/m^2	3.390 575 E - 02	ft^3/min	m^3/s	4.719 474 E - 04
lb/ft^2	kg/m^2	4.882 428 E + 00	ft^3/s	m^3/s	2.831 685 E - 02
			in.^{3}/min	m^3/s	2.731 177 E - 07
			Wavelength		
			A	nm	1.000 000 E - 01

THE GREEK ALPHABET

A, α - Alpha	I, ι - Iota	P, ρ - Rho
B, β - Beta	K, κ - Kappa	Σ, σ - Sigma
Γ, γ - Gamma	Λ, λ - Lambda	T, τ - Tau
Δ, δ - Delta	M, μ - Mu	Y, υ - Upsilon
E, ε - Epsilon	N, ν - Nu	Φ, φ - Phi
Z, ξ - Zeta	Ξ, ξ - Xi	X, χ - Chi
H, η - Eta	O, o - Omicron	Ψ, ψ - Psi
Θ, θ - Theta	Π, π - Pi	Ω, ϖ - Omega

SI PREFIXES

Prefix	Symbol	Exponential Expression	Multiplication Factor
exa	E	10^{18}	1 000 000 000 000 000 000
peta	P	10^{15}	1 000 000 000 000 000
tera	T	10^{12}	1 000 000 000 000
giga	G	10^{9}	1 000 000 000
mega	M	10^{6}	1 000 000
kilo	k	10^{3}	1 000
hecto	h	10^{2}	100
deka	da	10^{1}	10
Base Unit	---	10^{0}	1
deci	d	10^{-1}	0.1
centi	c	10^{-2}	0.01
milli	m	10^{-3}	0.001
micro		10^{-6}	0.000 001
nano	n	10^{-9}	0.000 000 001
pico	p	10^{-12}	0.000 000 000 001
femto	f	10^{-15}	0.000 000 000 000 001
atto	a	10^{-18}	0.000 000 000 000 000 001

Appendix

3

IMPERIAL UNIT DATA

DECIMAL EQUIVALENT OF FRACTIONS

Fraction (in.)	Decimal (in.)	Millimeter (mm)
1/64	0.015 625	0.396 875
1/32	0.031 250	0.793 750
3/64	0.046 875	1.190 625
1/16	0.062 500	1.587 500
5/64	0.078 125	1.984 375
3/32	0.093 750	2.381 250
7/64	0.109 375	2.778 125
1/8	0.125 000	3.175 000
9/64	0.140 625	3.571 875
5/32	0.156 250	3.968 750
11/64	0.171 875	4.365 625
3/16	0.187 500	4.762 500
13/64	0.203 125	5.159 375
7/32	0.218 750	5.556 250
15/64	0.234 375	5.953 125
1/4	0.250 000	6.350 000
17/64	0.265 625	6.746 875
9/32	0.281 250	7.143 750
19/64	0.296 875	7.540 625
15/16	0.312 500	7.937 500
21/64	0.328 125	8.334 375
11/32	0.343 750	8.731 250
23/64	0.359 375	9.128 125
3/8	0.375 000	9.525 000
25/64	0.390 625	9.921 875
13/32	0.406 250	10.318 750
27/64	0.421 875	10.715 625

DECIMAL EQUIVALENT OF FRACTIONS (Continued)

Fraction (in.)	Decimal (in.)	Millimeter (mm)
7/16	0.437 500	11.112 500
29/64	0.453 125	11.509 375
15/32	0.468 750	11.906 250
31/64	0.484 375	12.303 125
1/2	0.500 000	12.700 000
33/64	0.515 625	13.096 875
17/32	0.531 250	13.493 750
35/64	0.546 875	13.890 625
9/16	0.562 500	14.287 500
37/64	0.578 125	14.684 375
19/32	0.593 750	15.081 250
39/64	0.609 375	15.478 125
5/8	0.625 000	15.875 000
41/64	0.640 625	16.271 875
21/32	0.656 250	16.668 750
43/64	0.671 875	17.065 625
11/16	0.687 500	17.462 500
45/64	0.703 125	17.859 375
23/32	0.718 750	18.256 250
47/64	0.734 375	18.653 125
3/4	0.750 000	19.050 000
49/64	0.765 625	19.446 875
25/32	0.781 250	19.843 750
51/64	0.796 875	20.240 625
13/16	0.812 500	20.637 500
53/64	0.828 125	21.034 375
27/32	0.843 750	21.431 250

DECIMAL EQUIVALENT OF FRACTIONS (Continued)

Fraction (in.)	Decimal (in.)	Millimeter (mm)
55/64	0.859 375	21.828 125
7/8	0.875 000	22.225 000
57/64	0.890 625	22.621 875
29/32	0.906 250	23.018 750
59/64	0.921 875	23.415 625
15/16	0.937 500	23.812 500
61/64	0.953 125	24.209 375
31/12	0.968 750	24.606 250
63/64	0.984 375	25.003 125
1	1.000 000	25.400 000

SHEET METAL GAGE THICKNESS CONVERSIONS

Gage	in.	mm	Gage	in.	mm
30	0.0120	0.3048	16	0.0598	1.5189
29	0.0135	0.3429	15	0.0673	1.7094
28	0.0149	0.3785	14	0.0747	1.8974
27	0.0164	0.4166	13	0.0897	2.2784
26	0.0179	0.4547	12	0.1046	2.6568
25	0.0109	0.5309	11	0.1196	3.0378
24	0.0239	0.6071	10	0.1345	3.4163
23	0.0269	0.6833	9	0.1495	3.7973
22	0.0299	0.7595	8	0.1644	4.1758
21	0.0329	0.8357	7	0.1793	4.5542
20	0.0359	0.9119	6	0.1943	4.9352
19	0.0418	1.0617	5	0.2092	5.3137
18	0.0478	1.2141	4	0.2242	5.6947
17	0.0538	1.3665	3	0.2391	6.0731

WIRE GAGE DIAMETER CONVERSIONS

US Steel Wire Gage No.	Inches (in.)	Millimeters (mm)
7/0s	0.4900	12.447
6/0s	0.4615	11.7221
5/0s	0.4305	10.9347
4/0s	0.3938	10.0025
3/0s	0.3625	9.2075
2/0s	0.3310	8.4074
0	0.3065	7.7851
1	0.2830	7.1882
2	0.2625	6.6675
3	0.2437	6.1899
4	0.2253	5.7226
5	0.2070	5.2578
6	0.1920	4.8768
7	0.1770	4.4958
8	0.1620	4.1148
9	0.1483	3.7668
10	0.1350	3.429
11	0.1205	3.0607
12	0.1055	2.6797
13	0.0915	2.3241
14	0.0800	2.032
15	0.0720	1.8389
16	0.0625	1.5875
17	0.0540	1.3716
18	0.0475	1.2065
19	0.0410	1.0414
20	0.0348	0.8839

WIRE GAGE DIAMETER CONVERSIONS

US Steel Wire Gage No.	Inches (in.)	Millimeters (mm)
21	0.0317	0.8052
22	0.0286	0.7264
23	0.0258	0.6553
24	0.0230	0.5842
25	0.0204	0.5182
26	0.0181	0.4597
27	0.0173	0.4394
28	0.0162	0.4115
29	0.0150	0.381
30	0.0140	0.3556
31	0.0132	0.3353
32	0.0128	0.3251
33	0.0118	0.2997
34	0.0104	0.2642
35	0.0095	0.2413
36	0.0090	0.2286
37	0.0085	0.2159
38	0.0080	0.2032
39	0.0075	0.1905
40	0.0070	0.1778
41	0.0066	0.1678
42	0.0062	0.1575
43	0.0060	0.1524
44	0.0058	0.1473

AMERICAN PIPE DIMENSIONS

DIMENSIONS OF WELDED AND SEAMLESS PIPE

Nominal Pipe Size, in.	Outside Diameter	Nominal Wall Thickness (in.) For								
		Schedule 5S	Schedule 10S	Schedule 10	Schedule 20	Schedule 30	Schedule Standard	Schedule 40
1/8	0.405	---	0.049	---	---	---	0.068	0.068
1/4	0.540	---	0.065	---	---	---	0.088	0.088
3/8	0.675	---	0.065	---	---	---	0.091	0.091
1/2	0.840	0.065	0.083	---	---	---	0.109	0.109
3/4	1.050	0.065	0.083	---	---	---	0.113	0.113
1	1.315	0.065	0.109	---	---	---	0.133	0.133
1 1/4	1.660	0.065	0.109	---	---	---	0.140	0.140
1 1/2	1.900	0.065	0.109	---	---	---	0.145	0.145
2	2.375	0.065	0.109	---	---	---	0.154	0.154
2 1/2	2.875	0.083	0.120	---	---	---	0.203	0.203
3	3.5	0.083	0.120	---	---	---	0.216	0.216
3 1/2	4.0	0.083	0.120	---	---	---	0.226	0.226
4	4.5	0.083	0.120	---	---	---	0.237	0.237
5	5.563	0.109	0.134	---	---	---	0.258	0.258
6	6.625	0.109	0.134	---	---	---	0.280	0.280
8	8.625	0.109	0.148	---	0.250	0.277	0.322	0.322
10	10.75	0.134	0.165	---	0.250	0.307	0.365	0.365
12	12.75	0.156	0.180	---	0.250	0.330	0.375	0.406
14 O.D.	14.0	0.156	0.188	0.250	0.312	0.375	0.375	0.438
16 O.D.	16.0	0.165	0.188	0.250	0.312	0.375	0.375	0.500
18 O.D.	18.0	0.165	0.188	0.250	0.312	0.438	0.375	0.562
20 O.D.	20.0	0.188	0.218	0.250	0.375	0.500	0.375	0.594
22 O.D.	22.0	0.188	0.218	0.250	0.375	0.500	0.375	---
24 O.D.	24.0	0.218	0.250	0.250	0.375	0.562	0.375	0.688
26 O.D.	26.0	---	---	0.312	0.500	---	0.375	---

DIMENSIONS OF WELDED AND SEAMLESS PIPE (Continued)

Nominal Pipe Size, in.	Outside Diameter	Nominal Wall Thickness (in) For						
		Schedule 5S	Schedule 10S	Schedule 10	Schedule 20	Schedule 30	Schedule Standard	Schedule 40
28 O.D.	28.0	---	---	0.312	0.500	0.625	0.375	---
30 O.D.	30.0	0.250	0.312	0.312	0.500	0.625	0.375	---
32 O.D.	32.0	---	---	0.312	0.500	0.625	0.375	0.688
34 O.D.	34.0	---	---	0.312	0.500	0.625	0.375	0.688
36 O.D.	36.0	---	---	0.312	0.500	0.625	0.375	0.750
42 O.D.	42.0	---	---	---	---	---	0.375	---

See next table for heavier wall thicknesses.; all units are inches.

DIMENSIONS OF WELDED AND SEAMLESS PIPE

Nominal Pipe Size, in.	Outside Diameter	Nominal Wall Thickness (in.) For							
		Schedule 60	Extra Strong	Schedule 80	Schedule 100	Schedule 120	Schedule 140	Schedule 160	XX Strong
1/8	0.405	---	0.095	0.095	---	---	---	---	---
1/4	0.540	---	0.119	0.119	---	---	---	---	---
3/8	0.675	---	0.126	0.126	---	---	---	---	---
1/2	0.840	---	0.147	0.147	---	---	---	0.188	0.294
3/4	1.050	---	0.154	0.154	---	---	---	0.219	0.308
1	1.315	---	0.179	0.179	---	---	---	0.250	0.358
1 1/4	1.660	---	0.191	0.191	---	---	---	0.250	0.382
1 1/2	1.900	---	0.200	0.200	---	---	---	0.281	0.400
2	2.375	---	0.218	0.218	---	---	---	0.344	0.436
2 1/2	2.875	---	0.276	0.276	---	---	---	0.375	0.552
3	3.5	---	0.300	0.300	---	---	---	0.438	0.600
3 1/2	4.0	---	0.318	0.318	---	---	---	---	---

DIMENSIONS OF WELDED AND SEAMLESS PIPE (Continued)

Nominal Pipe Size in.	Outside Diameter	Nominal Wall Thickness (in.) For							
		Schedule 60	Extra Strong	Schedule 80	Schedule 100	Schedule 120	Schedule 140	Schedule 160	XX Strong
4	4.5	---	0.337	0.337	---	0.438	---	0.531	0.674
5	5.563	---	0.375	0.375	---	0.500	---	0.625	0.750
6	6.625	---	0.432	0.432	---	0.562	---	0.719	0.864
8	8.625	0.406	0.500	0.500	0.594	0.719	0.812	0.906	0.875
10	10.75	0.500	0.500	0.594	0.719	0.844	1.000	1.125	1.000
12	12.75	0.562	0.500	0.688	0.844	1.000	1.125	1.312	1.000
14 O.D.	14.0	0.594	0.500	0.750	0.938	1.094	1.250	1.406	---
16 O.D.	16.0	0.656	0.500	0.844	1.031	1.219	1.438	1.594	---
18 O.D.	18.0	0.750	0.500	0.938	1.156	1.375	1.562	1.781	---
20 O.D.	20.0	0.812	0.500	1.031	1.281	1.500	1.750	1.969	---
22 O.D.	22.0	0.875	0.500	1.125	1.375	1.625	1.875	2.125	---
24 O.D.	24.0	0.969	0.500	1.218	1.531	1.812	2.062	2.344	---
26 O.D.	26.0	---	0.500	---	---	---	---	---	---
28 O.D.	28.0	---	0.500	---	---	---	---	---	---
30 O.D.	30.0	---	0.500	---	---	---	---	---	---
32 O.D.	32.0	---	0.500	---	---	---	---	---	---
34 O.D.	34.0	---	0.500	---	---	---	---	---	---
36 O.D.	36.0	---	0.500	---	---	---	---	---	---
42 O.D.	42.0	---	0.500	---	---	---	---	---	---

All units are inches.

Appendix

5

DELTA FERRITE (FN) DIAGRAM

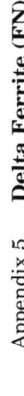

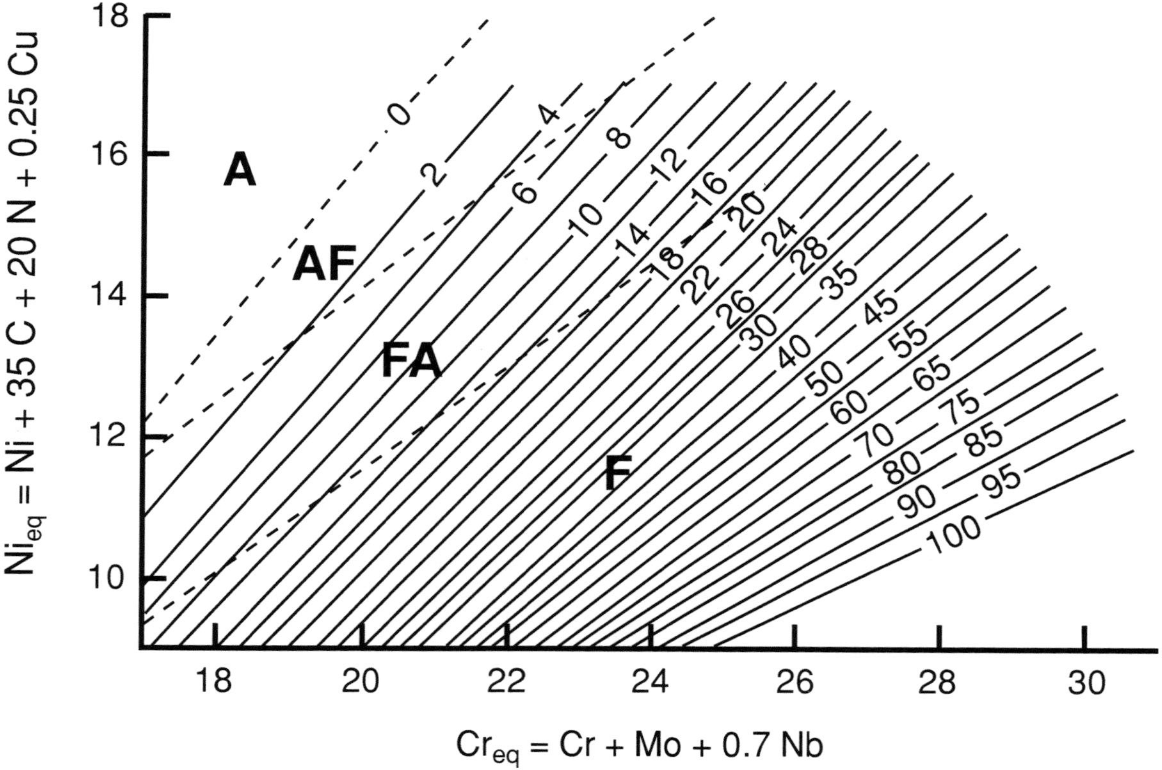
A
AF
FA
F
0
2
4
6
8
10
12
14
16
18
20
22
24
26
28
30
35
40
45
50
55
60
65
70
75
80
85
90
95
100
18
16
14
12
10
18
20
22
24
26
28
30
Ni_{eq} = Ni + 35 C + 20 N + 0.25 Cu
Cr_{eq} = Cr + Mo + 0.7 Nb

PERIODIC TABLE

Periodic Table of the Elements

Metals ──────── Nonmetals

Key to chart

Atomic Number → **50** +2 ← Oxidation States
Symbol → **Sn** +4
Atomic Weight → 118.69
−18−18−4 ← Electron Configuration

Transition Elements

Iᵃ	IIᵃ	IIIᵇ	IVᵇ	Vᵇ	VIᵇ	VIIᵇ		VIII		Iᵇ	IIᵇ	IIIᵃ	IVᵃ	Vᵃ	VIᵃ	VIIᵃ	O	Orbit
1 H +1 −1 1.0079 1																	**2** He 4.00260 2	K
3 Li +1 6.939 2−1	**4** Be +2 9.0122 2−2											**5** B +3 10.81 2−3	**6** C +2 +4 −4 12.011 2−4	**7** N +1 +2 +3 +4 +5 −3 14.0067 2−5	**8** O +1 −2 15.9994 2−6	**9** F −2 18.998403 2−7	**10** Ne 10.17₃ 2−8	K−L
11 Na +1 22.9898 2−8−1	**12** Mg +2 24.312 2−8−2											**13** Al +3 26.98154 2−8−3	**14** Si +2 +4 −4 28.08 2−8−4	**15** P +3 +5 −3 30.97376 2−8−5	**16** S +4 +6 −2 32.06 2−8−6	**17** Cl +1 +5 +7 −1 35.453 2−8−7	**18** Ar 39.948 2−8−8	K−L−M
19 K +1 39.09 −8−8−1	**20** Ca +2 40.08 −8−8−2	**21** Sc +3 44.955₈ −18−9−2	**22** Ti +2 +3 +4 47.9 −8−10−2	**23** V +2 +3 +4 +5 50.941 −8−11−2	**24** Cr +2 +3 +4 +5 +6 51.996 −8−13−1	**25** Mn +2 +3 +6 +7 54.9380 −8−13−2	**26** Fe +2 +3 +4 +7 55.847 −8−14−2	**27** Co +2 +3 58.9332 −8−15−2	**28** Ni +2 +3 58.71 −8−16−2	**29** Cu +2 +3 63.54 −18−1	**30** Zn +1 +2 65.38 −18−2	**31** Ga +2 69.72 −8−18−3	**32** Ge +2 +4 72.59 −8−18−4	**33** As +3 +4 74.9216 −8−18−5	**34** Se +4 +6 −3 78.96 −18−6	**35** Br +1 +5 −2 79.904 −8−18−7	**36** Kr 83.80 −8−18−8	K−L−M−N
37 Rb +1 85.467 −18−8−1	**38** Sr +2 87.62 −18−8−2	**39** Y +3 88.9059 −18−9−2	**40** Zr +4 91.22 −18−10−2	**41** Nb +3 +5 92.9064 −18−12−1	**42** Mo +6 95.94 −18−13−1	**43** Tc +6 +7 98.9062 −18−13−2	**44** Ru +6 +7 101.07 −18−15−1	**45** Rh +3 102.905 −18−16−1	**46** Pd +3 +4 106.4 −18−18−0	**47** Ag +1 +4 107.868 −18−18−1	**48** Cd +1 +2 112.40 −18−18−2	**49** In +3 114.82 −18−18−3	**50** Sn +2 +4 118.69 −18−18−4	**51** Sb +2 +4 −1 121.75 −18−18−5	**52** Te +3 +5 −2 127.60 −18−18−6	**53** I +4 +5 +7 −1 126.9045 −18−18−7	**54** Xe +1 +6 −3 131.30 −18−18−8	−L−M−N
55 Cs +1 132.9054 −18−8−1	**56** Ba +2 137.3 −18−8−2	**57*** La +2 +3 138.9055 −18−9−2	**72** Hf +3 178.49 −32−10−2	**73** Ta +5 180.948 −32−11−2	**74** W +6 183.85 −32−12−2	**75** Re +6 +7 186.207 −32−13−2	**76** Os +4 +6 +7 190.2 −32−14−2	**77** Ir +3 +4 192.9 −32−15−2	**78** Pt +2 +4 195.09 −32−16−2	**79** Au +2 +4 196.9665 −32−18−1	**80** Hg +1 +3 200.59 −32−18−2	**81** Tl +1 +2 204.37 −32−18−3	**82** Pb +2 +4 207.19 −32−18−4	**83** Bi +3 +5 208.980 −32−18−5	**84** Po +2 +4 (209) −32−18−6	**85** At (210) −32−18−7	**86** Rn (222) −32−18−8	−M−N−O
87 Fr +1 (223) −18−8−1	**88** Ra +2 226.0254 −18−8−2	**89**** Ac +3 (227) −18−9−2	**104** Rf +4 (261) −32−10−2	**105** Ha (262) −32−11−2	**106** (263) −32−12−2													−O−P−Q

The Metals Black Book – 3rd Edition

Appendix

7

INTERNATIONAL STANDARDS ORGANIZATIONS, TECHNICAL ASSOCIATIONS & SOCIETIES

International Standards Organizations

AENOR   Asociación Española de Normalización y Cetificación (Spain)
tel +34 1 310 48 51, fax +34 1 310 49 76
http://www.aenor.es/medioamb/

AFNOR   Association Française de Normalisation (France)
tel +33 1 42 91 55 55, fax +33 1 42 91 56 56
http://www.afnor.fr

ANSI   American National Standards Institute (USA)
tel +212 642 4900, fax +212 302 1286
http://web.ansi.org/default_js.htm

BSI   British Standards Institution (England)
tel +44 181 996 70 00, fax +44 181 996 70 01
http://www.bsi.org.uk/

CSA   Canadian Standards Association (Canada)
tel +416 747 4044, fax +416 747 2475
http://www.csa.ca/

CSCE   Canadian Society for Chemical Engineers (613) 526-4652

DIN   Deutches Institut für Normung e.V. (Germany)
tel +49 30 26 01 2260, fax +49 30 2601 1231
http://www.computerwoche.de/archiv/1978/39/7839c043.html

DS   Dansk Standard (Denmark)
tel +45 39 77 01 01, fax +45 39 77 02 02
http://www.diku.dk/users/khan/D/dansk.-0040.html

ELOT   Hellenic Organization for Standardization (Greece)
tel +30 1 201 50 25, fax +30 1 202 07 76
http://www.elot.gr/

International Standards Organizations (Continued)

IBN/BIN   Institut Belge de Normalisation/Belgisch Instituut voor Normalisatie (Belgium)
tel +32 2 738 01 11, fax +32 2 733 42 64

IPQ   Instituto Português da Qualidade (Potugal)
tel +351 1 294 81 00, fax +351 1 294 81 01
http://www.ipq.pt/homepage.html

ISO   International Organization for Standardization (Switzerland)
tel +41 22 749 01 11, fax +41 22 733 34 30
http://www.iso.ch/

ITM   Inspection du Travail et des Mines (Luxembourg)
tel +352 478 61 54, fax +352 49 14 47

JSA   Japanese Standards Association (Japan)
tel +03 3583 8074, fax + 033582 2390

NNI   Nederlands NormalisatieiInstituut (Netherlands)
tel +31 15 69 03 90, fax +31 15 69 01 90

NSAI   National Standards Authority of Ireland
tel +353 1 837 01 01, fax +353 1 836 98 21
http://www.nsai.ie/

NFS   Norges Standardiseringsforbund (Norway)
tel +47 22 46 60 94, fax +47 22 46 44 57
http://www.standard.no/nsf/

ON   Österreichisches Normungsindtitut (Austria)
tel +43 1 213 00, fax +43 1 213 00 650

SA   Standards Australia
tel +08 373 1540, fax +08 373 1051
http://www.standards.com.au/~sicsaa/

SCC   Standards Council of Canada
tel +800 267 8220, fax +613 995 4564
http://www.scc.ca/about/index.html

SIS   Standardiseringskommissione n i Sverige (Sweden)
tel +46 8 613 52 00, fax +46 8 411 70 35

SFS   Suomen Standardisoimisliitto r.y. (Findland)
tel +358 0 149 93 31, fax +358 0 146 49 25
http://www.ficc.fi/sfs.htm

SNV   Schweizerische Normen-Vereinigung (Switzerland)
tel +41 1 254 54 54, fax +41 1 254 54 74
http://www.snv.ch/

STRI   Technological Institute of Iceland
tel +354 587 70 02, fax +354 587 74 09
http://iti.is/orgchart.html

UNI   Ente Nazionale Italiano di Unificazione (Italy)
tel +39 2 70 02 41, fax +39 2 70 10 61 06

Technical Associations & Societies

AA	The Aluminum Association (202) 862-5100 http://www.aluminum.org/aa_indx.htm
AEE	The Association of Energy Engineers (404) 447-5083
AFS	American Foundrymen's Society (312) 824-0181 http://www.afsinc.org/
AISI	Association of Iron and Steel Engineers (412) 281-6323 http://www.aise.org/magazine/isead.htm
AlChE	American Institute of Chemical Engineers (212) 705-7338 http://www.aiche.org/, e-mail xpress@aiche.org
AMEC	Advanced Materials Engineering Centre (902) 425-4500
ASEE	American Society for Engineering Education (202) 331-3500 http://www.asee.org/
ASM	ASM International - The Materials Information Society (800) 336-5152 or (216) 338-5151 http://www.asm-intl.org/
ASME	American Society of Mechanical Engineers (212) 705-7722 http://www.asme.org/
ASNT	American Society for Nondestructive Testing (614) 274-6003 http://www.utexas.edu/ftp/depts/dos/cci/orgs/01771.html
ASQC	American Society for Quality Control (414) 272-8575 http://www.asqc.org/
ASTM	American Society for Testing and Materials (215) 299-5400 http://www.astm.org/
AWS	American Welding Society (305) 443-9353 or (800) 443-9353 http://www.aws.org/
CAIMF	Canadian Advanced Industrial Materials Forum (416) 798-8055
CASI	Canadian Aeronautic & Space Institute (613) 234-0191
CCA	Canadian Construction Association (613) 236-9455 http://www.cca-acc.com/
CCPE	Canadian Council of Professional Engineers (613) 232-2474 http://www.ccpe.ca/
CCS	Canadian Ceramics Society (416) 491-2886
CDA	Copper Development Association (212) 251-7200 http://www.copper.org/
CEN	European Committee for Standardization +32 2 550 08 11
CIE	Canadian Institute of Energy (403) 262-6969
CIM	Canadian Institute for Mining and Metallurgy (514) 939-2710 http://www.cim.org/
CMA	Canadian Manufacturing Association (416) 363-7261
CNS	Canadian Nuclear Society (416) 977-6152 http://www.cns-snc.ca/

Technical Associations & Societies (Continued)

CPI	Canadian Plastics Institute (416) 441-3222
CPIC	Canadian Professional Information Centre (905) 624-1058
CSEE	Canadian Society of Electronic Engineers (514) 651-6710
CSME	Canadian Society of Mechanical Engineers (514) 842-8121 http://www.freenet.edmonton.ab.ca/11/i/csme
CSNDT	Canadian Society for Nondestructive Testing (416) 676-0785 http://www.csndt.org/
EI	Engineering Information Inc. (212) 705-7600 http://www.ei.org/
FED	Federal & Military Standards (215) 697-2000
IEEE	Institute of Electrical & Electronic Engineers (212) 705-7900 http:// www.ieee.org
IES	Institute of Environmental Sciences (312) 255-1561 http://www.ei.org, e-mail ei@ei.org
IIE	Institute of Industrial Engineers (404) 449-0460 http://www.iienet.org/
IMMS	International Material Management Society (705) 525-4667
ISA	Instrument Society of America (919) 549-8411 http://www.isa.org/
ISS	Iron and Steel Society (412) 776-1535 http://www.issource.org/HomePg1.htm
ITI	International Technology Institute (412) 795-5300 http://www.iti.ca/welcome.html
ITRI	International Tin Research Institute (614) 424-6200
MSS	Manufactures Standardization Society of Valves & Fittings Industry (703) 281-6613
MTS	Marine Technology Society (202) 775-5966 http://www.cms.udel.edu/mts/
NACE	National Association of Corrosion Engineers (713) 492-0535 http://www.nace.org/
NAPE	National Association of Power Engineers (212) 298-0600 http://www.powerengineers.com/
NAPEGG	Association of Professional Engineers, Geologists and Geophysicists of the Northwest Territories (403) 920-4055 http://www.napegg.nt.ca/
NiDI	Nickel Development Institute (416) 591-7999
PIA	Plastics Institute of America (201) 420-5553
RIA	Robotic Industries Association (313) 994-6088 http://www.robotics.org/
SAE	Society of Automotive Engineers (412) 776-4841 http://www.sae.org/
SAME	Society of American Military Engineers (703) 549-3800 http://www.penfed.org/allies/al3same.htm

Technical Associations & Societies (Continued)

SAMPE   Society for the Advancement of Materials and Processing
        Engineering (818) 331-0616
        http://www.gatech.edu/studlife/soh/orgs/sampe.htm
SCC   Standards Council of Canada (800) 267-8220
        http:// www.scc.ca, e-mail info@sae.org
SCTE   Society of Carbide & Tool Engineers (216) 338 5151
SDCE   Society of Die Casting Engineers (312) 452-0700
SME   Society of Manufacturing Engineers (313) 271-1500
        http:// www.sme.org, e-mail krommar@sme.org
SPE   Society of Petroleum Engineers (214) 669-3377
        http://www.spe.org/
SSIUS   Specialty Steel Industry of the United States (202) 342-8630
SSPC   Steel Structures Painting Council (412) 268-3327
        http://www.horizonweb.com/pcn/90.htm
STC   Society for Technical Communications (202) 737-0035
        http://www.stc-va.org/
STLE   Society of Tribologists and Lubrication Engineers
        (312) 825-5536
TDA   Titanium Development Association (303) 443-7515
TMS   The Minerals, Metals, and Materials Society (412) 776-9000

For more web site addresses of other engineering related sites, visit our
Other Sites page at www.casti-publishing.com.

INDEX

A

B

C

D

F

I

J

M

T

W